BIOFUELS

BIOFUELS

ALTERNATIVE FEEDSTOCKS AND CONVERSION PROCESSES

Edited by

ASHOK PANDEY
National Institute for Interdisciplinary Science and Technology
Council of Scientific and Industrial Research
Trivandrum, India

CHRISTIAN LARROCHE
Biological Engineering Department
Chemical and Biochemical Engineering Laboratory
Polytech Clermont-Ferrand
Blaise Pascal University, France

STEVEN C RICKE
Steven C Ricke
Food Science Department
Division of Agriculture
University of Arkansas, USA

CLAUDE-GILLES DUSSAP
Laboratoire de Génie Chimique et Biochimique
Polytech Clermont-Ferrand
Blaise Pascal University, France

EDGARD GNANSOUNOU
Head, Bioenergy Group
École Polytechnique Fédérale de Lausanne
Lausanne, Switzerland

ELSEVIER

AMSTERDAM • BOSTON • HEIDELBERG • LONDON
NEW YORK • OXFORD • PARIS • SAN DIEGO
SAN FRANCISCO • SINGAPORE • SYDNEY • TOKYO
Academic Press is an imprint of Elsevier

Academic Press is an imprint of Elsevier

The Boulevard, Langford Lane, Kidlington, Oxford OX5 1GB, UK
Radarweg 29, PO Box 211, 1000 AE Amsterdam, The Netherlands
30 Corporate Drive, Suite 400, Burlington, MA 01803, USA
525 B Street, Suite 1900, San Diego, CA 92101-4495, USA

First edition **2011**

Library of Congress Cataloging-in-Publication Data
Biofuels : alternative feedstocks and conversion processes / edited by Ashok Pandey ... [et al.] — 1st ed.
p. cm.
ISBN 978-0-12-385099-7
1. Biomass energy. I. Pandey, Ashok.
TP339.B539 2011
333.95′39—dc22

2011005287

British Library Cataloguing in Publication Data
A catalogue record for this book is available from the British Library

ISBN: 978-0-12-385099-7

For information on all **Academic Press** publications
visit our web site at elsevierdirect.com

Printed and bound in USA
11 12 13 14 10 9 8 7 6 5 4 3 2 1

Contents

Preface vii
Contributors ix

I
GENERAL 1

1. Principles of Biorefining 3
2. Life-Cycle Assessment of Biofuels 25
3. Thermochemical Conversion of Biomass to Biofuels 51
4. Biomass-derived Syngas Fermentation into Biofuels 79

II
PRODUCTION OF BIOETHANOL FROM LIGNOCELLULOSIC FEEDSTOCKS 99

5. Lignocellulosic Bioethanol: Current Status and Future Perspectives 101
6. Technoeconomic Analysis of Lignocellulosic Ethanol 123
7. Pretreatment Technologies for Lignocellulose-to-Bioethanol Conversion 149
8. Production of Celluloytic Enzymes for the Hydrolysis of Lignocellulosic Biomass 177
9. Production of Hemicellulolytic Enzymes for Hydrolysis of Lignocellulosic Biomass 203
10. Hydrolysis of Lignocellulosic Biomass for Bioethanol Production 229
11. Production of Bioethanol from Agroindustrial Residues as Feedstocks 251
12. Fermentation Inhibitors in Ethanol Processes and Different Strategies to Reduce Their Effects 287

IIIA
PRODUCTION OF BIODIESEL FROM VEGETABLE OILS 313

13. Biotechnological Methods to Produce Biodiesel 315
14. Biodiesel Production in Supercritical Fluids 339
15. Production of Biodiesel Using Palm Oil 353
16. Biodiesel Production from Waste Oils 375

IIIB
PRODUCTION OF BIOFUELS FROM ALGAE 397

17. Production of Biodiesel from Algal Biomass: Current Perspectives and Future 399
18. Overview and Assessment of Algal Biofuels Production Technologies 415
19. Cultivation of Algae in Photobioreactors for Biodiesel Production 439

IV
PRODUCTION OF BIOHYDROGEN 465

20. Production of Biohydrogen: Current Perspectives and Future Prospects 467
21. Biohydrogen Production from Bio-oil 481
22. Biohydrogen Production from Industrial Effluents 499
23. Thermophilic Biohydrogen Production 525
24. Biohydrogen Production with High-Rate Bioreactors 537

V
PRODUCTION OF BIOBUUTANOL AND OTHER GREEN FUELS 569

25. Butanol Fuel from Biomass: Revisiting ABE Fermentation 571
26. Production of Green Liquid Hydrocarbon Fuels 587

Index 609

Preface

With the increasing demand of energy world over and depleting reserves of conventional fossil fuel, there has been growing global interest in developing alternative sources of energy. Also, there has been concern in growing economies with energy security. Biofuels offer much promise on these frontiers. In addition to above, they also offer benefits on environmental impact in comparison to fossil fuels. The present book provides state-of-the-art information on the status of the biofuel production and related aspects and also identifies the future R&D directions and perspectives.

The book has five sections. Section I is general and presents four chapters which deal with the principles of biorefineries, life cycle assessment of biofuels, thermochemical conversion of biomass to biofuels, and biomass-derived syngas fermentation into biofuels. Section II deals with different aspects of the production of second-generation bioethanol from lignocellulosic feedstocks. The first chapter in this section is introductory, giving state-of-the-art information on the status and perspectives; this is followed by a chapter on techno-economic analysis of lignocellulosic bioethanol. Subsequent chapters deal with the different aspects of bioconversion process such as the pretreatment of lignocellulosic biomass, production of cellulolytic and hemicellulolytic enzymes for the hydrolysis of lignocellulosic biomass, hydrolysis oflignocellulosic biomass, production of bioethanol from agro-industrial residues as feedstocks, and removal of inhibitory compounds from lignocellulosic hydrolyzates for bioethanol production. Section IIIA presents state-of-the-art information on the production of second-generation biodiesel from oilseeds. In this, the first chapter is introductory and presents current perspectives and future, followed by the biotechnological methods to produce biodiesel, biodiesel production in supercritical fluids, biodiesel production using palm oil, and biodiesel from waste oil. Section IIIB contains chapters dealing with the production of third-generation biofuels from algal sources. The first chapter in this section as usual presents the current perspectives and future, followed by life cycle assessment of algal biodiesel, and the cultivation of algae in photobioreactors.

Section IV is devoted on the fourth-generation biofuels, that is, biohydrogen. The section has five chapters and the first one gives general information with current perspectives and future. The other chapters are on biohydrogen production from bio-oils and industrial effluents, thermophilic biohydrogen production, and biohydrogen production with high-rate bioreactors. Section V provides two articles on the production of biobutanol and production of green liquid hydrocarbon fuels.

We thank the authors of all the chapters for their cooperation and also for their preparedness in revising the manuscripts in a time-framed manner. We also acknowledge the help from the reviewers, who in spite of their busy professional activities helped us by evaluating the manuscripts and gave their critical inputs to refine and improve the chapters. We warmly thank Dr. Marinakis Kostas and Dr. Anita Koch and the team of Elsevier for their cooperation and efforts in producing this book.

We sincerely hope that the current discourse on biofuels R&D would go a long way in bringing out the exciting technological possibilities and ushering the readers toward the frontiers of knowledge in the area of biofuels.

The text in all the chapters is supported by numerous clear, informative diagrams and tables. The book would be of great interest to the postgraduate students and researchers of applied biology, biotechnology, microbiology, biochemical, and chemical engineers working on biofuels.

Ashok Pandey
Christian Larroche
Steven Ricke
Claude-Gilles Dussap
Edgard Gnansounou
Editors

Contributors

Deepthy Alex Centre for Biofuels, Biotechnology Division, National Institute for Interdisciplinary Science and Technology, CSIR, Trivandrum 695 019, India

P. Alvira CIEMAT, Renewable Energy Division, Biofuels Unit Av. Complutese 22, 28040 Madrid

Irini Angelidaki Department of Environmental Engineering, Technical University of Denmark, Lyngby, Denmark

Amar Anumakonda UOP LLC, 25 E. Algonquin Road, Des Plaines, IL 60616, USA

M. Ballestero CIEMAT, Renewable Energy Division, Biofuels Unit Av. Complutese 22, 28040 Madrid

Thallada Bhaskar Bio-Fuels Division (BFD), Indian Institute of Petroleum (IIP), Council of Scientific and Industrial Research (CSIR), Dehradun 248005, India

Balagurumurthy Bhavya Bio-Fuels Division (BFD), Indian Institute of Petroleum (IIP), Council of Scientific and Industrial Research (CSIR), Dehradun 248005, India

Parameswaran Binod Centre for Biofuels, Biotechnology Division, National Institute for Interdisciplinary Science and Technology, CSIR, Trivandrum 695 019, India

Carlos A. Cardona Departamento de Ingeniería Química, Universidad Nacional de Colombia Sede Manizales, Cra. 27 No. 64-60, Manizales, Colombia

Elisa d'Avila Cavalcanti-Oliveira Universidade Federal do Rio de Janeiro, Instituto de Química, Av. Athos da Silveira Ramos, 149 - CT, Bloco A, lab. 549-1, CEP 21941-909 Rio de Janeiro, RJ, Brazil

Yi-Feng Chen School of Life Sciences, Tsinghua University, Beijing 100084, People's Republic of China

Francesco Cherubini Department of Energy and Process Engineering, Norwegian University of Science and Technology (NTNU), NO-7491 Trondheim, Norway

Arnaud Dauriat ENERS Energy Concept, P.O. Box 56, CH-1015 Lausanne, Switzerland

Joab Sampaio de Sousa Universidade Federal do Rio de Janeiro, Instituto de Química, Av. Athos da Silveira Ramos, 149 - CT, Bloco A, lab. 549-1, CEP 21941-909 Rio de Janeiro, RJ, Brazil

Vincenza Faraco Department of Organic Chemistry and Biochemistry, University of Naples "Federico II", Complesso Universitario Monte S. Angelo, via Cintia 4 80126, Naples, Italy

Edgard Gnansounou Bioenergy and Energy Planning Research Group (BPE), Ecole Polytechnique Fédérale de Lausanne (EPFL), CH-1015 Lausanne, Switzerland

Lalitha Devi Gottumukkala Centre for Biofuels, Biotechnology Division, National Institute for Interdisciplinary Science and Technology – CSIR, Trivandrum 695 019, India

Hari Bhagwan Goyal Bio-Fuels division (BFD), Indian Institute of Petroleum (IIP), Council of Scientific and Industrial Research (CSIR), Dehradun 248005, India

Denise Maria Guimarães Freire Universidade Federal do Rio de Janeiro, Instituto de Química, Av. Athos da Silveira Ramos, 149 - CT, Bloco A, lab. 549-1, CEP 21941-909 Rio de Janeiro, RJ, Brazil

Lien-Huong Huynh Department of Chemical Engineering, National Taiwan University of Science and Technology, 43 sec. 4, Keelung Road, Taipei 10607, Taiwan

K.U. Janu Centre for Biofuels, Biotechnology Division, National Institute for Interdisciplinary Science and Technology, CSIR, Trivandrum 695 019, India

Yi-Hsu Ju Department of Chemical Engineering, National Taiwan University of Science and Technology, 43 sec. 4, Keelung Road, Taipei 10607, Taiwan

Dimitar Karakashev Department of Environmental Engineering, Technical University of Denmark, Lyngby, Denmark

Keikhosro Karimi Chemical Engineering Department, Isfahan University of Technology, Iran

Susan Karp Department of Bioprocess Engineering and Biotechnology, Federal University of Parana, P.O. Box 19011, Curitiba, Brazil

Novy S. Kasim Department of Chemical Engineering, National Taiwan University of Science and Technology, 43 sec. 4, Keelung Road, Taipei 10607, Taiwan

Samir Kumar Khanal Department of Molecular Biosciences and Bioengineering (MBBE), University of Hawai'i at Mānoa, Agricultural Science Building 218, 1955 East-West Road, Honolulu, Hawaii 96822

Ajay Kumar Bio-Fuels Division (BFD), Indian Institute of Petroleum (IIP), Council of Scientific and Industrial Research (CSIR), Dehradun 248005, India

Amit Kumar Department of Mechanical Engineering, University of Alberta, Edmonton, Alberta, Canada T6G 2G8

Man Kee Lam School of Chemical Engineering, Universiti Sains Malaysia, Engineering Campus, Seri Ampangan, 14300 Nibong Tebal, Pulau Pinang, Malaysia

Duu-Jong Lee Department of Chemical Engineering, National Taiwan University, Taipei, Taiwan

Keat Teong Lee School of Chemical Engineering, Universiti Sains Malaysia, Engineering Campus, Seri Ampangan, 14300 Nibong Tebal, Pulau Pinang, Malaysia

Wen-Wei Li Department of Chemistry, University of Science and Technology of China, Hefei, 230026 China

S. Venkata Mohan Bioengineering and Environmental Centre (BEEC), Indian Institute of Chemical Technology (IICT), Hyderabad-500007, India

G. Mohanakrishna Bioengineering and Environmental Centre (BEEC), Indian Institute of Chemical Technology (IICT), Hyderabad-500007, India

Pradeep Chaminda Munasinghe Department of Molecular Biosciences and Bioengineering (MBBE), University of Hawai'i at Mānoa, Agricultural Science Building 218, 1955 East-West Road, Honolulu, Hawaii 96822

Ganti S. Murthy Biological and Ecological Engineering, Oregon State University, USA

Desavath Viswanath Naik Bio-Fuels Division (BFD), Indian Institute of Petroleum (IIP), Council of Scientific and Industrial Research (CSIR), Dehradun 248005, India

M.J. Negro CIEMAT, Renewable Energy Division, Biofuels Unit Av. Complutese 22, 28040 Madrid, Spain

Ashok Pandey Centre for Biofuels, Biotechnology Division, National Institute for Interdisciplinary Science and Technology, CSIR, Trivandrum 695 019, India

J. Pruvost GEPEA, Université de Nantes, CNRS, UMR6144, boulevard de l'Université, CRTT – BP 406, 44602 Saint-Nazaire Cedex, France

Julián A. Quintero Departamento de Ingeniería Química, Universidad Nacional de Colombia Sede Manizales, Cra. 27 No. 64-60, Manizales, Colombia

Kuniparambil Rajasree Centre for Biofuels, Biotechnology Division, National Institute for Interdisciplinary Science and Technology, CSIR, Industrial Estate PO, Trivandrum 695 019, India

Reeta Rani Singhania Biotechnology Division, National Institute for Interdisciplinary Science and Technology, CSIR, Trivandrum 695 019, India

Anjan Ray UOP India Pvt Ltd, 6th floor, Building 9B, Cyber City, DLF Phase III, Gurgaon-122002, India

Carlos Ricardo Soccol Department of Bioprocess Engineering and Biotechnology, Federal University of Parana, P.O. Box 19011, Curitiba, Brazil

Luis E. Rincón Departamento de Ingeniería Química, Universidad Nacional de Colombia Sede Manizales, Cra. 27 No. 64-60, Manizales, Colombia

Susanjib Sarkar Department of Mechanical Engineering, University of Alberta, Edmonton, Alberta, Canada T6G 2G8

Manju Sharma Department of Microbiology, Guru Nanak Dev University, Amritsar-143 005, India

Kuan-Yeow Show Department of Environmental Engineering, Faculty of Engineering and Green Technology, Universiti Tunku Abdul Rahman, Jalan University, Bandar Barat, 31900 Kampar, Perak, Malaysia

Ravindran Sindhu Centre for Biofuels, Biotechnology Division, National Institute for Interdisciplinary Science and Technology, CSIR, Trivandrum 695 019, India

Bhupinder Singh Chadha Department of Microbiology, Guru Nanak Dev University, Amritsar-143 005, India

Rawel Singh Bio-Fuels Division (BFD), Indian Institute of Petroleum (IIP), Council of Scientific and Industrial Research (CSIR), Dehradun 248005, India

S. Srikanth Bioengineering and Environmental Centre (BEEC), Indian Institute of Chemical Technology (IICT), Hyderabad-500007, India

Anders H. Strømman Department of Energy and Process Engineering, Norwegian University of Science and Technology (NTNU), NO-7491 Trondheim, Norway

Rajeev K. Sukumaran Centre for Biofuels, Biotechnology Division, National Institute for Interdisciplinary Science and Technology, CSIR, Trivandrum 695 019, India

Mohammad J. Taherzadeh School of Engineering, University of Borås, Sweden

Kok Tat Tan School of Chemical Engineering, Universiti Sains Malaysia, Engineering Campus, Seri Ampangan, 14300 Nibong Tebal, Pulau Pinang, Malaysia

Vanete Thomaz-Soccol Industrial Biotechnology Program, Positivo University, Curitiba, Brazil

E. Tomás-Pejó CIEMAT, Renewable Energy Division, Biofuels Unit Av. Complutese 22, 28040 Madrid, Spain

Luciana P.S. Vandenberghe Department of Bioprocess Engineering and Biotechnology, Federal University of Parana, P.O. Box 19011, Curitiba, Brazil

Adenise Woiciechowski Department of Bioprocess Engineering and Biotechnology, Federal University of Parana, P.O. Box 19011, Curitiba, Brazil

Qingyu Wu School of Life Sciences, Tsinghua University, Beijing 100084, People's Republic of China

Han-Qing Yu Department of Chemistry, University of Science and Technology of China, Hefei, 230026 China

Zhen-Peng Zhang Beijing Enterprises Water Group Limited, BLK 25, No. 3 Minzhuang Rd, Beijing, China

SECTION I

GENERAL

CHAPTER

1

Principles of Biorefining

*Francesco Cherubini**, *Anders H. Strømman*

Department of Energy and Process Engineering, Norwegian University of Science and Technology (NTNU), NO-7491 Trondheim, Norway

***Corresponding author: E-mail: francesco.cherubini@ntnu.no**

1 INTRODUCTION

1.1 Background

Driven by the increase in industrialization and population, the global demand for energy and material products is steadily growing. Since the world primary sources for energy and chemicals are fossil fuels, this growth raises important issues at environmental, economic, and social levels. Petroleum is exploited at a much faster rate than its natural regeneration through the planet C cycle, and the larger part of petroleum and natural gas reserves is located within a small group of countries. This production and consumption pattern is unsustainable because of equity and environmental issues that have far-reaching implications. In addition, there is a common increasing perception that the end of the cheap fossil era is around the corner, and prices for crude oil, transportation fuels, and petroleum-derived chemicals are likely to steadily increase in the years to come (Bentley et al., 2007; Greene, 2004). Climate experts widely agree that emissions of greenhouse gases (GHG), such as carbon dioxide (CO_2), methane (CH_4), and nitrous oxide (N_2O), arising from fossil fuel combustion and land-use change as a result of human activities, are perturbing the Earth's climate (Forster et al., 2007). Global warming and other issues can be mitigated by shifting from fossil sources to renewable energy resources, which are more evenly distributed than fossil resources and cause less environmental and social concerns.

Among the other energy sources, biomass resources are extremely promising since they are widespread and cheaply available in most of the countries. Today, biomass constitutes about 10% of the global primary energy demand, and it is mainly used in inefficient and traditional applications in developing countries (GBEP, 2007; IEO, 2009). Modern uses of biomass are restricted to developed countries to produce space heating, power, transportation biofuels (mainly bioethanol and biodiesel), and few chemical products. Given the variety of applications for biomass sources, it is extremely important to select the most promising options under environmental, economic and resource perspectives. Electricity and heat can be provided by several renewable alternatives (wind, sun, water, biomass, and so on), while biomass is very likely to be the only viable alternative to fossil resources for production of transportation fuels and chemicals. Today, more than 90% of the fossil carbon is used only for its energy content (Marquardt et al., 2010). This pattern is not likely to be followed in the future for biomass because of the lower efficiency in converting biomass into energy and the lower energy density of biomass than fossils. Stemming from these considerations, some authors convincingly argued that electricity should be produced by an increasing share of renewable sources, and the use of biomass be restricted to the production of transportation biofuels and carbon-based chemical products (Agrawal and Singh, 2010; Marquardt et al., 2010).

1.2 The Biorefinery Concept

The sustainable use of bio-based carbon suggests integrated manufacturing in biorefineries to selectively transform the variety of molecular structures available in biomass into a range of products including transportation biofuels, chemicals, polymers, pharmaceuticals, pulp and paper, food, or cattle feed (Cherubini, 2009,, 2010; Kamm et al., 2006a).

The biorefinery concept embraces a wide range of technologies able to separate biomass resources (wood, grasses, corn, etc.) into their building blocks (carbohydrates, proteins, fats, etc.) which can be converted to value-added products, biofuels, and chemicals. A biorefinery is a facility (or network of facilities) that integrates biomass conversion processes and equipment to produce transportation biofuels, power, and chemicals from biomass. Figure 1 gives an overview of the possible conversion pathways to produce the desired energy and material products from different biomass feedstocks, through jointly applied technological processes (Cherubini et al., 2009). The biorefinery concept is analogous to today's petroleum refinery, which produces multiple fuels and products from petroleum.

Biomass is constituted of an enormous variety of plant species with varying morphology and chemical composition. However, regardless of the phenotype, five main biomass components can be identified worldwide: lipids, starch, cellulose, hemicelluloses, lignin, and proteins. The average biomass available in the world is reported in Figure 2. It clearly appears that lignocellulosic biomass components such as cellulose, hemicelluloses, and lignin are by far the most abundant. Since they can be even gathered from waste streams (e.g., crop residues, paper and wood industries), or directly harvested from forests or biomass stands through sustainable management, their price tend to be lower than other biomass sources which need a dedicated agricultural plot. For this reason, this chapter has a special focus on the possibility to produce commodity chemicals from lignocellulosic sources, which have the largest chances for a massive market penetration in the near future.

2 FROM FOSSIL TO BIOMASS RAW MATERIALS

The elemental and chemical structure of biorefinery raw materials differs from that on which the current fossil refinery and chemical industry is based. Chemical and elemental composition of petroleum is compared with some lignocellulosic biomass feedstocks in

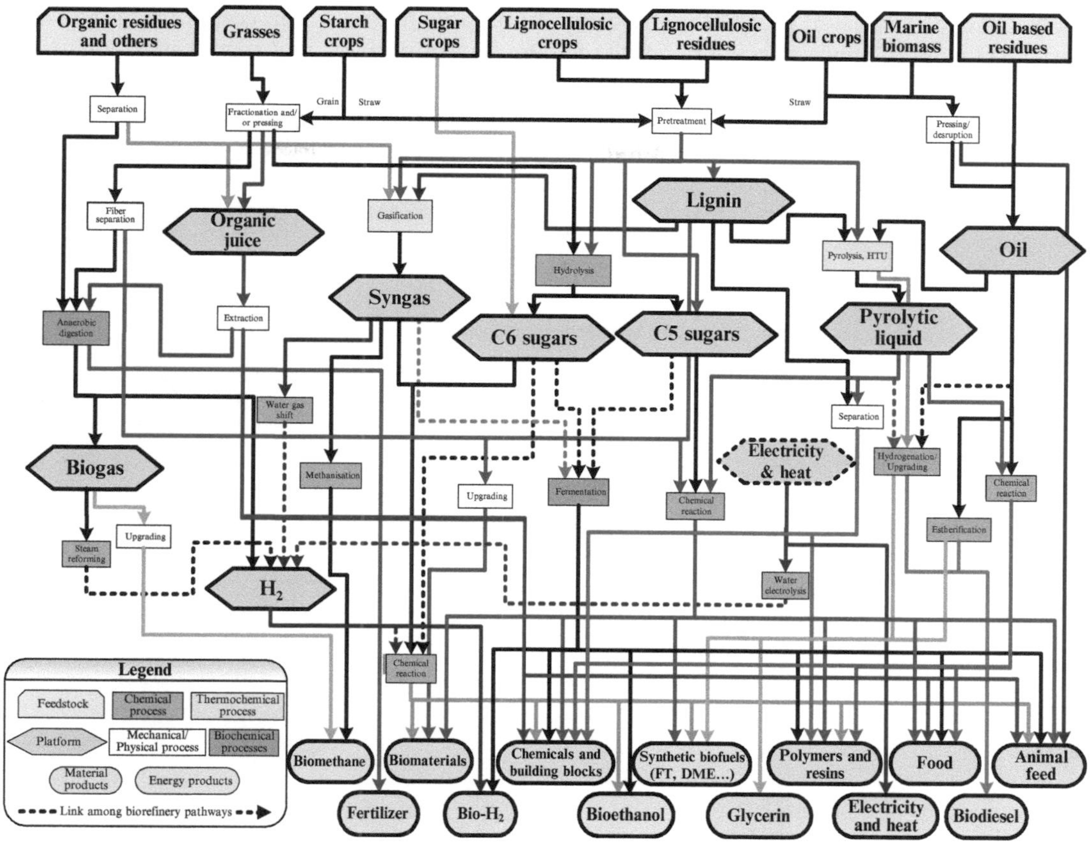

FIGURE 1 Main conversion routes for production of biofuels, energy, and chemicals from different biomass sources.

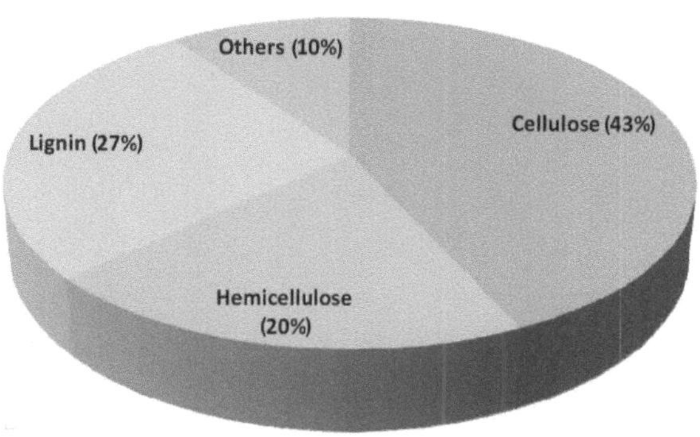

FIGURE 2 World average composition of the above ground standing biomass.

Table 1. Crude oil is a mixture of many different organic hydrocarbon compounds. The first step in oil refinery consists in the removal of water and impurities, and then distillation of the crude oil into its various fractions as gasoline, diesel fuel, naphtha, kerosene, lubricating oils, and asphalts is carried out. The relative volumes of the fractions formed depend on the processing conditions and the composition of the crude oil. The naphtha fraction is subsequently used as a feedstock for the production of just a few bulk chemicals from which all the major commodity chemicals are subsequently derived. An important characteristic of the naphtha

TABLE 1 Average Composition of Some Lignocellulosic Sources and Petroleum

Parameter	Unit (Dry)	Hardwood (Poplar)	Softwood (Pine)	Grass (Switchgrass)	Crop Residue (Corn Stover)	Petroleum
LHV	MJ/kg	19.5	19.6	17.1	16	42.7
Cellulose	%	42.9	44.5	32.0	37.7	
Glucan (C6)	%	42.9	44.5	32.0	37.7	
Hemicellulose	%	20.3	21.9	25.2	25.3	
Xylan (C5)	%	17.0	6.30	21.1	21.6	
Arabinan (C5)	%	1.20	1.60	2.84	2.42	
Galactan (C6)	%	0.70	2.56	0.95	0.87	
Mannan (C6)	%	1.42	11.4	0.30	0.38	
Lignin	%	26.6	27.7	18.1	18.6	
Acids	%	3.11	26.7	1.21	3	
Extractives	%	4.70	2.88	17.5	5.61	
Hydrocarbons	%					
Praffins	%	–	–	–	–	30
Naphthenes	%	–	–	–	–	49
Aromatics	%	–	–	–	–	15
Asphaltic	%	–	–	–	–	6
Elemental composition	%					
C	%	49.4	50.3	47.3	47	83-87
H	%	5.75	5.98	5.31	5.66	10-14
O	%	43.3	42.1	41.6	41.4	0.1-0.5
N	%	0.19	0.03	0.51	0.65	0.1-0.2
S	%	0.02	0.01	0.1	0.06	0.5-6
Minerals	%	2.43	0.32	5.95	10.1	0.1

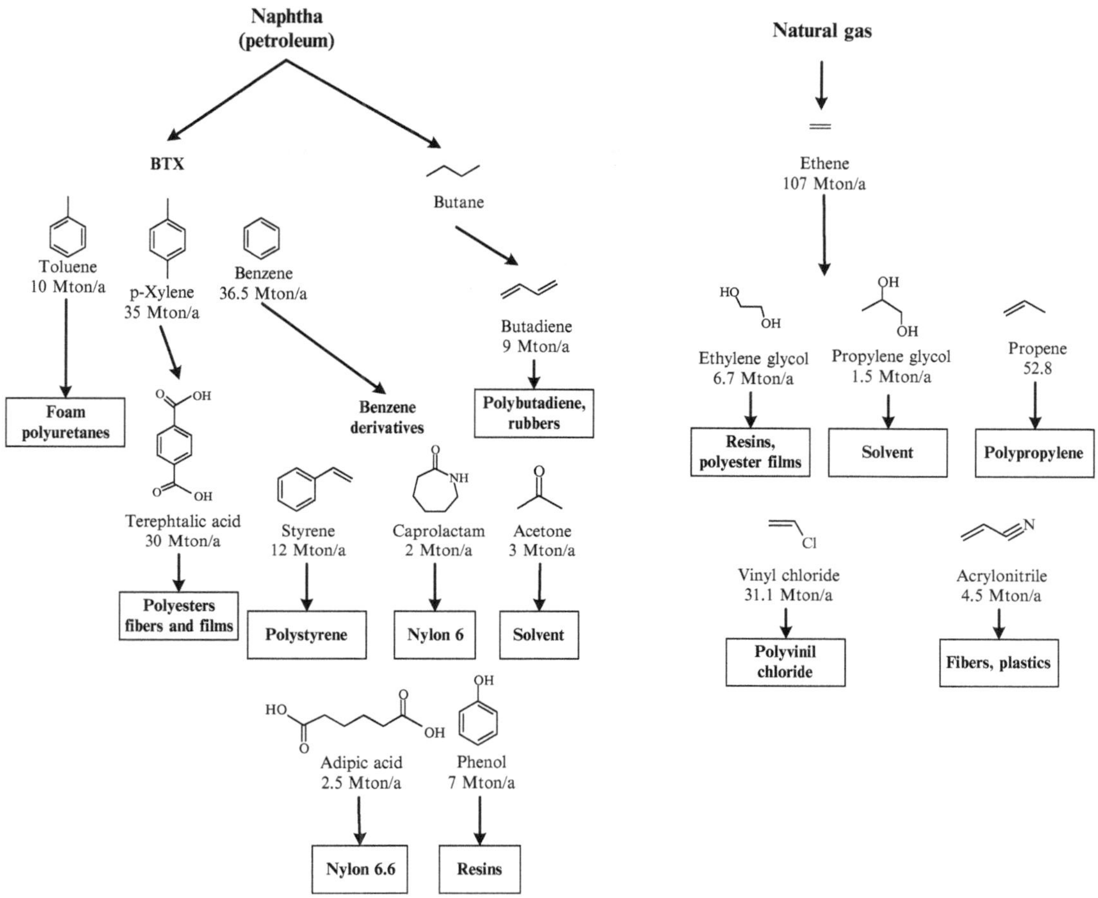

FIGURE 3 Schematic flow diagram of petrochemical production from fossils. The world market production is beneath the chemical name. The most common industrial applications for the specific chemical are even reported.

feedstock is that, unlike biomass, it is very low in oxygen content. The most important chemical products currently derived from oil and natural gas refinery are shown in Figure 3.

This figure shows that today's chemical industry processes fossil resources into a limited number of bulk chemicals from which a wide spectrum of secondary commodity chemicals are produced. These commodity chemicals have many applications in almost all the sectors of our society as textiles, plastics, resins, food and feed additives, and others. The bulk chemicals from which the majority of commodity chemicals can be produced are ethylene, propylene, batanes/butadiene, and the aromatic benzene, toluene, and xylene (BTX).

The composition of biomass is less homogeneous than petroleum. The share of biomass components in the feedstock can change and the elemental composition is a mixture of C, H, and O (plus other minor components such as N, S, and other mineral compounds). If compared to petroleum, biomass generally has less hydrogen, more oxygen, and a lower fraction of carbon. The compositional variety in biomass feedstocks is both an advantage and a drawback. An advantage is that biorefineries can make more classes of products than can petroleum refineries and can rely on a wider range of raw materials. A drawback is that a relatively larger range of processing technologies is needed, and most of these technologies are still at a precommercial stage (Dale and Kim, 2006). Another difference with petroleum resources concerns the seasonal changes which biomass suppliers have to face, since harvesting is usually not possible throughout the year. A switch from crude oil to biomass may require a change in the capacity of chemical industries, with a requirement to generate the materials and chemicals in a seasonal time frame. Alternatively, biomass may have to be stabilized prior to long-term storage in order to ensure continuous, year-round operation of the biorefinery (Clark et al., 2009).

More difficult is to adapt chemical processes to act on nonhomogeneous substrates, since the chemical industry has been built largely on the use of uniform and consistent raw materials (Hatti-Kaul, 2010). It is unlikely that this will change, so technologies will need to be developed to precondition biomass feedstocks to make their properties and reactivity patterns more stable, consistent, and uniform. One concept that may be of value is to separate the different biomass components early in biorefinery operations, so to make a distinction between those which are subject to energy uses (whose quality can be degraded) and those destined to chemical applications (which need high degree of purity and should be subject to milder process conditions to conserve the original structure).

3 BIOMASS PROCESSING IN BIOREFINERY

3.1 Basic Elemental Conversions in Biomass Processing

In order to be used for production of biofuels and chemicals, biomass needs to be depolymerized and deoxygenated. Deoxygenation is required because the presence of O in biofuels reduces the heat content of molecules and usually gives them high polarity, which hinders blending with existing fossil fuels (Lange, 2007). Chemical applications may require much less deoxygenation, since the presence of O often provides valuable physical and chemical properties to the product. Biomass feedstocks usually have an amount of carbon which must be retained throughout the value chain, few hydrogen, which must be added, and too much oxygen, which must be rejected along with other undesirable elements (such as nitrogen and sulfur). Hydrogen is usually added as water (H_2O), even if this implicates an addition of extra oxygen, which must be rejected. The addition of hydrogen as H_2 is more attractive and efficient (using proper metal catalysts) but underprivileged by the fact that elemental hydrogen is not present in nature and energy must be invested to produce it. Oxygen is rejected either as CO_2 or H_2O. In both cases, there are elemental issues: in the first case every mole of oxygen removes half a mole of carbon (thus reducing carbon efficiency), while in the second case 1 mol of oxygen removes 2 moles of hydrogen (which, contrarily, needs to be added). It would be most desirable to reject oxygen as O_2, but this is not a typical output of any biomass conversion process. The other undesired elements, sulfur and nitrogen, are usually rejected in their oxide forms (SO_2 and NO_2, respectively), thus contributing to rejection of excess oxygen.

3.2 Biomass Conversion Through Thermochemical or (Bio-)Chemical Processes

Biomass can be converted to chemicals through thermochemical or (bio-)chemical processes. The most promising thermochemical process is direct gasification of biomass, where the whole feedstock is kept at high temperature (>700 °C) with low oxygen levels to produce syngas, a mixture of H_2, CO, CO_2, and CH_4 (Gassner and Maréchal, 2009; van Vliet et al., 2009). These C-1 building blocks are then reassembled into the desired functional molecules. Other common thermochemical processes are pyrolysis and combustion for heat and power. These thermochemical approaches do not consider the complex molecular structures synthesized by nature, since they destroy the original biomass structure, which should be rather used to the maximum possible extent (Marquardt et al., 2010). Contrarily, the target of (bio-)chemical processes is to access the rich molecular structure already available in biomass without significant degradation of the basic components. For this purpose, the pretreatment step of lignocellulosic biomass is particularly important, since the three main biomass components must be efficiently separated into independent flows, lignin, cellulose, and hemicellulose, to be further processed (Fernandes et al., 2009; Kaparaju and Felby, 2010; Sun and Cheng, 2002). After pretreatment, these highly functionalized polymers have to be selectively depolymerized. Next, the resulting molecular structures need to be isolated and catalytically re-functionalized into target molecules. Such an advanced strategy offers the chance to establish conversion processes with higher carbon efficiency and lower entropic losses compared to conventional thermochemical processes. Although conceptually attractive, its implementation requires the tailoring of the industrial value chains to the nature of the bio-based raw materials. Preserving the natural molecular structures in the raw materials requires a shift from gas-phase reactions at high temperatures, prevalent in petroleum-based chemical engineering, to liquid-phase reactions at lower temperatures. Likewise, low-temperature separation technologies should be favored over classical distillation if possible. Higher viscosities of the process media and the management of large amounts of water are inevitable side effects offering their own challenges (Marquardt et al., 2010).

4 LIGNOCELLULOSIC MOLECULAR COMPONENTS AND THEIR DERIVATIVES

4.1 Lignin

The structure of lignin (see Figure 4) is complex and changes with the type of biomass source. Lignin is composed of phenylpropenyl (C9) randomly branched units. The phenylpropenyl building blocks, like guaiacols and syringols, are connected through carbon-carbon and carbon-oxygen (ether) bonds. Trifunctionally linked units provide numerous branching sites and alternate ring units (Holladay et al., 2007). Lignin offers a significant opportunity for enhancing the operation of a lignocellulosic biorefinery. Today, lignin is used as a source of heat and power for the processing plant (e.g., pulp and paper industry), but this approach seems to be shortsighted: lignin's native structure suggests that it could play a central role as a new chemical feedstock, particularly in the formation of supramolecular materials and aromatic chemicals. All current commercial nonenergy uses of lignin, except combustion and production of synthetic vanillin and dimethylsulfoxide (DMSO), take

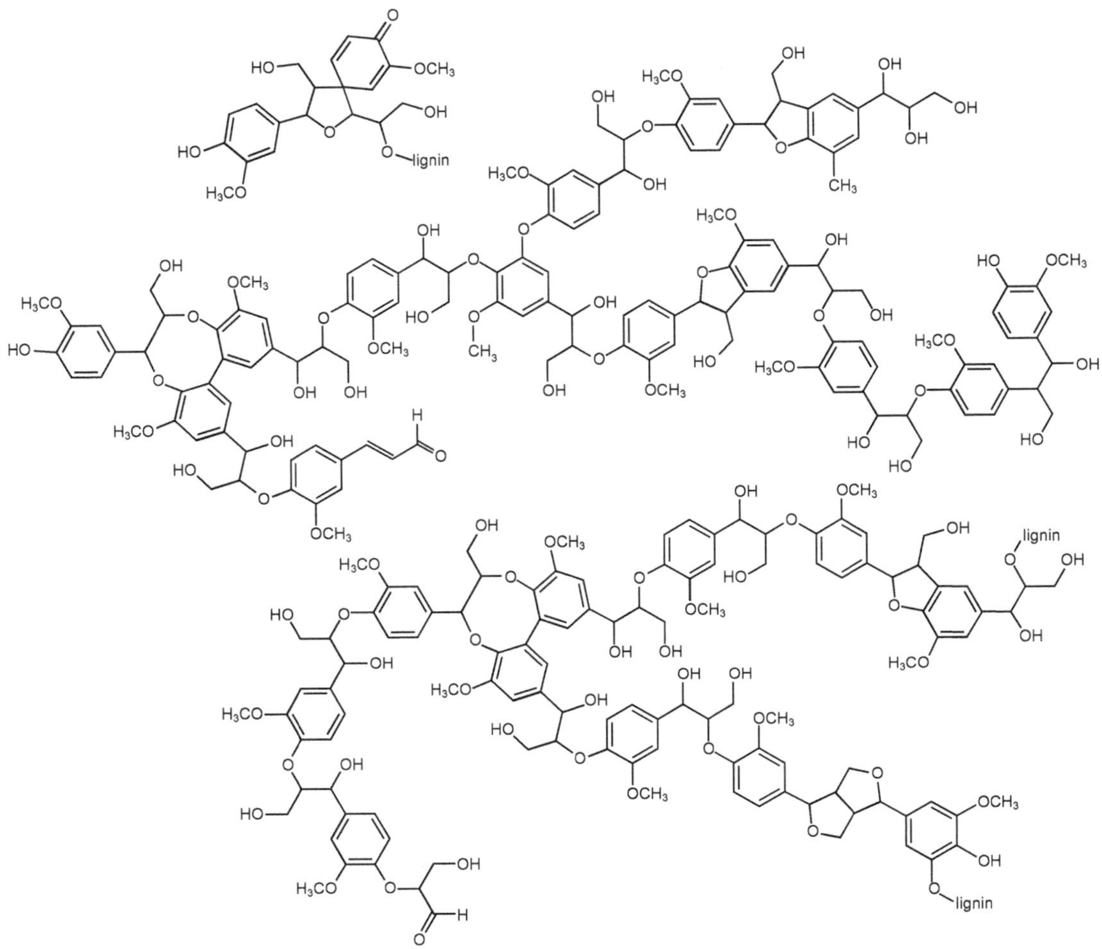

FIGURE 4 Chemical structure of softwood lignin.

advantage of lignin's polymer and polyelectrolyte properties. These are primarily applications targeted at dispersants, emulsifiers, binders, and sequestrants. Generally, lignin is used in these applications with little or no modification other than sulfonation or thio hydroxymethylation. These uses mainly represent relatively low value and limited volume growth applications. An economic study shows that when lignin is used for purposes other than power, the overall revenue improvement of a biorefinery concept is between \$12 and \$35 billion (Holladay et al., 2007). However, as will be shown hereinafter, significant technology developments are required to capture the lignin value benefit.

Besides the immediate opportunities for heat and power production, the specific types of products which can be produced from lignin can be grouped in two main categories:

1. Syngas-derived chemicals (near-term opportunity)
2. Aromatics (medium/long-term opportunity)

4.1.1 *Syngas-Derived Chemicals*

Gasification produces syngas, a mixture of H_2, CO, CH_4, and other light gases. Technology to produce methanol or dimethyl ether (DME) from syngas is well established (Li and Jiang, 1999; Peng, 2002; Sai Prasad et al., 2008). These products can be used directly or may be further converted to green gasoline via the methanol to gasoline process or to olefins via the methanol to olefins process (Cui et al., 2006; Lee, 1995). Because of the high degree of technology development in methanol and DME catalysts and processes, this conversion pathway is extremely promising. The main drawback for this technology is the purification of biomass-derived syngas, which is capital intensive, and demonstration that gasification can proceed smoothly with biorefinery lignin. Another promising use of syngas is the production of Fischer-Tropsch (FT) fuels (Wang et al., 2009). FT processes are well established but their application to biomass is still at a precommercial stage, due to the expensive purification of syngas streams and the need for catalyst and process improvements able to reduce unwanted side-products such as methane and higher molecular weight products such as waxes. Syngas can also be converted to mixed alcohols (like ethanol and other alcohol chemicals), but this technology has not been commercialized yet. Major challenges concern the catalyst and process improvements needed to increase the selectivity and consumption rate of the catalysts (Holladay et al., 2007). Finally, although syngas production via gasification is a well-developed technology for coal (and natural gas), there is continuing controversy over gasification economics at the scale needed for the lignocellulosic biorefinery.

4.1.2 *Aromatics*

Lignin is the most abundant renewable source which has aromatic units in its structure. As shown in Figure 3, the world demand for aromatics is consistent and increasing over the years. The possibility to establish a direct and efficient conversion of lignin to high-volume, low-molecular weight aromatic molecules is therefore extremely attractive. However, there are important technological barriers which must be overcome, given the resistant and robust lignin structure.

The basic chemical units of lignin shows very high potential for making BTX chemicals (Figure 5). Technologies able to efficiently depolymerize the polymer by breaking the C–C and C–O bonds are necessary. An aggressive, nonselective, depolymerization would bring to a mixture of BTX, phenols, and aliphatic fractions (C1-C3). These chemicals should be suitable for being directly used by the conventional petrochemical processes which convert the bulk aromatics into nylons, resins, polymers, and others. Development of the required aggressive and nonselective chemistries is part of the long-term opportunity but is likely to be achievable sooner than highly selective depolymerizations (presented below; Holladay et al., 2007).

A related technological challenge for the production of chemicals from lignin is the elaboration of proper separation techniques for the mixture intermediates from which the aromatic chemicals are to be isolated (Huang et al., 2008).

4.2 Cellulose and Hemicellulose

Carbohydrates are obtained from lignocellulosic resources after depolymerization of cellulose and hemicelluloses. Glucose (a sugar containing six carbons) is produced via hydrolysis of cellulose, whereas xylose and mannose are the main products obtained by hydrolysis of hemicellulose.

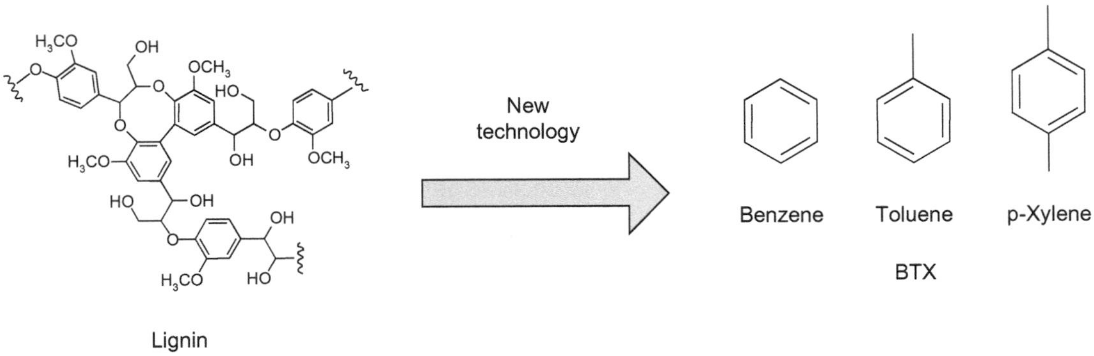

FIGURE 5 The production of BTX from lignin requires the development of a new technology.

Carbohydrates have the possibility to be converted to a wide spectrum of products by means of biochemical (e.g., fermentation) or chemical transformations. Fermentation of sugars to ethanol is already established in the market: nowadays more than 90% of the world ethanol production is derived from biomass feedstocks, while the remaining 10% is produced from oil or gas refinery (Patel, 2006). Further promising sugar derivatives through fermentation are organic acids like succinic, fumaric, malic, glutamic, aspartic, and others (Werpy and Petersen, 2004). Because of their functional groups, organic acids are extremely useful as starting materials for the chemical industry and may act as intermediate to production of fine chemicals. For many organic acids, the actual market is small, but an economical production process will create new markets by providing new opportunities for the chemical industry (Sauer et al., 2008). For example, succinic, fumaric, and malic acid could replace the petroleum-derived commodity chemical maleic anhydride in its applications. The market for maleic anhydride is huge, whereas the current market for the organic acids mentioned is small owing to price limitations. Once a competitive microbial production process for one of these acids is established, the market for that acid is expected to consistently increase. The technological barriers which keep these conversion routes at a precommercial stage concern microbial biocatalysts, which need to be improved to simultaneously reduce formation of byproducts and increase yields and selectivity. Issues of scale-up and system integration are also to be addressed (Werpy and Petersen, 2004).

In addition to microbial conversions, there are several catalytic transformations for carbohydrates, like oxidations, dehydration, hydrogenations, alkylations, among others, which are industrially feasible. Oxidation leads to valuable intermediates like gluconic acid, which is used for synthesis of pharmaceuticals, food additives, cleaning agents, and others (van Bekkum, 1998). Dehydration of sugars is a promising option for producing important platform chemicals like levulinic acid (LA; from glucose) and furfural (from xylose) which can be converted into a large portfolio of chemicals having many applications in the chemical industry and transportation sector (i.e., fuel additives; Bozell et al., 2000; Hayes et al., 2006). The technical barriers for this pathway concern the necessity to increase yields through more selective dehydration processes, perhaps supported by the development of new catalysts. Catalytic hydrogenation of sugars gives sugar alcohols, such as xylitol and sorbitol. Sorbitol is used as a sweetener as well as an intermediate for synthesis of vitamin C, food additives, and C4-C6 polyols for synthesis of alkyds (Blanc et al., 2000). Alkyds are polyesters formed via esterification between polyhydric alcohols and di- or poly-basic carboxylic acids or their anhydrides (Mäki-Arvela et al., 2007). These reaction pathways have a larger degree of development than fermentation routes, and some of them are already at a commercial stage. For instance, the production of sorbitol is practiced by several companies and has a production volume on the order of 0.1 million tons/years (Werpy and Petersen, 2004). These productions are usually based on batch technology, and the only technical development needed would be the use of a continuous process.

5 BIOREFINERY TO REPLACE EXISTING FOSSIL BULK CHEMICALS

Over the last decade, prices of fossil fuel feedstocks have increased, whereas prices of biomass resources have slowly and steadily decreased. This situation makes the possibility to produce the existing bulk chemicals from biomass rather than fossils an attractive option. In the following paragraphs, the current state of the art in the production of the bulk chemicals previously highlighted is investigated. The possible reaction pathways are summarized in Figure 6.

5.1 Ethylene

The production of this chemical from biomass sources can be achieved through dehydration of ethanol. This dehydration is favored at high temperatures (300-600 °C) and can be carried out over a wide variety of heterogeneous catalysts (Arenamnarta and Trakarnpruk, 2006; Takahara et al., 2005). There are no technological barriers to be faced for the production of ethene from ethanol at a commercial scale; this production is initially most likely to happen in regions with cheap and easy access to bioethanol (Haveren et al., 2008).

5.2 Propylene

Direct production of propene from sugars can be carried out via fermentation (Fukuda et al., 1987). Product yields are very low: the productivity needs to be improved by orders of magnitude to make this process economically viable (Haveren et al., 2008). An alternative production pathway consists in the dehydration of 2-propanol, which is produced by reduction of acetone. The latter can be obtained via the acetone, butanol, ethanol (ABE) fermentation process, which is largely studied in the scientific and industrial community (Ezeji et al., 2007). In addition, propene can be produced from dehydration of 1,2-propanediol (either called propene glycol). This glycol can be effectively produced from reduced sugars as sorbitol and xylitol or lactic acid, and such conversion routes have strong commercial potential (Haveren et al., 2008).

5.3 Butane and Butadiene

Starting from biomass, butadiene potentially can be produced from ethanol: ethanol is firstly dehydrogenated to acetaldehyde, which is then followed by aldol condensation and dehydration over a catalyst to form butadiene, with an overall yield of 70% (Weissermel and Arpe, 2003). Butadiene can subsequently be converted to butane by reduction.

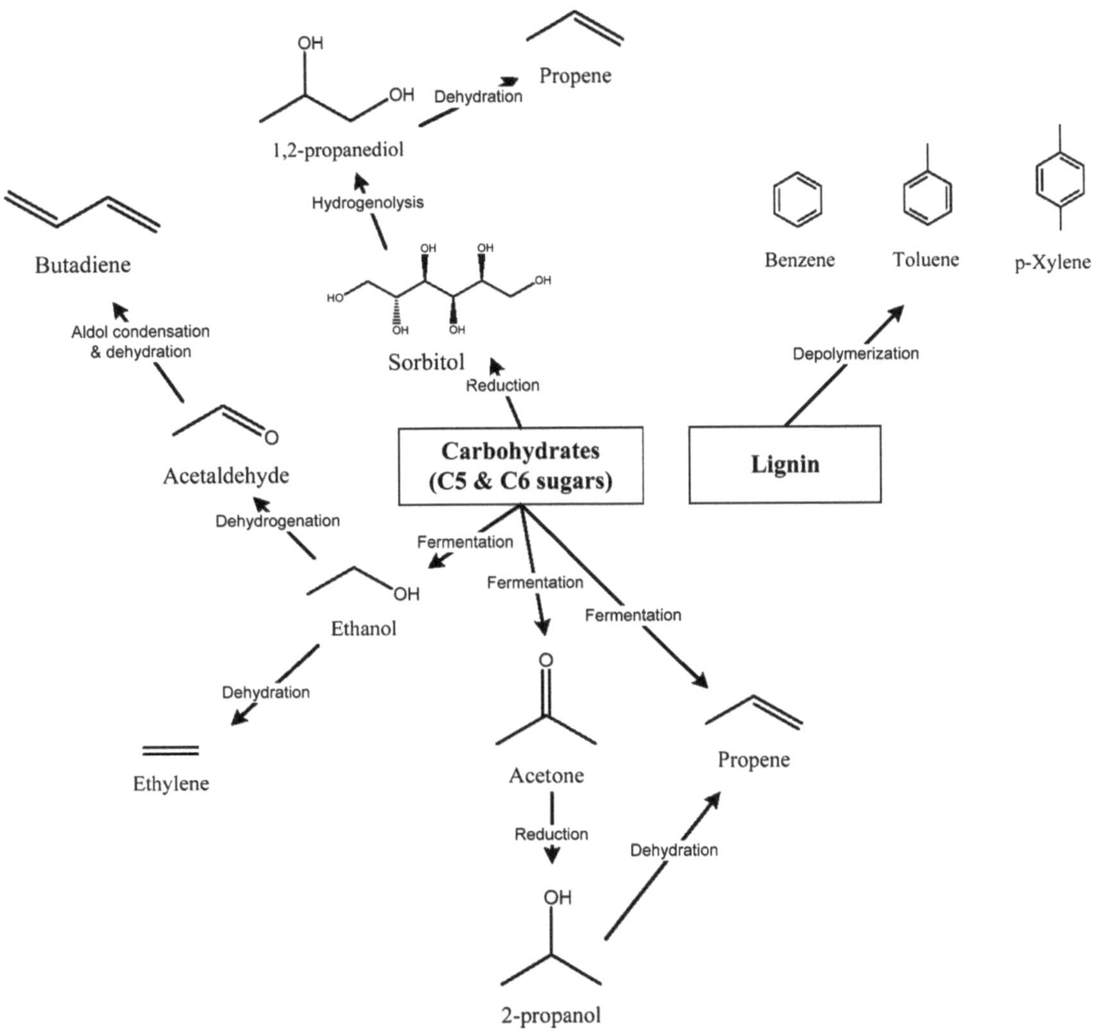

FIGURE 6 Main conversion pathways for producing the existing bulk chemicals in fossil refinery from lignocellulosic biomass.

5.4 Aromatics (BTX)

If the conversion of carbohydrates to oxygen-containing chemicals has been largely investigated, the replacement of bulk aromatic petrochemical compounds has received so far relatively little attention and limited success. Fermentation of glucose to a number of aromatic structures has been described in the patent literature. However, these aromatic structures themselves were neither bulk products nor the desired end product of the fermentation process (adipic acid; Haveren et al., 2008).

Utilization of specific terpenes could offer potential for the production of aromatic compounds such as, for example, substituted phenols or terephthalic acid and fine and specialty chemicals to be applied in the chemical or pharmaceutical industry (Costantino et al., 2009). However, current production volumes of terpenes are rather in the range of hundreds of thousands of tons instead of the million tons needed to substitute a significant amount of aromatics production.

Thanks to its original structure, the most promising feedstock for production of aromatics from biomass is lignin. The ideal conversion pathway would include the possibility to efficiently and selectively depolymerize lignin and separate from the resulting mixture the components of interest (e.g., BTX). Prior to be able to isolate aromatics and phenols from lignin, major technological improvements are needed. Another long term possibility to synthesize aromatics from biomass is the Diels-Alder cyclo-addiction of butadiene over a catalyst. Clearly, this route relies on an economic production pathway to butadiene prior being industrially taken into consideration.

5.5 N-containing Chemicals

The production of N-containing bulk chemicals from biomass is at a later stage of development than oxygenated chemicals. Genetically modified plants may produce elevated levels of amino acids, like lysine, which can be converted to caprolactam (a precursor of nylon), while fermentation of glucose can lead to N-containing compounds like glutamic acid and aspartic acid (see Figure 7). Other nitrogen-based chemicals could be produced by using protein waste streams from bioethanol and biodiesel production chains. Aspartic acid is an amino acid that can be produced by reaction of ammonia with fumaric acid, which can be theoretically produced from glucose fermentation. In order to be produced on a large scale, a direct fermentation route from glucose to aspartic acid is fundamental. Aspartic acid has

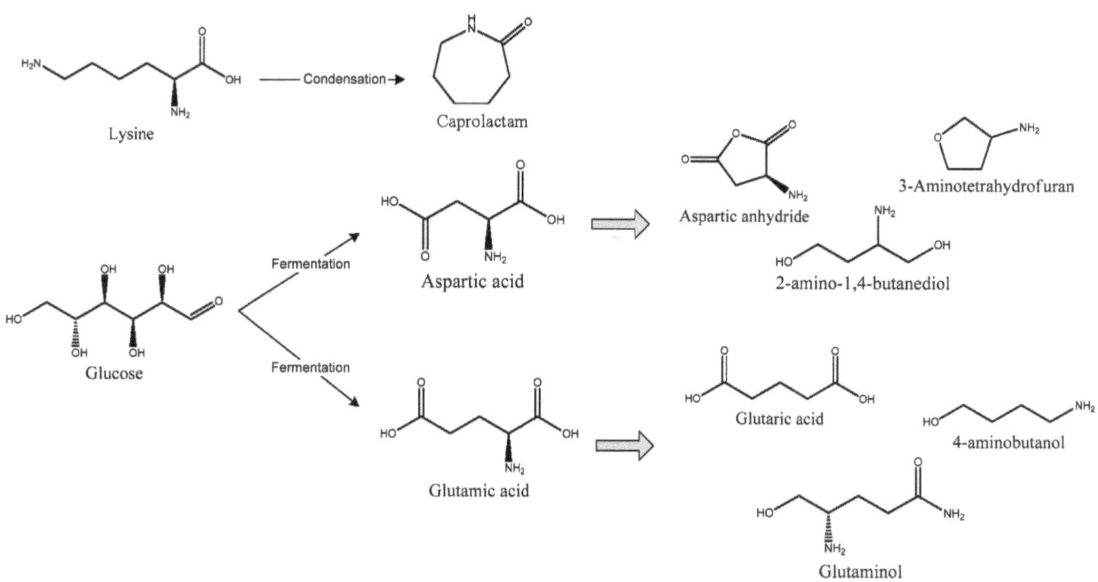

FIGURE 7 Schematic production of N-based chemicals from glucose.

large potential to be converted into a wide spectrum of N-containing chemicals (aspartic anhydride, pyrrolydone, and others).

Fermentation of sugars may even lead to the N-containing chemical glutamic acid. Glutamic acid is a five-carbon amino acid and has the potential to be a novel building block for five carbon polymers. The building block and its derivatives have the potential to build similar polymers but with new functionality to derivatives of the petrochemicals derived from maleic anhydride (Werpy and Petersen, 2004). These polymers could include polyesters and polyamides. The major technical hurdles for the development of glutamic acid as a building block include the development of very low-cost fermentation routes. There are currently several fermentation routes for the production of sodium glutamate. One of the major challenges for the development of a low-cost fermentation is to develop an organism that can produce glutamic acid as the free acid.

In general, there is a midterm potential for production of acrylic acid and other N-containing bulk chemicals like acrylonitrile, acrylamide, and caprolactam. The production of N-based chemicals from biomass is expected to become competitive in the market when large quantities of proteins (as a byproduct of biofuel production chains) will be available at affordable prices.

6 BIOREFINERY TO PRODUCE ALTERNATIVE PRODUCTS

In the previous section, the possibility to replace existing bulk chemicals from fossil refinery with the same bulk chemicals from oil refinery has been investigated. Unlike few cases, possible market penetration of biochemicals in the near term is limited and major technological barriers exist, especially in the production of aromatics. Rather than a head-to-head substitution of petrochemicals with biochemicals, biomass resources can be used to produce platform chemicals which better reflect the initial biomass composition and are easier to be achieved. At the same time, the products must ensure to meet the same functional properties expected by the consumers. The head-to-head substitution of petrochemicals with biochemicals is consistently disadvantaged by the presence of large quantity of oxygen in the biomass feedstock. Future product trees should accommodate the native oxygen content of biomass to reduce the need for deoxygenation. These considerations imply the need for a radical shift from petroleum-based to biomass-based chemical engineering aiming at new value chains with a new range of oxygenated products, novel production routes, and integrated biorefineries built from intensified unit operations which operate at moderate conditions (Marquardt et al., 2010).

6.1 New Bulk Chemicals from Lignin

Lignin has potential for a very selective depolymerization leading to a wide spectrum of oxygen-containing aromatics which are difficult to make via existing petrochemical routes (see Figure 8). These products preserve the lignin monomer structure and can be highly desirable if produced in reasonable quantity with an economic process. The major barrier of this conversion concerns the development of a technology that would allow highly selective bond scissions to maintain the monomeric lignin block structures (Holladay et al., 2007). In addition, proper markets and industrial applications for those aromatics which are related to the original lignin building blocks need to be established. Figure 9 shows the potential reaction

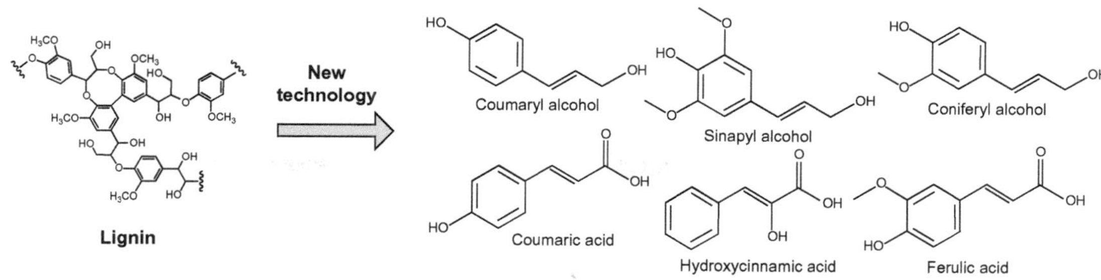

FIGURE 8 Products that preserve lignin original basic structure.

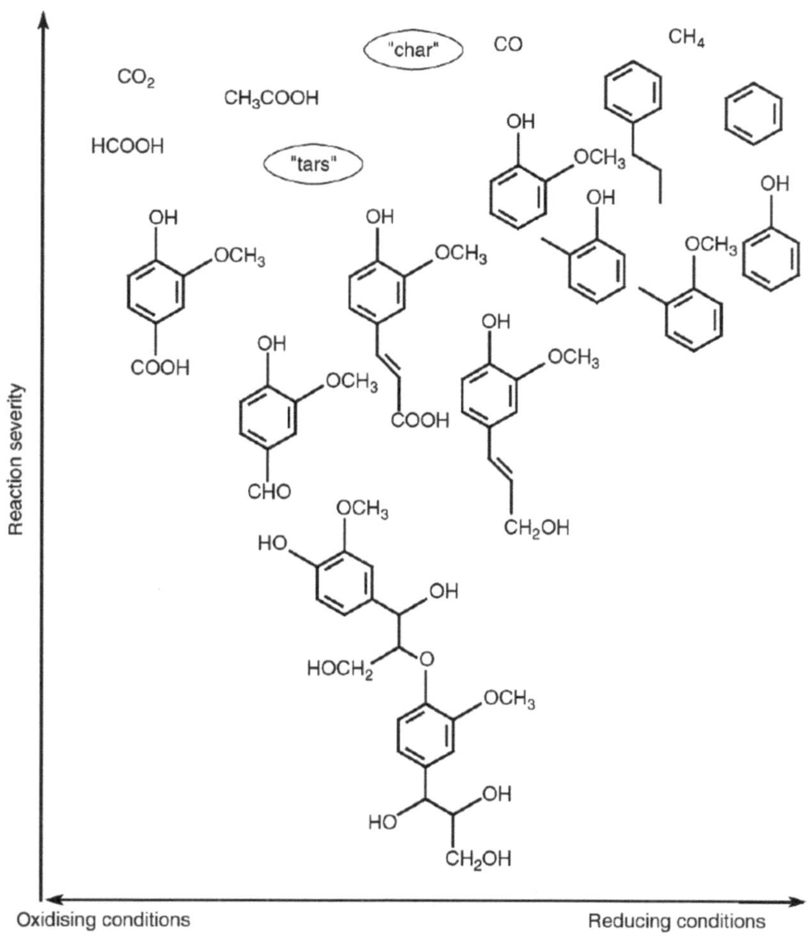

FIGURE 9 Potential reaction products from lignin decomposition at different reaction conditions (Haveren et al., 2008).

products from the decomposition of lignin via high temperature thermal processes (Haveren et al., 2008). This "cracking" of lignin results in a complex mixture of polyhydroxylated and alkylated phenol compounds, where the abundance and type of products are influenced by reaction conditions. Clearly, improved separation techniques for aromatic lignin monomers must be achieved.

6.2 New Bulk Chemicals from Carbohydrates

Figure 10 shows the selected new bulk chemicals and derivatives which can be produced from biomass. A total of 13 intermediates are identified as potential bulk chemicals from which a wide spectrum of products can be obtained. They are specified according to the number of C atoms:

- C2: ethanol
- C3: acetone, lactic acid, 3-Hydroxypropionic acid (HPA)
- C4: succinic acid
- C5: furfural, itaconic acid, xylytol, and LA.
- C6: sorbitol, HMF, 2,5-Furan dicarboxylic acid (FDCA), and gluconic acid.

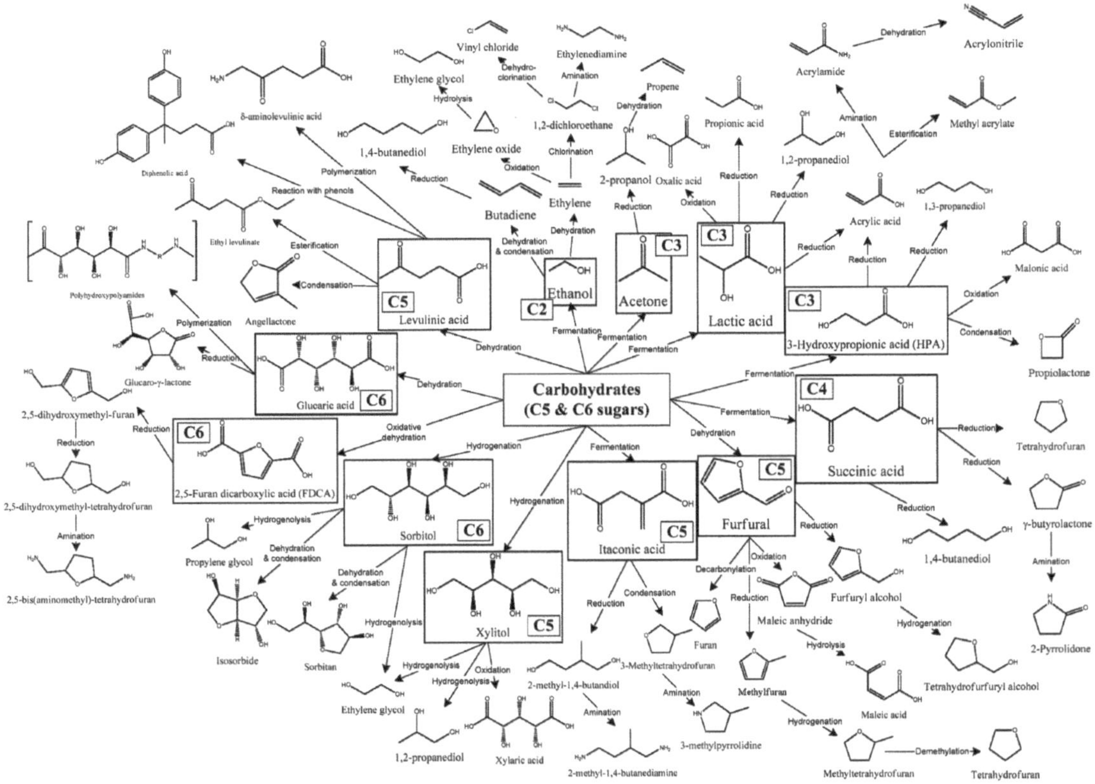

FIGURE 10 Scheme of the selected bulk chemicals obtained from carbohydrates and their main derivatives.

6.2.1 C2 Bulk Chemicals

Besides its uses as transportation biofuel, ethanol also has interesting applications as bulk chemical from which C2 derivatives can be produced. In particular, ethanol can be converted via dehydration to ethene, one of the bulk petrochemicals, which has a world production of 107 million tons/year. Once produced from bioethanol, ethene can be then used for the production of other important chemicals like 1,2-dichloroethane (world production of 20 million tons/year), vinyl chloride, butadiene, and others.

6.2.2 C3 Bulk Chemicals

Acetone is an important chemical compound with a market volume of 3 million tons/year. As already mentioned, it is possible to produce acetone via the ABE fermentation process. This process is widely studied and is expected to be competitive in the market within the next 5-10 years (Bos et al., 2010). Acetone can be a valuable bulk chemical for the production of propene, whose production from fossil refinery is large (50 million tons/year) due to its wide applications (mainly as polypropylene).

Lactic acid is a promising bulk chemical which can lead to many derivatives (in particular polymers), thanks to two reactive sites, the carboxylic group and the hydroxyl group. The production of lactic acid from biomass (fermentation of sugars) is already established in the market, with an annual production around 0.26 million tons and a 10% annual growth (Jem et al., 2010). Major applications are in the food sector, industrial uses, and personal care. Important derivatives which can be produced from lactic acid are acrylic acid via dehydration (current global market of 2 million tons/year) and 1,2-propanediol by reduction (1.5 million tons/year).

3-HPA has the potential to be a key bulk chemical for deriving both commodity and specialty chemicals. The basic chemistry of 3-HPA is not represented by a current petrochemically derived technology (Werpy and Petersen, 2004). Its production from biomass depends on the development of low-cost fermentation routes, since this conversion pathway should in principle have the same yields of that leading to lactic acid. The potential derivatives are similar to those produced from lactic acid, since they have identical reactive sites. In both cases, the development of new catalysts able to directly reduce the carboxylic acid groups to alcohols is required. The esterification of the carboxylic group to an ester, and then reduce the ester, is technically easier, but the process is more expensive. The dehydration of 3-HPA to acrylic acid and acrylamide will require the development of new acid catalyst systems that afford high selectivity (Werpy and Petersen, 2004).

6.2.3 C4 Bulk Chemicals

In fossil refinery, succinic acid is currently produced from butane/butadiene via maleic acid and has a production volume of 30-50 kilotons/year (Bos et al., 2010). This process is relatively expensive and the existing market for succinic acid is limited. However, if a more economic production route could be established, it has a potential market of hundreds to thousands tons, thanks to its many possible derivatives (Sauer et al., 2008). Succinic acid can be efficiently produced from fermentation of sugars, on condition that low-cost fermentation routes are established. The basic chemistry of succinic acid is similar to that of the petrochemically derived maleic acid/anhydride. These compounds can be converted via hydrogenation/reduction to butanediol, tetrahydrofuran, and gamma-butyrolactone.

In the case of succinic acid, the technical challenge is the development of catalysts that would not be affected by impurities in the fermentation. Noteworthy is the possibility to produce pyrrolidinones, so addressing a large solvent market (Werpy and Petersen, 2004).

6.2.4 C5 Bulk Chemicals

Furfural is the starting material for industrial production of furan compounds and today it is completely produced from biomass feedstocks rich in C5 sugars. The market volume is 0.2-0.3 million ton/year. It is obtained from hydrolysis of C5 sugars along with other degradation products. Removal of these impurities is expensive and industrial uses of furfural will benefit of an optimization of the furfural production process (Patel, 2006). Many valuable chemicals can be derived from furfural (e.g., maleic anhydride, furfuryl alcohol, etc.), and the chemistry for the conversions is well developed (Kamm et al., 2006b).

Itaconic acid has a chemistry similar to the fossil-derived chemicals maleic acid and maleic anhydride, which are used as monomers in the production of acrylate-based polymers and thermoset resins in oil refinery (Bos et al., 2010). Itaconic acid is currently produced via fungal fermentation and is used primarily as a specialty monomer. The major applications include the use as a copolymer with acrylic acid and in styrene-butadiene systems. The major technical hurdles for the development of itaconic acid as a bulk chemical include the development of very low-cost fermentation routes. The primary elements of improved fermentation include increasing the fermentation rate, improving the final titer, and potentially increasing the yield from sugar. Besides important chemical derivatives, itaconic acid can also undergo polymerization, but the properties of polyitaconic polymers need to be ascertained in order to evaluate its use as a polymer (Werpy and Petersen, 2004).

Xylitol is commercially produced from hydrogenation of xylose, the most abundant C5 sugar in hemicellulose. At the moment, there is limited commercial production of xylitol, but once a cheaper production route is established a large potential for production of ethylene glycol and 1,2-propanediol via hydrogenation is expected.

Another promising C5 bulk chemical is LA. It is produced from dehydration by means of acid treatment of C6 sugars like glucose and fructose. LA is one of the most important building blocks available from carbohydrates and has attracted interest from a number of large chemical industry firms: it has frequently been suggested as a starting material for a wide number of compounds (Bozell et al., 2000; Hayes et al., 2006; Kamm et al., 2006b; Werpy and Petersen, 2004). The technical barriers for this option include improvement of the process for LA production itself, even if the LA yield is already at 70% (Hayes et al., 2006). The family of chemical compounds available from LA is quite broad, and addresses a number of large volume chemical markets. Besides chemicals, LA shows promising efficiency in the conversion to methyltetrahydrofuran and ethyl levulinate, two fuel additives which can be blended up to 20% with gasoline and diesel (without requiring any modification of the engine).

6.2.5 C6 Bulk Chemicals

Sorbitol is produced by catalytic hydrogenation of glucose on a large industrial scale (1.1 million tons/year; Patel, 2006). Besides the food industry, it can be used for production of surfactants and polyurethanes. Sorbitol has potential for the production of isosorbide at low costs (if higher yields are achieved through optimization of process conditions and dehydration catalysts). Isosorbide is a very effective monomer for raising the glass transition temperature of polymers. The major applications are as a copolymer with PET for the use in bottle production. Hydrogenolysis of sorbitol leads to glycols, while direct polymerization forms polyesters for the resin market, whose characteristics need to be properly tested.

2,5-FDCA is formed by an oxidative dehydration of glucose, where side reactions still need to be minimized. FDCA has a large potential as a replacement for terephthalic acid, a widely used component in various polyesters, such as polyethylene terephthalate (PET) and polybutyleneterephthalate (PBT). This bulk chemical has high versatility in production of derivatives through simple chemical reactions: selective reduction leads to partially or fully hydrogenated products (with applications as new polyesters), combination with diamines produces new nylons, etc. (Werpy and Petersen, 2004). Like the other sugar-derived products, the primary technical barriers to production and use of FDCA include development of effective and selective dehydration processes for sugars.

Glucaric acid is the product of catalytic oxidation (with nitric acid, which should be replaced by oxygen) of glucose. Glucaric acid can serve as starting point for the production of a wide range of products with applicability in high volume markets, like new nylons (e.g., polyhydroxypolyamides) or new surfactants.

7 NEXT RESEARCH OUTLOOK

The success of the chemical industry in biomass conversion to chemical products is highly dependent on the development of new catalysts. Since the original molecular structure of biomass components is supposed to be preserved, the focus of catalysis research will have to shift from building functional structures out of simple building blocks to the re-functionalization of complex molecular structures (Marquardt et al., 2010). A crucial role is played by the next research achievements for basic chemical reactions like dehydration, condensation, hydrogenation, and so on, which require high selectivity to be implemented at commercial scale. Enzymatic or whole-cell biocatalysts are often high-performance alternatives resulting in high selectivity and yield (Stephanopoulos, 2007). Hybrid catalysts, combining enzymes with chemocatalysts in a complex molecular or nanoparticulate structure, constitute even more sophisticated options (Marquardt et al., 2010). In particular, the specific developments needed in the main conversion reactions are:

- Hydrogenation/reduction: this reaction is generally used to add hydrogen, e.g. to an acid functional group to form alcohols. Research developments should ensure the possibility to operate at milder conditions (pressure, temperature, etc.) giving high selectivity, by means of the improvements in catalyst performances. Catalysts should also improve their tolerance to inhibitory compounds and lifetime.
- Oxidation: this reaction oxidizes carbon and converts alcohols into acid functional groups. In future biorefineries, mineral oxidants like sulfuric acid and nitric acid should be replaced by air, molecular oxygen, dilute hydrogen peroxide, and others. Tolerance to inhibitory components of biomass processing streams should also be enhanced.
- Dehydration: this reaction removes oxygen from the substrate and it is fundamental for biomass processing. It requires improvements in the selectivity, needed to avoid side reactions. New heterogeneous catalysts (solid acid catalysts) are preferred over liquid catalysts.

- Fermentation: fermentation processes convert sugars into valuable products. In general, an improvement of microbial biocatalysts to reduce acetic acid coproducts and increase yields is needed. Lower costs to recover the products are necessary to scale-up.
- Polymerization: it is usually done through esterification to produce innovative polymers, whose applications need to be tested. Issues of selectivity and control of molecular weight and properties are still open.

The combination of new catalysts and new substrates offers innovative and largely unexplored opportunities to establish novel production pathways and novel innovative products with particular properties which must be still explored (Vennestrøm et al., 2010). The flexibility in tailoring the value chain, from feedstocks to the desired products (or vice versa), combined with the several possible uses of side streams, may lead to different options. These options must be systematically evaluated and screened to identify those with the best performances, including carbon efficiency, energy consumption, environmental impacts, and production cost. Ideally, such an evaluation should precede laboratory experiments in catalysis and production processes, in order to specifically focus research activities on the most promising alternatives.

References

Agrawal, R., Singh, N.R., 2010. Solar energy to biofuels. Ann. Rev. Chem. Biomol. Eng. 1, 343–364. doi: 10.1146/annurev-chembioeng-073009-100955.

Arenamnarta, S., Trakarnpruk, W., 2006. Ethanol conversion to ethylene using metal-mordenite catalysts. Int. J. Appl. Sci. Eng. 4, 21–32.

Bentley, R.W., Mannan, S.A., Wheeler, S.J., 2007. Assessing the date of the global oil peak: the need to use 2P reserves. Energy Pol. 35, 6364–6382.

Blanc, B., Bourrel, A., Gallezot, P., Haas, T., Taylor, P., 2000. Starch-derived polyols for polymer technologies: preparation by hydrogenolysis on metal catalysts. Green Chem. 2, 89–91.

Bos, H.L., Harmsen, P.F.H., Annevelink, E., 2010. Background information and biorefinery status, potential and sustainability. Star-COLIBRI project, p. 160.

Bozell, J.J., Moens, L., Elliott, D.C., Wang, Y., Neuenscwander, G.G., Fitzpatrick, S.W., et al., 2000. Production of levulinic acid and use as a platform chemical for derived products. Resour. Conserv. Recycl. 28, 227–239.

Cherubini, F., 2009. From oil refinery to biorefinery: production of energy and chemicals from biomass. Ital. Biol. 5, 46–61.

Cherubini, F., 2010. The biorefinery concept: using biomass instead of oil for producing energy and chemicals. Energy Convers. Manag. 51, 1412–1421.

Cherubini, F., Jungmeier, G., Wellisch, M., Willke, T., Skiadas, I., Ree, R.V., et al., 2009. Toward a common classification approach for biorefinery systems. Biofuels Bioprod. Bioref. 3, 534–546.

Clark, J., Deswarte, F., Farmer, T., 2009. The integration of green chemistry into future biorefineries. Biofuels Bioprod. Bioref. 3, 72–90. doi: 10.1002/bbb.119.

Costantino, V., Fattorusso, E., Mangoni, A., Perinu, C., Cirino, G., De Gruttola, L., et al., 2009. Tedanol: a potent anti-inflammatory ent-pimarane diterpene from the Caribbean Sponge Tedania ignis. Bioorg. Med. Chem. 17, 7542–7547.

Cui, Z.M., Liu, Q., Song, W.G., Wan, L.J., 2006. Insights into the mechanism of methanol-to-olefin conversion at zeolites with systematically selected framework structures. Angew. Chem. 118, 6662–6665. doi: 10.1002/ange.200602488.

Dale, B.E., Kim, S., 2006. Biomass refining global impact—the biobased economy of the 21st century. In: Kamm, B., et al. (Ed.), Biorefineries—Industrial Processes and Products (Status Quo and Future Directions). Wiley-VCH.

Ezeji, T., Qureshi, N., Blaschek, H.P., 2007. Production of acetone-butanol-ethanol (ABE) in a continuous flow bioreactor using degermed corn and *Clostridium beijerinckii*. Proc. Biochem. 42, 34–39.

Fernandes, T.V., Klaasse Bos, G.J., Zeeman, G., Sanders, J.P.M., van Lier, J.B., 2009. Effects of thermo-chemical pre-treatment on anaerobic biodegradability and hydrolysis of lignocellulosic biomass. Bioresour. Technol. 100, 2575–2579.

Forster, P., Ramaswamy, V., Artaxo, P., Berntsen, T., Betts, R., Fahey, D.W., et al., 2007. Changes in atmospheric constituents and in radiative forcing. In: Solomon, S. (Ed.), Climate Change 2007: The Physical Science Basis. Contribution of Working Group I to the Fourth Assessment Report of the Intergovernmental Panel on Climate Change, Cambridge, UK.

Fukuda, H., Kawaoke, Y., Fuji, T., Ogawa, T., 1987. Production of a gaseous saturated hydrocarbon mixture by Rhizopus japonicus under aerobic conditions. Agric. Biol. Chem. 51, 1529–1534.

Gassner, M., Maréchal, F., 2009. Thermo-economic process model for thermochemical production of Synthetic Natural Gas (SNG) from lignocellulosic biomass. Biomass Bioenergy 33, 1587–1604.

GBEP, 2007. A Review of the Current State of Bioenergy Development in G8 +5 Countries. GBEP Secretariat, Food and Agriculture Organization of the United Nations (FAO), Global Bioenergy Partnership, Rome.

Greene, N., 2004. Growing energy: how biofuels can help end america's oil dependence. Natural Resources Defense Council. New York, USA. p. 78.

Hatti-Kaul, R., 2010. Biorefineries—a path to sustainability? Crop Sci. 50, S-152–S-156. doi: 10.2135/cropsci2009.10.0563.

Haveren, J.v., Scott, E.L., Sanders, J., 2008. Bulk chemicals from biomass. Biofuels Bioprod. Bioref. 2, 41–57. doi: 10.1002/bbb.43.

Hayes, D.J., Fitzpatrick, S., Hayes, M.H.B., Ross, J.R.H., 2006. The biofine process—production of levulinic acid, furfural and formic acid from lignocellulosic feedstocks. In: Kamm, B. et al., (Ed.), Biorefienries—Industrial Processes and Products (Status Quo and Future Directions). Wiley-VCH.

Holladay, J.E., Bozell, J.J., White, J.F., Johnson, D., 2007. Results of Screening for Potential Candidates from Biorefinery Lignin, Top Value-Added Chemicals from Biomass. In: US Department of Energy, p. 87.

Huang, H.J., Ramaswamy, S., Tschirner, U.W., Ramarao, B.V., 2008. A review of separation technologies in current and future biorefineries. Separ. Purif. Technol. 62, 1–21.

IEO, 2009. International Energy Outlook, Energy Information Administration, Office of Integrated Analysis and Forecasting. U.S. Department of Energy, Washington, DC, p. 284.

Jem, K., van der Pol, J., de Vos, S., 2010. Microbial lactic acid, its polymer poly(lactic acid), and their industrial applications. In: Chen, G.G.Q. (Ed.), Plastics from Bacteria. Springer Berlin, Heidelberg, pp. 323–346.

Kamm, B., Kamm, M., Gruber, P.R., Kromus, S., 2006a. Biorefinery systems—an overview. In: Kamm, B. et al., (Ed.), Biorefineries—Industrial Processes and Products (Status Quo and Future Directions). Wiley-VCH.

Kamm, B., Kamm, M., Schmidt, M., Hirth, T., Schulze, M., 2006b. Lignocellulose based chemical products and product family trees. In: Kamm, B., et al. (Eds.), Biorefineries—Industrial Processes and Products (Status Quo and Future Directions). Wiley-VCH.

Kaparaju, P., Felby, C., 2010. Characterization of lignin during oxidative and hydrothermal pre-treatment processes of wheat straw and corn stover. Bioresour. Technol. 101, 3175–3181.

Lange, J.P., 2007. Lignocellulose conversion: an introduction to chemistry, process and economics. Biofuels Bioprod. Bioref. 1, 39–48.

Lee, S., 1995. Methanol-to-gasoline vs. DME-to-gasoline. II. Process comparison and analysis. Fuel science & technology international 13, pp. 1039–1057.

Li, K., Jiang, D., 1999. Methanol synthesis from syngas in the homogeneous system. J. Mol. Catal. A Chem. 147, 125–130.

Mäki-Arvela, P., Holmbom, B., Salmi, T., Murzin, D.Y., 2007. Recent progress in synthesis of fine and specialty chemicals from wood and other biomass by heterogeneous catalytic processes. Catal. Rev. Sci. Eng. 49, 197–340.

Marquardt, W., Harwardt, A., Hechinger, M., Kraemer, K., Viell, J., Voll, A., 2010. The biorenewables opportunity—toward next generation process and product systems. AIChE J. 56, 2228–2235. doi: 10.1002/aic.12380.

Patel, M., 2006. Medium and long-term opportunities and risks of the biotechnological production of bulk chemicals from renewable resources, The Brew Project. Utrecht University, Utrecht, The Netherlands, p. 474.

Peng, X.D., 2002. Kinetic Understanding of the Syngas-To-DME Reaction System and its Implications to Process and Economics. US Department of Energy, p. 53.

Sai Prasad, P.S., Bae, J.W., Kang, S.H., Lee, Y.J., Jun, K.W., 2008. Single-step synthesis of DME from syngas on Cu-ZnO-Al_2O_3/zeolite bifunctional catalysts: the superiority of ferrierite over the other zeolites. Fuel Proc. Technol. 89, 1281–1286.

Sauer, M., Porro, D., Mattanovich, D., Branduardi, P., 2008. Microbial production of organic acids: expanding the markets. Trends Biotechnol. 26, 100–108.

Stephanopoulos, G., 2007. Challenges in engineering microbes for biofuels production. Science 315, 801–804. doi: 10.1126/science.1139612.

Sun, Y., Cheng, J., 2002. Hydrolysis of lignocellulosic materials for ethanol production: a review. Bioresour. Technol 83, 1–11.

Takahara, I., Saito, M., Inaba, M., Murata, K., 2005. Dehydration of ethanol into ethylene over solid acid catalysts. Catal. Lett. 105, 249–252. doi: 10.1007/s10562-005-8698-1.

van Bekkum, H., 1998. Catalytic oxidation and carboxy-alkylation of carbohydrates, science and technology in catalysis. In: Hattori, H., Otsuka, K. (Eds.), Proceedings of the Third Conference on Advanced Catalytic Science and Technology, Tokyo, Japan, pp. 117–126.

van Vliet, O.P.R., Faaij, A.P.C., Turkenburg, W.C., 2009. Fischer-Tropsch diesel production in a well-to-wheel perspective: a carbon, energy flow and cost analysis. Energy Convers. Manag. 50, 855–876.

Vennestrøm, P.N.R., Christensen, C.H., Pedersen, S., Grunwaldt, J.D., Woodley, J.M., 2010. Next-generation catalysis for renewables: combining enzymatic with inorganic heterogeneous catalysis for bulk chemical production. ChemCatChem 2, 249–258. doi: 10.1002/cctc.200900248.

Wang, T., Wang, C., Zhang, Q., Wu, C., Ma, L., 2009. Catalytic reforming of biomass raw fuel gas to syngas for ft liquid fuels production. In: Goswami, D.Y., Zhao, Y. (Eds.), Proceedings of ISES World Congress 2007 (Vol. I-Vol. V). Springer, Berlin/Heidelberg, pp. 2366–2371.

Weissermel, K., Arpe, H.J., 2003. Industrial Organic Chemistry, fourth ed. Wiley VCH, Weinheim, Germany.

Werpy, T., Petersen, G., 2004. Top Value Added Chemicals from Biomass Volume I—Results of Screening for Potential Candidates from Sugars and Synthesis Gas. NREL. p. 76.

CHAPTER

2

Life-Cycle Assessment of Biofuels

Edgard Gnansounou[1]*, *Arnaud Dauriat*[2]

[1]Bioenergy and Energy Planning Research Group (BPE), Ecole Polytechnique Fédérale de Lausanne (EPFL), CH-1015 Lausanne, Switzerland

[2]ENERS Energy Concept, P.O. Box 56, CH-1015 Lausanne, Switzerland

*Corresponding author: Prof. Gnansounou; E-mail: edgard.gnansounou@epfl.ch

1 INTRODUCTION

During its earliest stage, the development of biofuel production in the industrialized countries was mostly driven by agricultural policies. The overproduction and low prices of crops called for diversification. Fuels derived from agricultural feedstocks were considered an ecologically valuable option for price stabilization in addition to fallowing. It was even seen as an alternative to the set-aside strategy. Two more motivations were highlighted. The perspective of oil depletion and concentration of petroleum resources in a limited number of regions which are politically instable increased the concerns about energy insecurity risks. Furthermore, due to global climate change, several industrialized countries committed to reduce their greenhouse gas emissions. The transportation sector was one of the priorities for public incentives.

Indeed, that sector is vulnerable to petroleum products which represent 98% of its final energy consumption worldwide. The high volatility of oil prices and the low competitiveness of sustainable biofuels when oil prices decrease under a certain threshold, make a claim for stable incentives in the early development stage of biofuel markets. However, the fast growth of biofuel production and the rise of the prices of agricultural commodities in 2008 fed some controversies about the sustainability of biofuels. In addition to the risks of competition with food and animal feed, the energy and greenhouse gas (GHG) balances of biofuels were debated. In response to the spread of reluctance to continue supporting publicly the utilization of biofuels, public authorities in several countries have imposed minimum sustainability targets for biofuels to be eligible for incentives (Escobar et al., 2008; Van Dam et al., 2008).

As an example, on 23 April 2009, the European Union issued a Directive on the promotion of the use of renewable energy with the requirement that each Member State shall ensure by 2020 at least 10% sustainable renewable energy in the final energy consumption of its transport sector. Article 17 of that directive notified the minimum sustainability criterion for biofuels (EC, 2009). For instance, concerning GHG emission reduction with respect to fossil fuels, increasing minimum targets were imposed: 35% in the year of entry into force of the Directive, 50% in 2017, and 60% for biofuels produced from plants that will start from 2017 onward.

The concerns about energy balances are related to both the life-cycle energy efficiency of biofuels and the saving of nonrenewable energy between biofuels and fossil fuels. The latter aspect is relevant with respect to the substitution efficiency of biofuels.

Monitoring the application of minimum targets on GHG emission reduction to biofuels, as well as estimating their substitution efficiency with respect to fossil fuels, is subject to significant uncertainty and inaccuracy associated with the methodology applied. Assessments of the environmental impact of biofuels (ADEME-DIREM-PWC, 2002; ADEME, 2010; Beer and Grant, 2007; CONCAWE-EUCAR-JRC, 2008; Elsayed et al., 2003; EMPA, 2007a; GM-LBST, 2002; Gnansounou and Dauriat, 2004, 2005; Kim and Dale, 2008; Macedo, 2004; Malça and Freire, 2006; Shapouri et al., 2002; VIEWLS, 2005; Wang, 2005) often significantly differ in methodological choices and consequently in their results. Table 1 shows an overview of the methodological choices in these studies.

For example, while some studies (Elsayed et al., 2003) are limited to a Well-to-Tank (WtT) approach (thereby excluding the utilization phase), other studies employ a Well-to-Wheel (WtW) approach. When included in the system, the utilization phase is taken into account either in a simplified way (usually by merely considering the difference in the LHV of fuels) or with more details (by considering the actual performance of fuels according to a specific engine technology and/or fuel blend). As far as the functional unit is concerned, the distance traveled (in km) is the unit of choice in most studies (in agreement with the principles of the WtW approach). Allocation by system expansion is the most widely used method, although in some studies a combination of methods is used. For instance, system expansion is combined with allocation by mass in ADEME-DIREM-PWC (2002). Economic allocation is the second most common approach. Fuel blends considered vary from one study to the other (usually between 5% vol. and 15% vol.), depending on the most frequent use of fuel-ethanol in the region of study. All the reviewed studies however also consider ethanol as a fuel component on its own, even though the way this is done does not depend on the actual fuel blend but rather on the difference in the LHV of fuels. Finally, land-use change is included with details in only a few studies (EMPA, 2007b; Elsayed et al., 2003; IFEU, 2004), based on IPCC (2003a) guidelines.

In the particular case of GHG balance, the magnitude of the discrepancy among the results is tremendously high. Farrel et al. (2006), based on a review of corn-based ethanol studies in the USA, attributed the main differences to the way coproducts are accounted for, the value of some input parameters, and some omission/inclusion of ambiguous inputs. Reijnders and Huijbregts (2003) focused on forest-based biofuels, analyzing the effect of the considered time frame on the emission factors, the choice of previous land use, the allocation of carbon sequestration and emissions during forest growth, and the fate of sequestered carbon after fuel wood harvesting. Börjesson (2009) focused on methodological choices and the influence of local conditions in wheat-based ethanol production in Sweden. He addressed the problem of coproducts allocation choices, the choice of the fuel used, and biogenic GHG emissions from cultivated soils.

In fact, quantitative investigations of the differences in the results from one study to the other are not straightforward due to the lack of information concerning the inventory data,

TABLE 1 Comparison of Methodological Choices in Reviewed Studies

Criteria	Subcriteria	Choice	ADEME-DIREM-PWC (2002)	ADEME (2010)	CONCAWE-EUCAR-JRC (2008)	GM-LBST (2002)	Macedo (2004)	Gnansounou and Dauriat (2004)	EMPA (2007b)	IFEU (2004)	CSIRO (2001)	VIEWLS (2005)	ANL-GM (2001)	Elsayed et al. (2003)
System definition and boundaries	Approach	Well-to-Tank (WtT)	x											X
		Well-to-Wheel (WtW)		x	x	x	x	x	x	x	x	x	x	
	Land use	Detailed		x					x	x				x
		Not included	x				x	x			x	x	x	
		Simplified			x									
	Blends	5			x		x	x	x	x				
		10		x			x				x			
		15							x		x		x	
		100	x		x	x	x	x	x	x	x	x	x	x
		Other		x				x						
	Use phase	Not included	x											x
		Simplified					x			x		x		
		Detailed		x	x	x		x	x		x		x	
Functional unit		1					x							
		MJ	x	x										x
		km (mile)		x	x	x		x	x	x	x	x	x	
		t of feedstock					x							
		ha								x				
Allocation methods		System expansion	x	x	x	x	x			x	x	x	x	x
		Mass	x											
		Energy		x							x			
		Economic						x	x		x		x	x

the assumptions made to complement unavailable data and modeling choices about system definition and boundaries, functional units, reference systems, and allocation methods. In the research presented in this chapter, an assessment platform was developed based on an extensive review of literature. Combinations of assumptions and modeling choices were defined to investigate the sensitivity of the results to several factors. The focus is mainly put on choices regarding the allocation method, the previous land use, the fuel blends, and the vehicle/fuel performance. This chapter aims to contribute to current discussions on how methodological choices and local conditions influence LCA results, addressing some important points often limitedly treated in the literature. The chapter considers wheat-based bioethanol production in Switzerland as a case study, with the aim of quantifying the variation in GHG emissions and nonrenewable energy use depending on methodological choices. The chapter is an updated version of a previous paper by Gnansounou et al. (2009).

2 THE CONCEPT OF LCA AND ITS APPLICATION TO BIOFUELS

Life-cycle analysis (LCA) or assessment is an internationally renowned methodology for evaluating the global environmental performance of a product along its partial or whole life cycle, considering the impacts generated from "cradle to grave." At its early age, the methodology was mainly dedicated to industrial products. Although the ISO 14040-series (ISO, 2006a,b) provides the standard for LCA, it was applied in a variety of ways and thus often leads to diverging results, especially in the case of biofuels. LCA of biofuels is often limited to energy and/or GHG balance. Several LCA studies (ADEME, 2010; ADEME-DIREM-PWC, 2002; Beer and Grant, 2007; CONCAWE-EUCAR-JRC, 2008; Elsayed et al., 2003; EMPA, 2007a; GM-LBST, 2002; Gnansounou and Dauriat, 2004; Macedo, 2004; Malça and Freire, 2006; Shapouri et al., 2002; VIEWLS, 2005; Wang, 2005) have been completed with various frameworks, scopes, accuracy, transparency and consistency levels, making it difficult to compare the results on a rational basis, even when addressing the same biofuel pathway (Panichelli et al., 2008).

The main assumptions found in the literature when estimating the reduction of GHG emissions of biofuels compared to fossil fuels are described in detail in a technical report by the Laboratory of Energy Systems (LASEN) of EPFL (Gnansounou et al., 2008a). Before introducing the general framework of the analyses made in this chapter, a short review of the most significant methodological issues of LCA is proposed, with a focus on the cases of biofuels.

2.1 System Definition and Boundaries

Depending on the goal and scope of the LCA, choices regarding system definition and boundaries are more or less accurate. The goal may be process design-, operation-, or policy-oriented. While the definition of the system is more detailed in case of design or operation improvement, the flowchart of biofuel pathways is simplified for policy-related LCA. In that latter case, the system boundaries are adapted to the purpose. For instance, if the intent is the comparison of various pathways of the same biofuel (e.g., bioethanol), a WtT LCA is appropriate because the pathways do not affect the performance of the fuel combustion in the vehicle's engine. The situation changes dramatically if the LCA intends to compare selected biofuels with their fossil substitutes, for example, bioethanol blends versus gasoline or more generally when different kinds of fuels and blends are compared. In these cases, the utilization stage plays a crucial role as the energy need in the vehicle tank for a given service (e.g., 100 veh.km) depends on the combustion performances that in turn vary from one blend to the other. Ignoring this important factor even for simplicity will lead to implicit assumptions on the combustion performances and therefore may induce inconsistency.

However, several authors used WtT boundaries while comparing GHG emissions of biofuels and fossil fuels (e.g., ADEME-DIREM-PWC, 2002; Elsayed et al., 2003). In other studies, the WtT was only a step for a complete WtWs assessment (e.g., Beer and Grant, 2007; CONCAWE-EUCAR-JRC, 2008; EMPA, 2007a; GM-LBST, 2002; Gnansounou and Dauriat, 2004; VIEWLS, 2005). Other aspects concerning the system definition and boundaries are the inclusion or not of land-use change and coproducts as part of the system. These two issues are addressed later on.

2.2 Functional Unit

When comparing biofuels with fossil fuels, it is of utmost importance to consider the same relevant service from the various systems. In case of motor fuels, as long as mobility is concerned, this service must be related to mechanical energy, in other words, to the distance travelled. Most studies however choose 1 MJ_{th} of fuel (in the tank) as the functional unit, regardless of the type of fuel (ADEME-DIREM-PWC, 2002; Elsayed et al., 2003; EMPA, 2007a; GM-LBST, 2002; Malça and Freire, 2006; Shapouri et al., 2002). This choice is not relevant as the mechanical efficiency can vary from one fuel or engine to the other.

For example, several tests (AEAT, 2002; EMPA, 2007b; IDIADA, 2003) have shown that the consumption of E5 (fuel blend consisting of 5% vol. bioethanol and 95% vol. gasoline) in liters is slightly less than the consumption of gasoline for the same service, that is, 100 km. In this specific case, it means that 1 MJ_{th} gasoline should be compared with less than 1 MJ_{th} E5. For simplicity, if one considers that the consumption of gasoline (in liters per 100 km) equals that of E5, then, 1 liter of fuel (gasoline or ethanol) should be a good functional unit for comparing ethanol with gasoline when the blend is E5. Using (even for simplicity) 1 MJ_{th} of fuel as the functional unit, when comparing gasoline to ethanol, means that one makes the implicit choice that 1 liter of gasoline should be compared with 1.5 liter of ethanol (given as the ratio of the LHV of gasoline, i.e., 31.9 MJ_{th}/l, to the LHV of bioethanol, i.e., 21.2 MJ_{th}/l). This choice between liter and MJ_{th} would be relevant however if ethanol were used as pure fuel (or at least as the main component of the fuel blend, e.g., E85) or in the case of heat applications.

ADEME (2010) has considered the effect of biofuels incorporation rate on the vehicle/fuel performance (in terms of liters per 100 km or MJ_{fuel}/km).

2.3 Reference System

In practice, two LCA methods can be distinguished. The attributional LCA is concerned with evaluation of a given product without any consideration of the interactions with a more global system such as the socioeconomic system or the agricultural system. Furthermore, this type of LCA is not in a comparative framework. For example, its purpose could be to improve a given pathway. In that case, the reference system is a baseline of the pathway. However, the public debate on biofuels is rather related to their "renewability" and carbon neutrality. For that reason, a more open methodology is required closed to the consequential LCA. Finally, it is proposed to include these two methodologies into a more general one based on system analysis. In the proposed methodology, the performance indicators are defined by comparing the studies system with a reference or baseline system. In most studies, the reference system is implicitly limited to a fossil fuel pathway (e.g., gasoline or diesel). In various cases, however, this picture is not complete: for example, when coproducts from the biofuel pathway replace an existing product whose performances are significantly different. In this situation, a reference substituted product should be defined. The same applies to the case when the production of feedstock for biofuels uses land that was previously storing carbon such as forests or grasslands. In this case, a "previous land-use" baseline should be included in order to determine the carbon emissions from this change of land use. When the same feedstock or the land was previously used for another purpose, an "alternative biomass use" or "alternative land use," respectively, may be included in the baseline in order to estimate the effects due to indirect land-use changes. The choice to include or not an "alternative use" depends on the assumption made concerning the substitution versus the addition of products. For example, if biofuels substitute overproduced food crops, there is no requirement for additional resources for replacing the substituted products. Conversely, in case of underproduction due to biofuels, additional resources of land or imported products will be required.

In the past, the land-use baseline was included (in a simplified way) in a very limited number of the LCA studies (e.g., ADEME, 2010; CONCAWE-EUCAR-JRC, 2008; GM-LBST, 2002; VIEWLS, 2005). In these three cases, the land used for growing energy crops was considered to be initially set aside (incl. extensive green crop cover with no farming inputs), and consequently no alternative use of land or biomass was assumed.

2.4 Land-Use Change

In their largely discussed work, Fargione et al. (2008) and Searchinger et al. (2008) showed the importance of including land-use change emissions in the GHG balance of biofuels. Righelato and Spracklen (2007) have even questioned biofuels production as a strategy to mitigate global warming. Direct land-use change concerns for example the case where production of energy crops for biofuels production leads to the conversion of land actually storing carbon (e.g., grassland, native ecosystems) to cultivated land for biofuels production. Missing to consider the previous storage of carbon will overestimate the reduction of GHG emissions of the biofuel chain. On the contrary, when the feedstock is produced on degraded soil, it can contribute to improve the soil carbon balance (Panichelli and Gnansounou, 2008). Consequently, the choice of the previous state of the land-use system can significantly affect the GHG balance of the biofuel. Direct land-use change is taken into account in a few recent studies (e.g., ADEME, 2010; CONCAWE-EUCAR-JRC, 2008; EMPA, 2007a). In the three studies, the recommendations of IPCC (2003a) are used for this purpose.

Taking into consideration indirect land-use change (land-use changes due to displaced activities or biomass use) is more complex as the indirect conversion of land is a global and dynamic issue that is difficult to relate accurately to biofuels production, more research works are needed for improving the methodologies on this aspect.

2.5 Allocation Methods

A high sensitivity to the allocation method has been reported for LCA results when evaluating carbon intensity and fossil energy consumption for bioethanol pathways (Beer and Grant, 2007; Kim and Dale, 2002; Malça and Freire, 2006). Allocation refers to the distribution of environmental burdens between coproducts in the LCA of a multifunctional system. The issue of allocation is one of the weaknesses of biofuels LCA. The ISO 14040-series (ISO, 2006a,b) recommends avoiding allocation whenever possible either through division of the whole process into subprocesses related to coproducts or by expanding the system limits to include the additional functions related to them (often referred to as system expansion or substitution and treated in the literature as an allocation method of its own). A complete subdivision is not possible for joint production processes due to the dependence between coproducts' flows. In fact, subdivision is only feasible when unit subprocesses are physically separate in space or time (combined production), so it is only on exceptional occasions that the allocation problem can be completely eliminated. Ekvall and Finnveden (2001), after screening a large sample of LCA case studies, did not find a case study where this was the case. In the present study, subdivision is applied to the stages downstream of the distillation process. According to Kim and Dale (2002), system expansion is based on the assumption that function-equivalent production systems have equal environmental impacts; that is rarely the case. Furthermore, this approach requires highly accurate data and can be subject to a high degree of uncertainty and/or inaccuracy; its implementation is difficult as the result depends significantly on the substitute that is chosen in the reference system. Finally, estimating the impact of this substitute may lead to another allocation problem. If "avoiding" allocation is not possible, then the ISO series recommends using a method that reflects the physical relationship between the environmental burdens and the coproducts. In that sense, allocation can be carried out by mass (wet or dry), carbon content, energy content or volume. Allocation on a weight basis relates products and coproducts using a physical property that is available and easy to interpret. But some researchers claim that it cannot be a good measure of energy functions (Malça and Freire, 2006; Shapouri et al., 2002). Energy allocation is mostly used in US biofuels studies by the US Department of Agriculture (Shapouri et al., 2002) and the Argonne National Laboratory (Wang, 2005). It is also the methodology chosen in the European Union (EC, 2009) and consequently applied by ADEME (2010). However, an objection can be made against this approach in the case where the coproducts are not meant for energy purposes. When physical properties are not appropriate, ISO recommends the use of other basis for allocation such as the economic value of the products. The rationale for economic allocation is that environmental burdens of a multifunctional process could be allocated according to the share on sales value, because demand is the main driving force of the production system. Price variation, subsidies, and market interferences could however imply difficulties in its implementation (Bergsma et al., 2006; Shapouri et al., 2002; Wang, 2005). In an LCA carried out in order to determine the net energy value (NEV) of bioethanol production, Shapouri et al. (2002) do not recommend this method because prices are determined for a number of market factors that are not related to the energy content. Guinée et al (2004) state that in spite of considerable price fluctuations, the shares on the total sales value remain quite constant, particularly in the longer term. According to some researchers (Weidema, 2003), attributional LCA requires market allocation, while consequential applications require system expansion. Allocation by mass is applied in ADEME-DIREM-PWC

(2002). System expansion (or substitution) is used in CONCAWE-EUCAR-JRC (2008), GM-LBST (2002), and VIEWLS (2005). The latter is also tested in ADEME (2010) by means of a sensitivity analysis. Kim and Dale (2002) investigated an expanded system including ethanol production from dry and wet milling, agricultural corn production, soybean oil and soybean meal from soybean milling as well as the urea production system for animal feed. Economic allocation was used in EMPA (2007a) and Gnansounou and Dauriat (2004). Elsayed et al. (2003) used alternatively economic allocation and substitution, depending on the biofuel pathway considered and the availability of data.

2.6 Life-Cycle Impact Assessment

The life-cycle inventory (LCI) is used to estimate the direct and indirect inputs and releases at each step of a biofuel pathway. The results are the use of resources (e.g., energy consumption) and the environmental emissions (e.g., carbon dioxide, sulfur oxides, nitrogen oxides). Through characterization factors, the outcomes of LCI are utilized to assess impact categories such as climate change, stratospheric ozone depletion, photo-oxidant formation, acidification, eutrophication, ecotoxicity, human toxicity, depletion of biotic resources, and depletion of abiotic resources. The impact categories describe environmental mechanisms which convert the outcomes of the LCI into environmental damages. Indicators can be derived from these mechanisms at intermediate levels (midpoints) or damages levels, (endpoints) after normalization and sometimes weighting approaches. The use of endpoint methods to derive a global indicator of impact is controversial. The proponents claim for simplicity in communication of the results of the LCA to nonscientific public. The opponents emphasize the subjective nature of the weighting process and on the reductionism related to that approach.

2.7 Variability of Results in LCA Studies

As mentioned earlier, LCA studies of biofuels (including project report, research papers, policy documents, etc.) are numerous and have become even more popular with the recent implementation of sustainability criteria in biofuels policy worldwide (especially in the European Union and the United States) and the increased research activities in advanced biofuel pathways (e.g., biofuels from microalgae). The results of biofuels LCA studies may vary significantly from one author to another for a variety of reasons: these include methodological choices as described in Table 1, but also the type of biofuel (including bioethanol, biodiesel, e.g., methyl esters of vegetable oils, so-called renewable diesel, e.g., from Fischer-Tropsch synthesis, biobutanol, etc.), the type of technology (including first-, second-, and third-generation technologies, biochemical or biothermal, dedicated biofuel production, or multioutput biorefinery concept, etc.) and their corresponding level of maturity, the type of feedstock considered for a given biofuel, and the conditions under which a given feedstock is produced (dedicated production, agricultural, forestry or industrial residues, wastes, etc.). In relation to the various aspects listed, the inventory data to characterize the production of biofuel may differ significantly depending on the level of detail (e.g., proven and long-lived technology, technology based on pilot/demonstration plant, laboratory-level experiments, physicochemical computer-based modeling, etc.). In addition, biofuels LCA studies will also vary with respect to the inventory database.

Many recent LCA studies have evaluated the environmental impact of biofuel production from microalgae (Batan et al., 2010; Campbell et al., 2011; Clarens et al., 2010; Collet et al., 2011; to cite only a few of the latest research papers). The large variety of research and development areas in this field (including biomass production, harvesting, separation, processing, transformation, and the possible products and applications) is detailed in Brennan and Owende (2010), Wijffels and Barbosa (2010), and Singh and Gu (2010). Wijffels and Barbosa (2010) conclude by emphasizing the need for life-cycle assessment of algal biofuel production processes, while Singh and Gu (2010) report: "An adequate LCA study is still not available which may help to present a clear picture of the situation. The reason is nonavailability of commercial plant data." A similar situation can be found concerning the production of cellulosic bioethanol (Singh et al., 2010; Spatari et al., 2010) and more generally with the concept of biorefinery and the coproduction of biofuels, biochemicals and/or biomaterials (thereby stressing even more the significance of the allocation method).

All the aspects listed make it very complicated to compare the results of various biofuels LCA studies, even for a given feedstock-technology-biofuel system, according to the hypotheses and methodological choices. This is illustrated in a detailed case study in the following sections of this chapter.

3 METHODOLOGY AND ASSUMPTIONS

The system studied in this chapter as for illustration is concerned with the production, distribution and use (WtW approach) of anhydrous fuel-bioethanol (99.7% wt.) as a transport fuel in Switzerland. Bioethanol is supposed to be produced from wheat also grown in Switzerland. The functional unit is 1 km.

The LCI is established by means of a spreadsheet model developed by ENERS Energy Concept and the Bioenergy and Energy Planning Research Group (BPE) of the Swiss Federal Institute of Technology, Lausanne (EPFL). The model is based on Microsoft® Excel and integrates the ecoinvent v2 database (ecoinvent, 2007; Frischknecht, 2004; Frischknecht et al., 2004), a reference in the field of LCA, developed by the Swiss Centre for Life-Cycle Inventories. The model is based on industrial data from the EU provided by the Swiss Alcohol Board, and was actually used in the implementation of various bioenergy datasets in the ecoinvent v2.0 database (Jungbluth et al., 2007; Nemecek and Kägi, 2007). The model reports on the consumption of resources (energy, chemicals, land, water, infrastructures) and emissions all along the production chain. The model was implemented in order to offer an extensive set of options regarding methodological choices.

The LCA carried out in this chapter complies with the ISO standard on LCA (ISO, 2006a,b). GHG emissions were assessed using the IPCC Global Warming Potentials (GWPs) with a time frame of 100 years. This method is most commonly used in the literature when dealing with global warming. The main greenhouse gases taken into account are carbon dioxide (CO_2), methane (CH_4), and nitrous oxide (N_2O), with respective global warming potentials (GWP) of 1, 23, and 296. Because biogenic CO_2 captured by photosynthesis during plant growth is eventually almost totally emitted as CO_2 during bioethanol production (fermentation) and utilization (combustion), only fossil CO_2 emissions are taken into account. Direct field emissions of N_2O are based on a model by Agroscope (Nemecek and Kägi, 2007), also included in the ecoinvent database.

3.1 Well-to-Tank System

The production of bioethanol from wheat grains gives rise to coproducts both at the agricultural stage (i.e., wheat straw) and at the industrial stage (i.e., wheat DDGS). Both coproducts may be used as animal feed or as fuel (Kaparaju et al., 2009). According to the most common practice in the European context, the reference use of the coproducts is considered to be animal feed. It is here considered that the land where the animal feed (baseline) was initially produced (now displaced by straw and DDGS) is turned into set-aside land. Similarly, it is considered in this reference framework that wheat is grown on land that was initially set aside (incl. green cover with no farming inputs). The corresponding systems are shown in Figure 1. When allocation is applied, the "from" (reference) and "to" (studied) systems are illustrated as in Figure 2 (showing the effect of allocation). When substitution is applied, the "from" and "to" systems are illustrated as in Figure 3, where the substituted products and the associated land use are included in the system studied with a negative impact in order to keep the reference system identical in all cases (i.e., limited to the production and use of gasoline). The effect of different allocation/substitution choices is investigated in the case study section.

The WtT GHG net GHG emissions of unleaded gasoline in the Swiss context are taken from ecoinvent and are equal to 0.018 kg $CO_{2\ eq,}/MJ_{th}$ (i.e., 0.782 kg $CO_{2\ eq.}$/kg or 0.586 kg $CO_{2\ eq.}$/l) at the service station.

3.2 Tank-to-Wheel System (TtW)

The fuels blends considered in the present article include E5, E10, and E85. The LHV, density, and biogenic carbon content of ethanol are taken as 26.8 MJ/kg, 0.790 kg/l, and

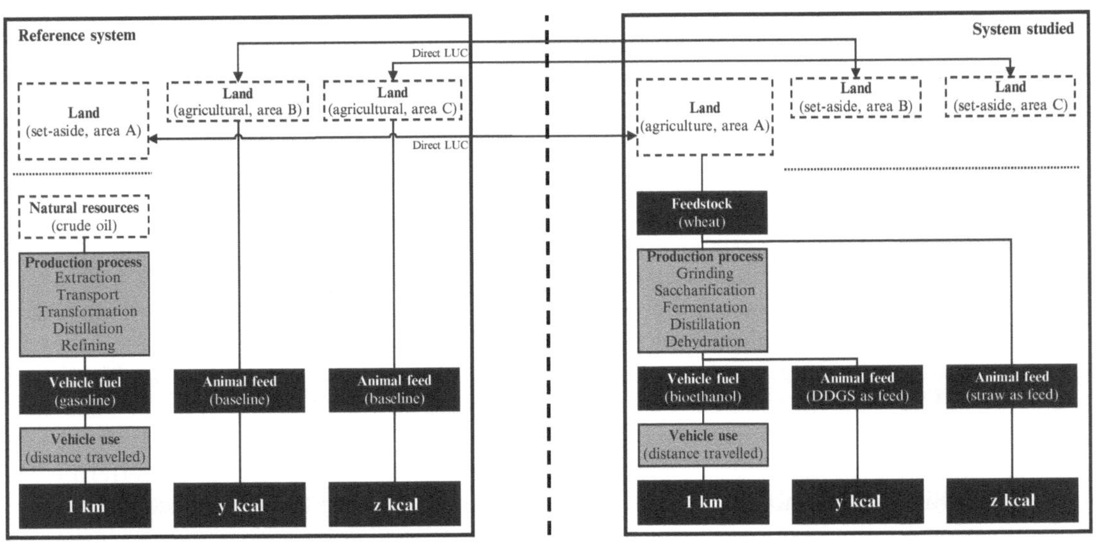

FIGURE 1 System definition and boundaries (from reference system to system studied).

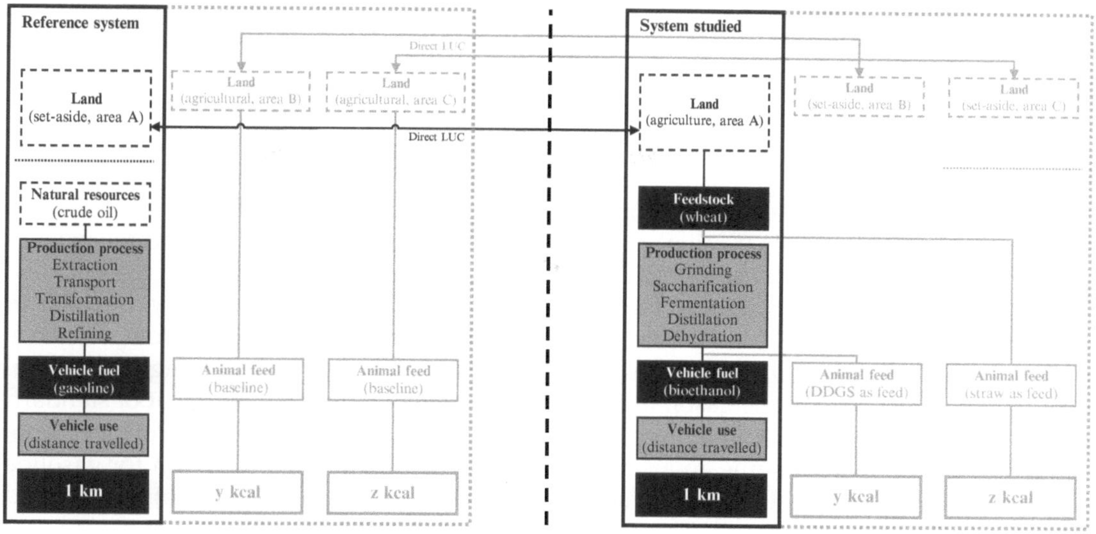

FIGURE 2 System definition and boundaries (in case of allocation by energy content, economic value, carbon content, or dry mass).

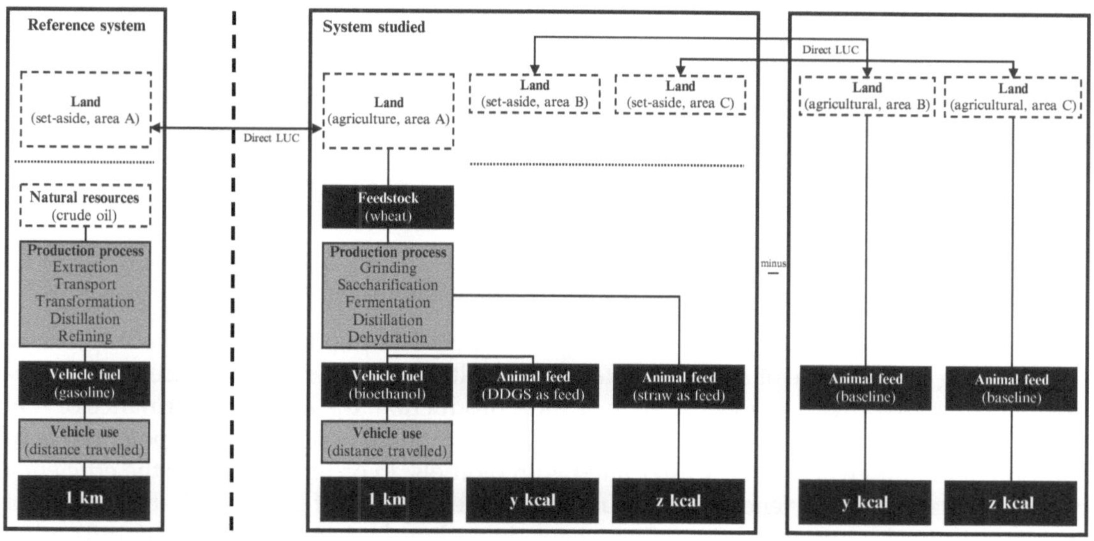

FIGURE 3 System definition and boundaries (in case of allocation by substitution, case of S-1, that is, DDGS and straw as animal feed).

0.520 kg C/kg. The LHV, density, and fossil carbon content of ethanol are taken as 42.5 MJ/kg, 0.750 kg/l, and 0.865 kg C/kg. The characteristics of the fuels blends are calculated according to the respective volume shares of ethanol and gasoline. The effect of considering different fuel blends and/or other hypotheses regarding fuel performance is investigated in the case study section.

As discussed previously, the performance of bioethanol as a vehicle fuel strongly depends on its rate of incorporation into gasoline. Indeed, although bioethanol shows a significantly lower LHV compared with gasoline (which leads to expect an increase in vehicle fuel consumption when ethanol is added to gasoline), many vehicle tests in the European context (AEAT, 2002; EMPA, 2002, 2007b; IDIADA, 2003) have demonstrated the improved efficiency (expressed in MJ_{th}/km) of gasoline-ethanol blends with respect to standard gasoline (Table 2). In 100% of the tests reported in these studies, gasoline-ethanol blends indeed show an improved efficiency (in MJ_{th}/km) compared with standard gasoline. On average, energy consumption per km is reduced by 2.7%, 7.5%, and 2.5% with E5, E10, and E85, respectively.

All these data, however, refer to fuel blends rather than ethanol specifically. In order to evaluate the WtW net GHG emissions of ethanol, it is necessary to define the fuel efficiency of the ethanol component in fuel blends. This is done in this chapter by assuming that the fuel efficiency (in km/MJ_{th}) of the gasoline component in fuel blends is equal to that of standard gasoline on its own, and that the difference is entirely explained by the presence of bioethanol in the fuel blend. If we assume an average fuel consumption of 2.564 MJ_{th}/km for gasoline (as reported in ecoinvent), the specific fuel consumption of ethanol (in MJ_{th}/km) is calculated according to the data in Table 2. The results are reported in Table 3.

3.3 Well-to-Wheel System

The WtW net GHG emissions of ethanol are calculated as the product of the WtT net GHG emissions and the specific fuel consumption of ethanol in the fuel blend (as reported in Table 3). The WtW net GHG emissions of ethanol (expressed in kg $CO_{2\ eq.}$/km) are then compared to those of unleaded gasoline.

3.4 Net Energy Use and Energy Substitution Efficiency

The net energy use of a fuel most often refers to the consumption of nonrenewable primary energy along the life cycle of a biofuel or fossil fuel. Although the energy balance is often limited to the comparative energy efficiency of fuels production, the actual performance of fuel blends must be taken into account in order to obtain a global picture of the potential substitution of nonrenewable energy associated with biofuels. In order to measure the efficiency of nonrenewable primary energy substitution over the life cycle of bioethanol, the concept of energy substitution efficiency is defined later, including both production and utilization of the biofuel.

According to the data in Table 3, the most efficient use of fuel-bioethanol is in the form of E10 (1.174 MJ_{th}/km compared to 1.413 MJ_{th}/km for bioethanol as E5 and 2.485 MJ_{th}/km for bioethanol as E85). The energy substitution efficiency is here defined as the ratio of the savings of nonrenewable primary energy of a given bioethanol system (incl. production and use) with respect to conventional gasoline to the savings of nonrenewable primary energy of an ideal bioethanol system (i.e., bioethanol with a zero nonrenewable primary energy consumption and utilization as E10).

TABLE 2 Effects of Ethanol on Vehicle Fuel Performance

				Variation of fuel consumption w.r.t gasoline[a]		
Testing body	**Fuel**	**Vehicle**	**Cycle**	**(l/km)**	**(kg/km)**	**(MJ_{th}/km)**
EMPA (2002)	E5	FORD Focus	NEFZ	−1.0%	−0.6%	−2.6%
			ECE	−1.9%	−1.6%	−3.5%
			EUDC	−0.7%	−0.4%	−2.4%
IDIADA (2003)	E5	RENAULT Megane	Stage III	−0.6%	−0.3%	−2.2%
AEAT (2002)	E10	TOYOTA Yaris	Cold ECE	−3.3%	−2.9%	−6.6%
			Cold EUDC	−1.6%	−1.2%	−4.9%
			WSL average	−1.1%	−0.6%	−4.4%
	E10	OPEL Omega	Cold ECE	−17.3%	−17.0%	−20.1%
			Cold EUDC	−14.5%	−14.1%	−17.3%
			WSL average	−6.4%	−6.0%	−9.5%
	E10	FIAT Punto	Cold ECE	−5.6%	−5.2%	−8.8%
			Cold EUDC	−12.5%	−12.2%	−15.5%
			WSL average	−3.0%	−2.5%	−6.2%
	E10	VW Golf	Cold ECE	−8.5%	−8.1%	−11.6%
			Cold EUDC	−3.8%	−3.4%	−7.1%
			WSL average	−4.3%	−3.9%	−7.6%
	E10	ROVER 416	Cold ECE	+1.1%	+1.6%	−2.3%
			Cold EUDC	−0.8%	−0.3%	−4.1%
			WSL average	−2.8%	−2.3%	−6.0%
EMPA (2007b)	E85	FORD Focus FFV	NEFZ	+35.0%	+41.8%	−2.5%
			ECE	+33.5%	+40.2%	−3.5%
			EUDC	+36.4%	+43.3%	−1.4%
Average	E5	–	–			−2.7%
Average[b]	E10	–	–			−7.5%
Average	E85	–	–			−2.5%

[a] *The variation of fuel consumption with respect to gasoline (in l/km, kg/km and MJ_{th}/km) is calculated according to the results presented in the various studies. The calculations are based on the actual characteristics and properties of the fuels as quoted in these studies, which may differ slightly from the data presented in Table 2. The average values at the bottom of the table are based on the variation in MJ_{th}/km.*

[b] *The average variation of fuel consumption for E10 is based on the complete set of results of the AEAT (2002) study. Only a part of these results are quoted in the table, which explains why the average calculated from the data given above may differ from the actual average of −7.5%.*

TABLE 3 Specific Fuel Efficiency/Consumption of Gasoline and Ethanol Components in Fuel Blends

	Fuel blend		Gasoline component			Ethanol component		
Fuel	(km/MJ_{th})	(MJ_{th}/km)	(% MJ/MJ)	(km/MJ_{th})	(MJ_{th}/km)	(% MJ/MJ)	(km/MJ_{th})	(MJ_{th}/km)
Gasoline	0.390	2.564	100.0%	0.390	2.564	0.0%	–	–
E5	0.401	2.496	96.6%	0.390	2.564	3.4%	0.708	*1.413*
E10	0.422	2.371	93.1%	0.390	2.564	6.9%	0.852	*1.174*
E85	0.400	2.501	21.0%	0.390	2.564	79.0%	0.402	*2.485*

4 CASE STUDY: BIOETHANOL FROM WHEAT

4.1 Definition of the System Under Study

The comparative analysis presented in this chapter is based on a reference case defined as follows: bioethanol is produced from wheat in a facility with a capacity of 40,000 t/yr (i.e., 134,000 t/yr of wheat). The corresponding agricultural area is of the order of 20,900 ha located in the region surrounding the ethanol plant. Beside fuel ethanol, the plant also produces about 48,600 t/yr of DDGS. Fuel ethanol is distributed over an average distance of 250 km (100 km by lorry and 150 km by train). The cultivation of wheat is carried out under the usual practice in Switzerland, with an average yield of 6.425 t/ha of grains (fresh matter at 15% wt. moisture) and 3.915 t/ha of straw (fresh matter at 15% wt. moisture). The grains are sent to the plant over a distance of 50 km (10 km by tractor and 40 km by lorry).

A simplified flow diagram of bioethanol production is presented in Figure 4. The stages common to both bioethanol and DDGS are shown in block A and include grinding,

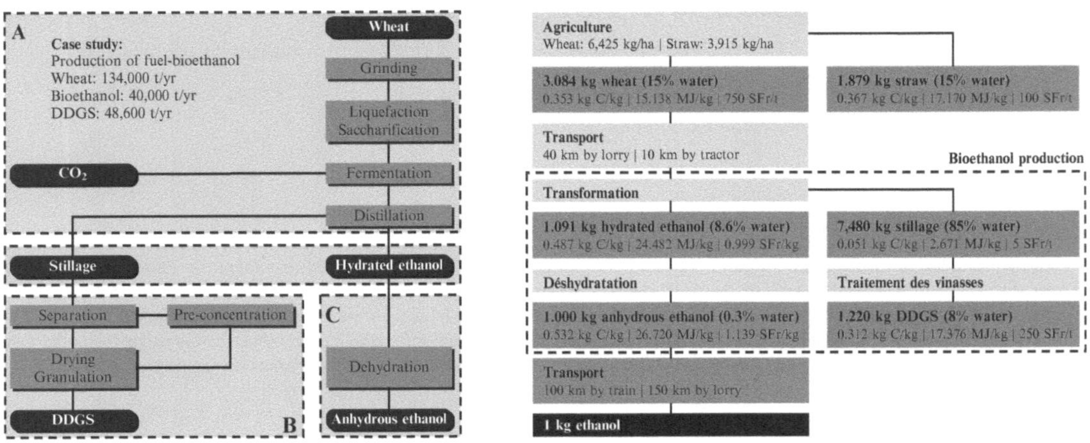

FIGURE 4 Simplified diagram of bioethanol production from wheat.

liquefaction, saccharification, fermentation, and distillation. Block B represents the stillage treatment, specific to DDGS. Block C is the dehydration stage, specific to bioethanol. When applicable, the allocation of impacts between bioethanol and DDGS is performed at the point of separation, after the distillation stage in this case (i.e., between hydrated ethanol and stillage). Allocation therefore applies to block A only (incl. production and delivery of wheat). The impacts associated to blocks B and C are fully allocated to DDGS and bioethanol, respectively.

Regarding the utilization phase, various blends are considered in this chapter. Each of these fuel blends has a specific performance (expressed in terms of fuel consumption per 100 km). The vehicle considered is a standard EURO 3 light-duty vehicle.

4.2 Definition of Scenarios (Sensitivity Analysis)

The effect of various methodological choices on the GHG and energy balance of bioethanol is evaluated, with an emphasis on allocation methods, land-use change, fuel blends, and vehicle/fuel performance. Each of these aspects is now explained in more detail.

4.2.1 Allocation Methods

Various allocation methods were investigated, including allocation by energy content, economic value, carbon content, dry mass, and substitution. The properties and prices of the coproducts are given in Table 4. Four various substitution scenarios are considered, namely: S-1) both straw and DDGS as animal feed, S-2) straw as animal feed and DDGS as fuel, S-3) straw as fuel and DDGS as animal feed, and S-4) both straw and DDGS as fuel. Each of these scenarios gives rise to a specific system definition (incl. the reference system). The "from" (reference) and "to" (studied) systems in the case of allocation (regardless of the method) and substitution (S-1) are illustrated in Figures 2 and 3, respectively.

4.2.2 Land-Use Change

The effect of choices regarding land-use change on the WtW net GHG emissions of bioethanol was investigated. The various land-use changes considered (and the corresponding annual soil carbon stock change) are summarized in Table 4. Land-use types are those defined by the IPCC (2003a). The annual soil carbon stock changes were derived from the IPCC tool, developed at the Colorado State University (IPCC, 2003b).

LUC-6, for instance, is concerned with the change from the so-called native ecosystem to long-term cultivated land. In this article, native ecosystems are considered to be forested areas, in accordance with the Swiss natural environment. In addition to the change of soil carbon stock, the removal of above ground biomass is also taken into account in this land-use change. In this land transformation process, it is considered (arbitrarily) that the wood is not valorized but simply cut down and burned on site. Therefore, the entire impact is allocated to the subsequent crop, in this case wheat for ethanol production. The corresponding loss of carbon stock (spread over 20 years) is 93.2 tons of carbon per hectare (i.e., 17.1 t $CO_{2\ eq.}$/ha.yr). It is important to bear in mind that this situation corresponds to the worst possible case and was selected on purpose in order to assess the magnitude of the land-use change effect in case of deforestation.

TABLE 4 Sets of Options Investigated in this chapter

			Agricultural stage		Industrial stage	
Ref.	**Method**	**Key**	**Wheat grains**	**Wheat straw**	**Bioethanol**	**Wheat DDGS**
(a) Allocation/substitution methods						
A-1	Allocation	Energy content	15.1 MJ_{th}/kg	17.2 MJ_{th}/kg	26.8 MJ_{th}/kg	16.4 MJ_{th}/kg
A-2	Allocation	Economic value	750 SFr/t	100 SFr/t	1,139 SFr/t	250 SFr/t
A-3	Allocation	Carbon content	0.353 kg C/kg	0.367 kg C/kg	0.520 kg C/kg	0.321 kg C/kg
A-4	Allocation	Dry mass	85% wt. dm	85% wt. dm	99.7% wt. dm	90% wt. dm
S-1	Substitution		–	Animal feed	–	Animal feed
S-2	Substitution		–	Fuel	–	Animal feed
S-3	Substitution		–	Animal feed	–	Fuel
S-4	Substitution		–	Fuel	–	Fuel

Ref.	**From**	**To**	**Annual soil carbon stock change (t C/ha yr)**
(b) Land-use change options and corresponding annual soil carbon stock changes			
LUC-1	Set aside	Long-term cultivated, reduced tillage, medium inputs	−0.22
LUC-2	Grassland, nondegraded		−1.07
LUC-3	Grassland, improved		−1.74
LUC-4	Grassland, moderately-degraded		−0.84
LUC-5	Grassland, severely-degraded		+0.35
LUC-6	Native ecosystem (forested land)		−1.07
LUC-7	Long-term cultivated, no tillage, medium inputs		−0.24
LUC-8	Long-term cultivated, reduced tillage, medium inputs		–
LUC-9	Long-term cultivated, full tillage, medium inputs		+0.30

			Variation of fuel consumption w.r.t gasoline			
Ref.	**Fuel**	**Basis**	**(l/km)**	**(kg/km)**	**(MJ_{th}/km)**	**Ethanol component (MJ_{th}/km)**
(c) Fuel blends and vehicle/fuel performance options						
E5-1	Ethanol, as E5	Actual tests	−1.0%	−0.7%	−2.7%	1.413
E10-1	Ethanol, as E10	Actual tests	−4.3%	−3.9%	−7.5%	1.174
E85-1	Ethanol, as E85	Actual tests	+34.9%	+41.8%	−2.5%	2.485
E-2	Ethanol	Volume basis	0.0%	–	–	1.703
E-3	Ethanol	Energy basis	–	–	0.0%	2.564

4.2.3 *Fuel Blends and Vehicle/Fuel Performance*

The effect of choices regarding fuels blends and vehicle/fuel performance was investigated. The fuel blends considered included E5, E10, and E85. Regarding fuel performance, various options were taken into account, namely: (1) fuel consumption data are based on actual vehicle tests in the European context; (2) fuel consumption of fuel blends (volume basis) is considered to be equal to that of standard gasoline; and (3) fuel consumption of fuel blends (energy basis) is considered to be equal to that of standard gasoline. The various options are summarized in Table 4. In each case, the corresponding specific fuel consumption of ethanol is given.

5 RESULTS

In order to present the results in a comprehensive way, the same structure as in the previous sections is used. Default methodological choices include A-1 (i.e., energy allocation) regarding allocation methods (based on the recommendations of the EC, 2009) and LUC-1 (i.e., set aside to cultivated land) regarding land-use change. The default choice regarding fuels blends and vehicle/fuel performance is E5-1 (i.e., ethanol used as E5, with fuel performance based on actual vehicle tests), according to the most common situation in the EU. The results showing the effect of methodological choices on the WtW net GHG emissions of bioethanol are given in Table 5. The same results are illustrated in Figure 5.

TABLE 5 WtW Net Emissions of GHG of Ethanol according to Selected Options

Allocation	LUC	Fuel	WtT ($kg\ CO_{2eq.}/MJ_{th}$)		TtW (MJ_{th}/km)	TtW ($kg\ CO_{2eq.}/km$)		WtW ($kg\ CO_{2eq.}/km$)	Index (–)
REF	REF	Gasoline	0.018	×	2.564	+ 0.190	=	0.237	100.0
A-1	LUC-1	E5-1	0.047	×	1.413		=	0.066	27.9
A-2	LUC-1	E5-1	0.106	×	1.413		=	0.150	63.4
A-3	LUC-1	E5-1	0.048	×	1.413		=	0.068	28.5
A-4	LUC-1	E5-1	0.041	×	1.413		=	0.057	24.2
S-1	LUC-1	E5-1	0.107	×	1.413		=	0.151	63.8
S-2	LUC-1	E5-1	0.012	×	1.413		=	0.017	7.0
S-3	LUC-1	E5-1	0.084		1.413		=	0.119	50.1
S-4	LUC-1	E5-1	−0.011	×	1.413		=	−0.016	−6.7
A-1	LUC-1	E5-1	0.047	×	1.413		=	0.066	27.9
A-1	LUC-2	E5-1	0.068	×	1.413		=	0.095	40.2

Continued

TABLE 5 WtW Net Emissions of GHG of Ethanol according to Selected Options—Cont'd

			WtT	TtW			WtW	Index
Allocation	**LUC**	**Fuel**	**(kg $CO_{2eq.}$/MJ_{th})**	**(MJ_{th}/km)**	**(kg $CO_{2eq.}$/km)**		**(kg $CO_{2eq.}$/km)**	**(–)**
A-1	LUC-3	E5-1	0.084	× 1.413		=	0.118	49.9
A-1	LUC-4	E5-1	0.062	× 1.413		=	0.087	36.9
A-1	LUC-5	E5-1	0.033	× 1.413		=	0.047	19.7
A-1	LUC-6	E5-1	0.177	× 1.413		=	0.249	104.9
A-1	LUC-7	E5–1	0.047	× 1.413		=	0.067	28.2
A-1	LUC-8	E5-1	0.042	× 1.413		=	0.059	24.7
A-1	LUC-9	E5-1	0.034	× 1.413		=	0.048	20.4
A-1	LUC-1	E5-1	0.047	× 1.413		=	0.066	27.9
A-1	LUC-1	E10-1	0.047	× 1.174		=	0.055	23.2
A-1	LUC-1	E85-1	0.047	× 2.485		=	0.116	49.1
A-1	LUC-1	E-2	0.047	× 1.703		=	0.080	33.7
A-1	LUC-1	E-3	0.047	× 2.564		=	0.120	50.7
A-2	LUC-1	E-3	0.106	× 2.564		=	0.273	115.0
A-2	LUC-2	E-3	0.161	× 2.564		=	0.412	173.7
A-2	LUC-3	E-3	0.204	× 2.564		=	0.522	220.1
A-2	LUC-4	E-3	0.146	× 2.564		=	0.374	157.8
A-2	LUC-5	E-3	0.070	× 2.564		=	0.179	75.6
A-2	LUC-6	E-3	0.447	× 2.564		=	1.146	383.2
A-2	LUC-7	E-3	0.108	× 2.564		=	0.276	116.4
A-2	LUC-8	E-3	0.092	× 2.564		=	0.237	99.8
A-2	LUC-9	E-3	0.073	× 2.564		=	0.188	79.0

The results are presented as net GHG emissions (as net energy use, respectively) of fuel-ethanol, expressed in kg $CO_{2\ eq.}$/km (in MJ_p/km, respectively). A positive value means that the system results in a net emission of GHG over the life cycle, whereas a negative value (only one case in the selected options below) means that the system is actually capturing GHG. These are then compared to gasoline in order to assess the actual balance and the potential for reducing GHG emissions and nonrenewable primary energy use. The net GHG emissions and net energy use of gasoline are 0.237 kg $CO_{2\ eq.}$/km and 3.493 MJ_p/km, respectively. Any smaller (larger, respectively) score for ethanol means that the system actually results in a reduction (an increase, respectively) of environmental impact with respect to gasoline.

Allocation	LUC	Fuel		IPCC Index
-	-	Gasoline		100.0
A-1	LUC-1	Bioethanol, as E5	E5-1	27.9
A-2	LUC-1	Bioethanol, as E5	E5-1	63.4
A-3	LUC-1	Bioethanol, as E5	E5-1	28.5
A-4	LUC-1	Bioethanol, as E5	E5-1	24.2
S-1	LUC-1	Bioethanol, as E5	E5-1	63.8
S-2	LUC-1	Bioethanol, as E5	E5-1	7.0
S-3	LUC-1	Bioethanol, as E5	E5-1	50.1
S-4	LUC-1	Bioethanol, as E5	E5-1	−6.7
A-1	LUC-1	Bioethanol, as E5	E5-1	27.9
A-1	LUC-2	Bioethanol, as E5	E5-1	40.2
A-1	LUC-3	Bioethanol, as E5	E5-1	49.9
A-1	LUC-4	Bioethanol, as E5	E5-1	36.9
A-1	LUC-5	Bioethanol, as E5	E5-1	19.7
A-1	LUC-6	Bioethanol, as E5	E5-1	104.9
A-1	LUC-7	Bioethanol, as E5	E5-1	28.2
A-1	LUC-8	Bioethanol, as E5	E5-1	24.7
A-1	LUC-9	Bioethanol, as E5	E5-1	20.4
A-1	LUC-1	Bioethanol, as E5	E5-1	27.9
A-1	LUC-1	Bioethanol, as E10	E10-1	23.2
A-1	LUC-1	Bioethanol, as E85	E85-1	49.1
A-1	LUC-1	Bioethanol	E-2	33.7
A-1	LUC-1	Bioethanol	E-3	50.7
A-2	LUC-1	Bioethanol	E-3	115.0
A-2	LUC-2	Bioethanol	E-3	173.7
A-2	LUC-3	Bioethanol	E-3	220.1
A-2	LUC-4	Bioethanol	E-3	157.8
A-2	LUC-5	Bioethanol	E-3	75.6
A-2	LUC-6	Bioethanol	E-3	483.2
A-2	LUC-7	Bioethanol	E-3	116.4
A-2	LUC-8	Bioethanol	E-3	99.8
A-2	LUC-9	Bioethanol	E-3	79.0

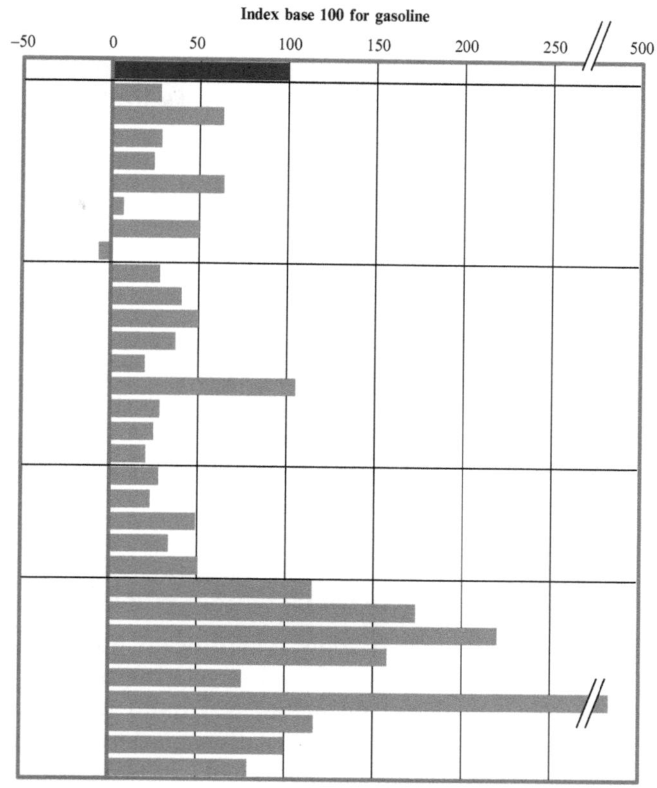

FIGURE 5 WtW net emissions of GHG of ethanol according to selected options.

5.1 Effect of Allocation Methods

The results indicate a strong influence of the choice of allocation method, with net GHG emissions ranging from -0.016 kg $CO_{2\ eq.}$/km (S-4, i.e., substitution with both straw and DDGS as fuel) to 0.151 kg $CO_{2\ eq.}$/km (S-1, i.e., substitution with both straw and DDGS as animal feed), that is, from -107% to -36% with respect to gasoline, respectively. In all cases, however, the net GHG emissions of bioethanol are lower than those of gasoline (0.237 kg $CO_{2\ eq.}$/km), with a percentage reduction of 36% in the "worst" case. The negative value in S-4 is explained by the fact that both the straw and the DDGS are replacing fossil energy agents for combined heat and power (CHP) applications. The electricity mix considered is that of Switzerland as reported in ecoinvent, while fuels for heat applications include 53% fuel oil and 47% natural gas (corresponding to the fuel mix in Switzerland). In S-1 and S-3 (i.e., substitution with straw as fuel and DDGS as animal feed), however, DDGS replace soybean meal (imported from Brazil and the US in equal shares, with a ratio of 0.82 kg of soybean meal per kg of DDGS based on dry weight protein content), where soybean oil is considered to be used as a feedstock for biodiesel production and to replace diesel fuel (substitution is applied over the global system). The consequence of using DDGS as animal feed in place of soybean meal is therefore unfavorable, showing on the net GHG emissions of bioethanol in S-1 and S-3.

As far as allocation methods are concerned, A-1 (energy), A-3 (carbon) and A-4 (dry mass) produce similar results, with net GHG emissions a lot more favorable than A-2 (economy). This is explained by the fact that wheat grains, straw, and DDGS have similar LHV and carbon contents. In this particular case of wheat to ethanol, the net GHG emissions in A-2 are not particularly sensitive to ethanol and grain prices. Increasing both prices by 50% results in an increase of the net GHG emissions by only 3% (allocation to grains with respect to straw and ethanol with respect to DDGS being already close to 100%).

5.2 Effect of Land-Use Change

The results show a strong influence of the land-use change considered, with net GHG emissions ranging from 0.047 kg $CO_{2\ eq.}$/km (LUC-5, i.e., grassland severely-degraded to cultivated land) to 0.249 kg $CO_{2\ eq.}$/km (LUC-6, i.e., forested land to cultivated land), that is, from -80% to +5% with respect to gasoline, respectively. It comes out that the net GHG emissions of bioethanol are lower than those of gasoline in all cases except when growing energy crops leads to deforestation.

5.3 Effect of Fuel Blends and Vehicle/Fuel Performance

The results again show a significant influence of the fuel blend and vehicle/fuel performance, with net GHG emissions ranging from 0.055 kg $CO_{2\ eq.}$/km (E10-1, i.e., ethanol used as E10 based on actual test data) to 0.120 kg $CO_{2\ eq.}$/km (E-3, i.e., energy basis), that is., from −77% to −49% with respect to gasoline, respectively. When taking into account actual fuel performance (from vehicle tests), E10 indeed appears to be the most favorable way of using fuel bioethanol as far as GHG emissions are concerned. This means that when considering a given volume of bioethanol to be introduced in a country, region or company, the most significant reduction of GHG emissions will be achieved by using the ethanol as E10. When comparing the net GHG emissions of fuel blends (and not only of the bioethanol component in the fuel blends), E85 leads to the most significant reduction of GHG emissions, before E10 and E5 in this order (in relation to the ethanol content in the fuel blend and the amount of gasoline displaced).

In case of lack of actual vehicle test data, option E-2 (equivalent to fossil reference on a volume basis) is undoubtedly the best choice for lower rates of ethanol incorporation (i.e., E5 to E20), while option E-3 (equivalent to fossil reference on an energy basis) is more appropriate for higher rates of incorporation (i.e., E85 to E100). In both situations however, options E-2 and E-3 lead to an underestimation of bioethanol merit (including the reduction of GHG emissions but also the reduction of nonrenewable energy consumption). In a very large majority of cases, E-3 (i.e., energy basis) is the choice adopted to take into account the utilization phase in a WtW approach.

When considering E5 (resp. E10) as the fuel blend, the error induced by choosing option 2 instead of option 1 is of the order of 20% (resp. 45%) to the disadvantage of ethanol. When considering E85 as the fuel blend, the error induced by choosing option 3 instead of option 1 is in the order of 3%, still to the disadvantage of ethanol. E-3 also leads to a small error when assessing the GHG balance of biodiesel, regardless of the incorporation rate.

5.4 Net Energy Use and Energy Substitution Efficiency

The energy substitution efficiency (ESE) was calculated for a number of the options investigated in this study. The calculations show the effect of allocation methods as well as fuel blends and vehicle/fuel performance on the WtW net energy use and the energy substitution efficiency. The results are given in Table 6 and illustrated in Figure 6.

TABLE 6 WtW Net NonRenewable Primary Energy Use and Energy Substitution Efficiency of Ethanol according to Selected Options

Allocation	LUC	Fuel	WtT (MJ_p/MJ_{th})		TtW (MJ_{th}/km)		WtW (MJ_p/km)	Index (–)	Energy substitution efficiency
REF	REF	Gasoline	1.362	×	2.564	=	3.493	100.0	–
A-1	LUC-1	E5-1	0.401	×	1.413	=	0.567	16.2	69.6%
A-2	LUC-1	E5-1	0.758	×	1.413	=	1.071	30.7	57.6%
A-3	LUC-1	E5-1	0.405	×	1.413	=	0.573	16.4	69.5%
A-4	LUC-1	E5-1	0.359	×	1.413	=	0.493	14.1	71.4%
S-1	LUC-1	E5-1	1.281	×	1.413	=	1.810	51.8	40.0%
S-2	LUC-1	E5-1	−0.220	×	1.413	=	−0.310	−8.9	90.5%
S-3	LUC-1	E5-1	0.450	×	1.413	=	0.636	18.2	68.0%
S-4	LUC-1	E5-1	−1.051	×	1.413	=	−1.485	−42.5	118.4%
A-1	LUC-1	E5-1	0.401	×	1.413	=	0.567	16.2	69.6%
A-1	LUC-1	E10-1	0.401	×	1.174	=	0.471	13.5	86.5%
A-1	LUC-1	E85-1	0.401	×	2.485	=	0.997	28.5	33.8%
A-1	LUC-1	E-2	0.401	×	1.703	=	0.684	19.6	55.4%
A-1	LUC-1	E-3	0.401	×	2.564	=	1.029	29.5	32.3%

Allocation	LUC	Fuel		Energy Index
-	-	Gasoline		100.0
A-1	LUC-1	Bioethanol, as E5	E5-1	16.2
A-2	LUC-1	Bioethanol, as E5	E5-1	30.7
A-3	LUC-1	Bioethanol, as E5	E5-1	16.4
A-4	LUC-1	Bioethanol, as E5	E5-1	14.1
S-1	LUC-1	Bioethanol, as E5	E5-1	51.8
S-2	LUC-1	Bioethanol, as E5	E5-1	-8.9
S-3	LUC-1	Bioethanol, as E5	E5-1	18.2
S-4	LUC-1	Bioethanol, as E5	E5-1	-42.5
A-1	LUC-1	Bioethanol, as E5	E5-1	16.2
A-1	LUC-1	Bioethanol, as E10	E10-1	13.5
A-1	LUC-1	Bioethanol, as E85	E85-1	28.5
A-1	LUC-1	Bioethanol	E-2	19.6
A-1	LUC-1	Bioethanol	E-3	29.5

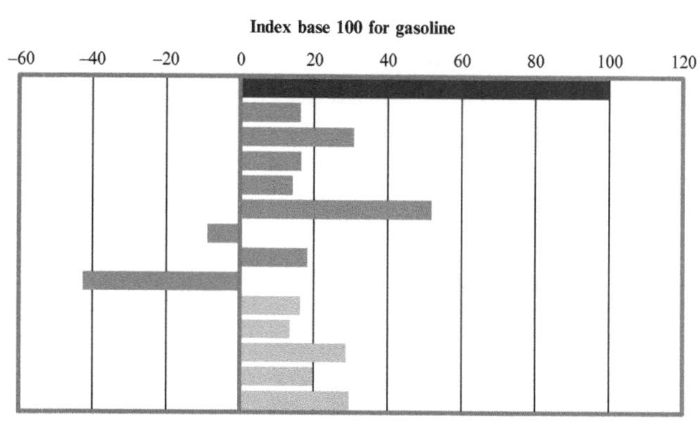

FIGURE 6 WtW net non-renewable primary energy use of ethanol according to selected options.

The results indicate that the choice of the allocation method has a significant impact on the WtW net energy use, with values ranging from -1.485 MJ_P/km (S-4, i.e., substitution with both straw and DDGS as fuel) to 1.810 MJ_P/km (S-1, i.e., substitution with both straw and DDGS as animal feed), that is, from -143% to -48% with respect to gasoline, with E5-1 as the option regarding fuels blend and vehicle/fuel performance. The effect of the fuel blend and vehicle/fuel performance is also significant, with net energy uses ranging from 0.471 MJ_P/km (E10-1, i.e., ethanol used as E10 based on actual test data) to 1.029 MJ_P/km (E-3, i.e., energy basis), that is, from -86% to -70% with respect to gasoline, respectively.

Both these methodological choices also significantly affect the energy substitution efficiency (ESE). For a given fuel blend and vehicle/fuel performance, the higher the nonrenewable primary energy use, the lower the ESE. For a given allocation method and bioethanol production pathway, the ESE is best when bioethanol is used in the form of E10. This notion is particularly useful when considering a given volume of bioethanol (at the scale of a country or a region for example). The results show how much more efficient it is to use this volume of bioethanol as E10 than to use it as E85 or even E5, for a given service (i.e., a given overall distance traveled). The situation is obviously different when considering a vehicle owner traveling a given distance every year. The best choice (in terms of both energy and GHG balance) for a specific consumer is obviously to use E85 (with a maximum volume of gasoline displaced), as long as the net energy use or net GHG emissions of the biofuel are better than those of gasoline.

5.5 Synthesis

It is reported in various publications and press articles that biofuels (incl. bioethanol and biodiesel) do actually contribute to global warming (Fargione et al., 2008; Reijnders and Huijbregts, 2008; Searchinger et al., 2008) under certain conditions. Other WtW studies, such as that of Beer and Grant (2007), found only marginal advantages for E10 blends (of the order of 4% reduction on GHG emissions). Looking at the results presented in this chapter so far, however, it does not seem to be the case for bioethanol from wheat produced in the Swiss context. This is not always true and is really the result of the default choices made in this chapter (which correspond to the most likely situation in the present European context). Let us assume A-2 as the allocation method (i.e., economy) and E-3 as the vehicle/fuel performance option (i.e., energy basis) and evaluate the net GHG emissions of bioethanol from wheat under various land-use change scenarios. This framework is among the most unfavorable set of options and actually corresponds to that of EMPA (2007a). Economic allocation is the default method in the ecoinvent database and the method chosen by the Swiss authorities to evaluate the sustainability of fuels in the frame of the Ordinance on the ecological balance of fuels. Vehicle/fuel performance based on the energy content of fuels is undoubtedly the most frequent hypothesis in LCA studies of biofuels (e.g., CONCAWE-EUCAR-JRC, 2008; EMPA, 2007a; GM-LBST, 2002; IFEU, 2004; VIEWLS, 2005). The results are presented in Table 5 and illustrated in Figure 5.

Under these assumptions, the variation of life-cycle GHG emissions with respect to gasoline ranges from −24% to +383%. Unless the land-use change leads to an improved annual carbon stock (i.e., LUC-5 LUC-8 and LUC-9 as shown in Table 4), the net GHG emissions of wheat to ethanol are indeed larger than those of gasoline. In the worst scenario (LUC-6, i.e., forested land to cultivated land), the net GHG emissions of bioethanol can be as large as 3.8 times that of gasoline.

6 CONCLUSIONS

The default case in this chapter for fuel-bioethanol production from wheat in the Swiss context considers allocation based on energy content (A-1), the switch from set-aside land to cultivated land (LUC-1) and vehicle/fuel performance based on actual vehicle test with fuel-ethanol used as E5 (E5-1). With net GHG emissions of 0.066 kg $CO_{2\ eq.}$/km (i.e., −72% with respect to gasoline) and a net energy use of 0.567 MJ_p/km (i.e., -84% with respect to gasoline), this default case may seem particularly advantageous compared to other similar studies. This default case, however, is realistic and corresponds to the most likely situation in the European context. Energy allocation is the methodology adopted by the European Union in its Directive on the promotion of the use of energy from renewable sources. Set-aside to long-term cultivated is a reasonable option when considering the production of biofuels from agricultural crops. Finally, fuel ethanol in the EU is mainly used as E5 at present. If the set of methodological choices as in EMPA (2007a) is applied to the same system, meaning economic allocation (A-2) and vehicle/fuel performance based on the energy content of fuels (E-3), the resulting net GHG emissions and net energy use are 0.273 kg $CO_{2eq.}$/km (i.e., +15% with respect to gasoline) and 1.944 MJ_p/km (i.e., −44% with respect to gasoline), respectively. These results are much more unfavorable and significantly different from those of the default case.

Various authors have demonstrated the significant effect of methodological choices on the GHG and energy balance of biofuels through review papers and other similar studies (Börjesson, 2009; Farrell and Sperling, 2007; Reijnders and Huijbregts, 2003). The present chapter quantifies these effects, based on a case study concerned with the production of fuel-ethanol from wheat in the Swiss context. In addition, it demonstrates and quantifies the effects of the fuel blend and the choices regarding vehicle/fuel performance.

The results presented in this chapter show a large variation of the net GHG emissions of wheat-based ethanol for transportation with a high sensitivity to the following factors: the method used to allocate the impacts between coproducts, the type of reference systems, the type of land-use change, and the type of fuel blend. Depending on the allocation method (energy content, economy, dry mass or carbon content), the net GHG emissions of ethanol may vary by a factor of up to 2.6 (with carbon content being the most favorable and economy the least favorable). When substitution is applied, the net GHG emissions of ethanol may even be negative when both straw and DDGS are used as fuels, thereby making the difference even more significant. Depending on the land-use change situation, the net GHG emissions of ethanol may vary by a factor of up to 6.4. Similarly, the hypotheses regarding actual fuel blends and vehicle/fuel performance may result in a variation of net GHG emissions by a factor of 2.2. Depending on the combinations of methodological choices and land-use change situations, the variation of life-cycle GHG emissions with respect to gasoline may range from −112% to +120% for the same ethanol production pathway.

In face of missing data and time stress, many studies use pragmatic approaches to evaluate the energy and GHG balance of biofuels. Thus, several studies are not transparent enough and methodological choices can turn a positive GHG balance into a negative one and vice versa. As policymakers will take decisions by using these results, it is important to establish the rationale of the evaluation methods. Some items need further research works, for example, rationale of allocation methods, indirect land-use change (Gnansounou et al., 2008b). Others are till now subject to low transparency and consistency requirements.

Especially concerning the boundaries of the system, the authors recommend to use a WtW approach. One should not mind if the implementation of the WtW should be simplified; utilization stage must be taken into account as long as comparison of different qualities of fuels is concerned, that is, fuels associate with different mechanical efficiencies. The functional unit must be appropriate, reflecting the fact that these fuels must be compared for the same service (e.g., the distance traveled). Finally, for transparency purpose, the reference system must be explicitly defined.

References

ADEME, 2010. Analyses de Cycle de Vie appliquées aux biocarburants de première génération consommés en France. Rapport final. Etude réalisée pour le compte de l'Agence de l'environnement et de la Maîtrise de l'Energie (ADEME), du Ministère de l'Ecologie, de l'Energie, du Développement Durable et de la Mer, du Ministère de l'Alimentation, de l'Agriculture et de la Pêche, et de France Agrimer par BIO Intelligence Service.

ADEME-DIREM-PWC, 2002. Bilans énergétiques et gaz à effet de serre des filières de production des biocarburants, rapport technique. ADEME-DIREM-PriceWaterhouseCoopers.

AEAT, 2002. Ethanol Emissions Testing. AEA Technology, prepared for the UK Department for Transport, UK.

ANL-GM (2001): GM, 2001. Well-to-wheels energy use, greenhouse gas emissions of advanced fuel/vehicle systems: North American analysis. Argonne National Laboratory, April 2001.

Batan, L., Quinn, J., Willson, B., Bradley, T., 2010. Net energy and greenhouse gas emission evaluation of biodiesel derived from microalgae. Environ. Sci. Technol 44 (20), 7975–7980.

Beer, T., Grant, T., 2007. Life-cycle analysis of emissions from fuel ethanol and blends in Australian heavy and light vehicles. J. Clean. Prod. 15 (8-9), 833–837.

Bergsma, G., Vroonhof, J., Dornburg, V., 2006. A Greenhouse Gas Calculation Methodology for Biomass-Based Electricity, Heat and Fuels—The view of the Cramer Commission. CE Delft.

Börjesson, P., 2009. Good or bad bioethanol from a greenhouse gas perspective—what determines this? Appl. Energy 86, 589–594.

Brennan, L., Owende, P., 2010. Biofuels from microalgae: a review of technologies for production, processing, and extractions of biofuels and co-products. Renew. Sustain. Energ. Rev. 14 (2), 557–577.

Campbell, P.K., Beer, T., Batten, D., 2011. Life cycle assessment of biodiesel production from microalgae in ponds. Bioresour. Technol. 102 (1), 50–56.

Clarens, A.F., Resurreccion, E.P., White, M.A., Colosi, L.M., 2010. Environmental life cycle comparison of algae to other bioenergy feedstocks. Environ. Sci. Technol 44 (5), 1813–1819.

Collet, P., Hélias, A., Lardon, L., Ras, M., Goy, R.A., Steyer, J.P., 2011. Life-cycle assessment of microalgae culture coupled to biogas production. Bioresour. Technol. 102 (1), 207–214.

CONCAWE-EUCAR-JRC, 2008. Well-to-wheels analysis of future automotive fuels and powertrains in the European context, Well-to-wheels report, version 2c. Joint study by CONCAWE, EUCAR and the Joint Research Centre of the European Commission.

CSIRO (2001): Beer, T., Grant, T., Morgan, G., Lapszewicz, J., Anyon, P., Edwards, J., Nelson, P., Watson, H., Williams, D., 2001. Comparison of transport fuels, Final report (EV45A/2/F3C) to the Australian Greenhouse Office on the Stage 2 study of Life-cycle emissions analysis of alternative fuels for heavy vehicles.

EC, 2009. Directive 2009/28/EC of the European parliament and the Council on the promotion of the use of energy from renewable source. L140/16 Official Journal of the European Union .

ecoinvent, 2007. ecoinvent Centre, ecoinvent data v2.0. CD-ROM, Swiss Centre for Life Cycle Inventories, Dübendorf, Switzerland.

Ekvall, T., Finnveden, G., 2001. Allocation in ISO 14041: a critical review. J. Clean. Prod. 9 (3), 197–208.

Elsayed, M.A., Matthews, R., Mortimer, N.D., 2003. Carbon and energy balances for a range of biofuels options. Sheffield Hallam University, Resources Research Unit, Project No. B/B6/00784/REP URN 03/836.

EMPA, 2002. Potential der Bio-Ethanol Beimischung im Benzin und Diesel-Treibstoff Untersuchungen hinsichtlich des Emissionsverhaltens an einem Nutzfahrzeug Motor und zwei Personenfahrzeugen. Bio-Ethanol Projekt, Untersuchungsbericht Nr. 202'672, prepared for Alcosuisse, Switzerland.

EMPA, 2007a. Ökobilanz von Energieprodukten: Ökologische Bewertung von Biotreibstoffen. EMPA, Materials Science and Technology.

EMPA, 2007b. Vergleich von Ethanol- (E85) und Benzinbetrieb an einen Ford Focus Flexifuel. Untersuchungsbericht Nr. 445'114, prepared for FORD Switzerland, Switzerland.

Escobar, J.C., Lora, E.S., Venturini, O.J., Yañez, E.E., Castillo, E.F., Almazan, O., 2008. Biofuels: environment, technology and food security. Renew. Sustain. Energ. Rev. in press, Corrected Proof.

Fargione, J., Hill, J., Tilman, D., Polasky, S., Hawthorne, P., 2008. Land Clearing and the Biofuel Carbon Debt. Science 319 (5867), 1235–1238.

Farrell, A.E., Plevin, R.J., Turner, B.T., Jones, A.D., O'Hare, M., Kammen, D.M., 2006. Ethanol can contribute to energy and environmental goals. Science 311 (5760), 506–508.

Farrell, A.E., Sperling, D., 2007. A Low-Carbon Fuel Standard for California, Parts 1 & 2. University of California.

Frischknecht, R., 2004. ecoinvent data v1.1. Int. J. LCA 10 (1), 1–2.

Frischknecht, R., Jungbluth, N., Althaus, H.J., Doka, G., Dones, R., Heck, T., et al., 2004. The ecoinvent database: overview and methodological framework. Int. J. LCA 10 (1), 3–9.

GM-LBST, 2002. Well-to-Wheel Analysis of Energy Use and Greenhouse Gas Emissions of Advanced Fuel/Vehicle Systems: A European Study. General Motors, L-B-Systemtechnik.

Gnansounou, E., Dauriat, A., 2004. Etude comparative de carburants par analyse de leur cycle de vie. Swiss Federal Institute of Technology of Lausanne, Laboratory of Energy Systems (LASEN), reports 415.105:2002 & 420.100:2004, prepared for Alcosuisse.

Gnansounou, E., Dauriat, A., 2005. Energy balance of bioethanol: a synthesis. Proc. of 14th European Biomass Conference & Exhibition, Paris, France.

Gnansounou, E., et al., 2008a. Estimating Energy and Greenhouse Gas Balances of Biofuels: Concepts and Methodologies. Swiss Federal Institute of Technology of Lausanne, Laboratory of Energy Systems (LASEN), working paper 437.100.

Gnansounou, E., et al., 2008b. Accounting for Indirect Land-Use Changes in GHG Balances of Biofuels: Review of Current Approaches. Swiss Federal Institute of Technology of Lausanne, Laboratory of Energy Systems (LASEN), working paper, 437.101.

Gnansounou, E., Dauriat, A., Villegas, J., Panichelli, L., 2009. Life cycle assessment of biofuels: energy and greenhouse gas balances. Bioresour. Technol. 100 (21), 4919–4930.

Guinée, J.B., Heijungs, R., Huppes, G., 2004. Economic allocation: examples and derived decision tree. Int. J. LCA 9 (1): 23–33.

IDIADA, 2003. Comparison of vehicle emissions at European Union annual average temperatures from E10 and E5 petrol. IDIADA Automotive Technology, Powertrain, Report LM030411, Spain.

IFEU (2004): Quirin, M., Gärtner, S.O., Pehnt, M., Reinhardt, G.A., 2004. CO2 Mitigation through biofuels in the transport sector. Status and perspectives. Institut für Energie- und Umweltforschung Heidelberg (IFEU), 2004, 55 pp.

IPCC, 2003a. Good Practice Guidance for Land Use, Land-Use Change and Forestry. Intergovernmental Panel on Climate Change (IPCC), National Greenhouse Gas Inventories Programme.

IPCC, 2003b. Tool for Estimation of Changes in Soil Carbon Stocks Associated with Management Changes in Cropland and Grazing Lands Based on IPCC Default Data. Developed at the Natural Resource Ecology Laboratory, Colorado State University.

ISO, 2006a. ISO Norm 14040:2006, Life Cycle Assessment: Principles and Framework. Environmental Management.

ISO, 2006b. ISO Norm 14044:2006, Life Cycle Assessment. Requirements and Guidelines. Environmental Management.

Jungbluth, N., Chudacoff, M., Dauriat, A., Dinkel, F., Doka, G., Faist Emmenegger, M., et al., 2007. Life Cycle Inventories of Bioenergy. Final report, ESU-services, Uster, CH.

Kaparaju, P., Serrano, M., Thomsen, A.B., Kongjan, P., Angelidaki, I., 2009. Bioethanol, biohydrogen and biogas production from wheat straw in a biorefinery concept. Bioresour. Technol. 100, 2562–2568.

Kim, S., Dale, B.E., 2002. Allocation procedure in ethanol production system from corn grain:I. System expansion. Int. J. LCA 7 (4), 237–243.

Kim, S., Dale, B.E., 2008. Life cycle assessment of fuel ethanol derived from corn grain via dry milling. Bioresour. Technol. 99, 5250–5260.

Macedo, I., 2004. Assessment of Greenhouse Gas Emissions in the Production and Use of Fuel Ethanol in Brazil. Government of the State of São Paulo, Brazil.

Malça, J., Freire, F., 2006. Renewability and life-cycle energy efficiency of bioethanol and bio-ethyl tertiary butyl ether (bioETBE): assessing the implications of allocation. Energy 31 (15), 3362–3380.

Nemecek, T., Kägi, T., 2007. Life Cycle Inventories of Swiss and European Agricultural Production Systems. Final report ecoinvent V2.0 No. 15a. Agroscope Reckenholz-Taenikon Research Station ART, Swiss Centre for Life Cycle Inventories, Zurich and Dübendorf, CH.

Panichelli, L., Dauriat, A., Gnansounou, E., 2008. Life cycle assessment of soybean-based biodiesel in Argentina for export. Int. J. LCA, online first.

Panichelli, L., Gnansounou, E., 2008. Estimating greenhouse gas emissions from indirect land-use change in biofuels production: concepts and exploratory analysis for soybean-based biodiesel production. J. Sci. Ind. Res. 67, 1017–1030.

Reijnders, L., Huijbregts, M.A.J., 2003. Choices in calculating life cycle emissions of carbon containing gases associated with forest derived biofuels. J. Clean. Prod. 11, 527–532.

Reijnders, L., Huijbregts, M.A.J., 2008. Palm oil and the emission of carbon-based greenhouse gases. J. Clean. Prod. 16, 477–482.

Righelato, R., Spracklen, D.V., 2007. Carbon mitigation by biofuels or by saving and restoring forests? Science 317 (5840), 902.

Searchinger, T., Heimlich, R., Houghton, R.A., Dong, F., Elobeid, A., Fabiosa, J., et al., 2008. Use of U.S. croplands for biofuels increases greenhouse gases through emissions from land use change. Science 319 (5867) 1238–1240.

Shapouri, H., Duffield, J., Wang, M., 2002. The energy balance of corn ethanol: an update. In USDA, Agricultural Economics Report No. 813.

Singh, J., Gu, S., 2010. Commercialization potential of microalgae for biofuels production. Renew. Sustain. Energ. Rev. 14 (9), 2596–2610.

Singh, A., Pant, D., Korres, N.E., Nizami, A.S., Prasad, S., Murphy, J.D., 2010. Key issues in life cycle assessment of ethanol production from lignocellulosic biomass: challenges and perspectives. Bioresour. Technol. 101 (13), 5003–5012.

Spatari, S., Bagley, D.M., MacLean, H.L., 2010. Life cycle evaluation of emerging lignocellulosic ethanol conversion technologies. Bioresour. Technol. 101 (2), 654–667.

Van Dam, J., Junginger, M., Faaij, A., Jurgens, I., Best, G., Fritsche, U., 2008. Overview of recent developments in sustainable biomass certification. Biomass Bioenergy 32, 749–780.

VIEWLS, 2005. Environmental and economic performance of biofuels, Volume I, Main report. VIEWLS Project, SenterNovem.

Wang, M., 2005. Energy and greenhouse gas emissions impacts of fuel ethanol. USDOE Argonne National Laboratory (ANL), Center for Transportation Research, presented at the NGCA Renewable Fuels Forum.

Weidema, B.P., 2003. Market Information in life cycle assessment. Prepared for the Danish Environmental Protection Agency.

Wijffels, R.H., Barbosa, M.J., 2010. An outlook on microalgal biofuels. Science 329, 796–799.

CHAPTER

3

Thermochemical Conversion of Biomass to Biofuels

Thallada Bhaskar, Balagurumurthy Bhavya, Rawel Singh, Desavath Viswanath Naik, Ajay Kumar, Hari Bhagwan Goyal*

Bio-Fuels division (BFD), Indian Institute of Petroleum (IIP), Council of Scientific and Industrial Research (CSIR), Dehradun 248005, India

***Corresponding author: Thallada Bhaskar; E-mail: tbhaskar@iip.res.in; thalladab@yahoo.com**

1 INTRODUCTION

The demand for energy sources to satiate human energy consumption continues to increase. Currently, the main energy source in the world is fossil fuels. Although it is not known how much fossil fuel is still available, it is generally accepted that it is being depleted and is nonrenewable. Prior to the use of fossil fuels, biomass was the primary source of energy for heat via combustion. With the introduction of fossil fuels in the forms of coal, petroleum, and natural gas, the world increasingly became dependent on these fossil fuel sources. Renewable energy is of growing importance in responding to concerns over the environment and the security of energy supplies. Given these circumstances, searching for other renewable forms of energy sources is reasonable. Other important consequences associated with fossil fuel uses include global warming. Also, fossil fuel resources are not distributed evenly around the globe, which makes many countries heavily dependent on imports.

Governments across the world are stimulating the utilization of renewable energies and resources such as solar, wind, hydroelectricity, and biomass. The three major forces that drive them are (i) secured access to energy; (ii) threat of climate change; (iii) develop/maintain agricultural activities (Lange, 2007). Agricultural economies could be supported by promoting the exploitation of local (bio) resources for food, energy, and material. Interestingly, each of these major drivers also represents one of the three dimensions of sustainability, namely, profitability (affordable energy), planet (climate change), and people (social stability).

Current use of fossil fuels is split, with about three-quarters for heat and power generation, about one-quarter for transportation fuel, and just a few percent for chemicals and materials (US Department of energy, 2006). The heat and power sector can be supplied with a variety of renewable sources, namely wind, solar, hydropower, and biomass. The transportation sector has a much more limited choice, however. At this time, biomass is the only resource that can provide renewable liquid fuels. Apart from the transportation sector, biomass is also a promising feedstock for the chemical industry due to the presence of a wide range of functionalities available with biomass, the natural polymer.

Biomass is unique in providing the only renewable source of fixed carbon, which is an essential ingredient in meeting many of our fuel and consumer goods requirements. Wood and annual crops and agricultural and forestry residues are some of the main renewable energy resources available (Bridgewater, 2006). Biofuel production has been growing rapidly in recent years.

Biomass, a renewable energy source, via photosynthesis, has provided energy for life for the longest period of existence. Industrial processes that take in biomass can be integrated with the natural photosynthesis/respiration cycle of vegetation. If used in this manner, biomass is a renewable energy source and by its utilization, much less CO_2 is added overall to the atmosphere compared with the fossil fuel counterpart processes. When combined with CO_2 sequestration, biomass-based processes can actually lower the CO_2 concentrated in the atmosphere (Van swaaij et al., 2004). Lignocellulosic biomass, which is not competing with the food chain, should be used for the production of fuels, chemicals, power, and heat. This competition can be avoided by first using the abundant residues from forests, agriculture, and subsequently energy crops. The potential of special energy crops is estimated to be in the range of 50-250 EJ/annum (Berndes et al., 2003).

Biomass combines solar energy and carbon dioxide into chemical energy in the form of carbohydrates via photosynthesis. The use of biomass as a fuel is a carbon neutral process since the carbon dioxide captured during photosynthesis is released during its combustion. Biomass includes agricultural and forestry residues, wood, byproducts from processing of biological materials, and organic parts of municipal and sludge wastes. Photosynthesis by plants captures around 4000 EJ/year in the form of energy in biomass and food (Kumar et al., 2009a).

The most important factor is that all fossil fuels are taken out from under the earth's surface, and its continuous excavation creates many geothermal disturbances. Biomass is grown and consumed only over the earth's surface and hence does not create such problems.

The events of the last few years have brought into sharp focus the need to develop sustainable green technologies for many of our most basic manufacturing and energy needs. Since the beginning of the new millennium, we have witnessed an ever-increasing merger of technical, economic, and societal demands for sustainable technologies. As such, this seeks to develop a new "carbohydrate-lignin economy" that will initially supplement today's petroleum economy and, as these nonrenewable resources are consumed, will become the primary resource for fuels, chemicals, and materials (Yunqiao et al., 2008).

2 FEEDSTOCKS FOR BIOFUELS

Biomass is harvested as part of a constantly replenished crop. This maintains a closed carbon cycle with no net increase in atmospheric CO_2 levels. There are five basic categories of material, that is, virgin wood, forestry materials, materials from arboricultural activities or from wood processing; energy crops: high-yield crops grown specifically for energy applications; agricultural residues: residues from agriculture harvesting or processing; food waste, from food and drink manufacture, preparation and processing, and postconsumer waste; industrial waste and coproducts from manufacturing and industrial processes.

Feedstocks that are used directly in a manner that is given to us by nature fall under the category of natural feedstocks. The first-generation biofuels use the edible biomass for producing biofuels. Some of them are sunflower seeds, jojoba oil, soya bean oil, safflower seeds for biodiesel production, and corn and sugar cane for producing ethanol. In contrast, the second-generation biofuels are produced from non edible feedstocks like lignocellulosic feedstocks which include agro residue (stalk, husk), forest residue (branch, twigs, bark, leaves), and several others.

In addition to growing currently available feedstocks on available land to produce biofuels, the realization of dedicated energy crops with enhanced characteristics would represent a significant step forward. The genetic sequences of a few key biomass feedstocks are already known, such as Poplar (Tuskan et al., 2006), and there are more in the sequencing pipeline. This genetic information gives scientists the knowledge required to develop strategies for engineering plants with far superior characteristics, such as diminished recalcitrance to conversion (Himmel et al., 2007).

Another area where genetic engineering could produce dramatically positive results is the development of perennial feedstocks that can reach high-energy densities over a short time with minimal fertilization and water consumption. By combining the known targeted climates and soil types present in the available conservation reserve program (CRP) and marginal lands with tailored feedstocks, it may be possible to develop grasses and short-rotation woody crops that maximize carbon and nitrogen fixation within these ecosystems. In addition to modifying the intrinsic polysaccharide/lignin composition and central metabolism of the feedstock itself, several research groups are attempting to express enzymes that are capable of breaking down cellulose into glucose directly within plants.

3 COMPOSITION OF LIGNOCELLULOSIC BIOMASS

Biomass is an organic material which stores sunlight in the form of chemical energy. It is available on a renewable basis. Here, we specifically mention the lignocellulosic biomass from plants and residues from various agricultural activities. Biomass is an organic material that is composed of polymers that have extensive chains of carbon atoms linked to macromolecules. The polymer back bone consists of chemical bonds linking carbon with carbon, or carbon with oxygen, or sometimes other elements such as nitrogen or sulfur. Instead of describing polymers in terms of the atomic structure of the chain, most can be viewed as assemblies of some larger molecular unit. In the case of cellulose, that unit is the glucan moiety, essentially a molecule of glucose with one molecule of water missing $(C_6H_{10}O_5)_n$. For hemicellulose, the unit is often a 5-carbon sugar, called xylose. However, hemicellulose polymers are not linear chains as in the cellulose polymer. Some are branched and other monomer units have side chains, with acetyl groups being very common. The lignin polymers are composed of phenyl propane subunits linked at various points on the monomer through C—C and C—O bonds. In addition, there are often side chain moieties such as methoxy groups. Wood-based biomass is available in large quantities and is cheap. It consists of three major components, that is, lignin, cellulose, and hemicellulose.

(i) *Cellulose*: It contains linear polysaccharides in the cell walls of wood fibers, consisting of D-glucose molecules bound together by β-1,4-glycoside linkages. Biomass comprises 40-50% cellulose.

(ii) *Hemicellulose*: It is an amorphous and heterogeneous group of branched polysaccharides (copolymer of any of the monomers of glucose, galactose, mannose, xylose, arabinose, and glucuronic acid). Hemicellulose surrounds the cellulose fibers and is a linkage between cellulose and lignin (15-30%). Hemicelluloses are heterogeneous polymers of pentoses (e.g., xylose, and arabinose), hexoses (e.g., mannose, glucose and galactose), and sugar acids. Unlike cellulose, hemicelluloses are not chemically homogeneous. Hemicelluloses are relatively easily hydrolyzed by acids to their monomer components consisting of glucose, mannose, galactose, xylose, arabinose, and small amounts of rhamnose, glucuronic acid, methylglucuronic acid, and galacturonic acid. Hardwood hemicelluloses contain mostly xylans, whereas softwood hemicelluloses contain mostly glucomannans. Xylans are the most abundant hemicelluloses. Xylans of many plant materials are heteropolysaccharides with homopolymeric backbone chains of 1, 4-linked β-D-xylopyranose units. Xylans from different sources, such as grasses, cereals, softwood, and hardwood, differ in composition. Besides xylose, xylans may contain arabinose, glucuronic acid, and acetic, ferulic and p-coumaric acids. The degree of polymerization of hardwood xylans (150-200) is higher than that of softwoods.

(iii) *Lignin*: It is a highly complex three-dimensional polymer of different phenylpropane units bound together by ether (C—O) and carbon-carbon (C—C) bonds. Lignin is concentrated between the outer layers of the fibers, leading to structural rigidity and holding the fibers of polysaccharides together (15-30%). Generally, softwoods contain more lignin than hardwoods. Lignins are divided into two classes, namely, guaiacyl lignins and guaiacyl-syringyl lignins. Although the principal structural elements in lignin have been largely clarified, many aspects of their chemistry remain unclear.

In addition, small amounts of extraneous organic compounds, that is, extractives, proteins, and inorganic constituents are found in lignocellulosic materials (about 4%; Stocker, 2008). Biomass residues like wheat straw, corn stover, or sugar cane bagasse contain much ash and N, S, Cl, and these quantities also depend on the geographical source.

4 LIGNOCELLULOSIC BIOMASS PRETREATMENT TECHNIQUES

Lignocellulosic biomass mainly consists of three components, namely, cellulose, hemicellulose, and lignin. Cellulose (major component) susceptibility to hydrolysis is restricted due to the rigid lignin and hemicellulose protection surrounding the cellulose micro fibrils. Therefore, an effective pretreatment is necessary to liberate the cellulose from the lignin-hemicellulose seal and also reduce cellulosic crystallinity. Some of the available pretreatment techniques include acid hydrolysis, steam explosion, ammonia fiber expansion (AFEX), alkaline wet oxidation, and hot water pretreatment. Besides reducing lignocellulosic recalcitrance, an ideal pretreatment must also minimize formation of degradation products that inhibit subsequent hydrolysis and fermentation. Pretreatment methods are subject to ongoing and intense research worldwide. Possible pretreatment methods can be classified as follows, although not all of them have been developed yet enough to be feasible for applications in large-scale processes (Taherzadeh and Karimi, 2008):

i. Physical pretreatments: milling (ball milling, two-roll milling, hammer milling, colloid milling, vibroenergy milling), irradiation (gamma ray, electron beam, microwave), others (hydrothermal, high-pressure steaming, expansion, extrusion, pyrolysis)
ii. Chemical and physicochemical pretreatment methods: explosion (steam explosion, ammonia fiber explosion, CO_2 explosion, SO_2 explosion),alkali treatment (treatment with sodium hydroxide, ammonia or ammonium sulfite), acid treatment (sulfuric acid, hydrochloric acid, phosphoric acid), gas treatment (chlorine dioxide, nitrogen dioxide, sulfur dioxide), addition of oxidizing agents (hydrogen peroxide, wet oxidation, ozone), solvent extraction of lignin (ethanol-water extraction, benzene-water extraction, ethylene glycol extraction, butanol-water extraction, swelling agents)
iii. Biological pretreatments (fungi and actinomycetes)

Mechanical comminuting reduces cellulose crystallinity, but power consumption is usually higher than inherent biomass energy. Steam explosion causes hemicellulose degradation and lignin transformation and is cost effective but destroys a portion of the xylan fraction, causes incomplete disruption of the lignin-carbohydrate matrix, and generates compounds inhibitory to microorganism. AFEX is an important pretreatment technology that utilizes both physical (high temperature and pressure) and chemical (ammonia) processes to achieve effective pretreatment. Besides increasing the surface accessibility for hydrolysis, AFEX promotes cellulose decrystallization and partial hemicellulose depolymerization and reduces the lignin recalcitrance in the treated biomass. This process is not efficient for biomass with high lignin content. CO_2 explosion increases accessible surface area; are cost effective and do not cause formation of inhibitory compounds but does not modify lignin or hemicelluloses. Ozonolysis reduces lignin content and do not produce toxic residues, but a large requirement of ozone makes it very expensive. Acid hydrolysis hydrolyzes hemicellulose to xylose and other sugars and alters lignin structure. Its disadvantages are high cost, equipment corrosion, and formation of toxic substances. Alkaline hydrolysis removes hemicelluloses and lignin and increases accessible surface area but long residence times are required, irrecoverable salts are formed and incorporated into biomass. Organosolv hydrolyzes lignin and hemicelluloses but solvents need to be drained from the reactor, evaporated, condensed, and recycled; hence, the process cost becomes high. Pulsed electrical field process is carried out in ambient conditions which disrupts plant cells and is simple equipment, but this process needs more research. Biological process involves degradation of lignin and hemicelluloses and has low-energy requirements, but the rate of hydrolysis is very low (Kumar et al., 2009b).

Lignocellulosic biomass has lignin, cellulose, and hemicelluloses with the complex structures with high molecular weight. The selective and effective lignocellulosic biomass conversion methods are highly desirable to produce the wide range of usable hydrocarbons as fuels, chemicals, and other products. The decomposition of complex structure can be performed by using biochemical and thermochemical methods using conventional and nonconventional energy sources.

5 BIOTECHNOLOGICAL CONVERSION

Following pretreatment, woody biomass can be converted into simple sugars by enzymatic deconstruction via a cellulase treatment. This remains the second most expensive component in the bioconversion of wood to bioethanol, despite the fact that research studies over the past decade have decreased cellulase costs by greater than a 10-fold basis. Numerous publications and reviews have highlighted the use of (i) separate hydrolysis and fermentation (SHF) and (ii) simultaneous saccharification and fermentation (SSF) to convert pretreated wood to ethanol (Wingren et al., 2003; Wyman, 1994). A process challenge in the conversion of wood to biofuels is the efficient conversion of all wood sugars (i.e., C_5 and C_6) to ethanol, especially for hardwoods which have greater amounts of pentoses.

One promising strategy has been to take a natural hexose ethanologen and add the pathways to convert other sugars (Helle et al., 2004; Lawford and Rousseau, 2002). An alternative approach to minimize the cost of cellulose deconstruction and conversion to ethanol is consolidated bioprocessing (CBP). CBP involves (i) bioproduction of cellulolytic enzymes from thermophilic anaerobic microbes, (ii) hydrolysis of plant polysaccharides to simple sugars and (iii) their subsequent fermentation to ethanol all in one stage (Lynd et al., 2005). This bioprocess is projected to reduce the cost of bioethanol by a factor of four over SSF, and these reduced costs and simplicity of operation have heightened research in this field.

6 THERMOCHEMICAL CONVERSION

The base of thermochemical conversion is the pyrolysis process in most cases. The products of conversion include water, charcoal (carbonaceous solid), biocrude, tars, and permanent gases including methane, hydrogen, carbon monoxide, and carbon dioxide depending upon the reaction parameters such as environment, reactors used, final temperature, rate of heating, and source of heat.

6.1 Combustion

Combustion is the sequence of exothermic chemical reactions between a fuel and an oxidant accompanied by the production of heat and conversion of chemical species. During the combustion of lignocellulosic biomass, the heat is generated due to oxidation reaction, where carbon, hydrogen, oxygen, combustible sulfur, and nitrogen contained in biomass react with air or oxygen. By far the most common means of converting biomass to usable heat energy is through straightforward combustion, and this account for around 90% of all energy attained from biomass (http://www.esru.strath.ac.uk/EandE/Web_sites/06-07/Biomass/HTML/combustion_technology.htm). It contributes over 97% of bioenergy production in the world. Combustion is a proven low-cost process, highly reliable technology, relatively well understood and commercially available. There are three main stages that occur during biomass combustion: drying, pyrolysis and reduction, and combustion of volatile gases and solid char.

Typically, the biomass contains high moisture and high oxygen content, which causes to have low heating values for biomass. The high moisture content is one of the most significant disadvantage features. Although the combustion reactions are exothermic, the evaporation of water is endothermic. As the moisture content increases, both the higher heating value (HHV) and lower heating value (LHV) decrease. HHV and LHV are used to describe the heat production of a unit quantity of fuel during its complete combustion. In determining the HHV and LHV values of a fuel, the liquid and vapor phases of water are selected as the reference states, respectively. The negative linear relationship exists between the moisture content and the heating value. Fouling (alkali and other elements) and corrosion (alkali, sulfur, chlorine, etc.) of the combustor are typical issues associated with biomass combustion. These are considered to be detrimental because of the resulting reduction in heat transfer in the combustor.

There are a number of combustion methods/technologies/reactors available for biomass combustion and the main ones can be categorized under two headings: Fixed-bed combustion systems and fluidized-bed combustion systems.

6.1.1 Fixed-Bed Combustion

There are two prominent types of fixed-bed combustion: underfeed stokers and grate firings. With these methods of combustion, air is primarily supplied through the grate from below, and initial combustion of solid fuel takes place on the grate and some gasification occurs. This allows for secondary combustion in another chamber above the first where secondary air is added. Generally, fixed-bed combustion is used in small-scale batch furnace for biomass containing little ash. Typical examples of fixed-bed systems are manual-fed systems, spreader-stoker systems, underscrew systems, throughscrew systems, static grates, and inclined grates.

6.1.1.1 UNDERFEED STOKERS

Generally suitable only for small-scale systems, underfeed stokers are a relatively cheap and safe option for biomass combustion. They have the advantage of being easier to control than other technologies, since load changes can be achieved quickly and with relative simplicity due to the fuel feed method. Fuel is fed into the furnace from below by a screw conveyor and then forced upward onto the grate where the combustion process begins. Underfeed stokers are limited in terms of fuel type to low ash content fuels such as wood chips. Due to ash removal problems, it is not feasible to burn ash-rich biomass as this can affect the air flow into the chamber and cause combustion conditions to become unstable.

6.1.1.2 GRATE FIRINGS

There are several different types of grate firing, with both fixed and moving grates commonplace. They have the distinct advantage over underfeed stokers in that they can accommodate fuels with high moisture and ash content as well as with varying fuel sizes. It is very important that fuel is spread evenly over the grate surface in order to ensure that air is distributed uniformly throughout the fuel and thus combustion is kept homogeneous and stable. There are a number of different types of grate firing including fixed grates, moving grates, rotating grates, horizontal/inclined grate, water cooling grate, dumping grate, and travelling grates.

The simplest fixed-bed system is composed of one combustion room with a grate. Generally, as soon as the new biomass feed is added into the furnace, it is pyrolyzed into volatile gases and chars. Primary and secondary air supplies are provided under and above the grate for the combustion of chars and volatile gases, respectively. The heat generated through the combustion of chars is responsible for providing enough heat for the pyrolysis of newly added biomass. Because of the high content of volatile matter in biomass fuels, a greater secondary air supply is required than the primary air supply. This is one of the major differences from the process of coal combustion. Recent developments have been made to enhance the combustion efficiency. One example is the cyclonic combustion system, which may be viewed as a modified fixed-bed system, suitable for the combustion of agricultural residues and particulate wood wastes at a high efficiency (Quaak et al., 1999).

6.1.2 Fluidized-Bed Combustion

Fluidized-bed furnaces operate in quite a different manner from fixed-bed furnaces and have a number of advantages associated with them. Fluidized-bed combustion uses silica sand (lime stone, dolomite, or other noncombustible materials) for bed material, keeps fuel and sand in furnace in boiling state with high-pressure combustion air, and burns through thermal storage and heat transmission effect of sand. It is suitable for high-moisture fuel or low-grade fuel. The typical operating temperatures are lower than fixed-bed systems. Depending on the blowing air velocity, fluidizing-bed systems can be further divided into Bubbling Fluidized-Bed (BFB) and Circulating Fluidized-Bed (CFB).

6.1.2.1 BUBBLING FLUIDIZED BED (BFB) COMBUSTION

The fundamental principle of a BFB furnace is that the fuel is dropped down a chute from above into the combustion chamber where a bed, usually of silica sand, sits on top of a nozzle distributor plate, through which air is fed into the chamber with a velocity of between 1 and 2.5 m/s (http://www.esru.strath.ac.uk/EandE/Web_sites/06-07/Biomass/HTML/combustion_technology.htm). The bed normally has a temperature of between 800 and 900 °C and the sand accounts for about 98% of the mixture, with the fuel then making up a small fraction of the fuel and bed material. BFBs have two main advantages in terms of fuel size and type over more traditional fixed-bed systems. First, they can cope with fuel of varying particle size and moisture content with little problem, and second, they can burn mixtures of different fuel types such as wood and straw. BFBs are only a practical option with larger plants with a nominal boiler capacity greater than 10 MWth.

6.1.2.2 CIRCULATING FLUIDIZED BED (CFB) COMBUSTION

If the air velocity is increased to 5-10 m/s then a CFB system can be achieved, where the sand is carried upward by the flue gases and a more thorough mixing of the bed material and fuel takes place. The sand is then separated from the gas in a hot cyclone or U beam separator at the top of the furnace and fed back into the combustion chamber where the whole process begins again. CFBs deliver very stable combustion conditions, but it involves higher cost. CFB systems exhibit several advantages, such as the adaptation to various fuels with different properties, sizes, shapes, and high moisture (up to 60%), and ash contents up to 50% (http://www.esru.strath.ac.uk/EandE/Web_sites/06-07/Biomass/HTML/combustion_technology.htm).

6.1.3 *Entrained Flow Combustion*

The fuel particles are transported into an externally heated silicon carbide (SiC) tube pneumatically through an insulated and water-cooled injector. Prior to the injection, the feeding stream, composed of air and fuel particles, has to pass through an agitation chamber for "disaggregation and filtering of pulses in the feeding." The feeding fuel is ignited by a natural gas/air burner at the reactor entrance (Jimenez and Ballester, 2006). There are three main stages that occur during biomass combustion: drying, pyrolysis and reduction, and combustion of volatile gases and solid char (IEA, International Energy Agency, Task 32: biomass combustion and co-firing: an overview. http://www.ieabioenergy.com/MediaItem.aspx?id=16).The combustion of volatile gases contributes to more than 70% of the overall heat generation. It takes place above the fuel bed and is generally evident by the presence of yellow flames.

Combined Heat and Power (CHP): Production of electricity and heat from one energy source at the same time is called CHP. In almost all cases, the production of electricity from biomass resources is most economical when the resulting waste heat is also captured and used as valuable thermal energy—known as CHP or cogeneration (http://www.epa.gov/chp/documents/biomass_fs.pdf). Biomass is most economical as a fuel source when the CHP system is located at or close to the biomass feed stock. In some cases, the availability of biomass in a location may prompt the search for an appropriate thermal host for a CHP application. In other circumstances, a site may be driven by a need for energy savings to search for biomass fuel within a reasonable radius of the facility (http://www.epa.gov/chp/basic/renewable.html).

Using biomass instead of fossil fuels to meet energy needs with CHP provides many potential environmental and economic benefits, which can include (i) reduced greenhouse gas and other emissions, (ii) reduced energy costs, (iii) improved local economic development, (iv) reduced waste, (v) expanded domestic fuel supply, (vi) reduced transmission and distribution losses. CHP offers distributed generation of electrical and/or mechanical power; waste heat recovery for heating, cooling, or process applications; and seamless system integration for a variety of technologies, thermal applications, and fuel types into existing building infrastructure. CHP systems typically achieve total system efficiencies of 60-80% for producing electricity and thermal energy (http://www.epa.gov/chp/documents/biomass_fs.pdf).

6.2 Carbonization

Biomass such as woody waste and food waste can be converted to a renewable energy source by means of carbonization processes. Carbonization processes for biomass is one of several technologies concerned with producing renewable energy sources and effectively reducing greenhouse gas production. Carbonization is done to obtain charcoal by heating solid biomass in the absence of air or oxygen. Carbonization is the term for the conversion of an organic substance into carbon or a carbon-containing residue through pyrolysis or destructive distillation. When biomaterial is exposed to sudden searing heat, it can be carbonized extremely quickly, turning it into solid carbon. From the point of view of waste, woody waste, food waste, and sewage sludge can be considered to contribute to biomass. The basic characteristics of woody waste and food waste, such as proximate analysis and heating value, are evaluated before carrying out carbonization tests. Medium-sized and small enterprises have been using carbonization technology for biomass, but the method is not used in large-scale operations because the production of carbonization residue by conventional technology is inefficient and uneconomical.

6.2.1 *Hydrothermal Carbonization (HTC)*

HTC is a thermochemical conversion process for biomass to yield a solid, coal-like product. It has been used for almost a century in different sciences, mainly to simulate natural coalification in the laboratory. Due to the need for efficient biomass conversion technologies, HTC has attracted some interest as a possible application for biomass in recent years, and R&D projects have been launched to assess its feasibility and discover additional possibilities for applications. HTC has been in use as a method for simulating natural coalification in coal petrology for nearly a century, and many experimental results have been published. It was introduced to this research field by Bergius as early as 1913 and was discussed controversially from then on. HTC is an exothermic process that lowers both the oxygen and hydrogen content of the feed (described by the molecular O/C and H/C ratio) by mainly dehydration and decarboxylation to raise its carbon content with the aim of achieving a higher calorific value. This is achieved by applying temperatures of 180-200 °C in a suspension of biomass and water at saturated pressure for several hours. With this conversion process, a lignite-like, easy-to-handle fuel with well-defined properties can be created from biomass residues, even with high moisture content. Thus, it may contribute to a wider application of biomass for energetic purposes (Behar and Hatcher, 1995; Funke and Ziegle, 2009; Mukherjee et al., 1996; Payne and Ortoleva, 2001; Ross et al., 1991; Siskin and Katritzky, 1991; Wolfs et al., 1960).

Many chemical reactions that might appear during HTC have been mentioned throughout the literature, but just few have been the focus of detailed investigations, for example, the hydrolysis of cellulose. It has been realized that the process is governed in sum by dehydration and decarboxylation, which means that it is exothermal. Simultaneously, functional groups are being eliminated to some extent. But the complex reaction network is not known in detail. So, for the time being, only a separate discussion of general reaction mechanisms that have been identified can provide useful information about possibilities of manipulating the reaction. These mechanisms include hydrolysis, dehydration, decarboxylation, condensation polymerization, and aromatization. They do not represent consecutive reaction steps but rather form a parallel network of different reaction paths. It is understood that the detailed nature of these mechanisms, as well as their relative significance during the course of reaction, primarily depends on the type of feed.

Although HTC has been known for nearly a century, it has received little attention in current biomass conversion research. Although it received great attention for biomass liquefaction and gasification, a technical implementation of HTC has only been developed with comparably low effort. This may be due to the fact that coal as an energy carrier is inferior to liquid or gaseous fuels. On the other hand, process requirements of HTC are comparably low while producing a fuel that is easier to handle and store because it is stable and nontoxic. Due to these facts, HTC may provide some advantages when considering small-scale, decentralized applications. Moreover, it might become a viable option for the production of functional carbonaceous materials.

The mildest reaction conditions in terms of temperature and pressure are employed in HTC. Lignocellulosic substrates have been extensively examined (Titirici et al., 2007) as reactants at temperatures from 170 to 250 °C over a period of a few hours to a day (Heilmann et al., 2010). Latest research on HTC focused on the preparation of functional carbonaceous materials and achieved interesting results for a future application to produce even more value-added materials. Low-value and widely available biomass can be converted into interesting carbon nanostructures using environment-friendly steps. These low-cost nanostructured carbon materials can then be designed for applications in crucial fields such as separation, energy conversion, and catalysis. Besides controlling the chemistry of carbonization (i.e., C—C linkage), two other important prerequisites for the achievement of useful properties are the control over morphology both at nano- and macroscale and the control over functionality by chemical means in HTC (Titirici and Antonietti, 2010).

6.2.2 Microwave-Assisted Hydrothermal Carbonization (MAHC)

The process uses microwave heating at 200 °C in acidic aqueous media to carbonize pine sawdust (Pinus sp.) and α-cellulose (Solucell®) at three different reaction times. Elemental analysis showed that the lignocellulosic samples subjected to MAHC yielded carbon-enriched material 50% higher than raw materials. In order to qualitatively evaluate the carbonization process, H/C and O/C were plotted using the van Krevelen (1950) diagram, which provides information about the changes in chemical structure after carbonization. These results showed that microwave-assisted HTC is an innovative approach to obtain carbonized lignocellulosic materials (Guiotoku et al., 2009).

6.3 Gasification

Gasification is the conversion of solid raw material into fuel gas or chemical feedstock gas otherwise called as synthesis gas, which can be upgraded to liquid fuels (diesel and gasoline) by Fischer-Tropsch synthesis. Biomass gasification is a process that converts carbonaceous biomass into combustible gases (e.g., H_2, CO, CO_2, and CH_4) with specific heating values in the presence of partial oxygen (O_2) supply (typically 35% of the O_2 demand for complete combustion) or suitable oxidants such as steam and CO_2.

When air or oxygen is employed, gasification is similar to combustion, but it is considered a partial combustion process. In general, combustion focuses on heat generation, whereas the purpose of gasification is to create valuable gaseous products that can be used directly for combustion, or be stored for other applications. In addition, gasification is considered to be more environmentally friendly because of the lower emissions of toxic gases into the atmosphere and the more versatile usage of the solid byproducts (Rezaiyan and Cheremisinoff, 2005).

Gasification can be viewed as a special form of pyrolysis, taking place at higher temperatures to achieve higher gas yields. Biomass gasification offers several advantages, such as reduced CO_2 emissions, compact equipment requirements with a relatively small footprint, accurate combustion control, and high thermal efficiency (Marsh et al., 2007;

Rezaiyan and Cheremisinoff, 2005). Gasification is normally carried out at temperatures over (727 °C)1000 K, but recently it has been demonstrated that H_2 and CO can be produced through the aqueous phase reforming of glycerol at lower temperatures <347 °C (<620 K) (Simonetti et al., 2007; Soares et al., 2006) at which integration of syngas production with FT upgrading is feasible. The ratio of CO/H_2 can be modified by the water gas shift reaction ($CO + H_2O \rightarrow CO_2 + H_2$).

The classification of gasification is based on several parameters such as types of gasifiers, gasification temperature, heating (direct or indirect), and gasification agent.

6.3.1 *Types of Gasifiers*

6.3.1.1 FIXED-BED GASIFIERS

Fixed-bed gasifiers generally produce low-heating-valued syngas. They are suitable for small or medium-scale thermal applications.

6.3.1.1.1 UPDRAFT (COUNTER-CURRENT) GASIFIERS The updraft gasifier is the simplest type of gasifier. The biomass is fed at the top while the air is injected at the bottom. Biomass and air move in a countercurrent direction. During its downward movement, biomass is firstly dried passing through a "drying zone." In the "distillation zone," biomass undergoes decomposition and is converted into volatile gases and solid char. The gases and char will be further converted into CO and H_2 as they pass through "reduction zone". Since some of the char settles down in the bottom of the reactor, heat is generated through its combustion in the "hearth zone" and is transported upward by the upflowing gas to maintain the pyrolysis and drying processes. In addition, CO_2 and H_2O vapor is also produced from char combustion.

Updraft gasifiers can accept biomass with relatively high moisture content (up to 60%). However, the resulting product gas has high tar content because the tar, newly formed during pyrolysis, does not have the opportunity to pass through the combustion zone.

6.3.1.1.2 DOWNDRAFT (CO-CURRENT) GASIFIERS The downdraft gasifier is currently one of the most widely used fixed-bed gasification systems. Different from the updraft gasifier, air in the downdraft gasifier is introduced into the reactor from the middle part. This design leads to the reversed order of the hearth zone and the reduction zone. In this gasifier, the injected air and biomass move cocurrently.

6.3.1.1.3 CROSS-FLOW GASIFIERS In a crossflow gasifier, biomass is added at the top of the reactor and moves downward. Air is introduced from one side of the reactor and the gas products are released from the other side of the reactor on the same horizontal level.

6.3.1.1.4 OPEN-CORE GASIFIERS Open-core gasifiers are generally employed to gasify biomass with low bulk density and high ash content. An example of this kind of biomass is rice husk. Instead of the narrow throat characteristic of other gasifiers, the open-core gasifier has a wide mouth for biomass injection to prevent fuel flow inhibition caused by bridging.

6.3.1.2 FLUIDIZED-BED GASIFIERS

Fluidized-bed reactors are widely employed as gasifiers. Fluidized-bed gasifiers can also be further classified into bubbling fluidized gasifiers and circulating fluidized gasifiers. In a bubbling fluidized gasifier, air is injected from the bottom of a grate, above which the moving bed is mixed with the biomass feed. The bed temperature is maintained at 700-900 °C. Biomass is pyrolyzed and cracked through contact with the hot bed material. In a circulating fluidized gasifier, the hot bed material is circulated between the reactor and a cyclone separator. During this circulation, bed materials and char go back to the reactor, while the ash is separated and removed from the system.

6.3.1.3 ENTRAINED FLOW GASIFIERS

In an entrained flow gasifier, the feed and air move cocurrently and the reactions occur in a dense cloud of very fine particles at high pressures, varying between 19.7 and 69.1 atm and very high temperatures >1000 °C. This type of gasifier has an elevated throughput of syngas (Zhang et al., 2010).

6.3.2 Low/High-Temperature Gasification

High-temperature gasification (typically above 1200 °C) results in a gas, which merely contains H_2 and CO as combustible components. At low-temperature however (typically below 1000 °C), also hydrocarbons are present in the gas. A CFB gasifier operated on biomass operated at 900 °C typically produces a gas containing 50% hydrocarbons (mainly methane, ethylene, and benzene) on energy basis (http://www.biosng.com/experimental-line-up/gasification-technology/).

6.3.3 Heating Source for Gasification

6.3.3.1 INDIRECT (OR ALLOTHERMAL) GASIFICATION

It is characterized by the separation of the processes of heat production and heat consumption. It therefore generally consists of two reactors connected by an energy flow. The biomass is gasified in the first reactor and the remaining solid residue (char) is combusted in the second reactor to produce the heat for the first process. Hot sand is circulated to transport the heat from the combustor to the gasifier. These indirect gasifiers theoretically are operated at an equilibrium based on the temperature dependence of the char yield in the gasifier. This means that at a low temperature, much char is remaining from the gasifier. Since this char is combusted to produce the heat, the temperature will rise until char yield matches the energy demand of the gasification (http://www.biosng.com/experimental-line-up/gasification-technology/).

6.3.3.2 PLASMA GASIFICATION

Plasma gasification is a gasification process that decomposes biomass into basic components, such as H_2, CO, and CO_2 in an oxygen-starved environment at an extremely high temperature. Plasma is regarded as the 4th state of matter; it is an ionized gas produced by electric discharges. A plasma torch is a tubular device that has two electrodes to produce an arc. It is an independent heat source that is neither affected by the feed characteristics nor the air/oxygen/steam supply. When electricity is fed, an arc is created, and the electricity is converted into heat through the resistance of the plasma. A plasma torch can heat the biomass feedstock to a temperature of 3000 °C or higher (up to 15,000 °C). Under such extremely elevated temperature, the injected biomass stream can be gasified within a few milliseconds without any intermediate reactions. The plasma technique has high destruction and reduction efficiencies. Any form of wastes, for example, liquid or solid, fine particles or bulk items, dry or wet, can be processed efficiently. In addition, it is a clean technique with little environmental impact. Plasma technique has great application potential for treating a wide range of hazardous wastes (Zhang et al., 2010).

6.3.3.3 CONCENTRATING-SOLAR BIOMASS GASIFICATION (CSBG)

The concept's key feature is the use of high-temperature heat from a solar-concentrating tower to drive the chemical process of converting biomass to a biofuel, obtaining a near-complete utilization of carbon atoms in the biomass. The aim of the concept is to obtain an easy to handle fuel with near-zero CO_2 emission and reduced land-use requirements compared to first- and second-generation biofuels. H_2 from water electrolysis with solar power is used for reverse water gas shift to avoid producing CO_2 during the process. The solar-driven third-generation biofuel requires only 33% of the biomass input and 38% of total land as the second-generation biofuel, while still exhibiting a CO_2-neutral fuel cycle. With CO_2 capture, second-generation biofuel would lead to the removal of 50% of the carbon in the biomass from the atmosphere. There is a trade-off between reduced biomass feed costs and the increased capital requirements for the solar-driven process; it is attractive at intermediate biomass and CO_2 prices (Hertwich and Zhang, 2009).

6.3.4 Gasification Agent

6.3.4.1 OXYGEN-BLOWN GASIFICATION

It is an alternative route for the production of a nitrogen-free product gas. To prevent local hotspots in the reactor, the oxygen is normally diluted with steam or CO_2. The methane content drops with increasing steam/O_2 and CO_2/O_2 ratio. The decrease on dry gas basis is mainly caused by the dilution by CO_2 or H_2 that is produced from steam by the CO shift reaction.

A low steam or CO_2/O_2 ratio produces a product gas with the highest CH_4 content, which is desired for synthetic natural gas (SNG) production. A low amount of CO_2 or steam also increases the gasifier efficiency, because less "inert" gas needs to be heated to the process temperature. A certain amount of oxygen dilution is required to prevent possible agglomeration of biomass (http://www.biosng.com/experimental-line-up/o2-blown-gasification/).

6.3.4.2 SUPER CRITICAL GASIFICATION OF BIOMASS

Super critical water gasification (SCWG) technology is suitable for wet biomasses and organic wastes. This technology takes advantage of the large amount of water in biomasses by using the water as a reaction medium, eliminating the costly feedstock-drying step. Supercritical water has a low dielectric constant close to that of organic compounds. The organic reactions under supercritical water, therefore, become more homogeneous, resulting in a higher reaction rate. The free radical condition of supercritical water also enhances the gas formation, leading to the high gas yield. As compared to conventional dry gasification,

SCWG produces a lower amount of tarry material and char as byproduct, due to the higher solubility and reactivity of the organic compounds in supercritical water. Nevertheless, because tar and char are difficult to gasify, they act as a drier to achieve complete gasification. The formation of tar and char also causes a reduction in the energy efficiency of the process by means of reactor plugging, heat exchanger fouling, and catalyst deactivation (Chuntanapum and Matsumura, 2010).

6.3.4.3 HYDROTHERMAL GASIFICATION OF BIOMASS

Hydrothermal gasification is the conversion of solid biomass into gaseous and/or liquid products in the presence of steam. Different hydrothermal biomass gasification processes are under development. In contrast to biomass gasification processes without water, biomass with the natural water content ("green biomass") can be converted completely and energetically efficiently to gases. Depending on the reaction conditions, methane or hydrogen is the burnable gas produced. Some processes use catalysts. In recent years, significant progress was achieved in the development of various hydrothermal biomass gasification processes. However, some challenges still exist and technical solutions are needed before large-scale production facilities can be built (Kruse, 2009).

6.4 Pyrolysis

Pyrolysis is the fundamental chemical reaction process that is the precursor of both the gasification and combustion of solid fuels, and is simply defined as the chemical changes occurring when heat is applied to a material in the absence of oxygen. The products of biomass pyrolysis include water, charcoal (carbonaceous solid), pyrolysis oils or tars, and permanent gases including methane, hydrogen, carbon monoxide, and carbon dioxide. The nature of the changes in pyrolysis depends on the material being pyrolyzed, the final temperature of the pyrolysis process, and the rate at which it is heated up. The pyrolysis process is a mildly endothermic reaction. The heat of vaporization of pure water is 2.26 KJ g^{-1} at 100 °C, while the chemical energy content of wood is only about 18.6 KJ g^{-1}. Most of the energy obtained from biomass goes in moisture removal. This reinforces the facts that lower the moisture content, greater is the energy obtained.

As typical lignocellulosic biomass materials such as wood, straws, and stalks are poor heat conductors, management of the rate of heating requires that the size of the particles being heated be quite small. Otherwise, in massive materials such as logs, the heating rate is very slow, and this determines the yield of pyrolysis products. Depending on the thermal environment and the final temperature, pyrolysis will yield mainly char at low temperatures, <450 °C, when the heating rate is quite slow, and mainly gases at high temperatures, >800 °C, with rapid heating rates. An intermediate temperature and under relatively high heating rates, the main product is a liquid bio-oil, a relatively recent discovery, which is just being turned to commercial applications. There are 3 stages in the pyrolysis process: The first stage, prepyrolysis, occurs between 120 and 200 °C with a slight observed weight loss, when some internal rearrangements, such as bond breakage, the appearance of free radicals, and the formation of carbonyl groups take place, with a corresponding release of small amounts of water (H_2O), carbon monoxide (CO), and CO_2. The second stage is the main pyrolysis process, during which solid decomposition occurs, accompanied by a significant weight loss from the initially fed biomass. The last stage is the continuous char devolatilization caused by the further cleavage of C—H and C—O bonds.

In reacting chemical systems, the term severity is used to capture the idea that both the duration of heating and the final temperature influence the chemical products of pyrolysis. Very-low-severity treatments of short duration to a maximum temperature of about 250 °C are sometimes called torrefaction and result in a product that has lost some H_2O and CO_2 from pyrolysis while retaining almost all of the heat value. Traditional charcoaling is a medium-severity process, while the production of bio-oils is a short-duration high-severity process, which, if the duration at high temperature is maintained, will go all the way to gas and soot.

Depending on the reaction temperature and residence time, pyrolysis can be divided into fast pyrolysis, intermediate pyrolysis, and slow pyrolysis. Typically, fast pyrolysis has an extremely short residence time (~1 s); the reaction temperature is approximately 100 °C higher than that of slow pyrolysis (e.g. ~500 °C vs. ~400 °C). Short reaction times combined with an elevated temperature generally result in a higher yield of liquid product. A conventional moderate or slow pyrolysis process, with a relatively long vapor residence time and low heating rate, has been used to produce charcoal for thousands of years (Zhang et al., 2010).

Among the short residence-time processes (0.5-5 s) under development are vacuum pyrolysis at about 300-400 °C and 0.3 atm (U. of Sherbrooke, Canada), flash pyrolysis at about 500-650 °C and 1 atm (U. of Waterloo, Canada), hydropyrolysis in an atmosphere of hydrogen at about 500-600 °C and 5-6 atm (HYFLEX TM, IGT), and flash pyrolysis in atmospheres of hydrogen or methane at 600-1000 °C and 1-70 atm (Brookhaven National Laboratory). An interesting report of a relatively long residence time (10-15 min heat up, several hours at temperature) pyrolysis study at reduced pressures of 0.0004-0.004 atm and temperatures of 250-320 °C of wild cherry wood seems to contrast with the results of several reports on flash pyrolysis (http://journeytoforever.org/biofuel_library/liquefaction.html).

6.4.1 *Slow Pyrolysis*

Heating of the lignocellulosic biomass in inert atmosphere for hours to a maximum temperature of 400-500 °C is called slow pyrolysis. The charcoal yield is 35-40% by weight. In general, the yield of liquid products would be less than the fast pyrolysis of biomass. Several types of catalysts can be employed for the pyrolysis of biomass and/or upgradation of the vapors produced from the thermal pyrolysis.

6.4.2 *Fast Pyrolysis*

The goal of fast pyrolysis is to produce liquid fuel from lignocellulosic biomass that can substitute for fuel oil in any application. The liquid can also be used to produce a range of specialty and commodity chemicals. The essential features of a fast pyrolysis process are very high heating and heat transfer rates, which often require a finely ground biomass feed. Carefully controlled reaction temperature of ca. 500 °C in the vapor phase and residence time of pyrolysis vapors in the reactor less than 1 s; and then quenching (rapid cooling) of the pyrolysis vapors to give the bio-oil product. The main product of fast pyrolysis is bio-oil, which is obtained in yields of up to 80 wt% of dry feed.

Fast pyrolysis is a promising process to produce transportable oil with a high volumetric energy density from bulky and inhomogeneous biomass. There are several applications foreseen for pyrolysis oil. It has been tested as a substitute for fuel oil or diesel in boilers, furnaces, engines, and turbines for heat and power generation and has been considered as a precursor for transportation fuels and chemicals. Water is the most abundant component in pyrolysis oil; typically, it is present in the range of 15-35 wt%. Probably all applications require different specifications with respect to the water content of pyrolysis oil. For fueling into a diesel engine, the water content should be below 30 wt% to decrease emissions of particles and to prevent ignition delay and phase separation. But there should also be a minimum amount of water present to limit NO_x emissions and to ensure a uniform temperature distribution in the cylinders. For cofeeding pyrolysis oil in a mineral oil refinery, nearly all water and most organically bound oxygen must be removed. Generally, less water is beneficial for the energy density, transportation costs, stability, and acidity of pyrolysis oil. Fast pyrolysis oil possesses many undesirable properties including a high total acid number (TAN ~200), low heating value (~6560 BTU/lb), high oxygen content (~40%), chemical instability, high water content (20%), and incompatibility with petroleum fractions. Inherent low-energy density makes pyrolysis oil expensive to transport, and the high TAN makes it metallurgically incompatible with conventional transport vessels and refinery hydroconversion equipment, both designed for feeds with TANs less than 2. In addition to these undesirable properties, pyrolysis oil is not miscible with petroleum fractions and if added into existing refinery equipment (hydrotreaters or hydrocrackers) will require a separate pyrolysis-oil feed system. Thus, pyrolysis oil needs effective pretreatment and upgradation before it is used as crude oil replacement. Depending on the reactors used, we have many kinds of fast pyrolysis processes.

6.4.2.1 ABLATIVE FAST PYROLYSIS

Ablative pyrolysis, in which much larger particle sizes can be employed than in other systems, as the heat is transferred from a hot surface to the biomass particle and the process, is limited by the rate of heat supply to the reactor rather than the rate of heat absorption by the pyrolyzing biomass. Ablative pyrolysis is fundamentally different from fluid bed processes from the mode of heat transfer through a molten layer at the hot reactor surface, use of large particles, and absence of a fluidizing gas.

6.4.2.2 VORTEX REACTOR

A vortex tube has certain advantages as a chemical reactor, especially if the reactions are endothermic, the reaction pathways are temperature dependent, and the products are temperature sensitive. With low-temperature differences, the vortex reactor can transmit enormous heat fluxes to a process stream containing entrained solids. This reactor has nearly plug flow and is ideally suited for the production of pyrolysis oils from biomass at low pressures and residence times to produce about 10 wt% char, 13% water, 7% gas, and 70% oxygenated primary oil vapors based on mass balances. This product distribution was verified by carbon, hydrogen, and oxygen elemental balances. The oil production appears to form by fragmenting all of the major constituents of the biomass. Cyclonic fast pyrolysis, also called vortex fast pyrolysis, separates the solids from the noncondensable gases and returns them to the mixer.

6.4.2.3 ROTATING CONE FAST PYROLYSIS: ROTATING CONE REACTOR

The rotating cone reactor is a novel reactor type for fast pyrolysis of biomass with negligible char formation, in which rapid heating and short residence time of the solids can be realized. Particles fed into the reactor first enter an impeller which is mounted in the base of the heated cone. After leaving the impeller, the particles flow outward over the conical surface and experience a high heat transfer rate due to their small distance from the heated surface. Biomass materials like wood, rice husks, or even olive stones can be pulverized and fed to the rotating cone reactor. Flash heating of the biomass will suppress coke-forming cracking reactions. Since no carrier gas is needed (cost reducing), the pyrolysis products will be formed at high concentrations. If additional thermal quenching of the gas outlet flow is applied, the amount of secondary tar decomposition reactions can be suppressed. In the rotating cone reactor, wood particles fed to the bottom of the rotating cone, together with an excess of inert heat carrier particles, are converted while being transported spirally upward along the cone wall. The cone geometry is specified by a top angle of $\pi/2$ radians and a maximum diameter of 650 mm. Products obtained from the flash pyrolysis of wood dust in a rotating cone reactor are noncondensable gases, bio-oil, and char. The biomass decomposes into 70% condensable gases with 15% noncondensable gases and 15% char.

6.4.2.4 BUBBLING FLUIDIZED BED (BFB) PYROLYSIS

A simple method for the rapid heating of biomass particles is to mix them with the moving sand particles of a high-temperature fluid bed. High heat transfer rates can be achieved, as the bed usually contains small sand particles, generally about 250 μm. The heat required is generated by combustion of the pyrolysis gases, and/or char, and eventually transferred to the fluid bed by heating coils. While the sand to biomass heat transfer is excellent (over 500 W/m^2 K), the heat transfer from the heating coils to the fluid bed will be low, due to the resistance inside the coils (gas to coil wall heat transfer estimated 100-200 W/m^2K), and the limiting driving force of around 300 °C as a maximum. In an optimistic case, at least 10-20 m^2 surface area is required per ton/h of biomass fed.

6.4.2.5 CIRCULATING FLUIDIZED BED (CFB) PYROLYSIS

CFB reactor has been widely used for the pyrolysis of lignocellulosic biomass into high yield of liquid products (Rapid Thermal Process, RTP; UOP).The CFB reactor has many advantages, for example, the simple structures, high production capacity, favorable conditions of heat and mass transfer, and the convenience of operation, etc., the CFB was used as the main reactor in this study. To reduce the operation cost, part of the pyrolysis gas was used as the carrier gas, while the rest and the pyrolysis char were recycled as heat.

The CFB could be divided into two zones corresponding to the main chemical processes. (i) pyrolysis zone: In this zone, feedstock was loaded into the bed and pyrolyzed very quickly. Since the feedstock particles were small and the heat exchanged rapidly, the heating rate was very high. For example, a small particle at 0.1-0.2 mm diameter could be heated at the rate of about 103 °C/s in an atmosphere at 1000 °C. In this zone, the main chemical process could be described as

$$\text{Biomass} \rightarrow \text{char} + \text{tar} + H_2O + \text{gas}\ (CO_2, CO, CH_4, C_nH_m, H_2).$$

Temperature was another essential factor affecting the pyrolysis besides heating rate. Because the relatively high temperature was favorable to form more noncondensable gas and decrease the tar yield, moderate and carefully controlled temperature was needed.

(ii) Reduction and cracking zone: Before the pyrolysis vapors were quenched by the condenser, further reactions had taken place; for example, the tar cracked and the char was reduced. These processes produced more noncondensable gas such as CO and H_2. Some C_nH_m also cracked at the same time. The main reactions could be expressed as

$$C + H_2O \rightarrow CO + H_2,$$

$$C + CO_2 \rightarrow 2CO,$$

$$CH_4 + H_2O \rightarrow CO + 3H_2,$$

$$CO + H_2O \rightarrow CO_2 + H_2,$$

$$\text{Tar} \rightarrow CH_4 + H_2O + C_nH_m + H_2.$$

Pyrolysis char contributed to secondary cracking by catalyzing secondary cracking in the vapor phase; rapid object of gasification is to get high-quality gas product. Thus, the high temperature of up to 900 °C is wanted to increase the gas product and decrease the tar, while the relatively long residence time contributes to the secondary reactions including char reduction, tar cracking, shift reaction, etc. So the amount of CO_2, CO, CH_4, and H_2 is far more, and the amount of C_nH_m is less in the gas product of gasification. By contrast, the objective of fast pyrolysis is to obtain more liquid product; it determines the operation conditions of moderate temperature and short residence time to increase the liquid production rate. Such operation conditions lead to the higher amount of C_nH_m and less amount of CO, CH_4, and H_2, which indicate that the degree of pyrolysis is not excessive.

6.4.2.6 AUGER (SCREW) REACTOR

The auger type of pyrolyzer has been identified as especially appealing for its potential to reduce operating costs associated with bio-oil production. This design may also be well suited for small, portable pyrolysis systems in a highly distributed or decentralized biomass processing scheme. The operating principle of this design is that biomass is continuously pyrolyzed by being brought into direct contact with a bulk solid heat transfer medium referred to as a "heat carrier." The heat carrier material, such as sand or steel shot, is heated independently before being metered into the reactor. On a gravimetric basis, thermodynamic calculations suggest a heat carrier feed rate 20 times the biomass feed rate. Two intermeshing, co-rotating 1inch augers quickly combine biomass and heat carrier in a shallow bed to effectively carry out the pyrolysis reactions. This mechanical mixing process, though not well understood, appears to be the essence of this alternative pyrolyzer design. Volatile vapors and aerosols exit at various ports, while char is transported axially through the 20 inch long reactor section and stored in a canister with the heat carrier (Brown, 2009).

6.4.3 *Ultra Fast Pyrolysis*

The ultra fast high-temperature pyrolysis will be carried out in a high-temperature fluid-wall reactor which can withstand working temperatures of up to 2200° C. The biomass is fed to the top of the reactor (rate of 1.0-1.8 kg/min). The feed falls and, at the same time is very quickly heated by radiation to the reaction temperature. The estimated heating rate is on the order of 10^6 °C/s for reactant surfaces. The fluid wall, produced by a nitrogen flow through the 30-cm diameter porous reactor core, prevents both reactants and products from reaching the reactor wall. The product distribution at the reactor exit has been determined for different operating conditions. The influence of reactor temperature, biomass feed rate, and biomass particle size on the product distribution and on the heating value of the exit gas has been investigated (Corella et al., 1988).

6.4.4 Hydropyrolysis

A better approach for biomass conversion is the integrated hydropyrolysis and hydroconversion of biomass to directly produce fungible gasoline and diesel fuel or blending components is carried out in two integrated stage. The first stage is a medium pressure, catalytically assisted, fast hydropyrolysis step completed in a fluid bed under moderate hydrogen pressure. Vapors from the first stage pass directly to a second-stage hydroconversion step where a hydrodeoxygenation catalyst removes all remaining oxygen and produces gasoline and diesel boiling range material. All the process steps are completed at essentially the same pressure, so that compression costs are minimized. A unique feature of this process is that all the hydrogen required for this process is produced by reforming the C_1—C_3 hydrocarbons, so no additional hydrogen is required. Pyrolysis is carried out in the presence of hydrogen at high pressure. The advantage of hydropyrolysis is the high quality of the products at the maximum liquid yield. The disadvantage is the high hydrogen consumption, which leads to high processing costs, but this is only a short-term economic consideration. If the H_2 from a carbon-free source becomes cost competitive, the hydropyrolysis can become commercially exploitable technique for the conversion of lignocellulosic biomass with complete utilization of carbon content (Agrawal and Singh, 2009; Marker et al., 2009).

6.4.5 Vacuum Pyrolysis

Vacuum or vacuum moving bed pyrolysis includes a combination of slow and fast pyrolysis conditions. Course solids are heated relatively slowly to temperatures higher than those of slow pyrolysis, while the gas is removed from the hot temperature zone relatively quickly by applying a reduced pressure of less than 0.20 atm in the process. Vacuum pyrolysis is not a rapid heating technique and is in the same thermal regime of time and temperature as slow pyrolysis (charcoal production); however, yields of liquid that are over 50% of the original biomass are achieved by removing the vapors as soon as they are formed by operating under a partial vacuum—essentially the converse of work at high pressures, in which the liquids are held in the charring mass to increase the yield of char.

6.5 Liquefaction

Hydrothermal liquefaction is the conversion of solid biomass into gaseous and/or liquid products in the presence of water. Liquefaction consists of the catalytic thermal decomposition of large molecules to unstable shorter species that polymerize again into a bio-oil.

Biomass is mixed with water and basic catalysts like sodium carbonate, and the process is carried out at lower temperatures than pyrolysis (252-472 °C) but higher pressures (50-150 atm) and longer residence times (5-30 min.). These factors combine to make liquefaction a more expensive process; however, the liquid product obtained contains less oxygen (12-14%) than the bio-oil produced by pyrolysis and typically requires less extensive processing (Elliott, 2007; Inoue et al., 1999; Karagoz et al., 2006; Kruse et al., 2003; Minowa et al., 1997).

6.5.1 Direct and Indirect Liquefaction

Currently, more research is being done on direct and indirect thermal liquefaction methods for biomass and wastes than on the other methods. Direct liquefaction is either reaction of biomass components with smaller molecules such as H_2 and CO (e.g., Pittsburg Energy Research Centre (PERC) and Lawrence Berkeley Laboratory, Berkeley, USA (LBL) processes) or short-term pyrolytic treatment, sometimes in the presence of gases such as H_2. Indirect liquefaction involves successive production of an intermediate, such as synthesis gas or ethylene, and its chemical conversion to liquid fuels, In 1983, after several years of laboratory and pilot-plant work on the PERC and LBL processes, which involve reaction of product oil or water slurries of wood particles with H_2 and CO at temperatures up to about 370 °C and pressures up to 272 atm in the presence of sodium carbonate catalyst, researchers concluded that neither process can be commercialized for liquid fuel production without substantial improvement. The most attractive approach to such improvement is believed to be a combination of solvolysis with a pyrolysis or reduction step. However, the oxygen content of the resulting complex liquid mixture is still high (6-10 wt%), and considerable processing is necessary to upgrade this material (http://journeytoforever.org/biofuel_library/liquefaction.html).

Direct liquefaction has some similarity with pyrolysis in terms of the target products (liquid products). However, they are different in terms of operational conditions. Specifically, direct liquefaction requires lower reaction temperatures but higher pressures than pyrolysis (0.5-2 atm for liquefaction vs. 0.01-0.05 atm for pyrolysis). In addition, drying of the feedstock is not a necessary step for direct liquefaction, but it is crucial for pyrolysis. Moreover, catalysts are always essential for liquefaction, whereas they are not as critical for pyrolysis. At the beginning of the liquefaction process, biomass undergoes depolymerization and is decomposed into monomer units. These monomer units, however, may be repolymerized or condensed into solid chars, which are undesirable (Zhang et al., 2010).

6.6 Co-processing

Investigation and large-scale application of co-gasification and co-pyrolysis of biomass and coal are becoming more common recently. In addition to the reduction of CO_2 emission, cogasification of biomass provides several advantages over biomass or coal gasification (Kumabe et al., 2007). One of the advantages is the reduction of sulfur and ash that cause equipment corrosion and environmental problems in coal gasification (Chmielniak and Sciazko, 2003; McLendon et al., 2004). It can also reduce the high cost of the feedstock and high tar generation in biomass gasification.

Likewise, co-pyrolysis has advantages over sole biomass or coal pyrolysis. Although pyrolysis of coal is a good method for producing liquid fuels, the yields of these products are limited because of the low H/C ratio in coal. The high H/C ratio in biomass renders biomass to act as a hydrogen donor in co-pyrolysis of biomass/coal blends. Moreover, the high thermochemical reactivity and high content of volatiles of biomass facilitate the conversion and the upgrading of the fuel. Therefore, it's considered promising to co-fire the two fuels as a step toward valid, sustainable utilization of coal and biomass and to minimize the impact on the environment.

Pyrolysis gas has a high heating value, 17 MJ/kg (http://www.nh.gov/oep/programs/energy/documents/biooil-nrel.pdf), and both pyrolysis oil and char can be gasified to produce syngas; it is a promising technique to further process the pyrolysis products through gasification to produce syngas more efficiently. Here, the process consists of the pyrolysis and subsequent gasification sections. In the first reactor, biomass is pyrolyzed with coal at 500-700 °C. The pyrolysis gas is quenched to produce liquid oil, and the char is flowed to the gasifier where steam and limited air are supplied to produce syngas.

6.7 Hydrolysis

Hydrolysis pathways are appropriate for lignocellulose processing if higher selectivity is desired in biomass utilization, for example, in the production of chemical intermediates or targeted hydrocarbons for transportation fuel. Selective transformations require isolation of sugar monomers, a step which is complex and expensive for lignocellulosic feedstocks. Once sugar monomers are isolated, however, they can be processed efficiently at relatively mild conditions by a variety of catalytic technologies (Alonso et al., 2010).

The ability to recover and use the major components of lignocellulosic biomass (cellulose, hemicellulose, lignin) is critical in developing economically viable bioproducts and biorefineries. This project focuses on the biomass pretreatment step of hemicellulose acid hydrolysis to recover the hemicellulose sugars and prepare the biomass for subsequent enzymatic or acid cellulose conversion. The ultimate goal is to identify promising routes to reduce the sugar production cost by 30% compared with established methods. Researchers are investigating three hydrolysis systems: water-rich hydrolysis, water-restricted, and near neutral pH. Using different reactor configurations (e.g., batch tube, Parr, flow through) with varying solids and pH levels, researchers have developed comprehensive data on the destructuring, disaggregation, and depolymerization of hemicellulose to sugars. Flow rate has been found to enhance hemicellulose removal, which is inconsistent with models typically applied to describe hemicelluloses hydrolysis. New models have been defined that reveal mass transfer could be important in explaining this anomaly. The flow through reactor experiments showed that lignin is modified as hemicellulose reacts, and the resulting disruption of lignin may play a significant role in enhancing cellulose digestion. In addition, researchers have shown that nonproductive adsorption on lignin can be reduced by prior treatment with low-cost proteins, thereby substantially cutting enzyme costs (Iranmahboob et al., 2002; Mosier et al., 2005; Patrick Lee et al., 1997; Wang et al., 2007; Yat et al., 2008).

The ideal process for cellulosic biomass conversion would be the production of liquid fuels from biomass in a single step at a short residence time. The liquid product produced in pyrolysis is called bio-oil, which is an acidic combustible liquid containing more than 300 compounds (Wang et al., 2008). Bio-oils are not compatible with existing liquid transportation fuels including gasoline and diesel. To use bio-oil as a conventional liquid transportation fuel, it must be catalytically upgraded (Carlson et al., 2008). Zeolite catalysts added into the pyrolysis process can convert oxygenated compounds generated by pyrolysis of the biomass into gasoline-range aromatics.

7 BIO-REFINERIES AND BIOFUELS

In addition to being a resource for energy generation, lignocellulosic biomass has potential to serve for multiple purposes. There is not necessarily a concurrence of various options. Highest value can be achieved by diverting individual components to optimum routes, thus aiming to achieve complete valorization of the material. Among possible target utilization options are to be mentioned in particular: (i) electricity and fuel generation; (ii) production of chemicals; (iii) precursors for industrial products such as biodegradable plastics; (iv) utilization as soil amendment.

The idea of so-called biorefineries is to process bioresources such as agricultural or forest biomass to produce energy and a wide variety of precursor chemicals and bio-based materials (Sigrid and Morar, 2009). Petroleum refineries are already built, and use of this existing infrastructure for the production of biofuels requires little capital investment (Marinangeli et al., 2006). Furthermore, the infrastructure for blending fuels as well as their testing and distribution is already in place at oil refineries. Three options are available for using petroleum refineries to convert biomass-derived feedstocks into fuels and chemicals: (i) fluid catalytic cracking (FCC), (ii) hydrotreating-hydrocracking, and (iii) utilization of biomass-derived synthesis gas (syngas) or hydrogen.

Cofeeding biomass-derived molecules into a petroleum refinery could rapidly decrease our dependence on petroleum feedstocks. Petroleum-derived feedstocks are chemically different than biomass-derived feedstocks; therefore a new paradigm in how to operate and manage a petroleum refinery is required. Another improvement toward the production of biofuels in a petroleum refinery would be if governments were to offer tax exemptions and subsidies to all types of biofuels, and not only for selected biofuels such as ethanol and biodiesel. As the price of petroleum continues to increase, we project that refining technology will be developed to allow the coproduction of bio- and petroleum-based fuels in the same (petroleum) refinery and even using the same reactors. A realistic practical scenario will be one in which both industries cooperate, with one producing the biofuel precursors and the other processing and converting them into valuable fuels (Huber and Corma, 2007).

8 TYPICAL ISSUES FOR LIFE-CYCLE ANALYSIS

(i) *Use of fossil fuel and raw materials to produce biofuels:* The whole life cycle of the production of biofuels involves the use of fossil fuel and raw materials to some extent. Whether the net gain balance out of the fossil input obtained in terms of low emissions is positive or not remains under discussion.

(ii) *Availability of land for fuel (food vs. fuel issue):* Current biofuel producers do not always have a secure access to raw materials due to limited grain reserves and the fact that the current costs of crude vegetable oil from "food crops" are variable. Bio-based energy industries are also currently in competition with food producers, and we perceive them as being a primary cause of the increase in food prices. In order to make biofuel production profitable and more sustainable, avoiding as much as possible competition with the food market, companies have to focus on second-generation biofuels made from alternative cheap feedstocks (e.g., (waste)-biomass, waste oils and fats, residues, etc.). (Ligno) cellulosic ethanol and biodiesel from waste oils, nonfood crops, or algae emerge as real alternatives to tackle this problem.

(iii) *Environmental impact:* Despite the fact that some studies carried out to date show first-generation biofuels may offer a low carbon balance, fossil fuel usage and GHG balance, further outputs and environmental indicators must be addressed. Water usage (in the growth of the crops), eutrophication (run off of lawn fertilizers into natural waters), and soil erosion are some of them. Second- and potential third-generation biofuels are more attractive in terms of crop economy.

(iv) *Socioeconomic impact:* Some sectors of the industry estimated that a robust global biofuel market will be fully established around 2012. The implementation of biofuels will also be highly dependent on the feasibility of the technologies employed for their production and the economics of the processes play a fundamental role in this regard.

At the moment, there is not yet a widely accepted definition of "sustainable biofuels," or a scheme for certification and labeling (Mol, 2007). Nevertheless, some agreement can be observed on four ecological issues that should be included in sustainability schemes such as GHG emissions, energy balance, biodiversity loss, specific environmental effects (i.e., soil condition and water use).The problem is that each feedstock is different and many crops produce their best yields in specific regions of the world or require certain soil or water conditions. These local differences demand specific attention and are not easily generalized. Furthermore, there is wide disagreement on the implementation of international conventions, while inclusion of social criteria is even more difficult (Oosterveer and Mol, 2010).

A holistic approach to valorization of lignocellulosic biomass needs to take into account sustainability of chosen options. If concepts are too heavily orientated toward energy production or industrial use, this can even be at the expense of environmental protection. If crop residues such as straw are no longer left on field, this will result in depletion of soil organic matter. While anaerobic digestion results in a digestate, which can be brought back to field to supply not only nutrients but also organic matter, thermal valorization and production of second-generation biofuels result in complete consumption of the biomass and consequently a lack of nutrients and organic matter. Soil requirements vary within a wide range and need to be assessed locally. Only lignocellulosic biomass which is in surplus of soil demand for organic matter should be considered for treatment options with complete consumption of the substrate. In regions with concern about declining organic content of soils, anaerobic digestion should be given special attention even if the net energy recovery is lower compared to that of alternative technologies with total consumption of the biomass (Sigrid and Morar, 2009).

9 PERSPECTIVES AND CHALLENGES

Several potential scenarios for biofuels can be foreseen in the future. The big hopes for the transport sector are second- and future third-generation biofuels, including biodiesel from microbial oil, the production of biobutanol (from nonedible feedstocks) as a more petrol-like fuel, and the preparation of biofuels from cellulosic and biomass nonedible feedstocks.

One of the critical factors that will influence the future prospects of biofuels is diversification. The future of biofuels as a sustainable (economic, social, and environmental definitions) technology is directly linked to the maximum use of byproducts that will make its production more cost effective.

Biology and synthetic biology would have the opportunity to design plants with special properties through genetic engineering to produce biomass feedstocks with requisite ratio and functionalities of lignin/cellulose (Luque et al., 2008). Such discoveries can significantly change the future of biofuels, as a major contribution to the GHG reduction through the regulation of the CO_2 fixed by plants in crops.

Acknowledgments

The authors thank The Director, Indian Institute of Petroleum, Dehradun, for his constant encouragement and support. RS thanks Council of Scientific and Industrial Research (CSIR), New Delhi, India, for providing Junior Research Fellowship (JRF).

References

Agrawal, R., Singh, N.R., 2009. Synergistic routes to liquid fuel for a petroleum-deprived future. AIChE 55 (7), 1898–1905.

Alonso, D.M., Bond, J.Q., Dumesic, J.A., 2010. Catalytic conversion of biomass to biofuels. Green Chem. 12, 1493–1513.

Behar, F., Hatcher, P.G., 1995. Artificial coalification of a fossil wood from brown coal by confined system pyrolysis. Energy Fuels 9, 984–994.

Berndes, G., Hoogwijk, M., Van den Broe, R., 2003. The contribution of biomass in the future global energy supply: a review of 17 studies. Biomass Bioenergy 25, 1–28.

Bridgewater, T., 2006. Biomass for energy. J. Sci. Food Agric. 86, 1755–1768.

Brown, J.N., *Development of a lab-scale auger reactor for biomass fast pyrolysis and process optimization using response surface methodology*. M.S. thesis, Iowa State University, Ames, 2009.

Carlson, T.R., Vispute, T.P., Huber, G.W., 2008. Green gasoline by catalytic fast pyrolysis of solid biomass derived compounds. Chem. Sus. Chem. 1, 397–400.

Chmielniak, T., Sciazko, M., 2003. Co-gasification of biomass and coal for methanol synthesis. Appl. Energy 74, 393–403.

Chuntanapum, A., Matsumura, Y., 2010. Char formation mechanism in supercritical water gasification process: a study of model compounds. Ind. Eng. Chem. Res. 49, 4055–4062.

Corella, J., Monzon, A., Santamaria, J., Gonzalez-Canibano, J., 1988. Ultra-fast biomass pyrolysis in a high-temperature (2200 °C), fluid-wall reactor. J. Sol. Energy Eng. 110 (1), 10–13.

Elliott, D.C., 2007. Energy Fuels 21, 1792–1815.

Funke, A., Ziegle, F., 2009. Hydrothermal carbonization of biomass: a literature survey focussing on its technical application and prospects. In: 17th European Biomass Conference and Exhibition, Hamburg, ETA Renewable Energies, Florence. pp. 1037–1050.

Guiotoku, M., Rambo, C.R., Hansel, F.A., Magalhaes, W.L.E., Hotza, D., 2009. Microwave-assisted hydrothermal carbonization of lignocellulosic materials. Mater. Lett. 63, 2707–2709.

Heilmann, S.M., Ted Davis, H., Jader, L.R., Lefebvre, P.A., Sadowsky, M.J., Schendel, F.J., et al., 2010. Hydrothermal carbonization of microalgae. Biomass Bioenergy 34, 875–882.

Helle, S.S., Murray, A., Lam, J., Cameron, D.R., Duff, S.J.B., 2004. Xylose fermentation by genetically modified *Saccharomyces cerevisiae* 259ST in spent sulfite liquor. Biores. Technol. 92, 163–171.

Hertwich, E.G., Zhang, X., 2009. Concentrating-solar biomass gasification process for a 3rd generation biofuel. Environ. Sci. Technol. 43, 4207–4212.

Himmel, M.E., Ding, S.Y., Johnson, D.K., Adney, W.S., Nimlos, M.R., Brady, J.W., et al., 2007. Biomass recalcitrance: engineering plants and enzymes for biofuels production. Science 315, 804–807.

Huber, G.W., Corma, A., 2007. Synergies between bio- and oil refineries for the production of fuels from biomass. Angew. Chem. Int. Ed. 46, 7184–7201.

Inoue, S., Okigawa, K., Minowa, T., Ogi, T., Biomass Bioenergy 16, 1999, 377–383.

Iranmahboob, J., Nadim, F., Monemi, S., 2002. Optimizing acid hydrolysis: a critical step for production of ethanol from mixed wood chips. Biomass Bioenergy 22, 401–404.

Jimenez, S., Ballester, J., 2006. Particulate matter formation and emission in the combustion of different pulverized biomass fuels. Combust. Sci. Technol. 178, 655–683.

Karagoz, S., Bhaskar, T., Muto, A., Sakata, Y., 2006. Chem. Eng. J. 108, 127–137.

Kruse, A., 2009. Hydrothermal biomass gasification. J. Supercrit. Fluids 47 (3), 391–399.

Kruse, A., Henningsen, T., Sinag, A., Pfeiffer, J., 2003. Ind. Eng. Chem. Res. 42, 3711–3717.

Kumabe, K., Hanaoka, T., Fujimoto, S., Minowa, T., Sakanishi, K., 2007. Co-gasification of woody biomass and coal with air and steam. Fuel 86, 684–689.

Kumar, A., Jones, D.D., Hanna, M.A., 2009a. Thermochemical biomass gasification: a review of the current status of the technology. Energies 2, 556–581.

Kumar, P., Barrett, D.M., Delwiche, M.J., Stroeve, P., 2009b. Methods for pretreatment of lignocellulosic biomass for efficient hydrolysis and biofuel production. Ind. Eng. Chem. Res 48, 3713–3729.

Lange, J.P., 2007. Lignocellulose conversion: an introduction to chemistry, process and economics. In: Centi, G., Van Santen, R. (Eds.), Catalysis for Renewables. Wiley-VCH, Weinheim.

Lawford, H.G., Rousseau, J.D., 2002. Performance testing of Zymomonas mobilis metabolically engineered for cofermentation of glucose, xylose, and arabinose. Appl. Biochem. Biotechnol. 98-100, 429–448.

Luque, R., Herrero-Davila, L., Campelo, J.M., Clark, J.H., Hidalgo, J.M., Luna, D., et al., 2008. Biofuels: a technological perspective. Energy Environ. Sci. 1, 542–564.

Lynd, L.R., van Zyl, W.H., McBride, J.E., Laser, M., 2005. Consolidated bioprocessing of cellulosic biomass: an update. Curr. Opin. Biotechnol. 16, 577–583.

Marinangeli, R., Marker, T., Petri, J., Kalnes, T., McCall, M., Mackowiak, D., et al., 2006. Opportunities for Biorenewables in Oil Refineries. Report No.DE-FG36-05GO15085, UOP.

Marker, T., Felix, L., Linck, M., 2009. Integrated hydropyrolysis and hydroconversion process for production of gasoline and diesel fuel from biomass. AIChE Extended Abstract.

Marsh, R., Hewlett, S., Griffiths, T., Williams, K., 2007. Advanced thermal treatment for solid waste—a wastemanager's guide. In: Proceeding of the 22nd International Conference on Solid Waste Management and Technology, Philadelphia (USA).

McLendon, T.R., Lui, A.P., Pineault, R.L., Beer, S.K., Richardson, S.W., 2004. High-pressure co-gasification of coal and biomass in a fluidized bed. Biomass Bioenergy 26, 377–388.

Minowa, T., Zhen, F., Ogi, T., Varhegyi, G.J., 1997. J. Chem. Eng. Jpn. 30, 186–190.

Mol, A.P.J., 2007. Boundless biofuels: between vulnerability and environmental sustainability. Sociol. Ruralis. 47, 297–315.

Mosier, N.S., Wyman, C., Dale, B., Elander, R., Lee, Y.Y., Holtzapple, M., et al., 2005. Features of promising technologies for pretreatment of lignocellulosic biomass. Bioresour. Technol. 96, 673–686.

Mukherjee, D.K., Sengupta, A.N., Choudhury, D.P., Sanyal, P.K., Rudra, S.R., 1996. Effect of hydrothermal treatment on caking propensity of coal. Fuel 75, 477–482.

Oosterveer, P., Mol, A.P.J., 2010. Biofuels, trade and sustainability: a review of perspectives for developing countries. Biofuels Bioprod. Bioref. 4, 66–76.

Patrick Lee, K.C., Bulls, M., Holmes, J., Barrier, J.W., 1997. Hybrid process for the conversion of lignocellulosic materials. Appl. Biochem. Biotechnol. 66, 1–23.

Payne, D.F., Ortoleva, P.J., 2001. A model for lignin alteration—Part I: a kinetic reaction-network model. Org. Geochem. 32, 1073–1085.

Quaak, P., Knoef, H., Stassen, H., 1999. Energy from Biomass, a Review of Combustion and Gasification Technologies. World bank technical paper no. 422. The International Bank for Reconstruction and Development, Washington (DC).

Rezaiyan, J., Cheremisinoff, N.P., 2005. Gasification Technologies—a Primer for Engineers and Scientists. CRC Press Taylor & Francis Groups, Boca Raton (FL).

Ross, D.S., Loo, B.H., Tse, D.S., Hirschon, A.S., 1991. Hydrothermal treatment and the oxygen functionalities in Wyodak coal. Fuel 70, 289–295.

Sigrid, K., Morar, M.V., 2009. Integration of lignocellulosic biomass into renewable energy generation concepts. ProEnvironment 2, 32–37.

Simonetti, D.A., Rass-Hansen, J., Kunkes, E.L., Soares, R.R., Dumesic, J.A., 2007. Coupling of glycerol processing with Fischer Tropsch synthesis for production of liquid fuels. Green Chem. 9, 1073–1083.

Siskin, M., Katritzky, A.K., 1991. Reactivity of organic compounds in hot water: geochemical and technological implications. Science 254, 231–237.

Soares, R.R., Simonetti, D.A., Dumesic, J.A., 2006. Glycerol as a source for fuels and chemicals by low temperature catalytic processing. Angew. Chem. Int. Ed. 45, 3982–3985.

Stocker, M., 2008. Biofuels and biomass-to-liquid fuels in the biorefinery: catalytic conversion of lignocellulosic biomass using porous materials. Angew. Chem. Int. Ed. 47, 9200–9211.

Taherzadeh, M.J., Karimi, K., 2008. Pretreatment of lignocellulosic wastes to improve ethanol and biogas production: a review. Int. J. Mol. Sci. 1621–1651.

Titirici, M.M., Antonietti, M., 2010. Chemistry and materials options of sustainable carbon materials made by hydrothermal carbonization. Chem. Soc. Rev. 39, 103–116.

Titirici, M.M., Thomas, A., Yu, S., Muller, J.O., Antonietti, M., 2007. A direct synthesis of mesoporous carbons with bicontinuous pore morphology from crude plant material by hydrothermal carbonization. Chem. Mater. 19, 4205–4212.

Tuskan, G.A., Difazio, S., Jansson, S., Bohlmann, J., Grigoriev, I., Hellsten, U., et al., 2006. The genome of black cottonwood, Populus trichocarpa (Torr & Gray). Science 313, 1596–1604.

US Department of energy, 2006. Energy Information Administration, International Energy Outlook. http://www.eia.doe.gov.

van Krevelen, D.W., 1950. Fuel 29, 269–284.

Van swaaij, W.P.M., Prins, W., Kersten, S.R.A., 2004. Strategies for the future of biomass for energy, industry and climate protection. In: van Swaaij, W.P.M., Fjallstroom, T., Helm, P., Grassi, A. (Eds.), Second World Biomass Conference: Biomass for Energy, Industry and climate Protection, ISBN: 88-89407-04-2. Published by ETA-Florence and WIP-Munich, Italy.

Wang, M., Wu, M., Huo, H., 2007. Life-cycle energy and greenhouse gas emission impacts of different corn ethanol plant types. Environ. Res. Lett. 2, 1–13.

Wang, L., Weller, C.L., Jones, D.D., Hanna, M., 2008. Contemporary issues in thermal gasification of biomass and its application to electricity and fuel production. Biomass Bioenergy 32, 573–581.

Wingren, A., Galbe, M., Zacchi, G., 2003. Techno-economic evaluation of producing ethanol from softwood: comparison of SSF and SHF and identification of bottlenecks. Biotechnol. Progr. 19, 1109–1117.

Wolfs, P.M.J., van Krevelen, D.W., Waterman, H.I., 1960. Chemical structure and properties of coal XXV—the carbonization of coal models. Fuel 39, 25–38.

Wyman, C.E., 1994. Ethanol from lignocellulosic biomass: technology, economics, and opportunities. Biores. Technol. 50, 3–15.

Yat, S.C., Berger, A., Shonnard, D.R., 2008. Kinetic characterization of dilute surface acid hydrolysis of timber varieties and switchgrass. Bioresour.Technol. 99, 3855–3863.

Yunqiao, P.u., Zhang, D., Singh, P.M., Ragauskas, A.J., 2008. The new forestry biofuels sector. Biofuels Bioprod. Bioref. 2, 58–73.

Zhang, L., Xu, C., Champagne, P., 2010. Overview of recent advances in thermo-chemical conversion of biomass. Energy Convers. Manage. 51, 969–982.

CHAPTER

4

Biomass-derived Syngas Fermentation into Biofuels

*Pradeep Chaminda Munasinghe, Samir Kumar Khanal**

Department of Molecular Biosciences and Bioengineering (MBBE), University of Hawai'i at Mānoa, Agricultural Science Building 218, 1955 East-West Road, Honolulu, Hawaii 96822.

***Corresponding author: E-mail: khanal@hawaii.edu**

1 BACKGROUND

Research on lignocellulosic biomass such as agri-residues (e.g., corn stover, wheat and barley straws, etc.), agri-processing byproducts (e.g., corn fiber, sugarcane bagasse, seed cake, etc.), and energy crops (e.g., switch grass, poplar, Napier grass, *Miscanthus*, etc.) has received considerable attention for bioenergy production, especially liquid transportation fuel in recent years. Lignocellulose is a renewable, nonfood feedstock with an annual availability of around 200×10^{12} kg (220 billion metric tons) globally. The United States alone has the potential of producing 1.3 billion dry tons of biomass annually, which could substitute more than 30% of the nation's petroleum consumption (United States Department of Agriculture (USDA) and United States Department of Energy (USDOE) Joint Report, 2005). Thus, lignocellulosic biomass could play an important role in the bio-based economy to produce a variety of biofuels and bio-based products. Lignocellulosic biomass consists of 40-50% cellulose, 20-40% hemicellulose, and 10-30% lignin. Although multiple conversion technologies are available for producing biofuels from biomass, there are two major pathways, namely biochemical and thermochemical. In biochemical conversion, the biomass is subjected to a combination of physical and chemical pretreatments to destruct the biomass structure. These pretreatments make the biomass accessible to enzymes. The pretreated biomass is subjected to enzyme hydrolysis to obtain fermentable sugars, which are then fermented to biofuels (Takara and Khanal, 2011). The biochemical route, however, has several drawbacks such as high pretreatment and enzymes costs, generation of inhibitory soluble compounds (acetic acid, furan derivatives, and various phenolic compounds), degradation of sugars, and low biomass to fuel conversion ratios (Lewis et al., 2010). On the other hand, in thermochemical conversion, the biomass is gasified to produce synthesis gas or syngas in short (a gas mixture predominantly consisting of CO, CO_2, and H_2). The syngas can be converted into liquid biofuels through Fischer-Tropsch (FT) synthesis (using metal catalysts) or direct microbial fermentation known as syngas fermentation (using microbial catalysts) (Henstra et al., 2007; Munasinghe and Khanal, 2010 (a)). The FT synthesis usually utilizes metal catalysts such as cobalt (Co), ferrous (Fe), copper (Cu), aluminum (Al), zinc (Zn), molybdenum (Mo), nickel (Ni), rubidium (Ru), and ruthenium (Rh) (Demirbas, 2007; Subramani and Gangwal, 2008). The major drawbacks of FT synthesis are the high costs of the metal catalyst, a fixed H_2:CO ratio (2:1), catalyst poisoning due to inert gases and contaminants containing sulfur, and high operating temperature and pressure (Phillips et al., 1994; Vega et al., 1990; Worden et al., 1991).

Syngas fermentation via biocatalysts (such as *Clostridium ljungdahlii, C. autoethanogenum, C. carboxydivorans, Butyribacterium methylotrophicum, Methanosarcina barkeri*, and *Rhodospirillum rubrum*) produces liquid/gaseous biofuels, and offers several advantages over the biochemical approach and the FT process. Some of the merits of syngas fermentation are the elimination of the need of expensive metal catalysts, a higher specificity of the biocatalysts, the independence of the H_2:CO ratio for bioconversion, the operation of bioreactors at ambient conditions, and the elimination of issues concerning noble metal poisoning (Bredwell et al., 1999; Klasson et al., 1990). Poor solubility of syngas in the aqueous phase and low product yield are the major limitations of syngas fermentation. These limitations have been the bottlenecks to commercialization of the syngas fermentation process.

This chapter critically reviews the existing literature on biomass-derived syngas fermentation into biofuels. Furthermore, relevant background information including pathways, microbial aspects, mass transfer, reactor design, and factors affecting syngas fermentation are also briefly discussed. In addition, the current developments, challenges, and future research directions in syngas fermentation to biofuels are also included.

2 FUNDAMENTAL ASPECTS OF SYNGAS FERMENTATION

2.1 Gasification of Biomass

Synthesis gas (syngas in brief) is a gas mixture of predominantly CO and H_2. Gasification of biomass feedstocks produces syngas through partial oxidation. Syngas quality largely depends on the compositions of biomass feedstock, gasifier types, and the gasifying agents. Other than the major constituents—CO and H_2, gasification of biomass also produces methane (CH_4), nitrogen (N_2), carbon dioxide (CO_2), water vapor, trace amounts of sulfur containing compounds, tar, higher hydrocarbons such as ethane (C_2H_6), ethylene (C_2H_4) and acetylene (C_2H_2), and particulate matter (Datar et al., 2004).

Gasifiers are mainly divided into two categories, namely, fixed-bed and fluidized-bed gasifiers. The fixed-bed gasifiers are characterized by the stationary reaction zone. Typically, in these gasifiers, biomass is fed from the top. Depending on the direction of biomass feeding and the oxidant employed, fixed-bed gasifiers are further divided into updraft and downdraft gasifiers. Fluidized-bed gasifiers use sand, ash, or char as moving media to increase the heat transfer and the gasification efficiency. Generally, the gasification of biomass takes place at high temperatures (fluidized bed: 750-900 °C and fixed bed: 1000-1200 °C). Table 1 shows the compositions of the produced gas mixtures from various gasification techniques and biomass feedstocks.

TABLE 1 Gas Compositions of Different Gasification Processes (Modified from Munasinghe and Khanal, 2010 (a))

Gasifier Type	Fluidized Bed	Updraft	Downdraft	Fluidized Bed	Fluidized Bed	Fluidized Bed	Fluidized Bed (Vernamo)	Fluidized Bed (Cyclone)
Biomass type	n/a	n/a	n/a	Switch grass	Bark	Coal	n/a	Sugarcane bagasse
Gasifying agent	Air	Air	Oxygen	Air	Air	Oxygen	O_2-Steam	Steam
N_2 (%)	50.0	53.0	3.0	56.8	42.9	1	<1	<1
CO (%)	14.0	24	48	14.7	19.6	67	12	17.4
CO_2 (%)	20	9	15	16.5	13.5	4	28.2	22.0
H_2 (%)	9	11	32	4.4	20.2	24	11.9	10.0
CH_4 (%)	7	3	2	4.2	3.8	0.02	8.2	3.4
H_2S (%)	n/a	n/a	n/a	n/a	Very low	1	n/a	n/a
Tars (g/m^3)	<10	>10	1	<1	<1	0	0.3	n/a
NH_3 (%)	n/a	n/a	n/a	n/a	n/a	0.04	n/a	n/a
H_2O (%)	n/a	n/a	n/a	n/a	Dry	3	38.1	45.0
Dust	High	Low	Low	n/a	n/a	n/a	n/a	n/a
C_2H_6, C_2H_4, and C_2H_2	n/a	n/a	n/a	3.2	n/a	n/a	1.3	2.2
H_2/CO	0.64	0.46	0.67	0.30	1.00	0.36	1.00	0.57
References	Bridgwater (1995)	Bridgwater (1995)	Bridgwater (1995)	Datar et al. (2004)	Subramani and Gangwal 2008)		Albertazzi et al. (2005)	Gabra et al. (2001)

n/a, not available.

2.2 Metabolic Pathways

Syngas-fermenting microorganisms such as *C. ljungdahlii* (Phillips et al., 1994), *C. carboxydivorans, C. autoethanogenum,* and *B. methylotrophicum* (Bredwell et al., 1999) follow the acetyl-CoA pathway (sometimes referred to as Wood-Ljungdahl Pathway) to produce biofuels (Henstra et al., 2007). Microorganisms that produce the intermediate acetyl-CoA from carbonyl or carboxyl precursors are known as acetogens (Brown, 2006). Though many acetogenic microbes produce acetate from alcohols and fatty acids, some are capable of producing organic acids and alcohols using CO_2 and H_2 (autotrophic acetogens) or CO (unicarbonotrophic acetogens) as their substrates.

Figure 1 shows the simplified acetyl-CoA pathway leading to the production of bio-based products such as ethanol, butanol, and butyrate and acetic acids from syngas. The essential reducing equivalents (–CO, –CoA, –Co–CH_3) are produced from H_2 and CO by hydrogenase and CO dehydrogenase (CODH) enzymes, respectively. In addition, the bifunctional CODH enzyme produces a carbonyl group from the reaction of carbon dioxide and water (Henstra et al., 2007). The produced reducing equivalents are then converted to acetyl-CoA by acetyl-CoA synthase (ASC) complex.

During the metabolic pathway, intermediate acetyl-CoA performs two major roles—it acts as a precursor for the cell macromolecule, and it serves as an energy source. It is essential to maintain a strict anaerobic environment during the acetyl-CoA pathway to avoid the consumption of reducing equivalents by other metabolic pathways (e.g., aerobic respiration). After several successive reactions, CO_2 is reduced to a methyl (–CH_3) group with the expense of 6 electrons and adenosine triphosphate (ATP). The produced methyl groups then react with the coenzyme and produce –Co–CH_3. During the later stage of the acetyl-CoA pathway,

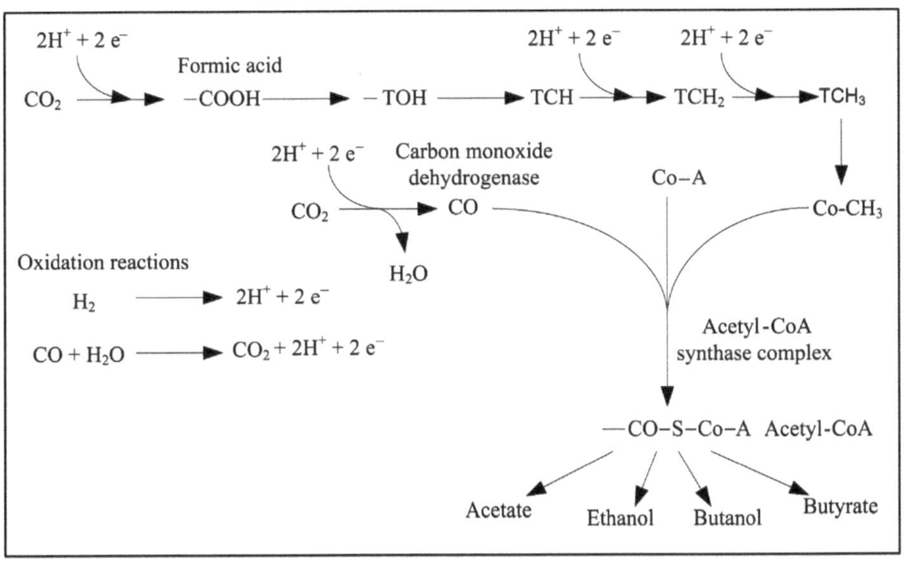

FIGURE 1 Modified acetyl-CoA pathway for converting syngas to biofuel. T, tetrahydrofolate; Co, a corrinoid protein (methyl group carrier).

the produced metabolites (–Co–CH_3, –CoA) react with CO to produce acetate. The enzyme complex—acetyl-CoA synthase enhances the reaction rate. This reaction recovers the metabolic energy invested during the early stages of the pathway. Acetate is further reduced to produce ethanol.

2.3 Biochemical Reactions

Acetic acid (CH_3COOH) and ethanol (C_2H_5OH) are the two major products from syngas fermentation. Equations (1)-(4) show the four basic reactions producing acetic acid and ethanol. In this case, the gaseous substrates CO and H_2 follow the acetyl-CoA pathway to produce acetic acid and ethanol under strict anaerobic conditions.

$$6CO + 3H_2O \rightarrow C_2H_5OH + 4CO_2 \quad \Delta G^\circ = -216.0 kJ/mol \tag{1}$$

$$6H_2 + 2\,CO_2 \rightarrow C_2H_5OH + 3H_2O \quad \Delta G^\circ = -97.1 kJ/mol \tag{2}$$

$$4CO + 2H_2O \rightarrow CH_3COOH + 2CO_2 \quad \Delta G^\circ = -135.0 kJ/mol \tag{3}$$

$$6H_2 + 2CO_2 \rightarrow CH_3COOH + 2H_2O \quad \Delta G^\circ = -54.8 kJ/mol \tag{4}$$

From Equation (1), it is clear that about one third of the carbon from CO is utilized in the product yield. The overall ethanol production, combining Equations (1) and (2), reveals that two thirds of the carbon from CO is converted to ethanol. During the acetyl-CoA pathway, hydrogen provides the required reducing equivalents and electrons when hydrogenase enzyme is present in the fermentation media (Equation (5)).

$$H_2 \rightarrow 2H^+ + 2e^- \tag{5}$$

If the hydrogenase enzyme is inhibited or hydrogen is not present in the fermentation broth, the required electrons are obtained from CO in the presence of CODH enzyme. In other words, CO is used in supplying electrons, rather than in the biofuel production. This obviously results in a drastic reduction in biofuel yields. It is therefore, important to maintain healthy concentrations of both hydrogen and CO during syngas fermentation.

Syngas-fermenting microorganisms are critical in biofuel production. Under optimum growth conditions, most of the known syngas-fermenting microbes tend to produce more acetate than alcohol products (e.g., ethanol, butanol, etc.). Vega et al. (1989) reported acetate to ethanol product ratio of 20:1. In order to improve the product formation from acidogenesis to solventogenesis, researchers investigated nutrient limitations, pH shifts, reducing agent addition (Klasson et al., 1992), and hydrogen addition.

3 MICROBIOLOGY OF SYNGAS FERMENTATION

Microorganisms can be categorized into two major types depending on their optimum growth temperature, namely, mesophilic and thermophilic. In general, optimum growth temperature for mesophilic microorganisms varies from 37 to 40 °C, whereas for thermophiles, the temperature range varies from 55 to 80 °C (Munasinghe and Khanal, 2010 (a)). Mesophilic microorganisms, for example, *Clostridium aceticum*, *Acetobacterium woodii*, *C. carboxydivorans*, and C. *ljungdahlii*, have been dominated in syngas fermentation with higher syngas to biofuel conversion efficiencies compared to the thermophilic counterpart (Henstra et al., 2007; Younesi et al., 2005).

C. ljungdahlii, one of the most widely used homoacetogenic microorganism, has an optimum growth temperature between 37 and 40 °C, and a pH of 5.8-6.0 (Tanner et al., 1993). However, *C. ljungdahlii* was reported producing an ethanol concentration as high as 48 g/L in a continuous stirred-tank reactor (CSTR) at a low pH of 4.0-4.5 in nutrient-limited culture media (Klasson et al., 1993). *C. carboxydivorans* (earlier referred to as bacterium strain *P7*) is another mesophilic organism which has an ability to grow on a gas mixture consists of CO, CO_2, H_2, and N_2, and produces mainly ethanol and acetic acid. The ethanol yield obtained was 0.16% (by weight) during a 10-day fermentation experiment at pH of 5.75 and temperature of 37 °C in a bubble column reactor (Rajagopalan et al., 2002). Later, the authors compared the ethanol yields of *C. carboxydivorans* and *C. ljungdahlii* and found out that the results were similar. In a separate study, Heiskanen et al. (2007) used *B. methylotrophicum* at 37 °C and a pH of 6.0-6.9 to convert syngas to acetic acid. The authors claimed a maximum acetic acid concentration of 1.3 g/L at a gas mixture of 40% H_2, 35% CO and 25% CO_2 after 144 h of fermentation.

The major advantage of thermophilic microbes in syngas fermentation is their capability of fermenting syngas at relatively high temperatures (around 60 °C) with higher conversion rates. Though the high temperature reduces the solubility of the component gases in the fermentation broth, benefit in product recovery improves the overall cost effectiveness of the process (Henstra et al., 2007). There were several attempts to utilize extreme thermophiles (optimum growth temperature >70 °C) in producing organic solvents. During the last decade, researchers were able to isolate several thermophilic microorganisms which were able to grow on CO as a substrate (Henstra et al., 2007). *Desulfotomaculum carboxydivorans*, *Carboxydocella sporoproducens*, *Moorella thermoacetica*, and *M. thermoautotrophica* are some examples of thermophiles with optimum temperature ranges of 55-58 °C.

4 SYNGAS CHARACTERISTICS

4.1 Syngas Composition

Gasification of biomass produces a gas mixture consisting of CO, CO_2, H_2, N_2, CH_4, trace amounts of NO_x and SO_x, tar, char, particulate matter, and higher hydrocarbons such as C_2H_2, C_2H_4, and C_2H_6. This gas mixture is sometimes referred to as a producer gas (Datar et al., 2004). The gas composition greatly depends on the type of feedstocks and the operating conditions of gasifier (Klasson et al., 1992). Table 1 summarizes the gas constituents from various types of gasifiers and feedstocks. Datar et al. (2004) used a producer gas consisting of N_2 (56.8%), CO (14.7%), CO_2 (16.5%), H_2 (4.4%), CH_4 (4.2%), C_2H_4 (2.4%), and C_2H_6 (0.8%) to determine the effects of the gas composition on cell growth, H_2 uptake, and acid and alcohol production. The authors found that there was a process alteration due to some trace species in the producer gas. Further, they suggested that the producer gas without purification could inhibit hydrogenase enzyme leading to low product formation.

4.2 Syngas Impurities

Biomass-derived syngas often contains additional constituents such as CH_4, some higher hydrocarbons (C_2H_2, C_2H_4 and C_2H_6), tar, ash, and char particles. Since most of the researchers use bottled synthetic gas mixtures for syngas fermentation studies, there are limited studies that examine the effects of impurities on syngas fermentation (Ahmed and Lewis, 2007). The authors reported the effects of NO on hydrogenase activity, cell growth, and product distribution using *C. carboxydivorans*. The authors further concluded that NO concentration below 40 ppm had no significant effect on syngas fermentation process. Ahmed et al. (2006) reported that tars could promote cell dormancy and product redistribution (ethanol and acetic acid) during syngas fermentation. Kundiyana et al. (2010) examined the ability of *Clostridium ragsdalei* ((ATCC BAA-622), previously known as strain *P11*) to grow and metabolize CO under microaerophilic condition (5% O_2) in a pilot-scale fermentor. Further, the growth of acetogens such as *M. thermoacetica* and *Clostridium magnum* in a medium supplemented with 21% O_2 was reported by Karnholz et al. (2002). These findings provide the prospects of scaling-up syngas fermentation for commercialization. Furthermore, the microbial catalysts used in syngas fermentation had higher tolerance to toxic gases and trace contaminants such as hydrogen sulfide (H_2S) and carbonyl sulfide (COS) than that of biochemical pathway (Worden et al., 1991; Younesi et al., 2005).

4.3 Syngas Cleaning

In general, the gasification of biomass is often followed by a gas clean-up and conditioning. The gas mixture is passed through a series of cyclones and filters to remove most of the undesirable pollutants (e.g., tar, particulate matter, and char). Datar et al. (2004) employed a condensation tower followed by acetone scrubbers to remove tar and moisture from the producer gas. A series of 0.025-μm filters were successfully used to clean producer gas mixture to prevent cell dormancy and product redistribution (Ahmed et al., 2006). Further, trace amounts of NO can be removed by using chemicals such as sodium hypochlorite, potassium permanganate, or sodium hydroxide. Syngas fermentation shows a high tolerance toward sulfur gases such as hydrogen sulfide (H_2S), and COS. Vega et al. (1990) found that CO-utilizing methanogenic microbes can grow in the presence of H_2S up to 2%.

5 CURRENT DEVELOPMENTS IN SYNGAS FERMENTATION

5.1 Biorefinery Concept

According to the Biomass Research and Development Technical Advisory Committee of the US Departments of Energy and Agriculture (2002) Report published by the U.S. Department of Energy and U.S. Department of Agriculture, the biorefinery is defined as "A processing and conversion facility that efficiently separates its biomass raw materials into individual components and converts these components into marketplace products including biofuels, biopower, and conventional and new bioproducts." Several papers discussed the major products and integrated biorefinery concept for syngas fermentation.

Ethanol is by far the most economical bio-product that is generated during syngas fermentation. Ethanol is currently being sold as a fuel additive to blend with gasoline. The existing gasoline engines can take up to 10% ethanol (known as E10) without modifying the engine. Biomass-derived syngas fermentation also produces other important bio-product such as acetic acid, butanol, and butyric acid (Datar et al., 2004). Acetic acid has numerous applications in chemical industries including synthesis of vinyl acetate and acetic anhydride. Butanol is considered as a better transportation fuel compared to ethanol due its high energy content and high vapor pressure. In addition, butanol is used in the production of butyl acetate and butyl acrylate which can be used as fuel additives to enhance the octane value of gasoline. Butyric acid is being used as a flavoring agent in the food processing industry.

Apart from the main products, organic acids and alcohols, the growth of anaerobic microbes also produces valuable biochemical such as polyester which serves as an energy storage unit for the organism (Brown, 2006). Most of the syngas-fermenting microorganisms produce these polyesters under stressed conditions such as nutrient imbalances. Polyhydroxyalkanoate (PHA) is one of the most known polyesters produced in the cells, and it is stored as a discrete granule. Polyester content of cell is as high as 80% (dry weight).

In conventional biochemical-based ethanol plants, lignin fraction of the biomass is considered as a low-value residue. Usually, 10 to 30% of biomass feedstock contains lignin which has a higher heating value of 9,111 Btu/lb. Therefore, the lignin recovered from the diverse feedstocks should be integrated into the process. Thermal cracking of lignin at high temperatures ranging from 250 to 600 °C showed the potential of producing low molecular weight gaseous feedstocks for further processing.

In an integrated biorefinery, the process is optimized to produce biofuel, along with other high-value products such as biopower and bio-based materials for a long-term sustainability.

5.2 Ethanol Fermentation

Ethanol is one of the major desirable products of syngas fermentation. Ethanol is commonly used as a direct additive to gasoline. It has an octane value of 129 and the energy content is about 70% of that of gasoline. Most of the syngas-fermenting microbes use acetyl-CoA pathway to produce ethanol. During that process, CO and H_2 are oxidized and produce electrons and H^+ ions necessary for the reactions, while CO_2 gets reduced to Co–CH_3 by accepting the electrons and H^+ ions. Toward the end of the pathway, Co–CH_3 and Co–A react with CO and produce acetyl-CoA under the influence of CODH and acetyl-CoA synthase enzymes. Acetyl-CoA acts as a building block for the production of a variety of biofuels including ethanol (Figure 1).

C. ljungdahlii is one of the most frequently used microorganisms in syngas fermentation to ethanol. Younesi et al. (2005) achieved an ethanol concentration of 0.6 g/L maintaining a syngas pressure of 1.8 atm in their bioreactor. The authors further reported that the high syngas pressure did not have a significant impact on acetic acid production, though it enhanced the ethanol yield. Klasson et al. (1990) reported a higher ethanol yield (3.0 g/L) by adding 0.02% yeast extract followed by cellobiose. The study further showed improvement in molar ratio of ethanol to acetate (>1.1) with the addition of 30 mg/L benzyl-viologen. Klasson et al. (1993) reported the highest ethanol concentration ever recorded (~48 g/L) with *C. ljungdahlii* at a pH of 4.0-4.5 in a completely stirred tank reactor under nutrient-limited condition during 560 h of fermentation.

5.3 Butanol Fermentation

B. methylotrophicum has the ability to convert syngas into acetic acid, butyric acid, and butanol. Shen et al. (1999) compared the physiological differences between the wild-type and the CO-adapted strains of *B. methylotrophicum,* and the production of both butyrate and butanol from CO. The authors found that the activity of the wild-type *B. methylotrophicum* was completely inhibited by the presence of CO. The study further reported that the CO-adapted strain produced significant amount of butyrate, while the wild-type produced only trace amounts of butyrate. The CO-adapted strain produced 0.33 g/L of butanol and 0.5 g/L ethanol at pH 6.0 from the microbes grown at 100% CO.

In a different study, Worden et al. (1989) studied the possibilities of ethanol and butanol production via syngas fermentation. The authors found an increase electron flow of 6-70% from CO into butyrate when the pH was lowered from 6.8 to 6.0. The high level of butyrate essentially increased the butanol yield in a two-stage fermentation process (Worden et al., 1991). During the two-stage process including acidogenic and solventogenic bioconversions, Worden et al. (1991) used two different biocatalysts, *B. methylotrophicum* and *Clostridium acetobutylicum,* in a two-stage process. The authors reported high butyrate (4 g/L) and acetate (8 g/L) concentrations while the biomass recirculation was maintained. The authors further reported a butanol concentration of 2.7 g/L from the continuous operation. Eqs (6) and (7) show the change in Gibbs free energy (ΔG°) for the reactions of CO bioconversion to butyric acid (C_3H_7COOH) and butanol (C_4H_9OH) (Worden et al., 1991).

$$10CO + 4H_2O \rightarrow C_3H_7COOH + 6CO_2 \quad \Delta G^\circ = -40.61\text{kJ/gmole CO} \tag{6}$$

$$12CO + 5H_2O \rightarrow C_4H_9OH + 8CO_2 \quad \Delta G^\circ = -37.68\text{kJ/gmole CO} \tag{7}$$

5.4 Methane Fermentation

There are several methane-fermenting microorganisms including *Methanobacterium thermoautotrophicum, Methanothermobacter thermoautotrophicus, M. barkeri, Methanosarcina acetivorans* strain C2A, *R. rubrum,* and *M. formicum* (O'Brien et al., 1984) that have been isolated for biomethane production from syngas. In syngas-to-methane fermentation, CO acts as an electron donor and CO_2 as an electron acceptor, which gets reduced to methane (CH_4). O'Brien et al. (1984) reported hydrogen production during the growth of *M. barkeri* on CO when the CO partial pressure exceeded 20 kPa. The authors further revealed a net consumption of hydrogen below CO partial pressure 20 kPa. Kluyver and Schnellen (1947) reported the production of intermediates such as H_2 and CO_2 in their suggested CO to methane pathway. Several studies reported the low growth rates of *M. barkeri* and *M. thermoautotrophicus* on CO compared to the growth on H_2 as the electron donor (O'Brien et al., 1984). The possible chemical reactions and the relevant Gibbs free energy contents of the conversion of CO to methane are given in Equations (8) and (9).

$$\text{From 100\% CO}, 4CO + 2H_2O \rightarrow CH_4 + 3CO_2 \quad \Delta G^\circ = -53.0\text{kJ/mole CO} \tag{8}$$

$$\text{From } H_2 \text{ and CO}, CO + 3H_2CH_4 + H_2O \quad \Delta G^\circ = -151.0\text{kJ/mole CO} \tag{9}$$

Sipma et al. (2003) reported the use of several granular anaerobic sludges to produce methane from CO at 30 and 55 °C. The authors found a significant increase in the CO to methane conversion efficiency (up to 90%). But the authors did not fully characterize the microbial communities in the sludge. According to some studies, methanogenesis is highly sensitive to CO concentration in the liquid phase (Klasson et al., 1990). However, successive transfers could enhance the ability of the microorganisms to grow on 100% CO (O'Brien et al., 1984). CO fermentation to methane opens up new area of syngas bioconversion to methane gas, which may overcome some of the challenges of syngas-to-ethanol fermentation.

5.5 Organic Acid Production

Bioconversion of syngas to organic acids (e.g., acetic and butyric acids) and alcohols (e.g., ethanol and butanol) follows the acetyl-CoA pathway (Henstra et al., 2007; Klasson et al., 1990; Phillips et al., 1994). The most common acidogenic microorganisms include *Clostridium thermoaceticum*, *C. ljungdahlii*, *Peptostreptococcus productus*, *A. woodii*, *Eubacterium limosum*, and *B. methylotrophicum*. Many of the reported fermentation studies have shown a high acetic acid production compared to the other organic acids. Younesi et al. (2005) reported an acetate concentration of 1.3 g/L at 1.4 atm pressure using *C. ljungdahlii*.

Butyrate is synthesized by the chemical intermediate acetyl-CoA reacting with butyryl-CoA (Brown, 2006). Acetic and butyric acid yields are highly dependent on the types of microbe and the substrate. Worden et al. (1989) reported that the production of butyrate was increased by 10-folds at the expense of acetate yield when the pH shift was from 6.8 to 6.0. Recovery of organic acids produced during syngas fermentation may provide opportunity for additional revenue generation from coproduct.

6 FACTORS AFFECTING SYNGAS FERMENTATION

6.1 Inhibitory Compounds

In general, the syngas mixture contains constituents such as ethylene (C_2H_4), ethane (C_2H_6), acetylene (C_2H_2), tar, ash, char particles, and gases containing sulfur and nitrogen (Ahmed et al., 2006). These impurities impair the fermentation process through scale formation in the pipes/joints, and inhibition of microbial catalysts and enzymes resulting in low cell growth and product yield. Datar et al. (2004) reported a cell dormancy, hydrogen uptake shutdowns, and a shift in metabolic pathways from acidogenesis to solventogenesis and vice versa, when the syngas was used without conditioning. Introduction of a 0.025-μm filter to remove tar, ash, and other particulate matter from the biomass-derived producer gas was able to overcome cell dormancy (Ahmed and Lewis, 2007). Further, the authors reported that nitrous oxide (NO) was found to be a potential inhibitor of hydrogenase enzyme activity, which reduced the available carbon for product formation. In order to eliminate the inhibitory effects of NO, some studies suggested to improve the gasification efficiency or to scavenge NO by chemicals including sodium hydroxide, potassium permanganate, or sodium hypochlorite. In a separate study, Klasson et al. (1993) reported that the growth of *C. ljungdahlii* was not significantly affected by H_2S concentrations as high as 5.2% (v/v).

It is evident that the biomass-generated syngas has inhibitory compounds that have adverse effects on syngas fermentation efficiency. Some of these impurities can be reduced by biomass pretreatment. Turn et al. (2003) reported that the fuel characteristics of sugarcane bagasse could be improved by pretreatments including milling and leaching. The authors reported a reduction of N, S, and Cl content of the sugarcane bagasse by 13%, 36%, and 62% from their ultimate analysis values (% dry basis) of 0.48%, 0.22%, and 0.65%, respectively. By combining the pretreatments, milling-leaching-milling, the authors reported a further reduction in N, S, and Cl contents by 27%, 82%, and 94%, respectively, from their initial contents. These pretreatments can be implemented to other lignocellulosic biomass feedstocks in order to reduce the production of nitrogen and sulfur compounds during gasification. Takara and Khanal (2011) introduced a new concept of wet or green processing of biomass for upfront juice extraction for coproduct generation, and the utilization of clean fiber for biofuel production. The authors reported the elimination of nitrogen compounds from the biomass to provide clean biomass feedstock for thermochemical conversion.

6.2 Mass Transfer

In microbial syngas fermentation, the gaseous substrates, such as CO and H_2, require transport from gas phase to the cell surface (Vega et al., 1990). In that case, the gaseous substrate is first absorbed at the gas-liquid interface and then diffused through the culture media to the cells. Microbes consume the diffused substrates as their carbon and energy sources and produce the metabolites such as biofuel and other byproducts.

There are several intermediate steps involved in transporting substrate gases into the microbial cells. These steps include the diffusion through the bulk gas to the gas-liquid interface, moving across the gas-liquid interface, transport into the bulk liquid surrounding the microbial cells, and the diffusive transport through the liquid-solid boundary. Out of these, the gas-liquid interface mass transfer is the major resistance for gaseous substrate diffusion (Klasson et al., 1992).

Poor solubility of a gaseous substrate in the culture media results in low substrate uptake by microbes and, thus, leads to low productivity. The volumetric mass transfer coefficient (k_La) is often used to quantify the solubility of a gas in the liquid phase. Klasson et al. (1992) proposed the following equation (Equation (10)) to calculate the mass transfer coefficient (k_L) in the liquid phase.

$$\frac{dN_S^G}{V_L dt} = \frac{k_L a}{H}(P_S^G - P_S^L) \tag{10}$$

where N_S^G(mol) is the molar substrate transferred from the gas phase, V_L (L) is the volume of the reactor, P_S^G and P_S^L (atm) are the partial pressures of the gaseous substrate in gas and the liquid phase, H (L·atm/mol) is Henry's law constant, and a (m^2/L) is the gas-liquid interfacial area for unit volume.

The difference in the partial pressures of the gaseous substrate ($P_S^G - P_S^L$) is the driving force for mass transfer and thus controls the solubility of the substrate. High-pressure operation improves the solubility of the gas in aqueous phase. However, at higher concentrations of gaseous substrates, especially CO, anaerobic microorganisms are inhibited. Therefore, the determination of a correlation between the substrate diffusion and the specific substrate uptake rate (q_S) (h^{-1}) is important in order to evaluate the process kinetics (Equation (11)).

$$q_S = \frac{q_m P_S^L}{K'_p + P_S^L + (P_S^L)^2/W'} \tag{11}$$

where q_m (h^{-1}), W' (atm) and K'_p(atm) are empirical constants. Furthermore, Q_S (mol/L h), the substrate uptake rate, can be written as $Q_S = q_S X$, where X (mol/L) is the microbial cell concentration. By comparing Equations (10) and (11), it is evident that the difference between the partial pressures of the substrate gases and the cell concentration of the reactor is directly proportional (Vega et al., 1990).

Different approaches such as high gas and liquid flow rates, large specific gas-liquid interfacial areas, increased pressure, different reactor configurations (Munasinghe and Khanal, 2010 (b)), innovative impeller designs, modified fluid flow patterns, varying mixing times and speeds, and the use of microbubble dispersers have been examined to enhance gas solubility in the liquid phase. Many of these approaches increase the agitator's power input to volume ratio which facilitates bubble breakup, and increases the interfacial surface area available for mass transfer. This approach, however, is not economically attractive for commercial syngas fermentation due to high energy costs. Additionally, higher power inputs can also damage the sensitive microorganisms in the culture media. In order to achieve energy efficient mass transfer, alternative bioreactor configurations such as trickling beds and airlift reactors were examined for syngas fermentation (Bredwell et al., 1999; Munasinghe and Khanal, 2010 (b)).

Yang et al. (2001) reported $k_L a$ values of 54 and 119 h^{-1} for H_2 and CO gases, respectively, in a slurry bubble column operated at temperature of 20 °C and pressure of 10 bar. In a separate study, Bredwell et al. (1999) reported a maximum $k_L a$ of 190 and 75 h^{-1} for H_2 gas in a stirred-tank reactor at a mixing speed of 300 rpm with and without microbubble sparging, respectively. The authors used a mixed culture of sulfate-reducing bacteria (SRB) in their study. Munasinghe and Khanal (2010 (b)) reported $k_L a$ values for CO, ranging from 0.4 to 91 L/h for eight different reactor configurations including a submerged composite hollow fiber membrane (CHFM) reactor. Some of the reported values of $k_L a$ for different reactor configurations under various hydrodynamic conditions are shown in Table 2.

6.3 Reactor Configuration

Bioreactor configuration is closely related to the effective gas-liquid mass transfer. Thus, reactor design plays an important role in syngas fermentation. High mass transfer rates, high cell densities, low operation and maintenance costs, and easy scale-up are some of the key parameters for designing an efficient bioreactor system. Similarly, the bioreactor size greatly depends on the rate of mass transfer for sparingly soluble gases (Klasson et al., 1992; Vega et al., 1990). Some of the commonly used reactor configurations are discussed in the following section.

6.3.1 Continuous Stirred-tank Reactor

CSTRs are the most commonly employed bioreactors in syngas fermentation. The CSTR has a continuous gas supply into the liquid phase, while agitator controls the gas-liquid mass transfer (Vega et al., 1990). Higher agitation speeds lead to a higher mass transfer rate between the substrate gases and the microbes. However, in industrial-scale fermentors, higher agitation speeds increase the agitator's power consumption, thus increasing the operational cost of the plant. Table 2 summarizes some of the performance parameters for various bioreactors.

TABLE 2 Volumetric Mass Transfer Coefficients (k_La) for Various Reactor Configurations and Hydrodynamic Conditions (Adopted from Munasinghe and Khanal, 2010 (b))

Reactor Configurations	N[a] (rpm)	Microorganisms	Gaseous Substrates	Volumetric Mass Transfer Coefficient (k_La) (h^{-1})	Reference
Trickle bed	n/a	n/a	Syngas	22	Cowger et al. (1992)
Continuous stirred tank	n/a	n/a	Syngas	38	
Continuous stirred tank	200	*B. methylotrophicum*	CO	14.2	Bredwell et al. (1999)
Continuous stirred tank	300	*SRB[b] mixed culture*	Syngas	31 for CO, 75 for H_2	
Continuous stirred tank	300	*C. ljungdahlii*	Syngas	35 for CO	
Continuous stirred tank	300	*R. rubrum*	Syngas	28.1 for CO	
Continuous stirred tank	450	*R. rubrum*	Syngas	101 for CO	
Stirred tank—microbubble sparger	200	*B. methylotrophicum*	CO	90.6	
Stirred tank—microbubble sparger	300	*SRB[b] mixed culture*	Syngas	104 for CO, 190 for H_2	
Packed bubble column	n/a	*R. rubrum*	Syngas	2.1	
Trickle bed	n/a	*R. rubrum*	Syngas	55.5	
Trickle bed	n/a	*SRB[b] mixed culture*	Syngas	121 for CO, 335 for H_2	
Trickle bed	n/a	*C. ljungdahlii*	Syngas	137 for CO	
Batch-stirred tank	n/a	*P. productus*	CO	7.15	Vega et al. (1990)
Stirred tank	300	*C. ljungdahlii*	CO	14.9	Klasson et al. (1993)
Stirred tank	400	*C. ljungdahlii*	CO	21.5	
Stirred tank	500	*C. ljungdahlii*	CO	22.8	
Stirred tank	600	*C. ljungdahlii*	CO	23.8	
Stirred tank	700	*C. ljungdahlii*	CO	35.5	
Bubble column	n/a	n/a	CO	72	Chang et al. (2001)
Stirred tank	400	n/a	CO	10.8-155	Riggs and Heindel (2006)

Continued

TABLE 2 Volumetric Mass Transfer Coefficients (k_La) for Various Reactor Configurations and Hydrodynamic Conditions (Adopted from Munasinghe and Khanal, 2010 (b))—Cont'd

Reactor Configurations	N[a] (rpm)	Microorganisms	Gaseous Substrates	Volumetric Mass Transfer Coefficient (k_La) (h^{-1})	Reference
Stirred tank	500	*R. rubrum*	Syngas	71.8	Younesi et al. (2008)
Column diffuser	n/a	n/a	CO	2.5-40.0	Munasinghe and Khanal (2010 (b))
20-μm bulb diffuser	n/a	n/a	CO	31.7-78.8	
Sparger only	n/a	n/a	CO	29.5-50.4	
Sparger with mechanical mixing	150	n/a	CO	33.5-53.3	
Sparger with mechanical mixing	300	n/a	CO	34.9-55.8	
Submerged composite hollow fiber membrane (CHFM) module	n/a	n/a	CO	0.4-1.1	
Air-lift combined with a 20-μm bulb diffuser	n/a	n/a	CO	49.0-91.1	
Air-lift reactor combined with a single-point gas entry	n/a	n/a	CO	16.6-45.0	

n/a, not applicable.
[a]*Agitation speed.*
[b]*Sulfate-reducing bacteria.*

6.3.2 *Bubble Columns*

The bubble column consists of a glass column mounted on a steel base (Datar et al., 2004). The substrate gases are introduced into the reactor through a fritted glass disk with a pore size of 4-6 microns. These reactors are mainly designed for industrial applications with large reactor volumes. High mass transfer rate and relatively low operational and maintenance cost are among the merits of this type of reactor, while back mixing and coalescence are common drawbacks of the system (Datar et al., 2004).

6.3.3 *Trickle-bed Reactors*

Trickle-bed reactor is a slender column with a packing media. The microbial culture media continuously flows down and the gaseous substrate flows either upward (countercurrent) or downward (cocurrent) direction depending on the application (Munasinghe and Khanal,

2010). The trickle-bed reactors are operated under the atmospheric pressure and no agitation is necessary. Therefore, trickle-bed reactors consume less energy than the conventional CSTR (Bredwell et al., 1999).

6.3.4 Microbubble Sparged Reactors

Microbubble sparged reactor is a combination of a CSTR reactor and a microbubble disperser. In this reactor configuration, the large specific interfacial area of the gas bubbles and the longer retention time stimulate the high gas-liquid mass transfer.

6.3.5 Membrane-based Reactors

CHFM have been proposed as a technologically and economically feasible method for syngas fermentation. Even though the technology is yet to be adopted for syngas fermentation, it has been widely studied in water and wastewater treatment. In CHFM reactors, the gas is introduced through the membrane fibers. The microbes grown as a biofilm on the surface of the membrane fibers utilize the CO and H_2 and produce biofuels. This novel CHFM system offers several advantages such as higher microbial cell retention, higher yield, and higher tolerance to toxic compounds (tar, acetylene, NO_x, etc.).

6.4 pH

pH is a critical parameter to obtain optimal microbial activity in the culture media. In general, acidogenic reactions are more favorable at higher pH values (6.0-7.0), whereas solventogenesis (alcohol production) requires low pH values (4.0-4.5) (Klasson et al., 1993). Researchers employed pH shift approach in order to achieve high alcohol yields. Low pH, however, inhibits the cell growth (Klasson et al., 1992). The optimum pH for most of the syngas-fermenting microbes varies between 4.5 and 7.3 depending on the species. For example, *C. ljungdahlii* has an optimum pH of 5.8-6.0.

6.5 Temperature

The temperature of the culture media affects the syngas fermentation in two ways. Firstly, it affects the growth kinetics and secondly, it affects the solubility of the syngas in aqueous medium. In most cases, the temperature of the culture media is decided based on the specific microorganism. The most favorable growth temperature for mesophilic microorganisms is 37-40 °C, while for thermophilic microbes, it is between 55 and 80 °C.

6.6 Types of Microorganism

Strict anaerobe, *C. ljungdahlii* is one of the most studied microorganisms in syngas fermentation. Apart from *C. ljungdahlii*, several other strict anaerobic mesophilic microbes have been identified that are capable of fermenting syngas into biofuels. Some of these microbes include *C. aceticum, A. woodii, C. autoethanogenum*, and *C. carboxydivorans* (Klasson et al., 1992; Rajagopalan et al., 2002; Younesi, et al., 2005). Significant efforts are being made to genetically engineer syngas-fermenting microbes to improve the yield.

6.7 Growth Culture Media

Growth media provides the microbes with all essential nutrients such as minerals, trace elements, vitamins, and reducing agents for their maximal growth. The selection of the growth media depends on the selected species and the targeted end products. Reducing agents (e.g., sodium thioglycolate, ascorbic acid, and benzyl-viologen) result in shift in the electron flow, thereby diverting the carbon flow from acid to alcohol production (Klasson et al., 1992). Researchers have developed their own protocols for media preparation, while some specific media are provided by American Type Culture Collection (ATCC) (e.g., ATCC culture 1754 for *C. ljungdahlii, acetobacterium* medium ATCC 1019).

7 INDUSTRIAL-SCALE SYNGAS FERMENTATION

Currently, there is not a single industrial-scale syngas fermentation-to-biofuel plant. Gas-to-liquid mass transfer is still considered as the major bottleneck for the commercialization of syngas fermentation technology. Keegan (2008) reported the construction of 40,000 gal per year pilot plant and a 40- to 100-million gal per year commercial-scale syngas fermentation facility. Kundiyana et al. (2010) reported a successful operation of a 100-L pilot scale syngas fermentation facility. Regardless of the recent developments in reactor designs, process optimizations, and microbial catalysts selection, the ethanol concentration from syngas fermentation is still just under 30 g/L. This leads to a high cost of ethanol recovery. For cost-effective ethanol recovery, its concentration should be around 15% (v/v). Therefore, in order to reduce the recovery cost, thus improving the overall economy of the process, industrial-scale syngas fermentation should focus on achieving higher ethanol concentration. This requires significant research and development in process microbiology.

8 CHALLENGES AND FUTURE RESEARCH DIRECTIONS

8.1 Syngas Quality

Gasification of biomass produces a gas mixture containing additional constituents such as tars, ash, particulate matter, higher hydrocarbons (e.g., C_2H_2, C_2H_4, and C_2H_6), and gaseous compounds containing sulfur and nitrogen other than CO, H_2, and CO_2. Many studies highlighted the adverse effects of these impurities on syngas fermentation including process upset, cell dormancies, and inhibition of enzymes. Therefore, syngas should be free from these impurities before entering into the fermenting process. The commonly adopted gas clean-up methods include cyclones, various types of filters and scrubbers, and rotating particle separators. Post gas clean-up operations always contribute to high operational and maintenance costs. Turn et al. (2003) proposed a pretreatment protocol combining milling and leaching to reduce S, N, and Cl compounds from biomass feedstocks from sugarcane family. Takara and Khanal (2011) reported the elimination of nitrogen compounds from biomass feedstocks by adopting wet or green processing through upfront juicing and clean fiber utilization for biofuel production.

8.2 Mass Transfer

Syngas fermentation to biofuels is often plagued by the limitations of gas-to-liquid mass transfer. Different approaches have been examined including advancements in impeller designs, fluid flow patterns, aerated power efficiency, mixing time, baffle design, and the use of microbubble dispersers to improve the mass transfer in the liquid phase. However, most of the suggested methods raise the concerns over long-term sustainability and economic feasibility issues when it comes to scale-up. The use of CHFM in syngas fermentation is an innovative approach which offers significant advantages over the conventional reactor configurations (Munasinghe and Khanal, 2010 (a)). More research effort is needed to develop innovative reactor design to obtain better mass transfer rate.

8.3 Product Recovery

Biofuel is one of the most desired products of syngas fermentation. Depending on the reactor design, process parameters, and microbial catalysts, the product stream may contain ethanol, acetic acid, butanol, butyrate, and other bio-based products and microbial cells (Lewis et al., 2010). Distillation is one of the traditional methods of recovering ethanol from the fermentation media. The main disadvantages of distillation process include high energy cost and the formation of azeotrope. Ultrasonic atomization, vapor recompression, vapor reuse and vacuum distillation, and selective adsorption of water are some of the alternative methods that have been examined in order to reduce the ethanol recovery cost. Liquid-liquid extraction is a widely used separation technique for acetic acid recovery. A suitable solvent can be used in order to extract a substantially pure acetic acid solution. Fockedey et al. (2008) proposed a novel extraction/re-extraction method using glycerol as one of the solvents to recover ethanol, acetic acid, and other byproducts.

8.4 Microbial Catalysts

Microbial catalysts play a major role in syngas fermentation by efficiently converting the gaseous substrates into biofuels. Even though the current syngas-fermenting microbes show high specificity, the product yields are comparatively low. Therefore, ideal microbial catalysts should tolerate fermentation inhibitors and process stresses such as product toxicity. The isolation of new microbial species which can effectively convert syngas into biofuels with higher product yields is one of the major challenges of scaling-up the syngas fermentation technology. Genetic engineering of microbial species capable of surviving in harsh conditions with a higher yield is another emerging research frontier in syngas fermentation.

8.5 Redirection of Metabolic Pathway

Several researchers highlighted the importance of pH shifts in syngas fermentation. In general, at high pH levels (6.0-7.0), acid production is more favorable. With acid production, the pH of the culture media drops and that might cause unfavorable environment for most of the solvent producing Clostridia. In this case, redirecting the metabolic pathway toward the solvent production by reducing the acid production might increase the ethanol yields.

9 SUMMARY

Biomass-derived syngas fermentation to biofuels is identified as a sustainable alternative for the fast depleting fossil-derived fuels. The process has several advantages including higher availability, low feedstock cost, and no competition with food and feed. The commercialization of syngas fermentation to biofuels is often plagued by the gas-to-liquid mass transfer limitations and low product yield. Innovative reactors designs and metabolic engineering aspects are being studied extensively in recent literature focusing on higher product yields. Use of CHFM in syngas fermentation is at its infant stage and needs extensive research to prove its economic and scale-up feasibilities. Similarly, research efforts should also be directed toward production of other biofuel such as butanol and gaseous fuel, methane.

References

Ahmed, A., Lewis, R.L., 2007. Fermentation of Biomass generated synthesis gas: effects of Nitric oxide. Biotechnol. Bioeng. 97 (5), 1080–1086.

Ahmed, A., Cateni, B.G., Huhnke, R.L., Lewis, S.R., 2006. Effects of biomass-generated producer gas constituents on cell growth, product distribution and hydrogenase activity of *Clostridium carboxidivorans* P7^T. Biomass Bioenergy 30, 665–672.

Albertazzi, S., Basile, F., Brandin, J., Einvall, J., Hultebrg, C., Fornasari, G., et al., 2005. The technical feasibility of biomass gasification for hydrogen production. Catal. Today 106, 297–300.

Biomass Research and Development Technical Advisory Committee of the US Departments of Energy and Agriculture, 2002. Roadmap for Biomass Technologies in the United States, Biomass Research and Development Technical Advisory Committee.

Bredwell, M.D., Srivastava, P., Worden, R.M., 1999. Reactor design issues for synthesis-gas fermentations. Biotechnol. Prog. 15, 834–844.

Bridgwater, A.V., 1995. The technical and economic feasibility of biomass gasification for power generation. Fuel 74, 631–653.

Brown, R.C., 2006. Biomass refineries based on hybrid thermochemical-biological processing—an overview. In: Kamm, B., Gruber, P., Kamm, M. (Eds.), Biorefineries—Industrial Processes and Products. Status Quo and Future Directions, vol. 1. Wiley-Vch Verlag GmbH & Co. KGaA, Weinheim.

Chang, I.S., Kim, B.H., Lovitt, R.W., Bang, J.S., 2001. Effect of CO partial pressure on cell-recycled continuous CO fermentations by *Eubacterium limosum* KIST612. Proc. Biochem. 37, 411–421.

Cowger, J.P., Klasson, K.T., Ackerson, M.D., Clausen, E.C., Gaddy, J.L., 1992. Mass-transfer and kinetic aspects in continuous bioreactors using *Rhodspirilium rubrum*. Appl. Biochem. Biotechnol. 34 (/35), 613–624.

Datar, R.P., Shenkman, R.M., Cateni, B.G., Huhnke, R.L., Lewis, R.S., 2004. Fermentation of biomass-generated producer gas to ethanol. Biotechnol. Bioeng. 86, 587–594.

Demirbas, A., 2007. Progress and recent trends in biofuels. Progress Energy Combust. Sci. 33, 1–18.

Fockedey, E., Wilde, J.D., Gerin, P.A., 2008. Separation of chemicals produced by acidogenic fermentation. In: The 2008 Annual Meeting Philadelphia, PA.

Gabra, M., Pettersson, E., Bachman, R., Kjellstrom, B., 2001. Evaluation of cyclone gasifier performances for gasification of sugar cane residue-part I: gasification of bagasse. Biomass Bioenergy 21, 351–369.

Heiskanen, H., Virkajarvi, I., Viikari, L., 2007. The effects of syngas composition on the growth and product formation of *Butyribacterium methylotrophicum*. Enzyme Microb. Technol. 41, 362–367.

Henstra, A.M., Sipma, J., Rinzema, A., Stams, A.J.M., 2007. Microbiology of synthesis gas fermentation for biofuel production. Curr. Opin. Biotechnol. 18, 200–206.

Karnholz, A., Kusel, K., Goner, A., Schramm, A., Drake, H.L., 2002. Tolerance and metabolic response of acetogenic bacteria toward oxygen. Appl. Environ. Microbiol. 68, 1005–1009.

Keegan, M., 2008. Making ethanol the Coskata way. Available from: http://thearticlewriter.com/autowriter (accessed 10.11.10).

Klasson, K.T., Ackerson, C.M.D., Clausen, E.C., Gaddy, J.L., 1990. Bioreactor design for synthesis gas fermentation. In: Biotechnology for the Production of Clean Fuels 28 August, Washington, USA.

Klasson, K.T., Ackerson, C.M.D., Clausen, E.C., Gaddy, J.L., 1992. Biological conversion of synthesis gases into fuels. Int. J. Hydrogen Energy 17 (4), 281–288.

Klasson, K.T., Ackerson, C.M.D., Clausen, E.C., Gaddy, J.L., 1993. Biological conversion of coal and coal-derived synthesis gas. Fuel 72, 1673–1678.

Kluyver, A.J., Schnellen, C., 1947. On the fermentation of carbon monoxide by pure cultures of methane bacteria. Arch. Biochem. 14 (1-2), 57–70.

Kundiyana, D.K., Huhnke, R.L., Wilkins, M.R., 2010. Syngas fermentation in a 100-L pilot scale fermentor: design and process considerations. J. Biosci. Bioeng. 109 (5), 492–498.

Lewis, R.S., Tree, D.R., Hu, P., Frankman, A., 2010. Syngas fermentation to ethanol: challengers and opportunities. In: Khanal, S.K., Surampalli, R.Y., Zhang, T.C., Lamsal, B.P., Tyagi, R.D., Kao, C.M. (Eds.), Bioenergy and Biofuel from Biowastes and Biomass. ASCE publication.

Munasinghe, P.C., Khanal, S.K., 2010 (a). Biomass-derived syngas fermentation into biofuels: opportunities and challenges. Bioresour. Technol. 101, 5013–5022.

Munasinghe, P.C., Khanal, S.K., 2010 (b). Syngas fermentation to biofuel: evaluation of carbon monoxide mass transfer coefficient (k_La) in different reactor configurations. Biotechnol. Prog. 26 (6), 1616–1621.

O'Brien, J.M., Wolkin, R.H., Moench, T.T., Morgan, J.B., Zeikus, J.G., Lupton, F.S., et al., 1984. Association of hydrogen metabolism with unitrophic or mixotrophic growth of *Methanosarcina barkeri* on carbon monoxide. J. Bacteriol. 158, 373–375.

Phillips, J.R., Clausen, E.C., Gaddy, J.L., 1994. Synthesis gas as substrate for the biological production of fuels and chemicals. Appl. Biochem. Biotechnol. 45-46, 145–157.

Rajagopalan, S., Datar, R.P., Lewis, R.S., 2002. Formation of ethanol from carbon monoxide via a new microbial catalyst. Biomass Bioenergy 23, 487–493.

Riggs, S.S., Heindel, T.J., 2006. Measuring carbon monoxide gas-liquid mass transfer in a stirred tank reactor. Biotechnol. Prog. 22, 903–906.

Shen, G.J., Shieh, J.S., Grethlein, A.J., Jain, M.K., Zeikus, J.H., 1999. Biochemical basis for carbon monoxide tolerance and butanol production by *Butyribacterium methylotrophicum*. Appl. Microbial. Biotechnol. 51, 827–832.

Sipma, J., Lens, P.N.L., Stams, A.J.M., Lattinga, G., 2003. Carbon monoxide conversion by araerobic bioreactor sludges. FEMS Microbiol. Ecol. 44, 271–277.

Subramani, V., Gangwal, S.K., 2008. A review of recent literature to search for an efficient catalytic process for the conversion of syngas to ethanol. Energy Fuels 22, 814–839.

Takara, D., Khanal, S.K., 2011. Green processing of tropical banagrass into biofuel and biobased products: an innovative biorefinery approach. Bioresour. Technol. 102 (2), 1582–1587.

Tanner, R.S., Miller, L.M., Yang, D., 1993. *Clostridium ljungdahlii* sp. nov., an acetogenic species in clostridial ribosomal-RNA homology group-I. Int. J. Syst. Bacteriol. 43, 232–236.

Turn, S.Q., Kinoshita, C.M., Jakeway, L.A., Jenkins, B.M., Baxter, L.L., Wu, B.C., Blevins, L.G., 2003. Fuel characteristics of processed, high-fiber sugarcane. Fuel Proc. Technol. 81, 35–55.

United States Department of Agriculture (USDA) and United States Department of Energy (USDOE) Joint Report, 2005. A billion-ton feed stock supply for bioenergy and bioproducts industry: technical feasibility of annually supplying 1 billion dry tons of biomass. Available from: http://www1.eere.energy.gov/biomass/pdfs/final_billionton_vision_report2.pdf (accessed on: 09.10.2010).

Vega, J.L., Prieto, S., Elmore, B.B., Clausen, E.C., Gaddy, J.L., 1989. The biological production of ethanol from synthesis gas. Appl. Biochem. Biotechnol. 20 (/21), 781–797.

Vega, J.L., Clausen, E.C., Gaddy, J.L., 1990. Design of bioreactors for coal synthesis gas fermentation. Conservat. Recycl. 3, 149–160.

Worden, R.M., Grethlein, A.J., Zeikus, J.G., Datta, R., 1989. Butyrate production from carbon monoxide by *Butyribacterium methylotrophicum*. Appl. Biochem. Biotechnol. 20/21, 687–698.

Worden, R.M., Grethlein, A.J., Jain, M.K., Datta, R., 1991. Production of butanol and ethanol from synthesis gas via fermentation. Fuel 70 (5), 615–619.

Yang, W., Wang, J., Jin, Y., 2001. Mass transfer characteristics of syngas components in slurry system at industrial conditions. Chem. Eng. Technol. 24 (6), 651–657.

Younesi, H., Najafpour, G., Mohameda, A.R., 2005. Ethanol and acetate production from synthesis gas via fermentation processes using anaerobic bacterium, *Clostridium ljungdahlii*. Biochem. Eng. J. 27, 110–119.

Younesi, H., Najafpour, G., Ismail, K.S.K., Mohamed, A.R., Kamaruddin, A.H., 2008. Biohydrogen production in a continuous stirred tank bioreactor from synthesis gas by anaerobic photosynthetic bacterium: *Rhodopirillum rubrum*. Bioresour. Technol. 99, 2612–2619.

SECTION II

PRODUCTION OF BIOETHANOL FROM LIGNOCELLULULOSIC FEEDSTOCKS

CHAPTER

5

Lignocellulosic Bioethanol: Current Status and Future Perspectives

Carlos Ricardo Soccol[1]*, *Vincenza Faraco*[2,3], *Susan Karp*[1], *Luciana P.S. Vandenberghe*[1], *Vanete Thomaz-Soccol*[4], *Adenise Woiciechowski*[1], *Ashok Pandey*[5]

[1]Department of Bioprocess Engineering and Biotechnology, Federal University of Parana, P.O. Box 19011, Curitiba, Brazil

[2]Department of Organic Chemistry and Biochemistry, University of Naples "Federico II", Complesso Universitario Monte S. Angelo, via Cintia 4 80126, Naples, Italy

[3]School of Biotechnological Sciences, University of Naples "Federico II", Naples, Italy

[4]Industrial Biotechnology Program, Positivo University, Curitiba, Brazil

[5]Centre for Biofuels, National Institute for Interdisciplinary Science and Technology, CSIR, Trivandrum, India

*Corresponding author: E-mail: soccol@ufpr.br

1 INTRODUCTION

The progressive depletion of fossil carbon sources has been causing increasing concern regarding both the security of their supply and greenhouse gas emission (GHG) and global warming. Biofuels can play a key role in solving these problems (Faaij, 2006). Among these, ethanol produced from corn starch in the United States, ethanol made from sugarcane in Brazil, and biodiesel produced mainly from rapeseed oil in Germany and France have been the most important commercial choices in recent years (Sagar and Kartha, 2007). Bioethanol has been recognized as a potential alternative to petroleum-derived transportation fuels (Oliveria et al., 2005), with several advantages such as high octane number, low cetane number, and high heat of vaporization (Balat, 2007).

A variety of biomass feedstock has been explored for ethanol production. According to Balat et al. (2008), bioethanol feedstock can be classified into three types: (i) lignocellulosic materials such as woody biomass, herbaceous perennials, and various wastes; (ii) starch-rich crops such as maize and grain sorghum; (iii) sucrose-rich crops such as sugarcane and sugarbeet.

In 2009, global production of bioethanol reached 19.5 billion gallons (agra-net.com, 2010), accounting for more than 94% of global biofuel production (International Risk Governance Council, 2007). Brazil and the US represent the world leaders, together accounting for about 90% of the world bioethanol production (Table 1).

1.1 Ethanol Production in the United Sates

In the United States, ethanol production achieved 12.1 billion gallons in 2010 (Service, 2010). In 2010, 204 biorefineries were operating and seven other biorefineries were under construction (Renewable Fuels Association—RFA, http://www.ethanolrfa.org/bio-refinery-locations/, accessed November 22, 2010). Ethanol is currently used as a gasoline additive in many states in the USA, where it has largely replaced methyl tertiary butyl ether in the last 6 years as a gasoline additive and oxygenate to reduce air pollution (Solomon, 2010). More than 95% of the ethanol produced for use in U.S. cars is currently manufactured from corn starch, and mostly used in 10% blends with gasoline (E10). All car engines produced in the United States after 1988 run on ethanol-alcohol E10 blends, and in most cases up to E20. Currently, over seven million cars in the United States have engines that can use an 85% ethanol blend (E85; U.S. Department of Energy, 2008).

TABLE 1 World Fuel Ethanol Production in 2008

Country	Millions of Gallons
USA	9000.0
Brazil	6472.2
European Union	733.6
China	501.9
Canada	237.7
Others	128.4
Thailand	89.8
Colombia	79.29
India	66.0
Australia	26.4
Total	17,335.2

Source: *Adapted from Renewable Fuels Association (http://www.ethanolrfa.org/pages/statistics#EIO, accessed November 22, 2010)*

1.2 Ethanol Production in Brazil

In Brazil, ethanol is produced exclusively from sugarcane. With an average productivity of about 6000 L/hectare (Soccol et al., 2010), the Brazilian sugarcane system of agroenergy now represents the most efficient system (Goldemberg, 2007). In 2008, ethanol production achieved around 6.6 billion gallons (Soccol et al., 2010) and the number of installed plants was estimated at 448 units (Udop, 2009). Currently, Brazil has more than 80% of its vehicles running with bioethanol and even small airplane engines are now being developed (Soccol et al., 2010).

Sugarcane was the chosen substrate for ethanol production due to its great adaptation to the Brazilian soil and weather conditions. Today, sugarcane ethanol production costs range from US$ 0.25–0.30 L^{-1} (Cerqueira Leite et al., 2009), and around 70% of these costs are due to feedstock purchase price (IBGE, 2008). The current trends are on the one hand to increase cane yield and on the other to raise the biomass fraction being transformed (Soccol et al., 2010). Today, the part of the sugarcane biomass used for bioenergy production corresponds to one-third of the plant, another one-third being bagasse, which is burnt for electricity production, whilst the remaining one-third is left in the field (Cortez et al., 2008). A significant increase in ethanol production yield would be possible by developing technologies for ethanol production from sugarcane bagasse, with a potential production of 600 million gallons of ethanol from 10 million tons of dry biomass (Soccol et al., 2010).

1.3 Cellulosic Ethanol Production

The use of sugar or starch as raw materials for fuel production competes with their use as foods (Pimentel et al., 2008), and the supply is not expected to be sufficient to face the increasing demand for ethanol fuel. Lignocellulosic biomass is an attractive alternative material for bioethanol fuel production. Lignocellulose is the most abundant renewable resource on Earth, and it constitutes a large component of the wastes originating from municipal, agricultural, forestry, and some industrial sources. The more widespread geographical distribution of lignocellulose sources, compared to fossil reserves, can provide security of supply by using domestic sources of energy. The use of lignocellulosic materials would minimize the conflict between land use for food (and feed) production and energy feedstock production. This raw material is less expensive than conventional agricultural feedstock and can be produced with lower input of fertilizers, pesticides, and energy. Biofuels from lignocellulose generate low net GHG emissions, reducing environmental impact, particularly on climate change.

The reduction of bioethanol cost depends mainly on the purchase price of feedstock and the cost of feedstock processing. At recent prices for corn, sugarcane, and cellulosic biomass, the latter feedstock is the least expensive (Lynd et al., 2009). However, order of conversion costs using current technology is the opposite of that of feedstock purchase cost: cellulosic biomass > corn > sugarcane. Taking both purchase price and current conversion technology into account, the near-term fuel cost is sugarcane ethanol < corn ethanol < cellulosic ethanol, whilst in the long term, incorporating advanced technological improvements (Lynd et al., 2009), the projected selling price of cellulosic ethanol is less than the purchase cost of the other feedstocks considered (Lynd et al., 2009). Cellulosic biomass is the best candidate for

TABLE 2 Comparison of Feedstock Categories with Respect to Several Social Objectives

	(A) Large-Scale Production		(B) Rural Economic Development		(C) Petroleum Displacement		(D) Fossil Fuel Displacement/ GHG Reduction		(E) Soil Fertility & Agricultural Ecology	(F) Low-Cost Fuels (Feedstock & Conversion)	
Feedstock Type	Per unit	Total	Now	Future	Per unit	Total	Per unit	Total		Now	Future
Cellulosic	••••	••••	••	•••••	•••••	•••••	•••••	•••••	•••••	••	•••••
Starch rich	••••	••	••••	•••	•••••	•••	•••	••	••	•••	•••
Sugar rich	•••	••	••••	•••	•••••	•••	•••••	••••	••••	•••	•••

Source: *Adapted from Lynd et al. (2009) with the author's permission.*
Ratings: ••••• Excellent; •••• Very good; ••• Good; •• Fair; • Poor.

large-scale energy production in the long term among the feedstock types reported in Table 2, for its potential in low fuel-production price, large-scale production, and environmental benefits. However, cellulosic biofuels are not produced at a competitive level yet, due to the high cost of processing with the currently available technologies (Lynd et al., 2008).

1.4 Cellulosic Ethanol Production Plants

Currently, some countries are producing ethanol from cellulosic feedstock at different development stages, and several public/private international projects have been developed in the biorenewable sector to promote a bio-based economy.

In Canada, Iogen's demonstration plant has been producing cellulosic ethanol since 2004, and in 2009 a production capacity of 1,464,978 L/year was achieved from wheat straw (http://www.iogen.ca/). This demonstration plant located in Ottawa is designed to process about 20-30 tons/day of feedstock (wheat, oat, and barley straw) and to produce approximately 5000-6000 L of cellulosic ethanol per day.

In summer 2005, Sekab's plant in Örnsköldsvik (Sweden) started to produce ethanol from sawdust (http://www.sekab.com). At present, Sekab sells ethanol to the 1400 E85 pumps throughout Sweden, and the number of fuel-flexible cars is about 147,000.

The U.S. Department of Energy and Abengoa Bioenergy New Technologies have signed a 4-year, $35.5 million (U.S.) contract in 2003 to develop the technology for Advanced Biorefining of Distillers Grain and Corn Stover Blends: Pre-Commercialization of Biomass-derived Process Technologies. In agreement with this contract, Abengoa Bioenergy (http://www.abengoabioenergy.com/corp/web/es/index.html) is developing pilot-scale processes for integrating lignocellulosic biomass ethanol with cereal ethanol production to achieve best overall economic results. The project involves conversion of residual starch, cellulose, and hemicellulose—mainly corn stover—to bioethanol. The first phase of this project, which has been completed at pilot plant testing phase, successfully demonstrated residual starch conversion and improved coproducts production. The 1.5 metric ton/day biomass refinery pilot plant, which has been in operation since September 2007, is being used for process optimization, generating engineering data for scaling up to commercial plant, and to generate large quantities of various coproducts for further development and evaluation.

Inbicon (http://www.inbicon.com/pages/index.aspx) is installing biorefineries to produce cellulosic ethanol in different sites. At Kalundborg (Denmark), their refinery can produce up to 1.4 million gallons/year of cellulosic ethanol from soft biomass such as bagasse from sugar production, miscanthus grass, and empty fruit bunches. In Malaysia, they are planning projects between 5 and 10 million gallons/year. In the United States, their first commercialization will produce 20 million gallons/year.

Poet (http://www.poet.com) have a network of 27 plants in seven states in the U.S., producing more than 1.6 billion gallons of ethanol annually. They invested over $40 million on research for construction of a commercial cellulosic ethanol plant (in Emmetsburg, Iowa) that should reach a production capacity of 25 million gallons/year, making use of corncobs.

2 FIRST-GENERATION FUEL ETHANOL PRODUCTION: THE FEEDSTOCK AND THE PROCESS AND THEIR CONSTRAINTS

First-generation biofuels are fuels produced from traditional agricultural crops, mainly represented by corn, sugarcane, and sugar beet.

Sugarcane and sugar beet are the most potential feedstocks for bioethanol production (United Nations Conference on Trade and Development, 2006), since they are sources of sucrose, a disaccharide composed of glucose and fructose linked by a β-1,2 bond. Sucrose is hydrolyzed by the enzyme invertase, which is produced by most species of *Saccharomyces*, so it is not necessary to prehydrolyze the substrate. This makes the ethanol production from sugar (sucrose) a very feasible process in comparison to corn ethanol. Two-thirds of the world sugar production is from sugarcane and one-third is from sugar beet (Linoj et al., 2006). Sugar cane is grown in tropical and subtropical countries, while sugar beet is grown only in countries with temperate climate.

Although there is a great number of starch sources, only a few of them have industrial importance: maize, cassava, potato, and wheat. Corn starch, for instance, is responsible for more than 80% of the worldwide starch market and the biggest production is in the United States (Jobling, 2004). In tropical countries, other starchy crops such as tubers (e.g., cassava) can be used for commercial production of fuel ethanol (Cardona and Sanchez, 2007). To produce ethanol from starch, it is necessary to break down the chains of this carbohydrate in order to obtain the glucose syrup, which can be converted into ethanol by *Saccharomyces cerevisiae*. This prehydrolysis is usually performed with starch-hydrolyzing enzymes and represents an additional cost in the process.

The production process of first-generation fuel ethanol comprises the following steps:

2.1 Pretreatment

The pretreatment of the raw material depends on the type of feedstock selected. However, some procedures are common in most of the cases, which: washing in order to remove impurities; grinding to reduce size; pressing to extract the juice and separate the solid fraction, usually called bagasse; physicochemical treatment of the extracted broth. This treatment includes separation of solids by sieving (alternatively, centrifugation), clarification by the action of pH and temperature, and subsequent decantation/sieving to separate colloids (Lima and Marcondes, 2002). Utilization of starchy raw materials necessarily demands a hydrolysis step, which can be chemical (acid) hydrolysis or enzymatic hydrolysis.

2.2 Fermentation

This bioprocess represents the conversion of fermentable sugars (hexoses) into ethanol by the metabolism of microorganisms, as represented by the following equation:

$$C_6H_{12}O_6 \rightarrow 2CH_3CH_2OH + 2CO_2$$

When sucrose is the substrate, the fermentation is usually performed by the yeast *Saccharomyces cerevisiae*, while the bacteria *Zymomonas mobilis* can be utilized to convert glucose into ethanol. The theoretical yield is 0.511 g of ethanol/g hexose. However, due to the formation of byproducts such as glycerol and organic acids and cellular maintenance, the maximum real yield is around 0.485 g ethanol/g hexose, which is called Pasteur yield (Lima and Marcondes, 2002).

During the fermentation process, several physicochemical factors such as temperature, dissolved oxygen, pH, and nutrients are essential for the appropriate development of the microorganism. For *Saccharomyces cerevisiae*, the temperature should be maintained at values below 32 °C, the ideal pH is between 4 and 5, and the sugar concentration should be no greater than 16°Brix. Once these parameters are adjusted, the fermentation starts and is usually conducted as a fed batch, in order to avoid the excessive formation of gas and foam. The fermentation tanks are usually open and have an internal or external temperature control system. After fermentation, the cells are recovered by centrifugation, treated with sulfuric acid and nutrients in order to improve viability and utilized as inoculum for the subsequent fermentation.

2.3 Distillation

The distillation device is composed of distillation columns, reboilers, located in the bottom of the columns, and condensers in the top of the columns. The fermented broth usually contains 7-7.5% (w/w) ethanol and enters the first column for a primary separation. The vapors of this column contain 35-45% (w/w) ethanol and are directed to a second column for a primary distillation. Then, the mixture feeds the third column, also called rectification column, which delivers ethanol at a concentration of 96.4°GL, the maximum possible concentration of the ethanol-water azeotropic mixture. Anhydrous ethanol, which is usually added to gasoline, is obtained by distillation in the presence of a solvent such as hexane or benzene, or by zeolite molecular sieves (Lima and Marcondes, 2002). The residue of distillation is called vinasse and represents an environmental problem, since 1 L of ethanol generates around 15 L of vinasse that must be disposed.

3 SECOND-GENERATION ETHANOL PRODUCTION

Second-generation biofuels are those produced from nonfood renewable resources. Some types of lignocellulosic biomasses such as straw, cornhusk, corn stover, sugarcane bagasse, and the organic fraction of municipal solid waste represent abundant, inexpensive, and readily available sources utilized for second-generation ethanol production. Currently, this readily available biomass is considered as a waste left at the field and is disposed of through agricultural burning after harvest (Dawson and Boopathy, 2007), or burnt at the industry to generate thermal energy.

3.1 Lignocellulose Structure

Lignocellulose is by far the major component of biomass, comprising around half of the plant matter produced by photosynthesis (also called photomass) and representing the most abundant renewable organic resource in soil. It consists of three types of polymers, cellulose, hemicellulose, and lignin, that are strongly intermeshed and chemically bonded by noncovalent forces and by covalent cross-linkages (Pérez et al., 2002; Sanchez, 2009). Compared with starchy biomass, it is considered as a quite recalcitrant material due to its highly lignified and crystalline structure. Only a small amount of the cellulose, hemicellulose, and lignin produced as byproducts in agriculture or forestry is used, the rest being considered waste.

Cellulose and hemicellulose are macromolecules constructed from different sugars, whereas lignin is an aromatic polymer synthesized from phenylpropanoid precursors. The composition and proportions of these compounds vary between plants (Sanchez, 2009).

Cellulose is a linear polymer that is composed of D-glucose subunits linked by β-1,4 glycosidic bonds forming the dimmer cellobiose. These form long chains (or elemental fibrils) linked together by hydrogen bonds and intra- and intermolecular van der Waals forces. This polymer usually is present as a crystalline form and a small amount of nonorganized cellulose chains forms amorphous cellulose. In the latter conformation, cellulose is more susceptible to enzymatic degradation (Pérez et al., 2002). Cellulose appears in nature to be associated with other plant compounds, mainly lignin, and this association may affect its biodegradation.

Hemicellulose is a polysaccharide with a lower molecular weight than cellulose. It is formed from D-xylose, D-mannose, D-galactose, D-glucose, L-arabinose, 4-O-methyl-glucuronic, D-galacturonic, and D-glucuronic acids, depending on the hemicellulose source. Sugars are linked together by β-1,4- and sometimes by β-1,3-glycosidic bonds.

Lignin is present in the cellular wall to give structural support, impermeability, and resistance against microbial attack and oxidative stress. It is an amorphous heteropolymer, non water-soluble, and optically inactive that is formed from phenylpropane units joined together by nonhydrolyzable linkages. This polymer is synthesized by the generation of free radicals, which are released in the peroxidase-mediated dehydrogenation of three phenyl propionic alcohols: coniferyl alcohol (guaiacyl propanol), coumaryl alcohol (p-hydroxyphenyl propanol), and sinapyl alcohol (syringyl propanol). This heterogeneous structure is linked by C—C and aryl ether linkages, with aryl-glycerol β-aryl ether being the predominant structure (Sanchez, 2009).

3.2 The Production Process: General Description of the Main Steps

The main challenge in second-generation bioethanol production from lignocellulosic biomass is the transformation of the complex polymers into simple sugars that can be assimilated by microorganisms during fermentation. These are the main steps in the conversion process:

3.2.1 *Mechanical Process*

The mechanical or physical process is usually the first step in biomass pretreatment and aims to reduce the size of the material, breaking the physical structure. It can be performed by milling, grinding, or chipping. The particle size after the physical treatment is an important parameter to be optimized. The use of very small particles may not be desirable due to higher energy consumption in milling stage as well as imposing negative effect on the subsequent pretreatment method (Talebnia et al., 2010). This process can demand a lot of energy since it requires equipments such as knife mill and hammer mill (Cadoche and Lopez, 1989).

3.2.2 *Physicochemical Pretreatments*

The aim of pretreatment is to promote the separation of the lignocellulose's components, that is, lignin, hemicellulose, and cellulose, reduce the crystallinity of cellulose, and disrupt the structure of biomass. This treatment should improve the formation of fermentable sugars, avoid the degradation of carbohydrates and, of course, be cost effective (Sun and Cheng, 2002). Fernandes et al. (2009) concluded that a physicochemical pretreatment utilizing alkali or acid enhanced the biodegradability of some lignocellulosic materials when the lignin content was high. To determine the best pretreatment process for a particular feedstock and product, rigorous technical and economic analysis is necessary (Aden et al., 2002).

3.2.2.1 STEAM EXPLOSION

Steam explosion is one of the most common methods for the pretreatment of lignocellulosic materials (McMillan, 1994). In this method, grinded biomass is treated with high-pressure saturated steam and then the pressure is quickly reduced, which makes the material undergo an explosive decompression. Steam explosion is typically initiated at a temperature of 160-260 °C (corresponding pressure of 0.69-4.83 MPa) for several seconds to a few minutes before the material is exposed to atmospheric pressure. The process causes the disrupting of the material's structure, the degradation of hemicellulose and lignin transformation due to the high temperature, thus facilitating the subsequent hydrolysis of cellulose (Öhgren et al., 2007). The steam treatment can be performed in the presence of a catalyst (acid or alkali). Many parameters such as temperature, residence time, catalyst concentration, time of presoaking, and moisture content can be optimized to improve steam pretreatment (Bruni et al., 2010).

3.2.2.2 ACID HYDROLYSIS

Acid hydrolysis can be employed in the pretreatment of cellulosic material, in order to cleave the interchain linkages in hemicelluose and cellulose. Concentrated acids such as H_2SO_4 and HCl are good hydrolysis agents but, on the other hand, they are very corrosive and hazardous (Sun and Cheng, 2002). Dilute acid hydrolysis has achieved high reaction rates and has improved cellulose hydrolysis significantly (Esteghlalian et al., 1997). During hot acid pretreatment, some of the polysaccharides are hydrolyzed, mostly hemicellulose. The resulting free sugars can degrade to furfural (from pentoses) and to 5-hydroxy-methyl-furfural or HMF (from hexoses). These compounds inhibit yeast cells and lead to decreased growth rate, ethanol production rate, and ethanol yield. In addition, their production means loss of fermentable sugars. Organic acids such as maleic and fumaric have been suggested as alternatives to avoid HMF formation (Kootstra et al., 2010).

3.2.2.3 ALKALINE HYDROLYSIS

Alkaline hydrolysis can also be used for pretreatment of lignocellulosic materials and the effect of the pretreatment depends on the lignin content of the materials (McMillan, 1994). The mechanism of alkaline hydrolysis is believed to be saponification of intermolecular ester bonds cross-linking xylan hemicelluloses and other components, for example, lignin and other hemicelluloses. Alkaline pretreatment processes utilize lower temperatures and pressures than other pretreatment technologies (Mosier et al., 2005). It can largely improve the cellulose digestibility and sugar degradation is less significant than in acid treatment; however, the application is hindered by high cost of alkalis (Talebnia et al., 2010). Dilute NaOH treatment of lignocellulosic material causes swelling, leading to an increase of internal surface area, a decrease in the degree of polymerization, a decrease in crystallinity, separation of structural linkages between lignin and carbohydrates, and disruption of the lignin structure (Fan et al., 1987).

3.2.2.4 LIQUID HOT WATER (LHW)

The treatment with LHW is still being developed and exhibits a great potential due to its simplicity, low generation of inhibiting byproducts, and high yields (Hamelinck et al., 2005). LHW is a hydrothermal pretreatment, where pressure is applied to maintain water in the liquid state at elevated temperature. Temperature in the range of 170-230 °C and pressure >5 MPa are commonly used (Talebnia et al., 2010). The principle of this process is a treatment of lignocellulose by subcritical pressurized water, eventually assisted by CO_2-enhanced hydrolysis. Its distinctly different behavior compared to water at ambient conditions is due to the dramatic changes in physical properties, namely, dielectric strength and ionic product, which in turn can easily be altered by changing temperature and pressure (Schacht et al., 2008). Higher xylan recovery suggesting lower generation of degradation products has already been demonstrated for the LHW treatment (Schacht et al., 2008).

3.2.3 *Enzymatic Hydrolysis*

Enzymatic hydrolysis converts cellulose to reducing sugars by the action of cellulases, so they can be fermented by yeasts or bacteria to ethanol (Sun and Cheng, 2002). The enzymatic hydrolysis is a multistep reaction that takes place in a heterogeneous system, in which insoluble cellulose is initially broken down at the solid-liquid interface via the synergistic action of endoglucanases (EC 3.2.1.4) and exoglucanases/cellobiohydrolases (EC 3.2.1.91). Subsequently, a liquid-phase hydrolysis of soluble intermediate products takes place, that is, short cellulo-oligosaccharides and cellobiose, that are catalytically cleaved to produce glucose by the action of β-glucosidase (EC 3.2.1.21) (Andric et al., 2010). Utility cost of enzymatic hydrolysis is low compared to acid or alkaline hydrolysis because it is usually conducted at mild conditions and does not cause corrosion problems (Duff and Murray, 1996). Both bacteria and fungi can produce cellulases for hydrolysis of lignocellulosic materials.

Enzymatic hydrolysis of cellulose consists of three steps: adsorption of cellulases to the surface of the cellulose, hydrolysis of cellulose to glucose, and desorption of cellulases. The noncellulose components—lignin and hemicellulose—and high crystallinity of cellulose make the adsorption of cellulase a rate-limiting step (Han and Chen, 2010). Substrate concentration is one of the main factors that affect the yield and initial rate of enzymatic hydrolysis of cellulose. At low substrate levels, an increase of substrate concentration normally results in an increase of yield and reaction rate of the hydrolysis (Cheung and Anderson, 1997). Several methods have been developed to reduce the inhibition of hydrolysis, including the use of high concentrations of enzymes, the supplementation of β-glucosidases during hydrolysis, and removal of sugars during hydrolysis by ultrafiltration or simultaneous saccharification and fermentation (SSF).

3.2.4 *Simultaneous Saccharification and Fermentation*

The SSF process has been extensively studied in order to reduce the inhibition of cellulases caused by end products of hydrolysis—glucose and short cellulose chains (Zheng et al., 1998). In the process, reducing sugars produced in cellulose hydrolysis or saccharification are simultaneously fermented to ethanol, which greatly reduces the product inhibition in hydrolysis. The SSF process increases the yields of ethanol by minimizing product inhibition as well as eliminates the need for separate reactors for saccharification and fermentation. The SSF process also showed to be superior to saccharification and subsequent fermentation due to the rapid assimilation of sugars by yeast during SSF (Krishna et al., 2001). The microorganisms used in the SSF are usually the fungus *Trichoderma reesei* and the yeast *S. cerevisiae* (Sun and Cheng, 2002). Hydrolysis is usually the rate-limiting process in SSF (Philippidis and Smith, 1995). Thermotolerant yeasts and bacteria have been used in the SSF to raise the temperature close to the optimal hydrolysis temperature (Prasad et al., 2007).

4 FEASIBILITY OF LIGNOCELLULOSIC ETHANOL PRODUCTION

4.1 Woody Biomass from Forestry

Forests cover about 9.5% of the Earth's surface, corresponding to around 32% of the land area and accounting for 89.3% of the total standing biomass and 42.9%, of the total annual world biomass production (Klass, 1998). Savanna and grasslands come second, accounting for 11% of total biomass production (Klass, 1998). Woody biomass as a feedstock has many advantages in terms of production, harvesting, storage, and transportation compared with herbaceous biomass (Zhu and Pan, 2010). Evaluation of the quantity of woody biomass available from forests and plantations has been reported by Perlack et al. (2005) and Smith et al. (2009). Woody biomass from forestlands comes from a number of different sources, including logging residues from harvest operations, fuel treatments (removing excess biomass), fuelwood, primary and secondary processing mill residues, and urban wood residues (Perlack et al., 2005).

4.1.1 *Ethanol Production from Wood-derived Lignocellulosic Substrates*

The two major species of woody biomass, hardwoods and softwoods, show differences affecting their processing for ethanol production, hardwood species showing lower recalcitrance and higher xylan and low mannan content than softwood species (Zhu and Pan, 2010).

Woody biomass pretreatment involves both physical and thermochemical processes (Zhu and Pan, 2010). Physical pretreatment of woody biomass provides size reduction, thus increasing its surface area and enhancing enzyme accessibility to cellulose. Different from herbaceous biomass, the size reduction of woody biomass is very energy intensive (Zhu et al., 2009b, 2010b).

Only few technologies were proven effective for pretreatment of woody biomass due to its high recalcitrance (Zhu et al., 2010c).

Alkaline-based pretreatments such as sodium hydroxide pretreatments (Zhao et al., 2008), lime pretreatment (Sierra et al., 2009), ammonia-based pretreatments (Gupta and Lee, 2009), and ionic liquid pretreatment (Sun et al., 2009) are not generally suitable for ethanol production from woody biomass since high alkali concentration and temperatures are required.

Diluted acid pretreatment at high temperatures allows efficient enzymatic saccharification of cellulose for certain hardwood species (Wyman et al. 2009). Pretreatment of size-reduced poplar wood of less than 6 mm at 190 °C with sulfuric acid (2% charge on wood) provided a recovery of 82.8% of total sugar (at an enzyme loading of 15 FPU/g cellulose), and a fermentation efficiency of the enzymatic hydrolysate of 81.4% by genetically modified *Saccharomyces cerevisiae* 424A(LNH-ST) (Wyman et al., 2009). Recently, Zhu et al. (2010b) reported the achievement of a substrate enzymatic digestibility (SED)—defined as the percentage of glucan on solid substrate converted to glucose enzymatically—of 80% for commercial-sized wood chips (6-38 mm) after pretreatment at 180 °C with sulfuric acid charge of 1.84%, followed by disk milling. The same conditions provided a SED of only about 40% when applied to softwood (Zhu et al., 2009a; Zhu et al., 2010b). Glucose recovery was increased at 80% when a two-stage dilute acid pretreatment at 190 and 210 °C was applied to size-reduced spruce wood of 2-10 mm (Monavari et al., 2009b).

Acid-catalyzed steam pretreatment of woody biomass has been largely investigated (Monavari et al., 2009a) and the results were recently reviewed by Zhu and Pan (2010). The acid-catalyzed steam explosion consists of acid-catalyzed steaming followed by a thermal flashing step, and it allows recovery of a high-concentration hemicellulose stream due to the low liquid-to-wood ratio. Efficient enzymatic saccharification was achieved for acid-catalyzed steam pretreated hardwood substrates, whilst a lower sugar recovery (about 65%) was obtained with softwood, even if it is improvable by two-step explosion (Monavari et al., 2009a). The main drawback of steam explosion is its energy consumption.

Sulfite pretreatment to overcome recalcitrance of lignocellulose (SPORL) (Wang et al., 2009; Zhu et al., 2009a, 2010a,b) represents an effective technology for pretreatment of woody biomass, including both hardwoods and softwoods. It consists of a diluted acid pretreatment in which sulfite or bisulfite is used as an additional catalyst, at typical acid and bisulfite loading on oven dry wood of about 0.5-1% and 1-3% for hardwood and 1-2% and 40-8% for softwood, respectively. SPORL is a mild pretreatment conducted at a temperature of 160-190 °C for a period of 10-30 min. The sulfite addition increases the pH, thus generating lower amounts of fermentation inhibitors, such as furfural and HMF

(Shuai et al., 2010; Wang et al., 2009; Zhu et al., 2009a), than dilute acid pretreatment. The partial sulfonation of lignin by sulfite provides wood softening, thus reducing energy consumption for size reduction. When compared with acid-catalyzed steam explosion, the pretreatment energy efficiency of SPORL is about 30 fold greater (Zhu and Pan, 2010). Moreover, sulfonation raises lignin hydrophilicity weakening the hydrophobic interaction between lignin and enzymes and thus facilitating cellulose saccharification. About 95% enzymatic saccharification of softwood substrates pretreated by SPORL was achieved within 48 h with enzyme loading of 15 FPU/substrate (Zhu et al., 2009a, 2010b). An overall sugar recovery of about 85% and an ethanol yield of 276 L/t were gained from lodgepole pine treated with the SPORL process (Zhu et al., 2010a). As a further economically relevant aspect of SPORL, dissolved lignosulfonate is recovered in pretreatment liquor representing a high-value marketable coproduct. Moreover and most importantly, since SPORL is based on sulfite pulping that is a commercially well-established process with low technological and environmental threats, this technology can be easily implemented adopting equipment largely practiced in the pulp and paper industry.

4.1.2 Suitability of Woody Biomass from Forestry as Raw Material for Ethanol Production

The estimation of current and potential energy production capacity from woody biomass is complicated by social issues such as the debate over shifting land uses and discussion on present and future productivity using conventional and new forest management options (Berndes et al., 2003). Removing forest residues can have ecological consequences on the ecosystem sustainability affecting soil health and quality and plant, animal, and insect communities, with detriment to biodiversity (Ares et al., 2007). Data and projections on average annual long-term harvest intensity are uncertain since they are scarce and dependent on changing degrees of reliability, and economical and ecological considerations determine the actual logging intensity.

Smeets and Faaij (2007) reported a projection of the energy production potential for woody biomass from forestry (woody biomass), including all products made from woody biomass and coming from the harvesting, processing, and use of woody biomass. The key factors considered in this study were the demand for woodfuel and industrial roundwood, plantation establishment rates and, especially, the supply of wood from forests, depending on the size of the forest area and the yield level. Their results showed that the global demand for woodfuel and industrial roundwood in 2050 can be met both with and without further deforestation, since woody biomass from forests, plantations, trees outside forests, and wood logging and processing residues can be a large source of bioenergy with a potential production of up to 98 EJ including deforestation and 111 EJ excluding deforestation, in 2050. However, economical and ecological factors may limit the supply of wood from forests. The total global bioenergy production potentials in 2050 were estimated to be 71, 64, 15, 0, and 8 EJ/year, taking into account the theoretical, technical, economical, ecological-economical, and ecological potentials of wood supplies from forests, respectively. The best candidates as woody biomass suppliers are the Caribbean and Latin America, the Commonwealth of Independent States and Baltic Stat and, in part, North America. Other regions with some potential included West Europe (mainly residues), East Asia (mainly residues), and Sub-Saharan Africa. Wood shortage was foreseen in 2050 for Japan, South Asia and the Middle East and North Africa.

Residues and waste may add an amount equivalent to 35 EJ roundwood, with a potential supply of bioenergy from wood logging residues and wood processing residues of 13-22 EJ in 2050 (Smeets and Faaij, 2007). Other studies reported values of 10-13 EJ in the year 2025 (Williams, 1995) and 11 EJ in 2050 (Williams, 1995).

4.2 Agricultural Crop Residues

Agricultural crop residues include both field and processing residues. Field residues consist of materials such as stalks and stubble (stems), leaves, and seed pods, left in the agricultural field after crop harvesting. Processing residues, including husks, seeds, bagasse and roots, are the materials left after the processing of the crop into a usable resource. Harvesting of cereals, vegetables, and fruits generates huge amounts of crop residues.

Among crop residues, sugarcane bagasse is a porous residue of cane stalks left over after the crushing and extraction of the juice from the sugarcane (Pandey et al., 2000a,b). It presents a great morphological heterogeneity and consists of fiber bundles and other structural elements such as vessels, parenchyma, and epithelial cells (Sanjuan et al., 2001). It is composed of 19-24% of lignin, 27-32% of hemicellulose, 32-44% of cellulose, and 4.5-9.0% of ashes. The remainder is mostly lignin plus lesser amounts of minerals, waxes, and other compounds (Jacobsen and Wyman, 2002). Sugar mills generate approximately 270-280 kg of bagasse (50% moisture) per metric ton of sugarcane (Rodrigues et al., 2003).

Cassava is another productive sector that generates large amount of residue. The cassava tubers processing for the large-scale production of starch result in solid and liquid wastes. An important residue is the bagasse, the waste material of the root containing part of the starch that was not previously extracted and fiber. The fibrous slurry constitutes about 15-20% of the processed cassava tuber, which contains around 50-70% starch on dry weight basis. Cassava bagasse is generally discarded to the environment without any treatment and causes serious concern about environmental pollution in areas where the starch industries are located (Jyothi et al., 2005). Table 3 shows the composition of cassava bagasse according to Vandenberghe et al. (1998).

Because of its low ash content, cassava bagasse could offer numerous advantages in comparison to other crop residues such as rice straw and wheat straw, which have 17.5% and 11.0% ashes, respectively, for usage in bioconversion processes using microbial cultures.

TABLE 3 Cassava Bagasse Constituents

Constituent	Percent Dry Basis
Moisture	11.20
Protein	1.61
Lipids	0.54
Fibers	21.10
Ash	1.44
Sugars (Starch)	63.00

In comparison to other agricultural residues, cassava bagasse can be considered as a rich solar energy reservoir due to cassava's easy regeneration capacity. When compared to sugarcane bagasse, it offers the advantage of being easily attacked by microorganisms without any pretreatment (Pandey et al., 2000a,b).

4.2.1 *Ethanol Production from Crop Residues*

Analyses of the effects of substrate composition, cellulose crystallinity, and particle size on the yields of ethanol production from bagasse and rice straw have shown that each type of feedstock requires a specific delignification pretreatment to optimize enzymatic hydrolysis (Rivers and Emert, 1988). Alkaline delignification of crop residues proved to be effective in separating cellulose, as a non-hydrolyzable product, from the lignin and hemicellulose, as hydrolyzable products, and the most efficient separation for bagasse and corn stover were obtained applying two alkaline hydrolysis cycles with 0.5 N KOH at 70 °C (Henderson et al., 2003). Relatively high yields of fermentable glucose were reported by enzymatic hydrolysis after alkaline delignification of crop residues (Li and Champagne, 2005a,b). At 40 °C, with an enzyme loading of 800 units/g of delignified substrate, the percentages of conversion to glucose in 24 h were 65.4 and 51.1% on a delignified dry biomass basis for KOH-treated corn stover and bagasse, respectively. These studies showed that physical and/or chemical pretreatments (grinding, drying, and phosphorylation) of non-hydrolyzable product have a great impact on the glucose yields and that the optimal pretreatment changes with the feedstock. Arvanitoyannis and Tserkezou (2008) recently reviewed methods and current and potential uses of corn and rice wastes. Among these, the production of bioethanol from corn stover using SSF was reported as an economically advantageous and environmentally friendly process. SSF of high dry matter content resulted in a high ethanol concentration in the fermented slurry, thereby decreasing the energy demand in the subsequent distillation step (Ohgren et al., 2006). Based on current technologies, dried cellulosic biomass from crop residues has been shown to be readily converted to bioethanol at a rate of 300 L of ethanol produced per ton of oven-dried biomass (Champagne, 2007).

4.2.2 *Suitability of Crop Residues as Raw Materials for Ethanol Production*

Energetic applications for crop residues may provide security of supply and mitigate climate change, and their use for ethanol production is strongly sustained in Brazil (Soccol et al., 2010), the USA (Fleming et al., 2006), and the EU (European Commission, 2006; Sticklen, 2006). In Canada, a much higher use of such residues to produce ethanol has been advocated by Champagne (2007), underlining that producing ethanol from crop residues presents important benefits such as the reduction in the potential air, water, and soil contamination associated with the land application of organic residuals. Champagne (2007) estimated that 5336 million liters of bioethanol could be produced from Canadian crop residues. However, crop residues available as raw materials for ethanol production should be evaluated considering their alternative possible applications, as pointed out by Reijnders (2008). Among the possible alternative uses of crop residues, especially important is their use for stabilizing and increasing the levels of soil organic carbon, with important effects on soil structure, limiting erosion, the provision of nutrients, counterbalancing acidification, and water-holding capacity of soils and soil fertility (Wilhelm et al., 2004). Because of the negative effects of removing crop residues from the soil, Lal (2008) suggested identifying alternate sources of biofuel feedstock, such as animal waste or municipal solid waste. On the other hand, Reijnders (2008) proposed a reduction in residue removal from the field, with a higher fraction removed only for residues from annual crops generating relatively large amounts of biomass. As an alternative, selection of residues that contain relatively high levels of available cellulose and hemicellulose for removal and ethanol production has also been proposed. In the case of corn stover, this fraction consists of cobs, leaves, and husks (Crofcheck and Montross, 2004). Another possible approach is returning the waste from processing crop residues—a residue rich in lignin and also containing unreacted cellulose and hemicellulose (Mosier et al., 2005)—to the field.

4.3 Municipal Solid Wastes

The cellulose content in MSW is mainly from paper wastes. The MSW fractions of office paper, coated paper, newsprint, and corrugated boxes contain 87, 42, 48, and 57% cellulose, respectively (Palmisano and Barlaz, 1996). Food waste contains a variable amount of cellulose, accounting for around 50% of the residue on average (Palmisano and Barlaz, 1996).

4.3.1 *Development of a Process for Bioethanol Production from MSW*

Only scarce information is available on the use of MSW as feedstock for bioethanol production. A dilute acid pretreatment (180 min with 3% w/w H_2SO_4) of residual corrugated cardboard showed to be effective in making the pretreated waste susceptible to enzymatic hydrolysis into hemicellulosic sugars and glucose by commercial enzymes-"Celluclast" cellulases (28 FPU/g of substrate) from *T. reesei* and "Novozym" β-glucosidase (360 IU/g of substrate) from *Aspergillus niger* provided by Novo Nordisk Bioindustrial (Yánez et al., 2004). A pretreatment with dilute strong acid followed by steam treatment at 120 °C for 15 min was performed on lignocellulosic solid wastes from selected sites in Tanzania and glucose concentration after enzymatic hydrolysis (with cellulase enzyme extracted from *T. reesei*, incubated at 55 °C for 6 h) of pretreated wastes was evaluated by Mtui and Nakamura (2005). They achieved a glucose concentration of 0.13 and 0.05 g/L (corresponding to 1 g of pretreated lignocellulosic material) from solid waste samples with high lignocellulose content (93%) and low lignocellulose content (14%), respectively.

Fifteen different pretreatments of selected biodegradable MSW fractions (carrot peelings and potato peelings typical of kitchen waste, grass typical of garden waste, and newspaper and scrap paper typical of paper/card fractions) to obtain the highest glucose yield for bioethanol production were compared by Li et al. (2007). Prehydrolysis treatments consisted of dilute acid (H_2SO_4, HNO_3 or HCl, 1 and 4%, 180 min, 60 °C), steam treatment (121 and 134 °C, 15 min), microwave treatment (700 W, 2 min), or a combination of two of these. Enzymatic hydrolysis was carried out with cellulases from *T. reesei* and *T. viride* (10 and 60 FPU/g of substrate) (Sigma). The highest glucose yield (73%) was obtained with a prehydrolysis treatment of 1% H_2SO_4 followed by steam treatment at 121 °C, and enzymatic hydrolysis with *T. viride* at 60 FPU/g substrate. The contributions of enzyme loading (49.39%) and acid concentration (47.70%) were significantly higher than the contribution of temperature during the steam treatment (0.13%) to the glucose yield.

Comparing hydrolysis of primary municipal wastewater sludge, secondary municipal wastewater sludge, and municipal biosolids, the highest fermentable glucose yield was found from the primary municipal wastewater sludge (Li and Champagne, 2005a,b). Both wet and dry substrates were submitted to different combinations of pretreatments, including drying, grinding, KOH, HCl, and HCl followed by KOH alkaline delignification at 40 °C for a period of 24 h. Results indicated that the cellulose in primary sludge is readily accessible to the enzymes. The KOH pretreatment was not particularly effective on the primary sludge, increasing its digestibility by only 4%. Similarly, when the primary sludge was treated with HCl, the glucose yield increased by 11.5% over that observed without acid and alkaline treatment (31.1%). The results implied that conversion of the cellulose contained in primary sludge into bioethanol might present a valuable waste-management alternative when employed as a wet feedstock, as drying and grinding are not necessary.

4.3.2 *Suitability of MSW as Raw Material for Ethanol Production*

MSW may be considered an alternative sustainable source for bioethanol and biogas production (Demirbas, 2006; Li et al., 2007). Ethanol production from MSW has environmental and economic benefits, even if, when compared with the use of MSW for biogas production, ethanol production may be less advantageous. However, in some countries lacking sufficient amounts of both agricultural and woody biomass, MSW has been identified as the only potential raw material for ethanol production (Faraco and Hadar, 2010).

In a recent study, Murphy and Power (2007) analyzed four scenarios for energy generation from newspaper: lignocellulosic biomass conversion to ethanol (transport fuel); codigestion with the organic fraction of MSW and production of CH_4-enriched biogas (transport fuel); cofiring with the MSW residue in an incinerator; gasification of newspaper as a sole fuel. Comparison of the profit/gate fee per ton of newspaper showed that the biogas scenario has a large economic advantage over the others, and the GHG analysis indicated that the biogas scenario generates the best net GHG savings.

Kalogo et al. (2007) modeled a facility for conversion of MSW into ethanol employing dilute acid hydrolysis and gravity pressure vessel technology, estimating its life-cycle energy use and air emissions. Results were compared with life-cycle assessments (LCAs) of vehicles fueled with gasoline, corn-ethanol, and energy crop cellulosic ethanol, assuming that the ethanol is utilized as E85 (blended with 15% gasoline) in a light-duty vehicle. MSW-ethanol production was also compared, as a waste-management alternative, with landfilling with gas-recovery options. For MSW-derived ethanol, the total energy use per vehicle per mile travelled proved to be less than that of corn-ethanol and cellulosic ethanol. Energy use from petroleum sources for MSW-ethanol was lower than for the other fuels. MSW-ethanol used in vehicles reduced net GHG emissions by 65% compared to gasoline, and by 58% compared to corn-ethanol. Relative GHG performance with respect to cellulosic ethanol depended on whether MSW classification was included or not. Thus, converting MSW into ethanol would result in a net fossil energy saving of 397-1830 MJ/million tons MSW compared to a net fossil energy consumption of 177-577 MJ/million tons MSW for landfilling. However, landfilling with gas recovery, either for flaring or for electricity production, would result in greater reductions in GHG emissions than the MSW-to-ethanol conversion. Stichnothe and Azapagic (2009) carried out an LCA to estimate the GHG emissions from bioethanol production using two alternative feedstocks, both derived from household waste: refuse-derived fuel (RDF) and biodegradable municipal waste (BMW). An integrated waste-management system was considered, taking into account recycling of materials and production of bioethanol in a combined gasification/biocatalytic process. For the functional unit defined as the "total amount of waste treated in the integrated waste management system," the best option was to produce bioethanol from RDF—this saved up to 196 kg CO_2 equiv. per ton of MSW, compared to the current waste management practice in the UK. However, if the functional unit was defined as "MJ of fuel equiv." and bioethanol was compared with petrol on an equivalent energy basis, the results showed that bioethanol from RDF offered no saving of GHG emissions compared to petrol, whereas bioethanol from BMW offered significant GHG saving potential over petrol. For a biogenic carbon content of 95%, the life-cycle GHG emissions from bioethanol were 6.1 g CO_2 equiv./MJ, which represents a saving of 92.5% compared to petrol. If the biogenic carbon of the BMW feedstock exceeded 97%, the bioethanol system became a carbon sequester. Compared to paper recycling, converting waste paper into bioethanol would save 460 kg CO_2 equiv./ton waste paper, or eight times more than recycling. Chester and Martin (2009) examined the major processes required to support a lignocellulosic MSW-to-ethanol infrastructure, computing cost, energy, and GHG effects for California. Their analysis was performed on MSW destined for landfills, for an ethanol plant employing a pretreatment by cocurrent dilute acid prehydrolysis, before enzymatic hydrolysis. Reductions in fossil energy consumption resulted primarily from displacement of gasoline and avoided emissions at the landfill (140 PJ/year). This was only partially offset by fossil energy increases in the plant and classification phases (32 PJ/year), with a resulting fossil energy reduction of 110 PJ/year. On the other hand, the authors found that ethanol production from MSW cannot be unequivocally justified from the perspective of net GHG avoidance. The avoided impact of diverting organic waste from the landfill presents the greatest system uncertainty. The net GHG impact is ultimately dependent on how well landfills control their emissions of decomposing organics. There is currently considerable uncertainty surrounding the recovery efficiency of landfill emission controls. A better understanding of carbon sequestration and methane capture performance within landfills is necessary before stronger conclusions can be drawn.

5 CONCLUDING REMARKS

Despite the technical and economic difficulties, lignocellulosic material is nowadays showing up as a very important alternative to produce biofuels—biogas and bioethanol. Its renewable characteristics contribute to the decrease of greenhouse gases, as far as this material is produced by photosynthesis, and the possibility to decrease the environmental damage generated with the final disposal of residues is also considerable.

5.1 Challenges for Lignocellulosic Ethanol Production

Recovering glucose from the cellulose molecule is not an easy task, first of all because the very peculiar glucose-glucose linkage of cellulose allows the "quasiplanar" molecular structure and the consequent very stable packaging among the cellulose molecules. Moreover, cellulose is found, in the lignocellulosic material, in a hard association with hemicellulose and lignin, which work out as a physical and chemical barrier to any physical or chemical agent. This structure is present in many types of feedstock, including agricultural and industrial residues, municipal solid waste, and pruning of trees, garden, and grass.

5.2 Perspectives for Lignocellulosic Ethanol Production

Great potentialities are observed for energy production from biomass. Biorenewable feedstocks can be converted into value-added chemicals and fuels with minimal waste and emissions. Thermochemical and biochemical conversion products from biorenewables are upgraded before ultimate refining processes. The upgrading includes fractionation for separation of primary products. The benefits of an integrated upgrading system are numerous because of the diversification in feedstocks and products. There are currently several different levels of integration in these systems which add to their sustainability, both economically and environmentally. Economic and production advantages increase with the level of integration in the system.

Depending on the feedstock considered, its availability for bioethanol production can vary considerably from season to season, and depending on geographic locations, could also pose difficulty in their supply. The changes in the price of feedstocks can highly affect the production costs of bioethanol. Because feedstocks typically account for greater than one-third of the production costs, maximizing bioethanol yield would be imperative (Balat et al., 2008).

Several agricultural residues such as corn stover, wheat and rice straw, residues from citrus processing, sugarcane, sugarbeet, coconut biomass, grasses and residues from the pulp and paper industry (paper mill sludge), from the extraction of castor and sunflower oil, residues from the wood industry as well as municipal cellulosic solid wastes, could eventually be used as raw materials to produce ethanol. However, the use of each source of biomass represents a technological challenge.

Each country must find the best and economical way to use their feedstocks and residues in order to produce biofuels. Brazilian bioethanol program is an example of the efficiency of sugarcane production and high technology bioethanol production. According to Petrobrás Biocombustíveis, the bioethanol production in Brazil may triplicate until 2020, passing from the actual 27.5 billion liters to 70 billion liters. The production of sugarcane, which is detonated from bioethanol production, occupies only 0.9% of areas that can be cultivated (excluding the areas of environmental protection). For food production, 15.98% of cultivable land is used. Thus, Brazil has sufficient territorial space to raise significantly the production of food and, also, the biofuels. However, in the years to come, the necessity to increase Brazilian biofuels production will probably be strongly attached to the use of biomass (sugarcane bagasse and leaves), which will not necessary demand new agricultable areas, but will certainly demand the development of proper technology, concerning technical and economic aspects.

Several novel markets for lignocellulosic residues have been identified recently. The use of fungi, reported by Sanchez (2009), in low-cost bioremediation projects might be attractive given their highly efficient lignocellulose hydrolysis enzyme machinery.

References

Aden, A., Ruth, M., Ibsen, K., Jechura, J., Neeves, K., Sheehan, J., et al., 2002. Lignocellulosic Biomass to Ethanol Process Design and Economics Utilizing Co-Current Dilute Acid Prehydrolysis and Enzymatic Hydrolysis of Corn Stover—NREL/TP-510-32438. Available from:http://butanol.com/docs/2002_NREL_Lignin.pdf. Accessed November 18, 2008.

agra-net.com, 2010. http://www.agra-net.com/portal2/home.jsp.

Andric, P., Meyer, A.S., Jensen, P.A., Dam-Johansen, K., 2010. Reactor design for minimizing product inhibition during enzymatic lignocellulose hydrolysis: I. Significance and mechanism of cellobiose and glucose inhibition on cellulolytic enzymes. Biotechnol. Adv. 28, 308–324.

Ares, A., Terry, T., Harrington, C., Devine, W., Peter, D., Bailey, J., 2007. Biomass removal, soil compaction, and vegetation control effects on five-year growth of Douglas-fir in coastal Washington. For. Sci. 53, 600–610.

Arvanitoyannis, I.S., Tserkezou, P., 2008. Corn and rice waste: a comparative and critical presentation of methods and current and potential uses of treated waste. Int. J. Food. Sci. Technol. 43, 958–988.

Balat, M., 2007. Global bio-fuel processing and production trends. Energy Explor. Exploit. 25, 195–218.

Balat, M., Balat, H., Cahide, O.Z., 2008. Progress in bioethanol processing. Prog. Energ. Combust. 34, 551–573.

Berndes, G., Hoogwijk, M., van den Broak, R., 2003. The contribution of biomass in the future global energy supply: a review of 17 studies. Biomass Bioenergy 25, 1–28.

Bruni, E., Jensen, A.P., Angelidaki, I., 2010. Steam treatment of digested biofibers for increasing biethanol production. Bioresour. Technol. 101, 7668–7671.

Cadoche, L., Lopez, G.D., 1989. Biol. Wastes 30, 153–157.

Cardona, C.A., Sanchez, O.J., 2007. Bioresour. Technol. 98, 2415–2457.

Cerqueira Leite, R.C., Leal, M.R.L.V.L., Cortez, L.A.B., Griffin, W.M., Scandiffio, M.I.G., 2009. Can Brazil replace 5% of the 2025 gasoline world demand with ethanol? Energy 34, 655–661.

Champagne, P., 2007. Feasibility of producing bio-ethanol from waste residues: a Canadian perspective. Canada Resour. Conserv. Recycling 50, 211–230.

Chester, M., Martin, E., 2009. Cellulosic ethanol from municipal solid waste: a case study of the economic, energy, and greenhouse gas impacts in California. Environ. Sci. Technol. 2788, doi:10.1021/es80.

Cheung, S.W., Anderson, B.C., 1997. Bioresour. Technol. 59, 81–96.

Cortez, L.A.B., Lora, E.E.S., Gómez, E.O., 2008. Biomassa Para Bioenergia. UNICAMP, Campinas.

Crofcheck, C.L., Montross, M.D., 2004. Effect of stover fraction on glucose production using enzymatic hydrolysis. Transact. Am. Soc. Agric. Eng. 47, 841–844.

Dawson, L., Boopathy, R., 2007. Bioresour. Technol. 98, 1695–1699.

Demirbas, A., 2006. Biogas production from the organic fraction of municipal solid waste, Energy sources, part A: recovery. Util. Environ. Eff. 28, 1127–1134.

Duff, S.J.B., Murray, W.D., 1996. Bioresour. Technol. 55, 1–33.

Esteghlalian, A., Hashimoto, A.G., Fenske, J.J., Penner, M.H., 1997. Bioresour. Technol. 59, 129–136.

European Commission, 2006. European Commission Presses for New Initiative to Further the Production of Biofuels. February 8, Available at:http://europa.eu.int/comm./agriculture/biomass/biofuel/.

Faaij, A.P.C., 2006. Modern biomass conversion technologies. Mitig. Adapt. Strat. Glob. Change 11, 343–375.

Fan, L.T., Gharpuray, M.M., Lee, Y.H., 1987. Cellulose hydrolysis, first ed Springer-Verlag, New York, NY.

Faraco, V., Hadar, Y., Renew. Sustain. Energy Rev. 2010. doi:10.1016/j.rser.2010.09.050.

Fernandes, T.V., Klaasse Bos, G.J., Zeeman, G., Sanders, J.P.M., van Lier, J.B., 2009. Effects of thermo-chemical pre-treatments on anaerobic biodegradability and hydrolysis of lignocellulosic biomass. Bioresour. Technol. 100, 2575–2579.

Fleming, J.S., Habibi, S., MacLean, H.L., 2006. Investigating the sustainability of lignocellulose-derived fuels for light-duty vehicles. Transport Res Part D 11, 146–159.

Goldemberg, J., 2007. Ethanol for a sustainable energy future. Science 315, 808–810.

Gupta, R., Lee, Y.Y., 2009. Pretreatment of hybrid poplar by aqueous ammonia. Biotechnol. Prog. 25, 357–364.

Hamelinck, C.N., van Hooijdonk, G., Faaij, A.P.C., 2005. Biomass Bioenergy 28, 384–410.

Han, Y.J., Chen, H.Z., 2010. Synergism between hydrophobic proteins of corn stover and cellulose in lignocellulose hydrolysis. Biochem. Eng. J. 48, 218–224.

Henderson, B., Champagne, P., Tudoret, M.J., Chemical separation of cellulose from lignin in sugarcane bagasse. Eighth Specialty Conf.: Environment & Sustainable Engineering & 31st Annual CSCE Congress. 2003. Proc. p. ENK-283.

IBGE, 2008. Brazilian Institute of Geography and Statistics [Instituto Brasileiro de Geografia e Estatística - IBGE]. Evolução da Produtividade da Cana-de-Açúcar no Brasil. Available at/www.ibge.gov.brS.

International Risk Governance Council, 2007. Governing the risks and opportunities of bioenergy. IRGC's Bioenergy Project, Geneva, Switzerland. August, Available at:www.irgc.org/irgc/IMG/pdf/IRGC ConceptNote Bioenergy 1408S.

Jacobsen, S.E., Wyman, C.E., 2002. Xylose monomer and oligomer yields for uncatalyzed hydrolysis of sugarcane bagasse hemicellulose at varying solids concentration. Ind. Eng. Chem. Res. 41, 1454–1461.

Jobling, S., 2004. Plant Biotechnol. 7, 210–218.

Jyothi, A.N., Sasikiran, K., Nambisan, B., Balagopalan, C., 2005. Proc. Biochem. 40, 3576–3579.

Kalogo, Y., Habibi, S., Maclean, H.L., Joshi, S.V., 2007. Environmental implications of municipal solid waste-derived ethanol. Environ. Sci. Technol. 41 (1), 35–41.

Klass, D.L., 1998. Biomass for Renewable Energy, Fuels, and Chemicals. Academic Press, New York.

Kootstra, A.M.J., Beeftink, H.H., Scott, E.L., Sanders, J.P.M., 2010. Comparison of dilute mineral and organic acid pretreatment for enzymatic hydrolysis of wheat straw. Biochem. Eng. J. 46, 126–131.

Krishna, S.H., Reddy, T.J., Chowdary, G.V., 2001. Simultaneous saccharification and fermentation of lignocellulosic wastes to ethanol using a thermotolerant yeast. Bioresour. Technol. 77, 193–196.

Lal, R., 2008. Crop residues as soil amendments and feedstock for bioethanol production. Waste Manage. 28, 747–758.

Li, C., Champagne, P., 2005a. Enzymatic hydrolysis of cellulose from various waste sources and their feasibility as feedstocks for ethanol production. Proceedings of the 20th international conference on solid waste technology and management.

Li, C., Champagne, P., 2005b. Feasibility of using waste materials as feedstocks for ethanol production. Int. J. Solid Waste Technol. Manage. 31 (2), 93–101.

Li, A., Antizar-Ladislao, B., Khraisheh, M., 2007. Bioconversion of municipal solid waste to glucose for bio-ethanol production. Bioprocess Biosyst. Eng. 30, 189–196.

Lima, L.R., Marcondes, A.A., 2002. Álcool Carburante: uma estratégia brasileira. Ed. UFPR, Curitiba. 248p.

Linoj, K.N.V., Dhavala, P., Goswami, A., Maithel, S., 2006. Liquid biofuels in South Asia: resources and technologies. Asian Biotechnol. Develop. Rev. 8, 31–49.

Lynd, L.R., Laser, M.S., Bransby, D., Dale, B.E., Davison, B., Hamilton, R., et al., 2008. How biotech can transform biofuels. Nat. Biotechnol. 26, 169–172.

Lynd, L.R., Larson, E., Greene, N., Laser, M., Sheehan, J., Dale, B.E., et al., 2009. The role of biomass in America's energy future: framing the analysis. Biofuels Bioprod. Bioref. 3, 113–123. doi: 10.1002/bbb.

McMillan, J.D., 1994. In: Himmel, M.E., Baker, J.O., Overend, R.P. (Eds.), Enzymatic Conversion of Biomass for Fuels Production, ACS Symposium Series, vol. 566. Pretreatment of lignocellulosic biomass. ACS, Washington, DC, pp. 292–324.

Monavari, S., Gable, M., Zacchi, G., 2009a. Impact of impregnation time and chip size on sugar yield in pretreatment of softwood for ethanol production. Bioresour. Technol. 100, 6312–6316.

Monavari, S., Gable, M., Zacchi, G., 2009b. The influence of solid/liquid separation techniques on the sugar yield in two-step dilute acid hydrolysis of softwood followed by enzymatic hydrolysis. Biotechnol. Biofuels 2, 1–9.

Mosier, N., Wyman, C., Dale, B., Elander, R., Lee, Y.Y., Holtzapple, M., et al., 2005. Features of promising technologies for pretreatment of lignocellulosic biomass. Bioresour. Technol. 96(6), 673–686.

Mtui, G., Nakamura, Y., 2005. Bioconversion of lignocellulosic waste from selected dumping sites in Dar es Salaam, Tanzania. Biodegradation 16, 493–499.

Murphy, J.D., Power, N., 2007. A technical, economic, and environmental analysis of energy production from newspaper in Ireland. Waste Manage. 27, 177–192.

Oliveria, M.E.D., Vaughan, B.E., Rykiel, E.J., 2005. Ethanol as fuel: energy, carbon 1239 dioxide balances, and ecological footprint. Bioscience 55, 593–602.

Ohgren, K., Rudolf, A., Galbe, M., Zacchi, G., 2006. Fuel ethanol production from steam pretreated corn stover using SSF at higher dry matter content. Biomass Bioenergy 30, 863–869.

Öhgren, K., Vehmaanperä, J., Siika-Aho, M., Galbe, M., Viikari, L., Zacchi, G., 2007. Enzyme Microb. Technol. 40 (4), 607–613.

Palmisano, A.C., Barlaz, M.A., 1996. Microbiology of Solid Waste. CRC Press, New York, p. 37.

Pandey, A., Soccol, C.R., Nigam, P., Soccol, V.T., 2000a. Biotechnological potential of agro-industrial residues. Part I. Sugarcane bagasse. Bioresour. Technol. 74, 69–80.

Pandey, A., Soccol, C.R., Nigam, P., Soccol, V.T., Vandenberghe, L.P.S., Mohan, R., 2000b. Bioresour. Technol. 74, 81–87.

Pérez, J., Muñoz-Dorado, J., De-la-Rubia, T., Martínez, J., 2002. Biodegradation and biological treatments of cellulose, hemicellulose and lignin: an overview. Int. Microbiol. 5, 53–63.

Perlack, R.D., Wright, L.L., Turhollow, A., Graham, R.L., Stokes, B., Erbach, D.C., 2005. Biomass as Feedstock for a Bioenergy and Bioproducts Industry: The Technical Feasibility of a Billion-Ton Annual Supply. Oak Ridge National Laboratory, US Dept. of Energy, Oak Ridge.

Philippidis, G.P., Smith, T.K., 1995. Appl. Biochem. Biotechnol. 51/52, 117–124.

Pimentel, D., Marklein, A., Toth, M.A., Karpoff, M., Paul, G.S., McCormack, R., et al., 2008. Biofuel impacts on world food supply: use of fossil fuel, land and water resources. Energies 1, 41–78.

Prasad, S., Singh, A., Joshi, H.C., 2007. Resour. Conserv. Recycl. 50, 1–39.

Reijnders, L., 2008. Ethanol production from crop residues and soil organic carbon. Resour. Conserv. Recycl. 52, 653–658.

Rivers, D.B., Emert, G.H., 1988. Factors affecting the enzymatic hydrolysis of bagasse and rice straw. Biol. Wastes 26, 85–95.

Rodrigues, R.C.L.B., Felipe, M.G.A., Sil, J.B.A., Vitolo, M., 2003. Response surface methodology for xylitol production from sugarcane bagasse hemicellulosic hydrolyzate using controlled vacuum evaporation process variables. Proc. Biochem. 38, 1231–1237.

Sagar, A.D., Kartha, S., 2007. Bioenergy and sustainable development? Annu. Rev. Env. Resour. 32, 131–167.

Sanchez, C., 2009. Lignocellulosic residues: Biodegradation and bioconversion by fungi. Biotechnol. Adv. 27, 185–194.

Sanjuan, R., Anzaldo, J., Vargas, J., Turrado, J., Patt, R., 2001. Morphological and chemical composition of pith and fibres from Mexican sugarcane bagasse. Holz als Roh-und Werkstoff 59, 447–450.

Schacht, C., Zetzl, C., Brunner, G., 2008. J. Supercrit. Fluids 46, 299–321.

Service, R.F., 2010. Is there a road ahead for cellulosic ethanol? Science Available at: www.sciencemag.org 329, 784–785.

Shuai, L., Yang, Q., Zhu, J.Y., Lu, F., Weimer, P., Ralph, J., et al., 2010. Comparative study of SPORL and dilute acid pretreatments of softwood spruce for cellulose ethanol production. Bioresour. Technol. 101, 3106–3114.

Sierra, R., Granda, C., Holtzaaple, M.T., 2009. Short term lime pretreatment of poplar wood. Biotechnol. Prog. 25, 323–332.

Smeets, E.M.W., Faaij, A.P.C., 2007. Bioenergy potentials from forestry in 2050 An assessment of the drivers that determine the potentials. Clim. Change 81, 353–390.

Smith, W., Miles, P.D., Perry, C.H., Pugh, S.A., 2009. Forest resources of the United States, 2007. General Technical Report WO-GTR-78. U.S. Department of Agriculture, Forest Service, Washington Office, Washington, DC, p. 336.

Soccol, C.R., Vandenberghe, L.P., Medeiros, A.B., Karp, S.G., Buckeridge, M., Ramos, L.P., et al., 2010. Bioethanol from lignocelluloses: status and perspectives in Brazil. Bioresour. Technol. 101 (13), 4820–4825.

Solomon, B.D., 2010. Biofuels and sustainability. Annals of the New York Academy of sciences issue: ecological economics reviews Ann. N.Y. Acad. Sci. 1185, 119–134.

Stichnothe, H., Azapagic, A., 2009. Bioethanol from waste: life cycle estimation of the greenhouse gas saving potential. Resour. Conserv. Recycl. doi:10.1016/j.resconrec.2009.04.012.

Sticklen, M., 2006. Plant genetic engineering to improve biomass characteristics for biofuels. Curr. Opin. Biotechnol. 17, 315–319.

Sun, Y., Cheng, J., 2002. Bioresour. Technol. 83, 1–11.

Sun, N., Rahman, M., Qin, Y., Maxim, M.L., Rodriguez, H., Rogers, R.D., 2009. Complete dissolution and partial delignification of wood in the ionic liquid 1-ethyl-3-methlimidazolium acetate. Green Chem. 11, 646–655.

Talebnia, F., Karakashev, D., Angelidaki, I., 2010. Production of bioethanol from wheat straw: an overview on pretreatment, hydrolysis and fermentation. Bioresour. Technol. 101, 4744–4753.

U.S. Department of Energy, 2008. Trend of total FFVs in use from 1998-2008, based on FFV production rates and life expectancy. Available from the Alternative Fuels & Advanced Vehicles Data Center. http://www.afdc.energy.gov/afdc/data/vehicles.html (accessed February 6, 2009).

Udop, 2009. União dos Produtores de Bioenergia. http://www.udop.com.br/index.php?item=alcool.

Wang, G.S., Pan, X.J., Zhu, J.Y., Gleisner, R., 2009. Sulfite pretreatment to overcome recalcitrance of lignocellulose (SPORL) for robust enzymatic saccharification of hardwoods. Biotechol. Prog. 25 (4), 1086–1093.

United Nations Conference on Trade and Development, 2006. Challenges and Opportunities for Developing Countries in Producing Biofuels. UNCTAD publication, UNCTAD/DITC/COM/2006/15, Geneva, November 27.

Vandenberghe, L.P.S., Soccol, C.R., Lebeault, J.M., Krieger, N., 1998. International Congress Biotech'98: Cassava wastes hydrolysate an alternative carbon source for citric acid production by *Candida lipolytica*. Portugal.

Wilhelm, W.W., Johnson, J.M.F., Hatfield, J.L., Voorhees, W.B., Linden, D.R., 2004. Crop and soil productivity response to corn residue removal: a literature review. Agron. J. 96, 1–17.

Williams, R.H., 1995. Variants of a Low CO_2 Emitting Energy Supply System (LESS) for the world. IPCC Second Working Group IIa Energy Supply Mitigation Options/Pacific Northwest Laboratories, Richland, Washington, USA. p. 39.

Wyman, C.E., Dale, B.E., Elander, R.T., Holtzapple, M., Ladisch, M.R., Lee, Y.Y., et al., 2009. Comparative sugar recovery and fermentation data following pretreatment of poplar wood by leading technologies. Biotechnol. Prog. 25, 333–339.

Yánez, R., Alonso, J.L., Parajó, J.C., 2004. Production of hemicellulosic sugars and glucose from residual corrugated cardboard. Process Biochem. 39, 1543–1551.

Zhao, Y.L., Wang, Y., Zhu, J.Y., Ragauskas, A., Deng, Y.L., 2008. Enhanced enzymatic hydrolysis of spruce by alkaline pretreatment at low temperature. Biotechnol. Bioeng. 99 (6), 1320–1328.

Zheng, Y.Z., Lin, H.M., Tsao, G.T., 1998. Biotechnol. Prog. 14, 890–896.

Zhu, J.Y., Pan, X.J., 2010. Woody biomass pretreatment for cellulosic ethanol production: technology and energy consumption evaluation. Bioresour. Technol. 100 (13), 4992–5002.

Zhu, J.Y., Pan, X.J., Wang, G.S., Gleisner, R., 2009a. Sulfite pretreatment (SPORL) for robust enzymatic saccharification of spruce and red pine. Bioresour. Technol. 100 (8), 2411–2418.

Zhu, J.Y., Wang, G.S., Pan, X.J., Gleisner, R., 2009b. Specific surface to evaluate the efficiencies of milling and pretreatment of wood for enzymatic saccharification. Chem. Eng. Sci. 64 (3), 474–485.

Zhu, J.Y., Zhu, W., OBryan, P., Dien, B.S., Tian, S., Gleisner, R., et al., 2010a. Ethanol production from SPORL-pretreated lodgepole pine: preliminary evaluation of mass balance and process energy efficiency. Appl. Microbiol. Biotechnol. 86 (5), 1355–1365.

Zhu, W., Zhu, J.Y., Gleisner, R., Pan, X.J., 2010b. On energy consumption for size-reduction and yield from subsequent enzymatic saccharification of pretreated lodgepole pine. Bioresour Technol 101 (8), 2782–2792.

Zhu, J.Y., Pan, X., Zalesny, R.S., Jr., 2010c. Pretreatment of woody biomass for biofuel production: energy efficiency, technologies, and recalcitrance. Appl. Microbiol. Biotechnol. 87, 847–857.

CHAPTER

6

Technoeconomic Analysis of Lignocellulosic Ethanol

Edgard Gnansounou[1]*, *Arnaud Dauriat*[2]

[1]Bioenergy and Energy Planning Research Group (BPE), Ecole Polytechnique Fédérale de Lausanne (EPFL), CH-1015 Lausanne, Switzerland

[2]ENERS Energy Concept, P.O. Box 56, CH-1015 Lausanne, Switzerland

*Corresponding author: Prof. Gnansounou; E-mail: edgard.gnansounou@epfl.ch

1 INTRODUCTION

Ethanol produced from lignocellulosic feedstock is expected to become mature in the space of 5-10 years and partly replace first-generation ethanol. Bioethanol demand is increasing rapidly in industrialized countries, particularly in the United States of America (USA) and in European countries, as a consequence of mandatory targets. Research is going on in several countries with the aim of improving the efficiency and economic performance of various pathways. The importance of lignocellulosic ethanol stems from the assumed possibility of using inexpensive feedstock, avoid direct and indirect competition with human food and animal feed, and reduce environmental risks, that is, soil degradation, and water and air pollution, which are associated with first-generation biofuels.

The necessity to monitor the research with the aim of concentrating the efforts on those steps that are more influential requires designing the process at the suitable level of detail and modeling the production cost using sets of relevant and consistent assumptions. Compared to technoeconomic analysis of the usual products, lignocellulosic ethanol shows such distinguished characteristics as significant variety of pathways, especially the possibility to use a large range of feedstock, high uncertainty about the economic drivers, large number of stakeholders involved in the pathways, and uncertainties related to their interactions.

Published works on lignocellulosic ethanol often simplify this complexity by focusing on limited pathways, a narrow range of feedstock, few choices of economic factors, and implicit assumptions with regard to the behavior of the stakeholders. These assumptions change significantly from one study to the other, thereby making it intractable to compare different technoeconomic evaluations. Existing reviews (for example, Galbe et al. (2007) highlight the variability of estimated ethanol production costs and find that the key drivers of those differences are feedstock cost and plant capacity.

During the last three decades, the amount of work on technoeconomic analyses of lignocellulosic ethanol has increased significantly with notable contributions of RD&D in the United States and to a lesser extent in Europe. This chapter reviews this work, focusing on the cases in the United States and Europe. The convergence and differences between the published results are pointed out. Finally, methodological issues are discussed, particularly with regard to how to tackle the value chain of biomass when performing a technoeconomic evaluation. This chapter provides an update version of Gnansounou and Dauriat (2010).

2 STATE OF THE ART

2.1 The U.S. Cases

Regarding the technoeconomic evaluation of lignocellulosic ethanol, the first detailed technical reports found in the literature concerning the U.S. cases date back to the mid-1980s. Especially in 1987, the U.S. National Renewable Energy Laboratory (NREL) received several technical reports delivered by subcontractors. Badger Engineers, Inc (1987) studied an acid hydrolysis-based ethanol plant using mixed hardwood chips as feedstock. Four design cases were analyzed (Table 1). The differences between them are related to the size of the plant, the type of hydrolysis, and the mode of electricity supply.

The main coproducts in all the analyzed cases are ethanol and furfural. The process description is based on eight unit areas, that is, feedstock handling, acid hydrolysis, fermentation, ethanol purification, furfural recovery, offsite tankage, waste treatment, and utilities. The economic evaluation is performed using Internal Rate of Return (IRR). In each case, the selling price of ethanol (after tax) required to reach a 15% IRR is estimated and results in a range of values from U.S. \$1.23 gallon^{-1} (U.S. \$0.32 l^{-1}) for the base case (design case I) to U.S. \$1.63 gallon^{-1} (U.S. \$0.43 l^{-1}) for the design case IV. The currency is for 1984.

TABLE 1 Early Design Cases of an Acid Hydrolysis Based Ethanol Plant (Badger Engineers, Inc, 1987)

Design	Unit	Base Case	Alternative Case	Small-Scale Plant (I)	Small-Scale Plant (II)
Production capacity	MM gal/yr	25	25	5	5
Number of hydrolysis stages	–	1	2	1	1
Wood feed rate	dry t/hr	73.8	66.0	14.8	14.8
Byproducts					
Furfural	MM l b/yr	130.2	93.1	26.0	26.0
Excess electricity	MW	22	–	4.4	–
Outside utilities required	–	No	No	No	Yes (4.1 MW)

Stone & Webster Engineering Corp. (1987), another subcontractor, studies the economic feasibility of an Enzyme-Based Ethanol Plant of 15 million gallons of ethanol per year using wood from cultivated eucalyptus tree farms. The plant is supposed to be located near Hilo, on the island of Hawaii. The description of the process includes feedstock handling, pretreatment by sulfuric acid impregnation and steam explosion, enzyme production, enzymatic hydrolysis, evaporation system to concentrate the glucose at the required level, fermentation, distillation, and anaerobic digestion. In the base case, only hexoses are fermented. The pentose fraction of the wood is utilized to produce biogas which is then burned with the lignin fraction to produce the steam required by the process. The economic evaluation is based on constant U.S. $ of 1984 and 15% IRR and results in a required ethanol selling price of U.S. $3.5 gallon^{-1} (U.S. $0.92 l^{-1}). The base case assumes 100% equity. The sensitivity analysis with 75% equity and 25% debt at a real interest of 8% reduces the required selling price to U.S. $3.04 gallon^{-1} (U.S. $0.80 l^{-1}).

A report on Economic feasibility of an enzymatic hydrolysis-based ethanol is also released by Chem Systems, Inc (1987). The size of the plant is 25 million gallons ethanol per year while the feedstock is supposed to be 80% hardwood (incl. 57% from Aspen forests) and 20% maples. The process is a separate hydrolysis and fermentation (SHF) with on-site enzyme production, carbon dioxide recovery, and furfural production. The pretreatment is dilute acid prehydrolysis. As for the case of Stone & Webster Engineering Corp., the sugar solution obtained after the saccharification step is concentrated using a multieffect evaporator. The economic feasibility analysis is performed based on IRR, and an ethanol-selling price of U.S. $2.06 gallon^{-1} (U.S. $0.54 l^{-1}) is found with the IRR set to 10%.

In addition to these feasibility studies, the technoeconomic evaluation of lignocellulosic ethanol owes much to two studies by NREL in association with other U.S. Research Institutes and Universities, that is, Wooley et al. (1999a) and Aden et al. (2002). Both studies are based on a detailed process design, mass and energy balance using ASPEN model and process economics evaluation. The former studies the simultaneous saccharification and cofermentation (SSCF) of yellow poplar wood. The size of the plant is 52.2 million gallons (198 million liters) of ethanol per year. The pretreatment is with dilute acid and the enzyme is produced onsite. The description of the process involves nine areas including SSCF, ethanol storage, cogeneration plant, and other utilities. The economic performance is estimated also based on 10% IRR. Five cases are evaluated: two cases represent the current state of technology (SOT) and the near-term best of industry; and three futuristic scenarios account for technology progress with ex ante snapshots of years 2005, 2010, and 2015. The economic performances of these cases are U.S. $1.44 gallon^{-1} ethanol, U.S. $1.16, U.S. $0.94, U.S. $0.82 and U.S. $0.76 (U.S. $ of 1997), respectively. In the case of 2015, the authors assume a 20% increase of carbohydrates due to biomass biotechnology improvements.

The second study (Aden et al., 2002) uses the same framework with the following main differences: (1) the feedstock is corn stover; the size of the plant is 69.3 million gallons ethanol per year; the onsite production of enzyme is removed and replaced by purchased enzymes. The levelized production cost based on 10% discount rate is U.S. $1.07 gallon^{-1} ethanol (U.S. $ of 2000). Updates of the technology model are provided in Aden (2008), Humbird and Aden (2008), and Aden and Foust (2009).

From 2002, the context of the technoeconomic evaluation of lignocellulosic ethanol had changed with the launch of the Biomass Program of the U.S. Department of Energy

(DOE). Since 2007, the design of this program has acquired a clear strategic goal with the aim of the public authorities to reduce the use of gasoline by 20% by 2017 and produce 35.10^9 l of renewable and alternative fuels in 2017. Concerning the RD&D in lignocellulosic bioethanol, a "Multi-Year Program Plan" (MYPP) is released and updated every 2 years, including so far 2005 (U.S. DOE 2005), 2007 (U.S. DOE 2007), and 2009 (U.S. DOE 2009). Two pathways are being studied, that is, thermochemical and biochemical.

In the framework of the "Biomass Program," Phillips et al. (2007) released a technical report on thermochemical ethanol with the goal to achieve economic competitiveness of lignocellulosic ethanol with starch-based ethanol by 2012. The feedstock is hybrid poplar wood chips. The process comprises seven main areas including feedstock handling and drying, gasification, gas cleanup and conditioning, alcohol synthesis, and alcohol separation. The economic evaluation is based on levelized production cost also termed Minimum Ethanol Selling Price (MESP) or Product Value (PV) in Kabir Kazi et al. (2010). Given a MESP of U.S. \$1.07 gallon^{-1} ethanol, the design case is such as to meet that target with a discount rate of 10%.

This approach is systematized in the MYPP (U.S. DOE, 2005; U.S. DOE, 2007; U.S. DOE, 2009; U.S. DOE, 2010), where a global Ethanol Programme Cost Target (EPCT) is fixed along with compatible cost targets for the different areas of the process. Furthermore, the EPCT (as well as the Ethanol Production Cost of the nth plant) changes from one MYPP to the other in order to reflect currency value, escalation factors, and the projected price of gasoline for the targeted year (Table 2).

As an example, the estimation of the EPCT in 2012 for the biochemical ethanol is based on the reference scenario by the Energy Information Administration (EIA, 2009) which forecasts the wholesale price of gasoline in 2012 at U.S. \$2.62 gallon^{-1} gasoline (U.S. \$ of 2007). Assuming a conversion factor of 0.67 gallon gasoline per gallon ethanol, the EPCT is set at U.S. \$1.76

TABLE 2 Ethanol Production Cost Breakdown According to U.S. MYPPs: 2012 Projections

		MYPP 2005	MYPP 2007	MYPP 2009
Currency (reference year)		U.S. \$ of 2002	U.S. \$ of 2007	U.S. \$ of 2007
Feedstock (total)	U.S. \$/dry ton	35.00	45.90	50.90
Ethanol yield	gal/dry ton	89.80	89.80	89.90
Supply chain areas				
Feedstock (total)	U.S. \$/gal	0.39	0.51	0.57
Prehydrolysis/treatment	U.S. \$/gal	0.21	0.25	0.26
Enzymes	U.S. \$/gal	0.10	0.10	0.12
Saccharification & Fermentation	U.S. \$/gal	0.09	0.10	0.12
Distillation & Solids recovery	U.S. \$/gal	0.13	0.15	0.16
Balance of plant	U.S. \$/gal	0.17	0.22	0.26
Ethanol production (total)	U.S. \$/gal	1.08	1.33	1.49

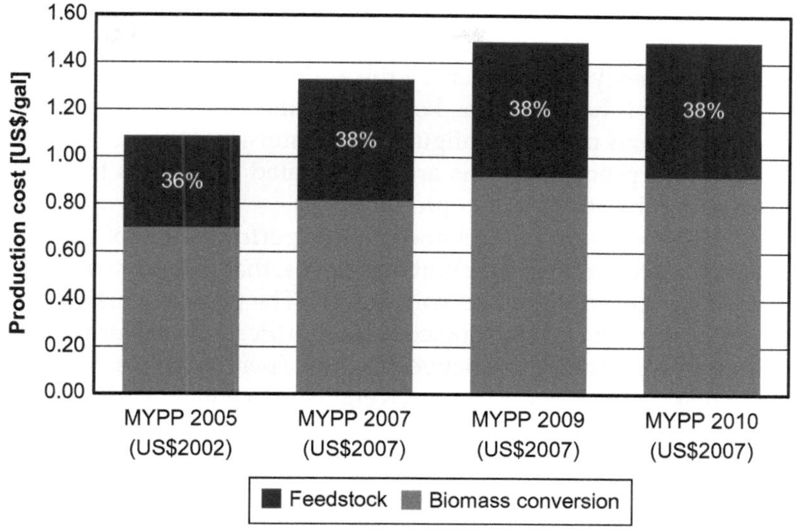

FIGURE 1 Ethanol production cost breakdown according to U.S. MYPPs: 2012 projections.

gallon^{-1} ethanol (U.S. $ of 2007). However, the Ethanol Cost Projection of the nth plant is at U.S. $1.49 gallon^{-1} ethanol (U.S. $ of 2007). That cost is very sensitive to economic assumptions. Kabir Kazi et al. (2010) found a levelized production cost of U.S. $3.40 gallon^{-1} ethanol (U.S. $ of 2007) for an nth plant supposed to be in operation in a 5-8 years' time frame. Compared to the MYPPs, the cost assumptions are higher, especially for the feedstock and enzymes. So is the case in Klein-Marcuschamer et al. (2010) where the MESP is estimated to U.S. $ 4.58 gallon^{-1} ethanol (U.S. $ of 2009) for the base case. Comparison between studies is not straightforward due to significant differences between the assumptions concerning the design and economical parameters. However, the series of MYPPs are more comparable.

The contribution made by feedstock production to the ethanol production cost increases from one MYPP to the other due to progress in understanding and estimating the feedstock production and logistics (Figure 1).

Other changes occur in the U.S. Energy policy which reinforces the role of biofuels. The Energy Independence and Security Act (EISA) of 2007 set a mandatory Renewable Fuel Standard (RFS). Accordingly, in 2022, the transportation fuel on U.S. market must contain a minimum of 36 billion gallon (136 billion liters) of renewable fuels. The objectives of the EISA with regard to biofuels are confirmed by the American Recovery and Reinvestment Act (ARRA) of 2009.

2.2 Other Technoeconomic Evaluation Cases

Besides the technoeconomic evaluations undertaken in the United States, significant contributions are brought by European Research institutions mainly in Sweden, the Netherlands, and Denmark. Galbe et al. (2007) present a review of the process economics of lignocellulosic ethanol published since 1996. They compare the production costs estimation of lignocellulosic ethanol of 15 studies undertaken in the United States and in Europe and point out the high variability of the results. However, that comparison is somewhat tricky as the year of U.S. $ currency is not given by the authors. They point out the ethanol yield and the energy demand of the process as key influencing factors of the ethanol production cost for given feedstock and process configuration. Water-insoluble solids (WIS) concentration and recirculation of process streams are investigated as options to reduce the energy demand and increasing the amount of coproducts.

Sassners et al. (2008) compare the technoeconomic performances of conversion of lignocellulosics to ethanol based on three different feedstocks, that is, a softwood (spruce), a hardwood (salix), and an agricultural residue (corn stover). The process consists of SO_2-catalyzed steam explosion pretreatment and simultaneous saccharification and fermentation (SSF). The feedstocks show significant differences between hexose/pentose ratios, that is, 7.4 for spruce, 2.9 for salix, and 1.6 for corn stover, based on weights' percentage of dry matters. However, for the percentage of C5 and C6 as a whole, corn stover is the first (68%) followed by spruce (67.5%) and salix (64.5%). The process capacity of the ethanol plant is supposed to be 200,000 dry tons of biomass per year. The process parameters are adjusted to experimental data and adapted to each feedstock. Enzymes are assumed to be purchased, while the yeasts are produced onsite. As an example, the temperature of steam pretreatment is 195, 190, and 205 °C and the yeast concentration is 3.0, 1.8, and 2.5 g/l, respectively for Salix, corn stover, and spruce.

Three base cases are evaluated, one for each feedstock where conversion factors for steam pretreatment and SSF are adapted from experimental and analytical research works at Lund University, Sweden. In the base cases, it is assumed that only the hexoses (glucan, galactan, and mannan) are converted to ethanol. Material and energy balances are evaluated using ASPEN PLUS. The overall ethanol yields—taking into account sugar consumption for yeast production and ethanol losses within the process—are estimated to be 239, 215, and 292 l/dry metric ton for Salix, corn stover, and spruce, respectively. Note that the corresponding ethanol yields from hexoses are 245, 302, and 426 l, respectively, per dry metric ton. Thus, the estimated yields correspond to 69.3%, 71.2%, and 68.5% of the potential yields for Salix, corn stover, and spruce, respectively. The lower value for spruce can be explained by the more severe pretreatment conditions which result in more degradation of sugars and higher level of inhibitors. Alternative cases where both hexoses and pentoses are converted to ethanol are evaluated. They result in overall yield of 314, 306, and 315 l/dry metric ton for Salix, corn stover, and spruce, respectively, these are 67.4%, 62.1%, 64.9% of the overall potential yield from hexoses and pentoses. Thus, compared to the base cases, the absolute yield (liters ethanol/dry metric ton of feedstock) increases with the conversion of pentoses into ethanol; however, the relative yields, that is (simulated yield with regard to assumed process condition)/(theoretical yield), decrease. These results suggest the need of a trade-off between, on one side severe pretreatment conditions which are favorable to a high digestibility of cellulose by enzymes but enhance the level of inhibitors and on the other side milder conditions that reduce the risk of hemicellulose sugars degradation and formation of inhibitors but decrease the digestibility of cellulose.

The authors define energy efficiency as the ratio between energy output (ethanol + solid fuel) and energy input (raw materials + electric power requirement). The raw materials, solid fuel (pellets), and ethanol are estimated using the higher heating value (HHV) and the efficiency for electricity generation is estimated to 30%. For the base cases, the authors find the following energy efficiencies for ethanol output only: 25 (Salix), 25 (corn stover), and 31% (spruce). These figures increase in case of the alternative cases and obviously when the outputs also consider solid fuel coproducts. In the latter case, the energy efficiency is in the range of 52-53% for Salix, 55% for corn stover, and 56% for spruce.

The economic evaluation consists in estimating annual production cost including annualized capital cost using 7% interest rate and 15 years' depreciation period, and annual operation costs. The costs are expressed in U.S. $. The authors do not indicate the year of the currency. For the base cases, the annual production costs (U.S. $) significantly vary, that is, U.S. $0.69 l^{-1} ethanol (spruce), 0.86 (corn stover), and 0.87 (Salix). For alternative cases, the costs become 0.66 (spruce), 0.67 (corn stover), and 0.72 (Salix).

Wingren et al. (2008) perform a technoeconomic evaluation of an SSF-based softwood to ethanol, with the objective to compare the impact of various downstream configurations, that is, after the SSF, on the ethanol production cost. The base case consists in conversion of wood chips of spruce into ethanol. The water content of the feedstock is 50% and the composition on a dry weight basis is as follows: 45.0% glucan, 12.6% mannan, 2.6% galactan, 7.1% pentosans, 28.1% lignin, and 4.6% acetyl groups, extractives, and ash. The conversion process is the same as in Sassners et al. (2008). The downstream process consists in distillation-rectification and evaporation. The unfiltered mash including ethanol, lignin, yeast, and water streaming from the SSF is preheated and distributed between the two distillation columns. The distillate is then sent to the rectifier while the stillage is processed in centrifuges for liquid-solid separation. The liquid is concentrated through an evaporator. The resulted syrup is blended to the stream with solid compounds and sent for drying. Part of the 85% dry matter resulting material is burned in the boiler to generate the primary process steam while the remainder is pelletized.

In the base case, the evaporator is composed of five effects. The alternative configurations analyzed by the authors include the following options: (1) increase the number of effects in the evaporator; (2) reduce the number of strippers from two to one and integrate it with the evaporator; (3) use a Mechanical Vapor Recompression (MVR) in order to increase the temperature of the latent heat leaving the last effect and use it to replace a significant part of the primary steam; the MVR requires however supplementary electrical energy; (4) finally, methanize the stillage and use the biogas to fuel the steam boiler while the produced sludge is burned in an incinerator. The economic evaluation uses the same approach as in Sassners et al. (2008). The interest rate, however, is 6%. The production cost in (U.S. $ per liter) varies between 0.546 for the MVR option to 0.591 for the base case. The case of anaerobic digestion results in 0.549 (U.S. $ per liter) production cost. That is close to the least cost of 0.546 U.S. $ per liter. The currency is supposed to be nominal U.S. $.

In the REFUEL project (2006-2008) funded by the European Commission under the Intelligent Energy Europe program, seven EU institutes have analyzed the prospects for biofuels in terms of resource potential, costs and impacts of different biofuels, including lignocellulosic ethanol. Although the project is rather focused on the cost and availability of resources within the European Union, the production cost of biofuels is taken into account. The data for bioethanol production from cellulosic materials based on enzymatic hydrolysis pathway are obtained from the Energy research Centre of the Netherlands (Kuijvenhoven, 2006) and the Copernicus Institute for Sustainable Development and Innovation of Utrecht

University (Hamelinck, 2004; Hamelinck et al., 2005). The economic evaluation (Londo et al., 2008) is based on constant of 2002 and results in a net production cost (including the sales of electricity as a byproduct) of 0.62 l^{-1} in 2010 (forest wood as feedstock, production capacity of 100,000 t ethanol per year), 0.59 l^{-1} in 2020 (200,000 t ethanol per year), and 0.50 l^{-1} in 2030 (400,000 t ethanol per year), given the learning curve based on expected global production and number of plants.

Seabra et al. (2010) compare the calculated technoeconomic performance for thermochemical and biochemical conversion of sugarcane residues, considering future conversion plants adjacent to sugarcane mills in Brazil. Process models developed by the NREL are adapted to reflect the Brazilian feedstock composition and used to estimate the cost and performance of these two conversion technologies. Like in previous works by the NREL, the technoeconomic performance in Seabra et al. (2010) is measured in terms of the MESP. The economic performance of the two technologies is quite similar in terms of the MESP, at U.S. \$0.318 l^{-1} (U.S. \$ of 2007) for biochemical conversion and U.S. \$0.329 l^{-1} for thermochemical conversion. The two figures refer to a sugar mill with a treatment capacity of 1000 t/h sugarcane.

3 KEY DRIVERS OF THE LIGNOCELLULOSIC ETHANOL PRODUCTION COST

The production cost of lignocellulosic ethanol is sensitive to key parameters such as the type, composition, and farm-gate price of the feedstock, the size of the ethanol plant, the conversion efficiency, and the level of investment costs. Some of these factors are illustrated in this section, in a harmonized framework. The same framework but in a different context is described in Gnansounou et al. (2005) for the production of ethanol from sweet sorghum bagasse.

The evaluation and analysis of bioethanol production cost is performed using an own spreadsheet model developed by the authors. The technology and process model is based on and follows closely the NREL design as reported in Wooley et al. (1999a). The model calculates all material and energy balances based on specified yields at each process step. Operating costs are calculated based on material flow and energy use, coupled with available cost information. Appropriate rates are used to size the equipment, and equipment costs are calculated based on NREL information for all the steps from feedstock handling and storage to manufacture of ethanol. The power law scale factors reported by NREL are used to estimate the change in cost of each equipment item with varying feedstock composition, feed capacity, yields, etc. The model is run initially at NREL conditions to ensure that it is correct and can duplicate the results from NREL. Changes are then made on various parameters to reflect the composition of selected feedstocks, yields, and specific costs. In particular, cost index values for plant capital, chemicals and materials, and labor are adapted according to the U.S. DOE's MYPP 2009, in order to match the present economic situation in the United States (+36% for plant cost, +38% for chemicals and materials, +24% for labor). Actual indices of 2007 are used in the present analysis. All the costs are expressed in U.S. \$ of 2007. The economic model applied in the spreadsheet is based on the one in NREL's ethanol process designs (Wooley et al., 1999a; Aden et al., 2002). The levelized production cost is evaluated based on a discount rate of 10%.

Four production options are analyzed, based on the type (and therefore composition) of feedstock, including (1) straw, (2) eucalyptus, (3) poplar, and (4) switchgrass. The composition of each feedstock is taken from the U.S. DOE's Biomass Feedstock Composition and Property Database (U.S. DOE, 2004) and is detailed in Table 3.

Again, the process design considered in the present analysis closely follows the one described by Wooley et al. (1999a). The feedstock is first crushed into chips before pretreatment with dilute sulfuric acid, where the hemicellulose is hydrolyzed. The resulting hydrolysate is detoxified in order to remove the acid as well as the inhibitors produced along the pretreatment. A portion of the detoxified hydrolysate is fed to a batch operation to produce cellulase enzymes by the fungus Trichoderma reesei. The bulk of the detoxified hydrolysate and the effluent from enzyme production are added to a reactor to release glucose from cellulose through enzymatic hydrolysis. In the same vessel and simultaneously, an organism ferments the sugars from hemicellulose plus the glucose released from cellulose to ethanol. This operation is referred to as SSCF for simultaneous saccharification and cofermentation (of C5 and C6 sugars). The fermented beer containing about 5% (vol.) ethanol passes on to distillation where it is concentrated to approximately 95% ethanol in the overhead. Molecular sieves then follow to recover fuel-grade ethanol (i.e., min. 99.7% wt. according to the European legislation). The solids, containing mostly lignin and solubles from distillation, are concentrated and burned to generate steam that can provide all of the heat and electricity for the process with some excess electricity left to export. Water is treated by anaerobic digestion and methane that results is also burned for steam generation. A schematic representation of the complete process is shown in Figure 2.

TABLE 3 Feedstock Proximate Analysis (Percentage Weight, Wet Basis)

Components	Straw (%)	Eucalyptus (%)	Poplar (%)	Switchgrass (%)
Moisture	15.0	30.0	50.0	50.0
Cellulose	27.7	34.0	21.3	16.8
Hemicellulose	24.9	18.1	28.7	27.6
Xylan	16.3	8.1	9.5	11.1
Arabinan	2.0	0.3	0.4	1.4
Mannan	0.3	0.7	2.0	0.2
Galactan	0.6	0.7	0.1	0.5
Acetate	1.9	3.0	2.3	0.7
Lignin	14.3	19.4	13.8	9.3
Ash	8.7	0.6	0.5	2.9
Other IS	2.1	2.0	0.0	1.2
Other SS	11.0	1.3	0.0	6.0
Total	100.0	100.0	100.0	100.0

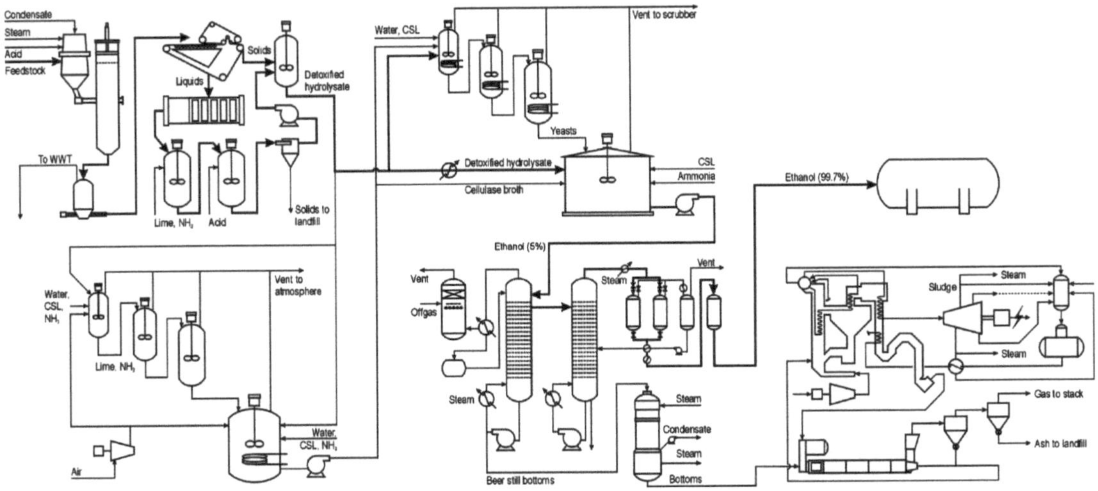

FIGURE 2 Schematic diagram of the ethanol production process *(adapted from Wooley et al., 1999a).*

Ethanol and possible excess electricity are the only two products according to the considered plant configuration. Other possible configurations mentioned in the preceding sections are not taken into consideration in the present illustration.

The reference ethanol production capacity is taken as 200 million liters per year (Ml/yr). The treatment capacity varies from 1600 to 2000 tons of dry matter (t DM) per day, according to the feedstock. Specific conversion yields of the prehydrolysis and fermentation reactions are taken from Aden et al. (2002).

The net production cost of ethanol is divided into (1) investment costs, (2) fixed operating costs (including salaries, general overhead, insurance, taxes, and maintenance), (3) variable operating costs (including purchase of consumables and sales of excess electricity), and (4) feedstock costs. Feedstock costs are separated from variable operating costs due to their large share of the net production cost.

Feedstock costs are divided into nontransport (farm gate) and transport costs and are calculated from the data in the European REFUEL project. Transport costs are divided into loading/unloading costs (U.S. \$0.19 ton^{-1}), fixed costs (U.S. \$2.57 ton^{-1}), and variable costs (U.S. \$0.10 ton^{-1}/km). Biomass is supposed to be collected within a circular area surrounding the ethanol plant with an availability factor of 10%. The collection radius is defined as the radius of half the collection area. Biomass yields are taken as 3.52, 12.60, 5.53, and 12.99 t DM per hectare per year, respectively, for straw (15% water), eucalyptus (30% water), poplar (50% water), and switchgrass (50% water).

Ethanol production costs as calculated by the spreadsheet model are given in Table 4. The main technical parameters including details of feedstock costs, ethanol yield, electricity production and consumption, project investment are also provided.

Feedstock costs vary from U.S. \$53 (eucalyptus) to U.S. \$123 t DM^{-1} (poplar). On a per-liter basis, feedstock costs vary from U.S. \$0.18 l^{-1} ethanol (eucalyptus) to U.S. \$0.42 l^{-1} ethanol (switchgrass). Total project investments vary from U.S. \$280 million (poplar) to U.S. \$310

TABLE 4 Ethanol Production Cost and Production Parameters as a Function of Feedstock

		Straw	Eucalyptus	Poplar	Switchgrass
General data					
Ethanol production capacity	Ml	200	200	200	200
Biomass treatment capacity	t DM/day	1 960	1 680	1 636	1 818
Total project investment	mio U.S. $	309	290	281	296
Feedstock					
Nontransport cost	U.S. $/t DM	97.30	52.81	123.33	118.01
Transport cost	U.S. $/t DM	11.65	9.57	17.06	13.29
Total cost	U.S. $/t DM	108.95	62.38	140.39	131.30
Yield	t DM/ha.yr	3.523	12.600	5.530	12.990
Average collection radius	km	55.7	27.3	40.6	27.9
Availability factor	ha/ha	10%	10%	10%	10%
Process					
Ethanol yield	l t DM^{-1}	291.3	339.9	349.0	314.1
Total electricity produced	MWh/yr	54.8	25.9	26.1	39.6
Net electricity consumed	MWh/yr	22.4	31.0	25.3	21.9
Excess electricity	MWh/yr	32.3	0.0	0.8	17.7
Electricity purchased	MWh/yr	0.0	5.1	0.0	0.0
Production cost					
Feedstock cost	U.S. $/l	0.37	0.18	0.40	0.42
Variable operating cost	U.S. $/l	0.02	0.07	0.05	0.03
Fixed operating cost	U.S. $/l	0.05	0.05	0.04	0.05
Investment cost	U.S. $/l	0.29	0.26	0.26	0.27
Total production cost	U.S. $/l	0.73	0.56	0.76	0.77
Total nonfeedstock cost	U.S. $/l	0.36	0.38	0.36	0.35

million (straw). Ethanol yields vary from 290 (straw) to 350 l/t DM (poplar). All feedstocks except eucalyptus lead to an excess of electricity, that is, production exceeds process requirements.

Ethanol production costs on a per-liter basis are largely dominated by feedstock and investment costs, while fixed and variable operating costs play a minor role. Apart from the case of eucalyptus which appears to be a cheaper feedstock, total production costs are composed of 50-55% feedstock costs, 35-40% investment costs, and 10% variable costs. If only nonfeedstock costs are taken into account, investment costs represent an average of 75%.

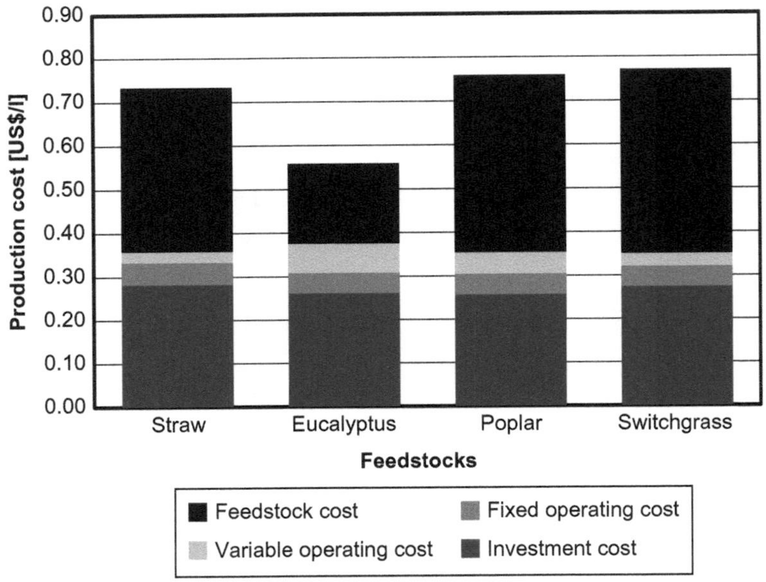

FIGURE 3 Ethanol production cost as a function of feedstock.

Second-generation ethanol is indeed heavier on investment than first-generation production pathways. Unless the selected feedstock for ethanol production turns out to be a waste in sufficient quantities at a reasonable distance from the plant, its cost on a per-liter basis is far from being negligible, even though it is less than for first-generation ethanol. These results show the importance of properly evaluating the availability and price of lignocellulosic feedstocks for ethanol production. The results regarding ethanol production costs on a per-liter basis are illustrated in Figure 3.

The sensitivity of the production cost with respect to parameters such as plant investment, feedstock cost, plant size, and ethanol yield is now evaluated.

3.1 Sensitivity of Ethanol Production Cost with Respect to Production Capacity

The analysis is performed for the case of ethanol production from straw. Similar results are obtained with other feedstocks. The production cost is calculated for ethanol plants with production capacities of 50, 100, 200, and 400 Ml/yr. The results are shown in Figure 4.

The choice of the production capacity has an effect not only on investment costs, but also on feedstock transport costs and fixed operating costs; salaries and maintenance costs depend on the size of the plant, but not linearly. According to the results in Figure 4, the larger the ethanol plant, the lower the production cost. It can be considered, due to the relatively low contribution of operating costs to the total production cost, that the effect of plant size on operating costs is almost negligible. The trade-off therefore is between investment costs and feedstock transport costs. On a per-liter basis, the larger the ethanol plant, the lower

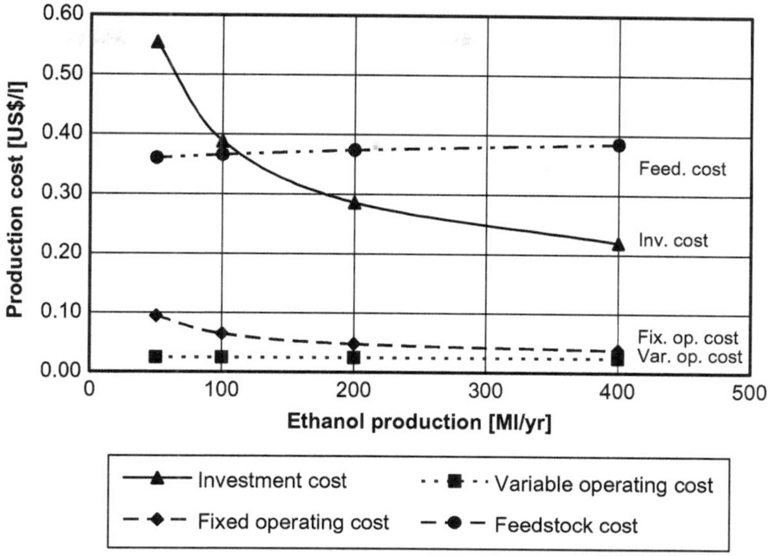

Production capacity	**Ml/yr**	**50**	**100**	**200**	**400**
Variable operating cost	US$/l	0.02	0.02	0.02	0.02
Fixed operating cost	US$/l	0.09	0.06	0.05	0.04
Investment cost	US$/l	0.55	0.39	0.29	0.22
Feedstock cost	US$/l	0.36	0.37	0.37	0.39
Total cost	**US$/l**	**1.03**	**0.84**	**0.73**	**0.67**

Type of feedstock		Straw	Straw	Straw	Straw
Conversion efficiencies		2002	2002	2002	2002
Ethanol production capacity	Ml/yr	50	100	200	400
Biomass treatment capacity	t DM/day	490	980	1960	3920
Total project investment	mio US$	152	211	309	469
Feedstock data					
Non-transport cost	US$/t DM	97.30	97.30	97.30	97.30
Transport cost	US$/t DM	7.65	9.30	11.65	14.96
Total cost	US$/t DM	104.95	106.60	108.94	112.25
Yield	t DM/ha.yr	3.523	3.523	3.523	3.523
Average collection radius	km	27.8	39.4	55.7	78.8
Availability	ha/ha	10%	10%	10%	10%
Process					
Ethanol yield	l/t DM	291.3	291.3	291.3	291.3
Total electricity produced	MWh/yr	13.7	27.4	54.8	109.5
Net electricity consumed	MWh/yr	5.6	11.2	22.4	44.9
Excess electricity	MWh/yr	8.1	16.2	32.3	64.6
Electricity purchased	MWh/yr	0.0	0.0	0.0	0.0

FIGURE 4 Sensitivity of ethanol production cost with respect to production capacity.

the investment cost due to economy of scale, but the larger the feedstock transport cost. The optimal size of an ethanol plant therefore largely depends on regional conditions and on the availability of feedstock. The latter will have an effect on feedstock transport costs, but may also have some on feedstock nontransport costs depending on local conditions. In the conditions described in the present analysis, a doubling of the production capacity (from 200 to 400 Ml/yr) results in a 10% reduction of the net production cost (from U.S. \$0.73 to U.S. \$0.67 l^{-1}). A halving of the production capacity (from 200 to 100 Ml/yr) results in a 15% increase of the net production cost (from U.S. \$0.73 to U.S. \$0.84 l^{-1}).

The trade-off between plant size and transport distance in favor of plant size in terms of production cost may be largely different when considering the environmental impact of ethanol production. The conversion infrastructure indeed is generally hardly significant when evaluating the energy or greenhouse gas (GHG) balance of biofuel production. Transport operations, however, especially biomass transport, are far from being negligible in terms of their environmental impact. Therefore, there might also be a trade-off between environmental impact and production cost in terms of plant size, with larger plants resulting in lower production cost but larger environmental impact due to more transport.

3.2 Sensitivity of Ethanol Production Cost with Respect to Ethanol Yield

Again, the analysis is performed for the case of ethanol production from straw. The production capacity is taken as 200 Ml/yr. The production cost is calculated according to four different sets of conversion efficiencies, including those of NREL's ethanol process designs (Wooley et al., 1999a; Aden et al., 2002). Two additional sets of conversion efficiencies are taken into account: one with conversion of only cellulose C6 sugars to ethanol with efficiencies as in Aden et al. (2002), referred to as "C6 only"; one corresponding to the theoretical maximum ethanol yield, referred to as "Max," The corresponding reaction-specific efficiencies are detailed in Table 5.

The corresponding ethanol yields are 189.2 ("C6 only"), 249.7 ("1999"), 291.3 ("2002"), and 340.4 ("Max"). The "C6 only" scenario optimizes the production and sales of excess electricity, while the "Max" corresponds to the maximum production of ethanol. The results are shown in Figure 5.

According to the results in Figure 5, the higher the ethanol yield, the lower the net production cost of ethanol. Higher ethanol yields also result in a lower electricity output. Ethanol production costs vary from U.S. \$1.06 ("C6 only") to U.S. \$0.73 l^{-1} ("2002), and could even be as low as U.S. \$0.65 l^{-1} under the "Max" scenario. The improvement of conversion efficiencies between the 1999 and the 2002 ethanol process designs by NREL results in an improved ethanol yield (+17%) and a reduced net production cost (−13%). The net production cost is largely dependent on the price of "renewable" electricity on the local market; U.S. \$0.02 kWh^{-1} in the present situation.

All cost components are affected by a change in ethanol yield, but at various degrees. Higher ethanol yields result in lower feedstock expenditures (less feedstock required per liter of ethanol output), but also in lower investment costs (lower treatment capacity for a given production capacity, and therefore smaller equipment and reduced investment), and lower fixed operating costs (in proportion somewhat to the investment cost).

TABLE 5 Conversion Rates Applied According to NREL (Aden et al., 2002; Wooley et al., 1999a)

		Conversion Rates			
Process Step	**Reactions**	*Low (%)*	*Wooley et al., 1999a (%)*	*Aden et al., 2002 (%)*	*Max (%)*
Pre-hydrolysis	Cellulose to glucose	5.0	6.5	7.0	10.0
	Xylan to xylose	70	75	90	100
	Mannan to mannose	70	75	90	100
	Galactan to galactose	70	75	90	100
	Arabinan to arabinose	70	75	90	100
	Acetate to acetic acid	100	100	100	100
Seed fermentation	Cellulose to glucose	20	20	20	20
	Glucose to ethanol	80	85	90	95
	Glucose to carbon dioxide	80	85	90	95
	Xylose to ethanol	75	80	80	90
	Xylose to carbon dioxide	75	80	80	90
Production fermentation	Cellulose to glucose	70	80	90	100
	Glucose to ethanol	90	92	95	95
	Glucose to carbon dioxide	90	92	95	95
	Xylose to ethanol	80	85	85	93
	Xylose to carbon dioxide	80	85	85	93

In the "C6 only" scenario where the hemicellulose is not converted to ethanol, unconverted solids are considered to be burned together with the lignin to produce heat and electricity. Depending on the process design, however, the hemicellulose may be converted to various value-added products. What is produced out of the various components of lignocellulosic biomass has a significant effect on the economics of cellulosic ethanol production, which is likely also to depend on local conditions.

3.3 Sensitivity of Ethanol Production Cost with Respect to Feedstock Nontransport Cost

Feedstock costs represent one of the most significant components of the production cost of ethanol. The sensitivity of ethanol production cost with respect to feedstock cost is analyzed for various values of nontransport feedstock costs, from U.S. \$25 to U.S. \$150 t DM^{-1}. The analysis is again performed for ethanol production from straw, in a facility with a production capacity of 200 Ml/yr. The results are shown in Figure 6.

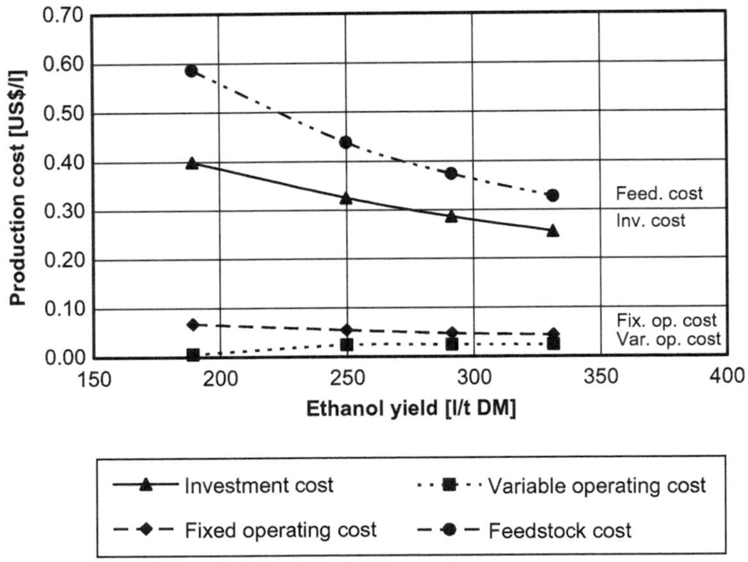

Ethanol yield	**l/t DM**	**189**	**250**	**291**	**331**
Variable operating cost	US$/l	0.01	0.02	0.02	0.02
Fixed operating cost	US$/l	0.07	0.05	0.05	0.04
Investment cost	US$/l	0.40	0.33	0.29	0.26
Feedstock cost	US$/l	0.59	0.44	0.37	0.33
Total cost	**US$/l**	**1.06**	**0.84**	**0.73**	**0.65**

Type of feedstock		Straw	Straw	Straw	Straw
Conversion efficiencies		C6 only	1999	2002	Max
Ethanol production capacity	Ml/yr	200	200	200	200
Biomass treatment capacity	t DM/day	3018	2287	1960	1724
Total project investment	mio US$	429	351	309	276
Feedstock data					
Non-transport cost	US$/t DM	97.30	97.30	97.30	97.30
Transport cost	US$/t DM	13.57	12.29	11.65	11.15
Total cost	US$/t DM	110.87	109.58	108.94	108.45
Yield	t DM/ha.yr	3.523	3.523	3.523	3.523
Average collection radius	km	69.1	60.2	55.7	52.2
Availability	ha/ha	10%	10%	10%	10%
Process					
Ethanol yield	l/t DM	189.2	249.7	291.3	331.3
Total electricity produced	MWh/yr	124.1	74.2	54.8	40.8
Net electricity consumed	MWh/yr	35.5	28.2	22.4	18.2
Excess electricity	MWh/yr	88.6	46.0	32.3	22.7
Electricity purchased	MWh/yr	0.0	0.0	0.0	0.0

FIGURE 5 Sensitivity of ethanol production cost with respect to ethanol yield.

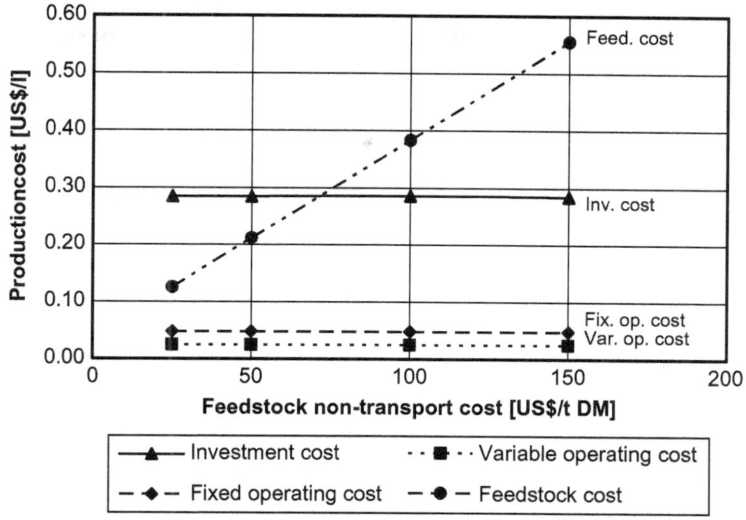

Feed. non-transport cost	**US$/t DM**	**25**	**50**	**100**	**150**
Variable operating cost	US$/l	0.02	0.02	0.02	0.02
Fixed operating cost	US$/l	0.05	0.05	0.05	0.05
Investment cost	US$/l	0.29	0.29	0.29	0.29
Feedstock cost	US$/l	0.13	0.21	0.38	0.55
Total cost	**US$/l**	**0.48**	**0.57**	**0.74**	**0.91**

Type of feedstock		Straw	Straw	Straw	Straw
Conversion efficiencies		2002	2002	2002	2002
Ethanol production capacity	Ml/yr	200	200	200	200
Biomass treatment capacity	t DM/day	1960	1960	1960	1960
Total project investment	mio US$	227	227	309	227
Feedstock data					
Non-transport cost	US$/t DM	25.00	50.00	100.00	150.00
Transport cost	US$/t DM	11.65	11.65	11.65	11.65
Total cost	US$/t DM	36.65	61.65	111.65	161.65
Yield	t DM/ha.yr	3.523	3.523	3.523	3.523
Average collection radius	km	55.7	55.7	55.7	55.7
Availability	ha/ha	10%	10%	10%	10%
Process					
Ethanol yield	l/t DM	291.3	291.3	291.3	291.3
Total electricity produced	MWh/yr	54.8	54.8	54.8	54.8
Net electricity consumed	MWh/yr	22.4	22.4	22.4	22.4
Excess electricity	MWh/yr	32.3	32.3	32.3	32.3
Electricity purchased	MWh/yr	0.0	0.0	0.0	0.0

FIGURE 6 Sensitivity of ethanol production cost with respect to feedstock non-transport cost.

Given the conversion efficiencies from Aden et al. (2002), each liter of ethanol requires 3.43 kg DM of straw. Given the hypotheses on biomass yield, that is, 3.52 t DM/ha in the case of straw and availability (10% in a circular area surrounding the ethanol plant), transport cost amounts to almost U.S. \$12 t DM^{-1} or U.S. \$0.04 l^{-1} ethanol.

The changes in nontransport feedstock costs only affect the feedstock cost components. None of the other cost components is affected by such changes. In case of a freely available feedstock (i.e., the only cost is the cost of collection), the net production cost of ethanol is found to be U.S. \$0.40 l^{-1}. However, due to the required amount of feedstock, such a condition would rarely appear. It comes out from the results in Figure 6 that total feedstock costs (including transport costs) exceed investment costs (on a per-liter basis) when nontransport feedstock costs exceed U.S. \$71 t DM^{-1}. The total cost of feedstock in the present analysis (on a per-liter basis) varies from U.S. \$0.13 (U.S. \$25 t DM^{-1} straw) to U.S. \$0.55 l^{-1} (U.S. \$150 t DM^{-1}). The average cost of straw according to the REFUEL project is considered to be U.S. \$97 t DM^{-1} (excluding transport), which corresponds to U.S. \$0.33 l^{-1} (or U.S. \$0.37 l^{-1} including transport costs). Although it is often considered that the availability and low cost of feedstock is one of the main advantages of second-generation biofuels, the results in the present analysis show that feedstock may still represent the largest cost component of cellulosic ethanol net production cost, depending on local and biomass market conditions.

The expected development of bioenergy in all forms (from heating to transportation purposes) and of nonenergy biomass applications is likely to modify the present notion of lignocellulosic waste. There might be situations where several facilities are in competition for a given biomass, which is likely to bring its price up. Therefore, cheap and largely available feedstock may often not be a reality, again depending on local conditions.

4 COST MANAGEMENT SYSTEM

Technoeconomic evaluation of lignocellulosic bioethanol is supposed to follow one of the three types of cost management system available in the literature of strategic cost management, that is, Value Engineering (VE), Target Costing (TC), and Combined Target Costing and Value Engineering (TC & VE). Each of these is described with emphasis on their application to lignocellulosic bioethanol.

4.1 Value Engineering

VE is a set of techniques which aim at reducing the production cost of a product or service by identifying the main cost reduction opportunities, generating cost improvement alternatives, and find out the best one (Ibusuki and Kaminski, 2007). In VE, each basic function in the system is specified and analyzed along with the interactions. The use of VE started during World War II when the shortage of resources forced to highly value creative and least cost designs. Nowadays, VE is used in order to design innovative products, increase the competitiveness, and access marketplace with low industrial and economic risks.

In the case of lignocellulosic ethanol, process design, modeling and cost analyses are included in VE (Wooley et al., 1999b). Cost reduction analyses dictate the detail level of the process design. Data collection and process flowsheeting allow a consistent design of each part of the process.

An alternative practice to VE is to only rely on designs made by external specialized engineering consultants with the risk to miss the overall consistency that requires an integration of knowledge. The complexity of technoeconomic evaluation of emerging technologies such as lignocellulosics to ethanol requires a pluridisciplinary approach only capable as long as the development of a morphological appraisal tool is concerned. Several issues are at stake along the process chain including the suitable choice and operation options of the feedstocks, pretreatment, enzymes production, saccharification, fermentation of most sugars, especially hexoses and pentoses, integration or not of the latter two segments, distillation, valorization of the stillage, and energy integration. The complementary use of Process Development Units (PDU) and sophisticated Process Simulators such as ASPEN PLUS has permitted significant progress during the last decades. VE allows to perform the best available estimates and the near-term expected states, that is, next 2 years of the lignocellulosic ethanol pathways.

The chosen feedstocks depend mostly on the availability and cost. In the United States, for example, two feedstocks are mainly considered by the NREL as base cases, that is, a hardwood (yellow poplar) and an agricultural residue (corn stover), while in a northern European forest country, as it is the case for Sweden, a softwood (spruce) is generally evaluated. There are significant differences between those three feedstocks that can impact the process design and the ethanol production cost. As an example, contrary to yellow poplar, the acetate levels in corn stover and in spruce are low, resulting in less costly detoxification step. The percentage of hexoses in spruce is also higher, thereby implying a higher potential yield in the current state of conversion efficiencies. However, the most significant feedstock impact on the ethanol production cost is the feedstock cost. In that sense, assumptions made in the United States earlier studies are often more optimistic than in European ones.

Although dilute acid and steam explosion are the two pretreatment processes mostly used in integrated assessments, other processes are under study and should deserve more attention especially liquid hot water, Ammonia Fiber Explosion (AFEX), and CO_2 Explosion which are more promising for meeting the following requirements: improve efficacy, reduce pretreatment costs, decrease inhibitors and toxic matters production, and enhance flexibility of feedstocks use and end coproducts valorization. Due to these challenges, pretreatment stage is considered as one of the most influencing stage for reducing the overall process cost.

Lignocellulosic feedstocks can be directly saccharified by acid hydrolysis. However, recycling the acid proves to be expensive. Enzymatic saccharification is then the alternative which is mostly studied in the reviewed papers. The major bottleneck of enzymatic saccharification is the cost of cellulases. Although they have been significantly reduced during the last decade, they remain high. Cellulases consist of at least three types of enzymes: endoglucanases weaken the structure of the cellulosic biomass by cutting randomly amorphous components of cellulose; exoglucanases attack the exposed ends and produce cellobiose units; and cellobiases hydrolyze the cellobiose into glucose. Trichoderma reesei, a mesophilic and filamentous fungus, is frequently used to produce cellulase complex. This organism produces abundant amounts of endoglucanases and exoglucanases but lesser cellobiases. Furthermore, cellulases are inhibited by cellobiose and glucose for certain concentration levels. Cellulases can be produced by solid-state fermentation (Pandey et al., 1999) or more usually by submerged fermentation (Tolan and Foody, 1999).

Besides more used cellulase-producing fungi, cellulase-producing bacteria are being considered for their biodiversity which allows isolating strains that can survive in harsh environments and produce enzymes which are stable even in extreme conditions (Maki et al., 2009). Another option for coping with inhibition of cellulases by end-products is to simultaneously produce and ferment glucose in the so-called SSF reactor. Besides potential improvement of the enzymes activity, SSF halves the number of reactors, decreases investment cost, and improves the overall production cost (Wingren et al., 2003). The main bottleneck for developing SSF is to cope with the difference between optimal temperatures of saccharification and fermentation. The most challenging in this way is to integrate the four steps, that is, enzyme production, saccharification, and fermentation of hexoses and pentoses. The consolidated bioprocessing CBP (Lynd et al, 2005) is the technology breakthrough that is expected for significantly reducing process costs. Direct Microbial conversion DMC (Lee, 1997) is one of the representatives of this concept.

Costing within VE consists in estimating the production cost of large-scale ethanol plant based on scale up of the demonstration plant, state-of-the-art technology and price quotes by process providers. Short- and medium-term costs are projected as well based on technological progress and learning curve. As an example, in the United States, the SOT report typically proceeds with VE-based costing. While short- and near-term maturing technologies are concerned with VE-based costing, futurist ones such as CBP should be excluded as the cost information is barely based on consolidated industrial data.

4.2 TC with or without VE

While costing within VE framework remains a standard "COST PLUS" approach, TC is rather a market-oriented method applied from the design stage. According to most of the production economics literature (Cooper and Slagmulder, 1997; Feil and Yook, 2004; Ibusuki and Kaminski, 2007; Kato 1993), TC originates from Japan where it is commonly used since the 1960s to manage production cost and gain competitiveness advantage. Few authors however investigate early adoption of TC in western countries. Wijewardena and de Zoysa (1999) perform a comparative analysis of cost management in Japan and Australia and find that several Australian companies apply TC as cost planning method. Dekker and Smidt (2003) survey the use of TC by Dutch firms, and Ellram (2006) investigates the TC practices in the United States and highlights the more frequent implementation of TC in R&D and supply chains contrary to assertions of previous works. Based on Ellram (1999), we derive the six-step application of TC to the design of lignocellulosic ethanol pathways (Figure 7).

4.2.1 *Step 1: Identify Desired Ethanol Characteristics*

The characteristics of lignocellulosic ethanol desired by the stakeholders are not only related to physical and chemical properties of the products as specified by technical standards but also such sustainability factors as environmental, social, and economic performances. These characteristics depend on several types of actors: public authorities define the minimum sustainability requirements if they setup mandates and develop incentives; potential intermediate purchasers may influence the sustainability characteristics beyond the minimum requirements level; consumers may express a willingness to pay for additional value; particular uses of the product may be prioritized by the consumers which result in certain

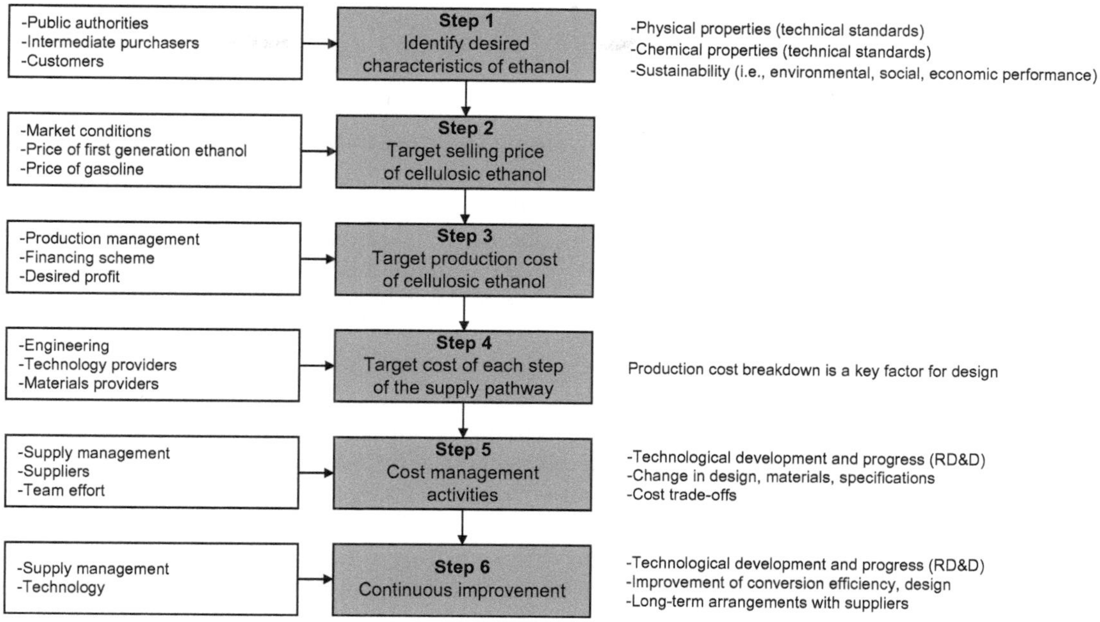

FIGURE 7 Target costing of lignocellulosic ethanol pathways *(modified from Ellram, 1999).*

values. These grounds may evolve in the future with the evolution of societal values and public regulation. In that sense, the comparison between lignocellulosic ethanol and gasoline must not be based only on heating values.

4.2.2 Step 2: Target Selling Price of Lignocellulosic Ethanol

With respect to step 1, the definition of the future selling price is not straightforward. A common practice is to consider as reference selling price either the market price of the first-generation bioethanol or the price of gasoline. If lignocellulosic bioethanol is considered as a distinct product compared to certain first-generation bioethanol types, the question whether it could be marketed as a distinct product is relevant. With the increase of the market share of ethanol, its price will be more and more correlated with the price of gasoline which in turn is volatile due the demand/supply of petroleum and refined products.

4.2.3 Step 3: Target Cost of Lignocellulosic Ethanol

Once the desired profit level is decided by the management, the overall allowable cost is estimated as price minus profit. The level of profit depends on the financing scheme. For technoeconomic evaluation, it is often assumed a 100% equity financing and a certain discount rate that results in a maximum allowable cost given the assumed price.

4.2.4 Step 4: Target Cost of Each Step of the Supply Pathway

Based on pieces of information gathered from engineering and potential materials and technology providers, the cost of each area is estimated. Apportioning the overall allowable cost into detailed costs of areas and subareas is the core of the TC approach. Each detailed cost is then a key factor for design, and material and equipment bill negotiation with the providers.

4.2.5 Step 5: Cost Management Activities

Distribution of the overall allowable cost among the areas and subareas in order to define target costs requires several cost management activities for targets to be robust enough. Long-term involvement of the stakeholders, particularly making the supply reliable and the suppliers faithful to the ethanol industry is one of the concerns of the cost management activities. Cost management at different areas and subareas in order to match the overall allowable cost is an integral part of the TC process. VE may be integrated in this step in order to conciliate cost allowance and cost targets.

4.2.6 Step 6: Continuous Improvement

In the course of the RD&D of lignocellulosic ethanol, information and knowledge are available with time. Development of new knowledge is liable to improve conversion efficiency and then reduce the process inputs for the same output. Efficient markets' structures of the technologies inputs and outputs, public accountability, long-term arrangements with the potential suppliers and customers, and new efficient designs are susceptible to reduce and stabilize cost and thus promote investment in the development of lignocellulosic ethanol.

5 CURRENT ECONOMIC EVALUATION OF LIGNOCELLULOSIC BIOETHANOL: SOME LIMITATIONS

Current practices of technoeconomic evaluation of lignocellulosic ethanol as they appear in scientific papers are rarely in full accordance with TC or VE approaches. So are practical cases of future commercial ethanol plants for which theoretical TC and VE are viewed as heavy processes. Even when applied, existing management cost systems show some drawbacks in the case of lignocellulosic ethanol where the resources are as important as the technology and values more relevant than market prices.

5.1 Accounting for the Competition Between Different Uses of Resources

Lignocellulosic ethanol is often treated as a product of an integrated system from feedstock production to the use of the produced ethanol. Therefore, the technical aspects of the supply chain are prioritized compared to the actors along the pathway. That way of assessment neglects the potential competition for resources. Complementary to VE, research on Value resources should be undertaken in order to identify the main uses which will compete with lignocellulosic ethanol for resources and how their markets would develop.

5.2 The Value of Lignocellulosic Resources

In the medium- to long-term, lignocellulosic resources can be used for energy production but also for chemicals and materials. Competition for resources is concerned with various conversion technologies including both energy and nonenergy uses. Using MARKAL, Gielien et al. (2000) study the optimal assessment of biomass uses in Western Europe for reducing GHG, by comparing energy production with materials applications. They conclude that the main substitution to fossil feedstocks will occur in transportation fuels, petrochemicals, and electricity generation. Although this approach mainly results in global scenarios which depend on the specifications of the objective function and constraints, competition between biomass applications will determine the delivery cost of biomass feedstocks. In a biomass-constrained case, facing several sales opportunities with different levels of willingness to pay, lignocellulosic feedstock producers will sell according to the expected maximum benefit based on opportunity cost. Thus, for a particular use, say lignocellulosic ethanol, the biomass procurement cost will not depend only on the cost of biomass activities but also on the comparative willingness to pay by biomass purchase competitors.

5.3 The Value of Lignocellulosic Ethanol and Coproducts

In the U.S. biomass program, the value of lignocellulosic ethanol is estimated as 65% of gasoline market price. Such modeling choice is acceptable as long as bioethanol is supposed to be used as pure ethanol or in a high blend rate with gasoline and providing that such characteristics as GHG emission reduction, renewability, absence of competition with food and feed, and lack of direct and indirect land-use impacts are not taken into account neither by the market nor by the public authorities. Full awareness of those characteristics by the customers implies a higher willingness to pay for sustainable lignocellulosic ethanol compared to another less sustainable ethanol. Public authorities can also use specific policy instruments such as feed-in tariff in order to stabilize the reference value of sustainable lignocellulosic ethanol and foster the investments. The issue of lignocellulosic ethanol value can be generalized to that of coproducts when established fossil-based markets exist. High value-added coproducts contribute to the competitiveness of bioethanol.

5.4 The Value of Intermediate Products such as Monosaccharides

Depending on existing markets, the value of intermediate products such as monosaccharides can be estimated based on the willingness to pay for various potential alternative products. That value is termed shadow price of intermediate products (Gnansounou et al., 2005). Estimation of the shadow prices allows evaluating the producer's willingness to sell lignocellulosic ethanol and his willingness to pay for feedstock.

5.5 Economic Evaluation Based on the Value Chain

Sustainable lignocellulosic ethanol is a specific product, the value of which should be estimated adequately. Given the nonintegration of feedstock delivery, conversion to ethanol, distribution and use segments, the supply chain must be evaluated using a value-based approach. Current market price projections cannot suitably consider the distinctive characteristics of sustainable lignocellulosic ethanol. Practical application of a value-based approach, however, needs to consider the specific environment of the ethanol plant.

6 CONCLUSION

While demonstration activities on cellulosic bioethanol in North America (i.e., USA and Canada) are concerned with both thermochemical and biochemical pathways, cellulosic bioethanol in Europe is mostly limited to the biochemical routes. Demonstration projects in Europe include those of Abengoa (Spain), BioGasol (Denmark), Inbicon (Denmark), M&G/Chemtex (Italy), Procethol 2G/Futurol (France), and SEKAB (Sweden). Other companies such as Novozymes, Danisco, or Syngenta are also supporting major efforts to develop cellulosic ethanol. Demonstration cellulosic bioethanol projects in the United States and Canada are even more numerous and varied. Most of the major actors have opted for the biochemical pathway, with either enzymatic hydrolysis (e.g., Abengoa, Inbicon, KL Energy, Mascoma, POET, QTeros, Verenium) or acid hydrolysis (e.g., BlueFire Ethanol). Thermochemical projects include those of Enerkem in Canada, Range Fuels and Coskata in the United States. Although cellulosic ethanol efforts are still in the research phase in other countries, significant work is underway (e.g., Praj Industries and Mission New Energy in India, Petrobras in Brazil). A notable R&D effort is also underway in Australia.

The review undertaken in this chapter raises the following issues and findings: the contribution of biomass cost to the overall production cost of lignocellulosic bioethanol proves to be one of the most significant; the standard production cost estimation should be replaced by an approach which makes use of VE, Value-resource, and TC; due to the complexity of the technoeconomic evaluation of lignocellulosic ethanol, the perceived risks by private investors will be high. Strategies to decrease these risks include promoting such projects as integration of second-generation with first-generation bioethanol and thus use existing residues and share equipments. Sugarcane bagasse is particularly concerned with such a strategy. Lignocellulosic biorefineries that aim at decreasing the production cost of bioethanol will be attractive only if the perceived risks by the investors are affordable. Low-risk profile biorefineries with stable product markets would be preferred to complex schemes with a high diversity of coproducts whose uncertainty would make profitability highly risky.

References

Aden, A., 2008. Biochemical production of ethanol from corn stover: 2007 state of technology model. National Renewable Energy Laboratory (NREL) NREL Report TP-510-43205.

Aden, A., Foust, T., 2009. Technoeconomic analysis of the dilute sulfuric acid and enzymatic hydrolysis process for the conversion of corn stover to ethanol. Cellulose 16 (4), 535–545.

Aden, A., Ruth, M., Ibsen, K., Jechura, J., Neeves, K., Sheehan, J., et al., 2002. Lignocellulosic biomass to ethanol process design and economics utilizing co-current dilute acid prehydrolysis and enzymatic hydrolysis for corn stover. National Renewable Energy Laboratory (NREL), NREL Report TP-510-32438.

Badger Engineers, Inc, 1987. Economic feasibility of an acid-hydrolysis-based ethanol plant. Subcontract Report SERI-STR-231-3142, prepared under subcontract ZX-3-03096-2, Cambridge, MA.

Chem Systems, Inc, 1987. Economic feasibility study of an enzymatic hydrolysis-based ethanol plant with prehydrolysis pretreatment. Subcontract Report SERI-STR-231-3135, prepared under subcontract XX-3-03097-1, Tarrytown, NY.

Cooper, R., Slagmulder, R., 1997. Target Costing and Value Engineering. Productivity Press.

Dekker, H., Smidt, P., 2003. A survey of the adoption and use of target costing in Dutch firms. Int. J. Prod. Econ. 84 (3), 293–305.

EIA, 2009. Annual Energy Outlook 2009. Energy Information Administration, Office of Integrated Analysis and Forecasting. U.S. Department of Energy, Washington, USA.

Ellram, L.M., 1999. The Role of Supply Management in Target Costing. Center for advanced Purchasing Studies, Tempe, AZ.

Ellram, L.M., 2006. The implementation of Target Costing in the United States: theory versus practice. J. Suppl. Chain Manag., 42 (1), 13–26, February 2006.

Feil, P., Yook, K.H., 2004. Japanese target costing: a historical perspective. Int. J. Strategic Cost Manag. (Spring 2004), 2 (4), 10–19.

Galbe, M., Sassner, P., Wingren, A., Zacchi, G., 2007. Process engineering economics of bioethanol production. Adv. Biochem. Eng./Biotechnology 108, 303–327.

Gielien, D.J., de Feber, M.A.P.C., Bos, A.J.M., Gerlagh, T., 2000. Biomass for energy or materials?: a Western European systems engineering perspective. Energy Policy 29 (4), 291–302.

Gnansounou, E., 2010. Production and use of lignocellulosic bioethanol in Europe: current situation and perspectives. Bioresour. Technol. 101 (13), 4842–4850.

Gnansounou, E., Dauriat, A., 2010. Techno-economic analysis of lignocellulosic ethanol: a review. Bioresour. Technol. 101 (13), 4980–4991.

Gnansounou, E., Dauriat, A., Wyman, C.E., 2005. Refining sweet sorghum to ethanol and sugar: economic trade-offs in the context of North China. Bioresour. Technol. 96 (9), 985–1002.

Hamelinck, C.N., 2004. Outlook for advanced biofuels. PhD thesis, Department of Science, Technology and Society and the Copernicus Institute for Sustainable Development and Innovation of Utrecht University.

Hamelinck, C.N., van Hooijdonk, G., Faaij, A.P.C., 2005. Ethanol from lignocellulosic biomass: techno-economic performance in short-, middle-, and long-term. Biomass Bioenergy 28, 384–410.

Humbird, D., Aden, A., 2008. Biochemical production of ethanol from corn stover: 2008 state of technology model. National Renewable Energy Laboratory (NREL), NREL Report TP-510-46214.

Ibusuki, U., Kaminski, P.C., 2007. Product development process with focus on value engineering and target-costing: a case study in an automotive company. Int. J. Prod. Econ. 105 (2), 459–474.

Kabir Kazi, F., Fortman, J., Anex, R., Kothandaraman, G., Hsu, D., Aden, A., et al., 2010. Techno-economic analysis of biochemical scenarios for production of cellulosic ethanol. National Renewable Energy Laboratory (NREL), NREL Report TP-6A2-46588.

Kato, Y., 1993. Target costing support system: lessons from leading Japanese companies. Manag. Account. Res. 4, 33–47.

Klein-Marcuschamer, D., Oleskowicz-Popiel, P., Simmons, B.A., Blanch, H.W., 2010. Technoeconomic analysis of biofuels: a wiki-based platform for lignocellulosic biorefineries. Biomass Bioenergy 34, 1914–1921.

Kuijvenhoven, J., 2006. Process design for an integrated lignocellulose to bioethanol production plant. In: Presentation at Netherlands Process Technology Symposium, October 25, 2006.

Lee, J., 1997. Biological conversion of lignocellulosic biomass to ethanol. J Biotechnol. 56 (1), 1–24.

Londo, H.M., Lensink, S.M., Deurwaarder, E.P., Wakker, A., de Wit, M.P., Junginger, H.M., et al., 2008. Biofuels development in the EU27+ until 2030: full-chain cost assessment and implications of policy options. REFUEL Project, WP4 final report.

Lynd, L.R., van Zil, W.H., McBride, J.E., Laser, M., 2005. Consolidated bioprocessing of cellulosic biomass: an update. Curr. Opin. Biotechnol. 16 (5), 577–583.

Maki, M., Leung, K.T., Qin, W., 2009. The prospects of cellulase-producing bacteria for bioconversion of lignocellulosic biomass. Int. J. Biol. Sci. 5, 500–516.

Pandey, A., Selvakumar, P., Soccol, C.R., Nigam, P., 1999. Solid-state fermentation for the production of industrial enzymes. Curr. Sci. 77, 149–152.

Phillips, S., Aden, A., Jechura, J., Dayton, D., Eggeman, T., 2007. Thermochemical ethanol via indirect gasification and mixed alcohol synthesis of lignocellulosic biomass. National Renewable Energy Laboratory (NREL), Technical Report NREL/TP-510-41168.

Sassners, P., Galbe, M., Zacchi, G., 2008. Techno-economic evaluation of bioethanol production from three different lignocellulosic materials. Biomass Bioenergy 32 (5), 422–430.

Seabra, J.E.A., Tao, L., Chum, H.L., Macedo, I.C., 2010. A techno-economic evaluation of the effects of centralized cellulosic ethanol and co-products refinery options with sugarcane mill clustering. Biomass and Bioenergy, 34 (8), 1065–1078.

Stone & Webster Engineering Corp., 1987. Economic feasibility of an enzyme-based ethanol plant. Subcontract Report SERI-STR-231-3138, prepared under subcontract ZX-3-03097-1, Boston, MA.

Tolan, J.S., Foody, B., 1999. Cellulase from submerged fermentation. Adv. Biochem. Eng./Biotechnology 65, 41–67.

U.S. DOE, 2004. Biomass Feedstock Composition and Property Database. Biomass Program, Energy Efficiency and Renewable Energy (EERE). U.S. Department of Energy (DOE) Accessed on Oct. 4, 2009.

U.S. DOE, 2005. Multi-Year Program Plan 2007-2012. Biomass and Biorefinery System R&D. Biomass Program, Energy Efficiency and Renewable Energy. U.S. Department of Energy.

U.S. DOE, 2007. Biomass Multi-Year Program Plan. Office of the Biomass Program, Energy Efficiency and Renewable Energy. U.S. Department of Energy.

U.S. DOE, 2009. Biomass Multi-Year Program Plan (MYPP). Office of the Biomass Program, Energy Efficiency and Renewable Energy. U.S. Department of Energy.

U.S. DOE, 2010. Biomass Multi-Year Program Plan (MYPP). Office of the Biomass Program, Energy Efficiency and Renewable Energy. U.S. Department of Energy.

Wijewardena, H., de Zoysa, A., 1999. A comparative analysis of management accounting practices in Australia and Japan: an empirical investigation. Int. J. Account. 34 (1), 49–70.

Wingren, A., Galbe, M., Zacchi, G., 2003. Techno-economic evaluation of producing ethanol from softwood: comparison of SSF and SHF and identification of bottlenecks. Biotechn. Progr. 19, 1109–1117.

Wingren, A., Galbe, M., Zacchi, G., 2008. Energy considerations for a SSF06-based softwood ethanol plant. Bioresour. Technol. 99 (7), 2121–2131.

Wooley, R., Ruth, M., Sheehan, J., Ibsen, K., Majdeski, H., Galvez, A., 1999a. Lignocellulosic biomass to ethanol process design and economics utilizing cocurrent dilute acid prehydrolysis and enzymatic hydrolysis: current and futuristic scenarios. National Renewable Energy Laboratory (NREL), NREL Report TP580-26157.

Wooley, R., Ruth, M., Glassner, D., Sheehan, J., 1999b. Process design and costing of bioethanol technology: a tool for determining the status and direction of research and development. Biotechnol. Progr. 15, 794–803.

CHAPTER

7

Pretreatment Technologies for Lignocellulose-to-Bioethanol Conversion

*E. Tomás-Pejó, P. Alvira, M. Ballesteros, M.J. Negro**

CIEMAT, Renewable Energy Division, Biofuels Unit Av. Complutese 22, 28040 Madrid

Tel: 0034913466056 Fax: 0034913460939

***Corresponding author: E-mail: mariajose.negro@ciemat.es**

1 INTRODUCTION

Nowadays, there is no doubt about the benefits of using renewable energies to diversify the energy sources and diminish petroleum dependence. Furthermore, renewable energies are inexhaustible, do not generate harmful residues, and are essential for reducing the greenhouse gases. Among other renewable energies, biomass shows additional benefits as it allows a certain grade of storage, favors the maintenance as well as development of agricultural and forest sectors, could imply energetic valorization of residues, and constitutes a realistic alternative for replacing fossil fuels in the transport sector.

About 98% of the fuels for transport come from petroleum, with negative consequences associated with supply security and CO_2 emissions (Gomez et al., 2008). During the last decades of the twentieth century, there has been an increasing interest in the production and use of liquid biofuels, either biodiesel (produced from oils and fats) or bioethanol (from sugar fermentation). These biofuels, obtained from biomass, are the only renewable products that can be easily integrated into the current fuel distributions systems, and they are one of the few alternatives for short-term diversification in the transportation sector.

Current production of fuel ethanol relies on bioethanol from sugars or starchy raw materials, but as those feedstocks are also employed for animal or human feed there has been much debate about its sustainability. In this context, lignocellulosic biomass is glimpsed as a key feedstock for bioethanol production because of its low cost, wide distribution, huge availability, and noncompetition with food crops.

Lignocellulosic materials, consisting mainly of cellulose, hemicellulose, and lignin, need to be hydrolyzed to monomeric sugars before being utilized by fermenting microorganisms. The preceding hydrolysis step can be performed through acid or enzymatic catalysts. In general, bioethanol production processes from lignocellulose based on enzymatic hydrolysis offer many more advantages than processes employing acids. While acid hydrolysis requires relatively high temperatures and implies corrosive operating conditions and generation of toxic compounds, enzymatic hydrolysis is advantageous due to its higher conversion efficiency and lower process energy requirements (Ballesteros, 2010). However, many physicochemical, structural, and compositional factors make the native lignocellulosic biomass recalcitrant and difficult to hydrolyze by enzymes. Thus, a previous pretreatment step is necessary to overcome these drawbacks and perform an efficient enzymatic hydrolysis. The aim of the pretreatment is to break down the lignin structure and disrupt the crystalline structure of cellulose to increase enzyme accessibility (Mosier et al., 2005). The mechanism for making the cellulose more accessible to enzymes depends on the pretreatment employed and nature of the raw material. While lignin is removed in ozonolysis, CO_2 explosion, and biological pretreatments, it is only redistributed in steam explosion and partially solubilized in liquid hot water (LHW). Hemicellulose is solubilized during wet oxidation and acid pretreatment; and mechanical comminution and ammonia fiber explosion (AFEX) have been shown to be good methods for reducing cellulose crystallinity.

Besides being considered an essential step in the biological conversion to ethanol, pretreatment has been described as one of the main economic costs in the process. In fact, it has been described as the second most expensive unit cost in the conversion of lignocellulose to ethanol based on enzymatic hydrolysis, preceded by feedstock cost (Merino and Cherry, 2007). It represents 33% of the total cost of the process which shows the necessity of developing efficient pretreatment technologies for reducing ethanol production cost (Lynd, 1996; Mosier et al., 2005).

The selection of an appropriate pretreatment determines the process configuration requirements for hydrolysis and fermentation as each step has a large impact on all subsequent stages. The chemistry of the pretreatment has a remarkable importance due to its impact on the global ethanol production process. Furthermore, pretreatment also affects the cost of the following operational steps, that is, downstream cost by determining fermentation toxicity, enzymatic hydrolysis rates, and enzyme loading as well as fermentation process variables. Figure 1 depicts the main interrelated factors of pretreatment, enzymatic hydrolysis, and fermentation in an ethanol production process from lignocellulose.

Focusing on the pretreatment step, sugar recovery yield, chip size required, and low energy demand have been described as decisive factors for an effective process (Banerjee et al., 2010; Yang and Wyman, 2008). These key properties necessary for a cost-effective pretreatment are included in Table 1.

It is widely known that harsh conditions during pretreatment lead to a partial hemicellulose and lignin degradation and generation of toxic compounds. Since these compounds are potentially inhibitors for yeasts, there are some strategies to diminish the impact of toxic compounds on the process: (i) removal of the inhibitors through some detoxification methods such as solvent extraction, anion exchange, overliming, and employment of zeolites or enzyme laccase; (ii) use of fermenting yeasts highly tolerant to the inhibitors or previously subjected to an adaptation procedure; and (iii) selecting an effective pretreatment minimizing

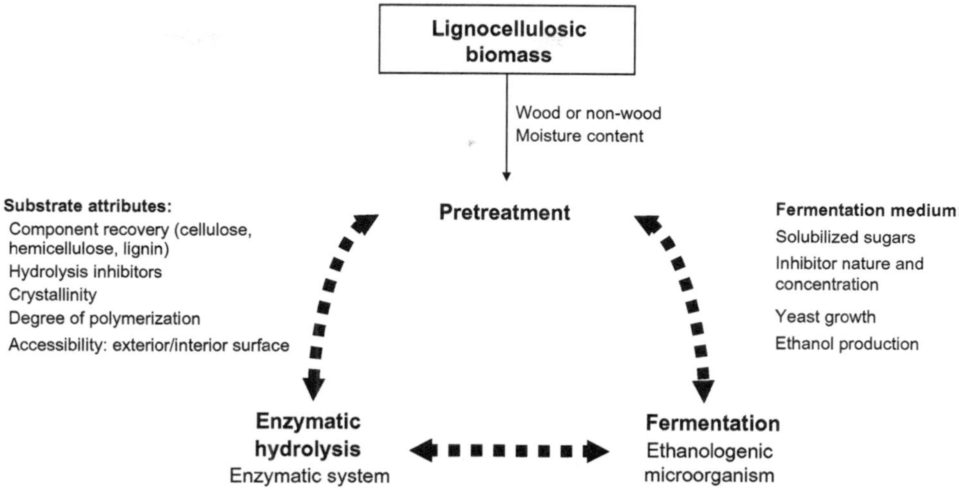

FIGURE 1 Interrelated factors between the main steps in an ethanol production process from lignocellulose.

TABLE 1 Key Factors for an Effective Pretreatment Method for Lignocellulosic Materials

Key Factors in an Effective Pretreatment
(1) High yields from multiple crops, sites ages and harvesting times.
(2) Solid fraction highly digestible
(3) No sugar degradation
(4) Low amount of toxic compounds
(5) Not requirement of size reduction
(6) Operation in reasonable size and moderate cost reactors
(7) Nonproduction of solid-waste residues
(8) Effectiveness at low moisture content
(9) Obtaining high sugar concentration
(10) Fermentation compatibility of the pretreated material
(11) Lignin recovery
(12) Minimum heat and power requirements

sugar degradation and inhibitor formation. Most detoxification methods only partially remove the toxic compounds and even imply a sugar loss. Furthermore, detoxification is an additional cost that can account for up to 22% of the total cost of the ethanol production (Von Sivers et al., 1994). Thus, an ideal pretreatment should increase the accessibility for enzymes while producing minimum concentration of inhibitory compounds that could affect the following hydrolysis and fermentation steps. The amount and nature of the inhibitory compounds is dependent on the raw material and on the chosen pretreatment which will be discussed in this chapter.

In the last years, pretreatment research has been focused on identifying, evaluating, developing, and demonstrating promising approaches that support the enzymatic hydrolysis of the pretreated biomass with lower enzyme dosages and shorter conversion times. Large number of pretreatment approaches have been investigated on a wide variety of feedstocks types, and there are several recent review articles which provide a general overview of the field (Alvira et al., 2010; Carvalheiro et al., 2008; Gírio et al., 2010; Hendriks and Zeeman, 2009; Taherzadeh and Karimi, 2008; Yang and Wyman, 2008).

This chapter summarizes the most novel and promising alternatives for an effective pretreatment of the lignocellulose for ethanol production.

2 TOXIC COMPOUNDS GENERATED DURING PRETREATMENT

As have been mentioned, severe conditions during pretreatment lead to generation of some toxic compounds that could affect the subsequent hydrolysis and fermentation steps.

The nature and concentration of the toxic compounds depend on the raw material (hardwood, softwood, herbaceous biomass, etc.), the pretreatment itself and conditions employed (temperature, residence time, pressure, pH, etc.) as well as the use of catalysts. Furthermore, owing to the variable nature of the raw materials and the different pretreatment methods, many degradation products cannot be identified accurately.

According to their origin, the degradation products can be divided into three groups: furan derivatives, weak acids, and phenolic compounds; all interfering in a different manner on enzymes and microorganisms. Their inhibition mechanism is not only based on the inhibitory effect caused by each compound individually but also on their interaction and synergy.

Since the generation of toxic compounds is closely related with the pretreatment technology, some pretreatments are known to release more inhibitors than others. Thus, the most common inhibitory compounds released from lignocellulose after different pretreatment technologies are depicted in Figure 2.

2.1 Furans

Among furan derivatives, 2-furaldehyde (furfural) and 5-hydroxymethylfurfural (HMF) constitute the main degradation compounds generated from pentoses and hexoses degradation, respectively. The concentration of these compounds depends mainly on the conditions employed for pretreatment. Thus, those pretreatments which employ acids as hydrolytic agents and utilize high temperature and time to reaction will produce furfural and HMF at higher levels (Wyman, 2007).

Most of the fermenting microorganisms are able to reduce furans to their corresponding less toxic alcohols. HMF is reduced to 2,5-*bis*-hydroxymethylfuran and furfural to furfuryl alcohol, and both could be also oxidized to formic acid under anaerobic conditions (Taherzadeh et al., 1999). If furans are present at high concentration, they exert an inhibitory effect interfering with glycolytic enzymes and synthesis of macromolecules provoking an enlarge of the lag phase and reducing the ethanol productivity (Almeida et al., 2007; Klinke et al., 2004). These effects depend on furan concentration but are highly related with the yeast strain.

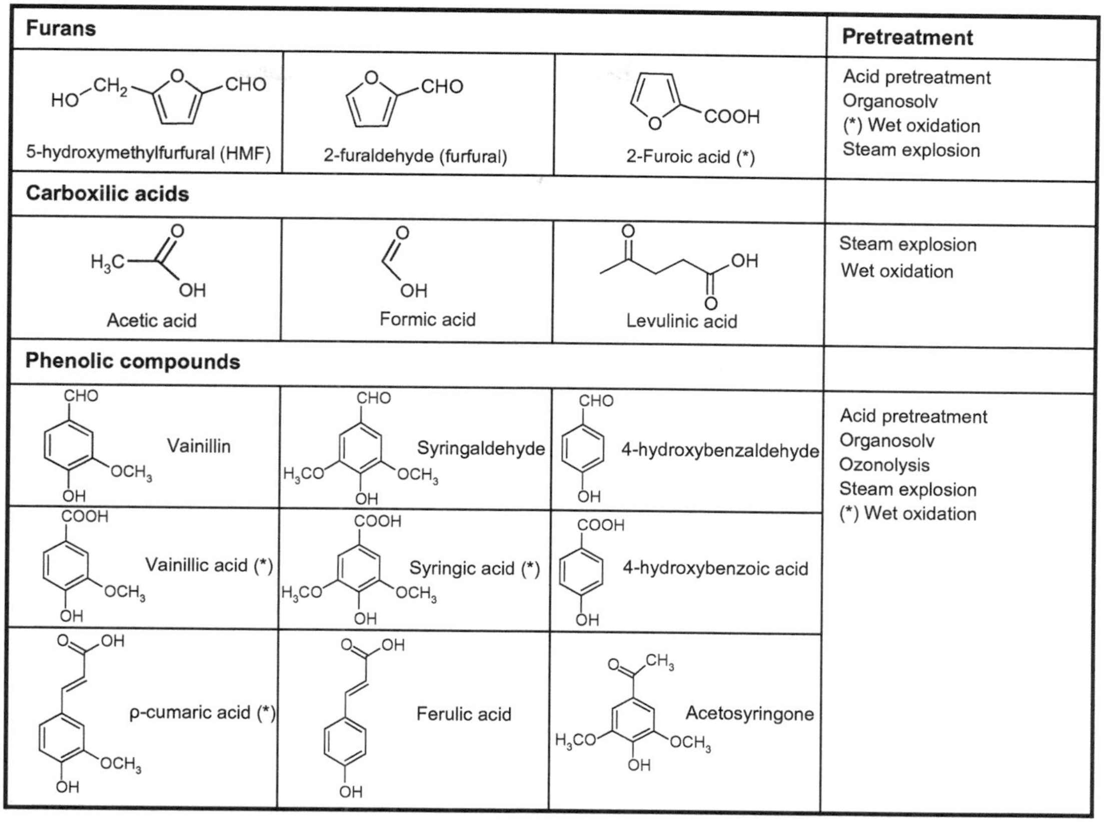

FIGURE 2 Main toxic compounds produced during different pretreatment technologies of lignocellulose.

2.2 Carboxylic Acids

Main carboxylic acids generated during pretreatment are acetic acid, produced from the acetyl groups in hemicelluloses, and formic acid, derived from furfural and HMF degradation. HMF could be also decomposed to levulinic acid being detected at lower concentration. Furthermore, hydroxycarboxylic acids such as glycolic acid and lactic acid are common degradation products from alkaline carbohydrate degradation (Klinke et al., 2004).

The undissociated form of weak acids can diffuse across the cell membrane and dissociate inside the cell due to the higher intracellular pH. This fact decreases intracellular pH which must be compensated by pumping protons out of the cell at expense of ATP. Thus, less ATP is available for biomass formation. Furthermore, if pumping capacity of the plasma membrane ATPase is overcome, acidification of cytoplasm and cellular death occur. Some studies have also reported that small amounts of acetic, levulinic, or formic acid could increase glucose consumption rates and ethanol yields because low concentration of acids stimulated the production of ATP(Almeida et al., 2007; Keating et al., 2006). The concentration of undissociated acids in lignocellulosic hydrolysates is dependent on the pH, and therefore pH control is necessary for minimizing acids toxicity.

2.3 Phenolic Compounds

Wide range of phenolic compounds derived from lignin decomposition is also generated during pretreatment. Identified phenols are monomers with an aliphatic substituent with different functional groups: aldehydes, ketones, or acids. Phenolic compounds are present in lower concentrations due to its minor solubilization. The concentration and type of phenolic compounds is highly dependent on the raw material since lignin content and chemical structure differ among the different lignocellulosic materials. The hydrolytic conditions during pretreatment are also very important for the functionality of the degradation products, that is, the phenolic aldehydes have been shown to be favored at oxidative acidic conditions (Klinke et al., 2002). After soda pulping wheat straw, phenols ρ-cumaric and ferulic acids are produced by the hydrolysis of esterified hemicellulose and lignin. Alkaline wet oxidation of wheat straw also produces cinnamic acid derivates. Furthermore, owing to oxidative cleavage of the conjugated double bonds, 4-hydroxybenzoic acid and vanillic acid are formed (Klinke et al., 2002).

Some other more abundant phenolic compounds are 4-hydroxybenzaldehyde, vanillin, synringaldehyde, syringic acid, and cathecol. These compounds are toxic because they affect the integrity of biological membranes (Almeida et al., 2007).

In general, it is accepted that there is a high amount of degradation products derived from lignin that remain unidentified.

3 PRETREATMENT PROCESSES

The pretreatment is a crucial step to alter structural characteristics of biomass increasing cellulose and hemicellulose accessibility to enzymes. The effectiveness of the pretreatment to improve the enzymatic hydrolysis has been attributed to a modification in the degree of polymerization and crystallinity index (Kumar and Wyman, 2010; Mansfield et al., 1999), to a disruption of the lignin-carbohydrate linkages (Laureano-Perez et al., 2005), to lignin and hemicelluloses removal (Pan et al., 2005) and to an increase of the porosity of the material (Chandra et al., 2007). Depending on pretreatment choice, the mechanism responsible for pretreatment effectiveness would be different.

During the last decades, a large number of diverse pretreatment technologies have been suggested. Those methods are usually classified into biological, physical, chemical, and physicochemical pretreatments.

3.1 Physical Pretreatments

3.1.1 Mechanical Comminution

The development of environment-friendly pretreatment such as milling or grinding that do not involve harmful residues and generation of degradation products has been widely studied.

Mechanical comminution pretreatment is used to reduce the particle size and crystallinity of lignocellulose in order to increase the specific surface area and reduce the degree of polymerization. This effect can be obtained by a combination of chipping, grinding, or milling depending on the final particle size of the material (10-30 mm after chipping and 0.2-2 mm after milling or grinding) (Sun and Cheng, 2002).

Different milling processes can be used to improve the enzymatic hydrolysis of lignocellulosic materials. For example, a new type of wet disk milling with lower energy consumption has been used for pretreating herbaceous biomass such as rice straw showing higher hydrolysis yields (glucose and xylose) than common dry milling besides (Hideno et al., 2009). Furthermore, mechanical pretreatments such as ball milling have been integrated in SSF processes for ethanol production from sugarcane bagasse with *Pichia stipitis* (Buaban et al., 2010).

The power requirement of this pretreatment is relatively high depending on the final particle size and the biomass characteristics. Particularly, the strong structure of forest biomass makes its size reduction very energy intensive and conducting some kind of chemical pretreatment prior to wood size reduction is appearing as an alternative (Zhu et al., 2010).

3.1.2 *Extrusion*

Size reduction is one of the most effective methods for increasing the enzymatic accessibility to lignocellulose. However, many of the physical methods for size reduction (milling, grinding, etc.) are not economically feasible because a very high-energy input is required. In this context, extrusion is a novel and promising physical pretreatment method for biomass conversion to ethanol production. In extrusion, materials are subjected to heating, mixing, and shearing, resulting in physical and chemical modifications during the passage through the extruder. The extruder has many advantages such as the ability to provide high shear, rapid heat transfer, and effective and rapid mixing (Karunanithy and Muthukumarappan, 2010a).

Screw speed and barrel temperature are believed to disrupt the lignocellulose structure causing defibrillation, and shortening of the fibers, and in the end, increasing accessibility of carbohydrates to enzymatic attack (Karunanithy et al., 2008a,b).

Because of its adaptability to many different process modifications such as the addition of chemicals or removal of materials, and the application of high pressure and expansion treatment (using steam or other solvents), extrusion has the potential to become an interesting option to pretreat lignocellulose. It has been recently employed for increasing the enzymatic hydrolysis yields of switchgrass (Karunanithy and Muthukumarappan, 2010a), corn stover (Karunanithy and Muthukumarappan, 2010b), wheat bran, and soybean hull (Lamsal et al., 2010).

3.2 Chemical Pretreatments

3.2.1 *Acid Pretreatment*

Acid pretreatments employ acids as catalysts which have stronger effect on hemicellulose and lignin than on crystalline cellulose. Its main objective is to solubilize the hemicellulosic fraction of the biomass making the cellulose more accessible to enzymes.

Acid catalyzed processes can be classified into two groups, treatments with concentrated acids or with diluted acids. However, the utilization of concentrated acids is less attractive for ethanol production due to the higher formation of inhibiting compounds. Furthermore, equipment corrosion problems and acid recovery are important drawbacks when using concentrated acid pretreatments. So, diluted acid pretreatment appears as a more favorable method than concentrated acid pretreatment for industrial applications and has been extensively studied for pretreating wide range of lignocellulosic raw materials. In this context, diluted acid pretreatment has been considered as candidate for large-scale bioethanol production. It can be performed at high temperature (e.g., 180 °C) during a short period of time, or at lower temperature (e.g., 120 °C) for longer retention time (30-90 min). It presents the advantage of solubilizing hemicellulose, mainly xylan, but also converting solubilized hemicellulose to fermentable sugars. Hemicellulose solubilization could avoid the addition of hemicellulases during the enzymatic hydrolysis.

Depending on the process temperature, some sugar degradation compounds such as furfural, HMF, and aromatic lignin degradation compounds are detected, and affect the microorganism metabolism in the fermentation step (Saha et al., 2005). Anyhow, this dilute acid pretreatment generates lower degradation products than concentrated acid pretreatments.

High enzymatic hydrolysis yields have been reported when pretreating lignocellulosic materials with diluted H_2SO_4 which is the most studied acid although HCl, H_3PO_4 and HNO_3 have also been tested (Mosier et al., 2005). Hydrolysis yield as high as 74% was shown when wheat straw was subjected to 0.75% v/v of H_2SO_4 at 121 °C for 1 h (Saha et al., 2005). Olive tree biomass was pretreated with 1.4% H_2SO_4 at 210 °C resulting in 76.5% of hydrolysis yields (Cara et al., 2008) and 92% of the theoretical maximum hydrolysis yield was obtained in enzymatic saccharification experiments from other woody biomass pretreated at 180 °C for 75 min with 2.75% H_2SO_4 (Ferreira et al., 2010). The improved enzymatic hydrolysis was reflected in an ethanol yield as high as 0.47 g/g glucose in fermentation tests with cashew apple bagasse pretreated with diluted H_2SO_4 at 121 °C for 15 min (Rocha et al., 2009).

Organic acids such as maleic, fumaric, or even acetic acid have been suggested as alternatives to inorganic acids. Organic acids do not promote degradation reactions that have been described in acid pretreatments, resulting in lower concentration of toxic compounds. Both maleic and fumaric acids have been compared with H_2SO_4 in enzymatic hydrolysis yields from wheat straw. Results showed than organic acids can pretreat wheat straw with high yields although fumaric acid was less effective than maleic acid. Furthermore, less amount of furfural was formed in the maleic and fumaric acid pretreatments than in H_2SO_4 pretreatment (Kootstra et al., 2009).

3.2.2 *Alkali Pretreatment*

The effect that some alkalis have on lignocellulosic biomass is the basis of alkaline pretreatments that can be performed at room temperature and residence times ranging from seconds to days. Alkali pretreatments increase cellulose digestibility and they are more effective for lignin solubilization, exhibiting less effect on cellulose and hemicellulose than acid or hydrothermal processes (Carvalheiro et al., 2008). This technology is effective depending on the lignin content of the biomass.

It is described to cause less sugar degradation than acid pretreatment, and it was shown to be more effective on agricultural residues than on wood materials (Kumar et al., 2009). Nevertheless, possible loss of fermentable sugars and some production of inhibitory compounds must be taken into consideration to optimize the pretreatment conditions.

NaOH, KOH, $Ca(OH)_2$, and NH_4OH are suitable alkaline pretreatments. NaOH causes swelling, increasing the internal surface of cellulose and decreasing the degree of polymerization and crystallinity, which also provokes lignin structure disruption (Taherzadeh and Karimi, 2008). NaOH has been reported to increase hardwood digestibility from 14% to 55% by reducing lignin content from 24-55% to 20% (Kumar et al., 2009). Furthermore, pretreated switchgrass revealed a great deal of pore formation in the NaOH pretreatment increasing the accessible surface area to the enzymes as well as decreasing lignin content (Nlewem and Thrash, 2010).

Normally, alkaline pretreatments are conducted at room or elevated temperatures, but recently cold NaOH solutions or NaOH/urea solutions have been employed. Raw plant fibers and cotton cellulose have been treated with NaOH at −5 °C and NaOH/urea at −20 °C, respectively. Furthermore, a novel approach in which ball-milled bamboo samples were subjected to ultrasound irradiation and NaOH/urea pretreatment at −12 °C showed an effective disruption of the recalcitrance of bamboo generating higher reactive cellulose (Li et al., 2010a).

$Ca(OH)_2$, known as lime, also removes acetyl groups from hemicellulose reducing steric hindrance of enzymes and enhancing cellulose digestibility (Mosier et al., 2005). This effect has been observed for enzymatic hydrolysis with corn stover (Kim and Holtzapple, 2006) or poplar wood (Chang et al., 2001) in which lime has been proven successfully at temperatures from 85 to 150 °C and for 3-13 h. To produce bioethanol with lime pretreatment, it is necessary to reduce pH as well as to separate the solid fraction to remove the alkali. However, solid fraction separation is not interesting owing to the significant amounts of fermentable sugars present in the liquid fraction. Novel lime pretreatment so-called calcium capturing by carbonation (CaCCO) in which lime is precipitated by carbonation has been studied (Mosier et al., 2005; Park et al., 2010). Pretreatment with lime has lower cost and less safety requirements compared to NaOH or KOH pretreatments.

Addition of an oxidant agent (oxygen/H_2O_2) to alkaline pretreatment (NaOH/$Ca(OH)_2$) can improve the performance by favoring lignin removal (Carvalheiro et al., 2008). Saccharification yields as high as 90-95% have been obtained in sorghum straw enzymatic hydrolysis (McIntosh and Vancov, 2010). Improvements on enzymatic hydrolysis have been also reflected in high ethanol production in simultaneous saccharification and cofermentation (SSCF) from wheat straw pretreated with diluted alkali (Saha and Cotta, 2006). Furthermore, it is remarkable the fact that no furfural or HMF are detected in hydrolysates obtained with alkaline peroxide pretreatment which favors the fermentation step in ethanol production processes (Taherzadeh and Karimi, 2008).

3.2.3 *Organosolv*

The organosolv pretreatment uses organic or aqueous solvents (ethanol, methanol, ethylene glycol, acetone, glycerol, tetrahydrofurfuryl alcohol, etc.) to extract lignin and provide more accessible cellulose. The organic solvent is mixed with water in various portions, added to the biomass and then heated to temperatures ranging 100-250 °C. Typically, acids (HCl, H_2SO_4, oxalic, or salicylic) can also be added as catalysts if the process is conducted at temperatures below 185-210 °C (Nahyun et al., 2010). Furthermore, in case of adding acid catalysts, the rate of delignification is increased and higher xylose yields are obtained (Zhao et al., 2009).

The main fractions obtained after pretreating biomass are: (i) cellulosic fibers; (ii) solid lignin, obtained after removal of the volatile solvent; and (iii) liquid solution of hemicellulosic sugars, mainly xylose.

Removal of solvents from the system is necessary using appropriate extraction and separation techniques, for example, evaporation and condensation. Solvents need to be separated because they might be inhibitory to enzymatic hydrolysis and fermentative microorganisms (Sun and Cheng, 2002). The high commercial price of solvents is another important factor to consider for industrial applications; thus, they should be recycled to reduce operational costs. For economic reasons, among all possible solvents, low-molecular weight alcohols with lower boiling points such as ethanol and methanol are favored.

Organosolv pretreatment produces a highly digestible cellulose substrate from almost all kind of raw materials, and lignin with the potential of high-value utilization can be recovered after pretreatment. Other benefit of organosolv pretreatment is that lignin removal minimizes the absorption problems of cellulolytic enzymes to lignin which is reflected in lower enzyme dosages requirements.

One of the drawbacks when employing organosolvents is related with the significant amount of furfural, HMF, and soluble phenols from lignin in the prehydrolysate obtained after pretreatment (Gírio et al., 2010; Zhu and Pan, 2010).

3.2.4 *Ozonolysis*

Ozone is an oxidizing agent that shows high delignification efficiency (Shatalov, 2008). Ozonolysis is usually performed at atmospheric conditions, room temperature, and normal pressure. Its effect is mainly limited to lignin, hemicellulose is slightly affected, and cellulose is not. Thus, the amount of degradation compounds derived from hemicellulose and cellulose is very low. Notwithstanding, ozone could react with lignin-based aromatic compounds generating some lignin-derived degradation products.

Ozone has been used to pretreat numerous lignocellulosic raw materials such as wheat straw and rye straw (García-Cubero et al., 2009), cotton straw (Silverstein et al., 2007), bagasse, and poplar among others (Kumar et al., 2009). Despite some interesting results, further research has to be performed regarding ethanol production from lignocellulosic materials pretreated with ozone. An important drawback to consider is the large amounts of ozone needed, which can make the process economically unviable.

3.2.5 *Ionic Liquids (ILs)*

The use of ILs as solvents for pretreatment of cellulosic biomass has received much attention during the last decade (Olivier-Bourbigou et al., 2010). They are capable to break down the extensive hydrogen-bonding network in the polysaccharides and promote its solubilization. ILs are salts, typically composed of large organic cations and small inorganic anions, which exist as liquids at relatively low temperatures, often at room temperature. Several imidazolium-based ILs were originally reported as good methods to dissolve large amounts of cellulose (Swatloski et al., 2002). The notable characteristics of ILs are their thermal and chemical stability, nonflammability, wide liquid temperature range, and good solvating properties for various types of materials (Hayes, 2009). Their solvent properties can be varied by adjusting the anion and the alkyl constituents of the cation. Since no toxic or explosive gases are formed, ILs are called "green" solvents.

Carbohydrates and lignin can be simultaneously dissolved in ILs with anion activity because ILs form hydrogen bonds between the nonhydrated chloride ions and the sugar hydroxyl protons in a 1:1 stoichiometry. As a result, the intricate network of noncovalent interactions among biomass polymers of cellulose, hemicellulose, and lignin is effectively disrupted while minimizing formation of degradation products. Although most available data showing the effectiveness of ILs have been developed using pure crystalline cellulose, recent studies have demonstrated that ionic ILs can be used to pretreat lignocellulosic biomass such as bagasse (Dadi et al., 2006), wheat straw (Li et al., 2009), or wood (Lee et al., 2009). Appropriated solvents for lignocellulosic material are 1-ethyl-3-methylimidazolium acetate and 1-allyl-3-metilimidazolium chloride (Mäki-Arvela et al., 2010).

Some authors reported that IL pretreatment of switchgrass significantly improved the enzymatic saccharification of both cellulose (96% glucose yield in 24 h) and xylan (63% xylose yield in 24 h) (Zhao et al., 2010). This improvement was attributed to the reduction in cellulose crystallinity and the delignification effect during dissolution-regeneration steps.

Other study showed a promising combined method for rice straw pretreatment using ILs and ammonia which recovered 82% of the cellulose with 97% of the glucose conversion, significantly higher than the individual ammonia or ILs treatments (Nguyen et al., 2010). The application of the synergic effect of ammonia and IL in the combined method significantly enhanced pretreatment efficiency by simplifying the sample communition, reducing the processing time for solubilization, using less enzyme amount for hydrolysis, and increasing the ILs recycling.

In a pretreatment method using 1-ethyl-3-methyl imidazolium diethyl phosphate, the yield of reducing sugars from wheat straw pretreated with this IL at 130 °C for 30 min was 54.8% after being enzymatically hydrolyzed for 12 h (Li et al., 2009). The fermentability of the hydrolysates obtained after enzymatic saccharification of the regenerated wheat straw was also evaluated showing no negative effect on the growth of *Saccharomyces cerevisiae* (Li et al., 2009).

For the large-scale application of ILs, development of energy-efficient recycling methods for ILs is a prerequisite and should be investigated in detail (Zavrel et al., 2009). Toxicity to enzymes and fermentative microorganisms must be also studied before ILs can be considered a real option for biomass pretreatment (Yang and Wyman, 2008; Zhao et al., 2010). Despite these current limitations, advanced research as potential synthesis of ILs from carbohydrates may play a role in reducing their cost. Development of ILs pretreatment could offer a great potential for future lignocellulose biorefinery processes.

3.3 Physicochemical Pretreatments

3.3.1 Wet Oxidation

Wet oxidation is an oxidative pretreatment method which employs oxygen or air as catalyst. When oxygen is not added, the process is similar to a hydrothermal pretreatment and comparable to the well-known steam explosion pretreatment.

Oxidative pretreatment is performed for 5-15 min at temperatures from 170 to 200 °C and at pressures from 10 to 12 bar O_2 (Kaparaju and Felby, 2010; Olsson et al., 2005). The addition of oxygen at temperatures above 170 °C makes the process exothermic reducing the total energy demand.

It has been proven to be an efficient method for solubilization of hemicelluloses and lignin. However, wet oxidation does not catalyze the hydrolysis of solubilized hemicellulose. In steam explosion and dilute acid pretreatments sugar monomers are produced, while in wet oxidation soluble sugars from hemicellulose are oligomers (Klinke et al., 2003).

Regarding toxic products generated during pretreatment, phenolic compounds are not end products during wet oxidation because they are further degraded to carboxylic acids, formic and acetic being the major degradation products. Phenol is more reactive than benzene due to the hydroxyl group that activates the aromatic ring by electron donation. Then, in wet oxidation, the phenol monomers are not end products but reaction intermediates. Furthermore, furfural and HMF production is lower during wet oxidation when compared to steam explosion or LHW methods. Addition of carbonate (Na_3CO_2) resulted in alkaline wet oxidation reducing even more the formation of toxics. Aldehydes are not stable under alkaline conditions where they undergo condensation reactions. Moreover, these are easily oxidized to carboxylic acids and only 2-furoic acid has been found among the furan derivatives in some wet oxidation pretreatments (Klinke et al., 2003).

An interesting feature of alkaline wet oxidations is that at alkaline conditions the formate ion is oxidized causing an increase in pH that helps to neutralize the carboxylic acids formed during the pretreatment and prevents the pH drop.

Pretreatment of wheat straw with Na_2CO_3, resulted in 96% recovery of the cellulose (65% converted to glucose) and 70% of hemicellulose (Klinke et al., 2002). High enzymatic hydrolysis yields have been also obtained after wet oxidation pretreatment of corn stover and spruce (Palonen et al., 2004). This technology has been widely used for ethanol production followed by simultaneous saccharification and fermentation (SSF) from corn stover (Varga et al., 2004), clover-ryegrass (Martin et al., 2008), or olive pulp (Haagensen et al., 2009).

Costs of oxygen and catalyst are considered one of the main disadvantages for wet oxidation development technologies.

3.3.2 *Microwave Pretreatment*

Microwave-based pretreatment combines both thermal and nonthermal effects generated in aqueous environment. The movement of ions and the vibration of polar molecules give rise to heat and extensive intermolecular collisions which accelerate chemical, physical, and biological processes. Compared to conduction/convection heating, which is based on supercritical heat transfer, microwave uses the ability of direct interaction between a heated object and an applied electromagnetic field to increase heat (Hu and Wen, 2008). Some of the advantages of employing microwave heating over conventional heating include reduction of process energy requirements, uniform and selective processing and capacity of starting and stopping the process instantaneously (Keshwani and Cheng, 2010). Furthermore, since the heat is generated internally via direct interaction between the electromagnetic field and components of the heated material, the heating is a faster process. When microwave is used to pretreat lignocellulose, it selectively heats the more polar part and this unique heating feature results in an improved disruption of the recalcitrant structures of lignocellulose. Regarding nonthermal effects, the electromagnetic field helps to accelerate the destruction of crystalline structures and changes the super molecular structure of lignocellulosic material improving its reactivity.

Microwave pretreatments were carried out at first by immersing the biomass in water, but recently the potential of different chemical reagents has been studied. Research evaluating different alkalis (Na_2CO_3, $Ca(OH)_2$, and NaOH) identified NaOH as the most effective reagent for switchgrass and coastal bermudagrass giving the highest total reducing sugars yields (Hu and Wen, 2008; Keshwani and Cheng, 2010). Alkali microwave pretreatment was also employed for pretreating rice straw and hulls in which case, results indicated a partially disruption of the lignin structure and more accessible cellulose to enzymes (Singh et al., 2010).

On the other hand, studies employing acetic and propionic acids for pretreating rice straw have shown those acids as good agents leading to swelling of cellulose, increasing the surface area and reducing its crystalline structure (Gong et al., 2010). Furthermore, when employing acids in combination with microwaves, hemicellulose degradation is enhanced.

The short length of the process as well as the low inhibitor production is reflected in high cost effectiveness. However, the feasibility of using a pretreatment method that involves microwave irradiation and chemicals in commercial scale is unknown, and it would be necessary to study the possibilities for performing the method in the future.

3.3.3 *Ultrasound Pretreatment*

Ultrasound, known as the mechanical waves at frequency above the hearing range for humans, has been employed in numerous biological and chemical processes. The effect of ultrasound on lignocellulosic biomass has been used for extracting hemicelluloses, cellulose, and lignin. It has been also concluded that ultrasound pretreatment may significantly increase the conversion of starch materials to glucose and therefore improve the ethanol yield in bioethanol production processes (Mielenz, 2001; Nikolić et al., 2010). However, less research has been addressed to study the hydrolysis performance of lignocellulosic materials pretreated with ultrasounds. In spite of the minor research, some researchers showed that saccharification corn stover and sugar cane bagasse were enhanced efficiently by ultrasonic pretreatment (Yachmenev et al., 2009).

Ultrasound waves produce cavitation and acoustic streaming in a liquid or slurry. Higher enzymatic hydrolysis yields after ultrasound pretreatment could be explained because cavitation effects caused by introduction of an ultrasound field into the enzyme processing solution greatly enhance the transport of enzyme macromolecules toward the substrate surface. Furthermore, mechanical impacts produced by the collapse of cavitation bubbles provide an important benefit of opening up the surface of solid substrates to the action of enzymes. In addition, the maximum effects of cavitation occur at 50 °C, which is the optimum temperature for many enzymes (Yachmenev et al., 2009).

3.3.4 *Liquid Hot Water*

LHW is a hydrothermal pretreatment that uses water at high pressures to maintain the liquid state at elevated temperatures (160-240 °C) and provoke alterations in the structure of the lignocellulose. It does not require any catalyst or chemical and usually involves temperatures of 150-230 °C for variable residence times from seconds to hours (Gírio et al., 2010; Hu et al., 2008). High variability on the pretreatment results is attributed to the different feedstocks.

During LHW pretreatment, most of the hemicellulose is solubilized, making the cellulose more accessible. Cellulose and lignin are not significantly affected and remain in the solid phase. Lignin is partially depolymerized and solubilized, but complete delignification is not possible by hot water alone, because of the recondensation of soluble components originated from lignin (Cara et al, 2007). Two-step pretreatment has been studied to optimize hemicellulosic sugars recovery and to enhance enzymatic hydrolysis yields. To avoid the formation of inhibitors, the pH should be kept between 4 and 7 because at this pH hemicellulosic sugars are retained in oligomeric form and monomers formation is minimized (Mosier et al., 2005).

LHW has been shown to remove up to 80% of the hemicellulose and to enhance the enzymatic digestibility of pretreated material in herbaceous feedstocks, such as corn stover (Mosier et al., 2001), sugarcane bagasse (Laser et al., 2002), and wheat straw (Pérez et al., 2007; Pérez et al., 2008).

In general, LHW pretreatments are attractive from a cost-savings potential: catalysts are not required and low corrosion allows the construction of low-cost reactors. It has also the major advantage that the solubilized hemicellulose and lignin products are present in lower concentrations, due to higher water input, and subsequently concentration of degradation products is reduced. In comparison to steam explosion, higher pentosan recovery and lower formation of inhibitors are obtained; however, water demanding in the process and energetic requirement are higher and it is not developed at commercial scale.

3.3.5 Ammonia Fiber Explosion

During the AFEX pretreatment, biomass is treated with liquid anhydrous ammonia at temperatures between 60 and 100 °C and high pressure for a variable period of time. After the residence time the pressure is released, vaporizing the ammonia and allowing its recovery and recycling. The ammonia has a marked effect on lignocellulose causing swelling and physical disruption of biomass fibers, partial decrystallization of cellulose, and breakdown of lignin-carbohydrates linkages (Chundawat et al., 2007; Laureano-Perez et al., 2005).

AFEX produces a solid pretreated material because during the pretreatment only a small amount of the material is solubilized and most of the biomass components remain in the solid fraction. Thus, since considerable hemicellulose is retained in the pretreated material, both cellulases and hemicellulases will be required in enzymatic hydrolysis process.

The AFEX pretreatment is more effective on agricultural residues and herbaceous crops, with limited effectiveness demonstrated on woody biomass and other high-lignin feedstocks (Wyman et al., 2005a). In general, early maturity grasses and agricultural residues require soft pretreatment conditions, while mature grasses and woody materials require more severe conditions (Balan et al., 2009; Bals et al., 2010).

At optimal conditions, AFEX can achieve more than 90% conversion of cellulose and hemicellulose to fermentable sugars. In fact, despite little removal of lignin or hemicellulose in the AFEX process, enzymatic digestion at low enzyme loadings is very high compared to other pretreatment alternatives (Wyman et al., 2005b). This may suggest that ammonia affects lignin and possibly hemicellulose differently than other chemicals, reducing the ability of lignin to adsorb enzyme and/or to make its access to cellulose more difficult.

Low formation of inhibitors for the downstream biological processes is one of the main advantages of the AFEX pretreatment, even though some phenolic fragments of lignin and other cell wall extractives may remain on the cellulosic surface.

Recently, AFEX pretreatment has been successfully used in SSCF processes with recombinant *S. cerevisiae* and *Escherichia coli* strains obtaining high ethanol yields from switchgrass and corn stover, respectively (Jin et al., 2010; Lau and Dale, 2010).

Another type of process utilizing ammonia is Ammonia Recycle Percolation (ARP) in which aqueous ammonia (5-15% wt) passes through a reactor packed with biomass. Temperature is normally fixed at 140-210 °C, reaction time up to 90 min, and percolation rate about 5 mL/min (Kim et al., 2008a). ARP can solubilize hemicellulose but cellulose remains intact. It leads to a short-chained cellulosic material with high glucan content (Yang and Wyman, 2008). An important challenge for ARP is to reduce liquid loading or process temperature to reduce energy cost. In this context, Soaking Aqueous Ammonia (SAA) appears as an interesting alternative since it is performed at lower temperature (30-75 °C) and is one of the few pretreatment methods where both glucan and xylan are retained in the solids. Due to that, it results in a pretreated material very interesting for being used when pentose-fermenting microorganisms are available. Furthermore, high xylose recovery at lower temperatures implies lower sugar degradation which is reflected in lower amount of inhibitory compounds. In this context, ethanol yields as high as 89.4% of the theoretical ethanol yield was shown from barley hull pretreated using SAA in an SSCF process using a recombinant *E. coli* KO11 (Kim et al., 2008b). Recently, a novel configuration so-called two-phase simultaneous saccharification and fermentation (TPSSF) has been studied to produce ethanol from corn stover pretreated with SAA obtaining ethanol yields of 84% (Li et al., 2010b). This process uses a single reactor to perform firstly SSF of xylan with *E. coli* KO11 and subsequently the SSF of glucan with *S. cerevisiae* D5A.

3.3.6 *Sulfite Pretreatment to Overcome Recalcitrance of Lignocellulose (SPORL)*

Most developed pretreatments, except organosolv, have low effectiveness on woody biomass due to the high recalcitrance caused by its physical and chemical properties. Special attention has been paid to the energy consumption for wood-size reduction before biomass pretreatment. However, some problems associated with the pretreatment of wood retained unresolved. In this context, a new pretreatment known as "SPORL" has been described. The objective of this pretreatment is to pretreat the wood chips in an aqueous sulfite (or/and bisulfite) solution followed by mechanical size reduction using disk refining (Zhu et al., 2009). The decrease of the strong recalcitrance of woody biomass by SPORL is achieved by combined effects of dissolution of hemicelluloses, depolymerization of cellulose, partial delignification, partial sulfonation of lignin as well as an increasing surface area through disk milling (Zhu and Pan, 2010).

Some results with spruce and pine have shown the effectiveness of employing SPORL for increasing the enzymatic cellulose conversion (Zhu et al., 2009; Zhu et al., 2010). Besides demonstrating the robust performance of SPORL pretreatment for producing susceptible substrates to be easily hydrolyzed, good ethanol yields have been obtained from lodgepole pine (Tian et al., 2010).

Degradation products (furfural and HMF) in the SPORL have been detected in lower concentration than in other pretreatments, which result very appropriate for the subsequent sugar fermentation.

3.3.7 *Supercritical Fluids*

The supercritical fluids are compounds that are in a gaseous form but are compressed at temperatures above their critical point to a liquid like density. Supercritical pretreatment conditions can effectively remove lignin increasing substrate digestibility; thus, lignin extraction has been studied by using supercritical fluids to pretreat lignocellulosic biomass. Furthermore, the addition of cosolvents such ethanol enhances lignin extraction.

A number of different supercritical fluids have been studied, water, carbon dioxide, and ammonia being some of the most common. Supercritical carbon dioxide (SC-CO_2) has been mostly used as an extraction solvent, but it is being considered for nonextractive purposes owing to its many advantages and potential benefits. CO_2 is nontoxic, noninflammable, leaves no harmful residues, and is inexpensive and readily available (Gao et al., 2010). In aqueous solution, CO_2 forms carbonic acid, which favors the polymers hydrolysis.

In a technology known as CO_2 explosion, this mechanism is facilitated by high pressure. After the explosive release of CO_2 pressure, disruption of cellulose and hemicellulose structure is observed and consequently accessible surface area of the substrate to enzymatic attack increases.

Operation at low temperatures compared to other methods prevents monosaccharides degradation, but in comparison to steam and ammonia explosion sugar yields obtained are lower. Nevertheless, a comparison of different pretreatment methods on several substrates showed that CO_2 explosion was more cost effective than ammonia explosion and formation of inhibitors was lower compared to steam explosion (Zheng et al., 1998). The improvement of enzymatic hydrolysis after CO_2 explosion was firstly reported with several woody raw materials such as southern yellow pine and aspen (Kim and Hong, 2001) but recently some studies have been performed with agricultural residues such rice straw (Gao et al., 2010).

Current efforts to develop these methods do not guarantee economic viability yet. A very-high-pressure requirement is specially a concerning issue. On the other hand, CO_2 utilization could be an attractive alternative to reduce costs because of its coproduction during ethanol fermentation.

3.3.8 *Steam Explosion*

Steam explosion is a physicochemical pretreatment previously used for deconstructing biomass for many purposes, that is, fiberboard building material. Nowadays, it is one of the most widely employed technologies for pretreating lignocellulose for bioethanol production.

It is a hydrothermal pretreatment in which the biomass is subjected to pressurized steam for a period of time ranging from seconds to several minutes, and then suddenly depressurized. As a physicochemical pretreatment, it combines mechanical forces and chemical effects due to the hydrolysis (autohydrolysis) of acetyl groups present in hemicellulose. Autohydrolysis takes place when high temperatures promote the formation of acetic acid from acetyl groups; furthermore, water can also act as an acid at high temperatures. The mechanical effects are caused because the pressure is suddenly reduced and fibers are separated due to the explosive decompression. In combination with the partial hemicellulose hydrolysis and solubilization, the lignin is redistributed and to some extent removed from the material. Hemicellulose removal increases enzyme accessibility to the cellulose microfibrils by exposing the cellulose surface. Steam explosion fractionates the biomass in two fractions: (i) a liquid fraction rich in monomeric and oligomeric sugars mainly from hemicelluloses solubilization, and; (ii) a solid fraction of digestible cellulose and lignin.

The most important factors affecting the effectiveness of steam explosion are particle size, temperature, residence time, and the combined effect of both temperature (*T*) and time (*t*), which is described by the severity factor ($\log R_0$) [$\log R_0 = \log(t^* e^{[T\text{-}100/14.75]})$] in which optimum value is highly dependent on the feedstock (Overend and Chornet, 1987). Higher severity results in an increased removal of hemicelluloses from the solid fraction and an enhanced cellulose digestibility but also promotes higher sugar degradation. The use of milder pretreatment conditions can minimize sugar degradation and generation of inhibitors. Different condition of process (time, temperature, and/or catalyst addition) on the same raw material gives rise to very different pretreated substrates. When severity increases, cellulose degree of polymerization decreases; furthermore, lignin content in the solid fraction increases owing to cellulose solubilization.

Steam explosion process offers several attractive features when compared to other pretreatment technologies. These include the potential for significantly lower environmental impact, lower capital investment, more potential for energy efficiency, less hazardous process chemicals and conditions, and complete sugar recovery (Avellar and Glasser, 1998). Among the main advantages, it is worth to mention the possibility of using high chip size, unnecessary addition of acid catalyst (except for softwoods), good hydrolysis yields in enzymatic hydrolysis, and its feasibility at industrial scale development. Although the possibility of avoiding acid catalysts has been stated as an advantage, the addition of an acid catalyst is a manner to increase cellulose digestibility and improve hemicellulose hydrolysis (Clark and Mackie, 1987; Sun and Cheng, 2002). In this context, many pretreatment approaches (SO_2 explosion) have included external acid addition (H_2SO_4) to catalyze the solubilization of the hemicellulose, lower the optimal pretreatment temperature, and give a partial hydrolysis of cellulose (Brownell et al., 1986; Tengborg et al., 1998). Notwithstanding, when using acids, the main drawbacks are related to equipment requirements and higher formation of degradation compounds (Mosier et al., 2005; Palmqvist and Hahn-Hägerdal, 2000). Since cost reduction and low-energy consumption are required for an effective pretreatment, high particle sizes as well as nonacid addition would be desirable to optimize the effectiveness of the process (Ballesteros et al., 2002; Hamelinck et al., 2005).

Steam explosion technology has been successfully proven for ethanol production from a wide range of raw materials as poplar (Oliva et al., 2003), eucalyptus (Ballesteros et al., 2004), olive residues (Cara et al., 2006), corn stover (Yang et al., 2010), wheat straw (Ballesteros et al., 2006), sugarcane bagasse (Martin et al., 2002), grasses (Viola et al., 2008), and hemp (Barta et al., 2010).

With the aim of maximizing sugar recoveries, some authors have suggested a two-step pretreatment (Monavari et al., 2009; Tengborg et al., 1998). In the first step, pretreatment is performed at low temperature to solubilize the hemicellulosic fraction, and the cellulose fraction is subjected to a second pretreatment step at temperatures higher than 210 °C. It offers some additional advantages such as higher ethanol yields and lower enzyme dosages during enzymatic hydrolysis (Söderström et al., 2002). Nevertheless, an economic evaluation is needed to determine the effectiveness of an additional steam explosion step. Furthermore, some authors have suggested the option to combine other methods with steam explosion to get an effective pretreatment. Therefore, alkaline peroxide pretreatment (Yang et al., 2010), IL treatment (Liu and Chen, 2006), organosolv extraction (Chen and Qiu, 2010), and superfine grinding pretreatment (Jin and Chen, 2007) coupled with steam explosion have appeared as interesting alternatives for pretreating lignocellulose.

One of the main drawbacks of steam explosion is the generation of some toxic compounds derived from sugar degradation during pretreatment that could affect the following hydrolysis and fermentation steps (Oliva et al., 2003; Zaldivar et al., 2001). Hence, it becomes necessary to use a robust strain in the subsequent fermentation step. The major inhibitors are furan derivatives, weak acids, and phenolic compounds. The main furan derivatives are furfural and HMF derived from pentoses and hexoses degradation, respectively.

Weak acids generated during pretreatment are mostly acetic acid, formed from the acetic groups present in the hemicellulosic fraction and formic and levulinic acids derived from further degradation of furfural and HMF. Wide ranges of phenolic compounds are generated due to the lignin breakdown varying widely between different raw materials. As the presence of toxic compounds is a significant obstacle for the development of large-scale ethanol production from lignocellulose, besides detoxification, several approaches such as genetic modification, evolutionary engineering, or adaptative strategies are nowadays appearing as promising alternatives to obtain more tolerant yeasts (Liu et al., 2005; Tomás-Pejó et al., 2010).

4 BIOLOGICAL PRETREATMENTS

The use of previously described pretreatments can lead in most cases to high-energy demand, some sugar degradation, and generation of toxic compounds. Fungal pretreatment has been previously explored to upgrade lignocellulosic materials for feed and paper applications (Camarero et al., 2001). Compared to the current leading pretreatment processes for bioethanol production (diluted acid, steam explosion, hydrothermal, and alkali extraction), fungal pretreatment of lignocellulose is considered an environment-friendly process with different advantages including no use of chemicals, reduced energy input, no requirement for pressurized and corrosion-resistant reactors, no waste stream generated and minimal inhibitors productions (Keller et al., 2003). Biological pretreatments employ microorganisms mainly brown, white and soft rot fungi which degrade lignin and hemicellulose and very little of cellulose, more resistant than the other components (Sánchez, 2009).

White rot fungi with selectivity to lignin degradation over cellulose can be successfully applied in microbial pretreatments. However, the patterns of cell wall deconstruction by white rot fungi vary among species and strains. Several white rot fungi such as *Phanerochaete chrysosporium*, *Ceriporia lacerata*, *Cyathus stercolerus*, *Ceriporiopsis subvermispora*, *Pycnoporus cinnarbarinus*, and *Pleurotus ostreaus* have been examined on different lignocellulosic biomass showing high delignification efficiency (Keller et al., 2003; Kumar et al., 2009; Shi et al., 2009).

(Wan and Li, 2010) found that *C. subvermispora* can effectively reduce recalcitrance of corn stover with high selectivity of lignin, high degradation rate, and minimal cellulose loss. In this case, when 5-mm corn stover was pretreated at 28 °C with 75% moisture content, overall glucose yields of 57.7%, 62.2%, and 66.6% were obtained after 18, 28, and 35 days of microbial pretreatment, respectively. Furthermore, for these conditions, the highest overall ethanol yield obtained was 57.8%.

Biological pretreatment by white rot fungi has been combined with organosolv pretreatment in an ethanol production process by SSF from beech wood chips (Itoh, 2003), *Pinus radiata* and *Acacia dealbata* (Muñoz et al., 2007). These experiments showed that both biological and chemical delignification processes may act synergistically, reducing the severity of the pretreatment and improving cellulose saccharification.

Brown fungal pretreatment has been recently pointed out as a good method for improving the enzymatic hydrolysis yields of *P. radiata* and *Pinus sylvestris* reaching saccharification yields around 70% (Ray et al., 2010). In this case, it was suggested that some organic acids secreted by the employed fungi *Caniophora puteana* reduced the pH and depolymerized it to some extent. Furthermore, combined brown rot decay-chemical delignification process from *P. radiata* wood chips resulted in an increase of ethanol production associated to both depolymerization of cellulose chains in wood and the selective delignification of organosolv pulp (Fissore et al., 2010).

Results from other studies have shown that fungal pretreatment of wheat straw for 10 days with a high lignin-degrading and low cellulose-degrading fungus (fungal isolate RCK-1) resulted in a reduction in acid loading for hydrolysis, an increase in the release of fermentable sugars, and a reduction in the concentration of fermentation inhibitors. Ethanol yield and volumetric productivity with *P. stipitis* were 0.48 g/g and 0.54 g/L.h, respectively (Kuhar et al., 2008).

In general, such processes offer advantages such as low capital cost, low energy, no chemicals requirement, mild environmental conditions, and no inhibitory compounds formation. The main drawback to develop biological methods is the low hydrolysis rate obtained in most biological materials compared to other technologies (Sun and Cheng, 2002). Several weeks to months are generally needed to obtain a high degree of lignin degradation with microbial pretreatment. However, when the microbial pretreatment is conducted concurrently with on-farm wet storage, the pretreatment time is no longer an issue (Wan and Li, 2010).

To move forward, for cost-competitive biological pretreatment of lignocellulose to improve the hydrolysis, and, eventually, improve ethanol yields, it is necessary to continue studying and testing more fungi for their ability to delignify the plant material quickly and efficiently.

5 CONCLUDING REMARKS

Different pretreatment methods to make the lignocellulose accessible to enzymes have been described and widely studied for improving ethanol production processes. The effects that some of the most studied pretreatments have on structure of lignocellulose are summarized in Table 2. The crystallinity of cellulose, its accessible surface area, and degree of polymerization as well as hemicellulose lignin disposal should be affected in a certain way during pretreatment which would affect in a different manner the subsequent enzymatic hydrolysis.

TABLE 2 Effect of Different Pretreatment Technologies on the Alteration of Lignocellulose

	Increases Accessible Surface Area	Cellulose Decrystallization	Hemicellulose Solubilization	Lignin Removal	Lignin Structure Alteration	Generation of Toxic Compounds
Mechanical comminution	H	H	0	0	0	0
Extrusion	H	H	0			
Acid	H	0	H	M	H	H
Alkali	H	H	M/H	H	H	L
Organosolv	M	–	H	M/H	M	M/L
Ozonolysis	M	M	M/H	H	M	L
Ionic liquids	M	H	H	M/H	M	M/L
Wet oxidation	H	–	H	M	H	L
Microwave	H	H	L	H	H	L
LHW	H	–	H	L	M	L
AFEX	H	M	M	L	H	L
SPORL	H	M	H	M	M	L
Supercritical fluids	M/H	–	M	H	M	M
Steam explosion	H	–	H	M	H	H
Biological	M	0	0	H		0/L

H, high effect; M, moderate effect; L, low effect; 0: no effect.

As shown in Table 3, each technology has advantages and disadvantages and an appropriate pretreatment will not only depend on the technology itself. While biological pretreatments are advantageous because of its low-energy consumption, mechanical comminution is very energy intensive. CO_2 explosion is shown as a cost-effective pretreatment; on the other hand, ozonolysis is not economically feasible due to the high cost of the large amount of ozone needed. Acid pretreatment generates high concentration of toxic compounds, but after wet oxidation only low amounts are detected. It is very difficult to conclude an ideal pretreatment and combination of different pretreatments could also be considered and might be interesting to obtain optimal fractionation of the different components and reach very high yields.

Pretreatment conditions and feedstocks would greatly affect the final pretreated material being, particle sizes, as well as time of harvesting and storage prior to pretreatment determinant for the effectiveness of process. Thus, the most appropriate pretreatment will depend on the nature of the feedstock and its recalcitrance.

TABLE 3 Advantages and Disadvantages with Different Methods for Pretreating Lignocellulosic Biomass

Pretreatment Method	Advantages	Disadvantages
Milling	Reduces cellulose crystallinity	High power and energy consumption
Concentrated acid	High glucose yield	High cost of acid and need to be recovered
		Reactor corrosion problems
	Ambient temperatures	Formation of inhibitors
Diluted Acid	Less corrosion problems than concentrated acid	Generation of degradation products
	Less formation of inhibitors	Low sugar concentration in exit stream
Alkaline	Effective lignin and hemicellulose solubilization	Requires alkali removal
Organosolv	Causes lignin and hemicellulose hydrolysis	High cost
		Solvents need to be drained and recycled.
Ozonolysis	Reduces lignin content	High cost of large amount of ozone needed
	Does not imply generation of toxic compounds	
Wet Oxidation	Efficient removal of lignin	High cost of oxygen and alkaline catalyst
	Low formation of inhibitors	
	Minimizes the energy demand (exothermic)	
LHW	Requires no catalyst and low-cost reactor	High water demanding
		High energy requirements
		Low-solids processing during pretreatment
AFEX	Increases accessible surface area	Not efficient for raw materials with high lignin content
	Low formation of inhibitors	High cost of large amount of ammonia
Steam explosion	Causes lignin transformation and hemicellulose solubilization	Generation of toxic compounds
	Cost-effective	
	Higher yield of cellulose and hemicellulose in the two-step method	Partial hemicellulose degradation
CO_2 Explosion	Increases accessible surface area	Does not affect lignin and hemicelluloses
	Cost-effective	
	Do not imply generation of toxic compounds	Very high pressure requirements
Biological	Degrades lignin and hemicellulose	Requires long incubation times
	Low energy consumption	Requires careful control of growth conditions

The present state of the lignocellulosic ethanol technology does not allow the production at commercial scale. Notwithstanding, several research laboratories and companies have scaled up (pilot and demonstration level) different pretreatment technologies for ethanol production processes from lignocellulose, but no commercial amounts of fuel are still produced (Table 4). Challenges in scaling up the technologies and reducing production costs need to be overcome.

Among others, Iogen Corporation has developed a demonstration plant applying a modified steam explosion for producing about 5000-6000 L of cellulosic ethanol per day.

TABLE 4 Companies Applying Different Pretreatment Technologies for Bioethanol Production from Lignocellulose

Company	Pretreatment	Country	Resource
Abengoa Bioenergy New Technologies	Steam Explosion	Spain	http://www.abengoabioenergy.com/corp/web/en/nuevas_tecnologias/tecnologias/hidrolisis/index.html
BioGasol Aps	Wet-explosion	Denmark	http://www.biogasol.com/Home-3.aspx
BlueFire Ethanol	Acid pretreatment	USA	http://bluefireethanol.com
Dupont Danisco Cellulosic Ethanol (DDCE)	NH_3 Steam recycled	USA	http://www.ddce.com/technology/index.html
Inbicom A/S	Hydrothermal	Denmark	http://www.inbicon.com/Technologies/Hydrothermal_pretreatment/Pages/Hydrothermal%20pretreatment.aspx
Iogen Corporation	Modified steam explosion	Canada	http://www.iogen.ca/cellulosic_ethanol/what_is_ethanol/process.html
Izumi Biorefinery	Acid Pretreatment	Japan	http://bluefireethanol.com/images/IZUMI_Status_2004_for_BlueFire_051606.pdf
KL Energy	Thermo-mechanical	USA	http://www.klenergycorp.com/technology-process.htm
Lignol Energy Corporation	Solvent pretreatment	Canada	http://www.lignol.ca/
Praj Industries	Thermo-chemical	India	http://www.praj.net/default.asp
Queensland University of Technology	Soda pulping and ionic liquid based pretreatments	Australia	http://www.scitech.qut.edu.au/news/news-event.jsp?news-event-id=32969
SEKAB	Acid pretreatment	Sweden	http://www.sekab.com
Terrabon Energy	Lime	USA	http://www.terrabon.com/mixalco_semiworksplant.php
Verenium	Acid pretreatment	USA	http://www.verenium.com/
Weyland AS	Acid pretreatment	Norway	http://www.weyland.no/teknologi

The Swedish company SEKAB has developed an industrial process for producing ethanol form lignocellulose pretreated with acid and Abengoa Bionergy has constructed a bioethanol pilot plant in Spain which can operate with steam exploded wheat straw.

Although the mechanism involved in converting lignocellulose to ethanol is well understood, much research has to be addressed to the fractionation of cellulose, hemicellulose, and lignin into pure fractions for making the whole process cost effective. It is necessary to determine the chemical and structural modifications that occur within the biomass during pretreatment to identify the limiting factors for different pretreatment technologies. Furthermore, technology bottlenecks in ethanol production processes from lignocellulose need to be overcome for commercial implementation.

References

Almeida, J.R.M., Modig, T., Petersson, A., Hahn-Hägerdal, B., Liden, G., Gorwa-Grauslund, M.F., 2007. Increased tolerance and conversion of inhibitors in lignocellulosic hydrolysates by *Saccharomyces cerevisiae*. J. Chem. Technol. Biotechnol. 82, 340–349.

Alvira, P., Tomás-Pejó, E., Ballesteros, M., Negro, M.J., 2010. Pretreatment technologies for an efficient bioethanol production process based on enzymatic hydrolysis: a review. Bioresour. Technol. 101, 4851–4861.

Avellar, B.K., Glasser, W.G., 1998. Steam-assisted biomass fractionation. I. Process considerations and economic evaluation. Biomass Bioenerg. 14, 205–218.

Balan, V., Sousa, L.D., Chundawat, S.P.S., Marshall, D., Sharma, L.N., Chambliss, C.K., et al., 2009. Enzymatic digestibility and pretreatment degradation products of AFEX-treated hardwoods (*Populus nigra*). Biotechnol. Prog. 25, 365–375.

Ballesteros, M., 2010. Enzymatic hydrolysis of lignocellulosic biomass. In: Waldron, K. (Ed.), Bioalcohol Production. Biochemical Conversion of Lignocellulosic Biomass. Woodhead Publishing. pp. 159–177.

Ballesteros, I., Oliva, J.M., Negro, M.J., Manzanares, P., Ballesteros, M., 2002. Enzymic hydrolysis of steam exploded herbaceous agricultural waste (*Brassica carinata*) at different particule sizes. Process Biochem. 38, 187–192.

Ballesteros, M., Oliva, J.M., Negro, M.J., Manzanares, P., Ballesteros, I., 2004. Ethanol from lignocellulosic materials by a simultaneous saccharification and fermentation process (SFS) with *Kluyveromyces marxianus* CECT 10875. Process Biochem. 39, 1843–1848.

Ballesteros, I., Negro, M., Oliva, J.M., Cabañas, A., Manzanares, P., Ballesteros, M., 2006. Ethanol production from steam-explosion pretreated wheat straw. Appl. Biochem. Biotechnol. 130, 496–508.

Bals, B., Wedding, C., Balan, V., Sendich, E., Dale, B.E., 2010. Evaluating the impact of ammonia fiber expansion (AFEX) pretreatment conditions on the cost of ethanol production. Bioresour. Technol. doi: 10.1016/j.biortech.2010.08.058.

Banerjee, S., Mudliar, S., Sen, R., Gir, B., Satpute, D., Chakrabarti, T., et al., 2010. Commercializating lignocellulosic bioethanol: technology bottlenecks and possible remedies. Biofuels Bioprod. Biorefining. 4, 77–93.

Barta, Z., Oliva, J.M., Ballesteros, I., Dienes, D., Ballesteros, M., Réczey, K., 2010. Refining hemp hurds into fermentable sugars or ethanol. Chem. Biochem. Eng. Q. 24, 331–339.

Brownell, H.H., Yu, E.K.C., Saddler, J.N., 1986. Steam-explosion Pretreatment of wood: effect of chip size, acid, moisture content and pressure drop. Biotechnol. Bioeng. 28, 729–801.

Buaban, B., Inoue, H., Yano, S., Tanapongpipat, S., Ruanglek, V., Champreda, V., et al., 2010. Bioethanol production from ball milled bagasse using an on-site produced fungal enzyme cocktail and xylose-fermenting *Pichia stipitis*. J. Biosci. Bioeng. 110, 18–25.

Camarero, S., Bocchini, P., Galletti, G.C., Martínez, M.J., Martínez, A.T., 2001. Compositional changes of wheat lignin by a fungal peroxidase analysed by pyrolysis-GC-MS. J. Anal. Appl. Pyrolysis. 58-59, 413–423.

Cara, C., Ruiz, E., Ballesteros, I., Negro, M.J., Castro, E., 2006. Enhanced enzymatic hydrolysis of olive tree wood by steam explosion and alkaline peroxide delignification. Process Biochem. 41, 423–429.

Cara, C., Romero, I., Oliva, J.M., Sáez, F., Castro, E., 2007. Liquid hot water of olive tree pruning residues. Appl. Biochen. Biotechnol. 136-140, 379–394.

Cara, C., Ruiz, E., Oliva, J.M., Sáez, F., Castro, E., 2008. Conversion of olive tree biomass into fermentable sugars by dilute acid pretreatment and enzymatic saccharification. Bioresour. Technol. 99, 1869–1876.

Carvalheiro, F., Duarte, L.C., Gírio, F.M., 2008. Hemicellulose biorefineries: a review on biomass pretreatments. J. Sci. Ind. Res. 67, 849–864.

Chandra, R.P., Bura, R., Mabee, W.E., Berlin, A., Pan, X., Saddler, J.N., 2007. Substrate pretreatment: the key to effective enzymatic hydrolysis of lignocellulosics? Adv. Biochem. Eng. Biot. 108, 67–93.

Chang, V.S., Nagwani, M., Holtzapple, M.T., 2001. Lime pretreatment of crop residues bagasse and newspaper. Appl. Biochemi. Biotechnol. 74, 1–28.

Chen, H., Qiu, W., 2010. Key technologies for bioethanol production from lignocellulose. Biotechnol. Adv. 28, 556–562.

Chundawat, S.P.S., Venkatesh, B., Dale, B.E., 2007. Effect of particle size based separation of milled corn stover on AFEX pretreatment and enzymatic digestibility. Biotechnol. Bioeng. 96, 219–231.

Clark, T.A., Mackie, K.L., 1987. Steam explosion of the softwood *Pinus radiata* with sulphur dioxide addition. I. Process optimization. J. Wood Chem. Technol. 7, 373–403.

Dadi, A.P., Varanasi, S., Schall, C.A., 2006. Enhancement of cellulose saccharification kinetics using an ionic liquid pretreatment step. Biotechnol. Bioeng. 95, 904–910.

Ferreira, S., Gil, N., Queiroz, J.A., Duarte, A.P., Domingues, F.C., 2010. Bioethanol from the Portuguese forest residue *Pterospartum tridentatum*—an evaluation of pretreatment strategy for enzymatic saccharification and sugars fermentation. Bioresour. Technol. 101, 7797–7803.

Fissore, A., Carrasco, L., Reyes, P., Rodríguez, J., Freer, J., Mendonça, R., 2010. Evaluation of a combined brown rot decay-chemical delignification process as a pretreatment for bioethanol production from *Pinus radiata* wood chips. J. Ind. Microbiol. Biotechnol. 37, 893–900.

Gao, M., Xu, F., Li, S., Ji, X., Chen, S., Zhang, D., 2010. Effect of SC-CO2 pretreatment in increasing rice straw biomass conversion. Biosyst. Eng. 106, 470–475.

García-Cubero, M.T., González-Benito, G., Indacoechea, I., Coca, M., Bolado, S., 2009. Effect of ozonolysis pretreatment on enzymatic digestibility of wheat and rye straw. Bioresour. Technol. 100, 1608–1613.

Gírio, F.M., Fonseca, C., Carvalheiro, F., Duarte, L.C., Marques, S., Bogel-Lukasik, R., 2010. Hemicelluloses for fuel ethanol: a review. Bioresour. Technol. 101, 4775–4800.

Gomez, L.D., Steele-King, C.G., Mc Queen-Mason, S.J., 2008. Sustainable liquid biofuels from biomass: the writing's on the walls. New Phytol. 178, 473–485.

Gong, G., Liu, D., Huang, Y., 2010. Microwave-assisted organic acid pretreatment for enzymatic hydrolysis of rice straw. Biosyst. Eng. 107, 67–73.

Haagensen, F., Skiadas, I.V., Gavala, H.N., Ahring, B.K., 2009. Pre-treatment and ethanol fermentation potential of olive pulp at different dry matter concentrations. Biomass Bioenerg. 33, 1643–1651.

Hamelinck, C.N., van Hooijdonk, G., Faaij, A.P.C., 2005. Ethanol from lignocellulosic biomass: techno-economic performance in short-, middle- and long-term. Biomass Bioenerg. 28, 384–410.

Hayes, D.J., 2009. An examination of biorefining processes, catalysts and challenges. Catal. Today 145, 138–151.

Hendriks, A.T.W.M., Zeeman, G., 2009. Pretreatments to enhance the digestibility of lignocellulosic biomass. Bioresour. Technol. 100, 10–18.

Hideno, A., Inoue, H., Tsukahara, K., Fujimoto, S., Minowa, T., Inoue, S., et al., 2009. Wet disk milling pretreatment without sulfuric acid for enzymatic hydrolysis of rice straw. Bioresour. Technol. 100, 2706–2711.

Hu, Z., Wen, Z., 2008. Enhancing enzymatic digestibility of switchgrass by microwave-assisted alkali pretreatment. Biochem. Eng. J. 38, 369–378.

Hu, G., Heitmann, J.A., Rojas, O.J., 2008. Feedstock pretreatment strategies for producing ethanol from wood, bark, and forest residues. Bioresources 3, 270–294.

Itoh, H., Wada, M., Honda, Y., Kuwahara, M., 2003. Bioorganosolve pretreatments for simultaneous saccharification and fermentation of beech wood by ethanolysis and white rot fungi. J. Biotechnol. 103, 273–280.

Jin, S., Chen, H., 2007. Superfine grinding of steam-exploded rice straw and its enzymatic hydrolysis. Biochem. Eng. J. 30, 225–230.

Jin, M., Lau, M.W., Balan, V., Dale, B., 2010. Two-step SSCF to convert AFEX-treated switchgrass to ethanol using commercial enzymes and *Saccharomyces cerevisiae* 424A(LNH-ST). Bioresour. Technol. 101, 8171–8178.

Kaparaju, P., Felby, C., 2010. Characterization of lignin during oxidative and hydrothermal pre-treatment processes of wheat straw and corn stover. Bioresour. Technol. 101, 3175–3181.

Karunanithy, C., Muthukumarappan, K., 2010a. Effect of Extruder Parameters and Moisture Content of Switchgrass, Prairie Cord Grass on Sugar Recovery from Enzymatic Hydrolysis. Appl. Biochem. Biotechnol. doi: 10.1007/s12010-010-8959-3.

Karunanithy, C., Muthukumarappan, K., 2010b. Influence of extruder temperature and screw speed on pretreatment of corn stover while varying enzymes and their ratios. Appl. Biochem. Biotechnol. 162, 264–279.

Karunanithy, C., Muthukumarappan, K., Julson, J.L., 2008a. A comparative study of acid and enzymatic hydrolysis of prairie cord grass pretreated using high shear bioreactor. Am. Soc. Agric. Biol. Eng. Ann. Int. Meet. 6, 3687–3696.

Karunanithy, C., Muthukumarappan, K., Julson, J.L., 2008b. Enzymatic hydrolysis of corn stover pretreated in high shear bioreactor. Am. Soc. Agric. Biol. Eng. Ann. Int. Meet. 6, 3603–3611.

Keating, J.D., Panganiban, C., Mansfield, S.D., 2006. Tolerance and adaptation of ethanologenic yeast to lignocellulosic inhibitory compounds. Biotechnol. Bioeng. 93, 1196–1206.

Keller, F.A., Hamilton, J., Nguyen, Q.A., 2003. Microbial pretreatment of biomass. Appl. Biotechnol. Bioeng. 105, 27–41.

Keshwani, D.R., Cheng, J.J., 2010. Microwave-based alkali pretreatment of switchgrass and coastal bermudagrass for bioethanol production. Biotechnol. Prog. 26, 644–652.

Kim, S., Holtzapple, M.T., 2006. Delignification kinetics of corn stover in lime pretreatment. Bioresour. Technol. 97, 778–785.

Kim, K., Hong, J., 2001. Supercritical CO_2 pretreatment of lignocellulose enhances enzymatic cellulose hydrolysis. Bioresour. Technol. 77, 139–144.

Kim, J.S., Kim, H., Lee, J.S., Lee, J.P., Park, S.C., 2008a. Pretreatment characteristics of waste oak wood by ammonia percolation. Appl. Biochem. Biotechnol. 148, 15–22.

Kim, T.H., Taylor, F., Hicks, K.B., 2008b. Bioethanol production from barley hull using SAA (soaking in aqueous ammonia) pretreatment. Bioresour. Technol. 99, 5694–5702.

Klinke, H.B., Ahring, B.K., Schmidt, A.S., Thomsen, A.B., 2002. Characterization of degradation products from alkaline wet oxidation of wheat straw. Bioresour. Technol. 82, 15–26.

Klinke, H.B., Olsson, L., Thomsen, A.B., Ahring, B.K., 2003. Potential Inhibitors From wet oxidation of wheat straw and their effect on ethanol production of *Saccharomyces cerevisiae:* wet oxidation and fermentation by yeast. Biotechnol. Bioeng. 81, 738–747.

Klinke, H.B., Thomsen, A.B., Ahring, B.K., 2004. Inhibition of ethanol-producing yeast and bacteria by degradation products produced during pre-treatment of biomass. Appl. Microbiol. Biotechnol. 66, 10–26.

Kootstra, A.M.J., Beeftink, H.H., Scott, E.L., Sanders, J.P.M., 2009. Comparison of dilute mineral and organic acid pretreatment for enzymatic hydrolysis of wheat straw. Biochem. Eng. J. 46, 126–131.

Kuhar, S., Nair, L.M., Kuhad, R.C., 2008. Pretreatment of lignocellulosic material with fungi capable of higher lignin degradation and lower carbohydrate degradation improves substrate acid hydrolysis and the eventual conversion to ethanol. Can. J. Microbiol. 54, 305–313.

Kumar, R., Wyman, C.E., 2010. Key features of pretreated lignocellulosic biomass solids and their impact on hydrolysis. In: Bioalcohol Production. Biochemical Conversion of Lignocellulosic Biomass. Woodhead Publishing. Series in Energy: Number 3. pp. 73–121.

Kumar, P., Barrett, D.M., Delwiche, M.J., Stroeve, P., 2009. Methods for pretreatment of lignocellulosic biomass for efficient hydrolysis and biofuel production. Ind. Eng. Chem. Res. 48, 3713–3729.

Lamsal, B., Yoo, J., Brijwani, K., Alavi, S., 2010. Extrusion as a thermo-mechanical pre-treatment for lignocellulosic ethanol. Biomass Bioenerg. 34, 1703–1710.

Laser, M., Schulman, D., Allen, S.G., Lichwa, J., Antal, M.J., Lynd, L.R., 2002. A comparison of liquid hot water and steam pretreatments of sugar cane bagasse for conversion to ethanol. Bioresour. Technol. 81, 33–44.

Lau, M.W., Dale, B.E., 2010. Effect of primary degradation-reaction products from Ammonia Fiber Expansion (AFEX)-treated corn stover on the growth and fermentation of *Escherichia coli* KO11. Bioresour. Technol. 101, 7849–7855.

Laureano-Perez, L., Teymouri, F., Alizadeh, H., Dale, B.E., 2005. Understanding factors that limit enzymatic hydrolysis of biomass. Appl. Biochem. Biotechnol. 121, 1081–1099.

Lee, S.H., Doherty, T.V., Linhardt, R.J., Dordick, J.S., 2009. Ionic liquid-mediated selective extraction of lignin from wood leading to enhanced enzymatic cellulose hydrolysis. Biotechnol. Bioeng. 102, 1368–1376.

Li, Q., He, Y.C., Xian, M., Jun, G., Xu, X., Yang, J.M., et al., 2009. Improving enzymatic hydrolysis of wheat straw using ionic liquid 1-ethyl-3-methyl imidazolium diethyl phosphate pretreatment. Bioresour. Technol. 100, 3570–3575.

Li, M.F., Fan, Y.M., Xu, F., Sun, R.C., Zhang, X.L., 2010a. Cold sodium hydroxide/urea based pretreatment of bamboo for bioethanol production: characterization of the cellulose rich fraction. Ind. Crop. Prod. 32, 551–559.

Li, X., Kim, T.H., Nghiem, N.P., 2010b. Bioethanol production from corn stover using aqueous ammonia pretreatment and two-phase simultaneous saccharification and fermentation (TPSSF). Bioresour. Technol. 101, 5910–5916.

Liu, L., Chen, H., 2006. Enzymatic hydrolysis of cellulose materials treated with ionic liquid [BMIM]Cl. Chin. Sci. Bull. 51, 2432–2436.

Liu, Z.L., Slininger, P.J., Gorsich, S.W., 2005. Enhanced Biotransformation of Furfural and Hydroxymethylfurfural by Newly Developed Ethanologenic Yeast Strains. Appl. Biochem. Biotechnol. 121-124, 451–460.

Lynd, L.R., 1996. Overview and evaluation of fuel ethanol from cellulosic biomass: technology, economics, the environment, and policy. Annu. Rev. Energ. Environ. 21, 403–465.

Mäki-Arvela, P., Anugwom, I., Virtanen, P., Sjöholm, R., Mikkola, J.P., 2010. Dissolution of lignocellulosic materials and its constituents using ionic liquids-A review. Ind. Crop. Prod. 32, 175–201.

Mansfield, S.D., Mooney, C., Saddler, J.N., 1999. Substrate and Enzyme Characteristics that limit Cellulose Hydrolysis. Biotechnol. Prog. 15, 804–816.

Martin, C., Galbe, M., Wahlbom, C.F., Hahn-Hägerdal, B., Jonsson, L.J., 2002. Ethanol production from enzymatic hydrolysates of sugarcane bagasse using recombinant xylose-utilising *Saccharomyces cerevisiae*. Enzyme Microb. Technol. 31, 274–282.

Martin, C., Thomsen, M.H., Hauggaard, H., Thomsen, A.B., 2008. Wet oxidation pretreatment, enzymatic hydrolysis and simultaneous saccharification and fermentation of clover-ruegrass mixtures. Bioresour. Technol. 99, 8777–8782.

McIntosh, S., Vancov, T., 2010. Enhanced enzyme saccharification of *Sorghum bicolor* straw using dilute alkali pretreatment. Bioresour. Technol. 101, 6718–6727.

Merino, S.T., Cherry, J., 2007. Progress and challenges in enzyme development for biomass utilization. Adv. Biochem. Eng. Biotechnol. 108, 95–120.

Mielenz, J.R., 2001. Ethanol production from biomass: technology and commercialization status. Curr. Opin. Microbiol. 4, 324–329.

Monavari, S., Galbe, M., Zacchi, G., 2009. The influence of solid/liquid separation techniques on the sugar yield in two-step dilute acid hydrolysis of softwood followed by enzymatic hydrolysis. Biotechnol. Biofuels 2, 6.

Mosier, N., Schulman, D., Ho, N., Antal, M.J., 2001. A comparison between hot liquid water pretreatment of corn stover. Bioresour. Technol. 40, 2934–2941.

Mosier, N., Wyman, C.E., Dale, B.D., Elander, R.T., Lee, Y.Y., Holtzapple, M., et al., 2005. Features of promising technologies for pretreatment of lignocellulosic biomass. Bioresour. Technol. 96, 673–686.

Muñoz, C., Mendoça, R., Baeza, J., Berlin, A., Saddler, J., Freer, J., 2007. Bioethanol production from bioorganosolv pulps of *Pinus radiata* and *Acacia dealbata*. J. Chem. Technol. Biotechnol. 82, 767–774.

Nahyun, P., Hye-Yun, K., Bon-Wook, K., Hwanmyeong, Y., In-Gyu, C., 2010. Organosolv pretreatment with various catalysts for enhancing enzymatic hydrolysis of pith pine (*Pinus rigida*). Bioresour. Technol. 101, 7046–7053.

Nguyen, T.A.D., Kim, K.R., Han, S.J., Cho, H.Y., Kim, J.W., Park, S.M., et al., 2010. Pretreatment of rice straw with ammonia and ionic liquid for lignocellulose conversion to fermentable sugars. Bioresour. Technol. 101, 7432–7438.

Nikolić, S., Mojović, L., Rakin, M., Pejin, D., Pejin, J., 2010. Ultrasound-assisted production of bioethanol by simultaneous saccharification and fermentation of corn meal. Food Chem. 122, 216–222.

Nlewem, K.C., Thrash, J., 2010. Comparison of different pretreatment methods based on residual lignin effect on the enzymatic hydrolysis of switchgrass. Bioresour. Technol. 101, 5426–5430.

Oliva, J.M., Sáez, F., Ballesteros, I., González, A., Negro, M.J., Manzanares, P., et al., 2003. Effect of lignocellulosic degradation compounds from steam explosion pretreatment on ethanol fermentation by thermotolerant yeast *Kluyveromyces marxianus*. Appl. Biochem. Biotechnol. 105, 141–154.

Olivier-Bourbigou, H., Magna, L., Morvan, D., 2010. Ionic liquids and catalysis: recent progress from knowledge to applications. Appl. Catal A-Gen. 373, 1–56.

Olsson, L., JØrgensen, H., Krogh, K.B.R., Roca, C., 2005. Bioethanol production from lignocellulosic material. In: Polysaccharides Structural Diversity and Functional Versatility. Marcel Dekker, New York, pp. 957–993.

Overend, R.P., Chornet, E., 1987. Fractionation of lignocellulosics by steam-aqueous pretreatments. Phil. Trans. R. Soc. Lond. 321, 523–536.

Palmqvist, E., Hahn-Hägerdal, B., 2000. Fermentation of lignocellulosic hydrolysates II: inhibitors and mechanism of inhibition. Bioresour. Technol. 74, 25–33.

Palonen, H., Thomsen, A.B., Tenkanen, M., Schmidt, A.S., Viikari, L., 2004. Evaluation of wet oxidation pretreatment for enzymatic hydrolysis of softwood. Appl. Biochem. Biotechnol. 117, 1–17.

Pan, X., Xie, D., Gilkes, N., Gregg, D.J., Saddler, J.N., 2005. Strategies to enhance the enzymatic hydrolysis of pretreated softwood with high residual lignin content. Appl. Biochem. Biotechnol. 124, 1069–1079.

Park, J.Y., Shiroma, R., Al Haq, M.I., Zhang, Y., Ike, M., Arai-Sanoh, Y., et al., 2010. A novel lime pretreatment for subsequent bioethanol production from rice straw—calcium capturing by carbonation (CaCCO) process. Bioresour. Technol. 101, 6805–6811.

Pérez, J.A., Gonzáles, A., Oliva, J.M., Ballesteros, I., Manzanares, P., 2007. Effect of process variables on liquid hot water pretreatment of wheat straw for bioconversion to fuel-ethanol in a batch reactor. J. Chem. Technol. Biotechnol. 82, 929–938.

Pérez, J.A., Ballesteros, I., Ballesteros, M., Sáez, F., Negro, M.J., Manzanares, P., 2008. Optimizing Liquid Hot Water pretreatment conditions to enhance sugar recovery from wheat straw for fuel-ethanol production. Fuel 87, 3640–3647.

Ray, M.J., Leak, D.J., Spanu, P.D., Murphy, R.J., 2010. Brown rot fungal early stage decay mechanism as a biological pretreatment for softwood biomass in biofuel production. Biomass Bioenerg. 34, 1257–1262.

Rocha, M.V., Rodrigues, T.H., de Macedo, G.R., Gonçalves, L.R., 2009. Enzymatic hydrolysis and fermentation of pretreated cashew apple bagasse with alkali and diluted sulfuric Acid for bioethanol production. Appl. Biochem. Biotechnol. 155, 407–417.

Saha, B.C., Cotta, M.A., 2006. Ethanol production from alkaline peroxide pretreated enzymatically saccharified wheat straw. Biotechnol. Prog. 22, 449–453.

Saha, B.C., Iten, L.B., Cotta, M.A., Wu, Y.V., 2005. Dilute acid pretreatment, enzymatic saccharification and fermentation of wheat straw to ethanol. Process Biochem. 40, 3693–3700.

Sánchez, C., 2009. Lignocellulosic residues: biodegradation and bioconversion by fungi. Biotechnol. Adv. 27, 185–194.

Shatalov, H.P., 2008. *Arundo donax* L. reed: new perspectives for pulping and bleaching. 5. Ozone-based TCF bleaching of organosolv pulps. Bioresour. Technol. 99, 472–478.

Shi, J., Sharma-Shivappa, R.R., Chinn, M., Howell, N., 2009. Effect of microbial pretreatment on enzymatic hydrolysis and fermentation of cotton stalks for ethanol production. Biomass Bioenerg. 33, 88–96.

Silverstein, R.A., Chen, Y., Sharma-Shivappa, R.R., Boyette, M.D., Osborne, J., 2007. A comparison of chemical pretreatment methods for improving saccharification of cotton stalks. Bioresour. Technol. 98, 3000–3011.

Singh, A., Tuteja, S., Singh, N., Bishnoi, N.R., 2010. Enhanced saccharification of rice straw and hull by microwave-alkali pretreatment and lignocellulolytic enzyme production. Bioresour. Technol. doi: 10.1016/j.biortech.2010.08.113.

Söderström, J., Pilcher, L., Galbe, M., Zacchi, G., 2002. Two-step steam pretreatment of softwood with SO_2 impregnation for ethanol production. Appl. Microbiol. Biotechnol. 98-100, 5–21.

Sun, Y., Cheng, J., 2002. Hydrolysis of lignocellulosic materials for ethanol production: a review. Bioresour. Technol. 83, 1–11.

Swatloski, R.P., Scott, K., Spear, J., Holbrey, D., Rogers, R.D., 2002. Dissolution of cellulose with ionic liquids. J. Am. Chem. Soc. 124, 4974–4975.

Taherzadeh, M.J., Karimi, K., 2008. Pretreatment of lignocellulosic wastes to improve ethanol and biogas production: a review. Int. J. Mol. Sci. 9, 1621–1651.

Taherzadeh, M.J., Gustafsson, L., Niklasson, C., Lidén, G., 1999. Conversion of furfural in aerobic and anaerobic batch fermentation of glucose by *Saccharomyces cerevisiae*. J. Biosci. Bioeng. 87, 169–174.

Tengborg, C., Stenberg, K., Galbe, M., Zacchi, G., Larsson, S., Palmqvist, E., et al., 1998. Comparison of SO_2 and H_2SO_4 impregnation of softwood prior to steam pretreatment on ethanol production. Appl. Microbiol. Biotechnol. 70-72, 3–15.

Tian, S., Luo, X.L., Yang, X.S., Zhu, J.Y., 2010. Robust cellulosic ethanol production from SPORL-pretreated lodgepole pine using an adapted strain *Saccharomyces cerevisiae* without detoxification. Bioresour. Technol. 101, 8678–8685.

Tomás-Pejó, E., Ballesteros, M., Oliva, J.M., Olsson, L., 2010. Adaptation of the xylose fermenting yeast *Saccharomyces cerevisiae* F12 for improving ethanol production in different fed-batch SSF processes. J. Ind. Microbiol. Biotechnol. 37, 1211–1220.

Varga, E., Klinke, H.B., Réczey, K., Thomsen, A.B., 2004. High solid simultaneous saccharification and fermentation of wet oxidized corn stover to ethanol. Biotechnol. Bioeng. 88, 567–574.

Viola, E., Cardinale, M., Santarcangelo, R., Villone, A., Zimbardi, F., 2008. Ethanol from eel grass via steam explosion and enzymatic hydrolysis. Biomass Bioenerg. 32, 613–618.

Von Sivers, M., Zacchi, G., Olsson, L., Hahn-Hägerdal, B., 1994. Cost analysis of ethanol production from willow using recombinant *Escherichia coli*. Biotechnol. Prog. 10, 555–560.

Wan, C., Li, Y., 2010. Microbial pretreatment of corn stover with *Ceriporiopsis subvermispora* for enzymatic hydrolysis and ethanol production. Bioresour. Technol. 101, 6398–6403.

Wyman, C.E., 2007. What is (and is not) vital to advancing cellulosic ethanol. Trends Biotechnol. 25, 153–157.

Wyman, C.E., Dale, B.E., Elander, R.T., Holtzapple, M., Ladisch, M.R., Lee, Y.Y., 2005a. Coordinated development of leading biomass pretreatment technologies. Bioresour. Technol. 96, 1959–1966.

Wyman, C.E., Dale, B.E., Elander, R.T., Holtzapple, M., Ladisch, M.R., Lee, Y.Y., 2005b. Comparative sugar recovery data from laboratory scale application of leading pretreatment technologies to corn stover. Bioresour. Technol. 96, 2026–2032.

Yachmenev, V., Condon, B., Klasson, T., Lambert, A., 2009. Acceleration of the enzymatic hydrolysis of corn stover and sugar cane bagasse celluloses by low intensity uniform ultrasound. J. Biobased Mater. Bioenergy. 3, 25–31.

Yang, B., Wyman, C.E., 2008. Pretreatment: the key to unlocking low-cost cellulosic ethanol. Biofuels Biop. Biorefining 2, 26–40.

Yang, M., Li, W., Liu, B., Li, Q., Xing, J., 2010. High-concentration sugars production from corn stover based on combined pretreatments and fed-batch process. Bioresour. Technol. 101, 4884–4888.

Zaldivar, J., Nielsen, J., Olsson, L., 2001. Fuel ethanol production from lignocellulose: a challenge for metabolic engineering and process integration. Appl. Microbiol. Biotechnol. 56, 17–34.

Zavrel, M., Bross, D., Funke, M., Büchs, J., Spiess, A.C., 2009. High-throughput screening for ionic liquids dissolving (ligno-)cellulose. Bioresour. Technol. 100, 2580–2587.

Zhao, X., Cheng, K., Liu, D., 2009. Organosolv pretreatment of lignocellulosic biomass for enzymatic hydrolysis. Appl. Microbiol. Biotechnol. 82, 815–827.

Zhao, H., Baker, G.A., Cowins, J.V., 2010. Fast enzymatic saccharification of switchgrass after pretreatment with ionic liquids. Biotechnol. Prog. 26, 127–133.

Zheng, Y., Lin, H.M., Tsao, G.T., 1998. Pretreatment for cellulose hydrolysis by carbon dioxide explosion. Biotechnol. Prog. 14, 890–896.

Zhu, J.Y., Pan, X.J., 2010. Woody biomass pretreatment for cellulosic ethanol production: technology and energy consumption evaluation. Bioresour. Technol. 101, 4992–5002.

Zhu, J.Y., Pan, X.J., Wang, G.S., Gleisner, R., 2009. Sulfite pretreatment (SPORL) for robust enzymatic saccharification of spruce and red pine. Bioresour. Technol. 100, 2411–2418.

Zhu, W., Zhu, J.Y., Gleisner, R., Pan, X.J., 2010. On energy consumption for size-reduction and yields from subsequent enzymatic saccharification of pretreated lodgepole pine. Bioresour. Technol. 101, 2782–2792.

Additional Resources

Abengoa Bionergy New Technologies, http://www.abengoabioenergy.com/corp/web/en/nuevas_tecnologias/tecnologias/hidrolisis/index.html.

BioGasol APS, http://www.biogasol.com/Home-3.aspx.

BlueFier Ethanol, http://bluefireethanol.com.

Dupont Danisco cellulosic ethanol, http://www.ddce.com/technology/index.html.

Inbicom A/S, http://www.inbicon.com/Technologies/Hydrothermal_pretreatment/Pages/Hydrothermal%20pretreatment.aspx.

Iogen corporation, http://www.iogen.ca/cellulosic_ethanol/what_is_ethanol/process.html.

Izumi Biorefinery, http://bluefireethanol.com/images/IZUMI_Status_2004_for_BlueFire_051606.pdf.

KL Energy, http://www.klenergycorp.com/technology-process.htm.

Lignol Energy Corporation, http://www.lignol.ca/.

Praj Industries, http://www.praj.net/default.asp.

Queensland University of Technology, http://www.scitech.qut.edu.au/news/news-event.jsp?news-event-id=32969.

SEKAB, http://www.sekab.com.

Terrabon Energy, http://www.terrabon.com/mixalco_semiworksplant.php.

Verenium, http://www.verenium.com/.

Weyland AS, http://www.weyland.no/teknologi.

CHAPTER

8

Production of Celluloytic Enzymes for the Hydrolysis of Lignocellulosic Biomass

*Reeta Rani Singhania**,†

Biotechnology Division, National Institute for Interdisciplinary Science and Technology, CSIR, Trivandrum-695019, India

***Corresponding author: E-mail: reetasinghania@gmail.com**

1 INTRODUCTION

In the last few decades of the twentieth century, there has been an increased demand for enzymes in many industrial applications due to the increasing concern for environmental safety and development of green processes to substitute several of the existing chemical processes. The demand for more stable, highly active, and specific enzymes is growing rapidly, and the projected world market for industrial enzymes is rapidly growing at an annual rate of about 7.6% and is estimated to be $6 billion in the year 2012 (World enzymes to 2011, Market study #2229 by Freedonia group. http://www.freedonia.com, 2007). Approximately 75% of the industrial enzymes are hydrolases, with carbohydrolases being the second largest group. The study of the biotechnology of cellulases and hemicellulases began in the early 1980s, initially in the animal feed industry and followed by food applications. Subsequently, these enzymes were used in the textile, laundry as well as the pulp and paper industries. The use of cellulases and hemicellulases has increased considerably over the last two decades, especially in the textile, food, brewery, and wine as well as the pulp and paper industries. Cellulases accounted for approximately 20% of the world enzyme market between 2005 ans 2010.

Cellulases are the second largest industrial enzyme by dollar volume, which is increasing with the increased demand for various industrial applications such as the detergent industry,

†Current address: Biological Engineering Department – Polytech Clermont-Ferrand, Université Blaise Pascal, 24 avenue des Landais, BP 206, F-63174 Aubière Cedex, France

textile industry, paper processing industry, animal feed industry and fruit juice industry. The commercial potential of using cellulases lies in its efficiency of converting lignocellulosic biomass into glucose through enzymatic hydrolysis, which can be utilized to generate a number of value-added products such as ethanol. There is a renewed interest in commercial utilization of lignocellulosic biomass to generate ethanol for the transport sector due to shortage of fossil fuels as well as environmental concern. Though there are several potential industrial applications of cellulases, the importance of lignocellulosic ethanol has brought cellulases to the main frontier. It is envisaged that cellulases may become the largest volume of industrial enzymes if ethanol from lignocellulosic biomass through the enzymatic route becomes a major transportation fuel. Lignocellulosic biomass is considered the only foreseeable source of energy (Lynd et al., 2002), and the future of humankind is predicted to be based on a carbohydrate-based economy directly dependent on biomass utilization. While moving toward a carbohydrate-based economy seems inevitable, there are also other issues to be addressed, such as the availability and sustainability of biomass for industry, possible scenario of monopolization, etc. Information is now available on the distribution of biomass availability on a regional basis (Pandey et al., 2009) which could be considered a milestone and could help to set up conversion technology plants, with a more feasible option for developing and underdeveloped countries, where cultivated land is dispersed. The lignocellulosic plant biomass is renewable and can be used for producing several compounds which are currently being sourced from petroleum. This potential has led to the development of a "biorefinery" concept where plant biomass is the raw material for generating fuel and chemicals. Lignocellulosic biomass is more attractive for the purpose as it does not compete with food availability, unlike starchy biomass. Cellulose is the most abundant and ubiquitous biopolymer on earth, considered to be an almost inexhaustible raw material. At the molecular level, it is a linear polymer of glucose composed of anhydroglucose units coupled to each other by β-1-4 glycosidic bonds. The number of glucose units in cellulose molecules varies from 250 to 10,000, depending on the source and pretreatment. Cellulose and hemicelluloses are the principal sources of fermentable sugars in lignocellulosic feedstock; however, nature has designed woody tissue for effective resistance to microbial attack. This is why crystalline cellulose is relatively impermeable not only to bigger molecules like protein but also to small molecules such as water in some cases. There are crystalline and amorphous regions in the polymer, and several types of surface irregularities exist. Due to the compact and stringent structure as well as its complex association with other components, very few reactive sites are available for enzyme attachment, which necessitates an appropriate pretreatment method (Pandey and Soccol, 2000). Suitable pretreatment methods disrupt lignin coating and make the fibers accessible to enzyme action.

2 CELLULASE: MODE OF ACTION

Cellulases are enzymes which hydrolyze the β-1, 4-D-glucan linkages in cellulose and produce as primary products glucose, cellobiose, and cello-oligosaccharides. Cellulases are produced by a number of microorganisms and comprise several different enzyme classifications. Three major types of cellulase enzymes are involved in the hydrolysis of native cellulose, namely, cellobiohydrolase (CBH), endo-β-1, 4-glucanase (EG), and β-glucosidase (BGL; Schulein, 1988). There are multiple enzymes within these classifications; for example,

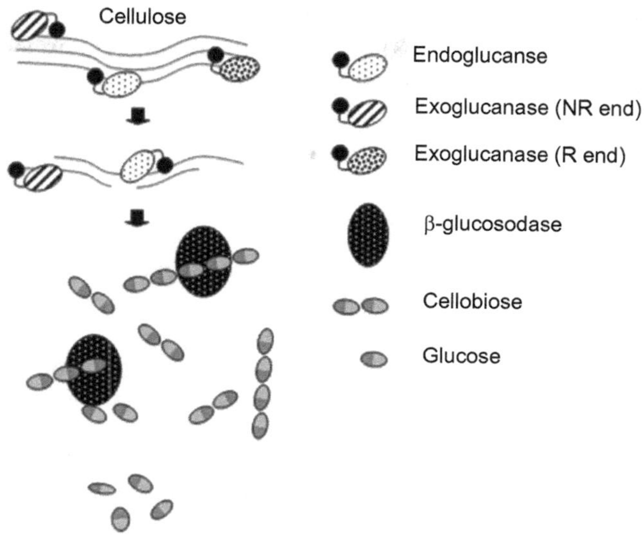

FIGURE 1 Mechanism of cellulase action.

the most studied fungus for cellulase production—*Trichoderma reesei*—produces two CBH components, not less than eight EG components and seven β-glucosidases (Aro et al., 2005). This is the most extensively studied multiple enzyme complexes comprising endoglucanase (EG), CBH, and β-glucosidases (BGL). For the complete hydrolysis of cellulose, the synergistic action of all the three enzyme components of cellulase is required. EG produces nicks in the cellulose polymer exposing reducing and nonreducing ends, exoglucanase (CBH) acts upon these reducing and nonreducing ends to liberate cello-oligosaccharides and cellobiose units, and β-glucosidases finally cleaves the cellobiose to liberate glucose completing the hydrolysis (Sukumaran et al., 2005). The complete cellulase system comprising EG, CBH, and BGL components thus acts synergistically to convert crystalline cellulose to glucose and has been depicted in Figure 1.

Majority of the cellulases have a characteristic two-domain structure with a catalytic domain (CD) and a cellulose-binding domain (CBD-also called carbohydrate-binding module, CBM) connected through a linker peptide (Ohmiya et al., 1997; Sakka et al., 2000). The core domain or the CD contains the catalytic site, whereas the CBDs help in binding of the enzyme to cellulose.

3 CELLULASE SYSTEMS AND THE CONTROL OF CELLULASE GENE EXPRESSION

Basic understanding of the cellulase systems and their regulation is imperative in the design of enzyme production and engineering strategies. Over several years of research, though the exact control mechanisms governing cellulase expression in microbes is not fully understood, considerable information is still available on this topic especially in the case of *T. reesei*.

Cellulase systems of microbes can be generally regarded as complexed or noncomplexed. Utilization of insoluble cellulose requires the production of extracellular cellulases by the organism. The cellulase systems consist of either secreted or cell associated enzymes belonging to different classes categorized based on their mode of action and structural properties. The three major type of cellulase activities recognized are

(1) EGs/1-4-β-D-glucanohydrolases/EG—(EC 3.2.1.4)
(2) Exoglucanases/1-4-β-D-glucan glucanohydrolases/Cellobiohydrolase/ CBH—(EC 3.2.1.74)
(3) β-Glucosidases/BG/BGL/β-glucoside glucohydrolases—(EC 3.2.1.21)

EGs cut at random at internal amorphous sites in the cellulose polysaccharide chain generating oligosaccarides and new chain ends. Exoglucanases act on the reducing and non-reducing ends of the cellulose chains liberating glucose, cellobiose, or cello-oligosaccharides as major products. β-Glucosidases hydrolyze soluble cellodextrins and cellobiose to glucose.

Noncomplexed cellulase systems from aerobic fungi and bacteria have the components of the cellulase system free and mostly secreted. Typical examples include the cellulase system from *T. reesei*. The fungus produces two exoglucanases—CBHI and CBHII, about eight EGs—EGI-EGVIII, and seven β-glucosidases—BGI-BGVII. The cellulase system of *Humicola insolens* is homologous to *T. reesei* and contains at least seven cellulases. An aerobic bacterium like Thermobifida also produces all components of the cellulolytic system including exo- and endoglucanases.

Complexed cellulase systems (Cellulosomes) on the other hand are native to anaerobic bacteria. Cellulosomes are protuberances on the cell wall of the bacteria which harbor stable enzyme complexes. The cellulolytic system of Clostridia has been studied in detail, and information on *Clostridium thermocellum* is by far the most comprehensive. In *C. thermocellum*, the cellulosome consists of a noncatalytic *cipA* protein which has different catalytic modules responsible for exo- and endoglucanase activities. Individual composition of the cellulosome varies with respect to the organism.

Cellulases are inducible enzymes and the regulation of the cellulase production is finely controlled by activation and repression mechanisms. Cellulase genes of *T. reesei* are coordinately regulated. The production of cellulolytic enzymes is induced only in presence of the substrate and is repressed when easily utilizable sugars are available. Natural inducers of cellulase systems have been proposed as early as 1962 (Mandels et al., 1962), and the disaccharide sophorose has since then been considered to be the most probable inducer of at least the *Trichoderma* cellulase system. It is proposed that the inducer is generated by the trans-glycosylation activity of basally expressed β-glucosidase. Cellobiose, δ-cellobiose-1-5 lactone and other oxidized products of cellulose hydrolysis can also act as inducers of cellulose. Lactose is another known inducer of cellulases and it is utilized in commercial production of the enzyme owing to economic considerations. Though the mechanism of lactose induction is not fully understood, it is believed that the intracellular galactose-1-phosphate levels might control the signaling. Glucose repression of cellulase system overrides its induction, and de-repression is believed to occur by an induction mechanism mediated by trans-glycosylation of glucose.

The promoter region of cellulases is found to harbor binding sites for the *CREI* catabolite repressor protein as well as sites for the transcriptional activators including Activator of Cellulase Expression proteins II (*ACE II*) besides the CCAAT sequence which binds general transcriptional activator complexes designated as "*HAP*" proteins, *ACEII* binds to the promoters of *cbh1* in *T. reesei*, and is believed to control the expression of *cbh1*, *cbh2*, *egl1*, and *egl2*. *Ace1* gene also produces a transcription factor similar to *ACEII* and has binding sites in the *cbh1* promoter, but it acts as a repressor of cellulase gene expression. Glucose repression of cellulase is supposed to be mediated through the carbon catabolite repressor protein *CRE1* in *T. reesei*. The promoter regions of *cbh1*, *cbh2*, *eg1*, and *eg2* genes of *T. reesei* have *CRE1*-binding sites indicating the fine control of these genes by carbon catabolite repression. Though this information gives better insight into the molecular biology of cellulase gene regulation, it is still unclear how the genes are coordinately regulated and what signals the activation of cellulase promoters by the transcriptional activators. Nevertheless, substantial research is being focused in this area and practical exploitation of the current knowledge can improve cellulase production by targeted interventions into the genetics of cellulolytic microbes.

4 CELLULASE PRODUCERS

Cellulolytic microbes are primarily carbohydrate degraders and are generally unable to use proteins or lipids as energy sources for growth (Lynd et al., 2002). There are wide variety of microorganisms involved in cellulase production including aerobic and anaerobic bacteria, white rot and soft rot fungi, and anaerobic fungi. In filamentous fungi, actinomycetes, and in aerobic bacteria, cellulases are mostly secreted as free molecules. Most of the bacteria are incapable of degrading crystalline cellulose since their cellulase system is incomplete. However, cellulolytic enzymes produced by filamentous fungi comprise all the three component of cellulase in different proportions and hence are capable of degrading cellulose completely. Most of the cellulases exploited for industrial applications are from filamentous fungi such as *Trichoderma*, *Penicillium*, *Fusarium*, *Humicola*, *Phanerochaete*, etc., where a large number of cellulases are encountered. Though the filamentous growth form causes difficulties in mass transfer compared to yeast or bacterial growth, efficient technologies have been developed for antibiotic, organic acid, and native enzyme production from filamentous fungi (Wiebe, 2003). *T. reesei* is one among the most potent cellulase producers studied in detail. It produces two CBHs (CBH I and CBH II) and the two EGs (EG1 and EG2), in a rough proportion of 60:20:10:10, which together can make up to 90% of the enzyme cocktail; while seven β-glucosidases-BGI-BGVII secreted by this fungus typically make up less than 1% (Lynd et al., 2002). Table 1 shows the commonly employed microorganism for cellulase production.

Even though *T. reesei*, *Penicillium*, *Aspergillus*, and *Humicola* can hydrolyze native cellulose, the reaction may be sometime very slow due to recalcitrance of biomass. Very rarely cellulose can be found in pure state in nature, as it usually is embedded in matrix of lignin and is bound with hemicelluloses. It is necessary to remove lignin from cellulose with proper pretreatment method to make cellulose accessible for the microorganisms. It is an important and a necessary step for commercial hydrolysis of lignocellulosic biomass.

5 PRETREATMENT

Cellulose and hemicelluloses are the principal source of C6 and C5 fermentable sugars in lignocellulosic feedstock; nature has designed woody tissue for effective resistance to microbial attack. It emphasizes on use of proper pretreatment method for making these accessible

TABLE 1 Major Microorganism Employed in Cellulase Production (Modified from Sukumaran et al., 2005)

	Microorganism	
Major Group	**Genus**	**Representative Species**
Fungi	*Trichoderma*	*T. reesei*
		T. longibrachiatum
		T. harzianum
	Humicola	*H. insolens*
		H. grisea
	Aspergillus	*A. niger*
		A. nidulans
		A. oryzae (recombinant)
	Penicillium	*P. brasilianum*
		P. occitanis
		P. decumbans
	Fusarium	*F. solani*
		F. oxysporum
	Melanocarpus	*M. albomyces*
	Phanerochaete	*P. chrysosporium*
Bacteria	*Bacillus*	*Bacillus* sp.
		Bacillus subtilis
	Pseudomonas	*P. cellulosa*
	Acidothermus	*A. cellulolyticus*
	Rhodothermus	*R. marinus*
	Clostridium	*C. acetobutylicum*
		C. thremocellum
Actinomycetes	*Thermononospora*	*T. fusca*
		T. curvata
	Cellulomonas	*C. fimi*
		C. bioazotea
		C. uda
	Streptomyces	*S. drozdowiczii*
		S. sp
		S. lividans

for microorganisms. Pretreatment of lignocellulosic biomass has been an actively researched field for several decades, and a wide variety of thermal, mechanical, chemical, and biological pretreatment approaches (and combinations thereof) have been investigated and reported in the scientific literature (McMillan, 1994). Pretreatment involves delignification of the feedstock in order to make cellulose more accessible during hydrolysis. It results in separation of lignin and hemicellulose components from cellulose, as well as enlarges the inner surface area of fibers thus paving a way for enhanced enzymatic hydrolysis. Steam explosion, alkali, and acid pretreatment are some of the common methods of pretreatment. Steam explosion is most commonly used and alkali pretreatment has been found to be better in lignin removal (Carrillo et al., 2005). Solid concentration is the key factor significantly affecting the process economics for a dilute acid pretreatment/enzymatic hydrolysis based process. Solid loading of 30% has been also investigated for dilute acid pretreatment. Still the relationship between enzymatic digestion and structural properties of pretreated material has to be explored for better understanding of the factors affecting cellulose hydrolysis. Active research has been carried out in this direction and several organic solvents as well as ionic liquids have been tested which shows promising results though have not reached to commercialization.

It is important that the selected pretreatment technology fulfill the following objectives:

1. Improve the enzymatic accessibility of the lignocellulosic compound
2. Result in the minimum loss of the potential sugars
3. Prevent the formation of molecules which are inhibitory to microbial degradation or enzymatic action
4. Pretreatment technology should be economically sound in order to make the overall process, that is, conversion of biomass to bioethanol a feasible technology

6 BIOPROCESSES FOR CELLULASE PRODUCTION

With the rejuvenated interest created due to their applications in lignocellulose conversion, several investigators worldwide are working on some or the other aspect of cellulase. Production of low titers of cellulase has always been a major concern, and thus several workers are trying to improve the production titers by adopting multifaceted approaches, which include the use of better bioprocess technologies, using cheaper or crude raw materials as substrates for enzyme production, bioengineering the microorganisms, etc. (Singhania, 2009). Bioprocess improvement strategies for enhancing the yield and specific activities of cellulases have also been well addressed by researchers worldwide. Majority of the reports on microbial production of cellulases utilizes the submerged fermentation technology (SmF) and the widely studied organism used in cellulase production—*T. reesei* has also been tested mostly in liquid media. However, in nature, the growth and cellulose utilization of aerobic microorganisms elaborating cellulases resembles solid-state fermentation (SSF) than a liquid culture (Hölker et al., 2004). During last two decades, SSF has regained interest due to the high titers of enzyme production employing fungal cultures. The lignocellulosic substrate type had the greatest impact on cellulase secretion. Some of the substrates significantly stimulated lignocellulolytic enzyme synthesis without supplementation of the culture medium with specific inducers (Elisashvili et al., 2009). Nevertheless, the advantages of better monitoring and handling are still associated with the submerged cultures.

Cellulase production in cultures is growth associated and is influenced by various parameters including the nature of the cellulosic substrate, pH of the medium, and nutrient availability; and a large-scale production of cellulases requires understanding and proper controlling of the growth and enzyme production capabilities of the producer. This is however extremely complicated since many factors and their interactions can affect cellulase productivity. Microbial cellulases are subject to induction and repression mechanisms and the process design and media formulation for cellulase production has to take care of these aspects. The media formulation for fermentation is of significant concern since no general composition can give the optimum growth and cellulase production. Also, the medium used is mostly specific for the organism concerned. In *T. reesei*, a basal medium after Mandels and Reese (1957) has been most frequently used with or without modifications. The carbon sources in majority of the commercial cellulase fermentations are cellulosic biomass including straw, spent hulls of cereals and pulses, rice or wheat bran, bagasse, paper industry waste, and various other lignocellulosic residues that induce the cellulase production. Majority of the cellulase production processes are batch processes, but fed batch or continuous mode helps to override the repression caused by the accumulation of reducing sugar. The major technical limitation in fermentative production of cellulases remains the increased fermentation times with a low productivity.

Information on the type of bioprocesses employed for cellulase production, microorganism employed as well as magnitude of production is available in reviews by Sukumaran et al. (2005) and Singhania et al. (2010).

6.1 Solid-State Fermentation

SSF is defined as the fermentation in absence or near absence of free water (Pandey, 1994). SSF for production of industrial enzymes is rapidly gaining interest as a cost-effective technology as the microorganisms, especially fungal cultures, produce comparatively high titers of metabolites due to the conditions of fermentation which shows similarity to the natural environment (Pandey et al., 1999, Singhania et al., 2009). Filamentous fungi as *T. reesei, A. niger, Penicillium* sp., etc. have been employed for cellulase production using SSF where a basal mineral salts medium was used for moistening the substrate. Figure 2 shows general steps for SSF process for the production of cellulase. Koji chamber can be used for large-scale production for the economic reason, though maintenance of sterile condition is difficult. Any cellulosic biomass could be employed as substrate. For cellulase production, inoculums can be prepared in stirred tank reactor and can be sprayed on to the sterile medium in the shallow tray. Either spores or mycelium can be used as inoculum in case of filamentous fungi. In this case, temperature and humidity are controlled inside the chamber, and incubation is allowed till 7 days or as per specified. Suitable buffer or distilled water with appropriate tween percentage is used as extraction liquid. Medium is homogenized with extraction liquid and centrifuged to remove the biomass and cell debris. Supernatant contains the extracellular cellulase which could be concentrated by acetone precipitation or salting out. For biomass hydrolysis, it could be used as such and for other application it depends on the degree of purity of cellulases required.

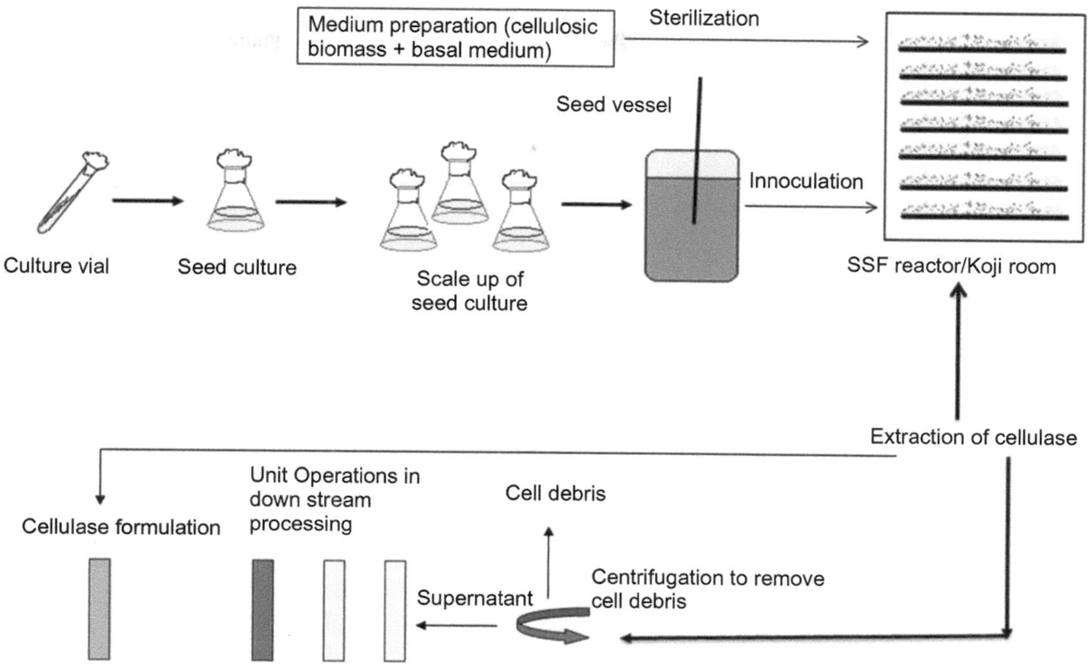

FIGURE 2 General outline of cellulase production employing Koji chamber as SSF bioreactor.

A well-designed solid-state fermentor should (1) have perfect control systems for temperature, air flow rate, and humidity; (2) have a well-designed system for preventing contamination; (3) be homogeneous in water activity, temperature, and composition so that microbes can grow uniformly; (4) be able to remove harmful metabolites, such as CO_2, quickly; and (5) be labor saving and easy to scale-up for handling solid medium. Till now, none of the available SSF bioreactors could satisfy all the points. Several bioreactors which were engineered for cellulase production to satisfy the discussed points and enable continuous monitoring are shallow tray fermentor, column fermentor, deep trough fermentor, rotating drum fermentor, stirred tank fermentor, rotating disk reactor, rocking drum reactor, and fluidized bed fermentor, though each of them had their own limitations (Cen and Xia, 1999).

Fermentors are engineered in a way to maintain growth and production conditions. There are several key factors which plays an important role in cellulase production via SSF such as pH, temperature, moisture content and water activity, aeration, and substrate composition. These operating conditions may differ with the organism and substrate used. For example, fungi prefer to grow at acidic pH, low moisture content (35-70%) compared to bacteria (70-90%) and usually grow well at 25-30 °C, whereas bacteria prefer neutral pH, high moisture content, and grow well at 37 °C. In case of SSF, lot of heat is generated due to vigorous metabolic activities of organisms. There is always limitation of heat transfer as it is relatively poor in solid layer and overheating occurs in substrate particle. This causes unfavorable conditions for spores germination, mycelia growth, and enzyme accumulation and secretion. Temperature control in the environment of the solid-state fermentor is relatively convenient to achieve, but temperature regulation within the solid substrate layer is relatively difficult. Few measures could be taken such as a proper thickness of the solid phase which could facilitate heat transfer and also aeration rate has to be controlled to supply oxygen and mass transfer for which convection could be employed. Controlling the moisture content in the medium is the key factor for cellulase production which is an essential component for growth of microorganisms. If the water content is high, the void space, as well as the gas-phase volume within the solid substrate, is reduced, which increases the mass transfer resistance of oxygen and carbon dioxide, as well as the possibility of contamination, whereas low water content is unfavorable to spore germination and substrate swelling. Substrate swelling is essential for fungi to attack and to digest the solid substrate. Water activity is even more important than the moisture content which is closely related moisture content but is not exactly equal to it. It gives the amount of unbounded water in the immediate surroundings. It is necessary to maintain the optimal value, but it tends to vary because of metabolism and evaporation. To a certain extent, it could be controlled by humidifier which could be incorporated into solid-state fermentor/bioreactor. Another important factor is pH which affects the growth of microorganism and hence the cellulase production. It is difficult to monitor the pH in solid substrate, but pH of the basal medium could be adjusted which usually contains nitrogen sources having buffering capacity. Solid cellulosic biomass has the buffering capacity which rule out the necessity to adjust the pH during SSF. Though there are several indirect methods for biomass measurement such as total protein estimation, fungal cell wall component measurement (*n*-acetyl glucosamine), etc., as well as direct methods such as CO_2 evolved and O_2 intake, but in case of SSF, measurement of biomass is difficult. So, it is not feasible to monitor the growth pattern of microorganisms which makes it difficult to develop suitable models for SSF. Nevertheless, solid substrate fermentation can be proposed as a better technology for commercial production of cellulases considering the low-cost input and ability to utilize naturally available sources of cellulose as substrate.

6.2 Submerged Fermentation

Submerged fermentation has been defined as fermentation in the presence of excess water. Almost all the large-scale enzyme producing facilities are using the proven technology of SmF due to better monitoring and ease of handling. Though bacteria and actinomycetes are also reported for cellulase production, the titers are very low to make the technology economically feasible. Most of the commercial cellulases are produced by the filamentous fungi—*T. reesei* or *A. niger*—under SmF. Cellulase production in cultures is highly influenced by various parameters including the nature of the cellulosic substrate, pH of the medium, nutrient availability, inducer supplementation, fermentation temperature, etc. Mostly, pure cellulose preparations like Solka-Floc and Avicell have been used in the liquid cultures of cellulolytic microbes for production of the enzymes, but while using soluble substrates, the break down products may hamper cellulase synthesis by promoting catabolite repression due to accumulation of free sugars. Increased production in fermentor may be achieved by a gradient feed of a suitable cellulose and maintenance of process conditions at their optimal. Large-scale production of cellulases requires understanding and proper controlling of the growth and enzyme production capabilities of the producer. Cellulases produced by compost organisms such as the filamentous fungi—*Trichoderma, Penicillium, Aspergillus, Humicola*, etc., can perform at diverse ranges of pH and temperature. Microbial cellulases are subject to induction and repression mechanisms and the process design and media formulation for cellulase production has to take care of these aspects. A two-stage continuous process for cellulase production could be employed in which the growth phase and production phase could be separated by different pH and temperature optima. This could help in overcoming the technical limitation of low productivity and long fermentation time for cellulase production. Repression by glucose and cellobiose is a known feature of cellulase systems, and several attempts have been directed toward development of mutants resistant to catabolite repression.

For submerged fermentation, huge bioreactors are available and also provide ease of control of various operating factors such as pH, temperature, aeration, etc., Figure 3 shows general steps involved in cellulase production via submerged fermentation. Till date, SmF is the most accepted technology for industrial production of primary and secondary metabolites. In submerged fermentation, all the parameters required for modeling can be monitored, and hence most of the modeling studies have been done for metabolites production is via SmF. List of fermentation technology adapted for cellulase production with the magnitude, microorganism employed as well as the amount of cellulase produced has been given by Sukumaran et al. (2005) as well as Singhania et al. (2010).

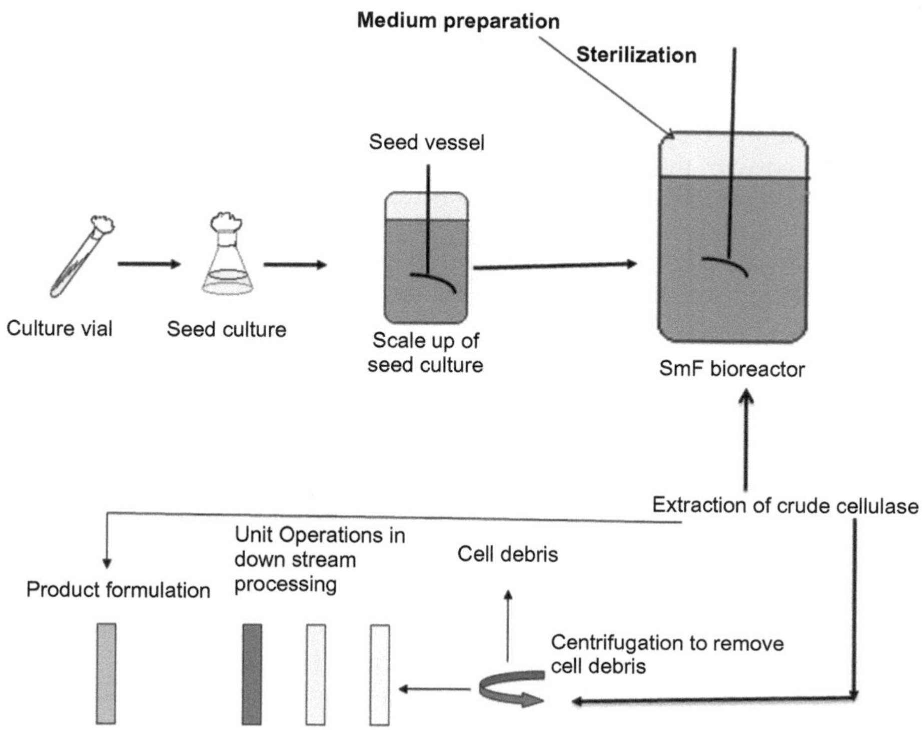

FIGURE 3 General steps involved in Cellulase production by SmF.

7 APPLICATIONS OF CELLULASES

Microbial cellulases find applications in a variety of industries where cellulases of varying degrees of purity are desired. Cellulases were initially investigated several decades back for the bioconversion of biomass which gave way to research in the industrial applications of the enzyme in animal feed, food, textiles, detergents, and in the paper industry. With the shortage of fossil fuels and the rising need to find alternative source for renewable energy and fuels, there is a renewal of interest in the bioconversion of lignocellulosic biomass using cellulases and other enzymes. In the other fields, however, the technologies and products using cellulases have reached the stage where these enzymes have become indispensable.

7.1 Textile Industry

Due to their ability in modifying the cellulosic fibers in a controlled and desired fashion so as to improve the quality of fabrics, cellulases have become the third largest group of enzymes used in the industry since their introduction only since a decade. Cellulases are used in the biostoning of denim garments for producing softness and the faded look of denim garments replacing the use of pumice stones which were traditionally employed in the industry. Cellulases act on the cellulose fiber to release the indigo dye used for coloring the fabric, producing the faded look of denim. The neutral/alkaline cellulases are the most preferred type of cellulases for the stonewash industry because they result in lower levels of back staining or redeposition and lower strength loss than acid cellulases. *H. insolens* cellulase is most commonly employed in the biostoning, though use of acidic cellulase from *Trichoderma* along with proteases is found to be equally good. Another important application of cellulases in the textile industry is for the biopolishing of fabric. Fuzz formation and pilling are common problems associated with the fabric using cotton or other natural fibers and cellulases are utilized for digesting off the small fiber ends protruding from the fabric resulting in a better finish. In addition to stone washing, the other textile applications in which cellulases have been used include softening and defibrillation. Cellulases have also been used in processes for providing localized variation in the color density of fibers.

7.2 Laundry and Detergents

Cotton or cotton blended garments tend to lose their color and become fluffy after several washings due to the altered microfibrillar structure. The fibrils are partially detached and form a layer over the surface of the fabric providing surface for reattachment of dirt from wash liquid. This also creates a dull look since the original color is masked by the thin layer of detached microfibrils harboring the dirt. Cellulases added to the detergents contribute to the primary washing performance, that is, the actual cleaning action, and also to the secondary washing performance of the detergent (ability to keep the dirt which has been removed from the fabric dissolved or suspended in the liquor, thus preventing it from being redeposited on the cleaned textile), and they have a finishing action which include the smoothing of textile by removing cellulose aggregates (antipilling), and they have a softening action that helps in color restoration. Cellulases, in particular EGIII and CBH I, are commonly used components of detergents for cleaning textiles. Several reports disclose EG III variants, in particular from *T. reesei*, suitable for the use in detergents. *T. viride* and *T. harzianum* are also industrially utilized natural sources of cellulases, as are Aspergillus, in particular *A. niger*. Cellulase preparations, mainly from species of *Humicola* active under mild alkaline conditions and at elevated temperatures, are commonly added in washing powders. *H. insolens* and *H. grisea* var. *thermoidea* cellulases for use in detergents are described which particularly deal with the fabric-softening effect.

7.3 Food and Animal Feed

In food industry, cellulases are used in extraction and clarification of fruit and vegetable juices, production of fruit nectars and purees, and in the extraction of olive oil. Glucanases are added to improve the malting of barley in beer manufacturing, and in wine industry, better maceration and color extraction is achieved by use of exogenous hemicellulases and glucanases. Cellulases are also used in carotenoid extraction in the production of food coloring agents. Animal feed industry is another major consumer of the cellulases in the processing of feed. *Trichoderma* cellulases, when used as a feed additive, improves the feed conversion ratio and/or increase the digestibility of a cereal-based feed.

7.4 Pulp and Paper Industry

In the pulp and paper industry, cellulases and hemicellulases have been employed for biomechanical pulping for modification of the coarse mechanical pulp and hand sheet strength properties, de-inking of recycled fibers, and for improving drainage and runnability of paper mills. The use of enzymes in wood pulping considerably reduces the energy requirement. As more and more importance is given to recycling of paper, the need for environment friendly de-inking of printed paper is also increasing. Cellulases are employed in the removing of inks, coating, and toners from paper. Biocharacterization of pulp fibers is another application where microbial cellulases are employed. Cellulases are also used in preparation of easily biodegradable cardboard. The enzyme in a stable formulation is added during the manufacturing process into the cardboard and gets activated once it contacts moisture. This helps accelerated degradation of the cardboard, making it a suitable biodegradable packaging material for several products. The enzyme is employed in the manufacture of soft paper including paper towels and sanitary paper, and preparations containing cellulases are used to remove adhered paper.

7.5 Biofuel

Perhaps the most important application currently being investigated actively is in the utilization of lignocellulosic wastes for the production of biofuel. The lignocellulosic residues represent the most abundant renewable resource available to mankind for effective utilization; their use is limited only by the lack of cost-effective technologies. A potential application of cellulase is the conversion of cellulosic materials to glucose and other fermentable sugars which in turn can be used as microbial substrates for the production of single cell proteins or a variety of fermentation products like ethanol. Organisms with cellulase systems and capable of converting biomass to alcohol directly are already reported in literature. Nevertheless, it is also noted that none of these systems described are effective alone to yield a commercially viable process. The strategy employed currently in bioethanol production from lignocellulosic residues is a multistep process involving pretreatment of the residue to remove lignin and hemicellulase fraction, cellulase treatment at 50 °C to hydrolyze the cellulosic residue to generate fermentable sugars, and finally use of a fermentative microorganism to produce alcohol from the hydrolyzed cellulosic material. The cellulase preparation needed for the bioethanol plant is prepared in the premises using the same lignocellulosic residue as substrate, and the organism employed is almost always *Trichoderma ressei*. In the effort to develop efficient technologies for biofuel production, significant research have been directed toward the identification of efficient cellulase systems and process conditions, besides studies directed at the biochemical and genetic improvement of the existing organisms utilized in the process. The use of pure enzymes in the conversion of biomass to ethanol or to fermentation products is currently uneconomical due to the high cost of commercial cellulases. Effective strategies are yet to resolve and active research has to be taken up in this direction. Overall, cellulosic biomass is an attractive resource that can serve as substrate for the production of value added metabolites and cellulases as such.

7.6 Cellulases for Bioconversion

Microbial cellulases find applications in a variety of industries where cellulases of varying degrees of purity are desired. Though cellulases were initially investigated several decades back for the bioconversion of biomass, this later became unattractive and the other industrial applications of the enzyme as in animal feed, food, textiles and detergents and in the paper industry were predominantly pursued. However, with the shortage of fossil fuels and the arising need to find alternative sources for renewable energy and fuels, there is a renewal of interest in the bioconversion of lignocellulosic biomass using cellulases and other lignocellulolytic enzymes.

Cellulases are available in the market under different names or trade mark for different applications which could be tried for biomass hydrolysis also. It would not be feasible to predict the efficiency of cellulases for bioconversion on the basis of standard assays as there are no clear relationships between cellulase activities on soluble substrates and those on insoluble substrates (Nieves et al., 2009). So, the soluble substrates should not be used to predict the efficiency of cellulases for processing relevant solid substrates, such as plant cell walls. The choice of the enzyme preparation for a particular biomass would be particularly more dependent on biomass characteristics rather than on standard enzyme activities measured (Kabel et al., 2005). Preparation having higher FPU activities are desirable for bioconversion as filter paper is seen as highly crystalline cellulose, the degradation of which, depends on the combination of activities of EG and CBH, where the EG create new chain ends for the CBH to split off cellobiose which further get attacked by BGL to give monomers as glucose. Preparations of cellulase from a single organism may not very efficient for hydrolysis of a particular feed stock. Though the filamentous fungi are the major source of cellulases and hemicellulases and the mutant strains of *Trichoderma* including *T. reesei*, *T. viride*, and

T. longibrachium are the best known producers of the enzyme, it is also well known that these species of *Trichoderma* have a low level of β-glucosidase activity resulting in an inefficient biomass hydrolysis. Cellulases for biomass conversion could be a blend or enzyme cocktail containing endo- and exocellulase, xylanase, β-glucosidase, pectinase, etc., which could vary for different biomass on the basis of their composition (Sukumaran et al., 2009). The hydrolytic efficiency of a multienzyme complex for lignocellulose saccharification depends both on properties of individual enzymes and their ratio in the multienzyme cocktail. The ideal cellulase complex must be highly active on the intended biomass feedstock, able to completely hydrolyze the biomass, operate well at mildly acidic pH, withstand process stress, and be cost effective. The success of any lignocellulosic ethanol project will depend on the ability to develop such cellulase systems. The key to developing cellulases those are effective toward a particular biomass feedstock is to artificially construct them either by enzyme assembly to form cocktails or to engineer the cellulase producers to express desired combination of cellulase enzymes. Both these approaches have been tried with success. Enzyme cocktails have been developed by mixing *T. reesei* cellulase with other enzymes including xylanases, pectinases, and β-glucosidases, and these cocktails were tried for hydrolysis of various feed stock. One of the recent examples of cocktails developed, include the multienzyme complex developed based on highly active *Chrysosporium lucknowense* cellulases (Gusakov et al., 2007).

With the enzyme majors Genencor and Novozymes already achieving their set targets of reducing enzyme cost for lignocellulosic ethanol production, and with still further improvements predicted, it becomes apparent that cost of enzymes may not be a major limiting factor in the biomass-ethanol process. Nevertheless, we have a long way to go in understanding the mechanisms of cellulase gene regulation and the structure to function relationships.

7.6.1 LCE (Lignocellulosic Ethanol) Employing Cellulases

The idea of generation of ethanol from lignocellulosic residues has been conceived by NREL (Northern Renewable Energy Laboratory) in USA. In order to make it competitive with gasoline by the turn of the century, an extensive program is going on with a strategy that will reduce the cost of bioconversion of biomass to biofuel ethanol, in countries like Canada, Denmark, and Brazil. It was proposed to be done in two steps, that is, hydrolysis of lignocellulosic material into their monomers and thereby its further conversion into ethanol by fermentation.

Due to the apparent advantages of ethanol having high octane rating and also being a renewable alternative to existing transport fuels, there is now an increased interest in commercializing technologies for its production from inexpensive biomass (Schell et al., 2004). Most of the fuel ethanol produced in the world is currently sourced from starchy biomass or sucrose (molasses or cane juice), but the technology for ethanol production from non-food-plant sources is being developed rapidly so that large-scale production will be a reality in the coming years. The process of converting low-value biomass to ethanol via fermentation depends on the development of economically viable cellulolytic enzyme to achieve effective depolymerization of the cellulosic content of the biomass. Reduction in cost of "biomass-ethanol" may also be achieved by efficient technologies for saccharification which includes the use of better "enzyme cocktails" and conditions for hydrolysis (Mathew et al., 2008). Cellulase preparation used in this process must hydrolyze crystalline cellulose completely, operate effectively at mild pH, withstand process stress, and they need not be derived from microbes that are generally regarded as safe (GRAS). The ability to engineer cellulase systems in anticipation of each application is key to successful optimization and commercialization. Agroresidues could be used as raw material for bioethanol production. A part of it can be used in the site itself for generating energy, but still a major part of it constitutes waste. Disposal of these residues itself is a major problem, causing pollution. Advances in industrial biotechnology offer ample opportunities on economic utilization of agroindustrial residues. According to Indian scenario, rice straw and sugarcane tops can be the probable feedstock for long-term motive (Pandey et al., 2009). Bioethanol production involves several steps starting from selection of proper feedstock, its pretreatment, cellulase production, hydrolysis of feedstock using cellulases, and finally fermentation of hydrolysate to obtain ethanol. This bioconversion of cellulose (enzymatic hydrolysis) is the costliest step in overall process which could be brought down by employing multifaceted approach as cheaper raw material for enzyme production, cheaper technology as SSF, appropriate feedstock for bioconversion as well as appropriate pretreatment method. Artificial cellulase preparation and engineering cellulases can help to modify cellulase to suit for the particular application. Expression cassettes, site-directed mutagenesis, and antisense technology have been successfully employed in designing cellulase. Potent cellulase gene from different filamentous fungi can be isolated, cloned, and expressed in the host organism to get better combination or synergism. Enzyme cocktail can be prepared using cellulases from different sources to achieve maximum efficiency which otherwise is not possible due to lack of one or the other component of native cellulase. Cellulase from *T. reesei* can be supplemented by β-glucosidase from *A. niger* to overcome repression and feed back inhibition of β-glucosidase in *T. reesei*.

Though the current applications of cellulases in industries such as food and textile themselves generate millions of dollars worth of economy, it is envisaged that the utilization of lignocellulosic biomass for biofuel production will be the major area where cellulases would be commercially exploited in the future. The greatest potential for ethanol production from biomass lies in the enzymatic hydrolysis of cellulose using cellulolytic enzymes (Singhania et al., 2008). Even after decades of research on these enzymes, the cost of cellulases still is high to be used economically in the bioconversion of biomass, and the major challenge for cellulosic ethanol is the cost reduction of enzymes. Large-scale applications of bioethanol in fuel blends will reduce the CO_2 and other emissions from transport sector. Approximately 17 million tons of fuel ethanol is currently being produced from sugar cane and starch crop residues in Brazil, USA, and some EU countries combined at the cost of about 0.5-0.7 $/l, which is about twice the price of gasoline. The USA and European market for bioethanol is projected to grow considerably in the coming years due to the policies taken to substitute at least a fraction of the fossil transport fuels by renewable biofuels. Lignocellulose to ethanol production technology has been extensively investigated in the USA, Canada, and some EU countries (Reith et al., 2002; Wooley et al., 1999).

8 CELLULASE MARKET SCENARIO

Current international players in the production of commercial cellulases include the enzyme-manufacturing giants Genencor and Novozymes. National Renewable Energy Laboratory of the United States have set their goals for reducing the cost of cellulases used in bioethanol production for which projects were initiated in 2000 with Genencor Corporation and Novozymes as contract partners. Genencor announced in 2004 that it has achieved an estimated cellulase cost in the range of $0.10-$0.20 per gallon of ethanol in NREL's cost model (Genencor press release 21 October 2004, Genencor celebrates major progress in the conversion of biomass to ethanol. http://genencor.com/cms/connect/genencor/media_relations/news/archive/2004/gen_211004_en.htm). Similarly, the collaborative subcontract between Novozymes and NREL has been able to reduce the cost of cellulases for biomass to ethanol to $0.10-0.18 gal^{-1} which is an almost 30 fold reduction from estimated cost in 2001. Novozymes predicts that their enzymes will make it possible to produce second-generation bioethanol by 2010. The company also has announced the setting up of an $80-100 million production facility in Nebraska for cellulase production (Novozymes Press Release, June 23, 2008. http://www.novozymes.com/en/MainStructure/PressAndPublications/PressRelease/2008/New+Facility+in+Nebraska.htm). The demand for cellulases is consistently on the rise due to its diverse applications. There are several other companies also involved in cellulase production for textile detergent, paper industries, and other industries. "Genencor" and "Novozyme" have played a significant role in bringing down the cost of cellulase several folds by their active research and are continuing to bring down the cost by adopting novel technologies. Recently, Genencor has launched Accelerase®1500, a cellulase complex intended specifically for lignocellulosic biomass processing industries. It is claimed to be more cost effective and efficient for bioethanol industries than the earlier Accelerase®1000. Accelerase®1500 is produced with a genetically modified strain of *T. reesei*. Accelerase®1500 is claimed to contain higher levels of β-glucosidase activity than all other commercial cellulases available today, so as to ensure almost complete conversion of cellobiose to glucose (http://www.genencor.com/wps/wcm/connect/genencor/genencor/products_and_services/business_development/biorefineries/products/accellerase_product_line_en.htm).

Genencor has also launched Accelerase® XY accessory xylanase enzyme complex that enhances both xylan (C5) and glucan (C6) conversion when blended with other Accelerase® enzyme products. Similarly, Accelerase® XC is an accessory xylanase/cellulase enzyme complex that contains a broad profile of hemicellulase and cellulase activities and enhances both xylan (C5) and glucan (C6) conversion when blended with other Accelerase® enzyme products. Also, Accelerase® BG is an accessory β-glucosidase enzyme that enhances glucan (C6) conversion when blended with cellulase products. There are several potential cellulases which may prove effective for biomass hydrolysis when supplemented with β-glucosidase, so indicating the importance of Accelerase® BG (http://www.genencor.com/wps/wcm/connect/genencor/genencor/products_and_services/business_development/biorefineries/products/accellerase_product_line_en.htm).

Novozyme has diverse range of cellulases available based on application such as Cellusoft®AP and Cellusoft®CR for bioblasting in textile mills, Carezyme® and Celluclean for laundry in detergent, Denimax® 6011 for stonewash industry at low temperature as well as many others specific for particular application (Novozymes Press Release, June 23, 2008. http://www.novozymes.com/en/MainStructure/PressAndPublications/PressRelease/2008/New+Facility+in+Nebraska.htm).

Novozyme also announced the availability of cellulase preparation specifically for biomass hydrolysis last year, though no information is available on the source of production as well as availability in the market. Amano Enzyme Inc. in Japan is also involved actively in enzyme production and is positioned among global enzyme producers. A majority of the world's total supply of industrial enzymes is produced in Europe, USA, and Japan.

Though most of the enzyme producing companies worldwide is involved in production and marketing of cellulases for diverse applications, there are very few of them who develop cellulases for biomass conversion, the most successful of them probably being Genencor and Novozyme. Table 2 shows the major players marketing cellulases with different trade mark and their source of origin, most of which may be genetically modified strains.

TABLE 2 Commercial Cellulases Produced by Companies and Their Sources

Enzyme Samples	Supplier	Source
Cellubrix (Celluclast)	Novozymes, Denmark	*T. longibrachiatum* and *A. niger*
Novozymes 188	Novozymes	*A. niger*
Cellulase 2000L	Rhodia-Danisco (Vinay, France)	*T. longibrachiatum/T. reesei*
Rohament CL	Rohm-AB Enzymes (Rajamaki, Finland)	*T. longibrachiatum/T. reesei*
Viscostar 150L	Dyadic (Jupiter, USA)	*T. longibrachiatum/T. reesei*
Multifect CL	Genencor Intl. (S. San Francisco, CA)	*T. reesei*
Bio-feed beta L	Novozymes	*T. longibrachiatum/T. reesei*
Energex L	Novozymes	*T. longibrachiatum/T. reesei*
Ultraflo L	Novozymes	*T. longibrachiatum/T. reesei*
Viscozyme L	Novozymes	*T. longibrachiatum/T. reesei*
Cellulyve	50L Lyven (Colombelles, France)	*T. longibrachiatum/T. reesei*
GC 440	Genencor-Danisco (Rochester, USA)	*T. longibrachiatum/T. reesei*
GC 880	Genencor	*T. longibrachiatum/T. reesei*
Spezyme CP	Genencor	*T. longibrachiatum/T. reesei*
GC 220	Genencor	*T. longibrachiatum/T. Reesei*
Accelerase®1500	Genencor	*T. Reesei*
Cellulase AP30K	Amano Enzyme	*A. niger*
Cellulase TRL	Solvay Enzymes (Elkhart, IN)	*T. reesei/T. Longibrachiatum*
Econase CE	Alko-EDC (New York, NY)	*T. reesei/T. Longibrachiatum*
Cellulase TAP106	Amano Enzyme (Troy, VA)	*T. viride*
Biocellulase TRI	Quest Intl. (Sarasota, FL)	*T. reesei/T. Longibrachiatum*
Biocellulase A	Quest Intl.	*A. niger*
Ultra-Low Microbial (ULM)	Iogen (Ottawa, Canada)	*T. reesei/T. Longibrachiatum*

9 ENGINEERED/ARTIFICIAL CELLULASES

Though several filamentous fungi are capable of cellulase production, the yield of the enzyme and the levels of individual cellulase components are not often satisfactory for commercialization. Improvements in cellulase titers as well as the ability to tailor the ratios of endo- and exoglucanases and β-glucosidase produced by organisms are highly desired for biomass conversion. Very relevant information related to cellulase gene regulation was revealed earlier, and now a study on the *T. reesei* genome revealed that the genome of the fungus contains fewer cellulases and hemicellulases than any other sequenced fungi despite being the best known producer of cellulases (Martinez et al., 2008). Genes coding for enzymes acting on carbohydrate polymers are distributed in clusters, and there are indications on the existence of numerous biosynthetic pathways for secondary metabolite production. However, the authors could not find any deep insight into the highly efficient protein secretion machinery in the fungus at least in the initial analysis. This work has tremendous implications in understanding the genetics of this important organism, which is used to produce cellulase enzymes and other important proteins. Also, such knowledge will enable improved production processes critical to reducing the cost of biomass conversion.

T. reesei and other filamentous fungi produce noncomplexed cellulases. Cellulase engineering for noncomplexed cellulase systems could be divided into three major research directions: (1) rational design for each cellulase, based on knowledge of the cellulase structure and the catalytic mechanism; (2) expression cassette and directed evolution for each cellulase, in which the improved enzymes or ones with new properties were selected after random mutagenesis and/or molecular recombination; and (3) the reconstitution of cellulase cocktails active on insoluble cellulosic substrates, yielding an improved hydrolysis rate or higher cellulose digestibility. Improvements in specific cellulase activities for noncomplexed cellulase mixtures can be implemented through cellulase engineering based on rational design or directed evolution for each component of cellulase, as well as its reconstitution.

Potent cellulase genes from filamentous fungi such as *Trichoderma* and *Aspergillus* can be isolated, cloned, and expressed in fungal hosts to get better combination or synergism. The cellobiohydrolase I (CBHI) promoter of *T. reesei* is a highly efficient known promoter with unusually high rate of expression under cellulase induction conditions and has been used to drive the expression of β-glucosidase and EG, thereby improving the cellulase profile of the host strain. The promoter has also been used to drive the expression of various homologous and heterologous proteins in *Trichoderma*. Glucose repression of cellulase genes has been addressed by using a truncated CBH I promoter lacking binding sites for the carbon catabolite repressor CRE1. Another major strategy employed for improving cellulase production in presence of glucose is to use promoters that are insensitive to glucose repression. For example, promoters of transcription elongation factors *1α* and *tef1*, and that of an unidentified cDNA (*cDNA1*) for driving the expression of EG and CBH in *T. reesei* could be used resulting into the de-repression of these enzymes (Nakari-Setala and Pentilla, 1995). These studies indicate that proper engineering of sequences to obtain expression of proteins from *cbh1* promoter and manipulations of the promoter to abolish repression can dramatically improve production of the cloned protein.

The cellulase system of *T. reesei* as well as of several other fungi is limited by the relatively lesser amount of β-glucosidase and its feed back inhibition by glucose. Beta-glucosidase which is insensitive or at least tolerant to glucose and cellobiose is highly desired for the conversion of cellulosic biomass to glucose as cellulase systems of several other fungi is limited by the relatively lesser amount of β-glucosidase and its feed back inhibition by glucose. Research on this line has yielded potential β-glucosidases from different microorganisms like *Candida peltata, Aspergillus oryzae,* and *A. niger*. One of the major approaches taken toward improving the cellulase complex for biomass hydrolysis is to increase the copy number of β-glucosidase gene and thus the amount of the BGL enzyme in the cellulase mixture produced by *T. reesei*, while other is to alter the cellulase profile of *T. reesei* by introducing glucose-tolerant BGL gene into the fungus.

Preparations of cellulase from a single organism may not be highly efficient for hydrolysis of different feed stock. Details have been given in Section 7.6.

Another interesting idea is the use of artificial cellulosomes generated by engineering cellulosome-bearing bacteria to express heterologous cellulases. Chimeric cellulosomes have been described for degradation of cellulosic substrates either by incorporating bacteria or fungal cellulases in cellulosomes by genetic engineering. The artificial cellulase complexes displayed enhanced activities compared to the corresponding free systems at least in the case of the bacterial enzymes. The benefits of developing heterologous cellulase expression systems in rapidly growing bacteria include substantial enhancement of enzyme stability and specific activity, the potential for greater cell densities using fed-batch cultures, a dramatic reduction in cell-growth time, and the potential for protein overproduction. The enhancement in activity could be due to the additional synergy induced by enzyme proximity within the complex and the effect of the cellulose-binding module offered by the chimeric scaffolding that anchors the whole complex at substrate surface.

The approaches discussed could be useful for developing cellulases for various specific applications, most importantly for bioconversion.

10 FUTURE PERSPECTIVES

Lignocellulose comprises a majority of the plant biomass produced on earth. This vast resource is the potential source of biofuels, biofertilizers, animal feed, and chemicals besides being the raw material for paper industry. Exploitation of this renewable resource needs either chemical or biological treatment of the material, and in the latter context cellulases have gained wide popularity over the past several decades. Research has shed light into the mechanisms of microbial cellulase production and has led to the development of technologies for production and applications of cellulose-degrading enzymes. However, there is no single process which is cost effective and efficient in the conversion of the natural lignocellulosic materials for production of value-added products. Use of the current commercial preparations of cellulase for bioconversion of lignocellulosic waste is economically not feasible.

The major goals for future cellulase research would be reduction in the cost of cellulase production, which could be attained either (1) by increasing the production level (2) using the cheaper raw material as a substrate for the production, (3) using the alternative cheaper production technologies such as SSF, (4) by improving the efficiency of cellulases. The former task may include such measures as optimizing growth conditions or processes, whereas the improvement in cellulase efficiency requires directed efforts in protein engineering and microbial genetics to improve the properties of the enzymes.

Optimization of growth conditions and processes has been attempted to a large extent in improving cellulase production. The section on fermentation production of cellulases describes many of these works basically dealing with empirical optimization of process variables to improve productivity. Many of the current commercial production technologies utilize submerged fermentation technology and employ hyperproducing mutants. In spite of several efforts directed at generating hyperproducers by directed evolution, the cost of enzymes has remained high. Alternative strategies thought of in cellulase production include mainly SSF on lignocellulosic biomass particularly by using host/substrate specific microorganisms. Filamentous fungi have been well exploited for the production of optimal enzyme complex for the degradation of host lignocellulose, as SSF imitates their natural survival conditions rather than generating an artificial habitat. It is also reported that the performance of enzyme complexes on lignocellulosic material is best when these complexes are prepared with the same lignocellulosic material as the host/substrate in fermentation. Another strategy is to use mixed culture in the production of enzyme. Mixed culture gives improved production and enzyme complexes with better hydrolytic activity. Thus, among the other strategies tried in production optimization and process developments for cellulase enzyme production, SSF may be considered as a cost-effective means for large-scale production of cellulases which probably would be several fold cheaper compared to the current commercial preparations (Singhania et al., 2007). But SSF has its own limitations, as it is still not feasible to monitor regularly as well as to provide controlled condition for the fermentation. Several large-scale SSF bioreactors have been engineered for cellulase production which has been discussed in detail in Cen and Xia (1999) review, focusing in the direction of overcoming these limitations.

Over several decades, the basic studies on cellulase have moved in the direction of understanding the enzymatic diversity. There is now a vast and diverse understanding of the regulation of enzyme production, but still we lack comprehensive and specific knowledge on the mechanism of induction of cellulase by any of the known inducers. No information is available on the nature of intracellular inducers, the possible signaling pathways, and the cofactors and transducers involved in the induction of cellulase. Recent reports have shown that cellulases are subject to regulation by various factors and some of the cis-acting promoter elements have been characterized. Active research in this field has led to genetic improvement of cellulase production by various methods including over expressing cellulases from the *cbh1* promoter of *T. reesei*, and generation of desired variation in the cellulase production profile of organism. The *cbh1* and *cbh2* promoters of *T. reesei* have also been exploited for expression of foreign proteins in *Trichoderma*. Feedback inhibition of cellulase biosynthesis by the end products—glucose and cellobiose generated by endogenous cellulolytic activity on the substrate is another major problem encountered in cellulase production. Cellobiose is an extremely potent inhibitor of the *CBH* and *EG* biosynthesis. *Trichoderma* and the other cellulase-producing microbes make very little β-glucosidase compared to other cellulolytic enzymes. The low amount of β-glucosidase results in a shortage of capacity to hydrolyze the cellobiose to glucose resulting in a feed back inhibition of enzyme production and in the case of biomass conversion applications—in the inhibition of cellulases. This issue has been addressed by various means like addition of exogenous β-glucosidases to remove the cellobiose and engineering β-glucosidase genes into the organism so that it is overproduced. More and more research is oriented in the direction of genetic manipulations of the cellulase producers for improving productivity. The developments in process design and medium formulations may be considered to have come to an age and the future definitely requires controlled genetic interventions into the physiology of cellulase producers to improve production and thereby make the cellulase production process more cost effective. The major tasks ahead include overriding the feed back control by glucose and development of integrated bioprocesses for the production of cellulases.

Improvements in cellulase activities or imparting of desired features to enzymes by protein engineering are probably other areas where cellulase research has to advance. Active site modifications can be imparted through site-directed mutagenesis and the mutant proteins can be used for understanding the mechanisms of action as well as for altering the substrate specificities or improving the activities. There are several reports of developments made in this direction. A mutant enzyme with EG-like features and improved activity by deleting the C terminal loop of *Clostridium fimi CELB* has been successfully generated (Meinke et al., 1995). Protein engineering has been successfully employed to improve the stability of a *Humicola* cellulase in presence of detergents, to improve the thermostability of an alkaline, mesophilic *endo*-1,4-ß-glucanase from alkaliphilic *Bacillus* sp., and for altering the pH profile of CBH and EG from *T. reesei*. Such modifications affecting the enzyme properties may be beneficial in improving the overall performance of the cellulases and a better understanding of their mode of action, which will enable better utilization of the enzymes in biomass conversion. More basic research is needed in this direction to be able to make designer enzymes suited for specific applications in the future.

11 CHALLENGES

Economic considerations are utmost important in case of cellulase as the final products are usually low-value products such as single-cell protein and ethanol. There are several steps involved for cellulase production either by SSF or SmF. Reduction or simplification of any of the step will ultimately leads to the economic feasibility of the technology. There are several challenges which have yet to be overcome, for example the recalcitrance of lignocellulosic biomass, which necessitates the pretreatment step to open up the fibers and decrease the crystallinity of cellulose, which again adds to the cost of lignocellulosic—value-added product technology such as bioethanol. Pretreatment methods also need to vary from biomass to biomass based on their compositional characteristic. For developing an economically feasible technology, the use of cheaper raw material as a substrate for cellulase production could bring down the production costs, where SSF seems promising. Also, eliminating the steps in downstream processing of the enzyme for bioconversion might help to bring down the cost of cellulases as would be other approaches like improving the specific activities, temperature, and low pH tolerance as well as engineering the organism for improved production. Catabolite repression is a subject of major concerned for cellulase production which could be overcome by continuous fermentation process or by coupling the enzymatic hydrolysis with the fermentation process as is favorable to eliminate product inhibition.

Most of the commercial cellulases available are produced from *T. reesei* and *A. niger*. But, *T. reesei* lack sufficient amount of β-glucosidase with glucose tolerance to undergo proper and complete hydrolysis. As observed, all components of the extracellular cellulase complex are essential for cellulose hydrolysis and in general, β-glucosidase that catalyzes cellobiose hydrolysis is either lacking or present in relatively small amounts in the extracellular cellulase complex of this fungus. Thus, the cellobiose, being not hydrolyzed completely due to lack of β-glucosidase, inhibits exo- and exoglucanases. β-glucosidases are also inhibited by their own product glucose. One way to solve this issue is to add a glucose-tolerant β-glucosidase to the reaction mixture containing other cellulase components and to employ this cocktail for the biomass hydrolysis which would increase the efficiency of hydrolysis. Enzyme cocktails have been employed successfully for biomass conversion.

12 CONCLUSION

Development of improved cellulases for bioconversion seems to help materialize the dream of developing eco-friendly lignocellulosic ethanol to a reality.

Petroleum resources are fast depleting and global warming is increasing at an alarming rate, signifying the need for alternative fuels that are less polluting, more energy efficient, and renewable. Biomass is the only renewable foreseeable source of energy which promises environmental sustainability. Technologies for biomass conversion are going to define the future economies, and biomass will be emerging as the energy currency. Worldwide, there is an explosion in interest on lignocellulose utilization which was earlier lost in oblivion. Though lignocellulose conversion technologies have not attained the state of maturity, the tremendous growth that has happened in this field in the recent years is an indicator of the world moving toward a carbohydrate-based economy. After decades of research on lignocellulose utilization, it is now a consensus opinion that enzyme-based technologies for biomass conversion are the most efficient, cost effective, and environment friendly. Presently, the cost of enzymes needed for biomass saccharification is the major hindrance to development of biomass conversion technologies. The leading enzyme companies claim and also have brought down the price of cellulases significantly. They have succeeded partly by developments in production technologies adopting multifaceted approaches such as adopting cheaper bioprocess technology, employing cheaper substrate, and employing engineered organisms and partly by developments of artificial/engineered cellulases and cocktails of enzyme. Although the commercial lignocellulosic ethanol production has just began in some parts of the world, still continuous research is needed to improve varied aspects on cellulase production (such as cost, specific activity, and substrate specificity) to achieve better technoeconomic feasibility. Artificial/engineered cellulases and enzyme cocktails rich in glucose-tolerant β-glucosidase have been proved successful for increasing the rate or efficiency of hydrolysis of biomass so as to prove the technology economically feasible.

Understanding of the microbial physiology and genetics of cellulase producers is still required. Recent report on the sequencing of *T. reesei* genome is a major step in this direction. Similar efforts will be needed in the case of other major cellulase producers also so that more information is built up on the molecular biology of cellulase producing fungi and their gene regulation. This information will be critical for future development of strains for cellulase production. With the current pace of research on cellulases, it can be asserted that more knowledge is generated in the near future that will aid our progress toward a greener and sustainable carbohydrate-based economy.

But the fact cannot be denied that despite several efforts, cellulase for bioconversion though available in the market is not easily accessible. It signifies the long way till to go.

References

Aro, N., Pakula, T., Penttila, M., 2005. Transcriptional regulation of plant cell wall degradation by filamentous fungi. FEMS Microbiol. Rev. 29, 719–739.

Carrillo, F., Lis, M.J., Colom, X., López-Mesas, M., Valldeperas, J., 2005. Effect of alkali pretreatment on cellulase hydrolysis of wheat straw: Kinetic study. Proc. Biochem. 40, 3360–3364.

Cen, P., Xia, L., 1999. Production of Cellulase by Solid-State Fermentation. Adv. Biochem. Eng. Biotechnol. 65, 69–92.

Elisashvili, V., Kachlishvili, E., Tsiklauri, N., Metreveli, E., Khardziani, T., Agathos, S.N., 2009. Lignocellulose-degrading enzyme production by white-rot Basidiomycetes isolated from the forests of Georgia. World J. Microb. Biot. 25, 331–339.

Gusakov, A.V., Salanovich, T.N., Antonov, A.I., Ustinov, B.B., Okunev, O.N., Burlingame, R., et al., 2007. Design of highly efficient cellulase mixtures for enzymatic hydrolysis of cellulose. Biotechnol. Bioengg. 97, 1028–1038.

Hölker, U., Höfer, M., Lenz, J., 2004. Biotechnological advantages of laboratory-scale solid-state fermentation with fungi. Appl. Microbiol. Biotechnol. 64, 175–186.

Kabel, M.A., van der Maarel, M.J.E.C., Gert, K., Alphons, G.J.V., Henk, A.S., 2005. Standard assays do not predict the efficiency of commercial cellulase preparations towards plant materials. Biotechnol. Bioeng. 93, 56–63.

Lynd, L.R., Weimer, P.J., van Zyl, W.H., 2002. Pretorius IS: microbial cellulose utilization: fundamentals and biotechnology. Microbiol. Mol. Biol. Rev. 66, 506–577.

Mandels, M., Reese, E.T., 1957. Induction of cellulase in *Trichoderma viride* as influenced by carbon sources and metals. J. Bacteriol. 73, 269–278.

Mandels, M., Parrish, F.W., Reese, E.T., 1962. Sophorose as an inducer of cellulase in *Trichoderma reesei*. J. Bacteriol. 83, 400–408.

Martinez, D., Berka, R.M., Henrissat, B., Saloheimo, M., Arvas, M., Baker, S.E., et al., 2008. Genome sequencing and analysis of the biomass-degrading fungus *Trichoderma reesei* (syn. *Hypocrea jecorina*). Nat. Biotech. 26, 553–560.

Mathew, G.M., Sukumaran, R.K., Singhania, R.R., Pandey, A., 2008. Progress in research on fungal cellulases for lignocellulose degradation. J. Sci. Ind. Res. 67, 898–907.

McMillan, J.D., 1994. Pretreatment of Lignocellulosic Biomass. In: Himmel, M.E., Baker, J.O., Overend, R.P., (Eds.), Enzymatic Conversion of Biomass for Fuels Production. ACS Symposium Series 566, American Chemical Society, Washington DC, pp. 292–324.

Meinke, A., Damude, H.G., Tomme, P., Kwan, E., Kilburn, D.G., Miller, R.C., Jr, et al., 1995. Enhancement of the Endo-beta-1,4-glucanase Activity of an Exocellobiohydrolase by Deletion of a Surface Loop. J. Biol. Chem. 270, 4383–4386.

Nakari-Setala, T., Pentilla, M., 1995. Production of *Trichoderma reesei* cellulases on glucose containing media. Appl. Environ Micorbiol. 61, 3650–3655.

Nieves, R.A., Ehrman, C.I., Adney, W.S., Elander, R.T., Himmel, M.E., 2009. Technical Communication: survey and analysis of commercial cellulase preparations suitable for biomass conversion to ethanol. World J. Microb. Biot. 14, 301–304.

Ohmiya, K., Sakka, K., Karita, S., Kimura, T., 1997. Structure of cellulases and their application. Gene. Rev. 14, 365–414.

Pandey, A., 1994. Solid-state fermentation: an overview. In: Pandey, A. (Ed.), Solid State Fermentation. Wiley Eastern Limited, New Delhi, India, pp. 3–10.

Pandey, A., Soccol, C.R., 2000. Economic utilization of crop residues for value addition: a futuristic approach. J. Sci. Ind. Res. 59, 12–22.

Pandey, A., Selvakumar, P., Soccol, C.R., Nigam, P., 1999. Solid state fermentation for the production of industrial enzymes. Curr. Sci. 77, 149–162.

Pandey, A., Biswas, S., Sukumaran, R.K., Kaushik, N., 2009. Study on Availability of Indian Biomass Resources for Exploitation. A report based on a nation-wise survey. TIFAC, New Delhi. p. 105.

Reith, J.H., den Uil, H., van Veen, H., de Laat, W.T.A.M., Niessen, J.J., de Jong, E., et al., 2002. Co-production of bioethanol, electricity and heat from biomass residues. 12th European Conference and Technology Exhibition on Biomass from Energy, Industry and Climate Protection, Amsterdam, The Netherlands.

Sakka, K., Kimura, T., Karita, S., Ohmiya, K., 2000. Molecular breeding of cellulolytic microbes, plants, and animals for biomass utilization. J. Biosci. Bioeng. 90, 227–233.

Schell, D.J., Riley, C.J., Dowe, N., Farmer, J., Ibsen, K.N., Ruth, M.F., et al., 2004. A bioethanol process development unit: initial operating experiences and results with a corn fiber feedstock. Biores. Technol. 91, 179–188.

Schulein, M., 1988. Cellulases of *Trichoderma reesei*. In: Wood, W.A., Abelson, J.N. (Eds.), Methods in Enzymology. Vol. 160, Academic Press, New York, pp. 234–242.

Singhania, R.R., 2009. Cellulolytic enzymes. In: Nigam, P., Pandey, A. (Eds.), Biotechnology for Agro-industrial residues utilization. Springer, USA, Ch 20, pp. 371–382.

Singhania, R.R., Sukumaran, R.K., Pandey, A., 2007. Improved cellulase production by *Trichoderma reesei* RUT C30 under SSF through process optimization. Appl. Biochem. Biotechnol. 142, 60–70.

Singhania, R.R., Binod, P., Pandey, A., 2008. Plant-based biofuels—an Introduction. In: Pandey, A. (Ed.), Handbook of Plant-Based Biofuels. Taylor & Francis, CRC Press, USA, Ch 1, pp. 1–10.

Singhania, R.R., Patel, A.K., Soccol, C.R., Pandey, A., 2009. Recent advances in solid-state fermentation. Biochem. Eng. J. 44, 13–18.

Singhania, R.R., Sukumaran, R.K., Patel, A.K., Larroche, C., Pandey, A., 2010. Advancement and comparative profiles in the production technologies using solid-state and submerged fermentation for microbial cellulases. Enzyme Microb. Technol. 46, 541–549.

Sukumaran, R.K., Singhania, R.R., Pandey, A., 2005. Microbial cellulases-Production, applications and challenges. J. Sci. Ind. Res. 64, 832–844.

Sukumaran, R.K., Singhania, R.R., Mathew, G.M., Pandey, A., 2009. Cellulase production using biomass feed stock and its application in lignocellulose saccharification for bioethanol production. Renew. Energ. 34, 421–424.

Wiebe, M.G., 2003. Stable production of recombinant proteins in filamentous fungi—problems and improvements. Mycologist. 17, 140–144.

Wooley, R., Ruth, M., Sheehan, J., Ibsen, K., 1999. Lignocellulosic biomass to ethanol process design and economics utilizing co-current dilute acid pre hydrolysis and enzymatic hydrolysis: current and futuristic scenarios. NREL Report, NREL/TP-580-26157.

CHAPTER

9

Production of Hemicellulolytic Enzymes for Hydrolysis of Lignocellulosic Biomass

*Sharma Manju, Bhupinder Singh Chadha**
Department of Microbiology, Guru Nanak Dev University, Amritsar-143 005, India
***Corresponding author: E-mail: chadhabs@yahoo.com**

1 INTRODUCTION

Lignocellulosics in the form of agroresidues and forestry biomass constitute a potentially enormous source of feedstock for bioconversion into biofuel, feed, and specialty chemicals (Kamm and Kamm, 2004; Ohara, 2003). Lignocellulosics are comprised of cellulose, hemicellulose, and lignin that are present as intertwined complex fibril macromolecular structure. The structural heterogeneity in terms of proportion of cellulose, hemicellulose, and lignin in different plant species, as well as the spatial distribution of the constituent molecules, is perhaps one of the major hindrances in developing universal enzyme-based bioconversion technologies for their optimal utilization (Sharma et al., 2010a,b). In this chapter, we focus on the technologies available for the utilization of hemicellulosic fraction.

2 STRUCTURE OF HEMICELLULOSE

The term hemicellulose refers to a group of homo- and heteropolymers consisting of xylopyranose, mannopyranose, glucopyranose, and galactopyranose main chains with a number of substituents resulting in structurally complex polymer (Girio et al., 2010; Zheng et al., 2009). The hemicelluloses derived from different plant sources also show significant differences in their composition and structure. Few of the recent reviews give a detailed account of the hemicellulose structure (Girio et al., 2010; Scheller and Ulvskov, 2010). β-1,4-xylans, the major components of hemicellulose, are the second most abundant polymer in nature, accounting for one-third of the renewable biomass available on earth and constitutes around 20-30% of the dry weight of tropical hardwood and annual plants (Dhiman et al., 2008). The homopolymeric backbone of β-1,4-linked D-xylopyranose units is substituted to varying degrees with 4-*O*-methylglucuronopyranosyl, α-L arabinofuranosyl, α-D-glucuronyl residues, acetyl, feruloyl, and/or *p*-coumaroyl side chain units (Sun et al., 2005). Xylan exists as *O*-acetyl-4-*O*-methylglucuronoxylan in hardwoods and as arabino-4-*O*-methylglucuronoxylan in softwoods, while xylans in grasses and annual plants are typically arabinoxylans consisting of β-1,4-linked backbone of D-xylopyranosyl residues to which α-L-arabinofuranosyl (Araf) residues are linked at C-3 and C-2 (Izydorczyk and Dexter, 2008). Arabinoxylan agroresidues such as straws have been identified in wheat, rye, barley, oat, rice, sorghum, corn fiber, rye grass, etc. (Polizeli et al., 2005). Arabinoxylans from rice, sorghum, finger millet, and maize bran are more complex than those from barley arabinoxylans. The former contain, in addition to arabinose residues, small amounts of xylopyranose, galactopyranose, and α-D-glucuronic acid or 4-*O*-methyl-α-D-glucuronic residues. One of the unique features of arabinoxylans is the presence of hydroxycinnamic acids, ferulic and p-coumaric, esterified to *O*-5 of Araf linked to *O*-3 of the xylose residues (Medina et al., 2010) where ferulate esters can dimerize via phenoxy radicals into dehydrodiferulate esters, which are responsible for covalent crosslinking between arabinoxylan chains and arabinoxylans and other cell wall constituents (Lazaridou et al., 2007). In addition, acetyl groups may be esterified at C-2 or C-3 of the xylose residues. The relative amount and the sequence of distribution of these structural elements vary depending on the source of arabinoxylans. The majority of arabinofuranosyl residues in arabinoxylans are present as monomeric substituents; however, a small proportion of oligomeric side-chains, consisting of two or more Araf residues, are linked via 1 → 2, 1 → 3, and 1 → 5 bonds (Wong, 2006).

In case of hardwood xylan, approximately seven out of 10 xylosyl residues carry α-*O*-methylglucuronyl residue at *O*-2. They are associated with the lignin via ester, ether, and glycosidic bonds in plant cell walls (Sun et al., 2005). A small percentage of hardwood is also composed of glucomannans which consist of β-(1-4) linked glucose and mannose units forming chains that are slightly branched. The ratio of mannose: glucose is about 1.5:1 or 2:1 in most hardwoods (Sande et al., 2009). At the C-2 position, D-galacturonic acid is linked with an L-rhamnose, whereas the L-rhamnose is connected to the xylose chain at its C-3 position (Vries and Visser, 2001). The differences in acetylation as well as the presence of *O*-2 substituted 4-*O*-methyl-α-D-glucuronic acid units in xylans, in addition to terminal methyl glucuronic acid units linked to the xylan backbone, have also been documented (Pinto et al., 2005).

Arabinoglucuronoxylan is a minor component of softwood hemicelluloses. The backbone of softwood xylan is made up of β-l, 4-xylose units, with branches at C-2 and C-3 position. For about every 10 units of xylose, there are two 4-*O*-methyl-α-D-glucuronic acid groups substituted at the C-2 position and one α-L-arabinose unit at C-3 position (Izydorczyk and Dexter, 2008). L-arabinose and 4-*O*-methyl-α-D-glucuronic acid groups help maintain the xylose backbone, which is otherwise degraded during base-catalyzed reaction (Peng et al., 2010). In softwoods, hemicelluloses are mainly in the form of galactoglucomannan that forms the backbone of linear or slightly branched chain of β-(1-4) linked D-mannopyranose and D-glucopyranose units. Galactoglucomannan can be roughly divided into two types: one with a low galactose content, sometimes referred to simply as glucomannan, and the other with a high galactose content. The ratios of galactose to glucose to mannose are 0.1-0.2:1:3-4 and

1:1:3 in the two types respectively. The hydroxyl groups at positions C-2 and C-3 in the backbone units are partly substituted by *O*-acetyl groups, on average one group per 3-4 hexose units (Girio et al., 2010); the degree of substitution varies with the source. In addition, arabinogalactan, xyloglucan, glucomannan, and other glucans are present (Albertsson et al., 2010; Laine, 2005; Scheller and Ulvskov, 2010).

Arabinogalactan is mainly known as a component of the heartwood of larches. The backbone consists of β-1-3-linked D-galactopyranose units and is highly branched at C-6 with side chains composed of β-1-6-linked D-galactose units, D-galactose, and L-arabinose units or single L-arabinose units and single D-glucuronic acid units. Other hemicelluloses include xyloglucan that are present mainly as a polysaccharide in the primary cell wall of higher plants and similar to structure of cellulose with β-(1 → 4)-linked D-glucosyl backbone containing α-D-Xylose-(1 → 6)-glucose substitutions. The xylosyl residues can be substituted at *O*-2 with β-Galactose, α-L-arabinose, or α-L-Fucose (Lopez et al., 2010). As well as β-glucans, there are linear homopolymers of D-glucopyranosyl (Glcp) residues linked mostly via two or three consecutive β-(1 → 4) linkages that are separated by a single β-(1 → 3) linkages (Lazaridou et al., 2007; Scheller and Ulvskov, 2010).

3 HEMICELLULASES

Due to the heterogeneity and complex chemical nature of hemicellulose, its hydrolysis into simpler constituents (monomers, dimers, or oligomers) requires the action of a wide spectrum of enzymes with diverse catalytic specificity and modes of action. Therefore, it is not surprising that microorganisms produce an arsenal of hemicellulolytic enzymes. Most important of these enzymes is endoxylanase (EC 3.2.1.8) that cleaves β-1,4-linked xylose backbone, while β-xylosidase (EC 3.2.1.37) cleaves xylose monomers from the nonreducing end of xylooligosaccharides and xylobiose. In addition, a variety of debranching enzymes, that is, α-arabinofuranosidase (EC 3.2.1.55), α-glucouronidase (EC 3.2.1.139), acetylxylan esterase (EC 3.1.1.72), α-galactosidases (EC 3.2.1.22), and β-mannosidases (EC 3.2.1.25), acetylxylan esterases (EC 3.1.1.72), ferulic acid esterases (EC 3.1.1.73), and ρ-coumaric acid esterases (EC 3.1.12) are required for efficient utilization (Figure 1) of hemicellulosic fraction (Shallom and Shoham, 2003).

4 ENDOXYLANASES

Xylanases-producing microrganisms have been isolated from diverse ecological niches like Southern Caucasus and Amazon forests, Antarctica, hot springs, composting soils, guts of earthworm, to name a few. Various bacterial and fungal cultures have been isolated and documented in several reviews (Maheshwari et al., 2000; Subramaniyan and Prema, 2002; Sunna and Antranikian, 1997; Vries and Visser, 2001). However, being an area of continued research, each passing year adds to the existing information about xylanases from different sources. A number of new species of microbes from diverse environments and ecological niches are being studied for the production of xylanolytic enzymes. Several new strains of thermophilic fungi, *Myceliophthora* sp., *Chrysosporium lucknowense*, *Malbranchea flava*, *Talaromyces thermophila*

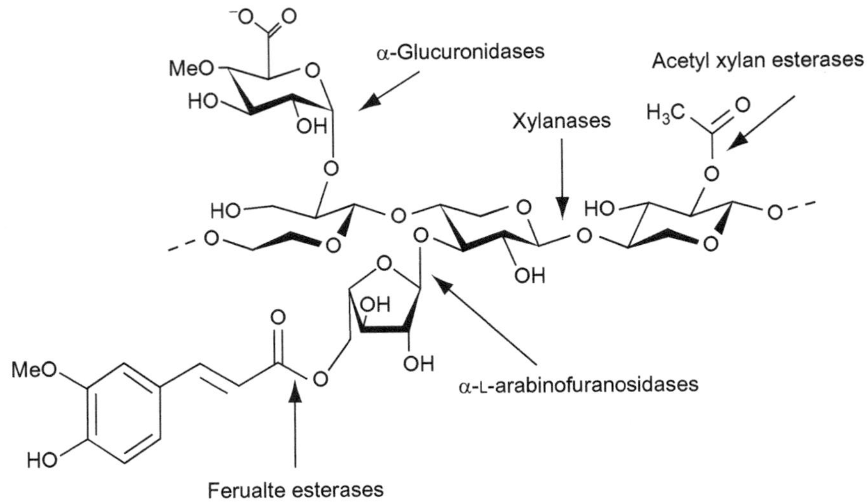

FIGURE 1 Enzymatic breakdown of arabinoxylan. Source: *www.google.com*.

(Chadha et al., 2004; Maalej et al., 2009; Sharma et al., 2008; Ustinov et al., 2008) have been isolated from composting piles where the temperature rises to ~70 °C reported to produce xylanolytic enzymes. Owing to their higher thermostability and other technical traits, xylanases from thermophilic strains of bacteria and fungi are important from biotechnological viewpoint. Some of the other novel xylanase-producing microrganisms reported in the recent past include basidiomycete *Cerioposis subvermisopora* (Magalhaes and Milagres, 2009); facultative anaerobe *Anoxybacillus pushchinoensis* A8 (Kacagan et al., 2008), *Alicyclobacillus* sp. A4 (Bai et al., 2010) as well as actinomycete strains of *Streptomyces thermonitrificans*, *S. thermocarboxydus* (Cheng et al., 2009; Kim et al., 2010). The isolation of genes from metagenomic library encoding for xylanases has also been reported in recent years. The environmental DNA library prepared from insect gut, manure waste, soil, and dairy cow rumen has yielded clones containing gene coding for xylanases (Brennan et al., 2004; Kim et al., 2008; Zhao et al., 2010). Novel xylanase showing 59% identity to endo-β-1-4-xylanase from *Cellulomonas pachnodae* was isolated from the soil metagenome (Kim et al., 2008). Clones harboring novel xylanases with two catalytic domains of family 43 and two CBD of family IV have been characterized (Zhao et al., 2010). The characterization of the crystal structure of CelM2, a bifunctional glucanase xylanase protein from the metagenome library, has revealed the metal effect and substrate-binding moiety (Nam et al., 2009).

Xylanases have been classified in families 5, 7, 8, 10, 11, and 43 on the basis of their amino acid sequences, structural folds, and mechanisms for catalysis (Collins et al., 2005; Cantarel et al., 2009). GH 10 and 11 xylanases represent the best studied xylanase families, and they differ in the number of subsites they possess, with GH 10 having four or five subsites and GH 11 having at least seven subsites (Dodd and Cann, 2009). While endoxylanases belonging to family 10 are characterized by high molecular weight (usually >30 kDa) and acidic pI, the members of family 11 have low molecular weight and basic pI, though exceptions do occur in some cases (Lagaert et al., 2009; Wong et al., 1988). The process of classifying xylanases in different families is supported by hydrophobic cluster analysis that predicts distinct protein-folding patterns as well as nucleotide sequences in these xylanases (Arora et al., 2009; Sapag et al., 2002). Compared to GH10 and GH11 endoxylanases, only a limited amount of data is available on the catalytic properties of xylanases from GH families 7, 8, and 43. The recent studies have reported the characterization of novel xylanases from *Trichoderma reesei* and *Erwinia chrysanthemi* belonging to glycoside hydrolase family 5 with exoacting mechanisms (Larson et al., 2003), and an endoacting xylanase from *Pseudoalteromonas haloplanktis* in family 8 (Collins et al., 2002). The GH5 enzyme from *E. chrysanthemi* is specialized for hydrolysis of 4-*O*-methyl-D-glucuronoxylan or its acetylated counterparts and does not attack other types of xylans, linear β-1,4-xylooligosaccharides, or esterified aldouronic acids (Vrsanska et al., 2007). However, a new bacterial xylanase belonging to GH 5 was found to be active on neutral, nonsubstituted xylooligosaccharides, showing a clear difference from other GH 5 xylanases characterized to date that show a requirement for methyl-glucuronic acid side chain for catalysis (Gallardo et al., 2010). The crystal structure of family 8 xylanase from an Antarctic bacterium, *P. haloplanktis*, showed that it appeared to have less salt bridges and increased number of hydrophobic residues that were exposed to the surroundings revealing their adaptation toward cold environment (Van Petegam et al., 2003). Pollet et al. (2010) evaluated the substrate preference and hydrolysis product profiles of different GH 8 xylanases in order to investigate their activities and substrate specificities. The findings of this study showed that GH 8 xylanases have narrow substrate specificities and the subtle amino acid changes in the glycon as well as the aglycon subsites probably form the basis of the observed differences between GH 8 xylanases. The GH 7 enzyme from *Trichoderma reesei* is considered as a nonspecific endo β-1,4-glucanase (Kleywegt et al., 1997), and the GH 43 enzyme from *Paenibacillus polymyxa* displays both xylanase and α-L-arabinofuranosidase activities (Gosalbes et al., 1991). Diverse physicochemical and functional characteristics, as well as folds and mechanisms of action of all the xylanase of different families, have been well discussed in an excellent review by Collins et al. (2005).

Catalytically, xylanases from families 10 and 11 can be differentiated on the basis of lower and higher substrate specificities, respectively. The lower substrate specificity of family 10 xylanases was demonstrated by their ability to catalyze the hydrolysis of cellulase substrate, pNP-cellobioside at a gluconic linkage, while the members of family 11 xylanase failed to recognize this as substrate (Biely et al., 1997; Collins et al., 2005). The substrate specificity of xylanases is reflected by the structural features of their active site. Each xylose is accommodated in a subsite (−) and (+), depending on whether it binds the glycone or aglycone regions of the substrate, respectively. Kinetic and structural investigations of GH11 xylanases indicate that their active sites potentially have upto three (−) subsites and three (+) subsites (Janis et al., 2005). In contrast, GH7 and GH10 xylanases have four to five subsites (Collins et al., 2005). Another feature that distinguishes GH10 and GH11 xylanases is the nature of the reaction products released from decorated xylans. GH11 xylanases produce substituted xylooligosaccharides both at the aglycone and glycone subsites (Maslen et al., 2007). The family 10 and 11 xylanases also differ in their action on 4-*O*-methyl-D-glucurono-D-xylan and rhodymenan, a β-1,3-β-1,4-xylan (Biely et al., 1997). A recent study assessed the activity of several GH10 and GH11 proteins with purified xylooligosaccharides substituted with MeGA and revealed that GH10 enzymes cleave xylan chains when MeGA is linked to xylose at the +1 subsite, whereas GH11 enzymes cleave when

MeGA is appended at the +2 subsite (Kolenova et al., 2006). Direct evidence for these results was reported in a recent study on the mass spectra of the products of hydrolysis for GH10 and 11 with arabinoxylan substrates (Maslen et al., 2007; Vardakou et al., 2008). These results suggest that EXs of family 10 are able to hydrolyze xylose linkages closer to side-chain residues and thus help to explain why these enzymes release shorter products than EXs of family 11 when incubated with arabinoglucuronoxylan substrates (Biely et al., 1997). This difference in substrate specificity for xylanases has important implications in the deconstruction of xylan (Dodd and Cann, 2009).

5 β-D-XYLOSIDASES

β-D-xylosidases (EC 3.2.1.37) are exotype glycosidases that hydrolyze short xylooligomers into single xylose units. An important role ascribed to β-xylosidases comes into play after the xylan has suffered a number of sequential hydrolyses by xylanase. This reaction leads to the accumulation of short oligomers of β-D-xylopyranosyl, which may inhibit the endoxylanase. β-xylosidase then hydrolyzes these products, removing the cause of inhibition, and increasing the efficiency of xylan hydrolysis (Zanoelo et al., 2004). Purified β-xylosidases usually do not hydrolyze xylan; their best substrate is xylobiose and their affinity for xylooligosaccharides is inversely proportionate to its degree of polymerization (Polizeli et al., 2005). β-xylosidases from filamentous fungi are usually liberated into the growth medium, that is, they are extracellular proteins. Although xylose is the end product inhibitor of β-xylosidases, it can act as inducer of xylanolytic gene expression. High yields of β-xylosidase on xylose were observed with *T. reesei* (Kristufek et al., 1995) and *A. versicolor* (Andrade et al., 2004). Recently, an extracellular xylose-tolerant β-xylosidase from *Paecilomyces thermophila* J18 was purified to homogeneity from the cell-free culture supernatant (Yan et al., 2008). However, cell-associated β-xylosidases have been reported from the cell extract of *Penicillium* sp., *Sclerotium* sp. grown on oat spelt xylan (OSX; Knob and Carmona, 2009). β-xylosidases from fungi are often monomeric glycoproteins, but some have been reported to possess two or three subunits (Polizeli et al., 2005; Xiong et al., 2007). They are grouped into five different families (GH3, GH39, GH43, GH52, and GH54) and their reaction mechanisms either result in inversion (GH43) or retention (GH3, GH39, GH52, and GH54) of stereochemical configuration at the anomeric carbon. The best characterized β-xylosidases are from GH3 and GH43 (Dodd and Cann, 2009). The crystal structures for two biochemically characterized GH43 β-xylosidases from *Selenomonas ruminantium* and *Geobacillus stearothermophilus* have revealed the presence of two domains, an N-terminal five bladed β-propeller domain and a C-terminal α/β-sandwich domain (Brunzelle et al., 2008). These enzymes possess two subsites for sugar binding and it is anticipated that only two xylose units will bind to the active site, thus extending the rest of the xylose units out into the solution. This prediction is further corroborated by biochemical analyses of GH43 β-xylosidases that reveal a decrease in catalytic efficiency (k_{cat}/K_M) when active on xylooligosaccharides longer than X_2, thus suggesting that these enzymes possess only two xylose-binding sites (Wagschal et al., 2009). Although most members of the GH43 family have bacterial origin, few filamentous fungi, namely, *Aspergillus oryzae*, *Penicillium herquei*, and *Cochliobolus carbonum* possess DNA sequences that encode putative family 43 β-xylosidases (Ito et al., 2003; Suzuki et al., 2010). Bravman et al. (2001a,b) reported the overexpression, purification, and biochemical characterization of a GH39 family β-xylosidase from *Bacillus stearothermophilus* T-6, and provided firm support for the assignment of Glu 160 as the acid-base catalyst of family 39 GHs. GH39 β-xylosidases have also been reported from *B. halodurans* (Muzard et al., 2009). GH3 represents a large group of glycosidic enzymes and possesses several distinct enzymatic activities including β-glucosidases, β-xylosidase, arabinofuranosidase, and N-acetyl-β-D-glucosaminidase activities (Faure, 2002). The spatial similarity between D-xylopyranose and L-arabinofuranose leads to bifunctional xylosidase/arabinosidase enzymes, found mainly in families 3, 43, and 54 (Mai et al., 2000). A bifunctional cell associated β-xylosidase belonging to GH3 family was purified from the cell extract of dimorphic fungus *Aureobasidium pullulans* strain ATCC 20254, grown on OSX (Ohta et al., 2010). The enzyme also showed some α-L-arabinofuranosidase activity (a novel mutant with AtBXL1 which encodes putative bifunctional β-D-xylosidase/α-L-arabinofuranosidase) has been identified in *Arabidopsis* mucilage secretory cells (Arsovski et al., 2009). The extensive structural and biophysical characterization of a family 52 β-xylosidase from *Geobacillus stearothermophilus* describes it as highly hydrated dimer protein whose active site was formed by the two promoters, and it probably involved aromatic residues (Contreras et al., 2008).

6 α-ARABINOFURANOSIDASES

α-Arabinofuranosidases (AFase) are accessory enzymes that hydrolyze the terminal, nonreducing α-L-arabinofuranosyl groups of arabinans, arabinoxylans, and arabinogalactans and act synergistically with other hemicellulases and pectic enzymes for the complete hydrolysis of hemicelluloses and pectins (Saha, 2000). Arabinan-degrading enzymes have been classified on the basis of their mode of action, that is, endoacting or exoacting. The arabinan-degrading enzymes that act in an endofashion are called endo-1,5-α-L-arabinanases (EC 3.2.1.99) and those that act in an exofashion are called α-L-arabinofuranosidases (EC 3.2.1.5). Exoacting α-L-arabinofuranosidases (EC 3.2.1.55) are active against *p*-nitrophenyl-α-L-arabinofuranoside and on branched arabinans, whereas endo-1,5-α-L-arabinofuranosidases (EC 3.2.1.99) are active only toward linear arabinans, and are not able to hydrolyze *p*-nitrophenyl-α-L-arabinofuranoside or arabic gum (Polizeli et al., 2005). Most of the arabinan-degrading enzymes reported in the literature are of the exoacting type. However, there are some reports of α-L-arabinofuranosidases capable of hydrolyzing both 1,3- and 1,5-α-L-arabinofuranosyl linkages in arabinoxylan (Corral and Ortega, 2006; Ichinose et al., 2008). Moreover, in some cases, α-L-AFases possessing β-xylosidase activity or xylanases with α-L-arabinofuranosidase activity also have been described (Arsovski et al., 2009). These enzymes expedite the hydrolysis of the glycosidic bonds by more than 10^{17} fold, making them one of the most efficient catalysts known. Arabinofuranosidases exist as monomers, but dimeric, tetrameric, and octameric forms have also been found (Panagiotou et al., 2003). They are classified into five GHs families, that is, GH3, GH43, GH51, GH54, and GH62 (Allgaier et al., 2010) and can hydrolyze glycosidic linkages at net inversion (GH43) or retention (GH51, 54) of stereochemical configuration at the anomeric carbon (Dodd and Cann, 2009; Carapito et al., 2009). AFs belonging to GH51 and 62 family release *O*-2 and *O*-3 linked arabinofuranosyl units from monosubstituted xylose.

A family 51 α-L-arabinofuranosidase from *Penicillium purpurogenum* was purified to homogeneity and characterized; the monomer with a molecular weight of 70 kDa exhibited low activity toward short arabinooligosaccharides and differed in some properties from other enzymes of this family (Fritz et al., 2008). Arabinofuranosidases of GH43 family result in the release of *O*-2 and *O*-3 substituted arabinose from monosubstituted xylose and display a variety of different substrate specificities. They are known to release *O*-3 linked arabinofuranosyl residues from double-substituted xylose (Hinz et al., 2009). A novel GH43 α-L-arabinofuranosidase from *Humicola insolens* that was cloned and expressed in *A. oryzae* was found to selectively hydrolyze arabinofuranosyl residues of doubly substituted xylopyranosyl residues in arabinoxylan. The synergistic action of two α-L-arabinofuranosidases from *H. insolens* belonging to GH51 along with the earlier-mentioned GH43 enzyme resulted in the removal of single sitting (1 → 2)-α-L-arabinofuranosyl units released after the GH43 enzyme had catalyzed the removal of (1 → 3)-α-L-arabinofuranosyl residues on doubly substituted xylopyranosyls in wheat arabinoxylan (Sorensen et al., 2006). Recently, crystal structures have been reported for GH43 arabinofuranosidase from *S. ruminantium* (Brunzelle et al., 2008) and *Bacillus subtilis* (Vandermarliere et al., 2009) which have the same N-terminal five bladed β-propeller fold common to GH43 enzymes but differ in the C-terminal domain. Due to this difference, these enzymes exhibit distinct substrate preferences with the *S. ruminantium* enzyme (SXA) having high activity on pNP-β-D-xylopyranoside followed by pNP-α-L arabinofuranoside and xylooligosaccharides (Jordan et al., 2007), whereas the *B. subtilis* showed highest activity on pNP-α-L arabinofuranoside and water extractable arabinoxylans (Bourgois et al., 2007).

7 ACETYLXYLAN ESTERASES

Acetylxylan esterases (3.1.1.72) are enzymes that are able to hydrolyze the ester linkage between acetyl and xylose residues in xylans. This deacetylation makes the xylopyranosyl units of the main xylan chain more accessible to degradation by endo-β-1,4-xylanases (EC 3.2.1.8). Acetylxylan esterases play an important role in the hydrolysis of xylan, as the acetyl side-groups can interfere with the approach of enzymes that cleave the backbone by steric hindrance, and their elimination thus facilitates the action of endoxylanases (Javier et al., 2007). The enzyme action on polysaccharide substrates creates new sites on the xylan main chain, suitable for productive binding with depolymerizing endoxylanases. The degradation of acetylxylan with endoxylanases proceeds faster and to a higher degree in the presence of acetylxylan esterases. They also deacetylate the partially acetylated xylooligosaccharides which makes the oligosaccharides fully susceptible to the action of β-xylosidases (Hinz et al., 2009). Two purified acetylxylan esterases from *C. lucknowense* were found to release all acetyl groups from acetylated xylan oligosaccharides except one, which was found to be located at the nonreducing end of the oligosaccharide suggesting that the esterases are able to cleave all ester linkages at the reducing end (Hinz et al., 2009). The production of acetylxylan esterases by various fungi and bacteria has been reported, but it has been important to distinguish between nonspecific acetyl esterase activity and acetylxylan esterases by using appropriate substrates (Li et al., 2008). They suggested that most of the esterases are serine type which attack on low molecular mass substrates such as 4-nitrophenyl acetate or 4-methylumbelliferyl acetate and employ a Ser-His-Asp (Glu) catalytic triad for catalysis. This mechanism involves the initial phase of acylation of the nucleophiles of serine residue followed by deacylation with water acting as a nucleophile (Taylor et al., 2006). However, the carbohydrate esterases of family 4, which also contain chitin deacetylases, do not operate on the earlier-mentioned aryl acetates and also do not possess Ser-His-Asp catalytic triad (Taylor et al., 2006). Cleavage of acetyl groups from the xylan is helpful in the removal of lignin. They may contribute to lignin solubilization by cleaving the ester linkages between lignin and hemicelluloses (Subramaniyan and Prema, 2002).

Feruloyl esterases (EC 3.1.1.73) are enzymes which hydrolyze the ester bond between the arabinose substitutions and ferulic acid. This later ester bond is involved in crosslinking xylan to lignin. Due to the ability of these residues to crosslink xylan and pectin polysaccharides to each other and to lignin, they are important for the structural integrity of the plant cell wall. Although some prokaryotic feruloyl esterases have been purified, the majority of these enzymes have been studied from eukaryotic systems. Feruloyl esterases can be divided into small monomeric enzymes, large dimeric enzymes, and monomeric enzymes based on molecular mass. On the basis of substrate specificity toward synthetic substrates and their capability to liberate diferuloyl bridges, these esterases can be divided into 4 groups, namely, A-D (Crepin et al., 2004). Benoit et al. (2008) introduced another classification of the ferulic acid esterases based on amino acid sequence homology and their activity toward methyl ferulate, methyl sinapate, and methyl caffeate. Most of the feruloyl esterases are extracellular and are active against xylan and xylan-derived oligosaccharides, from which they are able to release ferulic acid. Ferulic/coumaric acid esterases belong to the carbohydrate esterase (CE) family 1, whereas acetylxylan esterase activity has been described for members of CE 1-7, 12 and the recently discovered family 16 (Li et al., 2008).

8 α-D-GLUCURONIDASES

α-D-Glucuronidases (EC 3.2.1.131) are the enzymes that hydrolyze the α-1,2 linkages between glucuronic acid and xylose residues in glucuronoxylan. However, the substrate specificity varies with the microbial source, and some glucuronidases are able to hydrolyze the intact polymer (Wet and Prior, 2004). Acetyl groups close to the glucuronosyl substituents are known to partially hinder the α-glucuronidase activity. To date, all of the α-D-glucuronidases are classified as family 67 glycosidases, which catalyze the hydrolysis via the inverting mechanism (Shallom et al., 2004).

9 MANNANASES

Endo-1,4-β-D-mannanase (EC 3.2.1.78) catalyzes the random cleavage of β-D-1,4-mannopyranosyl linkages within the main chain of galactomannan, glucomannan, galactoglucomannan, and mannan. They liberate short-chain β-1,4-manno-oligomers, which can be further hydrolyzed to mannose by β-mannosidases (EC 3.2.1.25; Li et al., 2006). A variety of different organisms, including bacteria, fungi, higher plants, and animals, are known to produce mannanases (Chen et al., 2008; Li et al., 2006). Multiple extracellular mannanases have been reported among many fungi like *Trichoderma reesei, T. harzianum*, and *Aspergillus* sp. (Fattah et al., 2009). The interest in β-mannanase has recently increased, partly because of their potential pertinence in the food and paper and pulp industries (Dhawan and Kaur, 2007). Endo-1,4-β-D-mannanases are classified in GH families 5 and 26, whereas β-mannosidases are described in GH families 1, 2, and 5 (Cuong et al., 2009; Songsiriritthigul et al., 2010).

10 METHODS FOR ASSAY OF HEMICELLULOLYTIC ACTIVITY

Birchwood xylan (BWX) which is least substituted and contains 94% of carbohydrate as xylose (more than 90% is in the form of soluble xylan) is an ideal substrate for standardizing the activity of endoxylanase. Xylanase assay is usually done using BWX which is mainly present as methyl-glucouronoxylan as substrate and contains 90% xylan. The most widely used assay method that has been standardized after carrying out interlaboratory studies was suggested by Bailey et al., (1992). They found that given the nature of substrate and variation in batch to batch up to 17% standard deviation can be tolerated. Today, this method is most widely used as indicated by over 650 citations of the method. The use of arabinoxylans (wheat arabinoxylan; WAX/Rye arabinoxylan; RAX) shows high activity when compared to oatspelt xylan (OSX) and BWX. So, even though there are pitfalls in these methods, the hydrolysis of BWX using DNS method stands out as the most widely used method. Other methods involving the use of RBB (Remazol Brilliant Blue) dyed methyl glucuronoxylan which initiates the release of RBB have also been advocated; however, the high cost of this substrate is one of the limiting factors in its wide use. Megazyme, an Irish company, has also introduced azo-dyed xylan as substrate for xylanase activity. It has been observed that many authors bring about changes in the protocol which may lead to inaccurate assays and sometimes workers have erroneously reported the results where μg of xylose released instead of μmol of xylose released has been shown as enzyme units (Lakshmi et al., 2009). In this way, the xylanase activity is overestimated by 100 times. There are few reports where 4-nitrophenyl and 4-methylumbelliferyl glycosidases of xylobiose and xylotriose have been used as substrates for assay of endoxylanase activity (Ziser and Withers, 1994) which is considered to provide stable and linear hydrolysis over the period of assay when compared to xylan which show decrease in hydrolysis with time as the number of positions susceptible to hydrolysis decrease steadily. The use of fluorogenic substrates 6,8,-difluoro-4-methylumbelliferyl β-D-xylobioside for ultrasensitive continuous assay of xylanase has also been suggested. This HPLC-based method provides speed and sensitivity for measuring xylanase activity, as well as screening xylanase inhibitors in a high-throughput format (Ge et al., 2007). In yet another high-throughput screening (HTS) approach, multiplexed glycochip enzymatic assays based on a nanostructure initiator mass spectrometry (NIMS) have been developed by Northen and Coworkers at JBEI (Joint Bio Energy Institute, CA). In this NIMzyme assay, the enzyme substrates are immobilized on mass spectrometry surface using fluorous phase interactions (DOE Report, 2009).

The arabinofuranosidase activity is usually measured using pNp α-L-arabinofuranoside as substrate. However, cereal xylans are mono- and disubstituted with (1 → 2) and (1 → 3) linked α-L-arabinofuranosyl (α-L-Araf) residues. In addition, ferulic and ρ-coumaric acids are ester linked to arabinoxylans at *O*-5 of α-L-Araf units (Mastihubova and Biely, 2010; Pastel et al., 2009). In order to know the substrate specificity of α-L-arabinofuranosidase, the substrates β-D-xylp (1-2), α-L-araf (1-3), α-L-araf (1-3) mono, α-L-araf (1-3) di, and α-L-araf (1-2) di are employed, whereas pNP β-D xylopyranoside is substrate of choice for assay of β-xylosidase activity. Acetylxylan esterase activity is measured using pNp acetate or α- or β-naphthyl acetate or methylumbelliferyl acetate which though are nonspecific substrates for acetylxylan esterase activity but have been widely used in screening and identification of active fractions during purification (Blum et al., 1999). However, the most objective method uses hardwood acetylxylans where the amount of released acetic acid is determined either by HPLC- or enzyme-based assay (Megazyme) and few other commercial kits. Recent reports suggest using pNp ferulate as substrate for assay of feruloyl esterase activity; however, synthetic esters of cinnamic acid can be used as substrate where release of ferulic acid can be monitored using HPLC or determined spectrophotometrically at 340 nm (Ghatora et al., 2006; Mastihuba et al., 2002). Recent reports suggest the 4-nitrophenyl 5-*O*-transferuloyl α-L-arabinofuranoside and 4-nitrophenyl-2-*O*-transferuloyl α-L-arabinofuranoside as suitable substrates for determination as well as difference of FAE activity. α glucouronidase catalyzes the liberation of Me Glca and glcA from aldouronic acid on which MeGlca or Glca residues are linked to single xylopyranosyl residue or a non-reducing terminal xylopyranosyl residue of xylooligosaccharide. Therefore, glucouronoxylans can be used as substrate for assay of α-glucouronidase activity only in the presence of xylan depolymerizing enzyme (Puls and Schuseil, 1993). The most common substrate for α-glucuronidase activity is the aldouronic acids obtained by acidic/enzymatic hydrolysis of glucuronoxylan. An indirect method quantifies 4 nitrophenyl 2-*O*-(4-*O*-methyl-α-D-glucuronopyrnosyl) β-D-xylopyranose as substrate. Liberation of MeGlca from compounds yields an equivocal amount of pNP β-D-xylopyranoside which is hydrolyzed by β-xylosidase (Biely et al., 2000).

11 DOMAIN ORGANIZATION OF HEMICELLULASES

Most of the plant cell wall hydrolyzing enzymes typically comprise a catalytic module and one or more carbohydrate-binding modules (CBMs) that bind to a plant cell wall polysaccharide (Hachem et al., 2000). The primary function of CBMs is to increase the catalytic efficiency of the enzymes against soluble and/or insoluble substrates, and they do so by allowing inerrant alignment of the soluble enzyme with the insoluble polysaccharide. CBMs are also known to display some additional functions such as substrate disruption and sequestering and feeding of single polysaccharide chains into the active site of the catalytic modules (Subramaniyan and Prema, 2002). CBMs are located either at the N- or C-terminal, or both, and are classified into 61 different families in the CAZy database by sequence similarity and biochemical function (Coutinho and Henrissat, 1999). A wide variation exists in binding specificity within these types, for example, CBMs belonging to families 1, 2a, 3, 5, and 10 bind mainly to crystalline cellulose, whereas members of families 2b, 4, 6, 13, and 22 prefer xylan (Charnock et al., 2000). Three-dimensional structures of members of several CBM families have been elucidated and are now available from crystallographic as well as nuclear magnetic resonance (NMR) spectroscopic studies (Fujimoto et al., 2000).

Xylanases generally are known to have three types of domains, catalytic, noncatalytic (cellulose-binding domains), and thermostabilizing domains. Family 11 xylanases are found to contain a smaller catalytic domain than that of family 10 xylanases, and thus show lesser catalytic versatility (Biely et al., 1997; Sapag et al., 2002). Although xylanases contain a single catalytic domain, certain enzymes from *Neocalimastix patriciarum*, *Fibrobacter succinogenes*, and *N. frontalis* (Durand et al., 1996; Gilbert et al., 1992; Paradis et al., 1993) were found to contain two family 11 catalytic domains each. Structural analysis of both family 10 and 11 catalytic domains using X-ray crystallography revealed that family 10 has an eightfold β/α-barrel-shaped structure (Harris et al., 1996), while catalytic domains of family 11 xylanases fold into two β-sheets constituted mostly by antiparallel β-strands and one short α-helix (Gruber et al., 1998). Several studies have reported CBMs to potentiate the catalytic activity of cellulases against crystalline substrates, and xylanases against cellulose/xylan complexes. However, these domains do not potentiate the activity of GHs against soluble substrates (Ali et al., 2001). A family 2b CBM was found to increase the catalytic activity of a thermostable single domain family 10 xylanase (XynB) from *Thermotoga maritima* when fused at the C-terminus (Kittura et al., 2003). Similarly, Mangala et al. (2003) reported that the addition of a family 6 CBM to *B. halodurans* xylanase enhances its activity toward insoluble xylan. Araki et al. (2004) elucidated the essential role of the family-22 CBMs for β-1,3-1,4-glucanase activity of *Clostridium stercorarium* Xyn10B. Binding of CBMs to insoluble substrates was significantly enhanced by the presence of Na^+ and Ca^{2+} ions. Talabani et al. (2004) reported the structure determination of the xylan-binding CBM 36 domain of the *Paenibacillus polymyxa* xylanase 43A. The structural analysis revealed the molecular basis for its unique Ca^{2+}-dependent binding of xylooligosaccharides through coordination of the O_2 and O_3 hydroxyls, thus displaying its great potential for mapping the "glyco-architecture" of plant cells. In a recent study, the usefulness of synthetic xylan-binding modules as specific probes in analysis of hemicelluloses (xylan) in wood and fiber materials was demonstrated (Filonova et al., 2007). CBMs have also been used as affinity tags for purification of xylanases from *Myceliophthora* sp. (Badhan et al., 2007).

Recent studies report the characterization of a cellulose-binding domain from *Clostridium cellulovorans* endoglucanase-xylanase D and demonstrated that this domain can serve as a bifunctional fusion tag for solubilization of fusion partner as well as a domain for the immobilization, enrichment, and purification of molecules or cells on regenerated amorphous cellulose (Xu and Foong, 2008). The crystal structure of the family 31 CBM of β-1,3-xylanase from *Alcaligenes* sp. strain XY-234 (AlcCBM31) which shows affinity only with β-1,3-xylan was reported for the first time. The structure is based on typical immunoglobulin fold quite similar to CBM structures of families 34 and 9, which also adopt structures based on immunoglobulin folds (Hashimotoa et al., 2005).

CBDs have also been reported in other plant cell wall hydrolases such as mannanase (Stalbrand et al., 1995), acetylxylan esterase (Ferreira et al., 1993), and arabinofuranosidases (Black et al., 1996). Recently, a family 54 α-L arabinofuranosidase was reported to possess a CBM belonging to family 42 which specifically binds the arabinofuranose side chain of hemicellulose (Miyanaga et al., 2006).

12 MULTIPLICITY OF HEMICELLULASES

The production of a multienzyme system of xylanases, in which each enzyme has a special function, is one strategy for microorganisms to achieve effective hydrolysis of xylan. Most of the fungi-degrading lignocelluloses produce functionally diverse hemicellulases with many isoforms (Badhan et al., 2004; Wong et al., 1988). Perhaps the structural complexity of lignocelluloses has resulted in the need for these multiple forms. Various mechanisms have been suggested to account for the multiplicity of function and specificity of the xylan-degrading enzymes. Electrophoretically distinct xylanases could arise from post-translational modification (Martin et al., 2007) of a gene product such as differential glycosylation or proteolysis. The detection of minor xylanases may also be an artifact of the growth and/or purification conditions or these enzymes may have functions, which are not required in large amounts, for example, hydrolysis of linkages not found frequently (Wong and Saddler, 1992). Multiple endoxylanases can also be expressed by distinct alleles of one gene, or even by completely separate genes (Chavez et al., 2002; Lagaert et al., 2009). Heterogeneity of xylan substrates may be one of the reasons for the production of multiple forms of xylanases, and some of these isoforms may be substrate specific or may show wide specificity, while it may be a secondary activity for others (Wong et al., 1988). Many microorganisms are able to produce multiple endoxylanases in order to acclimatize to various plant structural polysaccharides. For example 2, 6, 10, and 12 types of xylanases are produced by *Bacillus firmus* and *M. flava* (Sharma et al., 2010a,b; Tseng et al., 2002), *C. lucknowense* (Ustinov et al., 2008), *Paenibacillus curdlanolyticus* B-6 (Pason et al., 2006), and a thermotolerant strain of *Myceliophthora* sp. (Badhan et al., 2007), respectively. Sharma et al. (2008) reported the molecular characterization of 16 different thermophilic/thermotolerant fungi isolated from composting materials capable of producing multiple xylanases (Figure 2a). Recently, two-dimensional electrophoresis approaches were employed to study the expression of multiple xylanases from *S. thermonitrificans* NTU-88 (Cheng et al., 2009). Presence of inducers or inhibitors in the medium also affects the production of enzymes, as expression of some of the genes may get induced or repressed by the presence of these agents. Expression of four *Cochliobolus carbonum* endo-1,4-β-xylanase genes (XYL1, XYL2, XYL3, and XYL4) and one exo-1,4-β-xylosidase gene (XYP1) was observed in the culture medium containing xylan; however, addition of glucose resulted in repression of all the four endoxylanases. The comparative analyses of the expression pattern of two genes from *P. purpurogenum*, *xynA* and *xynB* responsible for the production of endoxylanases XynA and XynB of families 10 and 11, respectively, were carried out under several induction and repression conditions. It was observed that the endoxylanase gene *xynB* was efficiently expressed with all the inducers (birch wood xylan, OSX, xylose, and xylitol), whereas *xynA* gene was expressed only in presence of OSX (Chavez et al., 2002). However, in case of production of multiple xylanases from thermophilic fungus *Myceliophthora* sp., it was observed that in addition to the type of carbon source, culture conditions also play an important role in multiplicity of xylanases, where rice straw induced expression of 3 and 5 xylanase isoforms under shake flask and solid-state fermentation (SSF), respectively (Badhan et al., 2004). Expression of multiple xylanases can also be induced by the positional isomers formed as a result of transglycosylation activity of enzymes produced at constitutive level (Saraswat and Bisaria, 1997). Multiple forms of enzymes may also result from horizontal gene transfer in the microorganisms living in similar ecological niche, and thereafter, evolving separately adapting to particular environmental conditions (Cpeljnik et al., 2004). The study of the functional importance of three xylanases from the saprophytic fungus *T. harzianum* showed a high degree of complementation of these xylanases in the hydrolysis of aspen xylan. Furthermore, the functional diversity of 10 xylanases from thermophilic fungus *Myceliophthora* sp. was analyzed using different types of xylan substrates, and it was concluded that xylanases are not redundant enzymes since each contributes

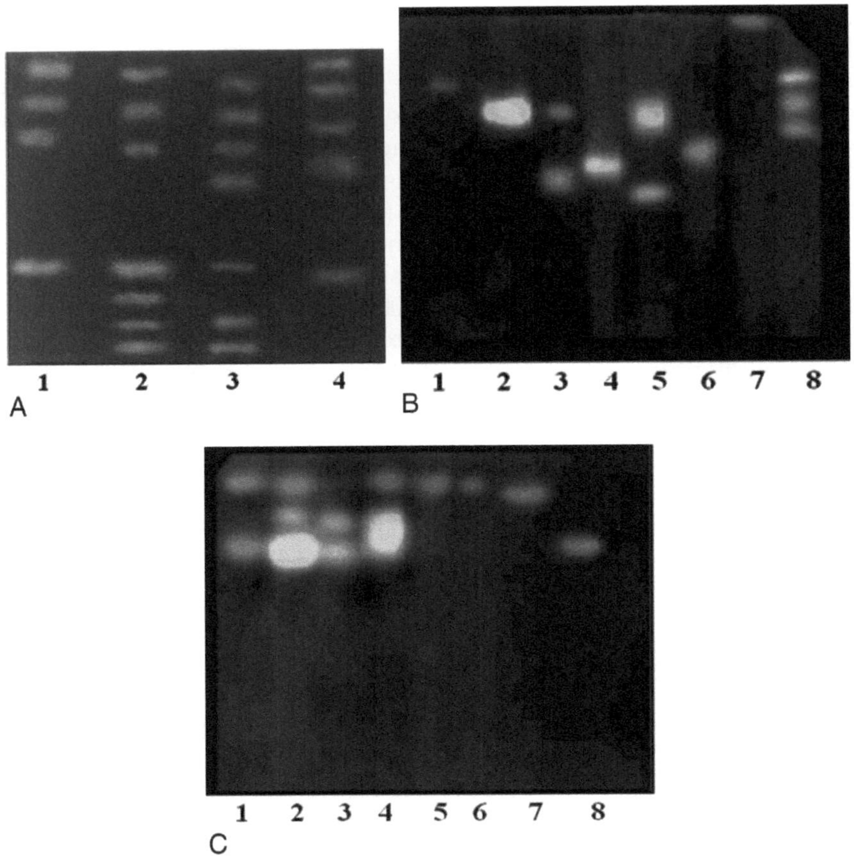

FIGURE 2 (a) Multiplicity of xylanase: Zymogram developed against PAGE lane 1, *Penicillium lagena;* 2, *Emericella nidulans;* lane 3, *Aspergillus terreus;* lane 4, *Humicola insolens.* (b) Multiplicity of arabinofuranosidase. Lanes 1-8: *Aspergillus niger, A. oryzae, A. awamori, A. tubingensis, A. terreus, A. niger, Penicillium oxalicum, P. janthenillum.* (c) Multiplicity of β-xylosidase. Lanes 1-8: *Penicillium janthenillum, P. oxalicum, A. niger, A. terreus, A. tubingensis, A. oryzae, A.niger.*

significantly and uniquely to the hydrolysis of the xylan. In spite of the fact that the multiform enzymes catalyze same reaction, they may differ in kinetic properties, regulatory characteristics, and/or stabilities (Naessens and Vandamme, 2003). Therefore, in order to elucidate the functional variations, the catalytic potential of each isoxylanase should be assayed against different substituted and unsubstituted xylan types (Ghatora et al., 2006; Wong et al., 1988).

Multiplicity has also been observed in β-xylosidases, α-L-arabinofuranosidases (AFs; Figure 2b and c), and acetylxylan esterases and feruloyl esterases (Ghatora et al., 2006; Vries and Visser, 2001). Two β-xylosidases liberated from the cell surface of *P. herquei* were purified and identified as GH43 enzymes (Ito et al., 2003). Three different forms of α-L-arabinofuranosidases from *P. purpurogenum* were separated by isoelectrofocusing and detected using the zymogram technique, out of which one arabinofuranosidase has been purified and identified as GH54 enzyme. Sinitsyna et al. (2003) isolated two arabinofuranosidases (AF-60 & AF-70) from among the major components of the xylanase system of *Penicillium canescens*. *B. subtilis* produces two α-L-arabinofuranosidases capable of releasing arabinosyl oligomers and L-arabinose from plant cell walls, both belonging to family 51 GHs but differing significantly in their substrate specificities (Inacio et al., 2008). Recently Hinz et al. (2009) reported the selective production, purification, and characterization of four arabinofuranosidases, two acetylxylan and ferulic acid esterases, and α-glucuronidase from the filamentous fungus *C. lucknowense*, thus demonstrating high potential of this fungus as a producer of hemicellulolytic enzymes. Thermophilic fungi including *H. insolens*, *Chaetomium thermophilum*, and *Melanocarpus* sp. were identified as prolific producers and expressed multiple esterases that were putatively classified as xylan acetyl esterase and feruloyl esterases on the basis of distinct preferential substrate specificities toward ρ-nitrophenyl acetate and ρ-nitrophenyl ferulate, respectively (Ghatora et al., 2006).

13 FUNCTIONAL GENOMICS APPROACH FOR STUDYING HEMICELLULASES

Functional genomics for system analysis of bacteria- and fungi-producing GHs have been important in profiling the expression of cellulases and hemicellulases predicting the functional strategy these fungi employ for degradation of plant cell wall. The analysis of the transcriptome and secretome datasets has been evaluated to identify the gene/proteins that are overexpressed in *Neurospora crassa* (Tian et al., 2009), *Postia placenta*, and *Phanerochaete chrysosporium* (Martinez et al., 2009; Wymelenberg et al., 2010). Viewed together with transcript profiles, *P. chrysosporium* employs an array of extracellular GHs to simultaneously attack cellulose and hemicelluloses. In contrast, under these same conditions, *P. placenta* secretes an array of hemicellulases but few potential cellulases (Wymelenberg et al., 2010). The studies reporting comparative secretomes of the fungal strains grown under submerged and SSF of *A. oryzae* (Oda et al., 2006), between two hypersecretory strains of *T. reesei* (Gimbert et al., 2008) or those grown in presence of different carbon sources, have also highlighted differential expression profiles and have also led to the identification of unreported putative arabinofuranosidases (Gimbert et al., 2008). Comparative studies have also highlighted differences in the relative abundance of proteases, cellulase/hemicellulase in the extracts of *T. reesei* Rut C-30, and commercial enzyme preparation Spezyme CP from the same organism (Nagendran et al., 2009). Quantitative iTRAQ secretome analysis of *A. niger* has revealed the presence of novel hydrolytic enzymes (Adav et al., 2010). The secretome of *A. fumigatus* has revealed the presence of variety of GHs that was found to be efficient in carrying out the saccharification of alkali-treated rice straw (Sharma et al., 2010a,b).

14 ENZYME PRODUCTION

A wide spectrum of cell wall-degrading enzymes including cellulases and hemicellulases (GHs) are produced by different fungi and bacteria. However, these microorganisms differ appreciably in their capability to produce these enzymes in terms of their activities as well as the spectrum of different GHs. Each microorganism differs in its genetic capacity

(Hinz et al., 2009) and consequently secretes a specific combination of GHs. However, most of the commercially important sources of hemicellulases are limited to fungi. Recent studies showed that the difference in the activity profiles of commercial cellulolytic/hemicellulolytic strains, for example, *T. reesei. A. niger* (Sorensen et al., 2005), and *C. lucknowense* (Emalfarb et al., 2003) could be related to their genetic capacity. Where *T. reesei* is known to be a good source of cellobiohydrolases and endoglucanases, *A. niger* is known to be a good source of β-glucosidase/xylosidase, whereas the *C. lucknowense* genetic system was found to be most elaborate for the expression of hemicellulases (Hinz et al., 2009).

Most of the studies on the production of hemicellulases are primarily focused on xylanases which are specifically required in the paper and pulp industry, for generating xylooligosaccharides (Pastel et al., 2009; Puchart and Biely, 2008; Sharma et al., 2010a,b). Most of the other applications however require the complete spectrum of hemicellulolytic enzymes, especially in the bioconversion process for converting hemicellulose fraction to monomeric sugars for further fermentation into biofuels and specialty chemicals (Ohara, 2003). Because of the differences in the structural composition of hemicellulose, defining the right balance of enzyme mixture is not easy. Alternative bioreactors such as the air-lift or bubble-column, which have a lower shear stress, seem to produce better results. For example, studies on xylanase and cellulase production by *A. niger* in various bioreactors showed that in general, better yield and productivity were obtained in a bubble-column and an air-loop air-lift than in the stirred-tank reactor. However, the relatively high cost of enzyme production has hindered the industrial application of the enzymatic process. Recent trends show that SSF which involves the growth of fungi on wet solids in the absence of free water is an attractive proposition because of the economic and engineering advantages (Rodrigues et al., 2007). The low moisture content results in low-energy consumption and prevents bacterial contamination and the problems caused by low gas distribution during submerged cultures, which make the SSF system good (Leite et al., 2007; Pandey et al., 2000). The hyphal mode of fungal growth and their good tolerance to low water activity and high osmotic pressure conditions make it efficient and competitive in natural microflora for bioconversion of solid substrates (Raimbault, 1998).

Some of the prolific producers of xylanase, *T. reesei* and *T. lanuginosus*, are known to produce >3000 U/mL under shake flask/submerged culture (Haapala et al., 1994; Singh et al., 2000), while *T. lanuginosus* has been reported to produce 48,000 (U/g substrate) under SSF (Sonia et al., 2005). There are several references in literature that suggest that fungi produce appreciably higher levels of xylanases when cultured under SSF in comparison to SmF. For example, *P. brasilinum, A. niger, Melanocarpus albomyces* produced almost 5-10 times higher activities under SSF as compared to SmF (Jorgensen et al., 2005; Narang et al., 2001; Thygeson et al., 2003). Similar observations have also been made on the production of arabinofuranosidase from *Arthrobacter* sp., where 0.1 (U/ml) was produced under shake flask conditions compared to 3.5 (U/g substrate) under SSF (Khandeparker et al., 2008). It has been estimated that SSF is 100 times more economical for cellulase production as compared to SmF (Antoine et al., 2010).

In order to produce a complete spectrum of hemicellulases, the nature and composition of the carbon sources used for induction of enzyme production plays a crucial role. Various carbon sources such as rice straw, wheat straw, wheat bran, corn cobs, bagasse, banana peels, etc. have been used for production hemicellulases (Sonia et al., 2005; Thygeson et al., 2003) The growth of cultures on different carbon sources has been shown to be associated with differential expression of functionally distinct xylanases (Badhan et al., 2007). It has been observed that not all the components of hemicellulases are produced in presence of one type of carbon source as *T. lanuginosus* produced maximal levels of xylanase and β-xylosidase in presence of corn cobs, whereas OSX was found to induce maximal levels of arabinofuranosidase, acetylxylan esterase, feruloyl esterase, and β-mannosidase which clearly suggests it is judicious to go for production of optimal level rather than maximal levels of production. Most of the work on optimization has been focused on endoxylanases, and there is dearth of work done where debranching has been considered during optimization. In a recent report, optimization of xylanases and debranching enzymes by thermophilic fungal strain *M. flava* grown on sorghum straw was optimized employing response surface methodology. Under optimal conditions, *M. flava* produced 16390, 9.49, 3.40, 69.8, and 2.25 (units/g substrate) of xylanase, β-xylosidase, arabinofuranosidase, acetyl esterase, and feruloyl esterase, respectively (Sharma and Chadha, 2010). In addition to carbon source, type of nitrogen sources, C:N ratio, initial medium pH, incubation temperature, inoculum level, inoculum age, initial moisture levels, etc. also play an important role in production of hemicellulases (Jatinder et al., 2006a; Sonia et al., 2005). The process optimization can be done by classical method that involves modification of one independent variable at a time, while all others are fixed at a certain level. The optimized conditions for production of hemicellulases have been reported for *Rhodothermus marinus* (Gomes et al., 2000), *Penicillium brasilianum* (Jorgensen et al., 2005), *Thermomyces lanuginosus* (Sonia et al., 2005), *Melanocarpus* sp. MTCC3922 (Jatinder et al., 2006a). Statistical approaches like response surface methodology, central composite design, multiple linear regression, back propagation neural network, and lazy learning algorithm have also been used for optimization of hemicellulases (Guerfali et al., 2010; Jatinder et al., 2006b, Meshram et al., 2008).

15 APPLICATIONS OF HEMICELLULASES

The xylanolytic enzymes used in the paper and pulp industry mainly for biobleaching and pectinolytic enzymes have been used for debarking; in addition to bleaching capability, xylanases have been found useful in other applications also, that is, clarification of juice and wine, starch separation and production of functional food ingredients, improving the quality of bakery products, in animal feed biotechnology, in debarking, deinking of recycled fibers, and in preparation of dissolving pulp (Beg et al., 2001; Polizeli et al., 2005; Techapun et al., 2003). The use of hemicellulases along with glucanases, cellulases, proteases, amylases, phytase, galactosidases, and lipases has become a common practice in the field of animal feed biotechnology. These enzymes bring about the breakdown of plant cell wall complex present in the ingredients of feed and reduce the viscosity of raw material. If xylanase is added to feed containing maize and sorghum, both of which are low-viscosity foods, it may improve the digestion of nutrients in the initial part of the digestive tract, resulting in a better use of energy (Polizeli et al., 2005). Addition of xylanases to rye-based diet of broiler chickens has been shown to increase weight of chicks (Bedford and Classen, 1992); moreover, the use of xylanases in combination with phytases has resulted in increase in egg and albumen weight from white and brown egg-laying hens (Silversides et al., 2006). Some of the family 11 xylanases produced by rumen bacteria of genera *Pseudobutyrivibrio* and *Butyrivibrio* show the cleavage of OSX into tetra or higher oligomers; these xylooligosaccharides could be helpful in promoting the proliferation of beneficial microflora (Craeyveld et al., 2008). Therefore, xylanases produced by these strains could be used as a feed additive for animals, and such strains can be used as probiotic for animals (Cpeljnik et al., 2004). Xylanase also play an important role in improving the quality of bread, breaking down hemicellulose in wheat flour, helping in the redistribution of water, and leaving the dough softer and easier to knead, resulting in increase in bread volumes and improved resistance to fermentation (Shah et al., 2006). Synergistic action of xylanases and related hemicellulases can be employed for generation of biofuel such as ethanol and xylitol from lignocellulosic biomass. Xylitol used as sweetener in food has odontological applications such as teeth hardening, and is used in chewing gum and toothpaste formulation (Beg et al., 2001). Xylanases with transglycosylation activities can also be used for designing the drugs and preparation of neoglycoproteins (Eneyaskaya et al., 2003). The use of xylanases in production of alkyl glycosides by hydrolysis of polysaccharides is a challenging opportunity. Xylanase purified from a strain of *A. pullulans* has been used for direct transglycosylation of xylan with 1-octanol and 2-ethylhexanol into octyl-β-D-xylobioside and 2-ethylhexyl-β-D-xylobioside, respectively (Matsumura et al., 1999). Xylan-debranching enzymes such as acetylxylan esterase and feruloyl esterases may enhance the process of solubilization of lignin-carbohydrate complex by removing substitutions and linkages between polymers during pulping (de Graaff et al., 2000). Acetylxylan esterase can be used in deinking of paper by aiding in the removal of substituents groups which hinder main-chain-degrading enzymes. Recently, esterases especially feruloyl esterases have been reported as being used for the bioconversion of lignocellulosic wastes, synthesis of esters in organic solvents, and isolation of phenolic acids as precursors of a variety of value-added chemicals (Garcia-Conesa et al., 2005). α-L-Arabinofuranosidases have been employed for aromatizing musks, wines and fruit juices, for delignification of paper pulp, for digestibility enhancement of animal feedstock, and for fractionation of sugar beet pulp into pectin, cellulose, and arabinose (Saha, 2000). A potential utilization of pectinases is in the treatment of softwoods, which has been shown to improve the efficiency of preservative treatment by rendering the wood more permeable for chemical preservatives (Gregorio et al., 2002).

References

Adav, S.S., Li, A.A., Manavalan, A., Punt, P., Sze, S.K., 2010. Quantitative iTRAQ secretome analysis of *Aspergillus niger* reveals novel hydrolytic enzymes. J. Proteome Res. 9, 3932–3940.

Albertsson, A.C., Voepel, J., Edlund, U., Dahlman, O., Lindblad, M.S.D., 2010. Design of Renewable Hydrogel Release Systems from Fiberboard Mill Wastewater. Biomacromolecules 11, 1406–1411.

Ali, M.K., Hayashi, H., Karita, S., Goto, M., Kimura, T., Sakka, K., et al., 2001. Importance of the carbohydrate-binding module of *Clostridium stercorarium* Xyn10B to *Xylan hydrolysis*. Biosci. Biotechnol. Biochem 65, 41–47.

Allgaier, M., Reddy, A., Park, J.I., Ivanova, N., D'haeseleer, P., Lowry, S., et al., 2010. Targeted discovery of glycoside hydrolases from a switchgrass-adapted compost community. PLoS ONE 5, e8812, doi:10.1371/journal.pone.0008812.

Andrade, S.V., Polizeli, M.L.T.M., Terenzi, H.F., Jorge, J.A., 2004. Effect of carbon source on the biochemical properties of the β-xylosidase produced by *Aspergillus versicolor*. Process. Biochem. 39, 1931–1938.

Antoine, A.A., Jacqueline, D., Thonart, P., 2010. Xylanase production by *Penicillium canescens* on soya oil cake in solid state fermentation. Appl. Biochem. Biotechnol. 160, 50–62.

Araki, R., Ali, M.K., Sakka, M., Kimura, T., Sakka, K., Ohmiya, K., 2004. Essential role of the family-22 carbohydrate-binding modules for β-1,3-1,4-glucanase activity of *Clostridium stercorarium* Xyn10B. FEBS Lett. 561, 155–158.

Arora, N., Banerjee, A.K., Mutyala, S., Murty, U.S., 2009. Comparative characterization of commercially important xylanase enzymes. Bioinformation 3, 446–453.

Arsovski, A.A., Popma, T.M., Haughn, G.W., Carpita, N.C., McCann, M.C., Western, T.L., 2009. AtBXL1 Encodes a Bifunctional β-d-Xylosidase/α-L-Arabinofuranosidase Required for Pectic Arabinan Modification in *Arabidopsis* Mucilage Secretory Cells1[C][W][OA]. Plant Physiol. 150, 1219–1234.

Badhan, A.K., Chadha, B.S., Sonia, K.G., Saini, H.S., Bhat, M.K., 2004. Functionally diverse multiple xylanases of thermophilic fungus *Myceliophthora* sp. IMI 387099. Enzyme Microb. Technol. 35, 460–466.

Badhan, A.K., Chadha, B.S., Saini, H.S., 2007. Purification of the alkalophilic xylanases from *Myceliophthora* sp. IMI 387099 using cellulose-binding domain as an affinity tag. World. J. Microbiol. Biotechnol. 24, 973–981.

Bai, Y., Wang, J., Yang, P., Shi, P., Luo, H., Meng, K., et al., 2010. A new xylanase from thermoacidophilic Alicyclobacillus sp. A4 with broad range pH activity and pH stability. J. Ind. Microbiol. Biotechnol. 37, 187–194.

Bailey, M., Biely, P., Poutanen, K., 1992. Interlaboratory testing methods for assay of xylanase activity. J. Biotechnol. 23, 257–270.

Bedford, M.R., Classen, H.L., 1992. The influence of dietary xylanase on intestinal viscosity and molecular weight distribution of carbohydrates in rye-fed broiler chick. In: Visser, J., Beldman, G., Van Somerren, M.A.K., Voragen, A.G.J. (Eds.), Xylan and Xylanases. Elsevier, Amsterdam, pp. 361–370.

Beg, Q.K., Kapoor, M., Mahajan, L., Hoondal, G.S., 2001. Microbial xylanases and their industrial applications: a review. Appl. Microbiol. Biotechnol. 56, 326–338.

Benoit, I., Danchin, E.G.J., Bleichrodt, R.J., De Vries, R.P., 2008. Biotechnological applications and potential of fungal feruloyl esterases based on prevalence, classification and biochemical diversity. Biotechnol. Lett. 30, 387–396.

Biely, P., Vršanská, M., Tenkanen, M., Kluepfel, D., 1997. Endo-β-1-4-xylanase families: differences in catalytic properties. J. Biotechnol. 57, 151–166.

Biely, P., de Vries, R.P., Vrsanska, M., Visser, J., 2000. Inverting character of α-glucuronidase A from *Aspergillus tubingensis*. Biochim. Biophys. Acta, 1474, 360–364.

Black, G.W., Rixon, J.E., Clarke, J.H., Hazlewood, G.P., Theodorou, M.K., Morris, P., et al., 1996. Evidence that linker sequences and cellulose-binding domains enhance the activity of hemicellulases against complex substrates. Biochem. J. 319, 515–525.

Blum, D.L., Li, X.L., Chen, H., Ljungdahl, L.G., 1999. Characterization of an acetyl xylan esterase from the anaerobic fungus *Orpinomyces* sp. strain PC-2. Appl. Environ. Microbiol. 65, 3990–3995.

Bourgois, T.M., Van Craeyveld, V., Van Campenhout, S., Courtin, C.M., Delcour, J.A., Robben, J., et al., 2007. Recombinant expression and characterization of XynD from *Bacillus subtilis* subsp. *Subtilis* ATCC 6051: a GH 43 arabinoxylan arabinofuranohydrolase. Appl. Microbiol. Biotechnol. 75, 1309–1317.

Bravman, T., Mechaly, A., Shulami, S., Belakhov, V., Baasov, T., Shoham, G., et al., 2001a. Glutamic acid 160 is the acid-base catalyst of β-xylosidase from Bacillus stearothermophilus T-6: a family 39 glycoside hydrolase. FEBS Lett. 495, 115–119.

Bravman, T., Zolotnitsky, G., Shulami, S., Belakhov, V., Solomon, D., Baasov, T., et al., 2001b. Stereochemistry of family 52 glycosyl hydrolases: β-xylosidase from Bacillus stearothermophilus T-6 is a retaining enzyme. FEBS Lett. 495, 39–43.

Brennan, Y.L., Callen, W.L., Christoffersen, L., Dupree, P., Goubet, F., Healey, S., et al., 2004. Unusual microbial xylanases from insect guts. Appl. Environ. Microbiol. 70, 3609–3617.

Brunzelle, J.S., Jordan, D.B., McCaslin, D.R., Olczak, A., Wawrzak, Z., 2008. Structure of the two-subsite beta-D-xylosidase from *Selenomonas ruminantium* in complex with 1,3 bis[tris(hydroxymethyl) methylamino]propane. Arch. Biochem. Biophys. 474, 157–166.

Cantarel, B.L., Coutinho, P.M., Rancurel, C., Bernard, T., Lombard, V., Henrissat, B., 2009. The carbohydrate-active enzymes database (Cazy): an expert resource for glycogenomics. Nucleic. Acids Res. 37 (Database issue), D233–D238.

Carapito, R., Imberty, A., Jeltsch, J.M., Byrns, S.C., Tam, P.H., Lowary, T.L., et al., 2009. Molecular basis of arabinobiohydrolase activity in phytopathogenic fungi: crystal structure and catalytic mechanism of *Fusarium graminearum* GH93 exo-α-L-Arabinanase. J. Biol. Chem. 284, 12285–12296.begin_of_the_skype_highlighting.

Chadha, B.S., Badhan, A.K., Mellon, F., Bhat, M.K., 2004. Two endoxylanases active and stable at alkaline pH from the newly isolated thermophilic fungus, *Myceliophthora* sp. IMI 387099. J. Biotechnol. 109, 227–237.

Charnock, S.J., Bolam, D.N., Turkenberg, J.P., Gilbert, H.J., Ferreira, L.M.A., Davies, G.J., et al., 2000. The X6 "thermostabilizing" domains of xylanases are carbohydrate-binding modules: structure and biochemistry of the Clostridium thermocellum X6b domain. Biochem 39, 5013–5021.

Chavez, R., Schachter, K., Navarro, C., Peirano, A., Aguirre, C., Bull, P., et al., 2002. Differences in expression of two endoxylanase genes (*xynA* and *xynB*) from *Penicillium purpurogenum*. Gene 293, 161–168.

Chen, X., Lu, W., Cao, Y., Li, D., 2008. Prokaryotic expression, purification and characterization of *Aspergillus sulphureus* beta mannanase and site-directed mutagenesis of the catalytic residues. Appl. Biochem. Biotechnol. 149, 139–144.

Cheng, H.L., Tsai, C.Y., Chen, H.J., Yang, S.S., Chen, Y.C., 2009. The identification, purification, and characterization of STXF10 expressed in *Streptomyces thermonitrificans* NTU-88. Appl. Microbiol. Biotechnol. 82, 681–689.

Collins, T., Meuwis, M., Stals, I., Claessens, M., Feller, G., Gerday, C., 2002. A novel family 8 xylanase: functional and physio-chemical characterization. J. Biol. Chem. 277, 35133–35139.

Collins, T., Gerday, C., Feller, G., 2005. Xylanases, xylanase families and extremophilic xylanases. FEMS Micbobiol. Rev. 29, 3–23.

Contreras, L.M., Gomez, J., Prieto, J., Clemente-Jimenez, J.M., Las Heras-Vazquez, F.J., Rodriguez-Vico, F., et al., 2008. The family 52 β-xylosidase from *Geobacillus stearothermophilus* is a dimer: structural and biophysical characterization of a glycoside hydrolase. Biochemica et Biophysica Acta 1784, 1924–1934.

Corral, O.C., Ortega, F.V., 2006. Xylanases. In: González, R.G., Pacheco, I.T. (Eds.), Advances in Agricultural and Food Biotechnology. Research Signpost. pp. 305–322.

Coutinho, P.M., Henrissat, B., 1999. Carbohydrate-active enzymes: an integrated database approach. In: Gilbert, H.J., Davies, G.J., Henrissat, B., Svensson, B. (Eds.), Recent Advances in Carbohydrate Bioengineering. Royal Society of Chemistry, Cambridge, pp. 3–12.

Cpeljnik, T., Krizaj, I., Marinsek-Logar, R., 2004. Isolation and characterization of the *Pseudobutyriovibrio xylanivorans* Mz5^T xylanase XynT—the first family 11 endoxylanase from rumen *Butyriovibrio*-related bacteria. Enzyme Microb. Technol. 34, 219–227.

Craeyveld, V.V., Swennen, K., Dornez, E., Wiele, T.V., Marzorati, M., Verstraete, W., et al., 2008. Structurally different wheat derived arabinoxylooligosaccharides have different prebiotic and fermentation properties in rats. J. Nutr. 138, 2348–2355.

Crepin, V.F., Faulds, C.B., Connerton, I.F., 2004. Functional classification of the microbial feruloyl esterases. Appl. Microbiol. Biotechnol. 63, 647–652.

Cuong, D.B., Thu, D.T., Berrin, J.G., Haltrich, D., Anh, T.K., Sigoillot, J.C., et al., 2009. Cloning, expression in *Pichia pastoris*, and characterization of a thermostable GH5 mannan endo-1,4-*β*-mannosidase from *Aspergillus niger* BK01. Microbial Cell Fact. 8, 59. doi:10.1186/1475-2859-8-59.

de Graaff, L.H., Visser, J., van den Broeck, H.C., Strozyk, F., Kormelink, F.J.M., Boonman, J.C., 2000. Method to alter the properties of acetylated xylan. United States Patent 6,010,892.

Dhawan, S., Kaur, J., 2007. Microbial mannanases: an overview of production and applications. Crit. Rev. Biotechnol. 27, 197–216.

Dhiman, S.S., Sharma, J., Battan, B., 2008. Industrial aspects and future prospects of microbial xylanases: a review. Bioresources 3, 1377–1402.

Dodd, D., Cann, I.K.O., 2009. Enzymatic deconstruction of xylan for biofuel production. GCB Bioenerg. 1, 2–17.

DOE Report, 2009. Bioenergy Research centres. An Overview of the Science. DOE/SC-01166.

Durand, R., Rascle, C., Fevre, M., 1996. Molecular characterization of xyn3, a member of the endoxylanase multigene family of the rumen anaerobic fungus *Neocalimastix frontalis*. Curr. Genet. 30, 531–540.

Emalfarb, M.A., Burlingame, R.P., Olson, P.T., Sinitsyn, A.P., Parriche, M., Bousson, J.C., et al., 2003. Transformation system in the field of filamentous fungal hosts. U.S. Patent No. 6,573,086.

Eneyaskaya, E.V., Brumer, H., Backinovsky, L.V., Ivanen, D.R., Kulminskaya, A.A., Shabalin, K.A., et al., 2003. Enzymatic synthesis of β-xylanase substrates: transglycosylation reactions of the β-xylosidase from *Aspergillus* sp. Carbohydr. Res. 338, 313–325.

Fattah, A.F.A., Hashem, A.M., Ismail, A.M.S., El-Refai, M.A., 2009. Purification and Some Properties of a-mannanase from *Aspergillus Oryzae* NRRL 3448. J. Appl. Sci. Res. 5, 2067–2073.

Faure, D., 2002. The family-3 glycoside hydrolases: from housekeeping functions to host-microbe interactions. Appl. Environ. Microbiol. 68, 1485–1490.

Ferreira, L.M.A., Wood, T.M., Williamson, G., Faulds, C., Hazlewood, G.P., Black, G.W., et al., 1993. A modular esterase from *Pseudomonas fluorescens* subsp. Cellulosa contains a non-catalytic cellulose-binding domain. Biochem. J. 294, 349–355.

Filonova, L., Gunnarsson, L.C., Daniel, G., Ohlin, M., 2007. Synthetic xylan-binding modules for mapping of pulp fibres and wood sections. BMC Plant Biol. 7, 54. doi:10.1186/1471-2229-7-54.

Fritz, M., Ravanal, M.C., Braet, C., Eyzaguirre, J., 2008. A family 51 α-L-arabinofuranosidase from *Penicillium purpurogenum*: purification, properties and amino acid sequence. Mycol. Res. 112, 933–942.

Fujimoto, Z., Kuno, A., Kaneko, S., Yoshida, S., Kobayashi, H., Kusakabe, I., et al., 2000. Crystal structure of *Streptomyces olivaceoviridis* E-86 β-xylanase containing xylan-binding domain. J. Mol. Biol. 300, 575–585.

Gallardo, O., Fernandez, M.F., Valls, C., Valenzuela, S.V., Roncero, M.B., Diaz, P., et al., 2010. Characterization of a family GhH5 xylanase with activity on neutral oligosaccharides and evaluation as a pulp bleaching aid. Appl. Environ. Microbiol. 76, 6290–6294.

Garcia-Conesa, M.T., Crepin, V.F., Goldson, A.J., Williamson, G., Cummings, N.J., Connerton, I.F., et al., 2005. The feruloyl esterase systems of *Talaromyces stipitatus*: production of three discrete feruloyl esterases, including a novel enzyme, TsFaeC, with a broad specificity. J. Biotechnol. 108, 227–241.

Ge, Y., Antoulinakis, E.G., Gee, K.R., Johnson, I., 2007. An ultrasensitive, continuous assay for xylanase using the fluorogenic substrate 6,8-difluoro-4-methylumbelliferyl β-D-xylobioside. Analytical. Biochem. 362, 63–68.

Ghatora, S.K., Chadha, B.S., Saini, H.S., Bhat, M.K., Faulds, C.B., 2006. Diversity of plant cell wall esterases in thermophilic and thermotolerant fungi. J. Biotechnol. 125, 434–445.

Gilbert, H.J., Hazlewood, G.P., Laurie, J.I., Orpin, C.G., Xue, G.P., 1992. Homologous catalytic domains in a rumen fungal xylanase: evidence for gene duplication and prokaryotic origin. Mol. Microbiol. 6, 2065–2072.

Gimbert, I.H., Margoet, A., Dolla, A., Jan, G., Molle, D., Lignon, S., et al., 2008. Comparative secretome analyses of two *Trichoderma reesei* RUT-C30 and CL847 hypersecretory strains. Biotechnol. Biofuels 1, 18. doi:10.1186/1754-6834-1-18.

Girio, F.M., Fonseca, C., Carvalheiro, F., Duarte, L.C., Maeques, S., Lukasik, R.B., 2010. Hemicelluloses for fuel ethanol: a review. Bioresour. Technol. 101, 4775–4800.

Gomes, J., Gomes, I., Terler, K., Gubala, N., Ditzelmuller, G., Steiner, W., 2000. Optimization of culture medium and conditions for α-L-arabinofuranosidase production by the extreme thermophilic eubacterium *Rhodothermus marinus*. Enzyme. Microb. Technol. 27, 414–419.

Gosalbes, M.J., Pérez-Gonzalez, J.A., Gonzalez, R., Navarro, A., 1991. Two β-glycanase genes are clustered in *Bacillus polymyxa*: molecular cloning, expression, and sequence analysis of genes encoding a xylanase and an endo-β-(1,3)-(1,4)-glucanase. J. Biotechnol. 173, 7705–7710.

Gregorio, A.D., Mandalari, G., Arena, N., Nucita, F., Tripodo, M.M., Curto, R.B.L., 2002. SCP and crude pectinase production by slurry-state fermentation of lemon pulps. Bioresour. Technol. 83, 89–94.

Gruber, K., Klintschar, G., Hayn, M., Schlacher, A., Steiner, W., Kratky, C., 1998. Thermophilic xylanase from *Thermomyces lanuginosus*: high resolution X-ray structure and modelling studies. Biochem 37, 13475–13485.

Guerfali, M., Chaabouni, M., Gargouri, A., Belghith, H., 2010. Improvement of α-L-arabinofuranosidase production by *Talaromyces thermophilus* and agro-industrial residues saccharification. Appl. Microbiol. Biotechnol. 85, 1361–1372.

Haapala, R., Linko, S., Parkkinen, E., Suominen, P., 1994. Production of endo-1,4-β-glucanase and xylanase by *Trichoderma reesei* immobilized on polyurethane foam. Biotechnol. Techn. 8, 401–406.

Hachem, M.A., Karlsson, E.N., Bartonek-Roxa, E., Raghothama, S., Simpson, P.J., Gilbert, H.J., et al., 2000. Carbohydrate-binding modules from a thermostable *Rodothermus marinus* xylanase: cloning, expression and binding studies. Biochem. J. 345, 53–60.

Harris, G.W., Jenkins, J.A., Connerton, I., Pikersgill, R.W., 1996. Refined crystal structure of the catalytic domain of xylanase A from *Psuedomonas fluorescens* at 1.8 A resolution. Acta Crystallogr. 52, 393–401.

Hashimotoa, H., Tamaib, Y., Okazakib, F., Tamarub, Y., Shimizua, T., Arakib, T., et al., 2005. The first crystal structure of a family 31 carbohydrate-binding module with affinity to β-1,3-xylan. FEBS Lett. 579, 4324–4328.

Hinz, S.W.A., Pouvreau, L., Joosten, R., Bartels, J., Jonathan, M.C., Wery, J., et al., 2009. Hemicellulase production in *Chrysosporium lucknowense* C1. J. Cereal. Sci. 50, 318–323.

Ichinose, H., Yoshida, M., Fujimoto, Z., Kaneko, S., 2008. Characterization of a modular enzyme of exo-1,5-α-L-arabinofuranosidase and arabinan binding module from *Streptomyces avermitilis* NBRC14893. Appl. Microbiol. Biotechnol. 80, 399–408.

Inacio, J.S., Correia, I.L., Nogueira, I.S., 2008. Two distinct arabinofuranosidases contribute to arabino-oligosaccharide degradation in *Bacillus subtilis*. Microbiol. 154, 2719–2729.

Ito, T., Yokoyama, E., Sato, H., Ujita, M., Funaguma, T., Furukawa, K., et al., 2003. Xylosidases associated with the cell surface of *Penicillium herquei* IFO 4674. J. Biosci. Bioeng. 96, 354–359.

Izydorczyk, M.S., Dexter, J.E., 2008. Barley β-glucans and arabinoxylans: molecular structure, physicochemical properties, and uses in food products. A review. Food. Res. Int. 41, 850–868.

Janis, J., Hakanpaa, J., Hakulinen, N., Ibatullin, F.M., Hoxha, A., Derrick, P.J., et al., 2005. Determination of thioxylo-oligosaccharide binding to family 11 xylanase using electrospray ionization fourier transform ion cyclotron resonance mass spectrometry and X-ray crystallography. FEBS. J 272, 2317–2333.

Jatinder, K., Chadha, B.S., Saini, H.S., 2006a. Optimization of medium components for production of cellulases by *Melanocarpus* sp. MTCC 3922 under solid-state fermentation. World J. Microbiol. Biotechnol. 22, 15–22.

Jatinder, K., Chadha, B.S., Saini, H.S., 2006b. Optimization of culture conditions for production of cellulases and xylanases by *Scytalidium thermophilum* using response surface methodology. World. J. Microbiol. Biotechnol 22, 169–176.

Javier, P.F.I., Oscar, G., Sanz-Aparicio, J., Díaz, P., 2007. Xylanases: molecular properties and applications. Ind. Enzymes. Section A, 65–82.

Jordan, D.B., Li, X.L., Dunlap, C.A., Whitehead, T.R., Cotta, M.A., 2007. Structure-function relationships of a catalytically efficient beta-D-xylosidase. Appl. Biochem. Biotechnol. 141, 51–76.

Jorgensen, H., Morkeberg, A., Kristian, B.R.K., Olsson, L., 2005. Production of cellulases and hemicellulases by three *Penicillium* species: effect of substrate and evaluation of cellulose adsorption by capillary electrophoresis. Enzyme Microb. Technol. 36, 42–48.

Kacagan, M., Canakci, S., Inan, K., Colak, D.N., Belduz, A.O., 2008. Characterization of a xylanase from thermophilic strain of *Anoxybacillus pushchioensis* A8 xylanase. Biologia 63, 599–606.

Kamm, B., Kamm, M., 2004. Biorefinery systems. Chem. Biochem. Eng. Q. 18, 1–6.

Khandeparker, R., Numan, M.T.H., Mukherjee, B., Satwekar, A., Bhosle, N.B., 2008. Purification and characterization of α-L-arabinofuranosidase from *Arthrobacter* sp. MTCC 5214 in solid-state fermentation. Process Biochem. 43, 707–712.

Kim, S.J., Lee, C.M., Han, B.R., Kim, M.Y., Yeo, Y.S., Yoon, S.H., et al., 2008. Characterization of a gene encoding cellulase from uncultured soil bacteria FEMS. Microbiol. Lett. 282, 44–51.

Kim, D.Y., Han, M.K., Oh, H.W., Park, D.S., Kim, S.J., Lee, S.G., et al., 2010. Catalytic properties of a GH10 endo-β-1, 4xylanase from *Streptomyces thermocarboxydus* Hy-15 isolated from the gut of *Eisenia fetida*. J. Mol. Catalysis. B: Enz. 62, 32–39.

Kittura, F.S., Mangalaa, S.L., Rus'da, A.A., Kitaokaa, M., Tsujibob, H., Hayashia, K., 2003. Fusion of family 2b carbohydrate-binding module increases the catalytic activity of a xylanase from *Thermotoga maritima* to soluble xylan. FEBS Lett. 549, 147–151.

Kleywegt, G.J., Zou, J.Y., Divne, C., Davies, G.J., Sinning, I., Stahlberg, J., et al., 1997. The crystal structure of the catalytic core of endoglucanase I from *Trichoderma reesei* at 3.6 Å resolution, and a comparison with related enzymes. J. Mol. Biol. 272, 383–397.

Knob, A., Carmona, E.C., 2009. Cell associated acid β-xylosidase production by *Penicillium sclerotiorum*. New Biotechnol. 26, 60–67.

Kolenova, K., Vrsanska, M., Biely, P., 2006. Mode of action of endo-β-1,4-xylanases of families 10 and 11 on acidic xylooligosaccharides. J. Biotechnol. 121, 338–345.

Kristufek, D., Zeilinger, S., Kubicek, C.P., 1995. Regulation of β-xylosidase formation by xylose in *Trichoderma reesei*. Appl. Microbiol. Biotechnol. 42, 413–417.

Lagaert, S., Belien, T., Volckaert, G., 2009. Plant cell walls: protecting the barrier from degradation by microbial enzymes. Semin. Cell Dev. Biol. 20, 1064–1073.

Laine, C., 2005. Structures of hemicelluloses and pectins in wood and pulp. Ph.D. Thesis. Helsinki University of Technology, KCL, Espoo, Finland.

Lakshmi, G.S., Rao, C.S., Rao, R.S., Hobbs, P.J., Prakasham, R.S., 2009. Enhanced production of xylanase by a newly isolated *Aspergillus terreus* under solid state fermentation using palm industrial waste: a statistical optimization. Biochem. Eng. J. 48, 51–57.

Larson, S.B., Day, J., Barba de la Rosa, A.P., Keen, N.T., McPherson, A., 2003. First crystallographic structure of a xylanase from glycoside hydrolase family 5: implications for catalysis. Biochem 42, 8411–8422.

Lazaridou, A., Biliaderis, C.G., Izydorczyk, M.S., 2007. Cereal β-glucans: structures, physical properties, and physiological functions. In: C.G. Biliaderis, M.S. Izydorczyk (Eds.), Functional Food Carbohydrates. CRC Press, Taylor & Francis Group, Boca Raton, pp. 1–72.

Leite, R.S., Daniela, A.B., Eduardo, D.S., Denis, S., Eleni, G., Roberto, S., 2007. Production of cellulolytic and hemicellulolytic enzymes from *Aureobasidium pulluans* on solid state fermentation. Appl. Biochem. Biotechnol. 137-140, 281–288.

Li, Y.N., Meng, K., Wang, Y.R., Yao, B., 2006. A beta-mannanase from *Bacillus subtilis B* 36: purification, properties, sequencing, gene cloning and expression in *Escherichia coli*. Z. Naturforsch C 61, 840–846.

Li, X.L., Skory, C.D., Cotta, M.A., Puchart, V., Biely, P., 2008. Novel family of carbohydrate esterases, based on identification of the *Hypocrea jecorina* acetyl esterase gene. Appl. Environ. Microbiol. 74, 7482–7489.

Lopez, M., Bizot, H., Chambat, G., Marais, M.F., Zykwinska, A., Ralet, M.C., et al., 2010. Enthalpic studies of xyloglucan-cellulose interactions. Biomacromolecules 11, 1417–1428.

Maalej, I., Behlaj, I., Masmoudi, N.F., Belghith, H., 2009. Highly thermostable xylanase of the thermophilic fungus Talaromyces thermophilus: purification and characterization. Appl. Biochem. Biotechnol. 158, 200–212.

Magalhaes, P.O., Milagres, A.M.F., 2009. Biochemical properties of a β-mannanase and a β-xylanase produced by *Ceriporiopsis subvermispora* during biopulping conditions. Int. Biodegrad. 63, 191–195.

Maheshwari, R., Bhardwaj, G., Bhat, M.K., 2000. Thermophilic fungi: their physiology and enzymes. Microbiol. Mol. Biol. Rev. 64, 461–488.

Mai, V., Wiegel, J., Lorenz, W.W., 2000. Cloning, sequencing, and characterization of the bifunctional xylosidase-arabinosidase from the anaerobic thermophile *Thermoanaerobacter ethanolicus*. Gene 247, 137–143.

Mangala, S.L., Kittur, F.S., Nishimoto, M., Sakka, K., Ohmiya, K., Kitaoka, M., et al., 2003. Fusion of family VI cellulose binding domains to *Bacillus halodurans* xylanase increases its catalytic activity and substrate-binding capacity to insoluble xylan. J. Mol. Catalysis B Enz. 21, 221–230.

Martinez, D., Challacombea, J., Morgensternc, I., Hibbettc, D., Schmoll, M., Kubicek, C.P., 2009. Genome, transcriptome, and secretome analysis of wood decay fungus *Postia placenta* supports unique mechanisms of lignocellulose conversion. PNAS 106, 1954–1959.

Martin, K., Mc Dougall, B.M., McIlroy, S., Chen, J., Seviour, R.J., 2007. Biochemistry and molecular biology of exocellular fungal β-(1,3)- and β-(1,6)-glucanases. FEMS. Microbiol. Rev. 31, 168–192.

Maslen, S.L., Goubet, F., Adam, A., Dupree, P., Stephens, E., 2007. Structural elucidation of arabinoxylan isomers by normal phase HPLC-MALDI-TOF/TOF-MS/MS. Carbohydr. Res. 342, 724–735.

Mastihuba, V., Kremnicky, L., Mastihubova, M., Willet, J.L., Cote, G.L., 2002. A spectrophotometric assay for feruloyl esterases. Anal. Biochem. 309, 96–101.

Mastihubova, M., Biely, P., 2010. Preparation of regioselectively feruloylated *p*-nitrophenyl α-L-arabinofuranosides and β-D-xylopyranosides—convenient substrates for study of feruloyl esterase specificity. Carbohydr. Res. 345, 1094–1098.

Matsumura, S., Sakiyama, K., Toshima, K., 1999. Preparation of octyl-β-D-xylobioside and xyloside by xylanase catalyzed direct transglycosylation reaction of xylan and octanol. Biotechnol. Lett. 21, 17–22.

Medina, G.N., Carvajal-Millan, E., Rascon-Chu, A., Marquez-Escalante, J.A., Guerrero, V., Salas-Munoz, E., 2010. Feruloylated arabinoxylans and arabinoxylan gels: structure, sources and applications. Phytochem. Rev. 9, 111–120.

Meshram, M., Kulkarni, A., Jayaraman, V.K., Kulkarni, B.D., Lele, S.S., 2008. Optimal xylanase production using *Penicillium janthinellum* NCIM 1169: a model based approach. Biochem. Eng. J. 40, 348–356.

Miyanaga, A., Koseki, T., Miwa, Y., Mese, Y., Nakamura, S., Kuno, A., et al., 2006. The family 42 carbohydrate-binding module of family 54 α-L-arabinofuranosidase specifically binds the arabinofuranose side chain of hemicellulose. Biochem. J. 399, 503–511.

Muzard, M., Aubry, N., Royon, R.P., O'Donohue, M., remond, C., 2009. Evaluation of the transglycosylation activities of a GH 39 β-D-xylosidase for the synthesis of xylose-based glycosides. J. Mol. Catalysis B Enz. 58, 1–5.

Naessens, M., Vandamme, E.J., 2003. Multiple forms of microbial enzymes. Biotechnol. Lett. 25, 1119–1124.

Nagendran, S., Hallen-Adams, H.E., Paper, J.M., Aslam, N., Walton, J.D., 2009. Reduced genomic potential for secreted plant cell-wall-degrading enzymes in the ectomycorrhizal fungus *Amanita bisporigera*, based on the secretome of *Trichoderma reesei*. Fungal Genet. Biol. 46, 427–435.

Nam, K.H., Kim, S.J., Hwang, K.Y., 2009. Crystal structure of a Celm2, bifunctional glucanase-xylanase protein from a metagenome library. Biochem. Biophys. Res. Commun. 383, 183–186.

Narang, S., Sahai, V., Bisaria, V.S., 2001. Optimization of xylanase production by *Melanocarpus albomyces* IIS68 in solid state fermentation using response surface methodology. J. Biosci. Bioengg. 91, 425–427.

Oda, K., Kakizono, D., Yamada, O., 2006. Proteomic analysis of extracellular protein from *Aspergillus oryzae* grown under submerged and solid state culture conditions. Appl. Environ. Microbiol. 72, 885–889.

Ohara, H., 2003. Biorefinery. Appl. Microbiol. Biotechnol. 62, 474–477.

Ohta, K., Fujimoto, H., Fujii, S., Wakiyama, M., 2010. Cell-associated β-xylosidase from *Aureobasidium pullulans* ATCC 20524: purification, properties, and characterization of the encoding gene. J. Biosci. Bioeng 110, 152–157.

Panagiotou, G., Topakas, E., Economou, L., Kekos, D., Macris, B.J., Christakopoulos, P., 2003. Induction, purification, and characterization of two extracellular α-L-arabinofuranosidases from *Fusarium oxysporum*. Can. J. Microbiol. 49, 639–644.

Pandey, A., Carlos, R.S., Mitchell, D., 2000. New development in solid-state fermentation: bioprocess and products. Process Biochem. 35, 1153–1169.

Paradis, F.W., Zhu, H., Krell, P.J., Phillips, J.P., Forsberg, C.W., 1993. The xynC gene from *Fibrobacter succinogenes* S85 codes for a xylanase with two similar catalytic domains. J. Bacteriol. 175, 7666–7672.

Pason, P., Kyu, K.L., Ratanakhanokchai, K., 2006. *Paenibacillus curdlanolyticus* strain B-6 xylanolytic-cellulolytic enzyme system that degrades insoluble polysaccharides. Appl. Environ. Microbiol. 72, 2483–2490.

Pastel, H., Virkki, L., Harju, E., Toumainen, P., Tenkanen, M., 2009. Presence of 1→3 linked 2-O-β-D-xylopyranosyl-α-L-arabinofuranosyl side chains in cereal arabinoxylan. Carbohydr Res. 344, 2480–2488.

Peng, F., Ren, J.L., Xu, F., Bian, J., Peng, P., Sun, R.C., 2010. Fractionation of alkali-solubilized hemicelluloses from delignified *Populus gansuensis*: structure and properties. J. Agric. Food Chem. 58, 5743–5750.

Pinto, P.C., Evtuguin, D.V., Pascoal Neto, C., 2005. Structure of hardwood glucuronoxylans: modifications and impact on pulp retention during wood kraft pulping. Carbohydr. Polymers. 60, 489–497.

Polizeli, M.L.T.M., Rizzatti, A.C.S., Monti, R., Terenzi, H.F., Jorge, J.A., Amorim, D.S., 2005. Xylanases from fungi: properties and industrial applications. Appl. Microbiol. Biotechnol. 67, 577–591.

Pollet, A., Schoepe, J., Dornez, E., Strelkov, S.V., Delcour, J.A., Courtin, C.M., 2010. Functional analysis of glycoside hydrolase family 8 xylanases shows narrow but distinct substrate specificities and biotechnological potential. Appl. Microbiol. Biotechnol. 87, 2125–2135.

Puchart, V., Biely, P., 2008. Simultaneous production of endo-β-1,4-xylanase and branched xylooligosaccharides by *Thermomyces lanuginosus*. J. Biotechnol. 137, 34–43.

Puls, J., Schuseil, J., 1993. Chemistry of hemicelluloses: relationship between hemicellulose structure and enzyme required for hydrolysis. In: Couhglan, M.P., Hazlewood, G.P. (Eds.), Hemicellulose and Hemicellulases. Portland Press, London, pp. 1–27.

Raimbault, M., 1998. General and microbiological aspects of solid substrate fermentation. Electronic J. Biotechnol. 1, 1–15.

Rodrigues, T.H.S., Dantas, M.A.A., Pinto, G.A.S., Goncalves, L.R.B., 2007. Tannase production by solid state fermentation of cashew apple bagasse. Appl. Biochem. Biotechnol. 136–140, 675–688.

Saha, B.C., 2000. α-L-Arabinofuranosidases: biochemistry, molecular biology and application in biotechnology. Biotechnol. Adv. 18, 403–423.

Sande, M.A., Teijeiro-Osorio, D., Remunan-Lopez, C., Alonso, M.J., 2009. Glucomannan, a promising polysaccharide for biopharmaceutical purposes. EJIPB. 72, 453–462.

Sapag, A., Wouters, J., Lambert, C., Ioannes, P., Eyzaguirre, J., Depiereux, E., 2002. The endoxylanases from family 11: computer analysis of protein sequences reveals important structural and phylogenetic relationships. J. Biotechnol. 95, 109–131.

Saraswat, V., Bisaria, V.S., 1997. Biosynthesis of xylanolytic and xylan debranching enzymes in *Melanocarpus albomyces* IIS 68. J. Ferment. Bioeng. 83, 352–357.

Scheller, H.V., Ulvskov, P., 2010. Hemicelluloses. Annu. Rev. Plant Biol. 61, 263–289.

Shah, A.R, Shah, R.K, Madamwar, D., 2006. Improvement of the quality of whole wheat bread by supplementation of xylanase from *Aspergillus foetidus*. Bioresour. Technol. 97, 2047–2053.

Shallom, D., Shoham, Y., 2003. Microbial hemicellulases. Curr. Opin. Microbiol. 6, 219–228.

Shallom, D., Golan, G., Shoham, G., Shoham, Y., 2004. Effect of dimer dissociation on activity and thermostability of the α-glucuronidase from *Geobacillus stearothermophilus*: dissecting the different oligomeric forms of family 67 glycoside hydrolases. J. Bacteriol. 6928–6937.

Sharma, M., Chadha, B.S., 2010. Characterization of functionally diverse alkaline active xylanases produced by thermophilic fungus *Malbranchea flava*. In: Presented at International conference on Genomic Sciences—Recent Trends. Madurai Nov 12-14, 2010.

Sharma, M., Chadha, B.S., Kaur, M., Ghatora, S.K., Saini, H.S., 2008. Molecular characterization of multiple xylanase producing thermophilic/thermotolerant fungi isolated fromcomposting materials. Lett. Appl. Microbiol. 46, 526–535.

Sharma, M., Chadha, B.S., Saini, H.S., 2010a. Purification and characterization of two thermostable xylanases from *Malbranchea flava* active under alkaline conditions. Bioresour. Technol. 101, 8834–8842.

Sharma, M., Soni, R., Nazir, A., Oberoi, H.S., Chadha, B.S., 2010b. Evaluation of glycosyl hydrolases in the secretome of *Aspergillus fumigatus* and saccharification of alkali treated rice straw. Appl. Biochem. Biotechnol. doi: 10.1007/s12010-010-9064-3.

Silversides, F.G, Scott, T.A, Korver, D.R, Afsharmanesh, M., Hruby, M., 2006. A study on the interaction of xylanase and phytase enzymes in wheat-based diets fed to commercial white and brown egg laying hens. Poult. Sci. 85, 297–305.

Singh, S., Pillay, B., Dilsook, V., Prior, B.A., 2000. Production and properties of hemicellulases by a *Thermomyces lanuginosus* strain. J. Appl. Microbiol. 88, 975–982.

Sinitsyna, O.A., Gusakov, A.V., Okunev, O.N., Serebryany, V.A., Vavilova, E.A., Vinetsky, U.P., et al., 2003. Recombinant endo-β-1,4-xylanase from *Penicillium canescens*. Biochem. (Moscow) 68, 1631–1638.

Songsiriritthigul, C., Buranabanyat, B., Haltrich, D., Yamabhai, Y., 2010. Efficient recombinant expression and secretion of a thermostable GH26 mannan endo-1,4-β-mannosidase from *Bacillus licheniformis* in *Escherichia coli*. Microbial Cell Fact. 9, 20. doi:10.1186/1475-2859-9-20.

Sonia, K.G., Chadha, B.S., Saini, H.S., 2005. Sorghum straw for xylanase hyper-production by *Thermomyces lanuginosus* (D_2W_3) under solid state fermentation. Bioresour. Technol. 96, 1561–1569.

Sorensen, H.R., Pedersen, S., Vikso-Nielsen, A., Meyer, A.S., 2005. Efficiencies of designed enzyme combinations in releasing arabinose and xylose from wheat arabinoxylan in an industrial ethanol fermentation residue. Enzyme Microb. Technol. 36, 773–784.

Sorensen, H.R., Jorgensen, C.T., Hansen, C.H., Jorgensen, C.I., Pedersen, S., Meyer, A.S., 2006. A novel GH43 α-L-arabinofuranosidase from *Humicola insolens*: mode of action and synergy with GH51 α-L-arabinofuranosidases on wheat arabinoxylan. Appl. Microbiol. Biotechnol. 73, 850–861.

Stalbrand, H., Saloheimo, A., Vehmaanpera, J., Henrissat, B., Penttila, M., 1995. Cloning and expression in *Saccharomyces cerevisae* of a *Trichoderma reesei* β-mannanase gene containing a cellulose binding domain. Appl. Environ. Microbiol. 61, 1090–1097.

Subramaniyan, S., Prema, P., 2002. Biotechnology of microbial xylanases: enzymology, molecular biology and application. Crit. Rev. Biotechnol. 22, 33–46.

Sunna, A., Antranikian, G., 1997. Xylanolytic enzyme from fungi and bacteria. Crit. Rev. Biotechnol. 17, 39–67.

Sun, X.F., Sun, R.C., Fowler, P., Baird, M.S., 2005. Extraction and characterization of original lignin and hemicelluloses from wheat straw. J. Agric. Food. Chem. 53, 860–870.

Suzuki, S., Fukuoka, M., Ookuchi, H., Sano, M., Ozeki, K., Nagayoshi, E., et al., 2010. Characterization of *Aspergillus oryzae* glycoside hydrolase family 43 β-xylosidase expressed in *Escherichia coli*. J. Biosci. Bioeng. 109 (2), 115–117.

Talabani, S.J., Boraston, A.B., Turkenburg, J.P., Tarbouriech, N., Ducros, V.M.A., Davies, G.J., 2004. Ab initio structure determination and functional characterization of CBM36: a new family of calcium-dependent carbohydrate binding modules. Structure 12, 1177–1187.

Taylor, E.J., Turkenburg, J.P., Gloster, T.M., Vincent, F., Brzozowski, A.M., Dupont, C., et al., 2006. Structure and activity of two metal-ion dependent acetyl xylan esterases involved in plant cell wall degradation reveals a close similarity to peptidoglycan deacetylases. J. Biol. Chem. 281: 10968–10975.

Techapun, C., Poosaran, N., Watanabe, M., Sasaki, K., 2003. Thermostable and alkaline-tolerant microbial cellulose-free xylanases produced from agricultural wastes and the properties required for use in pulp bleaching bioprocesses: a review. Process. Biochem. 38, 1327–1340.

Thygeson, A., Thomson, A.B., Schmidt, A.S., Jorgenson, H., Olsson, L., 2003. Production of cellulose and hemicellulose degrading enzymes by filamentous fungus cultivated on wet oxidised wheat straw. Enzyme. Microb. Technol. 32, 606–615.

Tian, C., Beeson, W.T., Iavarone, A.T., Sun, J., Marletta, M.A., Cate, J.H., et al., 2009. Systems analysis of plant cell wall degradation by the model filamentous fungus *Neurospora crassa*. PNAS. 106, 22157–22162.

Tseng, M.J., Yap, M.N., Ratanakhanokchai, K., Kyu, K.L., Chen, S.T., 2002. Purification and characterization of two cellulase free xylanases from an alkalophilic *Bacillus firmus*. Enzyme. Microb. Technol. 30, 590–595.

Ustinov, B.B., Gusakov, A.V., Antonov, A.I., Sinitsyn, A.P., 2008. Comparison of properties and mode of action of six secreted xylanases from *Chrysosporium lucknowense*. Enzyme Microbiol. Technol. 43, 56–65.

Vandermarliere, E., Bourgois, T.M., Winn, M.D., 2009. Structural analysis of a glycoside hydrolase family 43 arabinoxylan arabinofuranohydrolase in complex with xylotetroase reveals a different binding mechanism compared to other members of the same family. Biochem. J. 418, 39–47.

Van Petegam, F., Collins, T., Meuwis, M.A., Gerday, C., Feller, G., Van Beeuman, J., 2003. The structure of a cold-adapted family 8 xylanase at 1.3 Å resolution, structural adaptations to cold and investigations of the active site. J. Biol. Chem. 278, 7531–7539.

Vardakou, M., Dumon, C., Murray, J.W., 2008. Understanding the structural basis for substrate and inhibitor recognition in eukaryotic GH11 xylanases. J. Mol. Biol. 375, 1293–1305.

Vries, R.P.D., Visser, J., 2001. *Aspergillus* Enzymes Involved in Degradation of plant cell wall polysaccharides. Microbiol. Mol. Biol. Rev. 65 (4), 497–522.

Vrsanska, M., Kolenova, K., Puchart, V., Biely, P., 2007. Mode of action of glycoside hydrolase family 5 glucuronoxylan xylanohydrolase from *Erwinia chrysanthemi*. FEBS. J. 274, 1666–1677.

Wagschal, K., Heng, C., Lee, C.C., Wong, D.W., 2009. Biochemical characterization of a novel dual-function arabinofuranosidase/xylosidase isolated from a compost starter mixture. Appl. Microbiol. Biotechnol. 81, 855–863.

Wet, B.J.M., Prior, B.A., 2004. Microbial α-Glucuronidases. In: Lignocellulose Biodegradation Chapter 14. doi: 10.1021/bk-2004-0889. ch014 ACS Symposium Series, vol. 889. 241–254.

Wong, D.W.S., 2006. Feruloyl esterase a key enzyme in biomass degradation. Appl. Biochem. Biotechnol. 133, 87–122.

Wong, K.K.Y., Saddler, J.N., 1992. *Trichoderma* xylanases: their properties and applications. In: Visser, J. (Eds.), Xylans and Their Xylanases. Elsevier, Amsterdam, pp. 171–186.

Wong, K.K.Y., Tan, L.U.L., Saddler, J.N., 1988. Multiplicity of β-1,4-xylanases in microorganisms: functions and applications. Microbiol. Rev. 52, 305–317.

Wymelenberg, A.V, Gaskell, J., Mozuch, M., Sabat, G., Ralph, J., Skyba, O., et al., 2010. Comparative transcriptome and secretome analysis of wood decay fungi *Postia placenta* and *Phanerochaete chrysosporium*. Appl. Environ. Microbiol. 76: 3599–3610.

Xiong, J.S., Balland-Vanney, M., Xie, Z.P., Schultze, M., Kondorosi, A., Kondorosi, E., et al., 2007. Molecular cloning of a bifunctional b-xylosidase/a-Larabinosidase from alfalfa roots: heterologous expression in *Medicago truncatula* and substrate specificity of the purified enzyme. J. Exp. Bot. 58, 2799–2810.

Xu, Y., Foong, F.C., 2008. Characterization of a cellulose binding domain from *Clostridium cellulovorans* endoglucanase-xylanase D and its use as a fusion partner for soluble protein expression in *Escherichia coli*. J. Biotechnol. 135, 319–325.

Yan, Q.J., Wang, L., Jiang, Z.Q., Yang, S.Q., Zhu, H.F., Li, L.T., 2008. A xylose-tolerant β-xylosidase from *Paecilomyces thermophila*: characterization and its co-action with the endogenous xylanase. Bioresour. Technol. 99, 5402–5410.

Zanoelo, F.F., Polizeli, M.L.T.M., Terenzi, H.F., Jorge, J.A., 2004. Purification and biochemical properties of a thermostable xylose-tolerant β-D-xylosidase from *Scytalidium thermophilum*. J. Ind. Microbiol. Biotechnol. 31, 170–176.

Zhao, S., Wang, J., Bu, D., Liu, K., Zhu, Y., Dong, Z., et al., 2010. Novel glycoside hydrolases identified by screening a Chinese Holstein dairy cow rumen-derived metagenome library. Appl. Environ. Microbiol. 76, 6701–6705.

Zheng, Y., Pan, Z., Zhang, R., 2009. Overview of biomass pretreatment for cellulosic ethanol production. Int. J. Agric. Biol. Eng 2, 51–68.

Ziser, L., Withers, S.G., 1994. A short synthesis of β-xylobiosides. Carbohydr. Res. 265, 9–17.

CHAPTER

10

Hydrolysis of Lignocellulosic Biomass for Bioethanol Production

Parameswaran Binod, K.U. Janu, Raveendran Sindhu, Ashok Pandey*

Centre for Biofuels, Biotechnology Division, National Institute for Interdisciplinary Science and Technology, CSIR, Trivandrum - 695 019, India

***Corresponding author: E-mail: binodkannur@gmail.com**

1 INTRODUCTION

Production of ethanol from lignocellulosic biomass seems very attractive and sustainable due to several reasons, among which the renewable and ubiquitous nature of biomass and its noncompetitiveness with food crops are the major ones. Another significant factor which adds value as well as importance to lignocellulosic ethanol is the reduction in greenhouse gas emission. The utilization of lignocellulosic biomass for ethanol production necessitates the large-scale production technology to be cost effective and environmentally sustainable. Bioconversion of lignocellulosic materials into fermentable sugars is a biorefining area in which enormous research efforts have been invested, as it is a prerequisite for the subsequent production of bioethanol. Although extensive studies have been carried out to meet the future challenges of bioenergy generation, there is no self-sufficient process or technology available to convert the lignocellulosic biomass to bioethanol. The whole process primarily comprises the hydrolysis of lignocellulosic structure to fermentable sugars, followed by fermentation and finally distillation of the fermented broth. The hydrolysis of lignocellulosic material into fermentable sugars is a crucial stage, which mainly determines the overall process efficiency. Various methods are available for the generation of sugars from lignocellulosic biomass, of which the chemical and enzymatic methods have been proved to be more successful.

2 CHEMICAL HYDROLYSIS

Sugars are natural intermediates in the conversion of lignocellulosic biomass, but access to sugars is hindered by the recalcitrance of plant cell walls. Deriving sugars from this heterogeneous feedstock requires either physical or chemical disruption. Chemical hydrolysis is usually done by using acids. Concentrated mineral acids such as H_2SO_4 and HCl are commonly used for this process. Another method for deriving sugars from the biomass is to use less hazardous and more tractable cellulose solvents such as ionic liquids (ILs). These are salts with melting points near or below ambient temperature which can dissolve cellulose.

2.1 Acid Hydrolysis

The concentrated acid process for producing sugars from lignocellulosic biomass has a long history. The ability to dissolve and hydrolyze native cellulose in cotton using concentrated sulfuric acid followed by dilution with water was reported in the literature as early as 1883 (Harris, 1949). The concentrated acid disrupts the hydrogen bonding between cellulose chains, converting it to a completely amorphous state. Once the cellulose has been decrystallized, it forms a homogeneous gelatin with the acid. The cellulose is extremely susceptible to hydrolysis at this point. Thus, dilution with water at modest temperatures provides complete and rapid hydrolysis to glucose, with little degradation.

Most of the research on the concentrated acid hydrolysis processes has been done using corncobs. In 1918, researchers at the U.S. Department of Agriculture (USDA) proposed a process scheme for production of sugars and other products from corn cobs based on a two-stage process where the biomass is treated with dilute acid to remove the hemicellulose in the first stage, followed by decrystallization and hydrolysis of the cellulose fraction using concentrated acid in the second stage (LaForge and Hudson, 1918). In 1937, the Germans built and operated commercial concentrated acid hydrolysis plants using hydrochloric acid. Several such facilities were successfully operated. During World War II, researchers at USDA's Northern Regional Research Laboratory in Peoria, Illinois, further refined the concentrated sulfuric acid process for corncobs. They conducted process development studies on a continuous process that produced about 15-20% xylose sugar stream and 10-12% glucose sugar stream, with the lignin residue remaining as a byproduct. Separation of acid from the sugar stream after hydrolysis is a crucial factor. In 1948, a concentrated sulfuric acid hydrolysis process was commercialized in Japan where they used membranes to separate sugars and acid. Through this technique, they were able to achieve 80% recovery of acid (Wenzl, 1970). Further studies on hydrolysis resulted in the development of a process for improved recycling of sulfuric acid (Broder et al., 1992).

Arkenol Inc. USA developed concentrated acid hydrolysis technology to convert cellulosic materials into high-value chemicals and transportation fuels. The process includes a two-stage hydrolysis: in the first stage the biomass is treated with 90% sulfuric acid and in the second stage 30% sulfuric acid is used. The company owns several patents related to the development of this process, with the key patents related to acid-sugar separation and recovery. For sugar separation and recovery, a chromatography-based system, called a pseudomoving bed column, makes use of unique resins to preferentially retard the flow of one component of the stream to be separated. Resins may be anionic or cationic and will produce different results, separating the components of the feed into streams with unique concentration and purity. The simplified flow diagram of Arkenol process is shown in Figure 1. Using this technology, Arkenol has been able to take acid/sugar feed streams containing 12-15% sugar concentrations and produce a sugar stream with 98% purity. The recovered sulfuric acid is re-circulated and re-concentrated to the level required by the decrystallization and hydrolysis steps. The small quantity of acid remaining in the sugar is neutralized with lime to form hydrated gypsum, an insoluble precipitate that can be used in agriculture as a soil conditioner. The sugar stream, consisting of a mixture of C5 and C6 sugars, is mixed with nutrients and fermented with naturally occurring yeast specifically cultured by a proprietary method.

Although concentrated acid hydrolysis results in the release of fermentable sugars, they are toxic, corrosive, and hazardous and require reactors that are resistant to corrosion. This in turn makes the process very expensive. Hence, people are looking for more environment-friendly and economically feasible techniques for deriving sugars from lignocellulosic biomass. Dilute acid hydrolysis followed by enzymatic hydrolysis is one of them. Dilute acid hydrolysis has also been successfully developed for pretreatment, and it significantly improves the efficiency of the enzymatic hydrolysis step. Sulfuric acid concentration below 4% is generally used as it is comparatively inexpensive and helps in achieving high reaction

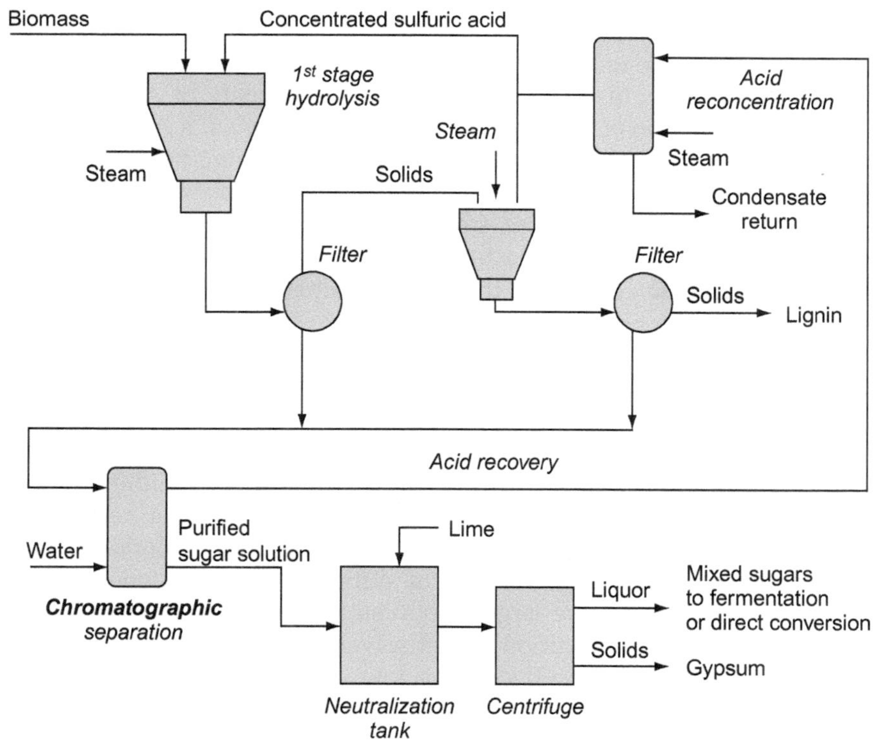

FIGURE 1 Simplified flow diagram of Arkenol process.

rates. Since sugar decomposition takes place at moderate temperature, this process requires a high temperature and neutralization of pH is also necessary for the downstream enzymatic hydrolysis or fermentation process. Apart from this, to make the process economically feasible, these acids must be recovered from the reaction mixture after hydrolysis.

2.2 Biomass Fractionation by ILs

The main challenge in lignocellulosic biomass to ethanol process is to separate lignin and cellulose, which appear as strongly bonded conglomerates in the lignocellulosic biomass. Suitable solvents allowing for the design of more cost-efficient and eco-friendly pulping processes would therefore be very helpful. The oldest method to dissolve cellulose, which was discovered in 1857, is dissolution in a mixture of copper (II) salts, ammonia, and sodium hydroxide. Although coagulation processes using this reagent performed fairly well, the challenge in this process was the necessity to recycle copper and ammonia from the dilute aqueous solutions of the coagulation bath. Therefore, this dissolution process was never realized on a large scale (Vagt, 2010).

In 1934, Charles Graenacher (Graenacher, 1934) proposed a concept for dissolving cellulose in molten organic salts. Using this method, he was able to dissolve cellulose in N-alkyl- or *N*-arylpyridinium chlorides in the presence of nitrogen-containing bases. At this time, the invention was treated probably as a novelty with little practical value, as molten salts were not readily available on a large scale. Another drawback might have been that the concentrations of cellulose obtained in these molten salts were rather low.

It was Robin Rogers with his research team at the University of Alabama who in 2002 applied ILs for the dissolution of cellulose (Swatloski et al., 2002). ILs are nonvolatile solvents under atmospheric conditions that are composed exclusively of ions held together by coulombic forces. IL-based pretreatment of lignocellulosic biomass offers an environment-friendly approach for the recovery of cellulose from lignocellulosic biomass. It is an emerging technique for pretreatment that significantly improves the digestibility of recalcitrant biomass under milder reaction conditions than conventional pretreatment processes such as dilute acid, alkali, ammonia fiber expansion, steam explosion, and organosolv pretreatment. In comparison to traditional solvents, ILs exhibit very interesting properties such as reasonable chemical inertness, production of no toxic or explosive gases during reaction, good thermal stability, low volatility, negligible vapor pressures, and unique solvation abilities that makes it an important candidate for lignocellulosic treatment. The combination of anion and cation affects their physical and chemical properties such as melting points, viscosity, hydrophobicity, and hydrolysis stability. Therefore, optimal ILs for certain applications can be designed. Cellulose-dissolving IL usually contains anions of chloride, formate, acetate or alkyl phosphonate, since these ions form strong hydrogen bonds with cellulose. Imidazolium-based ILs can dissolve large amounts of cellulose and the dissolved cellulose can be recovered back by the addition of antisolvents like water, ethanol, or methanol. Another interesting point regarding ILs is their low volatility which permits distillation of the volatile substances, thereby making IL recovery feasible. The ability to solubilize cellulose is useful for acid/base catalytic reactions in homogeneous solutions directly in the ILs or for direct enzymatic hydrolysis.

Fractionation of lignocelluloses using ILs faces some challenges to develop a feasible process: (i) the recovery and reuse of ILs, as the cost of IL is still high and (ii) the recovery of lignin and hemicellulose from the ILs after cellulose has been extracted. Due to the heterogeneous nature of lignocellulosic materials, it is necessary to screen a large variety of ILs to find a suitable one for a particular biomass.

Using 1-butyl-3-methylimidazolium chloride, Robin Rogers and co-workers were the first to be able to dissolve cellulose in technically useful concentrations by physical dissolution in an inert solvent without using any auxiliaries (Swatloski et al., 2002). The dissolution process of cellulose seems to be driven mainly by the anion of the IL. Anions such as halides, carboxylates, and phosphates are able to break very effectively intermolecular hydrogen bonds within the cellulose structures as they are not hydrated and are strong hydrogen bond acceptors. The presence of water decreases the solubility of cellulose through competitive hydrogen bonding processes. Cations with cyclic structures such as pyridinium, pyrazolium, the protonated diazabicycloundecene, and the most frequently used imidazolium cation showed the best results—leading to the suggestion that cations with a flatter molecular structure may support dissolution. The ability to dissolve cellulose decreases with increasing length of the alkyl chains on the cation. Overall, 1,3-dialkylimidazolium salts with no alkyl substitution in the 2-position are preferred as they show lower viscosities and allow cellulose concentrations as high as 20 wt% and more. 1-ethyl-3-methylimidazolium acetate [C2mim][OAc] turned out as the most preferred solvent for cellulose dissolution and processing as it is liquid at room temperature, offers relatively low viscosity (93 mPa s at 25 °C) and high dissolving power—even in the presence of up to 10 wt% of water. Concentrations of up to 25 wt% cellulose were achieved using [C2mim][OAc]. Furthermore, [C2mim][OAc] is not acutely toxic, shows no corrosion of stainless steel, and is highly miscible with water. The only limitation in using [C2mim][OAc] is the limited thermal stability of this IL. During processing of [C2mim][OAc], temperatures below 150 °C should be applied; otherwise, the decomposition of the imidazolium salt will lead to significant material loss (Vagt, 2010).

2.2.1 *Regeneration of the Cellulose and Recycling of the IL*

By adding water or any other solvent miscible with the IL, such as methanol, ethanol, or acetone, the dissolved cellulose is coagulated and can be regenerated quantitatively by centrifugation. The regenerated cellulose has almost the same degree of polymerization (DP) as the initial pulp, but the morphology changes significantly. The degree of crystallinity can be manipulated by exerting more or less stress on the regenerated material. During washing of the product with water, residual IL can easily be removed due to the very high affinity of this IL to water. After separation of the cellulose from the spin bath, a solution of the IL in water (or another solvent) is obtained. Both water and solvent can be removed by evaporation under reduced pressure, allowing the regeneration of the IL, which can then be reused for the dissolution step. Additional purification steps will be necessary after several regeneration cycles in order to remove impurities that are introduced into the process. These can be removed by filtration or, if necessary, by ion exchange. The recently discovered volatility of ILs offers an additional opportunity for further purification of the IL in the recycling step. At temperatures of 100-300 °C and under reduced pressure, the IL can be extensively purified (Vagt, 2010).

3 ENZYMATIC HYDROLYSIS

Enzymatic hydrolysis is carried out by cellulase enzymes which are highly specific, and the products of the hydrolysis are usually reducing sugars including glucose. Unlike chemical hydrolysis, enzymatic hydrolysis is conducted at mild conditions at a pH of 4.8 and temperature of 45-50 °C, which is optimum for the cellulase enzyme. The main advantage of enzymatic hydrolysis over chemical hydrolysis is that it does not create a corrosion problem (Duff and Murray, 1996). But the process takes several days whereas it is only a few minutes in the case of chemical hydrolysis. Moreover, the final product of enzymatic hydrolysis inhibits the enzyme and ultimately affects the process unless they are removed immediately after they are formed. Apart from this, a major bottleneck in lignocellulosic ethanol production, at present, is the cost of the enzymes.

3.1 Enzymes Involved in the Hydrolysis of Lignocellulosic Biomass

The first step in lignocellulosic ethanol production is chemical pretreatment to disrupt the lignin and expose the cellulose fraction. As the severity factor of the pretreatment process decreases, the sugar yield after enzymatic hydrolysis also decreases and there arises a requirement for different types of enzymes and their higher dosages to achieve maximum sugar yield from cellulose and hemicellulose fractions of the pretreated lignocellulosic biomass. Hence, the development of a cocktail of enzymes such as cellulases, hemicellulases, and other accessory enzymes is required for complete hydrolysis.

3.1.1 Cellulases

Cellulases distinguish themselves from most other classes of enzymes by being able to hydrolyze cellulose. According to the CAZy classification system (Carbohydrate-Active enzymes), these enzymes are classified in glycosyl hydrolase families based on their sequence homology and hydrophobic cluster analysis. Cellulose is enzymatically degraded to glucose by the synergistic action of three distinct classes of enzymes:

Endoglucanases (EGs) (EC 3.2.1.4), which hydrolyze internal β-1,4-glucosidic linkages randomly in the cellulose chain.

Cellobiohydrolases (CBHs, also known as exoglucanases) (EC 3.2.1.91), which progresses along the cellulose and cleave off cellobiose units from the ends.

β-glucosidases (BG also known as β-glucoside glucohydrolases) (EC 3.2.1.21), which hydrolyze cellobiose to glucose and also cleave off glucose units from cello-oligosaccharides.

Fungi are a good source for these enzymes. *Trichoderma reesei* produces two CBHs, five EGs, and two BGs. Several of these apparently redundant enzymes have been shown to exhibit synergy by either hydrolyzing different ends of the cellulose chain or exhibiting different affinities for different sites of attack. The whole hydrolysis process can be divided into two steps: primary hydrolysis and secondary hydrolysis. Primary hydrolysis involves EGs and exoglucanases and occurs on the surface of solid substrate releasing soluble sugars with a DP up to 6 into the liquid phase. This depolymerization step is the rate-limiting step for the whole cellulose hydrolysis process. Secondary hydrolysis occurs in the liquid phase involving primarily the hydrolysis of cellobiose to glucose by β-glucosidases. A schematic diagram of mechanism of cellulase action is shown in Figure 2.

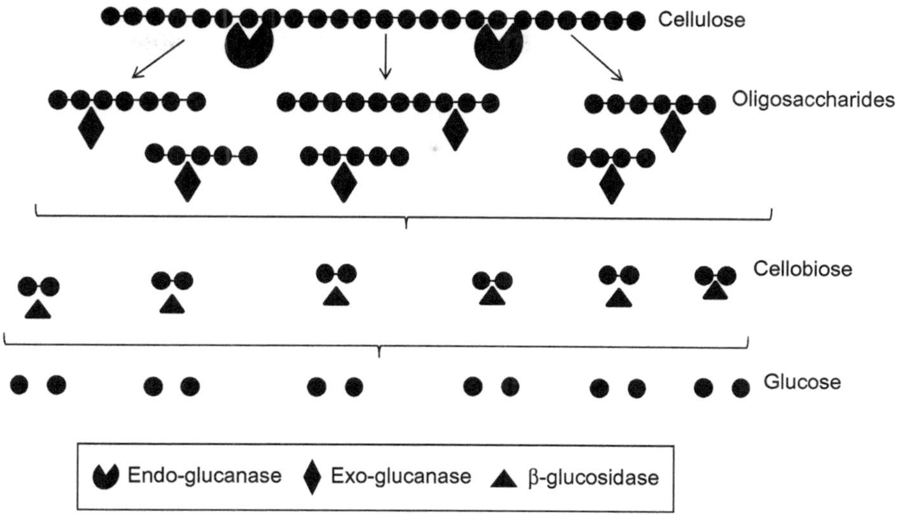

FIGURE 2 Mechanism of action of cellulase.

CBHs and EGs have a catalytic domain (CD) and a cellulose-binding domain (CBD). The function of the CBD is to bring the enzyme catalytic module in close contact with the substrate and ensure correct orientation. Removal of the CBD from the enzyme significantly impairs the hydrolysis of crystalline cellulose, demonstrating its importance. The CBD is connected to the CD with a glycosylated flexible linker, which help them to dock with and degrade crystalline cellulose. CBDs of CBHs are able to move laterally along the cellulose chain while the CD cleaves off cellobiose units. Only little is known about how the aromatic residues of the CBD interact with the cellulose crystal structure and how they desorb from the substrate and re-attach. Because of the insoluble nature of native cellulose and anchoring of CBDs, cellulases primarily work in a two-dimensional environment with the unidirectional movement of CBHs along the cellulose chain. Hence, the synergistic degradation of lignocellulose does not follow classic Michaelis-Menten kinetics. Moreover, factors like the heterogeneous nature of lignocellulose make understanding of hydrolysis mechanisms more complicated.

3.1.2 *Xylanases*

Another major component present in lignocellulosic biomass is xylan, which is the main carbohydrate present in hemicelluloses. These are polysaccharides made of xylose, a pentose sugar. Hydrolysis of xylan is carried out by a group of enzymes called xylanases. Removal of xylan from lignocelluloses using xylanases increases the accessibility of cellulose to enzymatic hydrolysis. Xylan does not form tightly packed crystalline structures like cellulose and is more susceptible to enzymatic hydrolysis. The complete hydrolysis of xylan requires the action of multiple xylanases with overlapping but different specificities and action. These enzymes consist of either a single domain or a number of domains, classified as catalytic and noncatalytic domains. *Aspergillus niger*, *T. reesei*, *Bacillus*, and *Humicola insolens* are some of the industrial sources of commercial xylanases, and their optimum temperature ranges from 40 to 60 °C. This enzyme system is composed of a repertoire of hydrolytic enzymes that act synergistically and convert xylan to its constituent sugars. Additional enzymes may also be needed depending on the hemicelluloses composition. The complete degradation of xylan requires the cooperative action of the following enzymes:

Endo-1, 4-β-xylanase (1, 4-β-d-xylan xylanohydrolases, EC 3.2.1.8) cleaves the glycosidic bonds in the xylan backbone releasing xylo-oligosaccharides.

β-xylosidase (1,4-β-d-xylan xylohydrolase, EC 3.2.1.37) acts upon the small oligosaccharides and cellobiose, generating β-d-xylopyranosyl residues from the nonreducing terminus.

α-arabinofuranosidase (EC 3.2.1.55) and *α-glucuronidase* (EC 3.2.1.139) remove the arabinose and 4-*O*-methyl glucuronic acid substituent, respectively, from the xylan backbone.

Esterases act upon the ester linkages between xylose units of the xylan and acetic acid (*Acetyl xylan esterase*, EC 3.1.1.72) or between arabinose side chain residues and phenolic acids such as ferulic acid (*Ferulic acid esterase*, EC 3.2.1.73) and p-coumaric acid (*p-coumaric acid esterase*). The mechanism of action of xylanase enzyme complex is schematically represented in Figure 3.

3.1.3 *Peroxidases*

Peroxidases are a group of enzymes involved in the degradation of lignin which is tightly bound to cellulose, making it inaccessible to the cellulase enzyme. Lignin peroxidase (LiP; also called ligninase [LiP], EC 1.11.1.7) and manganese peroxidase (also called Mn-dependent peroxidase [MnP], EC 1.11.1.7) are the two major components of the lignolytic enzyme system. These are heme-containing glycoproteins which require hydrogen peroxide as oxidant.

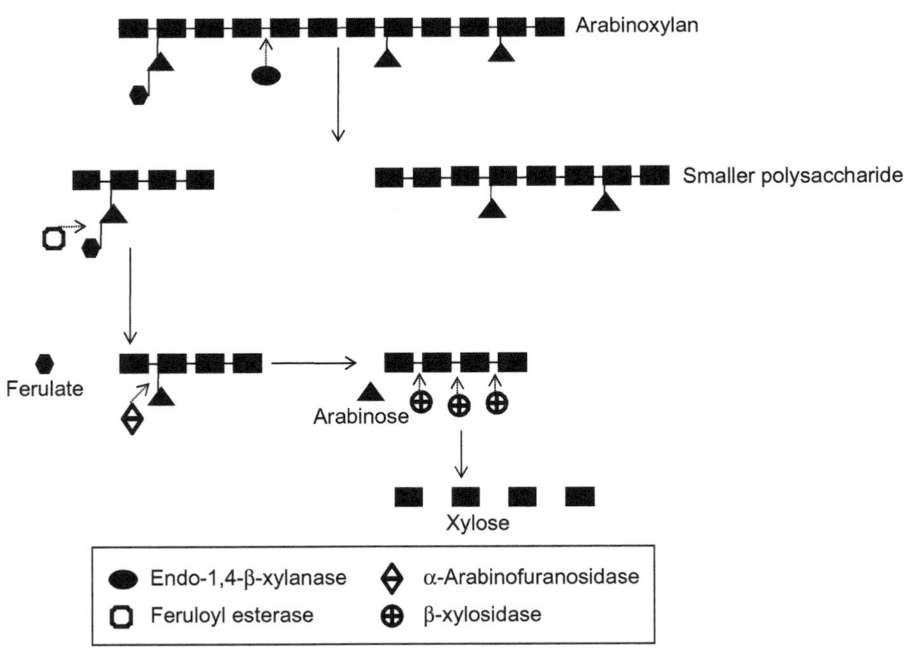

FIGURE 3 Mechanism of action of xylanases.

These enzymes were discovered in *Phanerochaete chrysosporium* and are called true ligninases due to their high redox potential.

LiP degrades nonphenolic lignin units (up to 90% of the polymer). The LiP isozymes are glycoproteins of 38-46 kDa, with pI values of 3.2-4.0. It has a distinctive property of an unusually low pH optimum near pH 3. The enzyme contains 1 mol of iron protoporphyrin IX per mole of protein. LiP oxidizes nonphenolic lignin substructures by abstracting one electron and generating cation radicals which are then decomposed chemically.

Schoemaker and Piontek (1996) described the mechanism of interaction of LiP with lignin polymer. Veratryl alcohol (valc), which is a secondary metabolite of white rot fungi, acts as a cofactor for the enzyme. It was observed that, in the depolymerization with fungal cultures, the presence of both LiP and valc stimulated the degradation of lignin:

$$\text{Lip} + H_2O_2 \rightarrow H_2O + \text{LiPI},$$

$$\text{LiPI} + \text{valc} \rightarrow \text{valc}^+ + \text{LiPII},$$

$$\text{LiPI} + 2H^+ \rightarrow \text{valc}^+ + H_2O + \text{LiP}.$$

In this process, LiP oxidizes the first molecule of valc to the corresponding radical cation (valc^+), which is liberated from the active site. Subsequently, the second substrate molecule is oxidized by LiPII to form a second valc^+. In the process, LiPII is converted to native enzyme.

MnP generates Mn3^+, which acts as a diffusible oxidizer on phenolic or nonphenolic lignin units through lipid peroxidation reactions. It oxidizes Mn(II) to Mn(III) which then oxidizes phenol rings to phenoxy radicals which lead to the decomposition of compounds.

$$2\text{Mn(II)} + 2H^+ + H_2O_2 \rightarrow 2\text{Mn(III)} + 2H_2O.$$

3.1.4 Laccases

Laccase (benzenediol: oxygen oxidoreductase, EC 1.10.3.2) is a copper-containing enzyme that belongs to the small group of enzymes called the blue copper proteins or the blue copper oxidases. These enzymes are also involved in the degradation of lignin. Laccase, alone or together with LiP lignin peroxidase and manganese peroxidase, has been demonstrated in a wide variety of white rot fungi and can completely mineralize this substrate. The presence of laccase in nonlignolytic fungi also has been demonstrated. Laccases may be constitutive or inducible enzymes. Several compounds like phenolic compounds, strictly related to lignin or lignin derivatives, have been shown to induce and improve laccase formation. However, nonlignin compounds and extracts from different origins are also found to be effective inducers of laccase production.

Laccases catalyze the oxidation of phenolic units in lignin and a number of phenolic compounds and aromatic amines to radicals, with molecular oxygen as the electron acceptor that is reduced to water. It shows a considerable diversity in molecular weight, pH optimum, and other properties. It has been shown that the ability of laccases to break down lignocellulose is increased by certain phenolic compounds (2,2 P-azino-bis-(3-ethylthiazoline-6-sulfonate (ABTS) or 3-hydroxyanthranilic acid (3-HAA) which act as mediators (Eggert et al., 1996). A mediator is a small molecule that acts as an "electron shuttle." Once it is oxidized by the enzyme, generating a strongly oxidizing intermediate, the comediator (oxidized mediator), it diffuses away from the enzymatic pocket and in turn oxidizes any substrate that, due to its size, could not directly enter into the active site. Due to this specificity for phenolic subunits in lignin and its restricted access to lignin in the fiber wall, laccase has a limited effect without these redox mediators. In an active holoenzyme form, the laccase molecule is a dimeric or tetrameric glycoprotein usually containing four copper atoms per monomer, bound to three redox sites (type 1, type 2, and type 3 Cu pair).

3.2 Other Helper Proteins in Hydrolysis

In the process of enzymatic hydrolysis of lignocellulosic materials, some proteins have been identified that are capable of nonhydrolytically loosening the packaging of cellulose fibril network, a process called amorphogenesis. These proteins act synergistically along with cellulases, thereby increasing the accessibility of cellulose to the enzymes. Hence, these helper proteins are called amorphogenesis-inducing agents. Swollenin is an example of such helper proteins, and is isolated from *T. reesei*. It comes under the category of expansin-like proteins which are proteins having a "loosening" effect on the cellulosic network within plant cell walls during growth. Swollenin contains an amino terminal fungal type cellulose-binding module linked to the plant expansin homologous module. It shows sequence similarity to the fibronectin (Fn) III-type repeats of mammalian titin proteins which have been shown to be able to unfold and refold easily, allowing the protein to stretch. Swollenin has been reported to disrupt the structure of cotton fibers without revealing any hydrolytic activity and formation of reducing sugars (Saloheimo et al., 2002). This indicates that the protein is involved in the swelling of the cellulosic network within the cell walls and is not active against the β-1,4-glycosidic bonds in cellulose. The protein increases the access of cellulases to cellulose chains by promoting the dispersion of cellulose aggregations and exposing individual cellulose chains to the enzyme. This ability makes it an important component in the enzyme mixture use for the hydrolysis of lignocellulosic biomass. There are several swollenin-like activities displayed by *T. reesei*, which differ in their modes of action but contribute synergistically to the efficient hydrolysis of the plant polysaccharides.

4 SEPARATE AND SIMULTANEOUS HYDROLYSIS

In the process of lignocellulosic ethanol production, two consecutive catalytic steps follow after pretreatment: enzymatic conversion of the cellulose to fermentable sugars in a process called saccharification or hydrolysis and conversion of these sugars to ethanol by fermentation. The hydrolysis and fermentation steps can be operated sequentially by Separate Hydrolysis and Fermentation (SHF) or concurrently by Simultaneous Saccharification and Fermentation (SSF).

In SHF, pretreated lignocellulosic materials are hydrolyzed to glucose and subsequently fermented to ethanol in separate reactors. Hence, both the hydrolysis and fermentation processes are performed at their optimum temperature, that is, 50 °C for hydrolysis and 37 °C for yeast fermentation. The drawback of the process is the accumulation of the hydrolysis products in the enzymatic reactor which causes feedback inhibition of the cellulolytic enzyme system. The cellulase activity is inhibited by the released sugars, mainly cellobiose and glucose. The inhibitory effect of glucose on cellulase is lower than that of cellobiose. Cellulase activity is reduced by 60% at a low cellobiose concentration of 6 g/l. But it has been reported that at a level of 3g/l of glucose, β-glucosidase activity is reduced by 75% (Philippidis and Smith, 1995). Another major problem in SHF is microbial contaminations due to the longer incubation time in hydrolysis. A possible source of contamination could be the enzymes and its sterilization is very difficult when in a large-scale operation.

SSF is a process where both hydrolysis and fermentation processes are carried out in a single reactor. In this process, glucose released by the hydrolyzing enzymes is consumed immediately by the fermenting microorganism present in the culture, and a low concentration of sugars is maintained in the media, thus reducing the problem of end product inhibition of cellulase. The optimal temperature for SSF is maintained around 38 °C, which is a compromise between the optimum temperature for hydrolysis (45-50 °C) and fermentation (30 °C). *T. reesei* and *Saccharomyces cerevisiae* are the microorganisms commonly used for SSF. Thermotolerant yeasts and bacteria have also been used to increase the temperature close to that of optimum hydrolysis temperature.

The following are the advantages of SSF.

(1) Increase of hydrolysis rate by reducing end product inhibition of cellulase
(2) Lower enzyme requirement
(3) Higher ethanol yield
(4) Lower requirement for sterile conditions
(5) Shorter process time
(6) Cost reductions by eliminating expensive reaction and separation equipment

The main disadvantage of SSF is the inhibition of cellulase enzyme by ethanol produced after fermentation, and ethanol inhibition may be a limiting factor in obtaining high ethanol yield. It is reported that 30 g/l ethanol reduces the enzyme activity by 25% (Wyman, 1996). Another major drawback is that the incomplete hydrolysis of the substrates at the end of the reaction which causes the close association of the yeast and adsorbed cellulases with the recalcitrant residue. This restricts the reuse of the high concentrations of yeasts that are necessary to ensure good ethanol production in the subsequent batch. As a result, much of the sugars released by cellulose hydrolysis are used to grow the yeast rather than fermenting the sugars to ethanol. Despite these disadvantages, SSF is the preferred method in many pilot-scale studies for ethanol production.

5 FACTORS AFFECTING ENZYMATIC HYDROLYSIS

The chemical and structural modifications occurring in the lignocellulosic biomass during pretreatment have a significant effect on sugar release patterns and subsequently the enzymes employed for enzymatic hydrolysis. Biomass composition plays a major role in determining the effectiveness of pretreatment and enzymatic hydrolysis. During enzymatic hydrolysis, cellulases tend to irreversibly bind to lignin through hydrophobic interactions that cause loss in enzyme activity. Hence, the amount and composition of lignin in the biomass used critically affects the formation of soluble sugars during enzymatic hydrolysis. Along with this, type of pretreatment employed, enzyme dosage and its efficiency for saccharification, etc. also have a great influence on biomass digestibility. Even though the individual impact of these factors on determining the efficiency of enzymatic hydrolysis has not been fully resolved, many of these factors are found to be interrelated during the saccharification process. The main factors that influence the enzymatic hydrolysis of lignocellulosic feed stocks can be divided into two groups: enzyme-related and substrate-related factors.

5.1 Enzyme-Related Factors

Several factors associated with the nature of the cellulase enzyme system have been suggested to be influential during the hydrolysis process. These include enzyme concentration, enzyme adsorption, synergism, end-product inhibition, mechanical deactivation (fluid shear stress or gas-liquid interface), thermal inactivation and irreversible (nonproductive) binding to lignin. In the process of enzymatic hydrolysis, the nature of the enzyme system employed, the mode of action (*endo*- vs. *exo*-enzymes), and their stereochemical mechanism of hydrolysis (inverting vs. retaining) are interrelated. In addition, the synergism between the enzymes can be of significant benefit in increasing the hydrolysis rates of complex substrate. Synergism is also substrate dependent, with some mixtures showing cooperative action on amorphous substrates, but not on microcrystalline cellulose. All these factors can collectively influence enzyme efficiency.

5.1.1 Incubation Temperature

Temperature has a profound effect on enzymatic conversion of lignocellulosic biomass. Temperature has been shown to also influence cellulase adsorption. A positive relationship between adsorption and saccharification of cellulosic substrate was observed at temperatures below 60 °C. The adsorption activities beyond 60 °C decreased, possibly because of the loss of enzyme configuration leading to denaturation of enzyme activity.

5.1.2 Effect of Surfactants

Surfactants are amphiphilic compounds that contain a hydrophilic head and a hydrophobic tail. They are capable of self-assembling into micelles and adsorb onto surfaces depending on the surfactant structure and the polarity of the surface. It has been shown that some surfactants have a positive effect on enzymatic hydrolysis. They increase hydrolysis efficiency significantly, allowing for either a faster hydrolysis rate or lower enzyme dosage (Helle et al., 1993). The addition of surfactants also facilitates efficient recycling of cellulases after saccharification, a process step that ideally needs to be considered to reduce the cost of lignocellulosic ethanol production.

Different mechanisms have been proposed for the positive effect of surfactant addition to the enzymatic hydrolysis of cellulose:

(1) Surfactants may cause a surface structure modification or disruption of the lignocellulose that increases enzyme accessibility to cellulose.
(2) Surfactants may affect enzyme-substrate interaction by preventing nonproductive adsorption of enzymes.
(3) Surfactants may act as enzyme stabilizers. They adsorb at the air-liquid interface and thus prevent enzyme denaturation during agitation in the hydrolysis mixture (Kim et al., 1982).

Lignin due to its ability to adsorb enzymes is known to have adverse effects on action of cellulases on lignocellulose. The CBD of CBHs have been shown to be the major contributing factor responsible for lignin adsorption, but both the structure and properties of the CBD as well as the CD are involved in the binding affinity. In lignin-containing substrates, the effect of surfactant addition is significant, resulting in almost doubling of the yield. The primary mechanism behind the increased hydrolysis efficiency is due to the hydrophobic interaction between lignin surfaces and surfactants. Based on kinetic analysis, Kaar and Holtzapple (1998) have found indications that surfactants could promote the availability of reaction sites through surface disruption, in turn increasing the hydrolysis rate. A number of surfactants have been examined for their ability to improve enzymatic hydrolysis. Nonionic surfactants are the most effective among them. Fatty acid esters of sorbitan polyethoxylates (tween80, tween20) and polyethylene glycol (PEG) are among the most effective surfactants reported for enhancing enzymatic hydrolysis. The hydrophilic portions of the bound surfactant protrude into the aqueous solution and prevent the non-productive adsorption of cellulases and thereby increase cellulose conversion. Addition of noncatalytic proteins such as bovine serum albumin (BSA) has a similar effect to the addition of nonionic surfactants. BSA is known to adsorb to surfaces, reducing unspecific binding by "filling up" adsorption sites on lignin surfaces. Although the use of surfactants imposes an additional cost to the ethanol production, significant benefits can be achieved by improving the efficiency of enzymatic hydrolysis that is the key process contributing to the cost of lignocellulosic ethanol production.

5.1.3 Inhibitors in Enzymatic Hydrolysis

Although the pretreatment process helps to improve the formation of sugars by enzymatic hydrolysis, it also leads to the degradation or loss of carbohydrates, which in turn leads to the formation of byproducts which are inhibitory to the hydrolysis and fermentation processes. The composition and concentration of the degradation products varies with certain pretreatment parameters like type of lignocellulosic biomass used, nature of the pretreatment process, temperature, time and pressure used for pretreatment. The main inhibitory compounds formed during pretreatment are as follows:

1. Organic acids—acetic acid, formic acid, and levulinic acid
2. Sugar degradation products—furfural and 5-hydroxymethylfurfural (5-HMF),
3. Lignin degradation products—vanillin, syringaldehyde, and 4-hydroxybenzaldehyde.

Acetic acid is released during the hydrolysis of hemicellulose in which the acetyl group of hemicellulose linked to the lignin is released and reacted in acid form; levulinic acid is the terminal product of oxidation of D-glucose and D-mannose; for formic acid, one is the terminal product of xylose oxidation, and another one is the byproduct of D-glucose and D-mannose oxidation to levulinic acid. Cellulases are found to be significantly inhibited by formic acid, whereas compounds such as vanillic acid, syringic acid, and syringylaldehyde, in addition to formic acid, cause significant inhibition of xylanases. A clear elucidation of the inhibitory effect of these degradation products will help in designing pretreatment technologies to release less strong inhibitors.

Hydrolysis is also found to be affected due to the end-product inhibition of cellulases. However, when working with insoluble substrate and kinetics that do not follow the

Michaelis-Menten model, it is difficult to determine the exact type of inhibition. Removal of end product is possible by using the SSF strategy. But in this case, inhibition of cellulases by fermentation products should also be considered. Ethanol is inhibitory to cellulases, although less compared to glucose (Chen and Jin, 2006). Hence, the effect of end products on cellulase has to be evaluated before selecting the hydrolysis and fermentation strategy.

5.2 Substrate-Related Factors

The rate of enzymatic hydrolysis of lignocellulose is profoundly affected by the structural features of cellulose (Fan et al., 1981) which include cellulose crystallinity, DP, available/accessible surface area, structural organization, that is, macrostructure (fiber) and microstructure (elementary microfibril), particle size, and presence of associated materials such as hemicellulose and lignin.

The typical time course of the enzymatic hydrolysis of the lignocellulosic material is characterized by the rapid initial rate of hydrolysis followed by slower and incomplete hydrolysis. Such a time course has been suggested to be due to the rapid hydrolysis of more easily available amorphous cellulose, with consequent increase of inherent degree of crystallinity, as the hydrolysis proceeds (Mansfield et al., 1999). The effect of substrate crystallinity has been shown to play a major role in limiting hydrolysis in some studies (Fan et al., 1981, 1980), while other studies have shown that, when all other substrate factors are similar, the degree of crystallinity of the substrate has no effect on hydrolysis (Puri, 1984).

The effect of the DP (number of glycosyl residues per cellulose chain) is essentially related to other substrate characteristics such as crystallinity. It has been shown that the depolymerization is largely a function of the nature of the cellulosic substrate being attacked. EGs preferentially attacking less ordered, inside regions of the cellulose chain contribute, thus, to a large extent to the rapid decrease of DP. On the contrary, *exo*glucanases (CBHs) hydrolyzing substrate from the chain ends releasing cellobiose as a product have little effect on the change of DP throughout the hydrolysis process. However, regardless of the substrate being hydrolyzed, there seems to be a "leveling off" of the cellulose DP, which is correlated with the increased recalcitrance of the residual (crystalline) cellulose.

Another major substrate characteristic influencing the hydrolysis process is accessibility of the substrate. Most often, accessibility is measured by the BET (Bennet-Emmit-Teller) method, which measures the surface area available to the nitrogen molecule (Masamune and Smith, 1964). The drawbacks of the method are that it involves the drying of the substrate, thus not allowing measurements on the material in its swollen state, and that the nitrogen molecule is substantially smaller in size compared to the enzyme molecule. As a consequence, Specific Surface Area (SSA) can be overestimated as small nitrogen molecules have access to pores and cavities on the fiber surface that cellulases cannot enter. External surface area is closely related to shape and particle size and, thus, a higher surface area-to-weight ratio should mean more available adsorption sites per mass of substrate. Consequently, substrate pretreatment methods often include cutting, that is, reduction in size, of the lignocellulosic material to increase SSA. Also, removal of lignin and hemicellulose by the pretreatment methods causes extensive changes in the structure and accessibility of cellulose (complementary to the desired effect of preventing enzyme loss by unproductive binding to lignin). Their removal leaves the cellulose more accessible and more open to swelling on contact with cellulases (Grethlein et al., 1984).

5.2.1 *Biomass Loading*

Substrate concentration is one of the main factors that affect the yield and initial rate of enzymatic hydrolysis of cellulose. Maintaining high solids concentrations throughout the conversion process from biomass to ethanol is important from an energy and economic viability viewpoint. High solids enzymatic hydrolysis takes place at solids levels where initially no significant amount of free water is present. This allows for a larger system capacity, less energy demand for heating and cooling of the slurry, and also less effluent discharge. Regarding the overall economic feasibility of lignocellulosic ethanol production, a high substrate concentration allows for the production of a concentrated sugar solution, which in turn is beneficial for the subsequent fermentation. By increasing the solids loading, the resulting sugar concentration and consequently ethanol concentration can be increased with significant effects on distillation. A sugar concentration of at least 8% (w/w) is required to achieve an ethanol yield of 4% (w/w) by which the energy required for the distillation can be significantly reduced. However, high substrate concentration can also cause substrate inhibition, which substantially lowers the rate of the hydrolysis. The extent of substrate inhibition depends on the ratio of total substrate to total enzyme (Penner and Liaw, 1994). The enzymatic conversion (percent of theoretical) is found to linearly decrease with increased solids concentration despite using a constant enzyme-to-substrate ratio. It may be explained by mass transfer limitations or nonproductive adsorption of enzymes. However, the specific mechanism behind the decreased hydrolytic efficiency is not fully studied.

Operating hydrolysis with high initial substrate concentration also faces the problem of product inhibition of especially the cellulolytic enzyme system. β-glucosidases from typical cellulase-producing microorganisms are to some extent inhibited by glucose. This results in accumulation of cellobiose, which in turn is a potent inhibitor of the CBHs, thereby affecting the saccharification efficiency. Another disadvantage of using high solid loadings is high slurry viscosity which causes insufficient mixing and also leads to excessive energy consumption. Moreover, water content in the hydrolysis slurry is important for the interaction between lignocellulose and cell wall-degrading enzymes. Thus, water content is essential for enzyme function, for enzyme transport mechanisms throughout the hydrolysis reaction, and for mass transfer of intermediates and end products (Felby et al., 2008).

The extent to which solid loading can be increased in hydrolysis varies with each lignocellulosic biomass. At both lab and industrial scale, 12-20% total solids is often considered the upper limit at which pretreated biomass can be mixed and hydrolyzed in conventional stirred tank reactors. Fed-batch operations can be employed to increase the final solid loadings.

6 RECYCLING OF ENZYMES

A key factor that prevents the commercialization of enzymatic cellulose hydrolysis is the high cost of cellulase enzymes. Enzyme cost is expected to account for more than 20% of ethanol production (Wooley et al., 1999). As much of it remains active after hydrolysis, recycling of cellulases makes the overall conversion process more economically feasible. Various methods have been used for recycling enzymes, which include sedimentation followed by ultra filtration or micro centrifugation, cation exchange chromatography, re-adsorption, and immobilization.

A simple method of recovering enzymes after hydrolysis by centrifugation was carried out by Moniruzzaman et al (1997). Their study shows that during initial stages of enzyme recycling, most of the initial enzyme activity could be recovered, but a gradual decrease in enzyme activity was observed at later stages of recycling, and this may be due to thermal or mechanical inactivation.

The ultra filtration method for enzyme recovery has proved to be an efficient way to recover cellulases as well as to continuously remove end products that are generated during hydrolysis that could potentially inhibit hydrolysis reactions (Tan et al., 1986). Mores et al (2001) reported cellulase recovery by a combined sedimentation and membrane filtration process. During the sedimentation step, the larger particles are removed so that they will not block the tubing or membrane filter. After sedimentation, the suspension is clarified using microfiltration. Ultra filtration membranes, made up of polysulfone or polyethersulfone, are used to separate sugars from cellulase. The enzymes are retained by the membranes while the water, sugars, ethanol, and other small molecules are removed. The retained cellulase can be reused for hydrolysis. The result indicates that 75% of the cellulase enzymes can be recovered in active form by membrane separation.

Another method for recycling enzymes is using amphiphilic lignin derivatives. The effect of amphiphilic lignin on cellulase recycling was investigated in a continuous multistage saccharification process of cellulosic materials using cellulase as catalyst. The results indicate that amphiphilic lignin is an excellent water-soluble polymeric carrier for immobilization of cellulase to preserve the hydrolytic activity for a long period (Uraki et al., 2001).

The potential economic benefit of surfactant addition on enzyme recycling was reported by Tu and Saddler (2010). Free cellulase re-adsorption on fresh steam exploded lodge pole pine and ethanol pretreated lodge pole pine was used to recover and recycle cellulase enzyme during hydrolysis. The economic analysis of enzyme cost versus surfactant cost suggests that a 66% reduction in total enzyme cost was achieved and tween80 was the most effective surfactant in enzyme recycling.

Reusability of enzymes by immobilization was carried out by Tu et al (2007). Their study evaluated the potential for immobilization of β-glucosidase on a methacrylamide polymer carrier, Eupergit C for lignocelluloses hydrolysis. The immobilization could facilitate enzyme recycling in sequential batchwise or semi-batchwise saccharification process. Eupergit C- immobilized β-glucosidase was examined for six successive rounds of lignocellulosic hydrolysis and exhibited relative stability during the subsequent five cycles.

7 METHODS FOR IMPROVING ENZYMATIC HYDROLYSIS

One strategy for achieving improved efficiency of enzymatic hydrolysis is to improve the specific activities of cellulases by genetic engineering. For lignocellulosic substrates, the nonproductive binding and inactivation of enzymes by the lignin component are the important factors limiting catalytic efficiency. Understanding the effect of these factors allows engineering of cellulases with improved activities (Berlin et al., 2005). The studies proved that naturally occurring cellulases with similar catalytic activity on a model cellulosic substrate can differ significantly in their affinities for lignin. Cellulases lacking CBDs have a high affinity for lignin which indicates the presence of lignin-binding sites on the CD.

Evolution of the cellulosome complex has led to colocalization of synergistic combinations of hydrolytic enzymes. This architectural feature has led to innovative molecular engineering approaches for diverse research and industrial applications (Nordon et al., 2009). The modular structure of the *T. reesei* endoglucanase IV (EG IV) was reconstructed by Liu et al (2006) by fusing EG IV with an additional catalytic module (EGIVCM). The genes were obtained through RT-PCR and gene fusion and expressed in recombinant *Pichia* strains. The results indicate that modification of the EGIV structure with an additional catalytic module in the C terminus improved specific activity of about fourfold.

Two strategies are used for improving the properties of individual cellulase components: (1) rational design and (2) directed evolution.

7.1 Rational Design

Rational design is the earliest approach to protein engineering. It was introduced after the development of recombinant DNA methods and site-directed mutagenesis. This strategy requires detailed knowledge of protein structure. The first step in rational design involves the selection of a suitable enzyme. In the next step, the amino acid site to be changed will be identified on the basis of a high-resolution crystallographic structure. Finally, the resultant mutant will be characterized.

The choice of a suitable enzyme for modification depends on the availability of data on the protein structure of an enzyme. Selection of a region of the protein to be modified requires the knowledge of the existing function of the region and also the desired modified function. Amino acid sequences can be modified through site-directed mutagenesis. The success is very difficult, because the information of structures and mechanisms is not available for a vast majority of enzymes. Even if the structure and catalysis mechanism of the target enzyme are well characterized, the molecular mutation basis for the desired function may not be achieved (Arnold et al., 2001). Large functional changes can be obtained with a few amino acid substitutions; it is difficult to discern the specific mutations responsible.

7.2 Directed Evolution

One of the advantages of directed evolution is that it is independent of enzyme structure and of the interactions between enzyme and the substrate (Zhang et al., 2006). The most important challenge of this method is developing tools to correctly evaluate the performance of mutants generated by recombinant DNA techniques. DNA shuffling is one of the methods to improve the properties of cellulases. Kim et al (2000) reported a fivefold increase in specific activity of *Bacillus subtilis* EG mutant generated by DNA shuffling. Murashima et al (2002) could enhance the thermostability of EG by sevenfold, using the family gene shuffling technique based on the parental *Clostridium* cellulosomal EGs.

8 KINETIC MODEL FOR ENZYMATIC HYDROLYSIS OF LIGNOCELLULOSES

A mathematical model is the general characterization of a process, object, or concept, in terms of mathematics, which enables the relatively simple manipulation of variables to be accomplished in order to determine how the process, object, or concept would behave in different situations. A model generally incorporates a number of parameters that are used to describe the desired process. The accuracy, to which the different parameters used in the model are experimentally determined, is usually an important issue. If parameters are difficult to determine, the introduction of errors in the model is inevitable. Thus, increasing the complexity of the model should be carefully evaluated as the uncertainty of the model can increase with increasing the number of parameters, as each parameter can introduce some additional variance into the system. Thus, the task of mathematical modeling of enzymatic degradation of cellulose is highly challenging as it is necessary to balance complex biological process with many variables, with the basic requirement of a model, that is, simplicity and robustness. It is therefore usually appropriate to make some approximations to reduce the model to a sensible size.

A simplified reaction scheme for modeling cellulose hydrolysis proposed by Kadam et al (2004) is shown in Figure 4.

Each enzymatic reaction is potentially inhibited by the sugar it generates or by the six sugars already present in the system, that is, glucose, cellobiose, galactose, mannose, xylose, and arabinose. To simplify model development, the sugar system is consolidated to three sugars: cellobiose, glucose, and xylose. The equations for Langmuir adsorption model, conversion of cellulose to cellobiose, conversion of cellobiose to glucose and mass balance equation used in this model are shown as follows (Kadam et al., 2004).

$$\text{Langmuir isotherm } \mathrm{E}_{i\mathrm{B}} = \frac{E_i \max K_{iad} E_{i\mathrm{F}} S}{1 + K_{iad} E_{i\mathrm{F}}}.$$

Cellulose-to-cellobiose reaction with competitive glucose, cellobiose, and xylose inhibition

$$r_1 = \frac{k_{1r} E_{1\mathrm{B}} R_S S}{1 + (\mathrm{G}_2/K_{1\mathrm{IG2}}) + (\mathrm{G}/K_{1\mathrm{IG}}) + (\mathrm{X}/K_{1\mathrm{IX}})}.$$

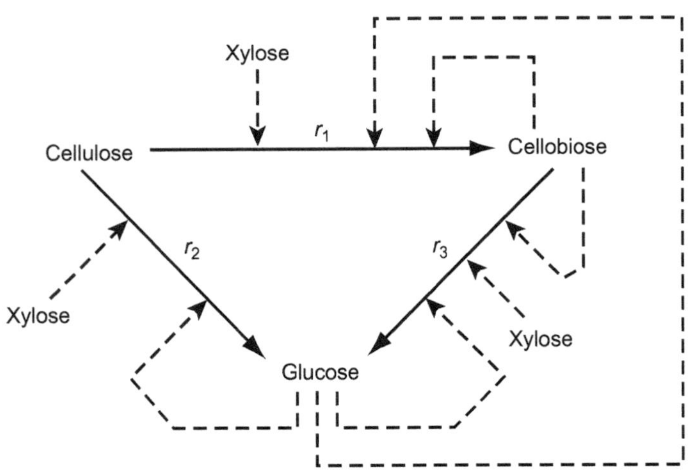

FIGURE 4 Reaction scheme for modeling cellulose hydrolysis. Enzymes involved in r1: *endo-β*-1, 4-glucanase and *exo-β*-1, 4-cellobiohydrolase. Enzymes-involved in r2: *exo-β*-1, 4-cellobiohydrolase and *exo-β*-1, 4-glucan glycohydrolase. Enzymes involved in r3: *β*-glucosidase *(Kadam et al., 2004).*

Cellulose-to-glucose reaction with competitive glucose, cellobiose, and xylose inhibition

$$r_2 = \frac{k_{2r}(E_{1B} + E_{2B})R_S S}{1 + (G_2/K_{2IG2}) + (G/K_{2IG}) + (X/K_{2IX})}.$$

Cellobiose-to-glucose reaction with competitive glucose and xylose inhibition

$$r_3 = \frac{k_{3r}E_{2F}G_2}{K_{3M}[1 + (G/K_{3IG}) + (X/K_{3IX})] + G_2}.$$

These rate equations assume that (1) enzyme adsorption follows a Langmuir-type isotherm with the first-order reactions (r_1 and r_2) occurring on the cellulose surface; (2) the cellulose matrix is uniform in terms of its susceptibility to enzymatic attack (i.e., no provision was made to include separately more reactive amorphous and more recalcitrant crystalline cellulose fractions); (3) enzyme activity remains constant; and (4) conversion of cellobiose to glucose occurs in solution and follows classical Michaelis-Menton kinetics (Kadam et al., 2004).

The kinetic model of enzymatic hydrolysis of cellulose by Wald et al (1984) incorporates enzyme adsorption, product inhibition, and a multiple enzyme system. This model considers enzyme adsorption as a function of available sorption sites and, thus, of accessible surface area via a Langmuir-type isotherm relationship. Although the model does not consider glucose end-product inhibition, it was capable of simulating saccharification of rice straw lignocellulose at high substrate (up to 333 g/l) and high enzyme (up to 9.2 FPU/ml) concentrations. Some of the kinetic models developed for enzymatic hydrolysis of lignocellulosic substrates are tabulated in Table 1.

TABLE 1 Kinetic Models Developed for Enzymatic Hydrolysis of Lignocellulosic Biomass

State of Substrate	Enzyme System	Kinetic Approach	Product Inhibition	Reference
Homogeneous material	Combined Endoglucanase and Cellobiohydrolase	Quasi-steady state	Competitive	Howell and Stuck (1975)
Homogeneous material	Combined Endoglucanase, Cellobiohydrolase; and β-glucosidase	Michaelis-Menten	Competitive	Huang (1975)
Degree of polymerization	Endoglucanase, Cellobiohydrolase; and β-glucosidase	Michaelis-Menten	Noncompetitive	Okazaki and Moo-Young (1978)
Homogeneous material	Combined Endoglucanase, Cellobiohydrolase; and β-glucosidase	Quasisteady state	Competitive	Howell and Mangat (1978)
Crystalline and amorphous	Combined Endoglucanase, Cellobiohydrolase; and β-glucosidase	Michaelis-Menten		Peiterson and Ross (1979)
Crystalline and amorphous	Combined Endoglucanase, Cellobiohydrolase; and β-glucosidase	Quasisteady state	Competitive	Ryu et al. (1982)
Homogenous material	Combined Endoglucanase and Cellobiohydrolase and β-glucosidase		Noncompetitive	Fan and Lee (1983)
Active and inert	Combined Endoglucanase, Cellobiohydrolase; and β-glucosidase	Quasisteady state	Competitive	Gan et al. (2003)

Adapted from Andersen (2007).

9 CONCLUSIONS

Conversion of lignocellulosic biomass into fermentable sugars is the key step in lignocellulosic ethanol production. Several challenges are involved in this process, which need to be addressed in order to improve process efficiency. Even though the conventional method of lignocellulosic hydrolysis using concentrated acids is an efficient process, there are several issues related to the environment, which makes one think of an alternative to replace this method with more environment-friendly processes. Using ILs for deriving cellulose from lignocellulosic materials seems to be a promising method, but there are several challenges that prevent this process being feasible. At present, the cost of ILs is too high and it is necessary to develop a technology to produce cheaper ILs. In addition, the recovery and reuse of ILs need to be addressed. Another challenge is to recover lignin and hemicelluloses from the ILs after cellulose has been extracted. So, there are immense opportunities for R&D in the area of IL-based processes for the production of lignocellulosic ethanol. Due to the heterogeneous nature of lignocellulosic materials, it is necessary to screen a large variety of ILs to find a suitable one for a particular biomass. Moreover, there occur wide possibilities for designing ILs based on the nature of lignocellulosic materials. Hydrolysis using enzymes is an attractive and environmentally safe alternative; still, there is a great deal of scope for research to improve the enzymatic conversion efficiency of lignocellulosic biomass to fermentable sugars by protein engineering approaches.

References

Andersen, N., 2007. Enzymatic Hydrolysis of Cellulose—Experimental and Modeling Studies BioCentrum-DTU. Technical University of Denmark, pp. 92.

Arnold, F.H., Wintrode, P.L., Miyazaki, K., Gershenson, A., 2001. How enzymes adapt: lessons from directed evolution. Trends Biochem. Sci. 26, 100–106.

Berlin, A., Gilkes, N., Kurabi, A., Bura, R., Tu, M., Kilburn, D., et al., 2005. Weak lignin-binding enzymes. Appl. Biochem. Biotechnol. 121–124.

Broder, J.D., Barrier, J.W., Lightsey, G.R., 1992. Conversion of cotton trash and other residues to liquid fuel from renewable resources. In: Cundiff, J.S. (Ed.), Proceedings of an alternative energy conference. American Society of Agricultural Engineers, St. Joseph, MI, pp. 189–200.

Chen, H., Jin, S., 2006. Effect of ethanol and yeast on cellulase activity and hydrolysis of crystalline cellulose. Enzyme Microb. Technol. 39, 1430–1432.

Duff, S.J.B., Murray, W.D., 1996. Bioconversion of forest products industry waste cellulosics to fuel ethanol: a Review. Bioresour. Technol. 55, 1–33.

Eggert, C., Temp, U., Dean, J.F.D., Eriksson, K.E.L., 1996. A fungal metabolite mediates degradation of non-lignin structures and synthetic lignin. FEBS Lett. 391, 144–148.

Fan, L.T., Lee, Y.H., 1983. Kinetic studies of enzymatic hydrolysis of insoluble cellulose: derivation of a mechanistic kinetic model. Biotechnol. Bioeng. 25, 2707–2733.

Fan, L.T., Lee, Y.H., Beardmore, D.H., 1980. Mechanism of the enzymatic hydrolysis of cellulose: effect of major structural features of cellulose on enzymatic hydrolysis. Biotechnol. Bioeng. 23, 177–199.

Fan, L.T., Lee, Y.H., Beardmore, D.H., 1981. The influence of major structural features of cellulose on rate of enzymatic hydrolysis. Biotechnol. Bioeng. 23, 419–424.

Felby, C., Thygesen, L.G., Kristensen, J.B., Jørgensen, H., Elder, T., 2008. Cellulose-water interactions during enzymatic hydrolysis as studied by Time Domain NMR. Cellulose 15, 703–710.

Gan, Q., Allen, S.J., Taylor, G., 2003. Kinetic dynamics in heterogeneous enzymatic hydrolysis of cellulose: an overview, an experimental study and mathematical modeling. Bioprocess Biotechnol. 38, 1003–1018.

Graenacher, C., 1934. Cellulose solution US Patent 1 943 176.

Grethlein, H.E., Allen, D.C., Converse, A.O., 1984. A comparative study of the enzymatic hydrolysis of acid pretreated white pine and mixed hardwood. Biotechnol. Bioeng. 26, 1498–1505.

Harris, E.E., 1949. Wood Saccharification Advances in Carbohydrate Chemistry. Academic Press, New York, pp. 153–188.

Helle, S.S., Duff, S.J.B., Cooper, D.G., 1993. Effect of surfactants on cellulose hydrolysis. Biotechnol. Bioeng. 42, 611–617.

Howell, J.A., Mangat, M., 1978. Enzyme deactivation during cellulose hydrolysis. Biotechnol. Bioeng. 20, 847–863.

Howell, J.A., Stuck, J.D., 1975. Kinetics of solka floc cellulose hydrolysis by *Trichoderma viride* cellulase. Biotechnol. Bioeng. 17, 873–893.

Huang, A.A., 1975. Kinetic studies on insoluble cellulose-cellulase system. Biotechnol. Bioeng. 17, 1421–1433.

Kaar, W.E., Holtzapple, M., 1998. Benefits from tween during enzymic hydrolysis of corn stover. Biotechnol. Bioeng. 59, 419–427.

Kadam, K.L., Rydholm, E.C., McMillan, J.D., 2004. Development and validation of a kinetic model for enzymatic saccharification of lignocellulosic biomass. Biotechnol. Prog. 20, 698–705.

Kim, M.H., Lee, S.B., Ryu, D.D.Y., 1982. Surface deactivation of cellulase and its prevention. Enzyme Microb. Technol. 4, 99–103.

Kim, Y.S., Jung, H.C., Pan, J.G., 2000. Bacterial cell surface display of an enzyme library for selective screening of improved cellulase variants. Appl. Environ. Microbiol. 66, 788–793.

LaForge, F.B., Hudson, C.S., 1918. The preparation of several useful substances from corn cobs. J. Ind. Eng. Chem. 10, 925–927.

Liu, G., Tang, X., Tian, S., Deng, X., Xing, M., 2006. Improvement of the cellulolytic activity of *Trichoderma reesei* Endoglucanase IV with an additional catalytic domain. World J. Microbiol. Biotechnol. 22, 1301–1305.

Mansfield, S.D., Mooney, C., Saddler, J.N., 1999. Substrate and enzyme characteristics that limit cellulose hydrolysis. Biotechnol. Proc. 15, 804–816.

Masamune, S., Smith, J.M., 1964. Adsorption rate studies—significance of pore diffusion. AIChE J. 10, 246–252.

Moniruzzaman, M., Dale, B.E., Hespell, R.B., Bothast, R.J., 1997. Enzymatic hydrolysis of high-moisture corn fiber pretreated by AFEX and recovery and recycling of the enzyme complex. Appl. Biochem. Biotechnol. 67, 113–126.

Mores, W.D., Knutsen, J.S., Davis, R.H., 2001. Cellulase recovery via membrane filtration. Appl. Biochem. Biotechnol. 91-93, 279–309.

Murashima, K., Chen, C.L., Kosugi, A., Tamaru, Y., Doi, R.H., Wong, S.L., 2002. Heterologous production of *Clostridium cellulovorans* Engb, using protease-deficient *Bacillus subtilis*, and preparation of active recombinant cellulosomes. J. Bacteriol. 184, 76–81.

Nordon, R.E., Craig, J.S., Foong, F.C., 2009. Molecular engineering of the cellulosome complex for affinity and bioenergy applications. Biotechnol. Lett. 31, 465–476.

Okazaki, M., Moo-Young, M., 1978. Kinetics of enzymatic hydrolysis of cellulose: analytical description of a mechanistic model. Biotechnol. Bioeng. 20, 637–663.

Peiterson, N., Ross, E.W., 1979. Mathematical model for enzymatic hydrolysis and fermentation of cellulose by *Trichoderma*. Biotechnol. Bioeng. 21, 997–1017.

Penner, M.H., Liaw, E.T., 1994. Kinetic consequences of high ratios of substrate to enzyme saccharification systems based on *Trichoderma* cellulase. In: Himmel, M.E., Baker, J.O., Overend, R.P. (Eds.), Enzymatic Conversion of Biomass for Fuels Production. American Chemical Society, Washington, DC, pp. 363–371.

Philippidis, G.P., Smith, T.K., 1995. Limiting factors in the simultaneous saccharification and fermentation process for conversion of cellulosic biomass to fuel ethanol. Appl. Biochem. Biotechnol. 51/52, 117–124.

Puri, V.P., 1984. Effect of crystallinity and degree of polymerization of cellulose on enzymatic saccharification. Biotechnol. Bioeng. 26, 1219–1222.

Ryu, D.D.Y., Lee, S.B., Tassinari, T., Macy, C., 1982. Effect on compression milling on cellulose structure and on enzyme hydrolysis kinetics. Biotechnol. Bioeng. 24, 1047–1067.

Saloheimo, M., Paloheimo, M., Hakola, S., Pere, J., Swanson, B., Nyyssönen, E., et al., 2002. Swollenin, a *Trichoderma reesei* protein with sequence similarity to the plant expansins, exhibits disruption activity on cellulosic materials. Eur. J. Biochem. 269, 4202–4211.

Schoemaker, H.E., Piontek, K., 1996. On the interaction of lignin peroxidase with lignin. Pure Appl. Chem. 68, 2089–2096.

Swatloski, R., Spear, S., Holbrey, J., Rogers, R., 2002. Dissolution of cellulose with ionic liquids. J. Am. Chem. Soc. 124, 4974.

Tan, T.K., Yeoh, H.H., Paul, K., 1986. Cellulolytic activities of *Trichoderma bamatum* grown on different carbon substrates. MIRCEN J. Appl. Microbiol. Biotechnol. 2, 467–472.

Tu, M., Saddler, J.N., 2010. Potential enzyme cost reduction with the addition of surfactant during the hydrolysis of pretreated softwood. Appl. Biochem. Biotechnol. 161, 274–287.

Tu, M.B., Chandra, R.P., Saddler, J., 2007. Evaluating the distribution of cellulases and the recycling of free cellulases during the hydrolysis of lignocellulosic substrates. Biotechnol. Prog. 23, 398–406.

Uraki, Y., Ishikawa, N., Nishida, M., Sano, Y., 2001. Preparation of amphiphilic lignin derivative as a cellulase stabilizer. J. Wood Sci. 47, 301–307.

Vagt, U., 2010. Cellulose dissolution and processing with ionic liquids. In: Wasserscheid, P., Stark, A. (Eds.), Handbook of Green Chemistry, Volume 6: Ionic Liquids. WILEY-VCH Verlag GmbH & Co. KGaA, Weinheim.

Wald, S., Wilke, C.R., Blanch, H.W., 1984. Kinetics of the enzymic hydrolysis of cellulose. Biotechnol. Bioeng. 26, 221–230.

Wenzl, H.F.J., 1970. Chapter IV: The Acid Hydrolysis of Wood the Chemical Technology of Wood. Academic Press, New York, pp. 157–252.

Wooley, R., Ruth, M., Glassner, D., Sheejan, J., 1999. Process design and costing of bioethanol technology: a tool for determining the status and direction of research and development. Biotechnol. Prog. 15, 794–803.

Wyman, C.E., 1996. Handbook on Bioethanol: Production and Utilization. Taylor & Francis, Washington, DC.

Zhang, P., Himmel, M.E., Mielenz, J.R., 2006. Outlook for cellulase improvement: screening and selection strategies. Biotechnol. Adv. 24, 452–481.

C H A P T E R

11

Production of Bioethanol from Agroindustrial Residues as Feedstocks

*Julián A. Quintero, Luis E. Rincón, Carlos A. Cardona**

Departamento de Ingeniería Química, Universidad Nacional de Colombia Sede Manizales, Cra. 27 No. 64-60, Manizales, Colombia

***Corresponding author: E-mail: ccardonaal@unal.edu.co**

1 INTRODUCTION

Worldwide high demand for energy, uncertainty of petroleum resources, and concern about global climatic changes have led to the resurgence in the development of alternative liquid fuels. Ethanol has always been considered a better choice as it reduces the dependence on crude oil and promises cleaner combustion leading to a healthier environment. Developing ethanol as fuel beyond its current role of fuel oxygenate would require lignocellulosics as a feedstock because of its renewable nature, abundance, and low cost (Saha et al., 2005).

Most of the fuel ethanol produced in the world is currently sourced from starchy biomass or sucrose (molasses or cane juice), but the technology for ethanol production from non-food plant sources is being developed rapidly such that large-scale production will be a reality in the coming years (Lin and Tanaka, 2006). Lignocellulosics mainly comprise cellulose, a polymer of six-carbon sugar, glucose; hemicellulose, a branched polymer comprising xylose; and other five-carbon sugars and lignin consisting of phenyl propane units. The presence of lignin limits the complete usage of cellulose and hemicellulose. Hemicelluloses are the most thermal-chemically sensitive. During thermal-chemical pretreatment, firstly the side groups of hemicellulose react, followed by the hemicellulose backbone (Sweet and Winandy, 1999). In biomass, cellulose is generally the largest fraction, about 40-50% by weight and hemicellulose about 20-40% (McKendry, 2002; Saxena et al., 2009). For example, the sugarcane bagasse contains 40-50% cellulose, 20-30% hemicellulose, 20-25% lignin and 1.5-3% ash. To convert these energy-rich molecules into simpler forms, it is necessary to remove the lignin from lignocellulosic materials.

The production of ethanol from lignocellulosic biomass involves different steps of pretreatment, hydrolysis (saccharification), fermentation, and ethanol recovery

(Van Zessen et al., 2003). A number of pretreatments such as concentrated acid hydrolysis (Liao et al., 2006), dilute acid hydrolysis (Cara et al., 2008), alkali treatment (Carrillo et al., 2005), sodium sulfite treatment (Kuhad et al., 1999; Kapoor et al., 2008), sodium chlorite treatment (Sun et al., 2004), steam explosion (Öhgren et al., 2005), ammonia fiber explosion (Teymouri et al., 2005) lime treatment (Kim and Holtzapple, 2005), and organic solvent treatment (Xu et al., 2006) have been used frequently to remove lignin and to improve the saccharification of the cell wall carbohydrates. The pretreatment is necessary to increase the rate of production and the total yield of monomeric sugars in the hydrolysis step.

Hydrolysis of biomass is essential for generation of fermentable sugars which are then converted to ethanol by microbial action. Acid and Enzimatic approaches, are primarily employed for biomass hydrolysis with varying efficiencies depending on treatment conditions, type of biomass, and the properties of the hydrolytic agents. The former is a mature technology but with the disadvantages of the generation of hazardous acidic waste and the technical difficulties in recovering sugar from the acid. The enzymatic method, however, is more efficient and proceeds under ambient conditions without generation of any toxic waste. The latter method which is under rapid development has immense potentials for improvement in cost and efficiency (Mishima et al., 2006). Commercialization of ethanol production from lignocellulosic biomass is hindered mainly by the high cost of the currently available cellulase preparations. Reduction in the cost of cellulases can be achieved only by concerted efforts which address several aspects of enzyme production from the raw material used for production to microbial strain improvement. Same situation is particularly observed as an analogy for the enzymes used in starch liquefaction and hydrolysis (Cardona and Sánchez, 2007) where the *in situ* production of different types of amylases was possible due to combined effects of scientists and industry reducing the costs significantly.

The produced monomeric hexoses (six carbon sugars) can be fermented to ethanol quite easily, while the fermentation of pentoses (five carbon sugars) is only done by a few strains. Volatile products are also not easily fermented to ethanol. A problem occurring during the fermentation is that the formed product ethanol is an inhibitor for the yeasts/bacteria that perform the fermentation. This puts a limit to the concentration of fermentable sugars (Hendriks and Zeeman, 2009). After fermentation, the ethanol has to be recovered from the fermentation broth by distillation (Mosier et al., 2005a). Furfural and other inhibitors like soluble lignin compounds also form a problem for the fermentation step, because such compounds can inhibit, or even stop the fermentation (Laser et al., 2002).

In the case of energy cogeneration, lignocellulosic biomass is also an attractive alternative; because it is largely available and able to recycle part of CO_2 emitted during its planting cycle. Biomass feedstocks have a reduced contribution to greenhouse effect compared to fossil fuels, at least if it is produced in a sustainable way no leading to any deforestation (Grassi and Allan, 2007). Interest for agroindustrial residues utilization as energy source is growing due to an understanding of its socioeconomic and political benefit effects (Grassi and Allan, 2007; Hall, 1997). In order to reduce GHG emissions and promote energy efficiency, substitution of fossil fuels with renewable sources helps to mitigate climate change as long as to generate renewable energy in a sustainable way (Shuit et al., 2009). Modern biomass utilization technologies, makes possible to add value to agroindustrial residues, using them as industrial energy source, by means of its combustion or gasification conversion, with less SO_2 and NOx content (Susta et al., 2003). Converting thus, this chemical energy into electricity in a process scheme known as bioelectricity.

Sources of biomass residues can be wood processing industry (sawdust, cut-offs, bark), agricultural industry (sugar cane bagasse, coconut shells, rice hulls, coffee husks, corn stover, oilseed cakes, wheat straw), food processing industry (organic waste animal manure and residues), wastewater and landfill municipal sewage (Susta et al., 2003; Ramírez et al., 2007). Different world economies usually have different uses for wood and agricultural residues: Non-commercial, for cooking and space heating in the poorer countries (Faaij, 2006), or commercial to produce electricity and/or district space heating in residential and commercial buildings, through direct combustion, gasification, anaerobic digestion, as well as, methanol and ethanol production (Faaij, 2006; Haq, 2010). However, fuels requirements of efficiency, low cost, and emissions will become more constraining, due to environmental regulation and legislation (Bram et al., 2005), making not all of biomass useful to be employed as fuel source. In this sense, residues from the biomass agroindustry can be successfully used in the energy industry due to their high availability and acceptable heating value. Among top used residues for bioelectricity production can be found: i) Rice Hulls obtained from paddy rice milling. It is used as energy source in large rice mills, through its direct combustion to produce the heat and power required in the operation of parboiling or rice noodles production. Also, it can be used in the charcoal production (Papong et al., 2004). ii) Sugar Cane Bagasse is the fibrous residue of juice removed in sugar cane milling and is one major biomass byproduct fuel, composed of trash, tops, and leaves of sugarcane plant (Larson et al., 2001). It can be used as fuel to produce heat and power for mills. Some facilities can produce an electricity excess able to be sold to the local grid (Papong et al., 2004; Coelho et al., 2000). iii) Oil palm residues are composed of empty fruit bunches (EFB) and fruit that contain crude palm oil, mesocarp fiber (MF), nuts, among others. Nut portion of the fruit can be processed to obtain crude palm kernel oil, among others. These residues can be used to produce heat and power, mainly EFB, allowing to satisfy mill requirements and sell electricity surplus to surrounding communities (Shuit et al., 2009; Papong et al., 2004).

However, scale and conversion efficiencies for biomass residues as fuel are still limited compared to fossil fuels (Bram et al., 2005). This makes them not economically attractive or unable to meet all the heat and power energy requirements of the process where it is applied. For this reason, it is usual to combine biomass with other fuels, such as natural gas or coal, in a configuration know as cofiring (Werther et al., 2000). In order to improve economy and efficiency of biomass-fired systems. This configuration has been already used in bioelectricity commercial applications in Finland, the Netherlands, and Belgium (Bram et al., 2005; Riccio and Chiaramonti, 2009).

2 LIGNOCELLULOSIC BIOMASS

For large-scale biological production of fuel ethanol, it is desirable to use cheaper and more abundant substrates. Lignocellulosic biomass is considered as an attractive feedstock fuel ethanol production because of its availability in large quantities at low cost (Cardona and Sánchez, 2007; Cheng et al., 2008) and its reduced competition with food but not necessarily with feed. To introduce ethanol as a large-scale transportation fuel, the production cost must be lowered to about the same level as oil and diesel. Today, the production cost of ethanol from lignocellulosics is still too high, which is the major reason why ethanol from this feedstock has not made its breakthrough yet. When producing ethanol from maize (made up from starch chains) or sugarcane (in the form of either cane juice or molasses), the raw material constitutes about 40-70% of the production cost (Sendelius, 2005; Quintero et al., 2008).

Many lignocellulosic materials have been tested for bioethanol production as was reviewed by Sánchez and Cardona (2008). In general, prospective lignocellulosic materials for fuel ethanol production can be divided into six main groups: crop residues (cane bagasse, corn stover, wheat straw, rice straw, rice hulls, barley straw, sweet sorghum bagasse, olive stones, and pulp), hardwood (aspen, poplar), softwood (pine, spruce), cellulose wastes (newsprint, waste office paper, recycled paper sludge), herbaceous biomass (alfalfa hay, switchgrass, reed canary grass, coastal Bermudagrass, thimothy grass, miscanthus grass), and municipal solid wastes (MSW) (see Table 1). Numerous studies for developing large-scale production of ethanol from lignocellulosics have been carried out in the world.

TABLE 1 Main Potential Lignocellulosic Materials for Fuel Ethanol Production

Raw Material	Pre-Treatment	Ref.
Almond shells	Autohydrolysis and dilute acid hydrolysis	Martinez et al. (1997)
Barley straw	Aqueous/steam fractionation	Belkacemi et al. (2001)
Coffee Cut	Hot liquid water and dilute acid	Quintero et al (2010)
Corncobs	Autohydrolysis	Garrote et al. (2008)
Corn fiber	Acid hydrolysis and hot liquid water	Kim and Lee (2005) and Allen et al. (2001b)
Corn stalks	Aqueous/steam fractionation	Belkacemi et al. (2001)
Empty fruit buches from palm oil	Dilute alkali	Piarpuzán et al (2010)
Pine pulp	Organosolv pretreatment	Kilpeläinen et al. (2007)
Pinus taeda	Dilute acid with different acids (HCl, H_2SO_4, HNO_3, and H_3PO_4).	Marzialetti et al. (2008)
Prosopis juliflora	Dilute acid (Sulfuric acid)	Gupta et al. (2009)
Ragi (*Eleusine coracana*)	Acid hydrolysis	Subba Rao and Muralikrishna (2006)
Rice straw and Rice hulls	Dilute acid and dilute alkali	Sukumaran et al. (2009)
Spent-Sawdust	Thermal dry treatment	Hideno et al. (2008)
Sugarcane bagasse	Dilute acid hydrolysis, dilute alkali and hot water	Quintero et al. (2011), Quintero et al. (2010), Cardona et al (2010), Sukumaran et al. (2009) and Han et al. (1983)
Willow	SO_2 with saturated steam	Von Sivers et al. (1994)
Woody slurry	Hot compressed water	Kobayashi et al. (2009)
Yellow poplar sawdust	Hot liquid water and dilute acid (sulfuric acid)	Allen et al. (2001a)

However, the main limiting factor is the higher degree of complexity inherent to the processing of this feedstock. This is related to the nature and composition of lignocellulosic biomass (which contain up to 75% of cellulose and hemicelluloses). Cellulose and hemicelluloses should be broken down into fermentable sugars in order to be converted into ethanol or other valuable products (xylans, xylitol, hydrogen, and enzymes). But this degradation process is complicated, energy consuming, and incompletely developed (Sánchez and Cardona, 2008). With the advent of modern genetics and other tools, the cost of producing sugars from these recalcitrant fractions and converting them into products such as ethanol can be significantly reduced in the future.

3 PRETREATMENT

3.1 Mechanical Pretreatment

Milling (cutting the lignocellulosic biomass into smaller pieces) is a mechanical pretreatment of the lignocellulosic biomass. The objective of a mechanical pretreatment is a reduction of particle size and crystallinity. The reduction in particle size leads to an increase of available specific surface and a reduction of the degree of polymerization (DP).

The increase in specific surface area, reduction of DP, and the shearing are all factors that increase the total hydrolysis yield of the lignocellulose in most cases by 5-25% (depends on kind of biomass, kind of milling, and duration of the milling), but also reduces the technical digestion time by 23-59% (thus an increase in hydrolysis rate) (Chang and Holtzapple, 2000). As no inhibitors (like furfural and HMF (hydroxymethylfurfural)) are produced, milling is suited for ethanol production. It has, however, a high-energy requirement (Cowling and Kirk, 1976; Pereira Ramos, 2003) and was found therefore not economically feasible as pretreatment. Taking into account the high-energy requirements of milling and the continuous rise of the energy prices, it is likely that milling is still not economically feasible.

3.2 Thermal Pretreatment

During this pretreatment, the lignocellulosic biomass is heated. If the temperature increases above 150-180 °C, parts of the lignocellulosic biomass, firstly the hemicelluloses and shortly after that lignin, will start to solubilize (Bobleter, 1994; Garrote et al., 1999). The composition of the hemicellulose backbone and the branching groups determine the thermal, acid, and alkali stability of the hemicellulose. From the two dominant components of hemicelluloses (xylan and glucomannan), the xylans are thermally the least stable, but the difference with the glucomannans is only small. Above 180 °C, an exothermal reaction (probably solubilization) of the hemicellulose starts (Domansky and Rendos, 1962). This temperature of 180 °C is probably just an indication of the temperature at which an exothermal reaction of the hemicellulose starts, because the thermal reactivity of lignocellulosic biomass depends largely on its composition (Hendriks and Zeeman, 2009).

During thermal processes, a part of the hemicellulose is hydrolyzed and forms acids. These acids are assumed to catalyze the further hydrolysis of the hemicellulose (Gregg and Saddler, 1996). The solubilization of lignin at temperatures above 160 °C produces phenolic compounds which have in many cases an inhibitory or toxic effect on bacteria and yeast

(Gossett et al., 1982). These soluble lignin compounds are very reactive and will, recondensate and precipitate on biomass if they are not removed quickly (Liu and Wyman, 2003).

3.2.1 *Steam Pretreatment/Steam Explosion (ST/SE)*

During steam pretreatment, the biomass is put in a large vessel and steamed at a high temperature, (temperatures up to 240 °C) and pressure is applied for a few minutes. After a set time, the steam is released and the biomass is quickly cooled. The objective of a steam pretreatment/steam explosion is to solubilize the hemicellulose and then to make the cellulose better accessible for enzymatic hydrolysis while formation of inhibitors is avoided. The difference between "steam" pretreatment and "steam explosion" pretreatment is the quick depressurization and cooling down of the biomass at the end of the steam explosion pretreatment, which causes the water in the biomass to "explode." During steam pretreatment, parts of the hemicellulose hydrolyze and form acids, which could catalyze the further hydrolysis of the hemicellulose. However, the role of the acids is probably not to catalyze the solubilization of the hemicellulose, but to catalyze the hydrolysis of the soluble hemicellulose oligomers (Mok and Antal, 1992). During steam pretreatment, the moisture content of the biomass influences the needed pretreatment time. The higher the moisture content, the longer the optimum steam pretreatment times (Brownell et al., 1986). The positive effect of steam pretreatment is mostly due to removal of a large part of the hemicellulose, causing an increase in cellulose fiber reactivity (Laser et al., 2002; Converse et al., 1989; Grohmann et al., 1986).

3.2.2 *Liquid Hot Water (LHW)*

In this case, liquid hot water (LHW) is used instead of steam. The objective of the liquid hot water is to solubilize mainly the hemicellulose to make the cellulose better accessible and to avoid the formation of inhibitors. To avoid the formation of inhibitors, the pH should be kept between 4 and 7 during the pretreatment (Mosier et al., 2005b; Weil et al., 1998). If catalytic degradation of sugars occurs, it results in a series of reactions that are difficult to control and result in undesirable side products. A difference between the LHW and steam pretreatment is the amount and concentration of solubilized products. In a LHW pretreatment, the amount of solubilized products is higher, while the concentration of these products is lower compared to steam pretreatment (Bobleter, 1994). This is probably caused by the higher water input in LHW pretreatment compared to steam pretreatment. The yield of solubilized (monomeric) xylan is generally also higher for LHW pretreatment; though this result diminishes when the solid concentration increases, because (monomeric) xylan is then further degraded by hydrolytic reactions to, xylose and furfural (Laser et al., 2002). At lower concentrations, the risk on degradation products like furfural and the condensation and precipitation of lignin compounds is reduced.

3.3 Chemical Treatment

3.3.1 *Acid Pretreatment*

Pretreatment of lignocellulose with acids at room temperature is done to enhance the anaerobic digestibility. The objective is to solubilize the hemicellulose, and to make the cellulose better accessible. The pretreatment can be done with dilute or strong acids. The main reaction that occurs during acid pretreatment is the hydrolysis of hemicellulose, especially xylan as glucomannan is relatively acid stable. Solubilized hemicelluloses (oligomers) can be subjected to hydrolytic reactions producing monomers, furfural, HMF, and other (volatile) products in acidic environments (Pereira Ramos, 2003). During acid pretreatment, solubilized lignin will quickly condensate and precipitate in acidic environments (Liu and Wyman, 2003; Shevchenko et al., 1999). The solubilization of hemicellulose and precipitation of solubilized lignin are more pronounced during strong acid pretreatment compared to dilute acid pretreatment.

3.3.2 *Alkaline Pretreatment*

During alkaline pretreatment, the first reactions taking place are solvation and saphonication. This causes a swollen state of the biomass and makes it more accessible for enzymes and bacteria. At "strong" alkali concentrations dissolution, "peeling" of end groups, alkaline hydrolysis, and degradation and decomposition of dissolved polysaccharides can take place. Loss of polysaccharides is mainly caused by peeling and hydrolytic reactions. This peeling is an advantage for later conversion, but lower molecular compounds are formed and the risk on degradation and loss of carbon, in the form of carbon dioxide, increases (Hendriks and Zeeman, 2009). An important aspect of alkali pretreatment is that the biomass on itself consumes some of the alkali. The residual alkali concentration after the alkali consumption by the biomass is the alkali concentration left over for the reaction (Gossett et al., 1982). Alkali extraction can also cause solubilization, redistribution, and condensation of lignin and modifications in the crystalline state of the cellulose. These effects can lower or counteract the positive effects of lignin removal and cellulose swelling (Gregg and Saddler, 1996). Alkaline pretreatment causes hemicellulose and parts of lignin to solubilize. The removal of hemicellulose has a positive effect on the degradability of cellulose. There is however often a loss of hemicellulose to degradation products and the solubilized lignin components often have an inhibitory effect. The loss of fermentable sugars and production of inhibitory compounds makes the alkaline pretreatment less attractive for the ethanol production.

3.3.3 *Oxidative Pretreatment*

An oxidative pretreatment consists of the addition of an oxidizing compound, like hydrogen peroxide or peracetic acid, to the biomass, which is suspended in water. The objective is to remove the hemicellulose and lignin to increase the accessibility of the cellulose. During oxidative pretreatment, several reactions can take place, like electrophilic substitution, displacement of side chains, cleavage of alkyl aryl ether linkages or the oxidative cleavage of aromatic nuclei (Hendriks and Zeeman, 2009). In many cases, the used oxidant is not selective and therefore losses of hemicellulose and cellulose can occur. A high risk on the formation of inhibitors exists, as lignin is oxidized and soluble aromatic compounds are formed.

3.4 Combinations

3.4.1 *Thermal Pretreatment in Combination with Acid Pretreatment*

A way to improve the effect of thermal steam or LHW pretreatment is to add an external acid. This addition of an external acid catalyzes the solubilization of the hemicellulose, lowers the optimal pretreatment temperature, and gives a better enzymatic hydrolyzable substrate

(Gregg and Saddler, 1996; Brownell et al., 1986). The lignocellulose is often impregnated (soaked) with SO_2 or H_2SO_4. During steam pretreatment, the SO_2 is converted to H_2SO_4 in the first 20 s of the process; after that, the catalytic hydrolyzation of the hemicellulose starts. Another important point is that gradual removal of hemicellulose and lignin can trigger reorientation of cellulose to a more crystalline form (Gregg and Saddler, 1996).

3.4.2 *Thermal Pretreatment in Combination with Alkaline Pretreatment*

Another way to improve the thermal pretreatment is to add an external alkali instead of an acid to the process. A very common alkaline thermal pretreatment is lime pretreatment. This pretreatment is usually carried out at temperatures of 100-150 °C with lime addition of approximately 0.1 g $Ca(OH)_2$ per g substrate (Chang et al., 2001a; Chang and Holtzapple (2000) attribute the effectiveness of lime pretreatment to the opening of the "acetyl valve" and partly opening the "lignin valve," making the substrate more accessible to hydrolysis. According to Kaar and Holtzapple (2000), lime pretreatment (with heating) is sufficient to increase the digestibility of low-lignin containing biomass, but not for high-lignin containing biomass. Chang et al. (2001a) mention that lime pretreatment of switchgrass and corn stover did not inhibit the enzymatic saccharification and fermentation steps.

3.4.3 *Thermal Pretreatment in Combination with Oxidative Pretreatment*

Wet oxidation is another oxidative pretreatment method, which uses oxygen as oxidator. The soluble sugars produced during wet-oxidation pretreatment are mainly polymers opposite to the monomers produced during steaming or acid hydrolysis as pretreatment. Phenolic monomers are no end products during wet oxidation but are further degraded to carboxylic acids. Also, the production of furfural and HMF can be low during wet oxidation, but part of the hemicellulose can be lost by reaction to carbon dioxide and water (Klinke et al., 2002)

3.4.4 *Thermal Pretreatment in Combination with Alkaline Oxidative Pretreatment*

According to Chang et al. (2001a), thermal lime pretreatment is not capable of removing enough lignin of high-lignin biomass to enhance the enzymatic digestibility, and therefore oxygen as oxidant must be included during the pretreatment. Low sugar degradation can be observed, probably as a result of the relative low temperature of 150 °C, applied during the pretreatment. The enzymatic digestibility of the treated biomass can increase up to 13 times compared to the untreated biomass (Chang et al., 2001b).

3.4.5 *Ammonia and Carbon Dioxide Pretreatment*

Other applied pretreatments are ammonia and carbon dioxide pretreatment. The ammonia pretreatment is conducted with ammonia loadings around 1:1 (kg ammonia/kg dw biomass) at temperatures ranging from ambient temperature with a duration of 10-60 days, to temperatures of up to 120 °C with a duration of several minutes (Kim and Lee, 2005; Alizadeh et al., 2005). Alizadeh et al. (2005) reported a sixfold increased enzymatic hydrolysis yield and a 2.5-fold ethanol yield after pretreatment. Bariska (1975) and Kim and Lee (2005) mention swelling of the cellulose and delignification as the responsible factors for the increased yield.

Carbon dioxide pretreatment is conducted with high-pressure carbon dioxide at high temperatures of up to 200 °C for several minutes. Explosive steam pretreatment with high-pressure carbon dioxide causes the liquid to be acidic and this acid hydrolyzes especially the hemicellulose (Puri and Mamers, 1983). Carbon dioxide is also applied as supercritical carbon dioxide (35 °C, 73 Bars), increasing the glucose yield of bagasse with 50-70% (Zheng et al., 1998), 14% for yellow pine, and 70% for aspen (Kim and Hong, 2001).

4 SACCHARIFICATION

4.1 Enzymatic

Microbial degradation of lignocellulosic waste is accomplished by the action of several enzymes, the most important of which are the cellulases. Three major types of cellulase activities are recognized (Lynd, 1996): (1) Endoglucanases (1,4-β-D-glucanohydrolases), (2) Exoglucanases, and (3) β-Glucosidases (β-glucoside glucohydrolases). Endoglucanases cut at random the internal amorphous sites in the cellulose polysaccharide chain generating oligosaccharides of various lengths, and consequently shorter chains appear. Exoglucanases act, in a progressive manner, on the reducing and non-reducing ends of the cellulose chains liberating either glucose (glucanohydrolases) or cellobiose (cellobiohydrolase) as major products. Exoglucanases can also act on microcrystalline cellulose peeling the chains from the microcrystalline structure (Sheehan and Himmel, 1999). β-Glucosidases hydrolyze soluble cellodextrins and cellobiose to glucose. The cellulase system of *Trichoderma reesei* consists of at least two exoglucanases, five endoglucanases, and two b-glucosidases. In addition to three major groups of cellulase enzymes, there are also a number of ancillary enzymes that attack hemicellulose, such as glucuronidase, acetylesterase, xylanase, β-xylosidase, galactomannanase, and glucomannanase (Duff and Murray, 1996). The enzymatic hydrolysis of lignocellulose is limited by several factors: crystallinity of cellulose, degree of polymerization (DP), moisture, available surface area, and lignin content (Chang and Holtzapple, 2000; Koullas et al., 1992; Laureano-Perez et al., 2005; Puri, 1984). Caulfield and Moore (1974) mentioned that decreasing particle size and increasing available surface rather than crystallinity affect the rate and extent of the hydrolysis. Other researchers (Grethlein, 1985; Grous et al., 1986; Thompson et al., 1992) concluded that the pore size of the substrate in relation to the size of the enzymes is the main limiting factor in the enzymatic hydrolysis of lignocellulosic biomass. Removal of hemicellulose increases the mean pore size of the substrate and therefore increases the probability of the cellulose to get hydrolyzed (Gregg and Saddler, 1996; Grethlein, 1985; Palonen et al., 2004). On the other hand, drying of pretreated lignocellulose can cause a collapse in pore structure, resulting in a decreased enzymatic hydrolyzability (Grous et al., 1986). Zhang and Lynd (2004) mention that cellulases can get trapped in the pores if the internal area is much larger than the external area, which is the case for many lignocellulosic biomasses. Lignin limits the rate and extent of enzymatic hydrolysis by acting as a shield, preventing the digestible parts of the substrate to be hydrolyzed (Chang and Holtzapple, 2000).

4.2 Dilute Acid

The dilute acid hydrolysis process is one of the oldest, simplest, and most efficient methods of producing ethanol from biomass. Dilute acid is used to hydrolyze the biomass to sugars. The first stage uses 0.7% sulfuric acid at 190 °C to hydrolyze the hemicelluloses present in the biomass. The second stage is optimized to yield the more resistant cellulose fraction.

This is achieved by using 0.4% sulfuric acid at 215 °C. The liquid hydrolyzates are then neutralized and toxic compounds are removed before fermentation of sugar solution (Brennan et al., 1986).

5 FERMENTATION

5.1 Microorganisms

Fungi, bacteria, and yeast microorganisms can be used for fermentation, specific yeast (*S. cerevisiae* also known as Baker's yeast) is frequently used to ferment glucose to ethanol. Theoretically, 100 g of glucose will produce 51.4 g of ethanol and 48.8 g of carbon dioxide. However, in practice, the microorganisms use some of the glucose for growth and the actual yield is less than 100%.

5.1.1 *Bacteria*

Ethanol-producing bacteria have attracted much attention in recent years because their growth rate is substantially higher than that of the *Saccharomyces* which is currently used for fuel ethanol production. With the recent advances in biotechnology, they have the potential to play a key role in making production of ethanol more economical (Dien et al., 2003). Among such ethanol-producing bacteria, *Z. mobilis* is a well-known organism used historically in tropical areas to make alcoholic beverages from plant sap (Skotnicki et al., 1983). The advantages of *Z. mobilis* are its high growth rate and specific ethanol production; unfortunately, its fermentable carbohydrates are limited to glucose, fructose, and sucrose. On the other hand, the Gram-negative strain *Zymobacter palmae,* which was isolated by Okamoto et al. (1993) using a broad range of carbohydrate substrates, is a facultative anaerobe that ferments hexoses, α-linked di- and tri-saccharides, and sugar alcohols (fructose, galactose, glucose, mannose, maltose, melibiose, sucrose, raffinose, mannitol, and sorbitol). This strain produces approximately 2 mol of ethanol per mole of glucose without accumulation of byproducts and shows productivity similar to that of *Z. mobilis* (Okamoto et al., 1993).

5.1.2 *Yeasts*

Metabolic pathway engineering is constrained by the thermodynamic and stoichiometric feasibility of enzymatic activities of introduced genes. Engineering of xylose metabolism in *S. cerevisiae* has focused on introducing genes for the initial xylose assimilation steps from *P. stipitis,* a xylose-fermenting yeast, into *S. cerevisiae,* a yeast traditionally used in ethanol production from hexose. However, recombinant *S. cerevisiae* created in several laboratories have used xylose oxidatively rather than in the fermentative manner that this yeast metabolizes glucose (Jin and Jeffries, 2004). D-Xylose is a major component of the hydrolyzate of hemicellulose from biomass. Therefore, ethanol production from xylose is essential for successful utilization of lignocellulose (Jeffries, 1985). Many bacteria, yeast, and fungi assimilate xylose, but only a few metabolize it to ethanol (Skoog and Hahn-Hagerdal, 1988).

Xylose-fermenting yeasts, such as *P. stipitis, Pachysolen tannophilus,* and *Candida shehatae* require precisely regulated oxygenation for maximal ethanol production (Skoog and Hahn-Hagerdal, 1988; Ligthelm et al., 1988) and detoxification of the hydrolyzate because they withstand the inhibitory environment of lignocellulose hydrolyzates poorly (Björling and Lindman, 1989; Hahn-Hägerdal et al., 1994; Sanchez and Bautista, 1988; van Zyl et al., 1991). These factors increase the cost of xylose fermentation. *S. cerevisiae* has an efficient anaerobic sugar metabolism, tolerates inhibitory industrial substrates better than other microorganisms (Olsson et al., 1992; Olsson and Hahn-Hägerdal, 1993), and ferments hexoses abundantly present in lignocellulosic hydrolyzates, such as glucose, mannose, and galactose with high yield and productivity.

5.1.3 *Fungi*

The filamentous fungus *Fusarium oxysporum* is known for its ability to produce ethanol by simultaneous saccharification and fermentation (SSF) of cellulose. However, the conversion rate is low and significant amounts of acetic acid are produced as a byproduct (Panagiotou et al., 2005). A few microbial species such as *Neurospora, Monilia, Paecilomyces,* and *Fusarium* have been reported to hold the ability to ferment cellulose directly to ethanol (Singh et al., 1992). *F. oxysporum* produces a broad range of cellulases and xylanases, which has been characterized earlier (Christakopoulos et al., 1996). Acetic acid was the major fermentation product of *Neocallimastix sp.*, another ethanol-producing fungus (Dijkerman et al., 1997).

5.2 Technological Configurations

The classic configuration employed for fermenting biomass hydrolyzates involves a sequential process where the hydrolysis of cellulose and the fermentation are carried out in different units. This configuration is known as separate hydrolysis and fermentation (SHF). In the alternative variant, the simultaneous saccharification and fermentation (SSF), the hydrolysis and fermentation are performed in a single unit. The most employed microorganism for fermenting lignocellulosic hydrolyzates is *S. cerevisiae,* which ferments the hexoses contained in the hydrolyzate but not the pentoses. Table 2 summarizes main intensification technologies that have been researched for improving fuel ethanol production feasibility.

5.2.1 *Separate Hydrolysis and Fermentation (SHF)*

When sequential process is utilized, solid fraction of pretreated lignocellulosic material undergoes hydrolysis (saccharification). This fraction contains the cellulose in an accessible form to acids or enzymes. Once hydrolysis is completed, the resulting cellulose hydrolyzate is fermented and converted into ethanol. One of the main features of the SHF process is that each step can be performed at its optimal operating conditions. The most important factors to be taken into account for saccharification step are reaction time, temperature, pH, enzyme dosage, and substrate load (Sánchez and Cardona, 2008).

5.2.2 *Simultaneous Saccharification and Fermentation (SSF)*

The SSF process has been extensively studied to reduce the inhibition of end products hydrolysis (Zheng et al., 1998; Saxena et al., 1992). In the process, reducing sugars produced in cellulose hydrolysis or saccharification are simultaneously fermented to ethanol, which greatly reduces the product inhibition to the hydrolysis. However, the need of employing more dilute media to reach suitable rheological properties makes the final product concentration to be low. In addition, this process operates at nonoptimal conditions for hydrolysis and requires higher enzyme dosage, which influences substrate conversion positively, but

TABLE 2 Process Integration Through Reaction-Reaction and Reaction-Separation Processes for Fuel Ethanol Production (Adapted from Cardona and Sánchez, 2007)

Technology	Bioagent	Substrate	Remarks	Ref.
Cofermentation (mixed culture)	*Saccharomyces cerevisiae* mutant + *Pichia stipitis* Respiratory deficient *S. diastaticus* + *P. stipitis*	Glucose and xylose Steam-exploded and enzymatically hydrolyzed aspen wood	Batch and continuous cultures; 100% glucose conversion and 69% xylose conversion. Continuous culture; EtOH conc. 13.5 g/L, yield 0.25 g/g, productivity 1.6 g/(L h); 100% conversion of glucose and xylose	Laplace et al. (1993)
Batch SSF (mixed culture)	*S. cerevisiae* + *Fusarium oxysporum*	Sweet sorghum stalks	Fungus produces cellulases and hemicellulases for hydrolysis process; formed sugars are converted into ethanol by concerted action of both microorganisms; 108-132% yield; EtOH conc. 35-49 g/L.	Mamma et al. (1995, 1996)
Batch SSF	Yeasts + *T. reesei* cellulases supplemented with β-glucosidase	Pretreated lignocellulosic biomass	3-7 d of cultivation; EtOH conc. 40-50 g/L for *S. cerevisiae*, 16-19 g/L for *Kluyveromyces marxianus*; 90-96% substrate conversion.	Ballesteros et al. (2004)
Semicontinuous SSF	*S. cerevisiae* + commercial cellulase supplemented with β-glucosidase	Paper sludge	Special design of solids-fed reactor; EtOH conc. 35-50 g/L; 0.466 g/g EtOH yield; 74-92% cellulose conversion; 1-4 months of operation	Fan et al. (2003)
Continuous SSF	*S. cerevisiae* + commercial cellulase supplemented with β-glucosidase	Dilute-acid pretreated hardwood	CSTR; residence time 2-3 d; 83% conversion; EtOH conc. 20.6 g/L	South et al. (1993)
Batch SSCF	Recombinant *Z. mobilis* + *T. reesei* cellulases	Dilute-acid pretreated yellow poplar	EtOH produced 17.6-32.2 g/L; yield 0.39 g/g; productivity 0.11-0.19 g/(L h)	McMillan et al. (1999)
Batch extractive cofermentation	*Z. mobilis*/*n*-dodecanol	Glucose and xylose	Modeling based on kinetic approach and liquid-liquid equilibrium; solvent is regenerated by flashing; productivity 2.2-3.0 g/(L h); solvent volume/aqueous volume ratio 1.33-3.0	Gutiérrez et al. (2005)
Continuous SSCF	Recombinant *Z. mobilis* + *T. reesei* cellulases	Dilute acid pretreated wood chips	Cascade of reactors; 92% glucose conversion, 85% xylose conversion	Wooley et al. (1999)

TABLE 2 Process Integration Through Reaction-Reaction and Reaction-Separation Processes for Fuel Ethanol Production (Adapted from Cardona and Sánchez, 2007)—Cont'd

Technology	Bioagent	Substrate	Remarks	Ref.
Continuous fermentation coupled with liquid-liquid extraction	Immobilized yeast/ n-dodecanol	Glucose-containing medium	18 d of operation; use of very concentrated feedstocks (10-48% w/w); 78% reduction of aqueous effluents	Gyamerah and Glover (1996)
Continuous extractive Fermentation	Immobilized *S. cerevisiae*/ n-dodecanol	Glucose	Pneumatically pulsed packed reactor; flowrates: solvent 1-2.55 L/h, medium 0.057-0.073 L/h; feed glucose conc. 261-409 g/L; EtOH conc. in solvent 3.37-10 g/L, in broth 9.4-33 g/L; yield 0.51; productivity 1.03 g/(L h)	Minier and Goma (1982)
	Clostridium thermohydrosulfuricum/ oleyl alcohol	Glucose	Flowrates: broth 0.15-0.55 L/h, solvent 0-18 L/h; feed glucose conc. 12.5-100 g/L; EtOH conc. in the broth <4.47 g/L, in the re-extraction water 3-14 g/L; 65 °C; productivity <0.128 g/(L h)	Weilnhammer and Blass (1994)
Fed-batch SSEF	*S. cerevisiae*/commercial cellulases/oleyl alcohol	Primary clarifier sludge from chemical pulping process/ cellulose	Reactor with up to 2.5% aqueous phase; 50% substrate conversion; 48-275 h cultivation; 65% increase in productivity compared to conventional fed-batch process	Moritz and Duff (1996)

EtOH conc., ethanol concentration at the end of batch culture or in the effluent for continuous processes; SSF, simultaneous saccharification and fermentation; SSCF, simultaneous saccharification and cofermentation; SSEF, simultaneous saccharification and extractive fermentation.

process costs negatively. Considering that enzymes account for an important part of production costs, it is necessary to find methods reducing the cellulases doses to be utilized (Sánchez and Cardona, 2008).

The microorganisms used in the SSF are usually the fungus *Trichoderma reesei* and *S. cerevisiae*. Hydrolysis is usually the rate-limiting process in SSF (Philippidis and Smith, 1995). Thermotolerant yeasts and bacteria have been used in the SSF to raise the temperature close to the optimal hydrolysis temperature. Ballesteros et al. (1991) have identified *Kluyveromyces marxianus* and *K. fragilis* that have the highest ethanol productivity at 42 °C from a number of yeast strains. *K. marxianus* has an ethanol yield of 0.5 g/g cellulose in 78 h using Solka Floc 200 as substrate at 42 °C (Ballesteros et al., 1991). SSF has some advantages such as increase of hydrolysis rate by conversion of sugars that inhibit the cellulase activity, lower enzyme requirement, higher product yield, lower requirements for sterile conditions since glucose is removed immediately and ethanol is produced, shorter process time and less reactor volume because a single reactor is used. However, ethanol may also exhibit inhibition to the cellulase activity in the SSF process. Some disadvantages of this configuration include incompatible temperature of hydrolysis and fermentation, ethanol tolerance of microbes, and inhibition of enzymes by ethanol.

5.2.3 *Pentoses Fermentation*

One of the main problems in bioethanol production from lignocellulosics is that *S. cerevisiae* can ferment only certain mono- and disaccharides like glucose, fructose, maltose, and sucrose. This microorganism is not able to assimilate cellulose and hemicellulose directly. In addition, pentoses obtained during hemicellulose hydrolysis (mainly xylose) cannot be assimilated by this yeast. A way to overcome this obstacle is through recombinant DNA technology (genetic engineering). Other approach to this problem is the use of pentose-fermenting microorganisms such as some species of yeasts and bacteria. In this case, configurations involving the separate fermentation of pentoses and hexoses have been proposed. Yeasts such as *Pichia stipitis*, *Candida shehatae*, and *Pachysolen tannophilus* can assimilate pentoses, but their ethanol production rate from glucose is at least five times less than that observed for *S. cerevisiae*. Moreover, their culture requires oxygen and ethanol tolerance is 2-4 times lower (Claassen et al., 1999). Pentose-fermenting yeasts require a careful control for maintaining low oxygen levels in the culture medium needed for their oxidative metabolism. Additionally, these yeasts successfully ferment pure xylose but not the aqueous hemicellulose streams generated during the biomass pretreatment, due to the presence of different inhibitors (Chandrakant and Bisaria, 1998).

5.2.4 *Simultaneous Saccharification and Cofermentation*

Other promising integration alternative is the inclusion of the pentose fermentation in the SSF, process called simultaneous saccharification and cofermentation (SSCF) (Cardona and Sánchez, 2007). In an initial stage, the cofermentation of mixed cultures was studied. For example, the coculture of *P. stipitis* and *Brettanomyces clausennii* has been utilized for the SSCF of aspen at 38 °C and pH of 4.8 yielding 369 L of ethanol per ton of aspen during 48-h batch process, as reported by Olsson and Hahn-Hägerdal (1996). In this configuration, it is necessary that both fermenting microorganisms be compatible in terms of operating pH and temperature. Chandrakant and Bisaria (1998) suggest that a combination of *C. shehatae* and *S. cerevisiae* is suitable for this kind of process. Similarly, a system including the isomerization of xylose and the fermentation with *S. cerevisiae* in a simultaneous way can be utilized. This system has been proven in nonpretreated spent sulfite liquor and in pretreated acid-hydrolyzed wheat straw (Linden and Hahn-Hagerdal, 1989) Some drawbacks of this configuration are the high byproduct formation in the form of CO_2 and xylitol, poor enzyme stability, incompatible pH and temperature (pH of 7.0 and 70 °C for the isomerization process), and the reversibility of the enzyme transformation (Chandrakant and Bisaria, 1998).

5.2.5 *Simultaneous Reaction and Separation*

The aforementioned types of integration allow the increase of process efficiency through the improvement of reaction processes. However, separation is the step where major costs are generated in process industry. Therefore, reaction-separation integration could have the highest impact on the overall process in comparison with homogeneous integration of processes (reaction-reaction, separation-separation) (Cardona and Sánchez, 2007; Cardona et al., 2009). This has been demonstrated in the case of reactive distillation processes (Pisarenko et al., 2001), particularly, in the lactic acid recovery by reactive distillation (Cardona et al., 2004), and in the production of allyl alcohol by allyl acetate butanolysis (Anokhina et al., 1996a,b).

One of the most promising technological configurations for fuel ethanol production from lignocellulosics is the system in which the removal of the product that causes the inhibition is accomplished through an extractive biocompatible agent (solvent). Migration of ethanol to solvent phase is known as extractive fermentation process. Minier and Goma (1982) showed that primary aliphatic alcohols with a chain length having less than twelve carbons inhibit the growth of yeast cells. They chose the fatty alcohol n-dodecanol as a solvent for *in situ* extraction of ethanol in an especially continuous pulse-packed column with immobilized cells of *S. cerevisiae*. This configuration allowed the utilization of very concentrated glucose feed due to the reduction of ethanol in the culture broth. In addition, immobilization seems to protect the cells against solvent toxicity (Aires Barros et al., 1987). Gyamerah and Glover (1996) implemented a process where the fermentation stage was coupled with an apparatus for liquid-liquid extraction in a continuous regime at pilot-scale level. They used n-dodecanol because of its very low toxicity for ethanol-producing microorganisms. However, this solvent has some drawbacks: it tends to form a stable emulsion with the culture broth, its melting point is relatively high (26 °C) considering fermentation conditions, and its distribution coefficient related to water is not very high (Kollerup and Daugulis, 1985). In addition, Kirbaşlar et al. (2001) experimentally showed that small amounts of water migrate to n-dodecanol in water-ethanol-n-dodecanol ternary systems. Weilnhammer and Blass (1994) proposed a simple model based on the mass balance of different components for the description of extractive fermentation with *C. thermohydrosulfuricum* using oleyl alcohol as a solvent; this model allowed the evaluation of the economy of the process with and without solvent based on production costs. Gutiérrez et al. (2005) modeled the batch extractive cofermentation from pretreated hydrolyzed lignocellulosic biomass coupling the equations representing the kinetics of the biological process with the equations describing the liquid-liquid equilibrium using different activity models. Liquid medium from this bioreactor is continuously removed in order to separate the cells and carry out the decantation of both phases. Aqueous phase is recycled back to the bioreactor and ethanol-rich solvent phase is flashed for the regeneration of solvent and the production of almost pure ethanol. L'Italien et al. (1989) proposed and tested a regime of fermentation using as a solvent supercritical carbon dioxide; for this, it was necessary to organize a cyclic process with periods of high-cell atmospheric fermentation followed by a period of hyperbaric conditions (7 MPa) for the rapid extraction of ethanol by CO_2; however, the complexity of the process and the loss of viability of the cells during prolonged intervals under high pressure makes this technology nonviable to date.

For the extractive fermentation process, the presence of microbial cells can reduce the rate of ethanol extraction. Crabbe et al. (1986) found that the yeast cells severely decreased the rate of extraction employing n-decanol as a solvent; the authors assumed that the studied effects can be extrapolated to n-dodecanol.

6 ENERGY PRODUCTION USING AGROINDUSTRIAL RESIDUES

Any ethanol process based on biomass should be compared to direct combustion for electricity or vapor. Last is the most used alternative in the world for lignocellulosic residues (Pandey et al., 2000a,b; Cardona et al., 2010). For example, industry prefers the sugarcane bagasse as a bioenergy source for cogeneration instead of alcohol (even if different studies consider this raw material as one of the best possibilities for second-generation alcohol). The maturity of alcohol technology from biomass is still low, and the profits from cogeneration are high. From this point of view, close to pretreatment and logistics problems, the real challenge for biomass ethanol production is the alternative use for cogeneration in many industries. Hereinafter, the conceptual elements for understanding the advantages or disadvantages of bioelectricity are discussed.

6.1 Bioelectricity Production

Bioelectricity production is a well-established technology, operating with a variety of feedstocks, able to replace coal-based thermoelectric stations that only convert about one-third of fuel energy into electricity. Usually in the last case, the remaining heat is lost, being a wasteful way to produce energy. This energy production technology is useful, where high investment costs for rural electrification, related to a construction and operation of long electricity grids connecting areas, are needed. In some cases, these costs make it impossible to bring electricity to these communities. So, local electric generation in decentralized plants represents a good alternative to provide energy for rural areas with difficult access. In fact, world economies such as United States, Brazil, Malaysia, Cuba use wood and agricultural residues for power generation in the electricity sector. For instance, the Energy Information Administration (EIA) from United States projects that biomass residues will generate 0.3% of the 5476 billion of kWh of electricity in this country (Haq, 2010). Among main agroindustrial residues used in bioelectricity production, can be counted, rice hulls (Papong et al., 2004), sugar cane bagasse (Papong et al., 2004; Coelho et al., 2000), as well as oil palm residues (Shuit et al., 2009; Papong et al., 2004). Bioelectricity is conventionally produced using a scheme of a boiler steam turbine (BST). Where biomass is burned on a grate or furnace, either fixed, moving, or fluidized, reacting with air, in different reaction schemes, with high reaction rates and high released heat, converting chemical energy stored in biomass into usable energy, such as mechanical power or electricity (Iakovou et al., 2010). For combustion or power generation, some important fuel properties can change according to chemical composition, ashes, moisture content (Grassi and Allan, 2007), and Low Heating Value (LHV) of biomass. In this way, combustion efficiencies can range 65-99% (Susta et al., 2003). Combustion is controlled with air flow rate, fuel flow rate, or a combination of both (Balat et al., 2009), producing hot gases around 1000-1500°K. Additionally, when biomass water content is >30%, a pretreatment process is required, using drying and milling processes. Heat produced during direct combustion in boilers is used to produce superheated steam, which moves the blades of a steam turbine generating electricity. Steam Turbine is one of the oldest machines, used to convert heat energy into useful mechanical or electrical energy by means of a steam flowing over blades, causing rotating motion and impelling the electricity generator. Quantum of the generated power is a function of two main factors: Steam flow and pressure drop through the turbine. These two parameters determine how much power can be produced with this device, for instance, high flow and pressure steam generated give high potential for power generation.

6.2 Biomass-Fired Cogeneration

Despite boiler steam turbine (BST) being a well-established technology, it still is a relatively high capital cost and low operating efficiency process with little capacity to be improved (Hall, 1997). For this reason, bioelectricity production must be modernized, fitting a sustainable path, where advanced conversion and energy utilization concepts can be employed (Haq, 2010). In this sense, biomass-fired cogeneration emerged as a thermodynamically efficient way of energy use, aimed both to satisfy heat and power local requirements or selling surplus electricity to public grid (Reith et al., 2002). Combined production of mechanical and thermal energy has remarkable cost and energy savings, operating also with a greater efficiency compared to systems which produce heat and power separately, converting four-fifths of fuel into usable energy. The use of a cogeneration configuration improves the energy utilization with economy and energy benefits (Prasad, 1995; Gupta et al., 2001; Tsuji et al., 2003; Biezma and Cristóbal, 2006; Uddin and Barreto, 2007).

Among fuels used in cogeneration that can be counted natural gas, biogas, coal, vegetable oil, and biomass. Preferential use of a fuel changes according to its price, availability, and technology. When biomass or its residues are used as fuel in cogeneration plants, this configuration is called biomass-fired cogeneration. Most of this type of plants are allocated attached to industrial sites. This guarantees a continuous supply of raw material avoiding also the logistic problem of gathering, transporting, and storing residues. Some examples are found in sugar and/or ethanol plants and paper mills (Haq, 2010; UN Foundation, 2007). Current available cogeneration technologies can be distinguished, according to its prime mover or its commercial use, into two main groups: (i) Highly commercially used, such as Steam turbine (ST) cogeneration system, Gas turbine (GT) cogeneration system, Combined cycle gas turbine (CCGT) cogeneration system, Internal Combustion Engine cogeneration system, Open cycle gas turbine (OCGT), and Steam Engine cogeneration system. (ii) Less commercially used Integrated Gasification Combined Cycle (IGCC), Microturbines, and Fuel Cells (Biezma and Cristóbal, 2006; Badami et al., 2008; Sanjay and Prasad, 2009). Combined cycle systems are considered as the most energy efficient among cogeneration plants.

6.3 Biomass-Integrated Gasification Combined Cycle (BIGCC)

BIGCC is the latest technology for energy conversion from biomass residues (Balat et al., 2009), holding the promise of an efficient, clean, and cost-effective power generation from biomass. Reducing investment costs, compared with conventional combustion and power generation in a steam Rankine cycle used in bioelectricity production (Reith et al., 2002; Gassner and Maréchal, 2010). In the last decade, different biomass gasification cogeneration projects have been initiated, such as Recovered Energy Resources project and United Technologies Research Center in United States of America, Skydraft in Sweden, ARBRE in United Kingdom, among others (HM Associates Inc, 2003; Resource Dynamics Corporation, 2004). Basic elements of BIGCC system include a biomass dryer, a gasifier chamber for converting biomass into a hot gas, a gas turbine generator fueled by combustion of the biomass-derived gas, a heat steam recovery generator (HRSG) to produce steam at different pressure levels with hot gas exhausted from gas turbine, and a steam turbine generator to produce additional energy (Larson et al., 2001). Main elements of BIGCC, Figure 1, will be explained as follows.

6.3.1 *Gasifier Chamber*

The Gasification is a thermochemical conversion technology of carbonaceous materials (coal, petroleum coke, and biomass) into a mixture of gaseous products (CO, CO_2, H_2O, H_2, CH_4) and small amounts of char and ash. Gas properties and composition change according to gasifying agent used (air, steam, steam-oxygen, oxygen-enriched air), gasification process, and biomass properties (Balat et al., 2009). Produced gas with this method is useful in a broader range of applications, including direct burning to produce heat and power

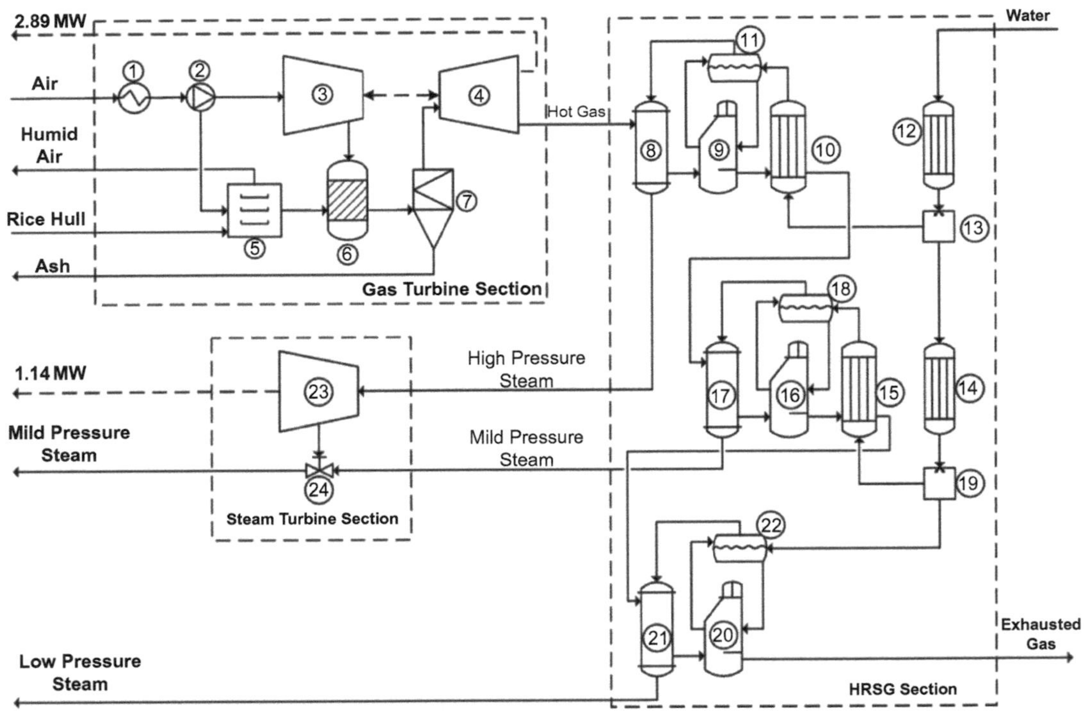

FIGURE 1 Flowsheet of Rice Hull fueled BIGCC plant. 1. Direct Use Heat Exchanger, 2. Air divisor, 3. Compressor, 4. Turbine, 5. Dryer, 6. Gasification and combustion chamber, 7. Cyclone, 8. High-Pressure Superheater, 9. High-Pressure Evaporator, 10. High-pressure Economizer, 11. High-pressure Drum, 12. Low-pressure Economizer1, 13. Flow Divisor, 14. Low-pressure Economizer2, 15. Intermediate-pressure Economizer, 16. Intermediate-pressure Evaporator, 17. Intermediate-pressure Superheater, 18. Intermediate-pressure Drum, 19. Flow Divisor, 20. Low-pressure Evaporator, 21. Low-pressure Superheater, 22. Low-pressure Drum, 23. Steam Turbine, 24. Flow Mixer.

or production of high-quality fuels or chemical products such as hydrogen or methanol (Faaij, 2006; UN Foundation, 2007). Gasification temperature usually ranges between 875 and 1275 K (Susta et al., 2003; Iakovou et al., 2010).

6.3.2 Gas Turbine

A gas turbine is a rotator engine that extracts energy from a flow combustion gas, it is able to produce power with an acceptable electrical efficiency, low emission and high reliability, composed of three main sections: compression (air pressure is increased, aimed to increase combustion efficiency), combustion and/or gasification (adiabatic reaction of air and fuel to convert chemical energy to heat), and expansion (obtained pressurized hot gas at high speed passing through a turbine generating mechanical work) (Badami et al., 2008; O'Brien and Bansal, 2000a; Branan, 2002). When a gas turbine operates using biomass feedstock, a previous thermochemical conversion process is required to convert biomass into a solid liquid or gaseous fuel (e.g., gasification, pyrolysis, charcoal among others) (Iakovou et al., 2010), different thermochemical process generates different configurations alternative for gas turbines: (i) pressurized gasification and direct combustion; (ii) pyrolysis and combustion of liquid fuel; (iii) direct combustion of converted biomass in biofuels (biogas, bioethanol, vegetable oil, or biodiesel); (iv) biomass pulverization and direct combustion and (v) indirect pressurized combustion of biomass and using hot gases for the turbine motion, replacing also the combustion chamber by a heat exchange. This configuration is also known as externally fired gas turbine (EFGT) (Bram et al., 2005; Prasad, 1995); here, any contact between the gas turbine internals and the biomass is avoided. In this way, the critical issue of severe gas cleaning required in internal combustion biomass systems is evaded (Bram et al., 2005).

6.3.3 Heat Recovery Steam Generator (HRSG)

Heat recovery steam generator is a high-efficiency steam boiler that uses hot gases from a gas turbine for reciprocating engine to generate steam in a thermodynamic Rankine Cycle. This system is able to generate steam at different pressure levels according to chemical process requirements (PGTHERMAL, 2009). HSRG system can use single, double, or even triple pressure levels, and choosing one configuration over other relies on process requirements. Furthermore, using high-pressure steam is an economical way to perform energy transportation (Branan, 2002). The HSRG system is integrated by the following units: Economizer, where cold water exits as saturated liquid; Evaporator, where saturated liquid is converted to steam and Superheater, where saturated steam is dried by overheating it beyond its saturation point (Sanjay and Prasad, 2009; PGTHERMAL, 2009; Jacek Topolski, 2010).

7 CASE STUDIES

In Colombia, 2.5 million of tons of paddy rice are produced per year, generating 500,000 tons of rice hulls. This waste is currently employed as floor covering in farms, for moisture retention in crops, and drying grains in furnaces. However, most of this residue remains unused becoming an environmental management problem (Ramírez et al., 2007). Because of that, use of rice hulls as fuel in biomass-fired cogeneration plants and for ethanol production has emerged as valuable alternative in Colombian energy sector, where energy from agroindustrial residues is attractive to provide electricity to rural sectors excluded from national central electric network.

7.1 Case Study 1: Ethanol Production from Rice Hulls

7.1.1 Process Description

Ethanol process scheme used in simulations is shown in Figure 2. Simulated pretreatment technologies were Liquid Hot Water (LHW) and Dilute Acid (DA). Both technologies present the same configuration with differences only in the acid additions and first hydrolysis conditions. Ethanol production process comprises five stages: pretreatment (first hydrolysis), saccharification (second hydrolysis), detoxification, fermentation, and separation. First operation unit is the crusher, which is used for size reduction. The final particle diameter expected in this unit is 1 mm. After milling, the raw material is subjected to acid hydrolysis (at 170 °C during 10 min) or liquid hot water pretreatment (at 220 °C during 10 min) which allows the conversion of hemicellulose into pentoses (mainly xylose). Furfural and HMF were considered as byproducts of this operation.

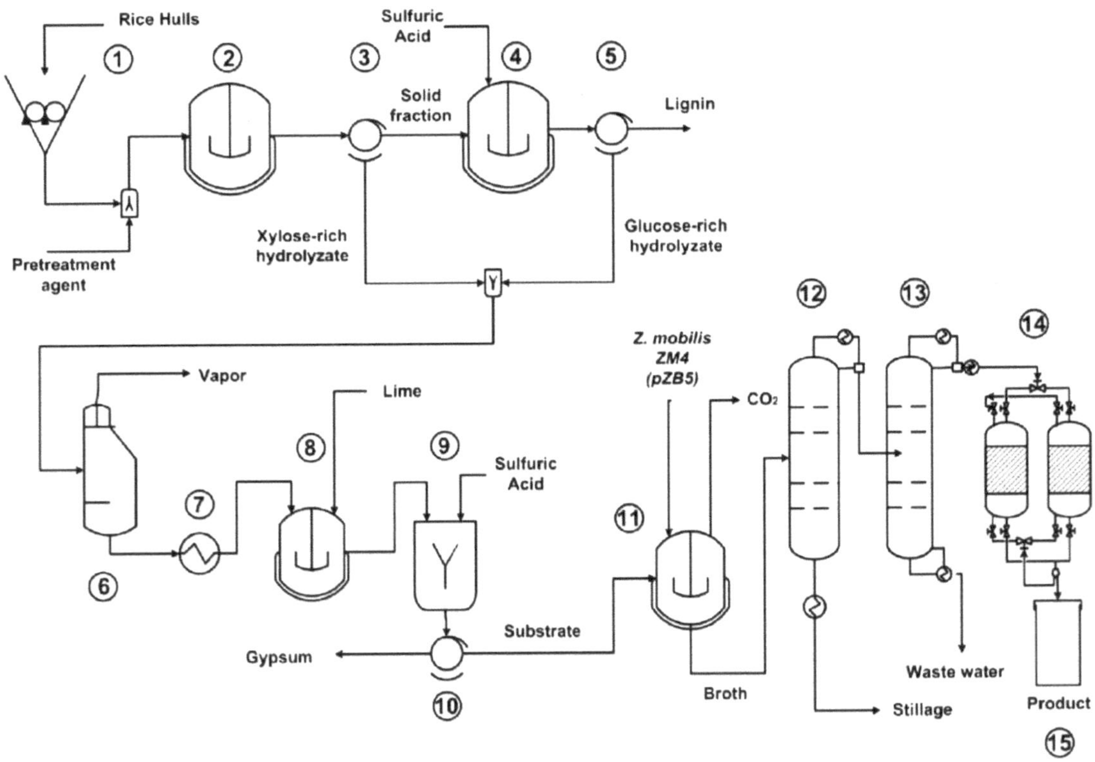

FIGURE 2 Flowsheet of fuel ethanol production from rice hulls using either LHW or DA pretreatment. 1. Crusher, 2. First stage hydrolysis, 3. Filter for separation of nonhydrolyzed fiber, 4. Second stage hydrolysis, 5. Filter for lignin separation, 6. Evaporator for sugars concentration, 7. Cooler, 8. Detoxificattion reactor, 9. Neutralization reactor, 10. Filter for gypsum separation, 11. Fermenter, 12. Concentration column, 13. Rectification column, 14. Molecular sieves for ethanol dehydration, 15. Product tank

After pretreatment, the solid fraction (cellulose and lignin) is separated from the reaction solution (hydrolyzate) and sent to the saccharification stage, where cellulose conversion is accomplished at 200 °C during 3 min. Liquid fractions from both hydrolyses are sent to the sugars concentration and detoxification. The aim of the concentration and detoxification is conditioning the substrate for fermentation. In detoxification, the liquid fraction is treated with calcium hydroxide at 60 °C during 30 min, in order to reduce the concentration of furfural and HMF which are inhibitory compounds for further fermentation. Then, the stream out of the detoxification is neutralized with sulfuric acid. The main stage of the whole process of ethanol production is the fermentation. At this point, the sugars streams coming from the previous stages are converted into ethanol by a recombinant bacteria *Zymomonas mobilis* ZM4 (pZB5) at 33 °C during 30 h. Produced ethanol is separated from the broth by continuous distillation and further rectification. Finally, molecular sieves columns are used for ethanol dehydration.

7.1.2 Simulation Procedure

The simulation of the technological configurations (ethanol production from rice hulls using DA and LHW) was carried out using Aspen plus (Aspen Technology Inc., USA). For the two processes, the simulation was started considering a plant capacity of about 100,000 L/day of anhydrous ethanol, and the required raw material was calculated using reported yields. Part of the physical property data of the components required for simulations were obtained from Wooley and Putsche (1996a, 1996b). The nonrandom two-liquid (NRTL) thermodynamic model was utilized to calculate the activity coefficients in the liquid phase and the Hayden-O'Conell equation of state was used to model the vapor phase.

The estimation of energy consumption was conducted based on the simulation data of thermal energy required by the heat exchangers, reboilers, and related units. DA and LHW were simulated using batch reactor with kinetic expressions. Detoxification and sugars fermentation were simulated including user subroutines by an Excel-Matlab interface for solving the mathematical model. The enzymatic hydrolysis was simulated based on a stoichiometric approach that considered the conversion of cellulose into glucose without kinetic models. The economic analysis was performed by using the Aspen Icarus Process Evaluator (Aspen Technology, Inc., USA) package. This analysis was estimated in US dollars for a 10-year period at an annual interest rate of 16.02% (typical for the Colombian economy), considering the straight line depreciation method and a 33% income tax. Prices and economic data used in this analysis correspond to Colombian conditions and were calculated at an exchange rate of 1950 Colombian pesos per US dollar. Rice hulls price of US$5/ton was estimated considering only the transport cost involved in this type of materials. Operator and supervisor labor costs were US$2.14/h and US$4.29/h, respectively. Electricity, potable water, and low steam pressure costs were US$0.03044/kWh, US$1.252/m^3, and US$8.18/ton. The aforementioned software estimates the capital costs of process units as well as the operating costs, among other valuable data, utilizing the design information provided by Aspen Plus and data introduced by the user for specific conditions as for example project location.

7.1.3 Simulation Results

Some simulations results of main streams for ethanol production from rice hulls are shown in Table 3. Both pretreatments achieved a complete hemicellulose conversion, but only with DA a significant cellulose conversion (31.20%) was accomplished. Due the cellulose

TABLE 3 Total Flow Rates and Compositions of Some Streams of Ethanol Production from Rice Hulls

	Raw Material	Dilute Acid Pretreatment				Liquid Hot Water Pretreatment			
Compounds	*Rice Hulls (wt%)*	*Xylose-Rich Hydrolyzate (wt%)*	*Glucose-Rich Hydrolyzate (wt%)*	*Substrate (wt%)*	*Product (wt%)*	*Xylose-Rich Hydrolyzate (wt%)*	*Glucose-Rich Hydrolyzate (wt%)*	*Substrate (wt%)*	*Product (wt%)*
Water	10.20	95.18	92.47	87.93	0.40	95.79	91.50	87.92	0.40
Cellulose	29.00	–	–	–	–	–	–	–	–
Hemicellulose	26.94	–	–	–	–	–	–	–	–
Lignin	15.00	–	–	–	–	–	–	–	–
Glucose	–	1.12	5.56	5.08	–	0.03	6.33	4.63	–
Xylose	–	3.40	0.06	5.95	–	4.05	0.07	6.06	–
Protein	–	–	–	–	–	–	–	–	–
Ash	16.61	–	–	–	–	–	–	–	–
Extractives	2.25	0.18	–	0.18	–	0.13	–	0.10	–
Furfural	–	0.01	–	–	–	Traces	–	–	–
HMF	–	–	1.43	0.80	–	–	1.63	1.18	–
Ethanol	–	–	–	–	99.60	–	–	–	99.60
Sulfuric acid	–	0.11	0.48	0.02	–	–	0.47	0.07	–
Calcium hydroxide	–	–	–	0.04	–	–	–	0.04	–
Total flow (kg/h)	10,000.00	82,287.65	28,967.74	51,327.46	2,739.84	75,093.72	36,663.31	50,528.91	2,673.51

conversion at pretreatment stage, DA showed higher sugars yield (40.61%) than LHW (30.87%). Sugars concentration in the substrate stream influenced the ethanol yield. In this way, when total sugars where concentrated up to 112 g/L (unit 6 in Figure 2), ethanol yield was 347.25 and 338.85 L/ton for DA and LHW, respectively. On the other hand, when the concentration unit was not included (total sugar concentration equal to 37 g/L), lower ethanol yields were obtained, 319.26 and 286.33 L/ton, for DA and LHW, respectively. Obtained yields from simulation were higher than that reported (236 L/ton) for ethanol from sugarcane bagasse (Botha and von Blottnitz, 2006). Plant capacity for ethanol production from rice hulls was around 80,000 L/day when sugar concentration was included, and 70,000 L/day with no concentration.

Higher furfural concentration (0.0044 g/L) was obtained when DA was used as pretreatment, due to the higher acids concentration. Sugars concentration helped in furfural and HMF reduction, because some amounts of these compounds are volatilized during this operation. When sugars concentration was not included, the higher furfural concentration was of 0.0161 g/L for DA. Although the furfural concentration was low compared to experimental values (0.7 g/L) reported by Laser et al. (2002) for SCB pretreated with LHW, the detoxification step was accomplished for furfural withdrawal. Detoxification extent for four evaluated cases was around 98%.

An important parameter involved in ethanol production from lignocellulosics is the energy consumption; this has been always one of the main disadvantages of this type of process. Table 4 shows the energy consumption of the evaluated cases, and Figure 3 shows the energy consumption by process stages. Highest energy consumptions were obtained when sugars concentrations were included. When pretreatment technologies were compared, LHW exhibited higher energy consumption than DA. For both evaluated pretreatment technologies, it was found that pretreatment and sugars concentration require most of the total energy of ethanol production, and separation is relevant only when sugars concentration was not considered.

Ethanol production cost using rice hulls for both evaluated pretreatment technologies is shown in Table 5. The highest production cost was obtained for DA with sugars concentration. Obtained costs are higher than those reported by Luo et al. (2009) (US$0.2051/L) using sugarcane bagasse for ethanol production and energy cogeneration from wastes.

TABLE 4 Energy Consumption of Ethanol Production Process from Rice Hulls

Pretreatment	Energy Consumption (MJ/L)
DA-SC[a]	86.75
DA[b]	46.29
LHW-SC[c]	96.75
LHW[d]	62.24

[a] *Dilute acid pretreatment with sugar concentration.*
[b] *Dilute acid pretreatment without sugar concentration.*
[c] *Liquid hot water pretreatment with sugar concentration.*
[d] *Liquid hot water pretreatment without sugar concentration.*

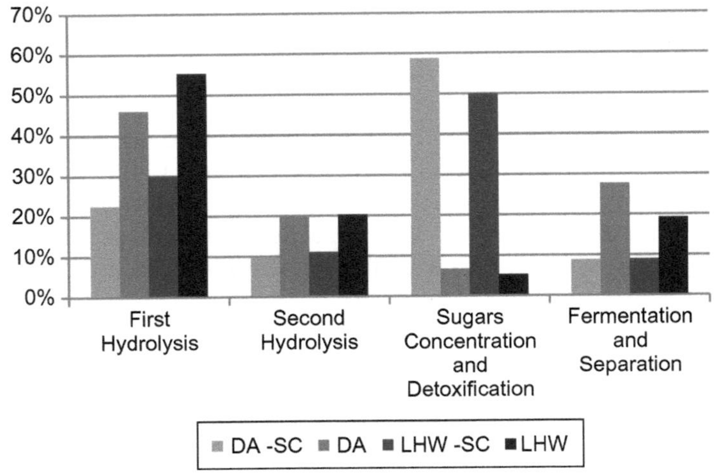

FIGURE 3 Energy requirements per stage in fuel ethanol production from rice hulls. DA-SC: Dilute acid pretreatment with sugar concentration. DA: Dilute acid pretreatment without sugar concentration. LHW-SC: Liquid hot water pretreatment with sugar concentration. LHW: Liquid hot water pretreatment without sugar concentration.

TABLE 5 Fuel Ethanol Production Cost from Rice Husk

Category	DA-SC[a] *US$/L*	DA[b] *US$/L*	LHW-SC[c] *US$/L*	LHW[d] *US$/L*
Raw materials[e]	0.0382	0.0416	0.0398	0.0490
Utilities[f]	0.2548	0.0823	0.2482	0.0728
Labor	0.0049	0.0047	0.0051	0.0052
Maintenance	0.0196	0.0385	0.0192	0.0244
Operating charges	0.0012	0.0012	0.0013	0.0013
Indirect plant expenses	0.0123	0.0216	0.0121	0.0148
General and administrative costs	0.0265	0.0152	0.0261	0.0134
Capital depreciation[g]	0.0843	0.1218	0.0849	0.0987
Total	0.4418	0.3269	0.4367	0.2797

[a] *Dilute acid pretreatment with sugars concentration.*
[b] *Dilute acid pretreatment without sugars concentration*
[c] *Liquid hot water pretreatment with sugars concentration*
[d] *Liquid hot water pretreatment without sugars concentration*
[e] *Price of rice hull US$ 0.005 kg^{-1}.*
[f] *Price of low-pressure steam US$ 8.18 ton^{-1}.*
[g] *Calculated using the straight line method.*

High ethanol production cost in the cases where sugars were concentrated is due to the high utilities cost that represents more than 56% of the total production cost. High utilities cost is a consequence of the high energy consumption of the pretreatment and sugars concentration steps (see Figure 3). When sugars concentration was not included, capital depreciation was the greatest contributing factor. As was suggested by Luo et al. (2009) and Cardona et al. (2010), energy cogeneration is required for energy supply of this type of process. In this sense, energy cogeneration is necessary for obtaining a lower production cost.

7.2 Study Case 2: Biomass-Integrated Gasification Combined Cycle (BIGCC) Fueled with Rice Hulls

7.2.1 Simulation Procedure

Simulation of Biomass-integrated gasification combined cycle (BIGCC) was carried out using Aspen Plus 2006.5 ® Software. (Aspen Technologies Inc., USA) All individual sections (Gas turbine, HSRG, and Steam Turbine) were modeled using a hierarchy block allowing to organize complex flowsheets in a hierarchical manner, containing simulation objects, such as streams, unit operation blocks, among others. The hierarchy appears as a single block on the flowsheet containing their own units and is interconnected with material streams. This simulation procedure facilitates a good global process understanding of interaction among its different sections. Biomass-integrated gasification combined cycle (BIGCC) system is designed using a rice hulls flow of 10,005 Kg/h, analyzing heat and power production capacity as well as its thermal efficiency, using mass and energy balances estimated via simulation. Considering that most of components are gases in this simulation and high-pressure levels can be used as base method, the SRK (Soave-Redlich-Kwong) equation of state to calculate water enthalpy with NBS steam tables. Gasification unit was modeled using a superstructure configuration, methodology that allows to simulate complex units (Smith, 2005), with a combination of available models in the simulator. In this case, breakdown reactions to produce flammable gas are modeled using an Aspen Plus RYield, using as input information rice hulls elemental analysis (see Table 6). Produced gases are then burned with pressurized air using an RGibbs model, producing high gases which pass over turbine blades. Model for Heat exchanger employed in Heat steam recovery generator section was HeatX, specifying a cold stream (water or steam) outlet condition. Compressor models were simulated using isentropic efficiency of GPSA Method of 0.72 and defining either discharge pressure or pressure decrease. This outlet specification was also employed for Turbines simulation. Rice hulls were introduced to simulator database as a nonconventional component according to its elemental and immediate analysis (see Table 6), while remaining elements were imported from simulator database.

7.2.2 Simulation Results

In gas turbine section, 10,005 Kg/h of rice hulls were gasified and then burned, producing 60021.88 Kg/h of combustion hot gases at 933.2 °C, generating 4 MW of power, corresponding to 0.2899 kWh of electricity per kg of Rice hulls. In Heat Steam Recovery Generator section, the hot gas stream at 933.2 °C was used to produce steam, with an exit gas stream at 132.53 °C, temperature lower than gas acid dew point. HRSG System produced steam at three

TABLE 6 Main Process Data for Simulation of BIGCC Using Rice Hulls as Fuel

Feedstock				
Rice Hulls[a]	Composition—elemental analysis	Carbon 36.6%, Hydrogen 5.83%, Nitrogen 3.31%, Oxygen 36.65%	Composition—immediate analysis	Moisture 9.3%, Fixed Carbon 15.4%, Volatile Matter 57.7%, Ash 17.6%
	Apparent Density (Kg/m^3)	389	Sphericity	0.49
	Mean Particle diameter (μm)	856	Flow (Kg/h)	10005
	Porosity	0.64	Mean particle diameter (μm)	856
	Composition		Flow (Kg/h)	
Air	Water 0.13%, Oxygen 23.1%, Nitrogen 75.4%, Carbon Dioxide 0.001%		55,000	
Water	Water 100%		20,000	

[a] *Ramírez et al. (2007)*

pressure levels: 6680.014 Kg/h at 130 bar (High Pressure), 8884.45 Kg/h at 30 bar (Intermediate Pressure), and 4435.54 Kg/h at 3 bar (Low Pressure). These conditions gave available steam utilities at 350.7 (130 bar), 253.9 °C (30 bar), and 153.6 °C (3 bar), useful to be employed for heating at different energy levels with a combined heating power of 16.189 MW corresponding to 1.6177 KJ of heating per Kg of rice hulls. Finally, in steam turbine section, high-pressure steam used to produce additional electricity power expanding this stream from 130 to 3 bar, generating additional 1.14 MW of power, corresponding to 0.1142 kWh of electricity per Kg of Rice hulls. Simulation results, of energy and mass balances, for gas turbine section are summarized in Tables 7 and 8, respectively.

TABLE 7 Mass Balance Results of BIGCC System

	Process Stream	
	Composition	***Flow (kg/h)***
HOT GAS	Oxygen 3.18%, Carbon Dioxide 22.57%, Nitrogen 65.94%, Carbon Monoxide 0.02%	6002.88
ASH	Ash	2321.16
MVAPHP	Steam at 130 bar and 350.7 °C	6680.014
MVAPMP	Steam at 30 bar 253.9 °C	88,844.502
MVAPLP	Steam at 3 bar, 123.9 °C	44,355.435

TABLE 8 Summary of Energy Produced and Consumed of BIGCC Plant

Energy Consumed (MW]		Net Energy Produced (MW]	
Pumping	1.19×10^{-2}	Gas Turbine	2.8019
Q Fired	19.3612	Steam Turbine	1.1424
Q Direct Use	1.2519	Q Heat Exchangers	17.4368

Operative performance of cogeneration system is evaluated using thermodynamic indexes, such as Energy Utilization Factor (EUF) (Feng et al., 1998; Yilmaz, 2004), Artificial Thermal Energy Efficiency (η_A), ΔFuel (O'Brien and Bansal, 2000a,b; Misa et al., 2007), and Fuel Energy Saving Ratio (FESR) (O'Brien and Bansal, 2000b; Misa et al., 2007; Ong'iro et al., 1996). For biomass-fired cogeneration plant fueled with rice hulls, operational performance results are summarized in Table 9. The EUF of 1.104, as an indicator of the plant combined capacity to produce energy, is reflected in a gain in its use by production of heat and power and related to energy consumption in gasification and syngas combustion from rice hulls. Cogeneration efficiency, measured by η_A, was 70%. Additionally, the fuel energy saving ratio (FESR) shown a 36.58% of fuel saved by the use of biomass-fired cogeneration system, with respect to consumption required by a direct Boiler steam Turbine system, employed for bioelectricity production. Considering that biodiesel process employs 0.0249 kWh of electricity and 0.0121 MJ of heating power per kilogram produced (Radich, 2006), and sugarcane-based ethanol production plant requires 0.2282 kWh of electricity and 25.419 MJ of heating per kilogram produced (Quintero et al., 2008). Total energy produced from 10,005 Kg/h of rice hulls is able to satisfy electricity requirements of this plant with a surplus able to be sold to grid or provide light to close small community. However, produced heating power is not high enough to meet productive plants heating requirements, making it necessary either to increase rice hulls feed or use a cofiring scheme. Combining biomass fuel with natural gas or coal, the selection of one method or other is conditioned by agroindustrial residue availability and technical and economical aspects (Riccio and Chiaramonti, 2009).

TABLE 9 Cogeneration Performance Summary

EUF	1.104	Total heating potential (kJ per kg of Rice Husk]	1.618
Artificial Thermal Energy Efficiency (η_A)	0.709	Total electricity potential (kWh per kg of Rice Husk]	0.404
ΔFUEL	11.166	Local electricity selling prices (USD/MW]	$ 105.208
FESR	0.366	Potential income by electricity selling (USD/h]	$ 336.033

8 CONCLUSIONS

In recent years, it has been suggested that instead of traditional feedstocks, cellulosic biomass (cellulose and hemicellulose), including agricultural and forestry residues, waste paper, and industrial wastes, could be used as an ideally inexpensive and abundantly available source of sugar for fermentation into transportation fuel ethanol. The efficiency of biomass conversion to ethanol depends upon the ability of the microorganism used in the process to utilize these diverse carbon sources and amount of fraction present in biomass. The potentiality of sugar crops, agro, and urban/industrial residues feedstocks for production of ethanol as an alternative fuel and energy sources was shown, which is renewable, sustainable, efficient, and safe for environment. The cost of ethanol production from lignocellulosic material is relatively high based on current technologies, and the main challenges are low yield and high cost of hydrolysis. There is need of process optimization for detoxification and maximize conversion of agroindustrial residues feedstocks for production of ethanol as a cheaper substrate like molasses and other directly fermentable materials. Although bioethanol production has been greatly improved by new technologies, there are still challenges that need further investigations. These challenges include maintaining a stable performance of the genetically engineered microorganisms in commercial-scale fermentation operations and developing more efficient pretreatment technologies for the lignocellulosic biomass and integrating the optimal components into economic ethanol production systems.

A reduction of the cost of ethanol production can be achieved by reducing the cost of either the raw materials or the cellulase enzymes. Reducing the cost of cellulase enzyme production is a key issue in the enzymatic hydrolysis of lignocellulosic materials. Energy requirement for pretreatment is a strong criterion in the evaluation of ethanol production process from lignocellulosic biomass because it represents the main section of energy demand. The simulation presented in this work showed that DA and LHW are efficient processes, but if energy cogeneration is not implemented, these pretreatment methods are not applicable for industry given the large amount of energy required. Ethanol production cost from rice hulls highly depends on energy consumption. At current context and market prices, a project of ethanol production from these wastes is not feasible in Colombia.

Biomass-fired cogeneration is an efficient way to produce electricity with an increased capacity to generate heat and power using fuel efficiently. In this sense, utilization of agroindustrial residues has the potential of being a cost-effective way to produce bioelectricity, in countries whose electric generation is highly seasonal, complementing electricity production in dry seasons. Replacing, also the use of fossil-based thermoelectric power plants, as well as its emissions, offering both electricity security and an environmentally safe waste management and disposal of these residues. In order to increase competitiveness of production of bioethanol from agroindustrial residues; integrated use of biomass-fired cogeneration is valuable alternative addressed to increase project incoming and viability. Although with the amount of rice hulls considered in the Study Case 2 the system did not generate an enough heating potential to meet a standard biofuel productive plant requirements; this potential can be effectively improved increasing the amount of burned biomass or when this alternative is not cost-effective using a cofiring system. Where biomass waste burning could be combined with fossil fuels, that is, coal or preferably natural gas in a co-firing scheme as a cost-effective way to improve energy production. Being a useful scheme employed in cases where the same agroindustrial waste is used both in a productive plant and to meet its energy requirements through cogeneration; and is neither technically nor economically efficient to use a more feedstock amount in energy production than to obtain main product.

Acknowledgments

The authors express their acknowledgments to the Colombian Institute for Development of Science and Technology (Colciencias) and Universidad Nacional de Colombia sede Manizales, for the financial support to this work.

References

Aires Barros, M.R., Cabral, J.M.S., Novais, J.M., 1987. Production of ethanol by immobilized Saccharomyces bayanus in an extractive fermentation system. Biotechnol. Bioeng. 29 (9), 1097–1104.

Alizadeh, H., et al., 2005. Pretreatment of switchgrass by ammonia fiber explosion (AFEX). Appl. Biochem. Biotechnol. 124 (1), 1133–1141.

Allen, S.G., et al., 2001a. A comparison between hot liquid water and steam fractionation of corn fiber. Ind. Eng. Chem. Res. 40 (13), 2934–2941.

Allen, S.G., et al., 2001b. A comparison of aqueous and dilute-acid single-temperature pretreatment of yellow poplar sawdust. Ind. Eng. Chem. Res. 40 (10), 2352–2361.

Anokhina, E.A., Cardona, C.A., Pisarenko, Y.u.A., Ponomarev, E.L., 1996a. Basic design stages of combined processes by the example of the production of allyl alcohol by allyl acetate butanolysis. Part 1: the selection of a combined process for the production of allyl alcohol. Chem. Ind. Russ. 9, 537.

Anokhina, E.A., Cardona, C.A., Pisarenko, Y.u.A., Ponomarev, E.L., 1996b. Basic design stages of combined processes by the example of the production of allyl alcohol by allyl acetate butanolysis. Part 2: separation of a reaction mixture removed from the reaction-rectification column for allyl alcohol synthesis. Chem. Ind. Russ. 11, 689.

Badami, M., et al., 2008. Design and performance evaluation of an innovative small scale combined cycle cogeneration system. Energy 33 (8), 1264–1276.

Balat, M., et al., 2009. Main routes for the thermo-conversion of biomass into fuels and chemicals. Part 2: gasification systems. Energy Convers. Manage. 50 (12), 3158–3168.

Ballesteros, I., et al., 1991. Selection of thermotolerant yeasts for simultaneous saccharification and fermentation (SSF) of cellulose to ethanol. Appl. Biochem. Biotechnol. 28-29 (1), 307–315.

Ballesteros, M., et al., 2004. Ethanol from lignocellulosic materials by a simultaneous saccharification and fermentation process (SFS) with *Kluyveromyces marxianus* CECT 10875. Pro. Biochem. 39 (12), 1843–1848.

Bariska, M., 1975. Collapse phenomena in beechwood during and after NH_3 impregnation. Wood Sci. Technol. 9 (4), 293–306.

Belkacemi, K., Turcotte, G., Savoie, P., 2001. Aqueous/steam-fractionated agricultural residues as substrates for ethanol production. Ind. Eng. Chem. Res. 41 (2), 173–179.

Biezma, M.V., Cristóbal, J.R.S., 2006. Investment criteria for the selection of cogeneration plants—a state of the art review. Appl. Therm. Eng. 26 (5-6), 583–588.

Björling, T., Lindman, B., 1989. Evaluation of xylose-fermenting yeasts for ethanol production from spent sulfite liquor. Enzyme. Microb. Technol. 11 (4), 240–246.

Bobleter, O., 1994. Hydrothermal degradation of polymers derived from plants. Progr. Polym. Sci. (Oxford) 19 (5), 797–841.

Botha, T., von Blottnitz, H., 2006. A comparison of the environmental benefits of bagasse-derived electricity and fuel ethanol on a life-cycle basis. Energy Pol. 34 (17), 2654–2661.

Bram, S., De Ruyck, J., Novak-Zdravkovic, A., 2005. Status of external firing of biomass in gas turbines. Proc Inst. Mech. Eng. J. Power Energy 219 (2), 137–145.

Branan, C., 2002. Rules of Thumb for Chemical Enginners: A Manual of Quick, Accurate Solutions to Everyday Process Engineering Problems. In: Gulf Professional Publishing, p. 426.

Brennan, A.H., Hoagland, W., Schell, D.J., 1986. High temperature acid hydrolysis of biomass using an engineering scale plug flow reactor: results of low solids testing. Biotechnol. Bioeng. 17, 53–70.

Brownell, H.H., Yu, E.K.C., Saddler, J.N., 1986. Steam-explosion pretreatment of wood: Effect of chip size, acid, moisture content and pressure drop. Biotechnol. Bioeng. 28 (6), 792–801.

Cara, C., et al., 2008. Conversion of olive tree biomass into fermentable sugars by dilute acid pretreatment and enzymatic saccharification. Bioresour. Technol. 99 (6), 1869–1876.

Cardona, C.A., Sánchez, Ó.J., 2007. Fuel ethanol production: Process design trends and integration opportunities. Bioresour. Technol. 98 (12), 2415–2457.

Cardona, C.A., Marulanda, V.F., Young, D., 2004. Analysis of the environmental impact of butylacetate process through the WAR algorithm. Chem. Eng. Sci. 59, 5839–5845.

Cardona, C.A., Matallana, L.G., Gutierrez, L.F., 2009. Environmental performance of reactive separation processes. In: 8th World Congress of Chemical Engineering. Montreal, Canada.

Cardona, C.A., Quintero, J.A., Paz, I.C., 2010. Production of bioethanol from sugarcane bagasse: Status and perspectives. Bioresour. Technol. 101 (13), 4754–4766.

Carrillo, F., et al., 2005. Effect of alkali pretreatment on cellulase hydrolysis of wheat straw: Kinetic study. Pro. Biochem. 40 (10), 3360–3364.

Caulfield, D., Moore, W.E., 1974. Effect of varying crystallinity of cellulose on enzymic hydrolysis. Wood Sci. 6 (4), 375–379.

Chandrakant, P., Bisaria, V.S., 1998. Simultaneous bioconversion of cellulose and hemicellulose to ethanol. Crit. Rev. Biotechnol. 18, 295–331.

Chang, V., Holtzapple, M., 2000. Fundamental factors affecting biomass enzymatic reactivity. Appl. Biochem. Biotechnol. 84-86 (1), 5–37.

Chang, V., et al., 2001a. Oxidative lime pretreatment of high-lignin biomass. Appl. Biochem. Biotechnol. 94 (1), 1–28.

Chang, V.S., et al., 2001b. Simultaneous saccharification and fermentation of lime-treated biomass. Biotechnol. Lett. 23 (16), 1327–1333.

Cheng, K.K., et al., 2008. Sugarcane bagasse hemicellulose hydrolysate for ethanol production by acid recovery process. Biochem. Eng. J. 38 (1), 105–109.

Christakopoulos, P., et al., 1996. Purification and characterisation of a major xylanase with cellulase and transferase activities from *Fusarium oxysporum*. Carbohydr. Res. 289, 91–104.

Claassen, P.A.M., et al., 1999. Utilisation of biomass for the supply of energy carriers. Appl. Microbiol. Biotechnol. 52 (6), 741–755.

Coelho, S.T., et al., 2000. Evaluation of externalities in sugarcane-origin cogeneration process in Brazil. In: First World Conference and Exhibition for Energy and Industry. Sevilla, Spain.

Converse, A., et al., 1989. Kinetics of thermochemical pretreatment of lignocellulosic materials. Appl. Biochem. Biotechnol. 20-21 (1), 63–78.

Cowling, E.B., Kirk, T.K., 1976. Properties of cellulose and lignocellulosic materials as substrates for enzymatic conversion processes. Biotechnol. Bioeng. Symp. (6), 95–123.

Crabbe, P.G., Tse, C.W., Munro, P.A., 1986. Effect of microorganisms on rate of liquid extraction of ethanol from fermentation broths. Biotechnol. Bioeng. 28 (7), 939–943.

Dien, B.S., Cotta, M.A., Jeffries, T.W., 2003. Bacteria engineered for fuel ethanol production: current status. Appl. Microbiol. Biotechnol. 63 (3), 258–266.

Dijkerman, R., et al., 1997. The anaerobic fungus *Neocallimastix* sp. Strain L_2: growth and production of (hemi) cellulolytic enzymes on a range of carbohydrate substrates. Curr. Microbiol. 34 (2), 91–96.

Domansky, R., Rendos, F., 1962. On the pyrolysis of wood and its components. Holz Roh-Werkst 20, 473–476.

Duff, S.J.B., Murray, W.D., 1996. Bioconversion of forest products industry waste cellulosics to fuel ethanol: a review. Bioresour. Technol. 55 (1), 33.

Faaij, A., 2006. Modern biomass conversion technologies. Mitig. Adapt. Strat. Glob. Change 11 (2), 335–367.

Fan, Z., et al., 2003. Conversion of paper sludge to ethanol in a semicontinuous solids-fed reactor. Bioproc. Biosyst. Eng. 26 (2), 93–101.

Feng, X., Cai, Y.N., Qian, L.L., 1998. A new performance criterion for cogeneration system. Energy Convers. Manage. 39 (15), 1607–1609.

Garrote, G., Domínguez, H., Parajó, J.C., 1999. Hydrothermal processing of lignocellulosic materials. Eur. J. Wood Wood Prod. 57 (3), 191–202.

Garrote, G., et al., 2008. Coproduction of oligosaccharides and glucose from corncobs by hydrothermal processing and enzymatic hydrolysis. Ind. Eng. Chem. Res. 47 (4), 1336–1345.

Gassner, M., Maréchal, F., 2010. Increasing conversion efficiency in fuel ethanol production from lignocellulosic biomass by polygeneration—and a paradoxon between energy and exergy in process integration. In: 23rd International Conference on Efficiency, Cost, Optimization, Simulation and Environmental Impact of Energy Systems (ECOS 2010). Lausanne, Switzerland.

Gossett, J.M., et al., 1982. Heat treatment and anaerobic digestion of refuse. J. Environ. Eng. Div. ASCE 108 (EE3), 437–454.

Grassi, G., Allan, A., 2007. Bioelectricity and cogeneration. In: Ranalli, P. (Ed.), Improvement of Crop Plants for Industrial End Uses. Springer.

Gregg, D., Saddler, J.N., 1996. A techno-economic assessment of the pretreatment and fractionation steps of a biomass-to-ethanol process. Appl. Biochem. Biotechnol. A Enzym. Eng. Biotechnol. 57-58, 711–727.

Grethlein, H.E., 1985. The effect of pore size distribution on the rate of enzymatic hydrolysis of cellulosic substrates. Biotechnology 3 (2), 155–160.

Grohmann, K., Torget, R., Himmel, M., 1986. Dilute acid pretreatment of biomass at high solids concentrations. Biotechnol. Bioeng. Symp. 59, 80.

Grous, W.R., Converse, A.O., Grethlein, H.E., 1986. Effect of steam explosion pretreatment on pore size and enzymatic hydrolysis of poplar. Enzyme. Microb. Technol. 8 (5), 274–280.

Gupta, A.K., Lewis, M.J., Daurer, M., 2001. Swirl effects on combustion characteristics of premixed flames. J. Eng. Gas Turb. Power 123 (3), 619–626.

Gupta, R., Sharma, K.K., Kuhad, R.C., 2009. Separate hydrolysis and fermentation (SHF) of *Prosopis juliflora*, a woody substrate, for the production of cellulosic ethanol by *Saccharomyces cerevisiae* and *Pichia stipitis*-NCIM 3498. Bioresour. Technol. 100 (3), 1214–1220.

Gutiérrez, L.F., Sánchez, Ó.J., Cardona, C.A., 2005. Modeling of batch extractive fermentation for the fuel ethanol production. In: 8th Conference on Process Integration, Modelling and Optimisation for Energy Saving and Pollution Reduction. Naxos, Italy.

Gyamerah, M., Glover, J., 1996. Production of ethanol by continuous fermentation and liquid-liquid extraction. J. Chem. Technol. Biotechnol. 66 (2), 145–152.

Hahn-Hägerdal, B., et al., 1994. An interlaboratory comparison of the performance of ethanol-producing microorganisms in a xylose-rich acid hydroysate. Appl. Microbiol. Biotechnol. 41 (1), 62–72.

Hall, D.O., 1997. Biomass energy in industrialised countries—a view of the future. For. Ecol. Manage. 91 (1), 17–45.

Han, Y.W., Catalano, E.A., Ciegler, A., 1983. Chemical and physical properties of sugarcane bagasse irradiated with. gamma. rays. J. Agric. Food Chem. 31 (1), 34–38.

Haq, Z., 2010. Biomass for electricity generation. Available from: www.eia.doe.gov/oiaf/analysispaper/biomass/pdf/biomass.pdf.

Hendriks, A.T.W.M., Zeeman, G., 2009. Pretreatments to enhance the digestibility of lignocellulosic biomass. Bioresour. Technol. 100 (1), 10–18.

Hideno, A., et al., 2008. Optimization of the thermal dry treatment to enhance the enzymatic hydrolysis of a spent-sawdust matrix used for Grifola frondosa cultivation. Energy and Fuels 22 (1), 120–122.

HM Associates Inc, 2003. Assessment of the Commercial Potential for Small Gasification Combined Cycle and Fuel Cell Systems Phase II Final Draft Report. U.S. Department of Energy, Office of Fossil Energy, Office of Coal and Power Systems.

Iakovou, E., et al., 2010. Waste biomass-to-energy supply chain management: a critical synthesis. Waste Manag. 30 (10), 1860–1870.

Jacek Topolski, J.B., 2010. Efficiency of HRSG within a Combined Cycle with Gasification and Sequential Combustion at GT26 Turbine, in Reacting Flows Department. Institute of Fluid-Flow Machinery Polish Academy of Sciences.

Jeffries, T.W., 1985. Emerging technology for fermenting D-xylose. Trends Biotechnol. 3 (8), 208–212.

Jin, Y.S., Jeffries, T.W., 2004. Stoichiometric network constraints on xylose metabolism by recombinant *Saccharomyces cerevisiae*. Metab. Eng. 6 (3), 229–238.

Kaar, W.E., Holtzapple, M.T., 2000. Using lime pretreatment to facilitate the enzymic hydrolysis of corn stover. Biomass Bioenergy 18 (3), 189–199.

Kapoor, M., Nair, L.M., Kuhad, R.C., 2008. Cost-effective xylanase production from free and immobilized *Bacillus pumilus* strain MK001 and its application in saccharification of *Prosopis juliflora*. Biochem. Eng. J. 38 (1), 88–97.

Kilpeläinen, I., Xie, H., King, A., Granstrom, M., Heikkinen, S., Argyropoulos, D.S., 2007. Dissolution of wood in ionic liquids. 55 J. Agric. Food Chem (22), 7.

Kim, K.H., Hong, J., 2001. Supercritical CO_2 pretreatment of lignocellulose enhances enzymatic cellulose hydrolysis. Bioresour. Technol. 77 (2), 139–144.

Kim, S., Holtzapple, M.T., 2005. Lime pretreatment and enzymatic hydrolysis of corn stover. Bioresour. Technol. 96 (18 Spec. Iss.), 1994–2006.

Kim, T., Lee, Y., 2005. Pretreatment of corn stover by soaking in aqueous ammonia. Appl. Biochem. Biotechnol. 124 (1), 1119–1131.

Kirbaşlar, Ş.I., et al., 2001. Equilibrium data on water-ethanol-1-dodecanol ternary system. Su-Etanol-1-Dodekanol Üçlü Sisteminin Denge Verileri 25 (2), 111–115.

Klinke, H.B., et al., 2002. Characterization of degradation products from alkaline wet oxidation of wheat straw. Bioresour. Technol. 82 (1), 15–26.

Kobayashi, N., et al., 2009. Characteristics of solid residues obtained from hot-compressed-water treatment of woody biomass. Ind. Eng. Chem. Res. 48 (1), 373–379.

Kollerup, F., Daugulis, A.J., 1985. A mathematical model for ethanol production by extractive fermentation in a continuous stirred tank fermentor. Biotechnol. Bioeng. 27 (9), 1335–1346.

Koullas, D.P., et al., 1992. Correlating the effect of pretreatment on the enzymatic hydrolysis of straw. Biotechnol. Bioeng. 39 (1), 113–116.

Kuhad, R.C., Manchanda, M., Singh, A., 1999. Hydrolytic potential of extracellular enzymes from a mutant strain of *Fusarium oxysporum*. Bioprocess Eng. 20 (2), 133–135.

L'Italien, Y., Thibault, J., LeDuy, A., 1989. Improvement of ethanol fermentation under hyperbaric conditions. Biotechnol. Bioeng. 33 (4), 471–476.

Laplace, J.M., et al., 1993. Cofermentation of glucose and xylose to ethanol by a respiratory-deficient mutant of *Saccharomyces cerevisiae* co-cultivated with a xylose-fermenting yeast. J. Ferment. Bioeng. 75 (3), 207–212.

Larson, E.D., Williams, R.H., Leal, M.R.L.V., 2001. A review of biomass integrated-gasifier/gas turbine combined cycle technology and its application in sugarcane industries, with an analysis for Cuba. Energy Sustain. Dev. 5 (1), 54–76.

Laser, M., et al., 2002. A comparison of liquid hot water and steam pretreatments of sugar cane bagasse for bioconversion to ethanol. Bioresour. Technol. 81 (1), 33–44.

Laureano-Perez, L., et al., 2005. Understanding factors that limit enzymatic hydrolysis of biomass: Characterization of pretreated corn stover. Appl. Biochem. Biotechnol. A Enzym. Eng. Biotechnol. 124 (1-3), 1081–1099.

Liao, W., et al., 2006. Acid hydrolysis of fibers from dairy manure. Bioresour. Technol. 97 (14), 1687–1695.

Ligthelm, M.E., Prior, B.A., du Preez, J.C., 1988. The oxygen requirements of yeasts for the fermentation of d-xylose and d-glucose to ethanol. Appl. Microbiol. Biotechnol. 28 (1), 63–68.

Linden, T., Hahn-Hagerdal, B., 1989. Fermentation of lignocellulose hydrolysates with yeasts and xylose isomerase. Enzyme. Microb. Technol. 11 (9), 583–589.

Lin, Y., Tanaka, S., 2006. Ethanol fermentation from biomass resources: current state and prospects. Appl. Microbiol. Biotechnol. 69 (6), 627–642.

Liu, C., Wyman, C.E., 2003. The effect of flow rate of very dilute sulfuric acid on xylan, lignin, and total mass removal from corn stover. Ind. Eng. Chem. Res. 43 (11), 2781–2788.

Luo, L., van der Voet, E., Huppes, G., 2009. Life cycle assessment and life cycle costing of bioethanol from sugarcane in Brazil. Renew. Sustain. Energy Rev. 13 (6-7), 1613–1619.

Lynd, L.R., 1996. Overview and evaluation of fuel ethanol from cellulosic biomass: Technology, economics, the environment, and policy. Ann. Rev. Energy Environ. 21 (1), 403–465.

Mamma, D., et al., 1995. An alternative approach to the bioconversion of sweet sorghum carbohydrates to ethanol. Biomass Bioenergy 8 (2), 99–103.

Mamma, D., et al., 1996. Bioethanol from sweet sorghum: simultaneous saccharification and fermentation of carbohydrates by a mixed microbial culture. Pro. Biochem. 31 (4), 377–381.

Martinez, J.M., et al., 1997. Hydrolytic pretreatment of softwood and almond shells. Degree of polymerization and enzymatic digestibility of the cellulose fraction. Ind. Eng. Chem. Res. 36 (3), 688–696.

Marzialetti, T., et al., 2008. Dilute acid hydrolysis of loblolly pine: a comprehensive approach. Ind. Eng. Chem. Res. 47 (19), 7131–7140.

McKendry, P., 2002. Energy production from biomass (part 1): overview of biomass. Bioresour. Technol. 83 (1), 37–46.

McMillan, J.D., et al., 1999. Simultaneous saccharification and cofermentation of dilute-acid pretreated yellow poplar hardwood to ethanol using xylose-fermenting *Zymomonas mobilis*. Appl. Biochem. Biotechnol. A Enzym. Eng. Biotechnol. 77-79, 649–665.

Minier, M., Goma, G., 1982. Ethanol production by extractive fermentation. Biotechnol. Bioeng. 24 (7), 1565–1579.

Misa, T., et al., 2007. Thermodynamic analysis of an in-situ cogeneration plant. In: Proceedings of the Institution of Mechanical Engineers—Part A—Power & Energy. Professional Engineering Publishing, pp. 591–602.

Mishima, D., et al., 2006. Comparative study on chemical pretreatments to accelerate enzymatic hydrolysis of aquatic macrophyte biomass used in water purification processes. Bioresour. Technol. 97 (16), 2166–2172.

Mok, W.S.L., Antal Jr, M.J., 1992. Uncatalyzed solvolysis of whole biomass hemicellulose by hot compressed liquid water. Ind. Eng. Chem. Res. 31 (4), 1157–1161.

Moritz, J.W., Duff, S.J.B., 1996. Simultaneous saccharification and extractive fermentation of cellulosic substrates. Biotechnol. Bioeng. 49 (5), 504–511.

Mosier, N., et al., 2005a. Features of promising technologies for pretreatment of lignocellulosic biomass. Bioresour. Technol. 96 (6), 673–686.

Mosier, N., et al., 2005b. Optimization of pH controlled liquid hot water pretreatment of corn stover. Bioresour. Technol. 96 (18 Spec. Iss.), 1986–1993.

O'Brien, J.M., Bansal, P.K., 2000a. Modelling of cogeneration systems part 2: development of a Quasi-static cogeneration model (steam turbine cogeneration analysis). In: Proceedings of the Institution of Mechanical Engineers—Part A—Power & Energy. Professional Engineering Publishing, pp. 125–143.

O'Brien, J.M., Bansal, P.K., 2000b. Modelling of cogeneration systems part 3: application of steam turbine cogeneration analysis to auckland hospital cogeneration utility system. In: Proceedings of the Institution of Mechanical Engineers—Part A—Power & Energy. Professional Engineering Publishing, pp. 227–241.

Öhgren, K., Galbe, M., Zacchi, G., 2005. Optimization of Steam Pretreatment of SO_2-impregnated corn stover for fuel ethanol production. In: Davison, B.H. et al., (Ed.), Twenty-Sixth Symposium on Biotechnology for Fuels and Chemicals. Humana Press, pp. 1055–1067.

Okamoto, T., et al., 1993. Zymobacter palmae gen. nov., sp. nov., a new ethanol-fermenting peritrichous bacterium isolated from palm sap. Arch. Microbiol. 160 (5), 333–337.

Olsson, L., Hahn-Hagerdal, B., 1996. Fermentation of lignocellulosic hydrolysates for ethanol production. Enzyme Microbiol. Technol. 8, 312–331.

Olsson, L., Hahn-Hägerdal, B., 1993. Fermentative performance of bacteria and yeasts in lignocellulose hydrolysates. Pro. Biochem. 28 (4), 249–257.

Olsson, L., Lindén, T., Hahn-HäGerdal, B., 1992. Performance of microorganisms in spent sulfite liquor and enzymatic hydrolysate of steam-pretreated Salix. Appl. Biochem. Biotechnol. 34-35 (1), 359–368.

Ong'iro, A., et al., 1996. Thermodynamic simulation and evaluation of a steam CHP plant using ASPEN Plus. Appl. Therm. Eng. 16 (3), 263–271.

Palonen, H., et al., 2004. Evaluation of wet oxidation pretreatment for enzymatic hydrolysis of softwood. Appl. Biochem. Biotechnol. A Enzym. Eng. Biotechnol. 117 (1), 1–17.

Panagiotou, G., et al., 2005. Intracellular metabolite profiling of *Fusarium oxysporum* converting glucose to ethanol. J Biotechnol 115 (4), 425–434.

Pandey, A., et al., 2000a. Biotechnological potential of agro-industrial residues. I: Sugarcane bagasse. Bioresour. Technol. 74 (1), 69–80.

Pandey, A., et al., 2000b. Biotechnological potential of coffee pulp and coffee husk for bioprocesses. Biochem. Eng. J. 6 (2), 153–162.

Papong, S., et al., 2004. Overview of Biomass Utilization in Thailand. Cleaner Technology Advancement Program, National Metal and Materials Technology Center, National Science and Technology Development Agency, Ministry of Science and Technology, Bangkok, Thailand.

Pereira Ramos, L., 2003. The chemistry involved in the steam treatment of lignocellulosic materials. Quimica Nova 26 (6), 863–871.

PGTHERMAL, 2009. Introduction to HRSG design. Available from: http://www.hrsgdesign.com/design0.htm.

Philippidis, G.P., Smith, T.K., 1995. Limiting factors in the simultaneous saccharification and fermentation process for conversion of cellulosic biomass to fuel ethanol. Appl. Biochem. Biotechnol. 51-52 (1), 117–124.

Piarpuzán, D., Quintero, J.A., Cardona, C.A., 2011. Empty fruit bunches from oil palm as a potential raw material for fuel ethanol production. Biomass and Bioenergy 35 (3), 1130–1137.

Pisarenko, Y.A., et al., 2001. Reactive distillation design: analysis of the process statics. Rev. Chem. Eng. 17 (4), 253–327.

Prasad, S.B., 1995. Biomass-fired steam power cogeneration system: a theoretical study. Energy Convers. Manage. 36 (1), 65–77.

Puri, V.P., 1984. Effect of crystallinity and degree of polymerization of cellulose on enzymatic saccharification. Biotechnol. Bioeng. 26 (10), 1219–1222.

Puri, V.P., Mamers, H., 1983. Explosive pretreatment of lignocellulosic residues with high-pressure carbon dioxide for the production of fermentation substrates. Biotechnol. Bioeng. 25 (12), 3149–3161.

Quintero, J.A., et al., 2008. Fuel ethanol production from sugarcane and corn: comparative analysis for a Colombian case. Energy 33 (3), 385–399.

Quintero, J.A., Cardona, C.A., Triana, C., 2010. Analysis of Sugarcane Bagasse and Coffee Cut. in 32nd Symposium on Biotechnology for Fuels and Chemicals. Clearwater Beach, USA.

Quintero, J.A., Cardona, C.A., 2011. Process Simulation of Fuel Ethanol Production from Lignocellulosics using Aspen Plus. Industrial & Engineering Chemistry Research. In Press, Corrected Proof.

Radich, A., 2006. Biodiesel performance, costs, and use. [cited 2006 23-10-10]; Available from: http://www.eia.doe.gov/oiaf/analysispaper/biodiesel/.

Ramírez, J.J., Martínez, J.D., Petro, S.l., 2007. Basic design of fluidized bed gasifier for rice husk on a pilot scale. Latin Am. Res. 37, 299–306.

Reith, J.H., et al., 2002. Co-production of bio-ethanol, electricity and heat from biomass residues. In: 12th European Conference and Technology Exhibition on Biomass for Energy, Industry and Climate Protection. Amsterdam, the Netherlands.

Resource Dynamics Corporation, 2004. Combined Heat and Power Market Potential for Opportunity Fuels. U.S. Department of Energy.

Riccio, G., Chiaramonti, D., 2009. Design and simulation of a small polygeneration plant cofiring biomass and natural gas in a dual combustion micro gas turbine (BIO_MGT). Biomass Bioenergy 33 (11), 1520–1531.

Saha, B.C., et al., 2005. Dilute acid pretreatment, enzymatic saccharification and fermentation of wheat straw to ethanol. Pro. Biochem. 40 (12), 3693–3700.

Sanchez, B., Bautista, J., 1988. Effects of furfural and 5-hydroxymethylfurfural on the fermentation of *Saccharomyces cerevisiae* and biomass production from Candida guilliermondii. Enzyme. Microb. Technol. 10 (5), 315–318.

Sánchez, Ó.J., Cardona, C.A., 2008. Trends in biotechnological production of fuel ethanol from different feedstocks. Bioresour. Technol. 99 (13), 5270–5295.

Sanjay, O., Prasad, B.N., 2009. Comparative performance analysis of cogeneration gas turbine cycle for different blade cooling means. Int. J. Therm. Sci. 48 (7), 1432–1440.

Saxena, A., Garg, S.K., Verma, J., 1992. Simultaneous saccharification and fermentation of waste newspaper to ethanol. Bioresour. Technol. 42 (1), 13–15.

Saxena, R.C., Adhikari, D.K., Goyal, H.B., 2009. Biomass-based energy fuel through biochemical routes: a review. Renew. Sustain. Energy Rev. 13 (1), 167–178.

Sendelius, J., 2005. Steam Pretreatment Optimisation for Sugarcane Bagasse in Bioethanol Production. Lund University, Sweden.

Sheehan, J., Himmel, M., 1999. Enzymes, energy, and the environment: a strategic perspective on the U.S. Department of Energy's Research and Development Activities for Bioethanol. Biotechnol. Prog. 15 (5), 817–827.

Shevchenko, S.M., Beatson, R.P., Saddler, J.N., 1999. The nature of lignin from steam explosion/enzymatic hydrolysis of softwood: Structural features and possible uses. Appl. Biochem. Biotechnol. A Enzym. Eng. Biotechnol. 77-79, 867–876.

Shuit, S.H., et al., 2009. Oil palm biomass as a sustainable energy source: a Malaysian case study. Energy 34 (9), 1225–1235.

Singh, A., Kumar, P., Schügerl, K., 1992. Bioconversion of Cellulosic Materials to Ethanol by Filamentous Fungi, in Enzymes and Products from Bacteria Fungi and Plant Cells. Berlin/Heidelberg, 29–55.

Skoog, K., Hahn-Hagerdal, B., 1988. Xylose fermentation. Enzyme. Microb. Technol. 10 (2), 66–80.

Skotnicki, M.L., et al., 1983. High-productivity alcohol fermentations using *Zymomonas mobilis*. Biochem. Soc. Symp. 48, 53–86.

Smith, R., 2005. Chemical Process: Design and Integration. Wiley.

South, C.R., Hogsett, D.A., Lynd, L.R., 1993. Continuous fermentation of cellulosic biomass to ethanol. Appl. Biochem. Biotechnol. 39-40 (1), 587–600.

Subba Rao, M.V.S.S.T., Muralikrishna, G., 2006. Hemicelluloses of ragi (*Finger millet, Eleusine coracana*, indaf-15): isolation and purification of an alkali-extractable arabinoxylan from native and malted hemicellulose B. J. Agric. Food Chem. 54 (6), 2342–2349.

Sukumaran, R.K., et al., 2009. Cellulase production using biomass feed stock and its application in lignocellulose saccharification for bio-ethanol production. Renew. Energy 34 (2), 421–424.

Sun, J.X., et al., 2004. Isolation and characterization of cellulose from sugarcane bagasse. Polym. Degrad. Stabil. 84 (2), 331–339.

Susta, M.R., Luby, P., Mat, S.B., 2003. Biomass Energy Utilization & Environment Protection—Commercial Reality and Outlook. Power-Gen Asia.

Sweet, M.S., Winandy, J.E., 1999. Influence of degree of polymerization of cellulose and hemicellulose on strength loss in fire-retardant-treated southern pine. Holzforschung 53 (3), 311–317.

Teymouri, F., et al., 2005. Optimization of the ammonia fiber explosion (AFEX) treatment parameters for enzymatic hydrolysis of corn stover. Bioresour. Technol. 96 (18), 2014–2018.

Thompson, D.N., Chen, H.C., Grethlein, H.E., 1992. Comparison of pretreatment methods on the basis of available surface area. Bioresour. Technol. 39 (2), 155–163.

Tsuji, H., et al., 2003. High Temperature Air Combustion from Energy Conservation to Pollution Reduction. In: CRC Press, p. 401.

Uddin, S.N., Barreto, L., 2007. Biomass-fired cogeneration systems with CO_2 capture and storage. Renew. Energy 32 (6), 1006–1019.

UN Foundation, 2007. Biomass conversion technologies. Available from: http://www.globalproblems-globalsolutions-files.org/gpgs_files/pdf/UNF_Bioenergy/UNF_Bioenergy_5.pdf.

Van Zessen, E., et al., 2003. Lignocellulosic ethanol, a second opinion. Report 2GAVE-03.11, [cited 2008 04 11]; Available from: http://library.wur.nl/WebQuery/wurpubs/lang/339809.

van Zyl, C., Prior, B.A., du Preez, J.C., 1991. Acetic acid inhibition of d-xylose fermentation by *Pichia stipitis*. Enzyme. Microb. Technol. 13 (1), 82–86.

Von Sivers, M., et al., 1994. Cost analysis of ethanol production from willow using recombinant Escherichia coli. Biotechnol. Prog. 10 (5), 555–560.

Weil, J., et al., 1998. Continuous pH monitoring during pretreatment of yellow poplar wood sawdust pressure cooking in water. Appl. Biochem. Biotechnol. A Enzym. Eng. Biotechnol. 70-72, 99–111.

Weilnhammer, C., Blass, E., 1994. Continuous fermentation with product recovery by in-situ extraction. Chem. Eng. Technol. 17 (6), 365–373.

Werther, J., et al., 2000. Combustion of agricultural residues. Progress Energy Combust. Sci. 26 (1), 1–27.

Wooley, R., Putsche, V., 1996a. Development of an ASPEN PLUS physical property database for biofuels components. Report No. NREL/MP-425-20685. National Renewable Energy Laboratory, Golden, CO, USA.

Wooley, R., Putsche, V., 1996b. Report NREL/MP-425-20685. National Renewable Energy Laboratory, Golden, CO, USA, p. 38.

Wooley, R., Ruth, M., Sheehan, J., Ibsen, K., Majdeski, H., Galvez, A., 1999. Lignocellulosic biomass to ethanol process design and economics utilizing co-current dilute acid prehydrolysis and enzymatic hydrolysis. Current and futuristic scenarios. Technical Report NREL/TP-580-26157. National Renewable Energy Laboratory, Golden, CO, USA, p. 123.

Xu, F., et al., 2006. Comparative study of alkali- and acidic organic solvent-soluble hemicellulosic polysaccharides from sugarcane bagasse. Carbohydr. Res. 341 (2), 253–261.

Yilmaz, T., 2004. Optimization of cogeneration systems under alternative performance criteria. Energy Convers. Manage. 45 (6), 939–945.

Zhang, Y.H.P., Lynd, L.R., 2004. Toward an aggregated understanding of enzymatic hydrolysis of cellulose: Noncomplexed cellulase systems. Biotechnol. Bioeng. 88 (7), 797–824.

Zheng, Y., Lin, H.M., Tsao, G.T., 1998. Pretreatment for cellulose hydrolysis by carbon dioxide explosion. Biotechnol Prog 14 (6), 890–896.

CHAPTER

12

Fermentation Inhibitors in Ethanol Processes and Different Strategies to Reduce Their Effects

Mohammad J. Taherzadeh[1,*], *Keikhosro Karimi*[2]

[1]School of Engineering, University of Borås, Sweden

[2]Chemical Engineering Department, Isfahan University of Technology, Iran

*Corresponding author: E-mail: mohammad.taherzadeh@hb.se

1 INTRODUCTION

Fermentation is the heart of an ethanol process, in which the sugars are converted to ethanol by a variety of microorganisms. The raw materials or substrate for the fermentation is generally a solution containing natural sugars such as sugarcane or beet sugar juices, "molasses," the byproduct of the sugar industry or any other residual or low-value products such as fruit juice byproducts and residuals. Another category of raw materials for fermentation is the sugar solution produced from a prior hydrolysis process of, for example, grains or lignocelluloses (Figure 1). These hydrolysis steps are generally carried out by enzymes or acids. However, to have an efficient hydrolysis, a pretreatment is usually necessary, which could be by physical, chemical, thermal, or biological means (Figure 1). In summary, the raw materials used for ethanol production, as well as the steps prior to fermentation, can contain or create some chemical compounds that reduce the microorganism's ability of ethanol production. These "inhibitors" may reduce the yield or productivity of ethanol, reduce the viability of the microorganisms, or completely stop the fermentation. Furthermore, the carbon source of the fermentation process, that is, sugars, or the major product of the fermentation, that is, ethanol, can also act as an inhibitor. In this chapter, the major inhibitors in different ethanol processes, as well as different methods to avoid the inhibition effects or remove the inhibitors, are discussed.

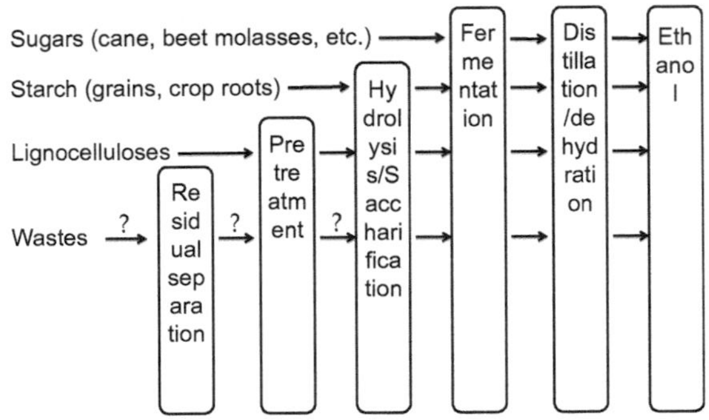

FIGURE 1 An overview of the ethanol process using different raw materials.

2 COMMON INHIBITORS FOR ETHANOL PRODUCTION

2.1 Ethanol as an Inhibitor for "Ethanol" Production

Ethanol is a fermentation product as well as a well-known toxic compound for organisms, even ethanol-producing microorganisms. It is an important antimicrobial compound with a wide-ranging and nonspecific mode of action (John, 1998). Ethanol can easily diffuse through the cell membranes and reduce glucose metabolisms (Ricci et al., 2004). Ethanol can reduce the activity and even denature the glycolytic enzymes. The reason for the inhibitory effects of ethanol is most likely related to reduction of water activity (a_w), ratio of vapor pressure of water in the substance, and the vapor pressure of pure water at the same temperature. Pure water has a_w of 1.0, and most of bacteria and fungi need a_w of higher than 0.91 and 0.7, respectively, and prefer an a_w of 0.99 to grow. *Saccharomyces cerevisiae* can grow only in a medium with a_w between 0.9 and 1.0, with optimal value between 0.975 and 0.999 (John, 1998). The effects of a_w on growth also depend on temperature, pH, nutrient availability, and other factors. High temperature is a source of stress that is synergistic to water stress. Water activity influences the stability of enzymes and proteins significantly.

Addition of ethanol to water can sharply reduce water activity. For instance, 20% ethanol (w/v) can reduce the a_w to 0.895. The presence of 5% ethanol was reported to reduce intracellular pH and affect hydrated cell components such as lipid-lipid bonding in cell membrane, membrane-associated proteins, and glycolytic enzymes, which are vulnerable to the effects of water stress (Jones and Greenfield, 1986; Rose, 1993). Furthermore, fermentation medium compounds, for example, sugars and other metabolites, could further decrease the a_w. Therefore, there is a synergic effect in ethanol and other medium components, and they have negative effects on metabolism and growth.

Industrial strains for ethanol production should have high ethanol tolerance. At low alcohol concentrations, for example, less than 2%, the inhibitory effect is generally negligible, but it could be increased rapidly at higher concentrations (Maiorella et al., 1983). The current ethanol plants produce ethanol with a concentration of about 9–10%, while the concentration for ethanol from lignocelluloses might be in the order of 4–5% (Gnansounou and Dauriat, 2010). There are few microorganisms that can tolerate ethanol at more than 11% concentration (Breisha, 2010).

The metabolic response of the cells to high ethanol concentrations is by synthesizing compatible solutes such as glycerol and trehalose that help the cells to protect against the effects of water stress and hydrogen-bond disruption by ethanol (John, 1998; Ogawa et al., 2000; Taherzadeh et al., 2002). Production of glycerol, an unwanted byproduct of ethanol production, can reduce the a_w problem by protecting hydrated cell components from the effects of water stress. Intracellular trehalose is related to tolerance of yeast cells to ethanol. Trehalose can protect the cells against the water stress induced by ethanol by decreasing the membrane permeability in the presence of ethanol. Trehalose has a protective role against the different types of water stress such as heat shock, freezing, and dehydration (Devantier et al., 2005; John, 1998). Furthermore, ethanol at high concentrations, plays a direct role in the control of gene expression.

A correlation between ethanol and glycogen in the yeast cell has been reported (Dake et al., 2010). The presence of moderate amounts of ethanol (e.g., 2-8%) increases the glycogen and other carbohydrates in the cells, whereas higher concentrations of ethanol (e.g., 10-12%) deplete the glycogen and carbohydrate content along with decreasing cell weight. The plasma membrane is a prime target for ethanol action, while membrane-bound insoluble glycogen might play a protective role in combating ethanol stress (Dake et al., 2010).

Several methods have been suggested to selectively remove the ethanol produced from the fermenting broth to eliminate ethanol inhibition effects (Maiorella et al., 1983). System with vacuum (Ghose et al., 1984; Maiorella et al., 1983), membranes (Escobar et al., 2001; Mori and Inaba, 1990; Shabtai et al., 1991), and extractive fermentation (Daugulis, 1994; Kapucu and Mehmetoglu, 1998; Minier and Goma, 1982) are among the investigated methods.

2.2 Sugar Inhibition

2.2.1 *Catabolite or Sugar Inhibition*

High concentrations of sugars can decrease ethanol yield and productivity. It is referred to as catabolite or sugar inhibition, which can reduce the activity of the enzymes in the fermentative pathway. The sugar inhibition depends on the strain and typically starts at a concentration of 150 g/l glucose. However, there are strains that can grow well on higher concentrations of, for example, 250 g/l (Osho, 2005) or up to 500 g/l sugar (Ok and Hashinaga, 1997). On the other hand, the sugar concentrations of more than 500 g/l result in too low a_w and a very harsh condition for the microorganisms to grow. An industrial application of this fact is molasses, which is a concentrated residual of the sugar industries with about 50% sugar concentration. Molasses can be stored in an open environment for even years without biological degradation.

At a high concentration of sugars, most of the ethanol-producing microorganisms produce ethanol even under aerobic condition rather than producing biomass (crabtree effect). In this condition, production of the oxidative enzymes is inhibited and the cells are forced to fermentative metabolism. However, for industrial application, the strains without catabolite repression are more desirable (Maiorella, 1983).

2.2.2 *Low Sugar Concentration*

The rate of ethanol production under anaerobic conditions is related to the sugar concentration by Monod equation (Maiorella, 1983):

$$v = \frac{v_m C_s}{K_m + C_s},$$

where v is the specific ethanol productivity (g ethanol/g biomass·h), C_s is the sugar substrate concentration (g/l). K_m, saturation constant, is typically between 0.2 and 0.4 g/l, which is much lower than C_s for industrial applications. Therefore, ethanol productivity is very close to v_m and independent of sugar concentration in typical fermentation processes. However, very low sugar concentrations, for example, less than about 3 g/l, can starve the ethanol-producing microorganisms and reduce the ethanol productivity (Maiorella, 1983).

2.2.3 *Interaction of Mixture Sugar Fermentation*

When a mixture of sugars, for example, hemicellulosic hydrolyzate that generally contains glucose, xylose, mannose, and galactose is fermented, the catabolite repression can reduce ethanol production. Catabolite repression, which has been observed for most microorganisms, allows them to adapt quickly to a preferred sugar and metabolize it rapidly. On the other hand, this phenomenon could inhibit synthesis of enzymes involved in catabolism of the sugars other than the preferred one. For instance, when *E. coli* bacteria are cultivated on a mixture of glucose and lactose, they assimilate glucose first and then lactose (Deutscher, 2008; Stulke and Hillen, 1999). Most microorganisms prefer glucose and fructose, and they cannot start assimilating other sugars while these sugars are available in appreciable amounts (Gancedo, 1998). Therefore, when there is a cellulosic hydrolyzate that contains mainly glucose, the rate of ethanol production might be higher than hemicellulosic hydrolyzates, even if it mainly contains hexoses.

2.3 Inhibition by High Salt Concentration

High concentrations of alkali and heavy metal salts, available in many substrates, for example, lignocellulosic hydrolyzates and molasses, can inhibit the fermentation process. In pretreatment and hydrolysis, the metals corrosion could be a main source of inorganic salts in the solution. Furthermore, salts could also be formed by neutralization in some of the pretreatment methods, such as dilute-acid or alkaline processes. In other processes salts are part of the processes, in which the carbon source is produced; such as sugar industries, in which molasses with high concentration of calcium and other salts are produced. This high concentration of salts results in high osmotic stress of the culture, which has negative effects for ethanol fermentation. Baker's yeast and other microorganisms try to reduce this effect by producing, for example, glycerol (Adler et al., 1985; André et al., 1988; Taherzadeh et al., 2002). Several studies have been carried out to obtain *S. cerevisiae* strains with greater salt tolerance (Sanchez and Cardona, 2008).

One of the strategies in industrial ethanol processes for reducing fresh water and nutrient consumption, and decreasing the amount of wastewater produced, is recycling the stillage, which is the major part of the wastewater of ethanol distilleries (Castro et al., 2010; Larsson et al., 1997; Maiorella et al., 1984). However, this strategy results in accumulation of salts in the process. High concentrations of cations such as Ca^{2+}, Mg^{2+}, K^+, Na^+ and anions, for example, Cl^- and $SO4^{2-}$ could severely inhibit the yeast growth and decrease the productivity of ethanol (Klinke et al., 2004). High mineral salt availability in the medium has more inhibitory effects than high ethanol concentration in, for example, continuous fermentation of xylose by *Thermoanaerobacter thermosaccharolyticum* (Lynd et al., 2001).

2.4 Other Inhibitors Following the Raw Materials

A number of inhibitors are present in the different substrates used for industrial processes. Molasses, the byproduct of sugar industries, is one of the major industrial sources of ethanol production. It contains a variety of inhibitors such as furfural, hydroxymethylfurfural, phenol, vanillin, vanillic acid, hexanol, and heptanol (Gough et al., 1997; Zauner et al., 1979). High osmotic pressure as a result of salt build up in sweet sorghum juice and molasses can also affect their fermentation (Gibbons and Westby, 1989; Maiorella et al., 1983). Proper dilution and precipitation was suggested to improve the yield of ethanol from molasses and the juice.

Spent sulfite liquor (SSL) is a byproduct of pulp mills by sulfite process and widely used as a substrate for ethanol production. It contains different sugar degradation products, for example, furfural, HMF, carboxylic acids, for example, acetic acid, lignosulfonates, and assorted salts (Keating et al., 2006; Nigam, 2001; Taherzadeh et al., 2003). Hydroxycarboxylic acids, for example, glycolic acid and lactic acid, are common degradation products of wood in pulping of wood (Alén et al., 1991). Evaporation, overliming with $Ca(OH)_2$, and microorganisms adaptation are suggested pretreatments for a successful fermentation of SSL (Nigam, 2001).

Over 88 million tons of citrus fruits are produced globally per year. Half of these fruits end in juice factories and are squeezed and the remainder, including segment membranes, peel, and other byproducts, are called citrus wastes. These wastes contain appreciable amounts of carbohydrate polymers, which are suitable for ethanol production. However, this waste contains a native inhibitor, D-limonene, which is a very toxic compound (Pourbafrani et al., 2007). It is a severe inhibitor of microorganisms in ethanol production. Removal of limonene from citrus processing waste to less than 0.1%, by, for example, steam purging, is necessary for efficient bioconversion of the waste (Widmer et al., 2010).

The substrates containing Maillard- and caramelization-reactions products contain inhibitors. These components can reduce the formation of ethanol up to 80% (Tauer et al., 2004). The Maillard reaction is a chemical reaction between an amino acid and a reducing sugar at elevated temperatures that can occur in many processes, such as concentration of sugars or malt processing for brewing, as the natural amino acids as well as the sugars are available (Tauer et al., 2004). In caramelization, the reducing carbohydrates undergo enolization, dehydration, and cyclization reactions (Rufian-Henares and de la Cueva, 2008). Both Maillard and caramelization reactions could result in the formation of inhibitory components such as HMF.

3 FORMATION OF INHIBITORS IN LIGNOCELLULOSES HYDROLYZATES

The choice of pretreatment and/or acid hydrolysis of lignocelluloses may result in a variety of inhibitors (Jeon et al., 2010; Taherzadeh and Karimi, 2007; Taherzadeh and Karimi, 2008). Furans, carboxylic acids, and phenolic compounds are the prominant and most investigated inhibitors (Almeida et al., 2009a; Himmel, 2008):

Hemicellulose → Xylan → Xylose → Furfural → Carboxylic acids
Hemicellulose → Mannan → Mannose → HMF → Carboxylic acids
Hemicellulose → Galactan → Galactose → HMF → Carboxylic acids
Hemicellulose → Arabinan → Arabinose → Furfural → Carboxylic acids
Cellulose → Glucan → Glucose → HMF → Carboxylicacids
Lignin → Phenolic compounds
Acetyl groups → Acetic acid.

These chemicals can affect the microorganisms by different mechanisms that are described here in brief. There have been many attempts to identify some key inhibitors by different methods such as microtox assay (Rivard et al., 1996), but the results cannot generally reflect the fermentability of different hydrolyzates. The concept of "severity factor, Log (R_0)" was another factor presented in 1987 by Overend as (Palmqvist and Hahn-Hägerdal, 2000b):

$$\mathrm{Log}(R_0) = \log\left[t \times \exp\left(\frac{T - T_{\mathrm{ref}}}{14.75}\right)\right]$$

where t is the residence time of pretreatment or hydrolysis, T is the reaction temperature, and T_{ref} is the reference temperature, set at 100 °C. A higher severity factor means a more inhibiting hydrolyzate. This equation is useful, when heat treatment in combination with, for example, acids is applied. However, there are many other factors as well as some synergistic effects between different inhibitors, which make the application of this equation very limited or misleading in predicting the fermentability of the hydrolyzates.

3.1 Furan Derivatives

Inhibition effects of the furan derivatives on microorganisms have been extensively studied (Almeida et al., 2009a). In this section, origin, nature, and inhibition mechanism of the well-known furans, that is, furfural and hydroxymethyl furfural are discussed.

3.1.1 Furfural

Furfural is known as one of the most important fermentation inhibitors of many microorganisms (Baek et al., 2008; Duarte et al., 2005; Gutierrez et al., 2002; Keating et al., 2006; Kelly et al., 2008; Lu et al., 2007; Miller et al., 2009a). In the presence of acids at high temperatures, five carbon sugars such as xylose can undergo dehydration and lose three water molecules to form furfural (Zeitsch, 2000):

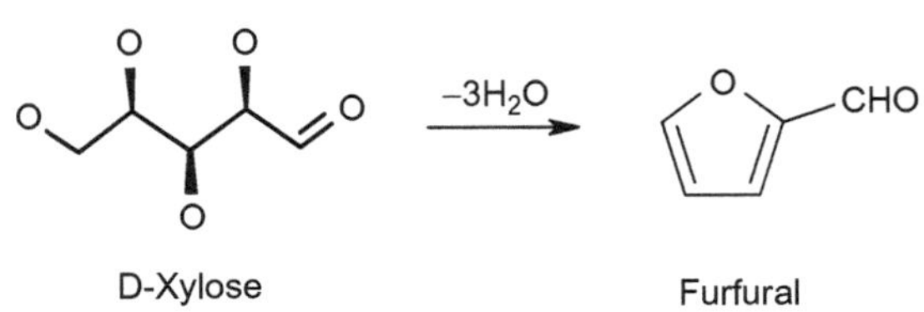

It is therefore impossible to completely avoid furfural formation in acid treatments, particularly in dilute-acid hydrolysis, which is among the well-developed methods for hydrolysis and pretreatment of lignocelluloses.

The effects of furfural on fermentation by different microorganisms have been the subject of several investigations, particularly in the fermentation of the acid hydrolyzates of lignocelluloses (Azhar et al., 1981; Banerjee et al., 1981b; Brovenko and Gusel'nikova, 1993; Horvath et al., 2001; Karimi et al., 2005; Palmqvist and Hahn-Hägerdal, 2000a; Palmqvist et al., 1999). While furfural concentration is in the order of 1.0 g/l and above, it has clear negative effects on many bacteria, yeasts, and filamentous fungi on the vitality, viability, specific growth rate, lag phase, ethanol yield, and ethanol productivity (Almeida et al., 2007; Azhar et al., 1982; Nilvebrant et al., 2001; Taherzadeh et al., 2000a). However, the inhibitory effects of furfural depend on its concentration and the strain used. Furfural decreases the rate of ethanol production but typically does not affect the final ethanol yield (Gutierrez et al., 2002; Taherzadeh et al., 1999a). Furthermore, there is a synergistic effect between furfural and other inhibitors which are available in hydrolyzates (Almeida et al., 2007).

The inhibitory effects of furfural on the performance of microorganisms have been explained through different mechanisms. Several intercellular enzymes which are involved in growth and fermentation, particularly dehydrogenases, have shown to be sensitive to furfural (Modig et al., 2002). It can also affect the energy metabolism by affecting the glycolytic and TCA enzymes. The following enzymes, which are very sensitive to furfural, are among the most important enzymes involved in ethanol production (Figure 2):

- *Hexokinase* (enzyme 1 in Figure 2; Taherzadeh, 1999), which is responsible in phosphorylation of six-carbon sugars for breakdown of glucose, can be inhibited by furfural,
- *Glyceraldehyde 3-phosphate dehydrogenase* (GPD, enzyme 6 Figure 2) can be inhibited by furfural. GPD is probably one of the key inhibited enzymes by furfural in the glycolysis pathway.
- *Alcohol dehydrogenase* (ADH, enzyme 14 in Figure 2) is responsible for transformation of acetaldehyde to ethanol and vice versa (Modig et al., 2002). Therefore, furfural can inhibit formation of ethanol by inhibiting ADH.
- *Pyruvate dehydrogenase* (PDH, enzyme 28 in Figure 2) is responsible for transformation of pyruvate into acetyl-CoA. The acetyl-CoA is an intermediate, which link the glycolysis to the TCA cycle. Therefore, inhibition of PDH (Modig et al., 2002) can inhibit the cellular respiration and consequently the growth of the microorganisms.

Besides the effects of furfural on these enzymes, furfural may damage the vacuole and mitochondrial membranes, chromatin, and actin (Almeida et al., 2007). In conclusion, furfural can redirect the cell energy by intracellular ATP and NAD(P)H levels reduction by enzymatic inhibition, consumption of cofactors, and damaging of membrane, genetic materials, and some proteins (Almeida et al., 2007). However, the cells can tolerate low concentrations of furfural and convert them to less toxic compounds such as furfuryl alcohol and furoic acid (Taherzadeh et al., 1999a). Furfuryl alcohol, which is the reduced form of furfural, is produced from furfural using ADH (Horvath et al., 2001), while furoic acid, the oxidized product of furfural, is produced using aldehyde dehydrogenase (Modig et al., 2002; Sarvari Horvath et al., 2003).

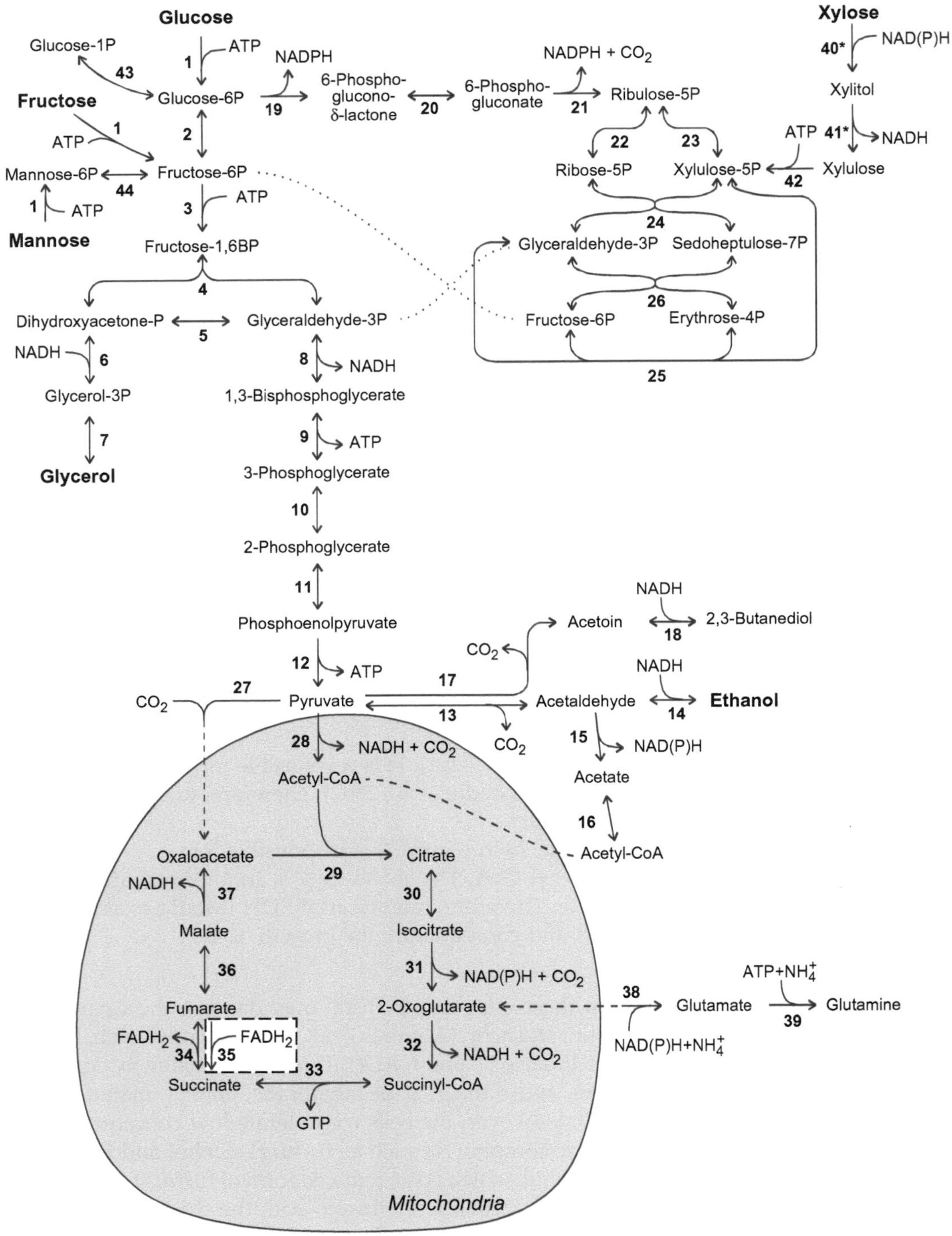

FIGURE 2 The central metabolic pathways (EMP, PPP, TCA cycle, NH_4^+, and sugar assimilation) in yeast *(adapted from Taherzadeh, 1999)*. Dashed lines indicate transport across mitochondrial membrane. $NAD(P)^+$, ADP, H^+,

FIGURE 2—CONT'D P_i, and CoA are not included in the picture for clarity. The enzymes marked with "*" are not present in wild-type *S. cerevisiae*. Enzyme notations are:

1	Glucokinase (GLK) and Hexokinase (HXK)	23	Ribulose-5-phosphate epimerase (RPE)
2	Phosphoglucose isomerase (PGI)	24	Transketolase (TKL)
3	Phosphofructokinase (PFK)	25	Transketolase (TKL)
4	Aldolase (ALD)	26	Transaldolase (TAL)
5	Triosephosphate isomerase (TPI)	27	Pyruvate carboxylase (PYC)
6	Glycerol-3-phosphate dehydrogenase (GPD)	28	Pyruvate dehydrogenase (PDH)
7	Glycerol-3-phosphatase (GPP)	29	Citrate synthase (CS)
8	Triosephosphate dehydrogenase (TDH)	30	Aconitase
9	Phosphoglycerate kinase (PGK)	31	Isocitrate dehydrogenase (IDH, IDP1, IDP2)
10	Phosphoglycerate mutase	32	2-Oxoglutarate dehydrogenase (OGDH)
11	Enolase (ENO)	33	Succinyl-CoA synthetase
12	Pyruvate kinase (PYK)	34	Succinate dehydrogenase
13	Pyruvate decarboxylase (PDC)	35	Fumarate reductase (cytosolic, FRDS)
14	Alcohol dehydrogenase (ADH)	36	Fumarase
15	Aldehyde dehydrogenase (AlDH)	37	Malate dehydrogenase
16	Acetyl-CoA synthetase (ACS)	38	Glutamate dehydrogenase (GDH)
17	Three pathways via acetaldehyde-TPP complex	39	Glutamine synthetase (GS)
18	Butanediol dehydrogenase	40*	Xylose reductase (XR)
19	Glucose-6-phosphate dehydrogenase	41*	Xylitol dehydrogenase (XDH)
20	Lactonase	42	Xylulokinase
21	6-Phospho-gluconate dehydrogenase	43	Phosphoglucomutase (PGM)
22	Ribose-5-phosphate ketol-isomerase (RKI)	44	Phosphomannoisomerase (PMI)

3.1.2 Hydroxymethylfurfural

Hydroxymethylfurfural (HMF) is a product of dehydration of hexose sugars and carbohydrates:

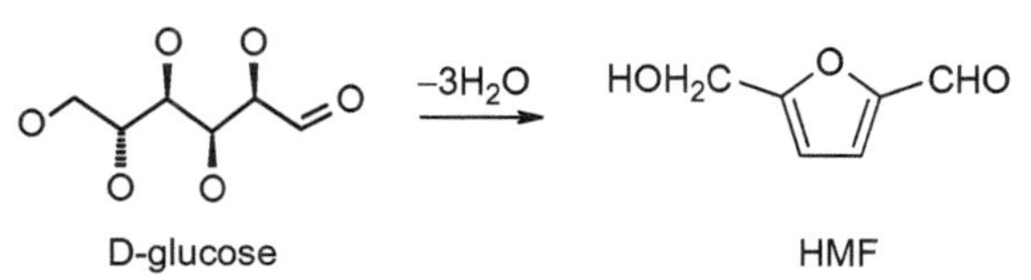

HMF has been identified in a wide variety of heat-processed substrate containing hexose sugars such as fruit juices and molasses, and polycarbohydrates, for example, cellulosic materials. HMF is available in many substrates for ethanol production, since it is an unavoidable product of acid hydrolysis of lignocelluloses, as well as Maillard reaction and caramelization of sugars (Rufian-Henares and de la Cueva, 2008; Tauer et al., 2004).

HMF is considered a potential feedstock for renewable liquid fuels and chemicals. However, it was reported to be a severe inhibitor for many microorganisms. The effects of HMF on ethanol production by *S. cerevisiae* have been deeply investigated in several studies (Laadan et al., 2008; Palmqvist and Hahn-Hägerdal, 2000b; Taherzadeh et al., 2000b). The negative effects on growth rate and fermentation rate were reported for HMF in concentrations of more than 1.0 g/l (Banerjee and Vishwanathan, 1974; Banerjee et al., 1981a; Taherzadeh et al., 2000b). The bioconversion rate of HMF is slower than furfural; however, it is a less severe inhibitor for many microorganisms than furfural (Sanchez and Bautista, 1988). When a high concentration of HMF is available in the medium, no cell growth occurs. Prolongation of the lag phase is the main effect of HMF on fermentation. Similar to furfural, HMF can inhibit the activity of several enzymes, for example, ADH, PDH, and ALDH. HMF, similar to furfural, is converted to less toxic compounds in respirative and fermentative conditions (Taherzadeh et al., 2000b):

$$\text{HMFurfuryl alcohol} \xleftarrow[+\text{NAD(P)H}]{\text{reduction}} \text{HMF} \xrightarrow[-\text{NAD(P)H}]{\text{oxidation}} \text{HMfuroicacid}.$$

3.2 Carboxylic Acids

Acetic acid, formed by deacetylation of hemicellulosic part of biomass, is the most common carboxylic acid available in the lignocellulosic hydrolyzates. Levulinic acid is a product of HMF breakdown in acidic and elevated temperatures conditions. Formic acid can be formed from furfural and HMF degradation under acidic and high temperature conditions. However, unlike furans and phenolic compounds, no synergistic effects were shown among the three acids (Larsson et al., 1999a).

The inhibitory effects of carboxylic acids are related to accumulation of their intracellular anions. Only the molecular (undissociated) form of carboxylic acids can diffuse through the membranes of microorganisms; thus, the toxicity of the acids is pH dependent (e.g., Figure 3;

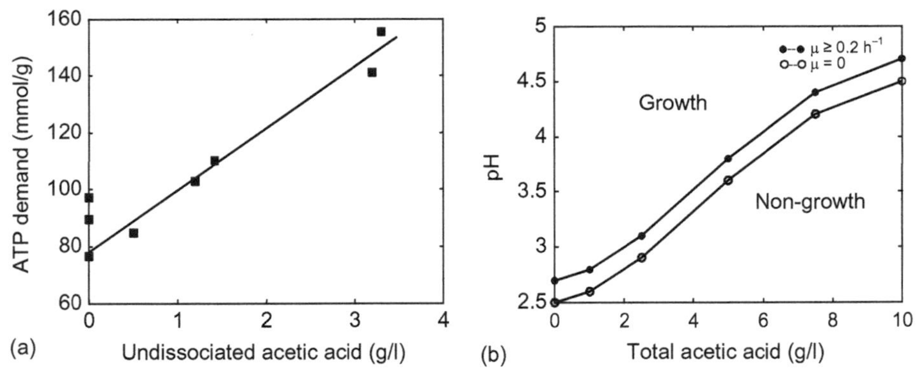

FIGURE 3 (a) The ATP demand for the growth of *S. cerevisiae* in media containing different concentrations of undissociated acetic acid, and (b) region of anaerobic growth of *S. cerevisiae* as a function of medium pH and total acetic acid concentration *(adapted from Taherzadeh, 1999).*

Almeida et al., 2007; Gottschalk, 1987; Taherzadeh et al., 1997b). The acid molecules can dissociate in the cells, since the intercellular pH (≈7) is higher than that of the medium (typically pH 4.5-5.5 for the yeasts). It results in decreasing the cytosolic pH. Therefore, the cells try to pump out the protons across the cell membrane. This pumping needs energy in the form of ATP (Figure 3). Therefore, high amounts of the acid showed negative effects on biomass and ethanol production. However, low concentrations of the acids can improve the production of ethanol by stimulating ATP production and lower yield of biomass formation (Taherzadeh et al., 1997b).

There are some efforts to replace glycerol formation by reducing acetate to ethanol as the mechanism for NADH reoxidation by metabolic engineering (cf. Figure 2). Glycerol is an unavoidable byproduct of ethanol production by natural ethanol-producing microorganism under anaerobic conditions. Formation of glycerol is essential for reoxidizing NADH produced in anaerobic fermentation. As a strategy to improve ethanol yields (Taherzadeh et al., 1996), glycerol elimination by partial addition of acetate was suggested. This could be a successful strategy to consume acetic acid, which is an inhibitor, and consume less carbon source for glycerol production (Guadalupe Medina et al., 2010).

The yeast genes conferring protection against acetic acid have been recently identified (Mira et al., 2010; Zhang et al., 2011), and several genes for genetic engineering were recommended to obtain more robust yeast strains against acetic acid toxicity. These genes are involved in many cellular activities such as transcription, internal pH homeostasis, carbohydrate metabolism, cell wall assembly, biogenesis of mitochondria, ribosome and vacuole, and in the sensing, signaling, and uptake of various nutrients, in particular iron, potassium, and amino acids. Furthermore, it was shown that the higher concentration of K^+ can improve the tolerance to acetic acid (Mira et al., 2010).

The concentration of acetic acid is usually more than formic and levulinic acid in lignocellulosic hydrolyzates. However, acetic acid showed less inhibitory effects than levulinic acid. Formic is more inhibitory than levulinic and acetic acid. The more inhibitory effects of formic acid is related to its lower p*Ka* value. It leads to availability of its undissociated form in lower concentrations as well as its lower molecule size. Levulinic acid is highly hydrophobic and can diffuse into the cells easier than acetic acid, thus showing more inhibitory effects than acetic acid (Almeida et al., 2007).

3.3 Phenolic Compounds

Lignin breakdown as well as carbohydrate degradation during acid hydrolysis can produce a wide range of phenolic compounds. The types of these compounds depend on the treatment and lignin structure in the biomass (Almeida et al., 2007). A mixture of phenolics such as syringaldehyde, 4-hydroxybenzaldehyde, catechol, vanillin, 4-hydroxybenzoic acid, dihydroconiferyl alcohol, coniferyl aldehyde, and syringic acid was reported as lignin degradation products. These components are divided into three groups 4-hydroxybenzyl (H), guaiacyl (G), and syringyl (S). Softwood lignins mainly produce G phenols, whereas hardwoods and agricultural residue produce all the three types (Himmel, 2008). It should be noticed that some of these phenolics are thought to originate from extractive components rather than lignin (Klinke et al., 2004). Among the different phenolic compounds, the lower molecular weight compounds are more toxic to microorganisms.

The phenolic compounds can decrease the rate of ethanol production, but usually not ethanol yield. They could also decrease the growth rate and yields by affecting the cell membrane integrity and activities. The resistance of microorganisms to these types of inhibitors depends on the composition of the membrane, particularly fatty-acid composition, phospholipid headgroups, and in the protein content of cell membrane (Heipieper et al., 1994). It is difficult to find the mechanisms of inhibition of the phenolic compounds on microorganisms. Furthermore, these compounds have generally low concentration in the culture and are sometimes difficult to analyze by traditional methods. These phonelic compounds can also affect the membrane activity of the cells (Almeida et al., 2007).

3.4 Glycolaldehyde

Formation of glycolaldehyde from lignocellulosic materials was recently investigated, and it was presented as a severe inhibitor (Jayakody et al., 2011). Glycolaldehyde ($HOCH_2$–CH=O) is the simplest possible sugar, which is the only possible 2-carbon monosaccharide. It was first detected when lignocelluloses were treated with pressurized hot water. In this process, glycolaldehyde was formed in ranges from 1 to 24 mM, while 1-10 mM is enough for severely inhibiting cell cultivation. This component is believed to be one of the main substances responsible for inhibiting fermentation after pressurized hot water degradation of lignocelluloses. The genes encoding ADH, methylglyoxal reductase, polysomes, and the ubiquitin ligase complex are required for glycolaldehyde tolerance (Jayakody et al., 2011). However, more investigation is necessary to study the formation of glycolaldehyde during the other treatment, for example, acid hydrolysis.

4 STRATEGIES FOR MINIMIZING THE EFFECTS OF INHIBITORS

There are several different strategies to deal with the inhibitors for ethanol production. In this section, these strategies are briefly discussed.

4.1 Substrate Concentration

The concentrations of sugars, ethanol, and salts should remain less than the threshold of the tolerance of the microorganism used. Therefore, the following points should be considered:

(a) *The suitable concentration of sugars is high enough to stimulate maximum assimilation rate of sugars and less than the corresponding inhibiting concentration. If too high a concentration is going to be used, fed-batch or continuous cultivations would be suitable, to keep low sugar concentrations in the culture.*
(b) *Avoid high concentration of ethanol by choosing a suitable sugar concentration, or removing the produced ethanol by, for example, vacuum evaporation or membrane separation.*
(c) *If the substrate creates a very high osmotic stress because of high sugars or salts (e.g., molasses), it should be diluted to an acceptable level suitable for the microorganism used.*
(d) *When the stillage or wastewater is reused, its blending with the fresh water should be in an appropriate ratio to avoid the accumulation of inhibitors.*

4.2 Avoiding the Formation of Inhibitors During Pretreatments

One of the points in pretreatment or hydrolysis of substrate, especially in complex substrates such as starchy and lignocellulosic materials, is to avoid the formation of inhibitors. For this purpose, one should keep the following points in mind:

(a) *Try to avoid destroying the carbohydrate by choosing suitable pretreatment or hydrolysis conditions.*

(b) *It is better to avoid inhibitory chemicals and reagents for pretreating lignocellulosic materials.*

(c) *Separate the inhibitors after the pretreatment, by, for example, washing the solid pretreated materials.*

(d) *Enzymatic hydrolysis does not usually create high amounts of inhibitors for fermentation. However, it is difficult to keep the enzymes sterile and keep the hydrolysis process uncontaminated, and some inhibitory components might be produced by contamination.*

4.3 *In Situ* Detoxification by Fermenting Microorganisms

It is possible for the fermenting microorganism to detoxify some inhibitors. Conversion of toxic compounds to non- or less-toxic compounds by fermenting organisms is probably the easiest way to get rid of inhibitors, even if it makes a lag phase or reduces the ethanol yield up to a reasonable extent. The *in situ* detoxification of the severe inhibitors is discussed in this section.

As long as furfural is available in the media, no growth can occur for most of the microorganisms. However, the microorganisms can convert furfural to less inhibitory alcohol under aerobic and anaerobic conditions (Taherzadeh et al., 1999a). The bioconversion of furfural resulted in depletion of NAD(P)H, which may be the reason for increasing the levels of acetaldehyde in the media (Palmqvist et al., 1999; Taherzadeh et al., 1999a):

$$\text{Furfuryl alcohol} \xleftarrow[+\text{NAD(P)H}]{\text{anaerobic reduction}} \text{Furfural} \xrightarrow[+\text{NAD(P)}^+]{\text{aerobic oxidation}} \text{Furoic acid},$$

$$\text{Ethanol} \xleftarrow[+\text{NAD(P)H}]{\text{anaerobic reduction}} \text{Acetaldehyde} \xrightarrow[+\text{NAD(P)}^+]{\text{aerobic oxidation}} \text{Acetic acid}.$$

Furoic acid can be formed only under aerobic condition, while furfuryl alcohol is the product of furfural detoxification under both aerobic and anaerobic conditions. The bioconversion reactions are shown to be coupled with oxidation of NADH (Bowman et al., 2010). Higher expression levels of reductases in yeasts and bacteria are related to their tolerance to furfural and HMF (Gutiérrez et al., 2006; Petersson et al., 2006).

Furfural reductase, a type of alcohol-aldehyde oxido-reductase, catalyzes the essentially irreversible reduction of furfural with NADPH (as cofactor) and is relatively specific for the reduction of furfural (Gutierrez et al., 2006). Overexpression of the gene encoding the furfural reductase was suggested for development of the furfural-tolerant strain for conversion of lignocellulosic biomass to alternative fuels and biobased products (Gutierrez et al., 2006).

Miller et al. (2009b) reported genes encoding 12 oxidoreductases to vary in response to furfural challenges, based on mRNA expression levels in a parent and mutant strain of an ethanologenic *E. coli*. When expressed, eight genes increased furfural tolerance in the parent strain.

Expression of three of the silenced genes was reported to decrease furfural tolerance compared to that in the parent bacteria. It is also possible to obtain furfural tolerant strains, for example, *S. cerevisiae* NRRL Y-50049, which has the ability to convert biotransform furfural to furan methanol (Liu et al., 2008).

The medium composition can also affect the resistance of microorganisms. For instance, amino acid enrichment of the substrate enhanced the ability of *Pichia stipitis* to resist furfural and HMF. The addition of excessive amounts of $MgSO_4$ makes the cells more susceptible to inhibitor exposure (Slininger et al., 2009).

Most microorganisms are able to convert HMF to less-toxic compounds. HMF can be reduced mainly to Hydroxymethylfurfuryl alcohol (HMF alcohol or HMFAL) and less than 4% to hydroxymethyl furan carboxylic acid (HMFCA) by HMF reductases either under aerobic or anaerobic conditions (Nemirovskii and Kostenko, 1991; Nemirovskii et al., 1989; Taherzadeh et al., 2000b). Furthermore, some other HMF degradation products were also reported as the result of condensation between HMF and acetaldehyde (Taherzadeh et al., 2000b). Its reduction is coupled with both NADH and NADPH oxidation. It was shown (Almeida et al., 2009b) that NADH-dependent HMF reductase influenced the production of ethanol more than NADPH-dependent HMF reduction.

4.4 Microorganisms Modification

4.4.1 *Microorganisms Adaptation*

Adaptation of microorganisms to inhibiting hydrolyzates is suggested as an alternative or improvement to detoxification. The adaptation can increase the fermentation rate and yield of ethanol production from an inhibitory media (Taherzadeh and Karimi, 2007; Tomas-Pejo et al., 2010). In some cases, outstanding improvements have been reported. For instance, the 90-h lag phase was reduced to 17 h, when a strain of *S. cerevisiae* adapted to a media supplemented with 17 mM furfural after 300 generations (Heer and Sauer, 2008). Adaptation may also be used to improve simultaneous utilization of different sugars. For instance, several strains of *S. cerevisiae* cannot utilize galactose in the presence of glucose. However, the adapted strain may be able to utilize glucose and galactose simultaneously in the presence of acetic acid (Olsson and HahnHagerdal, 1996).

The best method for adaptation of fermentative organisms is probably the continuous fermentation process. This process could naturally adapt the microorganisms to the inhibitors, besides providing other advantages such as higher ethanol productivity, lower labor costs and time for cleaning and filling, and microbial contamination (Brethauer and Wyman, 2010). Weber et al. (2010) reviewed and compared the properties of various microorganisms and current research efforts to develop a reliable microorganism for efficient conversion of lignocelluloses to ethanol.

4.4.2 *Genetic Engineering*

Metabolic engineering can be applied for solving the inhibition problems. Overexpressing genes encoding enzymes for resistance to inhibitors such as cloning of the laccase genes and altering cofactor balances is among the strategies for fermentation of toxic hydrolyzates (Parawira and Tekere, 2011). Overexpressing genes encoding furfural reductase, laccase, and phenylacrylic acid decarboxylase are among the most often recommended methods

(Li et al., 2009). To construct a strain for efficient conversion of hydrolyzates, the strain should ferment the sugar mixture to ethanol and resist/metabolize inhibitors in lignocellulosic hydrolyzates. The strain should preferably not produce lactic acid and xylitol as byproducts. Yan et al. (2009) used the biotechnology of inactive intergeneric fusion between *S. cerevisiae* and *Pachysolen tannophilis* for this propose. The strain results showed good ethanol production yields, ethanol tolerance, and resistance to the inhibitors in dilute-acid hydrolyzate. However, it has xylitol as a byproduct.

4.5 Fermentation Strategy to Overcome Inhibitions

4.5.1 Batch, Fed-Batch, or Continuous Operations

The fermentation process can be carried out in batch, fed-batch, or continuous modes of operations. While all the materials are fed once to the reactor in the batch process, in the fed-batch and continuous processes they are fed continuously with a small amount of the media. Therefore, the sugars as well as the inhibitors are in high concentrations at the beginning of the fermentation in a batch reactor, while their concentrations can be kept low in the fed-batch and continuous modes. A combination of these facts, as well as the ability of converting some of the inhibitors "at their low concentrations" by the fermenting microorganisms, results in the success of the fed-batch and continuous fermentation, while the batch fermentations of the same hydrolyzates might fail (Taherzadeh et al., 1999b).

If the hydrolyzate is severely inhibiting, the cells will not grow in batch cultivation and no fermentation will occur (Taherzadeh et al., 1997a). However, if the culture is slightly toxic, *in situ* detoxification can partly occur, but might not be enough to complete the fermentation. Chung and Lee (1985) tried to ferment acid hydrolyzates by continuous cultivation. Although they succeeded in the fermentation, they observed 90% decrease in the viable cell number after three residence times. Reducing the cell viability results in lowering the dilution rate, and consequently, the productivity of the cultivation is reduced and eventually washout of the cells occurs. Although the cells can be retained in the reactor by, for example, filtration (Brandberg et al., 2007; Lee et al., 1996), the cells have difficulties with toxic substrates and lose their viability. In such cases, fed-batch cultivation could be an alternative.

Fed-batch combines part of the advantages of both batch and continuous cultivations. In fed-batch cultivation, there is no washout of biomass, and concentration of inhibitors in the media is kept low if *in situ* detoxification can occur. In this case, inhibitor concentrations and cell viability are functions of the dilution rate (Taherzadeh et al., 1999b). When a high dilution rate of an inhibiting hydrolyzate is applied, both ethanol production and cell growth can be stopped (Figure 4a). At a low dilution rate, the cells continue to produce ethanol and convert the inhibitors, but the growth rate is very close to zero when the initial glucose has been assimilated (Figure 4b). Consequently, there is an optimum (or maximum) dilution rate for the fed-batch cultivation of each hydrolyzate.

If the hydrolyzate is easily fermentable, one can apply a relatively high dilution rate and minimize the total fermentation time. On the other hand, for a severely inhibiting hydrolyzate, a low dilution rate should be applied. The low dilution rate provides enough time for the cells to convert part of the inhibitors, which results in a less inhibiting medium. These facts were the basis of adaptive control of dilution rates in fed-batch processes (Nilsson et al., 2001; Taherzadeh et al., 2000c)

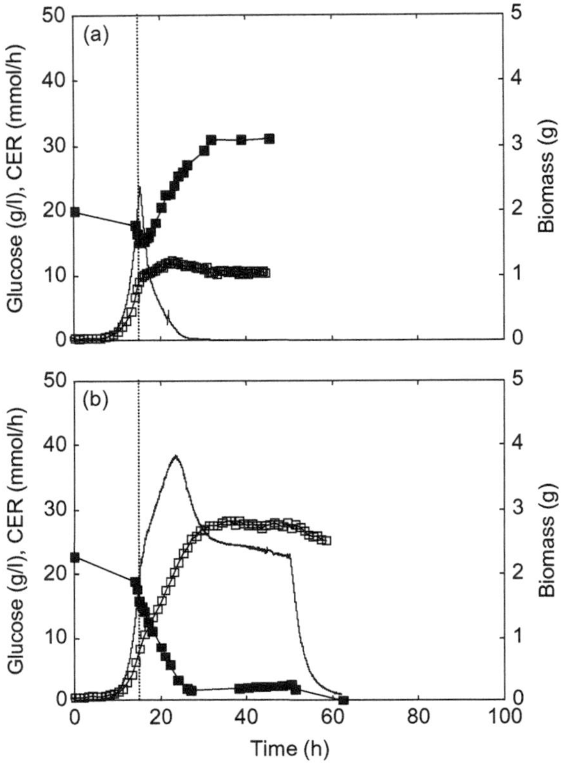

FIGURE 4 Effects of feeding rate in fed-batch cultivation on growth and sugar consumption. The figure shows carbon dioxide evolution rate (solid line), total biomass (□), and glucose concentration (▪) during anaerobic fed-batch fermentation by *S. cerevisiae* of a severely inhibiting spruce hydrolyzates. The reactor contained initially 1.0 l of defined synthetic medium (20 g/l glucose), and the final volume was 2.7-2.8 l. Feed rates (a) 100 ml/h, (b) 50 ml/h. Starting of feed is indicated by the dashed line.

4.5.2 Operation with High Cell Density

When the cells are exposed to the inhibitors, they are able to convert some of the inhibitors e.g. furfural and HMF to less toxic compounds (Taherzadeh et al., 1999a). This ability is reported to be dependent on the cell concentration in the inhibiting culture (Navarro, 1994). It means that when higher cell density is applied, the cells have a better chance to survive in the toxic environment. This concept was applied in several attempts to increase the cell concentrations (Figure 5) by, for example, cell recycling via filtration, sedimentation or

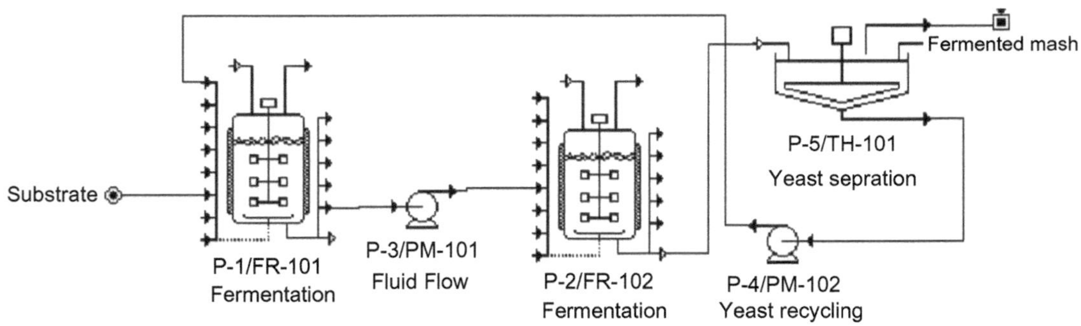

FIGURE 5 Schematic diagram of a two-reactor high-cell density fermentation system with yeast recycling.

immobilization (Brandberg et al., 2007), or using flocculating yeast strains (Dehkhoda et al., 2009; Purwadi et al., 2007). These high-cell density systems, particularly with flocculating yeast, were reported as a successful method for fermenting toxic hydrolyzates.

4.5.3 *Encapsulation*

Another strategy to overcome the inhibition effect is to use encapsulated cells (Park and Chang, 2000). In addition to providing high cell density, the cells are captured in this method inside a capsule, in which the microenvironment created enables the cells to tolerate toxic media (Talebnia et al., 2005). Continuous fermentation of a toxic dilute-acid hydrolyzate with free cells resulted in reducing cell viability to 25% at dilution rate (D) 0.1 h^{-1} and complete failure of the fermentation at D 0.2 h^{-1}. However, continuous cultivation of the same hydrolyzate with encapsulated baker's yeast even at D 0.5 h^{-1} was successful and the cells retained their viability at 85% (Talebnia and Taherzadeh, 2006). The theory behind this success is that as the cells located next to the membranes are exposed to the inhibitors and convert them, the back layers of the cells receive only sugars and grow and eventually replace the first layers (Talebnia and Taherzadeh, 2007). Despite the success of encapsulation in fermenting the toxic hydrolyzates, the capsules still need further development before commercialization.

4.6 Detoxification of the Substrates

Different methods have been recommended for the detoxification of lignocellulosic hydrolyzates. However, the need for detoxification and the method used depend both on the type of inhibitors and tolerance of the fermenting microorganism. The detoxification is typically accompanied by additional equipment, production of additional wastes, and loss of sugars. Therefore, detoxification should be considered only if the fermentation cannot succeed without it.

4.6.1 *Biological Detoxifications*

Specific enzymes or microorganisms can be used for detoxification of toxic compounds in the hydrolyzates in a process called bioabatement. Some specific enzymes such as lignolytic enzymes, for example laccase, can detoxify the hydrolyzates by oxidative polymerization of low-molecular-weight phenolic compounds (Mussatto and Roberto, 2004). Many microorganisms such as *Trichoderma reesei*, *Coniochaeta ligniaria*, *Trametes versicolor*, *Candida guilliermondii*, *Pseudomonas putida*, *Amorphotheca resinae*, and *Ureibacillus thermosphaericus* were used for detoxification of hydrolyzates (Parawira and Tekere, 2011; Zhang et al., 2010a). White-rot fungi such as *T. versicolor* produce laccase and peroxidase enzymes that can detoxify wood hydrolyzates by detoxification of acid and phenolic compounds (Jönsson et al., 1998). Furthermore, some microorganisms can selectively remove the inhibitors. For instance, a mutant species of *S. cerevisiae* is able to remove acetic acid effectively from 6.8 to less than 0.4 g/l in wood hydrolyzate (Schneider, 1996).

Although biological detoxification is efficient in eliminating fermentation lag times associated with inhibitory compounds, the prolongation of incubations with the bioabating microorganisms results in the consumption of some glucose and reduced production of ethanol (Nichols et al., 2010).

4.6.2 Physical Detoxifications

In these methods, the toxic compounds are removed from hydrolyzates without addition of enzymes or chemicals. Vacuum evaporation (Dehkhoda et al., 2009; Zhu et al., 2009) for reducing the volatile compounds such as furfural, acetic acid, and vanillin is probably the most commonly applied process among the physical methods. Evaporation has two advantages for fermentation of hydrolyzates:

(a) *Reducing the contents of volatile compounds,*
(b) *Concentrating sugars in the media, as the acid hydrolyzates are usually dilute and the concentration can help the process to produce a higher concentration of ethanol.*

Improvements of fermentation after vacuum evaporation were observed in some cases (Converti et al., 2000). However, the process suffers from a serious drawback, which is concentrating the nonvolatile toxic compounds such as lignin derivatives. It may even result in reducing the fermentability of the hydrolyzates (Parajo et al., 1997; Silva and Roberto, 1999). A high amount of evaporation may even make the hydrolyzates unfermentable for microorganisms (Walton et al., 2010).

4.6.3 Chemical Detoxification

Treatment of hydrolyzates with some chemicals could change the degree of toxicity through precipitation and ionization of some inhibitors. It is possible to adsorb the toxic compounds on ion-exchange resins, activated charcoal, or diatomaceous earths (Carvalho et al., 2006; Fargues et al., 2010; Rodrigues et al., 2001). However, overliming or treatment with $Ca(OH)_2$ is among the most common chemical detoxification methods which could partially remove the phenolic compounds, furfural, and HMF, and significantly enhance hydrolyzate fermentability (Martinez et al., 2001; Millati et al., 2002). In this method, the pH is increased to 10-12 by the addition of lime and remains for a period of time before reducing to the fermenting pH. However, the process may result in the loss of a part of the sugars in hydrolyzates. Nevertheless, choosing a correct pH and retention time can result in removing furans, while the sugars remain unaffected (Purwadi et al., 2004).

Larsson et al. (1999b) compared detoxification with sodium or calcium hydroxide; sulfite; evaporation; anion exchange; enzymatic detoxification with phenoloxidase laccase; and detoxification with the fungus *T. reesei*. Anion exchange, treatment with laccase, treatment with calcium hydroxide, and treatment with *T. reesei* were the most efficient detoxification methods reported, whereas evaporation and treatment with sulfite were the least efficient methods. Anion exchange at pH 10 was the best method for removing all three major groups of the inhibitory compounds but it resulted in partial degradation of fermentable sugars as well.

Addition of reducing agents such as dithionite and sulfite to enzymatic hydrolyzates of spruce wood or sugarcane bagasse showed improvement in both SHF (simultaneous hydrolysis and fermentation) and SSF (simultaneous saccharification and fermentation) processes (Alriksson et al., 2011). These chemicals can be added directly, *in situ*, to the fermentation vessel. This means that the treatment can be performed at a temperature and pH suitable for fermentation without inhibition of enzymes and microorganisms and degradation of fermentable sugars.

4.6.4 Combined Detoxification

Combinations of different detoxification methods might be used to detoxify lignocellulosic hydrolyzates. Different combinations of chemicals with or without adsorption can be applied. For instance, treatment with $Ca(OH)_2$ and then with activated charcoal (Converti et al., 2000; Kuhad et al., 2010) or adsorption onto ion-exchange resins (Canilha et al., 2010) are probably among the most efficient methods.

A single detoxification process might not be enough to make the acid hydrolyzate fermentable for microorganisms. For instance, the toxicity of dilute-acid hydrolyzate of hazelnut shell, which is a high-lignin lignocellulose, cannot be overcome by several single detoxification methods. The most effective method for this propose is pretreatment of the shells with 3% NaOH for delignification followed by detoxification by overliming, and then by two times treatment with charcoal. This could reduce 97% in furans and 88.4% in phenolic compounds and efficiently increase the yields and productivity of the subsequent ethanol production. However, in addition to the costs of detoxification, this treatment resulted in significant loss of sugars (Arslan and Eken-Saracoglu, 2010).

Recently, microbial fuel cell has been reported to reduce the levels of inhibitors (furfural, HMF, vanillic acid, 4-hydroxybenzaldehyde, and 4-hydroxyacetophenone) from stillage with almost complete removal of the inhibitors, while simultaneously producing electricity. This could be an interesting idea for water recycling in lignocellulosic biorefineries (Borole et al., 2009).

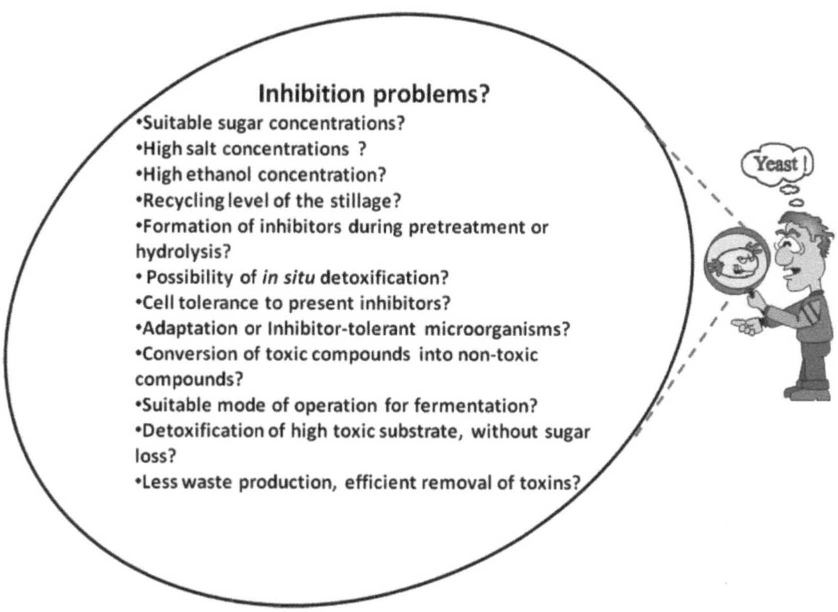

FIGURE 6 A summary of the reasons for the inhibition of fermentation and possible action to remove or reduce the inhibition effects.

5 CONCLUSION

Inhibition in ethanol production may be caused by many factors, including the high concentrations of sugars, ethanol, and salts, as well as the raw materials such as limonene in citrus wastes, or the materials formed in the pretreatment/hydrolysis, such as carboxylic acids, furans, and phenolic compounds. The inhibition effects can be reduced or removed by choosing a suitable level of the components in the substrates and choosing the correct methods and optimizing the pretreatment and hydrolysis steps to reduce the inhibitors, or doing a detoxification prior to the fermentation. However, if the inhibitors enter the bioreactors, choosing the right strategy for the fermentation mode of operation, a tolerant organism, or an organism that can convert the inhibitors might be a suitable method. Figure 6 summarizes these conclusions.

References

Adler, L., Blomberg, A., Nilsson, A., 1985. Glycerol metabolism and osmoregulation in the salt-tolerant yeast Debaryomyces hansenii. J. Bacteriol. 162, 300–306.

Alén, R., Sjöström, E., Suominen, S., 1991. Application of ion-exclusion chromatography to alkaline pulping liquors; separation of hydroxy carboxylic acids from inorganic solids. J. Chemical Technol. Biotechnol. 51, 225–233.

Almeida, J.R., Modig, T., Petersson, A., Hähn-Hägerdal, B., Lidén, G., Gorwa-Grauslund, M.F., 2007. Increased tolerance and conversion of inhibitors in lignocellulosic hydrolysates by *Saccharomyces cerevisiae*. J. Chemical Technol. Biotechnol. 82, 340–349.

Almeida, J.R., Bertilsson, M., Gorwa-Grauslund, M.F., Gorsich, S., Liden, G., 2009a. Metabolic effects of furaldehydes and impacts on biotechnological processes. Appl. Microbiol. Biotechnol. 82, 625–638.

Almeida, J.R., Bertilsson, M., Hahn-Hagerdal, B., Liden, G., Gorwa-Grauslund, M.F., 2009b. Carbon fluxes of xylose-consuming *Saccharomyces cerevisiae* strains are affected differently by NADH and NADPH usage in HMF reduction. Appl. Microbiol. Biotechnol. 84, 751–761.

Alriksson, B., Cavka, A., Jonsson, L.J., 2011. Improving the fermentability of enzymatic hydrolysates of lignocellulose through chemical in-situ detoxification with reducing agents. Bioresour. Technol. 102, 1254–1263.

André, L., Nilsson, A., Adler, L., 1988. The role of glycerol in osmotolerance of the yeast Debaryomyces hansenii. J. Gen. Microbiol. 134, 669–677.

Arslan, Y., Eken-Saracoglu, N., 2010. Effects of pretreatment methods for hazelnut shell hydrolysate fermentation with *Pichia stipitis* to ethanol. Bioresour. Technol. 101, 8664–8670.

Azhar, A.F., Bery, M.K., Colcord, A.R., Roberts, R.S., Corbitt, G.V., 1981. Factors affecting alcohol fermentation of wood acid hydrolyzate. Biotechnol. Bioeng. Symp. 11, 293–300.

Azhar, A.F., Bery, M.K., Colcord, A.R., Roberts, R.S., 1982. Development of a yeast strain for the efficient ethanol fermentation of wood acid hydrolyzate. Dev. Ind. Microbiol. 23, 351–360.

Baek, S.W., Kim, J.S., Park, Y.K., Kim, Y.S., Oh, K.K., 2008. The effect of sugar decomposed on the ethanol fermentation and decomposition reactions of sugars. Biotechnol. Bioprocess Eng. 13, 332–341.

Banerjee, N., Vishwanathan, L., 1974. Effect of furfural and 5-hydroxymethyl furfural on the growth and alcohol production by yeasts. In: Proc. Annu. Conv. Sugar Technol. Assoc. India 40, G1–G4.

Banerjee, N., Bhatnagar, R., Viswanathan, L., 1981a. Development of resistance in *Saccharomyces cerevisiae* against inhibitory effects of Browning reaction products. Enzyme Microb. Technol. 3, 24–28.

Banerjee, N., Bhatnagar, R., Viswanathan, L., 1981b. Inhibition of glycolysis by furfural in *Saccharomyces cerevisiae*. Eur. J. Appl. Microbiol. Biotechnol. 11, 224–228.

Borole, A.P., Mielenz, J.R., Vishnivetskaya, T.A., Hamilton, C.Y., 2009. Controlling accumulation of fermentation inhibitors in biorefinery recycle water using microbial fuel cells. Biotechnol. Biofuels 2, 1–14.

Bowman, M.J., Jordan, D.B., Vermillion, K.E., Braker, J.D., Moon, J., Liu, Z.L., 2010. Stereochemistry of furfural reduction by a *Saccharomyces cerevisiae* aldehyde reductase that contributes to in situ furfural detoxification. Appl. Environ. Microbiol. 76, 4926–4932.

Brandberg, T., Karimi, K., Taherzadeh, M.J., Franzen, C.J., Gustafsson, L., 2007. Continuous fermentation of wheat-supplemented lignocellulose hydrolysate with different types of cell retention. Biotechnol. Bioeng. 98, 80–90.

Breisha, G.Z., 2010. Production of 16% ethanol from 35% sucrose. Biomass Bioenerg. 34, 1243–1249.

Brethauer, S., Wyman, C.E., 2010. Review: continuous hydrolysis and fermentation for cellulosic ethanol production. Bioresour. Technol. 101, 4862–4874.

Brovenko, G., Gusel'nikova, T., 1993. Chemical composition of wood hydrolyzates as a substrate for microbial protein synthesis. 1. Furfural and hydroxymethylfurfural. Gidroliz. Lesokhim. Prom-st. 1, 6–10.

Canilha, L., Carvalho, W., Felipe, M.G., Silva, J.B., Giulietti, M., 2010. Ethanol production from sugarcane bagasse hydrolysate using *Pichia stipitis*. Appl. Biochem. Biotechnol. 161, 84–92.

Carvalho, G.B.M., Mussatto, S.I., Candido, E.J., Silva, J., 2006. Comparison of different procedures for the detoxification of eucalyptus hemicellulosic hydrolysate for use in fermentative processes. J. Chem. Technol. Biotechnol. 81, 152–157.

Castro, G.A., Caicedo, L.A., Almeciga-Diaz, C.J., Sanchez, O.F., 2010. Solvent extraction of organic acids from stillage for its re-use in ethanol production process. Waste Manag. Res. 28, 533–538.

Chung, I.S., Lee, Y.Y., 1985. Ethanol fermentation of crude acid hydrolyzate of cellulose using high-level yeast inocula. Biotechnol. Bioeng. 27, 308–315.

Converti, A., Dominguez, J.M., Perego, P., da Silva, S.S., Zilli, M., 2000. Wood hydrolysis and hydrolysate detoxification for subsequent xylitol production. Chem. Eng. Technol. 23, 1013–1020.

Dake, M.S., Jadhv, J.P., Patil, N.B., 2010. Variations of two pools of glycogen and carbohydrate in *Saccharomyces cerevisiae* grown with various ethanol concentrations. J. Ind. Microbiol. Biotechnol. 37, 701–706.

Daugulis, A.J., 1994. Integrated fermentation and recovery processes. Curr. Opin. Biotechnol. 5, 192–195.

Dehkhoda, A., Brandberg, T., Taherzadeh, M.J., 2009. Comparison of vacuum and high pressure evaporated wood hydrolyzate for ethanol production by repeated fed-batch using flocculating *Saccharomyces Cerevisiae*. Bioresources 4, 309–320.

Deutscher, J., 2008. The mechanisms of carbon catabolite repression in bacteria. Curr. Opin. Microbiol. 11, 87–93.

Devantier, R., Scheithauer, B., Villas-Boas, S.G., Pedersen, S., Olsson, L., 2005. Metabolite profiling for analysis of yeast stress response during very high gravity ethanol fermentations. Biotechnol. Bioeng. 90, 703–714.

Duarte, L.C., Carvalheiro, F., Neves, I., Girio, F.M., 2005. Effects of aliphatic acids, furfural, and phenolic compounds on Debaryomyces hansenii CCMI 941. Appl. Biochem. Biotechnol. 121-124, 413–425.

Escobar, J.M., Rane, K.D., Cheryan, M., 2001. Ethanol production in a membrane bioreactor: pilot-scale trials in a corn wet mill. Appl. Biochem. Biotechnol. 91-93, 283–296.

Fargues, C., Lewandowski, R., Lameloise, M.L., 2010. Evaluation of ion-exchange and adsorbent resins for the detoxification of beet distillery effluents. Ind. Eng. Chem. Res. 49, 9248–9257.

Gancedo, J.M., 1998. Yeast carbon catabolite repression. Microbiol. Mol. Biol. Rev. 62, 334–361.

Ghose, T.K., Roychoudhury, P.K., Ghosh, P., 1984. Simultaneous saccharification and fermentation (SSF) of lignocellulosics to ethanol under vacuum cycling and step feeding. Biotechnol. Bioeng. 26, 377–381.

Gibbons, W.R., Westby, C.A., 1989. Cofermentation of sweet sorghum juice and grain for production of fuel ethanol and distillers' wet grain. Biomass 18, 43–57.

Gnansounou, E., Dauriat, A., 2010. Techno-economic analysis of lignocellulosic ethanol: a review. Bioresour. Technol. 101, 4980–4991.

Gottschalk, G., 1987. Control of product formation in anaerobes. Dechema Monographs 105, 43–53.

Gough, S., Brady, D., Nigam, P., Marchant, R., McHale, A.P., 1997. Production of ethanol from molasses at 45 °C using alginate-immobilized *Kluyveromyces marxianus* imb3. Bioprocess Eng. 16, 389–392.

Guadalupe Medina, V., Almering, M.J., van Maris, A.J., Pronk, J.T., 2010. Elimination of glycerol production in anaerobic cultures of a *Saccharomyces cerevisiae* strain engineered to use acetic acid as an electron acceptor. Appl. Environ. Microbiol. 76, 190–195.

Gutierrez, T., Buszko, M.L., Ingram, L.O., Preston, J.F., 2002. Reduction of furfural to furfuryl alcohol by ethanologenic strains of bacteria and its effect on ethanol production from xylose. Appl. Biochem. Biotechnol. 98-100, 327–340.

Gutierrez, T., Ingram, L.O., Preston, J.F., 2006. Purification and characterization of a furfural reductase (FFR) from *Escherichia coli* strain LYO1—an enzyme important in the detoxification of furfural during ethanol production. J. Biotechnol. 121, 154–164.

Gutiérrez, T., Ingram, L.O., Preston, J.F., 2006. Purification and characterization of a furfural reductase (FFR) from *Escherichia coli* strain LYO1—an enzyme important in the detoxification of furfural during ethanol production. J. Biotechnol. 121, 154–164.

Heer, D., Sauer, U., 2008. Identification of furfural as a key toxin in lignocellulosic hydrolysates and evolution of a tolerant yeast strain. Microbial. Biotechnol. 6, 497–506.

Heipieper, H.J., Weber, F.J., Sikkema, J., Keweloh, H., de Bont, J.A.M., 1994. Mechanisms of resistance of whole cells to toxic organic solvents. Trends Biotechnol. 12, 409–415.

Himmel, M.E., 2008. Biomass Recalcitrance: Deconstructing the Plant Cell Wall for Bioenergy. Blackwell, Oxford.

Horvath, I.S., Taherzadeh, M.J., Niklasson, C., Liden, G., 2001. Effects of furfural on anaerobic continuous cultivation of *Saccharomyces cerevisiae*. Biotechnol. Bioeng. 75, 540–549.

Jayakody, L.N., Hayashi, N., Kitagaki, H., 2011. Identification of glycolaldehyde as the key inhibitor of bioethanol fermentation by yeast and genome-wide analysis of its toxicity. Biotechnol. Lett 33, 285–292.

Jeon, Y.J., Xun, Z., Rogers, P.L., 2010. Comparative evaluations of cellulosic raw materials for second generation bioethanol production. Lett. Appl. Microbiol. 51, 518–524.

John, E.H., 1998. Ethanol-induced water stress in yeast. J. Ferment. Bioeng. 85, 125–137.

Jones, R.P., Greenfield, P.F., 1986. Role of water activity in ethanol fermentations. Biotechnol. Bioeng. 28, 29–40.

Jönsson, L.J., Palmqvist, E., Nilvebrant, N.O., Hahn-Hagerdal, B., 1998. Detoxification of wood hydrolysates with laccase and peroxidase from the white-rot fungus *Trametes versicolor*. Appl. Microbiol. Biotechnol. 49, 691–697.

Kapucu, H., Mehmetoglu, U., 1998. Strategies for reducing solvent toxicity in extractive ethanol fermentation. Appl. Biochem. Biotechnol. 75, 205–214.

Karimi, K., Brandberg, T., Edebo, L., Taherzadeh, M.J., 2005. Fed-batch cultivation of Mucor indicus in dilute-acid lignocellulosic hydrolyzate for ethanol production. Biotechnol. Lett. 27, 1395–1400.

Keating, J.D., Panganiban, C., Mansfield, S.D., 2006. Tolerance and adaptation of ethanologenic yeasts to lignocellulosic inhibitory compounds. Biotechnol. Bioeng. 93, 1196–1206.

Kelly, C., Jones, O., Barnhart, C., Lajoie, C., 2008. Effect of furfural, vanillin and syringaldehyde on *Candida guilliermondii* growth and xylitol biosynthesis. Appl. Biochem. Biotechnol. 148, 97–108.

Klinke, H.B., Thomsen, A.B., Ahring, B.K., 2004. Inhibition of ethanol-producing yeast and bacteria by degradation products produced during pre-treatment of biomass. Appl. Microbiol. Biotechnol. 66, 10–26.

Kuhad, R.C., Gupta, R., Khasa, Y.P., Singh, A., 2010. Bioethanol production from Lantanacamara (red sage): pretreatment, saccharification and fermentation. Bioresour. Technol. 101, 8348–8354.

Laadan, B., Almeida, J.R., Radstrom, P., Hahn-Hagerdal, B., Gorwa-Grauslund, M., 2008. Identification of an NADH-dependent 5-hydroxymethylfurfural-reducing alcohol dehydrogenase in *Saccharomyces cerevisiae*. Yeast 25, 191–198.

Larsson, M., Galbe, M., Zacchi, G., 1997. Recirculation of process water in the production of ethanol from softwood. Bioresour. Technol. 60, 143–151.

Larsson, S., Palmqvist, E., Hahn-Hägerdal, B., Tengborg, C., Stenberg, K., Zacchi, G., et al., 1999a. The generation of fermentation inhibitors during dilute acid hydrolysis of softwood. Enzyme Microb. Technol. 24, 151–159.

Larsson, S., Reimann, A., Nilvebrant, N.O., Jonsson, L.J., 1999b. Comparison of different methods for the detoxification of lignocellulose hydrolyzates of spruce. Appl. Biochem. Biotechnol. 77-9, 91–103.

Lee, W., Lee, J., Park, B., Kim, M., Park, S., Chang, H., 1996. Continuous ethanol production from wood hydrolysate by chemostat and total cell retention culture. Korean J. Chem. Eng. 13, 453–456.

Li, H., Zhang, X., Shen, Y., Dong, Y., Bao, X., 2009. Inhibitors and their effects on *Saccharomyces cerevisiae* and relevant countermeasures in bioprocess of ethanol production from lignocellulose—a review. Sheng Wu Gong Cheng Xue Bao 25, 1321–1328.

Liu, Z.L., Moon, J., Andersh, B.J., Slininger, P.J., Weber, S., 2008. Multiple gene-mediated NAD(P)H-dependent aldehyde reduction is a mechanism of in situ detoxification of furfural and 5-hydroxymethylfurfural by *Saccharomyces cerevisiae*. Appl. Microbiol. Biotechnol. 81, 743–753.

Lu, P., Chen, L.J., Li, G.X., Shen, S.H., Wang, L.L., Jiang, Q.Y., et al., 2007. Influence of furfural concentration on growth and ethanol yield of Saccharomyces kluyveri. J. Environ. Sci. (China) 19, 1528–1532.

Lynd, L.R., Baskaran, S., Casten, S., 2001. Salt accumulation resulting from base added for pH control, and not ethanol, limits growth of *Thermoanaerobacterium thermosaccharolyticum* HG-8 at elevated feed xylose concentrations in continuous culture. Biotechnol. Prog. 17, 118–125.

Maiorella, B.L., 1983. Ethanol. In: Young, M. (Ed.), Industrial Chemicals, Biochemicals and Fuels, V. 3, Comprehensive Biotechnology. Pergamon Press, pp. 861–914.

Maiorella, B., Blanch, H.W., Wilke, C.R., 1983. By-product inhibition effects on ethanolic fermentation by *Saccharomyces cerevisiae*. Biotechnol. Bioeng. 25, 103–121.

Maiorella, B.L., Blanch, H.W., Wilke, C.R., 1984. Feed component inhibition in ethanolic fermentation by *Saccharomyces cerevisiae*. Biotechnol. Bioeng. 26, 1155–1166.

Martinez, A., Rodriguez, M.E., Wells, M.L., York, S.W., Preston, J.F., Ingram, L.O., 2001. Detoxification of dilute acid hydrolysates of lignocellulose with lime. Biotechnol. Prog. 17, 287–293.

Millati, R., Niklasson, C., Taherzadeh, M.J., 2002. Effect of pH, time and temperature of overliming on detoxification of dilute-acid hydrolyzates for fermentation by *Saccharomyces cerevisiae*. Process Biochem. 38, 515–522.

Miller, E.N., Jarboe, L.R., Turner, P.C., Pharkya, P., Yomano, L.P., York, S.W., et al., 2009a. Furfural inhibits growth by limiting sulfur assimilation in ethanologenic *Escherichia coli* strain LY180. Appl. Environ. Microbiol. 75, 6132–6141.

Miller, E.N., Jarboe, L.R., Yomano, L.P., York, S.W., Shanmugam, K.T., Ingram, L.O., 2009b. Silencing of NADPH-dependent oxidoreductase genes (yqhD and dkgA) in furfural-resistant ethanologenic *Escherichia coli*. Appl. Environ. Microbiol. 75, 4315–4323.

Minier, M., Goma, G., 1982. Ethanol production by extractive fermentation. Biotechnol. Bioeng. 24, 1565–1579.

Mira, N.P., Palma, M., Guerreiro, J.F., Sa-Correia, I., 2010. Genome-wide identification of *Saccharomyces cerevisiae* genes required for tolerance to acetic acid. Microb. Cell Fact. 9, 79.

Modig, T., Liden, G., Taherzadeh, M.J., 2002. Inhibition effects of furfural on alcohol dehydrogenase, aldehyde dehydrogenase and pyruvate dehydrogenase. Biochem. J. 363, 769–776.

Mori, Y., Inaba, T., 1990. Ethanol production from starch in a pervaporation membrane bioreactor using Clostridium thermohydrosulfuricum. Biotechnol. Bioeng. 36, 849–853.

Mussatto, S.I., Roberto, I.C., 2004. Alternatives for detoxification of diluted-acid lignocellulosic hydrolyzates for use in fermentative processes: a review. Bioresour. Technol. 93, 1–10.

Navarro, A.R., 1994. Effects of furfural on ethanol fermentation by *Saccharomyces cerevisiae*—mathematical-models. Curr. Microbiol. 29, 87–90.

Nemirovskii, V., Kostenko, V., 1991. Transformation of yeast growth inhibitors, which occurs during biochemical processing of wood hydrolyzates. Gidroliz. Lesokhim. Prom-st. 1, 16–17.

Nemirovskii, V.D., Gusarova, L.A., Rakhmilevich, Y.D., Sizov, A.I., Kostenko, V.G., 1989. Furfurol and hydroxymethylfurfurol transformation routes during culturing of nutrient yeasts. Biotekhnologiya 5, 285–289.

Nichols, N.N., Dien, B.S., Cotta, M.A., 2010. Fermentation of bioenergy crops into ethanol using biological abatement for removal of inhibitors. Bioresour. Technol. 101, 7545–7550.

Nigam, J.N., 2001. Ethanol production from hardwood spent sulfite liquor using an adapted strain of *Pichia stipitis*. J. Ind. Microbiol. Biotechnol. 26, 145–150.

Nilsson, A., Taherzadeh, M.J., Lidén, G., 2001. Use of dynamic step response for control of fed-batch conversion of lignocellulosic hydrolyzates to ethanol. J. Biotechnol. 89, 41–53.

Nilvebrant, N.O., Reimann, A., Larsson, S., Jonsson, L.J., 2001. Detoxification of lignocellulose hydrolysates with ion-exchange resins. Appl. Biochem. Biotechnol. 91-3, 35–49.

Ogawa, Y., Nitta, A., Uchiyama, H., Imamura, T., Shimoi, H., Ito, K., 2000. Tolerance mechanism of the ethanol-tolerant mutant of sake yeast. J. Biosci. Bioeng. 90, 313–320.

Ok, T., Hashinaga, F., 1997. Identification of sugar-tolerant yeasts isolated from high-sugar fermented vegetable extracts. J.Gen. Appl. Microb. 43, 39–47.

Olsson, L., HahnHagerdal, B., 1996. Fermentation of lignocellulosic hydrolysates for ethanol production. Enzyme Microb. Technol. 18, 312–331.

Osho, A., 2005. Ethanol and sugar tolerance of wine yeasts isolated from fermenting cashew apple juice. Afr. J. Biotechnol. 4, 660–662.

Palmqvist, E., Hahn-Hägerdal, B., 2000a. Fermentation of lignocellulosic hydrolysates. I: inhibition and detoxification. Bioresour. Technol. 74, 17–24.

Palmqvist, E., Hahn-Hägerdal, B., 2000b. Fermentation of lignocellulosic hydrolysates. II: inhibitors and mechanisms of inhibition. Bioresour. Technol. 74, 25–33.

Palmqvist, E., Almeida, J., Hahn-Hägerdal, B., 1999. Influence of furfural on anaerobic glycolytic kinetics of *Saccharomyces cerevisiae* in batch culture. Biotechnol. Bioeng. 62, 447–457.

Parajo, J.C., Dominguez, H., Dominguez, J.M., 1997. Improved xylitol production with Debaryomyces hansenii Y-7426 from raw or detoxified wood hydrolysates. Enzyme Microb. Technol. 21, 18–24.

Parawira, W., Tekere, M., 2011. Biotechnological strategies to overcome inhibitors in lignocellulose hydrolysates for ethanol production: review. Crit. Rev. Biotechnol. 31, 20–31.

Park, J., Chang, H., 2000. Microencapsulation of microbial cells. Biotechnol. Adv. 18, 303–319.

Petersson, A., Almeida, J.R., Modig, T., Karhumaa, K., Hahn-Hagerdal, B., Gorwa-Grauslund, M.F., et al., 2006. A 5-hydroxymethyl furfural reducing enzyme encoded by the *Saccharomyces cerevisiae* ADH6 gene conveys HMF tolerance. Yeast 23, 455–464.

Pourbafrani, M., Talebnia, F., Niklasson, C., Taherzadeh, M.J., 2007. Protective effect of encapsulation in fermentation of limonene-contained media and orange peel hydrolyzate. Int. J. Mol. Sci. 8, 777–787.

Purwadi, R., Niklasson, C., Taherzadeh, M.J., 2004. Kinetic study of detoxification of dilute-acid hydrolyzates by Ca(OH)2. J. Biotechnol. 114, 187–198.

Purwadi, R., Brandberg, T., Taherzadeh, M.J., 2007. A possible industrial solution to ferment lignocellulosic hydrolyzate to ethanol: continuous cultivation with flocculating yeast. Int. J. Mol. Sci. 8, 920–932.

Ricci, M., Martini, S., Bonechi, C., Trabalzini, L., Santucci, A., Rossi, C., 2004. Inhibition effects of ethanol on the kinetics of glucose metabolism by *S. cerevisiae*: NMR and modelling study. Chem. Phys. Lett. 387, 377–382.

Rivard, C.J., Engel, R.E., Hayward, T.K., Nagle, N.J., Hatzis, C., Philippidis, G.P., 1996. Measurement of the inhibitory potential and detoxification of biomass pretreatment hydrolysate for ethanol production. Appl. Biochem. Biotechnol. 57-8, 183–191.

Rodrigues, R., Felipe, M.G.A., Silva, J., Vitolo, M., Gomez, P.V., 2001. The influence of pH, temperature and hydrolyzate concentration on the removal of volatile and nonvolatile compounds from sugarcane bagasse hemicellulosic hydrolyzate treated with activated charcoal before or after vacuum evaporation. Braz. J. Chem. Eng. 18, 299–311.

Rose, A.H., 1993. Composition of the envelope layers of *Saccharomyces cerevisiae* in relation to flocculation and ethanol tolerance. J. Appl. Bacteriol. 74 (Suppl.), 110S–118S.

Rufian-Henares, J.A., de la Cueva, S.P., 2008. Assessment of hydroxymethylfurfural intake in the Spanish diet. Food Addit. Contam. Part A Chem. Anal. Control Expo. Risk Assess 25, 1306–1312.

Sanchez, B., Bautista, J., 1988. Effects of furfural and 5-hydroxymethylfurfural on the fermentation of *Saccharomyces cerevisiae* and biomass production from *Candida guilliermondii*. Enzyme Microb. Technol. 10, 315–318.

Sanchez, O.J., Cardona, C.A., 2008. Trends in biotechnological production of fuel ethanol from different feedstocks. Bioresour. Technol. 99, 5270–5295.

Sarvari Horvath, I., Franzen, C.J., Taherzadeh, M.J., Niklasson, C., Liden, G., 2003. Effects of furfural on the respiratory metabolism of *Saccharomyces cerevisiae* in glucose-limited chemostats. Appl. Environ. Microbiol. 69, 4076–4086.

Schneider, H., 1996. Selective removal of acetic acid from hardwood-spent sulfite liquor using a mutant yeast. Enzyme Microb. Technol. 19, 94–98.

Shabtai, Y., Chaimovitz, S., Freeman, A., Katchalski-Katzir, E., Linder, C., Nemas, M., et al., 1991. Continuous ethanol production by immobilized yeast reactor coupled with membrane pervaporation unit. Biotechnol. Bioeng. 38, 869–876.

Silva, C.J.S.M., Roberto, I.e.C., 1999. Statistical screening method for selection of important variables on xylitol biosynthesis from rice straw hydrolysate by *Candida guilliermondii* FTI 20037. Biotechnol. Techniques 13, 743–747.

Slininger, P.J., Gorsich, S.W., Liu, Z.L., 2009. Culture nutrition and physiology impact the inhibitor tolerance of the yeast *Pichia stipitis* NRRL Y-7124. Biotechnol. Bioeng. 102, 778–790.

Stulke, J., Hillen, W., 1999. Carbon catabolite repression in bacteria. Curr. Opin. Microbiol. 2, 195–201.

Taherzadeh, M.J., 1999. Ethanol from lignocellulose: physiological effects of inhibitors and fermentation strategies. PhD Thesis, Chalmers University of Technology, Göteborg, Sweden.

Taherzadeh, M.J., Karimi, K., 2007. Enzymatic-based hydrolysis processes for ethanol from lignocellulosic materials: a review. Bioresources 2, 707–738.

Taherzadeh, M.J., Karimi, K., 2008. Pretreatment of lignocellulosic wastes to improve ethanol and biogas production: a review. Int. J. Mol. Sci. 9, 1621–1651.

Taherzadeh, M.J., Lidén, G., Gustafsson, L., Niklasson, C., 1996. The effect of pantothenate deficiency and acetate addition on anaerobic batch fermentation of glucose by *Saccharomyces cerevisiae*. Appl. Microbiol. Biotechnol. 46, 176–182.

Taherzadeh, M.J., Eklund, R., Gustafsson, L., Niklasson, C., Lidén, G., 1997a. Characterization and fermentation of dilute-acid hydrolyzates from wood. Ind. Eng. Chem. Res. 36, 4659–4665.

Taherzadeh, M.J., Niklasson, C., Lidén, G., 1997b. Acetic acid—friend or foe in anaerobic batch conversion of glucose to ethanol by *Saccharomyces cerevisiae*? Chem. Eng. Sci. 52, 2653–2659.

Taherzadeh, M.J., Gustafsson, L., Niklasson, C., Liden, G., 1999a. Conversion of furfural in aerobic and anaerobic batch fermentation of glucose by *Saccharomyces cerevisiae*. J. Biosci. Bioeng. 87, 169–174.

Taherzadeh, M.J., Niklasson, C., Lidén, G., 1999b. Conversion of dilute-acid hydrolyzates of spruce and birch to ethanol by fed-batch fermentation. Bioresour. Technol. 69, 59–66.

Taherzadeh, M.J., Gustafsson, L., Niklasson, C., Lidén, G., 2000a. Inhibition effects of furfural on aerobic batch cultivation of *Saccharomyces cerevisiae* growing on ethanol and/or acetic acid. J. Biosci. Bioeng. 90, 374–380.

Taherzadeh, M.J., Gustafsson, L., Niklasson, C., Lidén, G., 2000b. Physiological effects of 5-hydroxymethylfurfural on *Saccharomyces cerevisiae*. Appl. Microbiol. Biotechnol. 53, 701–708.

Taherzadeh, M.J., Niklasson, C., Liden, G., 2000c. On-line control of fed-batch fermentation of dilute-acid hydrolyzates. Biotechnol. Bioeng. 69, 330–338.

Taherzadeh, M.J., Adler, L., Lidén, G., 2002. Strategies for enhancing fermentative production of glycerol—a review. Enzyme Microb. Technol. 31, 53–66.

Taherzadeh, M.J., Fox, M., Hjorth, H., Edebo, L., 2003. Production of mycelium biomass and ethanol from paper pulp sulfite liquor by *Rhizopus oryzae*. Bioresour. Technol. 88, 167–177.

Talebnia, F., Taherzadeh, M.J., 2006. In situ detoxification and continuous cultivation of dilute-acid hydrolyzate to ethanol by encapsulated *S. cerevisiae*. J. Biotechnol. 125, 377–384.

Talebnia, F., Taherzadeh, M.J., 2007. Physiological and morphological study of encapsulated *Saccharomyces cerevisiae*. Enzyme Microb. Technol. 41, 683–688.

Talebnia, F., Niklasson, C., Taherzadeh, M.J., 2005. Ethanol production from glucose and dilute-acid hydrolyzates by encapsulated *S. cerevisiae*. Biotechnol. Bioeng. 90, 345–353.

Tauer, A., Elss, S., Frischmann, M., Tellez, P., Pischetsrieder, M., 2004. Influence of thermally processed carbohydrate/amino acid mixtures on the fermentation by *Saccharomyces cerevisiae*. J. Agric. Food Chem. 52, 2042–2046.

Tomas-Pejo, E., Ballesteros, M., Oliva, J.M., Olsson, L., 2010. Adaptation of the xylose fermenting yeast *Saccharomyces cerevisiae* F12 for improving ethanol production in different fed-batch SSF processes. J. Ind. Microbiol. Biotechnol. 37, 1211–1220.

Walton, S., van Heiningen, A., van Walsum, P., 2010. Inhibition effects on fermentation of hardwood extracted hemicelluloses by acetic acid and sodium. Bioresour. Technol. 101, 1935–1940.

Weber, C., Farwick, A., Benisch, F., Brat, D., Dietz, H., Subtil, T., et al., 2010. Trends and challenges in the microbial production of lignocellulosic bioalcohol fuels. Appl. Microbiol. Biotechnol. 87, 1303–1315.

Widmer, W., Zhou, W., Grohmann, K., 2010. Pretreatment effects on orange processing waste for making ethanol by simultaneous saccharification and fermentation. Bioresour. Technol. 101, 5242–5249.

Yan, F., Bai, F., Tian, S., Zhang, J., Zhang, Z., Yang, X., 2009. Strain construction for ethanol production from dilute-acid lignocellulosic hydrolysate. Appl. Biochem. Biotechnol. 157, 473–482.

Zauner, E., Bronn, W.K., Dellweg, H., Tressl, R., 1979. Inhibitory activity of impurities of beet molasses and plant protectants on yeast (*Saccharomyces cerevisiae*) respiration and fermentation. Branntweinwirtschaft 119, 158–160, 163.

Zeitsch, K.J., 2000. The Chemistry and Technology of Furfural and Its Many By-Products, first ed Elsevier.

Zhang, J., Chu, D., Yu, Z., Zhang, X., Deng, H., Wang, X., et al., 2010a. Process strategy for ethanol production from lignocellulose feedstock under extremely low water usage and high solids loading conditions. Sheng Wu Gong Cheng Xue Bao 26, 950–959.

Zhang, J.G., Liu, X.Y., He, X.P., Guo, X.N., Lu, Y., Zhang, B.R., 2011. Improvement of acetic acid tolerance and fermentation performance of *Saccharomyces cerevisiae* by disruption of the FPS1 aquaglyceroporin gene. Biotechnol. Lett. 33, 277–284.

Zhu, J.J., Yong, Q., Xu, Y., Yu, S.Y., 2009. Comparative detoxification of vacuum evaporation/steam stripping combined with overliming on corn stover prehydrolyzate. In: 2009 International Conference on Energy and Environment Technology, Vol. 3, Proceedings, pp. 240–243.

SECTION IIIA

PRODUCTION OF BIODIESEL FROM VEGETABLE OILS

CHAPTER

13

Biotechnological Methods to Produce Biodiesel

Denise Maria Guimarães Freire, Joab Sampaio de Sousa, Elisa d'Avila Cavalcanti-Oliveira*

Universidade Federal do Rio de Janeiro, Instituto de Química, Av. Athos da Silveira Ramos, 149 - CT, Bloco A, lab. 549-1, CEP 21941-909 Rio de Janeiro, RJ, Brazil

***Corresponding author: E-mail: freire@iq.ufrj.br**

1 INTRODUCTION

The process currently being used for industrial-scale biodiesel production makes use of an alkali catalyst (usually NaOH, KOH, or sodium methoxide) in the transesterification of triacylglycerol (TAG) with methanol. Industry has favored this process because of the high conversion obtained in a short time and the low cost of the catalysts. However, there are some drawbacks in the process that have encouraged researchers and business people to look into different biodiesel production methods. The choice of catalyst is fundamental, as this determines the characteristics of the raw material, the reaction conditions, and the purification steps in the process, as shown in Table 1.

The use of enzymes (lipases) as catalysts in biodiesel production overcomes the problems inherent to alkali catalysts. Lipases are a group of enzymes that are initially described by their capacity to catalyze the hydrolysis of ester bonds in long-chain TAGs, producing free fatty acids (FFAs) and glycerol. For this reason, they have been defined as glycerol ester hydrolases (E.C. 3.1.1.3). These enzymes not only catalyze the cleaving of carboxyl-ester bonds (hydrolysis), but can also catalyze the reverse reaction (esterification and transesterification) in water-restricted systems.

However, despite the advantages of using enzymes (table 1), biodiesel production plants using lipases are not yet an industrial-scale reality. The reason for this is that there are some challenges that are yet to be overcome before biocatalysts can be made feasible for biodiesel production, such as their higher cost, biodiesel productivity, and inhibition by reactants and products.

TABLE 1 General Considerations About the Different Types of Processes for Biodiesel Production (Al-Zuhair, 2007; Balat and Balat, 2010; Fijerbaek et al., 2009; Loreto et al., 2005; Marchetti et al., 2007; Nielsen et al., 2008; Zhang et al., 2003a,b)

Type of Process	Composition of Raw Material	Reaction Conditions	Purification Step
Alkaline catalyst	• The raw material must be of high quality in order maintain the yield of the process: free of FFAs (<0.5%) and water (<0.1–0.3%); • FFAs react with the catalyst and form soap; • The water promotes hydrolysis of the alkyl esters to FFAs; • The raw material of high quality can represent 70–95% of the final cost of biodiesel	• High conversion (~99%); • Short reaction time (90 mim); • Temperature around 60°C; • Molar ratio methanol/oil 6:1; • 1% of catalyst based on the mass of oil; • Homogeneous and low-cost catalyst	• If there is soap formation, the separation step become unfeasible; • The catalyst is usually homogeneous and can not be reused; • The catalyst has to be removed from the product and a large volume of alkaline wastewater is generated and must be properly treated; • Glycerol (co-product) is contaminated with salts of catalyst neutralization, exhibiting low sale value
Acid catalyst	• Low-quality low-price raw materials can be employed; • FFAs of raw material are esterified to biodiesel, but the presence of water may diminish the reaction conversion	• High conversion (>90%); • The transesterification of TAGs is slow, only the esterification of FFAs is rapid; • Temperature of 60-120°C; • Homogeneous and low-cost catalyst; • The acid catalyst is corrosive to the equipment	• The catalyst is usually homogeneous and can not be reused; • Neutralization and removal of the catalyst;
Enzymatic catalyst	• Low-quality low-price raw materials can be employed; • FFAs are converted into biodiesel, without loss of raw material; • For some lipases, the water does not negatively interfere in reaction conversion;	• High conversion (>95%); • Mild reaction conditions (30–40 ° C and atmospheric pressure); • Long reaction time (8-72h); • Low energy consumption; • High costs of the biocatalyst	• Easy separation of biodiesel and biocatalyst by filtration; • Easy separation of biodiesel and glycerol by decanting; • Fewer steps process; • The immobilized enzyme can be reused; • Glycerol is of high quality and has a high sale value; • Enzymes are biodegradable

TABLE 1 General Considerations About the Different Types of Processes for Biodiesel Production (Al-Zuhair, 2007; Balat and Balat, 2010; Fijerbaek et al., 2009; Loreto et al., 2005; Marchetti et al., 2007; Nielsen et al., 2008; Zhang et al., 2003a,b)—Cont'd

Type of Process	Composition of Raw Material	Reaction Conditions	Purification Step
Non-catalytic (supercritical)	• Low-quality low-price raw materials can be employed; • FFAs are converted into biodiesel, without loss of raw material	• High conversion (98%); • Short reaction time (7–15 min); • High temperature (250-300°C) and pressure (10-25 MPa); • High energy consumption; • Possible generation of thermal degradation products; • No catalyst cost	• No catalyst separation step; • Easy separation of biodiesel and glycerol by decanting

2 ENZYMATIC TRANSESTERIFICATION

Transesterification is a term that is widely used to describe an important class of organic reactions, where one ester is converted into another. This transfer of an acyl group can happen between an ester and an acid (acidolysis), one ester and another ester (interesterification) or between an ester and an alcohol (alcoholysis; Gunstone and Herslöf, 2004).

In broad terms, the transesterification reaction between TAGs and alcohol to produce biodiesel is a sequence of three consecutive and reversible reactions, by which DAG and MAG are formed as intermediates.

There are some factors that influence conversion by enzymatic transesterification, such as the substrate used (TAG and alcohol), the molar ratio between the substrates, the water content in the reaction medium, whether a solvent is used, the temperature, whether the enzyme is free or immobilized, the lipase concentration, and others.

Despite the many reports in the literature describing biodiesel synthesis using different lipases, it is hard to make any generalizations about the optimal reaction conditions because lipases from different sources tend to respond differently to changes in the reaction medium (Bajaj et al., 2010; Jothiramalingam and Wang, 2009).

2.1 Alcohol

Biodiesel can be produced using primary short-chain alcohols like methanol, ethanol, propanol, and butanol, as well as secondary alcohols like isopropanol and 2-butanol (Fijerbaek et al., 2009). The prerequisites for selecting the alcohol for industrial-scale biodiesel production are that it must be cheap and in plentiful supply. Currently, only methanol and ethanol meet these two requirements, and of these two substances, methanol is the more widely used as it is cheaper and more readily available in most countries although ethanol has the dual advantages of being renewable and less toxic (Antczak et al., 2009; Deng et al., 2005; Demirbas, 2007). Meanwhile, in enzyme catalysis, it is generally the case that the shorter the alcohol chain, the more likely it is to deactivate the lipase. It is believed that this is because it strips off the microlayer of water surrounding the lipase, which is essential for the optimal conformation of the enzyme (Fijerbaek et al., 2009; Zheng et al., 2009).

Shimada et al. (1999) studied the transesterification of vegetable oil with methanol catalyzed by an immobilized lipase from *Candida antarctica* (Novozym 435) in a solvent-free system. What they found was that when the methanol concentration exceeded its solubility level, it deactivated the lipase irreversibly. To prevent lipase deactivation and keep the stoichiometric ratio required for the total conversion of the TAG, the reaction involved adding methanol in three steps. This three-step process converted 98.4% of the oil to its corresponding methyl esters in 48 h, and the immobilized lipase was able to be reused 50 times. In another study, Watanabe et al. (2000) used a two-step strategy for adding the alcohol. They started by adding one-third of the alcohol at the beginning of the reaction, but then found that after it had been converted into biodiesel (10 h reaction time), they could add the rest of the alcohol in a single step, since its solubility was raised by the presence of the biodiesel. As such, the reaction time was reduced to 36 h and the enzyme was reused 70 times, achieving >95% conversion. This stepwise addition of a short-chain alcohol was adopted by researchers investigating other lipases, such as *Candida sp.* (Lu et al., 2007), *Pseudomonas fluorescens* (Soumanou and Bornscheuer, 2003), *Rhizopus oryzae* (Chen et al., 2006). Shimada et al. (2002) explained the lower enzyme deactivation by longer-chain alcohols (>3 carbons) by the fact that they are more apolar and more soluble in oil at the stoichiometric ratio. However, as already mentioned, each lipase has different properties. With *P. fluorescens*, high conversion (>90%) was possible with 4.5 molar equivalent of methanol added at the beginning of the reaction (Soumanou and Bornscheuer, 2003). In another study (Salis et al., 2008) the use of two lipases, from *P. fluorescens* and *Pseudomonas cepacia* (now *Burkholderia cepacia*), resulted in 58% and 37% conversion, respectively, in the presence of 1:8 oil/methanol molar ratio in a solvent-free system, while another six lipases tested were completely inactive under these conditions.

The excess alcohol above and beyond the stoichiometric ratio increases the reaction rate, but may also deactivate the enzyme, compromising the number of times the enzyme can be reused or even the conversion of the reaction when enzyme deactivation is more severe (Antczak et al., 2009). There are also some arguments against using excess alcohol in industrial-scale processes, such as higher energy consumption, larger equipment requirements, and the need to treat the unreacted alcohol (Fijerbaek et al., 2009).

To prevent the alcohol deactivating the enzyme, many researchers have used organic solvents in the reaction medium in a bid to increase the solubility of the alcohol and reduce its concentration. (Iso et al., 2001; Mittelbach, 1990; Nelson et al., 1996; Ranganathan et al., 2008; Royon et al., 2007).

2.2 Water Content

It is known that the water content in nonaqueous media affects the activity of enzymes, reducing their rigidity and consequently enhancing their activity. When biodiesel production is catalyzed by lipases, if the water content exceeds the optimal concentration, biodiesel conversion is affected because a competing inhibition reaction takes place that enables the hydrolysis of the TAGs, DAGs, MAGs, and alkyl esters (Shah et al., 2004).

The ideal water content in the reaction medium varies greatly depending on the enzyme and the reaction medium, and so must be studied on a case-by-case basis. For example, Kaieda et al. (2001) found that the water concentrations that resulted in the best conversions were 8-20% for *Candida rugosa* lipase, 4-20% for *P. fluorescens* lipase, and 1-2% for *P. cepacia* lipase. Deng et al. (2005) tested several immobilized commercially available lipases and found that with the exception of *C. antarctica*, the conversion obtained from the transesterification reaction with all the others (*Thermomyces lanuginosus, Rhizomucor miehei, P. cepacia,* and *P. fluorescens*) was higher when anhydrous ethanol was replaced with hydrous ethanol (4% water). Kaieda et al. (1999), who used a *R. oryzae* lipase and the stepwise addition of methanol, observed that the addition of 4-30% water in proportion to the substrate mass resulted in higher conversions.

It is also very important to take account of the water present in the reagents and even in the enzyme in order to design appropriate reaction medium. Studies of lipase reutilization at different water concentrations have to be carried out since water can influence enzyme stability, making it crucially important for designing an economically feasible process (Deng et al., 2005; Triantafyllou et al., 1995).

Some authors have noted that adding water to the reaction medium can protect lipases against deactivation in the presence of short-chain alcohols (Kaieda et al., 1999; Kaieda et al., 2001; Noureddini et al., 2005; Pizarro and Park, 2003).

Those lipases that respond well to reaction media with a higher water content are of interest for use with raw materials containing water, as this would rule out the need for a dehydration pretreatment stage. For example, exchanging anhydrous ethanol for ethanol containing 5% water, which is cheaper, had no impact on the transesterification reaction catalyzed by a *P. cepacia* lipase (Shah and Gupta, 2007). However, the water content in the biodiesel must be kept within the specifications required by law. Thus, unless the raw material already contains water, it is preferable to maintain a low water concentration in the reaction medium (Deng et al., 2005).

2.3 Organic Solvent Use

Organic solvents are used in the enzymatic production of biodiesel to obtain a homogeneous reaction medium by ensuring greater solubility of both the hydrophobic compounds (like TAG and biodiesel) and the hydrophilic compounds (e.g., alcohol and glycerol). Solvents also serve to reduce the viscosity of the reaction medium, enabling a higher diffusion rate to be achieved and reducing mass transfer problems (Fijerbaek et al., 2009). A suitable solvent must therefore be found, which both enhances the catalytic activity of the enzyme and keeps it stable.

Soumanou and Bornscheuer (2003) studied methanolysis using six different solvents and found that for the apolar solvents (hexane, cyclohexane, *n*-heptane, isooctane, and petroleum ether) the three lipases being studied (*P. fluorescens, T. lanuginosus,* and *R. miehei*) achieved good conversions (60-80%); yet when acetone was the solvent, conversion into biodiesel was low for all the lipases (< 20%). Kojima et al. (2004) assessed the lipase activity of *C. cylindracea* (now *C. rugosa*) after incubation for 72 h in different solvents and found the same behavior: the polar solvents reduced enzyme activity, while the hydrophobic solvents kept it stable. Polar solvents may alter the native conformation of the enzymes by disrupting hydrogen bonding and hydrophobic interactions, leading to a very low alcoholysis rate

(Soumanou and Bornscheuer, 2003). Also, polar solvents tend to strip the water molecules present on the surface of the enzyme, causing a reduction in its catalytic activity (Gorman and Dordick, 1991 cited in Lara and Park, 2004).

One important organic solvent is tert-butanol, which is relatively hydrophilic and has been used successfully as a novel reaction medium for the lipase-mediated methanolysis of biodiesel production. Due to steric hindrance, this alcohol is not accepted by the lipases as a substrate, and as a solvent it has the ability to dissolve oil, methanol, and glycerol, leading the authors to believe that the negative effects caused by methanol and glycerol on lipase performance could be entirely eliminated (Li et al., 2006; Royon et al., 2007). Using a combination of immobilized lipases from *T. lanuginosus* and *C. antarctica* in a system containing tert-butanol, 95% conversion into biodiesel was obtained and the enzymes were reused in over 200 cycles without any obvious loss in lipase activity (Li et al., 2006). Using lipase from *C. antarctica* lipase, 95% conversion was obtained in the reaction with *t*-butanol, while the lipase itself was used for 500 h without any loss of activity (Royon et al., 2007).

One interesting option is to use fossil diesel as a solvent in enzymatic transesterification reactions, as this way the solvent does not have to be separated from the product at the end of the reaction. The group that investigated this solvent (Kojima et al., 2004; Park et al., 2008) studied biodiesel using waste activated bleaching earth as a substrate, and obtained 97-100% conversion.

The use of solvents resolves several problems, but their use on an industrial scale is not desirable because of the cost of the solvent itself and the cost of recovering it at the end of the reaction. The use of solvents also makes it necessary to use larger reactors, since it occupies a large volume, and also raises operational risks because of solvents are toxic and flammable.

Although similar conversions have been obtained with and without solvents (Kumari et al., 2007; Soumanou and Bornscheuer, 2003), solvent-free enzymatic biodiesel production is characterized by lower reaction rates than when solvents are used (Mittelbach, 1990), which is something that must be improved to make it feasible (Fijerbaek et al., 2009).

2.4 Types of Biocatalysts

Free and immobilized lipases have been studied for biodiesel production, including, more recently, as whole-cell immobilized lipases. Each type of biocatalyst has its strengths and weaknesses when it comes to reducing the contribution of the biocatalyst in the final cost of the biodiesel.

2.4.1 Free Biocatalysts

Free enzymes are far cheaper than immobilized lipases. They can be purchased in an aqueous solution composed of the enzyme solution plus nothing more than a stabilizer to prevent enzyme denaturation (e.g., glycerol or sorbitol) and a preservative to inhibit microbial growth (e.g., benzoate; Nielsen et al., 2008). Several studies have obtained high biodiesel conversions (>90%) using soluble lipases from *C. rugosa*, *P. cepacia*, and *P. fluorescens* (Kaieda et al., 2001), *R. oryzae* (Kaieda et al., 1999), and *C. cylindracea* (now *C. rugosa*; Park et al., 2008).

To prevent the addition of water to the reaction medium, the solution containing the free lipase can be freeze-dried. However, this combined freezing and drying process sometimes reduces enzyme activity. It has been reported that pH tuning (when an enzyme solution is freeze-dried in a buffer whose pH is the same as the optimal pH of the enzyme in an aqueous medium) may protect enzymes from deactivation (Roy and Gupta, 2004). However, Nielsen et al. (2008) strongly recommend that this kind of nonformulated enzyme preparation be used with care and on a small scale, because the powder containing the enzyme is allergenic if inhaled.

Even free lipases are considerably more expensive than the chemical catalysts currently used in biodiesel production. If they are to be economically feasible, lipases must be reusable. When free lipases are used in biodiesel production, they can be partially recovered in the aqueous phase. However, their indefinite reuse is restricted by the build-up of glycerol (Nielsen et al., 2008). Furthermore, most enzyme molecules are insoluble in anhydrous media, and tend to clump together, which reduces the surface area of the biocatalyst. One way of getting round both these problems is to immobilize the enzyme (Bisen et al., 2010; Shah et al., 2003).

2.4.2 *Immobilized Biocatalysts*

Enzyme reutilization is an important key to making the production of a commodity like biodiesel possible by enzymatic means (Hsu et al., 2001). The longer an enzyme can be reused, the lower its contribution to the overall price of the product. Immobilized lipases have attracted most interest by researchers for their potential use in biodiesel production, since immobilization serves not only to enable recovery and reutilization, but also to enhance enzyme stability (Ranganathan et al., 2008). The reason immobilized enzymes are more stable is because their molecular mobility is lower. This helps prevent denaturation, which can be caused by chemicals or high temperatures, while they are also protected from mechanical damage inside the support (Ranganathan et al., 2008). The immobilized biocatalysts can be recovered at the end of the reaction by filtration alone, or can be packed in columns for use in a continuous-flow process (Nielsen et al., 2008).

Different immobilization techniques have been tried for lipases for biodiesel production: adsorption, covalent attachment, entrapment, and cross-linkage (Ganesan et al., 2009).

Adsorption, the simplest and most widely used technique for immobilizing lipases, consists of bonding the lipase to the support surface through weak forces such as van der Waals or hydrophobic interactions. However, given the low bond strength between the enzyme and the support, it may be desorbed throughout the reaction (Jegannathan et al., 2008 cited by Tan et al., 2010).

The most widely studied lipase for biodiesel production is the *C. antarctica* lipase immobilized by adsorption on acrylic resin (Novozym 435—manufactured by Novozymes). This lipase has been reported to obtain over 95% conversion into biodiesel (Shimada et al., 1999; Watanabe et al., 2000) and has been reused 70 times without any reduction in the conversion (Watanabe et al., 2000).

Some studies have used lipases immobilized by covalent bond onto a support matrix for biodiesel production. *T. lanuginosus* lipase immobilized by covalent attachment onto polyglutaraldehyde activated styrene-divinylbenzene copolymer catalyzed the conversion of 97% rapeseed oil into biodiesel in 24 h, and was reused for 10 reactions with no loss of activity (Dizge et al., 2009).

A relatively new immobilization method involves cross-linked enzyme aggregates (CLEAs) and protein-coated microcrystals (PCMCs). These have been tested for production of biodiesel with a *P. cepacia* lipase in solvent-free conditions, obtaining 92% conversion with CLEAs and 99% with PCMCs, both after 2.5 h (Kumari et al., 2007).

On the other hand, the use of immobilized lipases can incur some problems, since large molecules (TAG, FAME) have to diffuse through small pores to access the enzyme, while low-solubility reagents (methanol) have to penetrate the oil-filled pores in the support (internal diffusion). There are also external restrictions on mass transport, with the possibility of a film being produced around the support. The formation of a layer of reagents or products around the immobilized enzyme (external diffusion) can usually be minimized by increasing the agitation rate in the reactor, or increasing the flow rate if it is a fixed-bed reactor (Ranganathan et al., 2008). At the start of a transesterification reaction catalyzed by immobilized lipases, the reaction system is made up of three immiscible phases (TAG, alcohol, and immobilized enzyme), but as the alkyl esters are formed, they operate as a solvent for the substrate and the reaction becomes a two-phase (liquid and solid) reaction, ameliorating the diffusion-related problems in the system (Noureddini et al., 2005).

2.4.3 *Whole-Cell Biocatalysts*

Whole-cell immobilized lipases have been studied for biodiesel production. This kind of biocatalyst should be cheaper to produce because it does not require many of the steps in the downstream process, such as the isolation and purification of the enzyme after fermentation (Ban et al., 2001; Li et al., 2007a; Zeng et al., 2006).

Qin et al. (2008) used lyophilized free whole cells of *R. chinensis* for biodiesel production, obtaining yields of 86% FAME. Torres et al. (2003) used whole-cell lipase of *Aspergillus flavus* to catalyze methanolysis combined with oil extraction. The authors obtained 92% conversion after 96 h of reaction.

However, the free cells in the reaction mixture are difficult to reuse, so cell immobilization could solve this problem. Ban et al. (2001) did just this, immobilizing a whole-cell biocatalyst of *R. oryzae* on reticulated polyurethane foam. The fungal biomass was immobilized on the support spontaneously during fermentation and a high conversion of 90% on biodiesel could be achieved.

In a later work, Ban et al. (2002) showed that when this same biocatalyst was treated with a solution of glutaraldehyde (cross-linking), it enhanced the stability of the intracellular lipase from *R. oryzae* and yielded 80% conversion into methyl esters in a solvent-free system (Tamalampudi et al., 2008). Ying and Chen (2007) studied the cells of lipase-producing *Bacillus subtilis* encapsulated within a net of hydrophobic carrier with magnetic particles. This biocatalyst was recoverable by magnetic separation. When methanolysis was carried out using waste cooking oil, the proportion of methyl esters in the reaction mixture reached about 90% after 72 h in a solvent-free system.

Matsumoto et al. (2001) constructed a strain of *S. cerevisiae* with high-level expression of intracellular *R. oryzae* lipase. They obtained 71% conversion into methyl esters after 165 h with permeabilized cells. The yeast *S. cerevisiae* cell surface display system for the lipase from *R. oryzae* was developed by Matsumoto et al. (2002), who obtained 78% methanolysis after 72 h in a solvent-free system. These differences in yield and conversion rate of methyl esters might be attributed to the easier access of molecules from the substrate to the cell surface displayed lipase, which did not need to be permeabilized for methanolysis to occur.

3 ENZYMATIC ESTERIFICATION

Some basic differences have been identified between the transesterification of TAGs and the esterification of FFAs. The former is a sequence of three reactions (DAG and MAG are formes as intermediates), while esterification involves parallel reactions of FFAs to make biodiesel, which is quicker (Marchetti et al., 2008). The greater polarity of FFAs than TAGs makes the short-chain alcohols more soluble in the reaction medium (Du et al., 2007). Also, water is one of the products of esterification reaction and shifts the equilibrium toward hydrolysis when the concentration exceeds optimal levels.

A variety of waste oils and fats have been used for enzymatic biodiesel production (Fijerbaek et al., 2009; Table 2); indeed, the use of low-quality raw materials with a high FFA concentration (low aggregate value) is one way of reducing the overall cost of producing biodiesel, helping make the use of more expensive catalysts like lipases economically feasible.

The effect of the presence of FFAs on the tolerance of lipases to alcohol (methanol) has been studied using a *C. antarctica* lipase as a model. In a previous work, Shimada et al. (1999) showed that when refined oil (100% TAG) was used as a substrate, the maximum methanol concentration that could be employed without deactivating the lipase was a 1.5:1 methanol/oil molar ratio, which is the solubility limit of methanol for this system. Based on this finding, Du et al. (2007) tested different proportions of oleic acid and refined oil as substrates, varying the concentration of oleic acid to refined oil mass from 5% to 100%. They observed that the tolerance of the *C. antarctica* lipase to the methanol was higher when

TABLE 2 Examples of Enzymatic Biodiesel Production Employing Low-Cost Raw Materials in Solvent-Free System

Raw Material	Lipase	Conversion (%)	Alcohol Type	Reaction Time (h)	Reuse of Lipase (cycles)	Reference
Soybean oil deodorizer distillate (28% FFA)	*C. antarctica*	95	Methanol	10	–	Du et al. (2007)
Acid oil (78% FFA) + refined oil	*C. antarctica*	97 (two steps)	Methanol	24 each step	>100 each step	Watanabe et al. (2007a)
Acid oil hydrolysate (92% FFA)	*C. antarctica*	99 (two steps)	Methanol	24 each step	40 each step	Watanabe et al. (2007b)
Waste fatty acids from tuna oil (100% FFA)	*C. antarctica*	98 (two steps)	Methanol	24 each step	>45	Watanabe et al. (2002a)
Rice bran oil dewaxed/degummed (85% FFA)	*C. antarctica* or *R. miehei*	96	Methanol	6	14	Lai et al. (2005)
Madhuca indica acid oil (~20% FFA)	*P. cepacia*	99	Ethanol	2.5	–	Kumari et al. (2007)
Grease (8.5% FFA)	*P. cepacia*	96	Ethanol	18	–	Hsu et al. (2001)

the oleic acid concentration was higher. For instance, when the substrate was 100% oleic acid, a 30:1 methanol/oil molar ratio yielded around 90% biodiesel, while a substrate with just 30% oleic acid deactivated the lipase and no biodiesel was produced. The improved enzyme stability was probably brought about by the greater solubility of the methanol in the presence of the FFAs (Du et al., 2007). It has also been reported that the esterification of FFAs with methanol catalyzed by a *C. antarctica* lipase is quicker than the methanolysis of TAG and requires a lower lipase concentration (Lai et al., 2005; Shimada et al., 1999; Watanabe et al., 2002a,b; Watanabe et al., 2005). It can be concluded that a reaction system that consists primarily of the esterification of FFAs may reduce total lipase costs and reaction times.

The importance of removing water (a product of esterification) during biodiesel production has been demonstrated by some authors using a *C. antarctica* lipase and methanol in a solvent-free system. When Du et al. (2007) used soybean oil deodorizer distillate (25-35% FFAs) as a substrate, they obtained a higher conversion into biodiesel (95%) when they added an adsorbent to the reaction medium to control the water content. Meanwhile, Watanabe et al. (2005) used mixtures with 50-90% FFA in the TAG as a substrate, observing that the FFAs from the mixture were efficiently esterified with methanol, but the water produced by this process significantly inhibited the methanolysis of the TAGs, when a 1:1 methanol/FA molar ratio was used.

The use of large quantities of methanol may be one way of overcoming methanolysis inhibition by water, and high biodiesel conversions may be obtained when the reaction equilibrium is shifted toward the production of methyl esters. Thus, the presence of water has less of an impact on esterification than on transesterification (Watanabe et al., 2005). Watanabe et al. (2007b) used glycerol to absorb the water produced during the esterification of the acid oil hydrolysate (92% FFA). The glycerol removed the water from the medium, resulting in a higher FAME yield, without any increase in the partial glyceride content being detected during the reaction.

It has been noted that the yield of biodiesel from low-cost raw materials is usually lower than it is from refined materials. This could be caused by the other compounds in these materials, such as phospholipids found in crude oils which often inhibit lipase activity. When crude rice bran oil (20% FFAs) was used as a substrate, 56% FAME was obtained after 12 h. This concentration rose to 88% when dewaxed/degummed rice bran oil (20% FFAs) was used as the substrate (Lai et al., 2005). Watanabe et al. (2007b) detected low lipase stability in a mixture of acid oil hydrolysate with 1-2 mol methanol, which was assumed to have been caused by some inhibitors contained in the acid oil hydrolysate, since in a previous work by the same group (Watanabe et al., 2005) the lipase had been found to be stable in a mixture of pure FFAs with 1-2 mol methanol. This inhibition was controlled by the addition of 5-7 mol methanol. The authors related two hypotheses: one where the concentration of inhibitors was reduced by dilution with methanol, the other where the inhibitors adhered to the lipase in the presence of 1-2 mol methanol (low polarity) were released in the presence of 5-7 mol (high polarity; Watanabe et al., 2007b).

Studies of lipases other than the *C. antarctica* lipase have been carried out using a solvent in the reaction medium. The *Penicillium expansum* lipase was used in FAME production from waste oil (~20% FFA) in the presence of tert-amyl alcohol. The water produced during the reaction was removed by the addition of adsorbents, resulting in a higher conversion into FAME (93%; Li et al., 2009). Likewise, the use of adsorbents by Deng et al. (2003) in the esterification of oleic acid and methanol catalyzed by the *Candida sp.* lipase in the presence of petroleum ether resulted in over 90% conversion. Wang et al. (2006) used a combination of lipases from *C. antarctica* and *T. lanuginosus* to catalyze biodiesel production from soybean oil deodorizer distillate in a medium with tert-butanol. Both the FFAs and the glycerides were converted into biodiesel simultaneously and reached a 97% conversion with the addition of an adsorbent with no obvious loss in lipase activity even after 120 cycles. Li et al. (2007b) used whole cells of *R. oryzae* and tert-butanol as a solvent, observing that the increase from 0% to 20% in FFAs in the oil resulted in higher conversion into biodiesel.

4 HYDROESTERIFICATION

Hydroesterification is a process that combines two basic processes, hydrolysis and esterification, in sequential reactions in order to produce biodiesel. This methodology allows the use of raw materials with high concentrations of free fatty acids and water (as normally occurs with waste raw materials) without pre-treatment, since water is one of the reagents and high concentrations of fatty acids is the expected product of the hydrolysis reaction.

4.1 Enzymatic Hydrolysis

The hydrolysis of oil and fat is an important industrial process. The products (FFAs and glycerol) are basic raw materials for a whole host of applications.

Noor et al. (2003) studied the hydrolysis of palm oil in a stirred tank bioreactor by lipase-SP398, produced by Novo Nordisk S/A. Almost all the palm oil was hydrolyzed in 90 min, and the addition of gum arabic, which operated as a surfactant, increased the hydrolysis rate.

Meanwhile, Talukder et al. (2010a) studied the hydrolysis of crude (unrefined) palm oil by the *C. rugosa* lipase, followed by the esterification of the FFAs from this oil with methanol by the *C. antarctica* lipase. The oil was completely hydrolyzed in 4 h in the presence of isooctane. The biocatalysts were reused for up to 10 cycles in hydrolysis and 50 cycles in esterification, with no significant loss of activity.

Watanabe et al. (2007b) studied the enzymatic conversion of acid oils (byproduct of the refining of vegetable oils) into FFAs catalyzed by *C. rugosa* lipase and obtained an oil with 92% FFAs. The second step encompassed the esterification reaction catalyzed by *C. antarctica* lipase that obtained conversion of 96% after 24 h reaction. The final product contained 91 wt% methyl esters. Both steps could be repeated for 40 cycles without reduction of reaction conversion.

Pugazhenthi and Kumar (2004) studied the hydrolysis of olive oil by the pancreatic lipase immobilized on poly methyl methacrylate-ethylene glycol dimethacrylate. In this study, the immobilized enzyme was used in reactions for over 50 h during 25 days.

In a study by Gan et al. (1998), sunflower oil was completely hydrolyzed by a *C. cylindracea* lipase in an integrated reaction system involving an agitated tank reactor coupled to an ultrafiltration system, which provided the simultaneous separation of the product during the enzymatic hydrolysis of the oil. They found that the continuous separation of the reaction product (glycerol) and the recirculation of the free lipase in the system enhanced the production of FFAs.

There is also increasing interest in the use of membrane technology to combine reactions involving lipases with separation systems in the processing of oils and fats for use in lipid refining (Koike et al., 1992), separation (Raman et al., 1996), discoloration (Reddy et al., 1996), and decontamination (Vavra and Koseuglu, 1994).

4.2 Hybrid Catalysis

Saifuddin et al. (2009) developed a hybrid catalysis process for biodiesel production using waste cooking oil with high acidity (low quality) as a raw material. The lipase used, from *Candida rugosa*, hydrolyzed 88% of the cooking oil in 5 h at 40 °C. The hydrolysate was then used in an esterification reaction catalyzed by sulfuric acid (2.5%) at a 1:15 raw material/methanol molar ratio, yielding up to 83% biodiesel in 1 h.

Ting et al. (2008) studied the use of a *C. rugosa* lipase immobilized on chitosan in the hydrolysis step. The authors obtained 88% of the soybean oil hydrolysis after 5 h of reaction. The hydrolysate was esterified with methanol at a 1:15 molar ratio by acid catalysis (2.5% sulfuric acid), obtaining 99% conversion into biodiesel after 12 h at 50 °C.

Talukder et al. (2010b) studied the use of cooking oil for biodiesel production by enzymatic hydrolysis accompanied by chemical esterification. The *C. rugosa* lipase completely hydrolyzed the oil after 10 h. The FFAs were converted into biodiesel by chemical esterification using Amberlyst 15 (acidic styrene divinylbenzene) and a 99% conversion into biodiesel was obtained after 2 h. In this work, there was a loss of enzyme activity and the hydrolysis yield fell to 92% after five runs.

Cavalcanti-Oliveira et al. (2011) studied the use of a *T. lanuginosus* lipase (TL 100 L) in the hydrolysis of soybean oil in a hydroesterification process. The lipase hydrolyzed 89% of the oil after 48 h. This stage was followed by the esterification of the FFAs with methanol, which was catalyzed by niobic acid in pellets. They obtained 92% conversion of the FFAs into FAMEs after 1 h.

Sousa et al. (2010) studied the Physic nut lipase (*Jatropha curcas* L.) for the hydrolysis of different raw materials for biodiesel production using hydroesterification. The best conversions were obtained using soybean oil and jatropha oil, obtaining up to 98% FFA after 2 h. The esterification of the FFAs from the jatropha oil with methanol was catalyzed by niobic acid in pellets, obtaining up to 97% conversion into biodiesel after 2 h. The biodiesel obtained from this process fulfilled all the legal requirements for its commercial use.

5 REACTOR CONFIGURATIONS

One of the problems to be overcome when biocatalysis is used for obtaining biodiesel on a large scale is the right setup and operation of the bioreactor, given that both factors, as well as the form of the biocatalyst (whether free or immobilized), have a direct impact on the stability of the enzyme and whether it can be reused, which are crucial for reducing costs in enzyme catalysis. The set-up of the bioreactor for free or immobilized enzyme preparations should take into account how the biocatalyst will be reused from the product stream (biodiesel and glycerol). When free biocatalysts are used, ultrafiltration or centrifugation units may be coupled to the system. However, there are more process options and bioreactors to choose from if the biocatalyst is immobilized.

The most widely used reactors for enzymatic biodiesel production are packed-bed reactors (PBRs) and stirred-tank reactors (STRs). However, as biodiesel is a chemical commodity, its production in continuous-flow systems would certainly reduce the operational costs of its production. As a result, these reactors are the most widely used in continuous operations with heterogeneous catalysts, such as immobilized enzymes. Even so, there are other reactor set-ups that are worth investigating, such as fluidized-bed reactors (FBRs), expanded bed reactors and membrane reactors (Fijerbaek et al., 2009).

Several authors have investigated the use of PBRs operating continuously and in batches using enzymes immobilized on supports or whole cells as a biocatalyst. Table 3 summarizes the main works in the literature that use PBRs to obtain biodiesel using enzymes.

Generally speaking, biodiesel production using continuous-flow PBRs has attained good enzyme stability and conversions, both with and without the use of solvents. Solvents may add to the overall production cost, but on a commercial scale, the absence of solvents may incur a marked drop in pressure on the bed, causing serious operational problems. PBRs should operate at low flow rates or using larger biocatalyst particle sizes to minimize such a drop in pressure. Fijerbaek et al. (2009) noted a drop in the effectiveness factor (η) as the particle size of commercial biocatalysts increased. This was the equivalent of a 34% drop in the reaction rate due to the increased particle diameter and the correspondingly larger pore diffusion distance. These factors should therefore be taken into account when the bioreactor/system is being chosen. FBRs have certain features that help overcome these problems, but imply in designing equipment that efficiently separates and recovers the biocatalyst. Another problem to be overcome in continuous-flow PBRs is the adsorption of the glycerol formed during the reaction on the immobilized biocatalyst bed, causing the inhibition of enzyme activity. Jachmanián et al. (2010), in their investigation of the composition of the substrate, adjusted the ratio between the oil, alcohol, and solvent in such a way as to prevent the separation of alcohol from the substrate and/or glycerol from the product mixture, leading to optimal enzyme performance and productivity in continuous PBRs. Other procedures may help keep up enzyme activity, such as adding silica gel to the bed, using solvents, or using semicontinuous-flow processes that provide the opportunity for the biocatalyst to be washed periodically.

The use of STRs has also been investigated on a smaller scale. They generally yield high conversion rates to begin with because of the high dispersion rate of the alcohol in the oil. However, some problems arising from physical damage to the biocatalyst caused by shear stress have been reported. Hama et al. (2007) noted that biodiesel production in batch-stirred-tank reactor (BSTR) (150 rpm) using whole *R. oryzae* cells immobilized in polyurethane foam initially resulted in similar conversions to those obtained from PBRs. However, after 10 operation cycles, the conversion dropped to less than 10% of the initial value because of cell exfoliation. Meanwhile, Ognjanovic et al. (2009) obtained high conversions (transesterification of sunflower oil and methanol) using a commercial enzyme, Novozyme 435, in a BSTR equipped with a six-blade turbine impeller. This system provided good dispersion

TABLE 3 Enzymatic Biodiesel Production in Packed-Bed Reactors (PBR)

Oil/Fat Source Alcohol	Lipase Source	Conditions	Conversion Ratio	Stability/ Operation Time	Reference
Waste oil and methanol	*Candida antarctica* (Novozyme 435)	3 bioreactors continuously operated in series with addition of 1 molar equivalent alcohol for each bed ($\tau_R = \approx 2.7$ h each bed, $T = 30$ °C). Cosolvent-free system and glycerol removal	90%	100 d	Watanabe et al. (2001)
Vegetable oil and methanol	*Candida antarctica* (Novozyme 435)	3 bioreactors in series with addition of 1 molar equivalent alcohol for each bed ($T = 30$ °C). Cosolvent-free system and glycerol removal	90%	100 d	Shimada et al. (2002)
Cotton seed oil and methanol	*Candida antarctica* (Novozyme 435)	1 bioreactor continuously operated with 4.2:1 oil/alcohol ratio. 32,5 vol% tert-butanol as cosolvent.	95%	500 h	Royon et al. (2007)
Soybean oil and methanol	*Candida antarctica* (Novozyme 435)	1 bioreactor continuously operated ($\tau_R = 30$-40 min, $T = 52$ °C) with 4.3:1 oil/alcohol ratio. *n*-hexane:tert-butanol (9:1, v/v) as cosolvent.	75%	–	Shaw et al. (2008)
Soy bean oil and ethanol	*Candida antarctica* (Novozyme 435)	1 bioreactor continuously operated ($\tau_R = 6$ h). 1:12 oil/alcohol ratio at 70 °C. Pressurized propane (60 bar, 7:1 ratio propane/oil) as cosolvent.	70-75%	24 h	Rosa et al. (2009)
Sun flour oil and methyl methanol	*Candida antarctica* (Novozyme 435)	1 bioreactor batch operated ($\tau_{RT} = 8$-10 h, $T = 45$ °C) with 3:1 alcohol/grease molar ratio. Cosolvent-free system	93-96%	72 h	Ognjanovic et al. (2009)
Waste cooking palm oil and methanol	*Candida antarctica* (Novozyme 435)	2 bioreactors operated in series. 1:4 oil/alcohol ratio ($\tau_{RT} = 4$ h, $T = 40$ °C). Tert-butanol (1/1 v/v of oil) as cosolvent.	80%	120 h	Halim et al. (2009)
Soy bean oil and isopropanol	*Candida antarctica* (Novozyme 435)	1 bioreactor continuously operated ($\tau_R = \approx 1$ h, $T = 51.5$ °C). 1:4 oil/alcohol ratio. Cosolvent-free system.	75%	168 h	Chang et al. (2009)

Canola oil and methanol	*Candida antarctica* (Novozyme 435) plus pieces of loofa	1 bioreactor batch operated ($\tau_{RT} = 72$ h, $T = 38$ °C). 1:4.3 oil/alcohol. Cosolvent-free system	97%	432 h	Hajar et al. (2009)
Sunflower oil and isopropanol	*Candida antarctica* (Novozyme 435)	1 bioreactor continuously operated ($T = 50$ °C) with oil/alcohol/isopropyl ester weight ratio of 35:35:30.	90%	210 h	Jachmanián et al. (2010)
Waste oil and methanol	*Candida sp* immobilized lipase in Cotton membrane	3 bioreactors continuously operated in series with 3 stepwise additions of alcohol (24 h) ($\tau_R = \approx 100$ min, $T = 40$ °C). Petroleum ether (3/2, v/v of oil) as cosolvent and glycerol removal by hydrocyclone	92%	500 h (32% conversion ratio)	Nie et al. (2006)
Waste cooking oil and methanol	Candida sp immobilized lipase in textile cloth	3 bioreactors in series with addition of 1 molar equivalent alcohol for each bed ($T = 45$ °C). Hexane/oil weight ratio of 15:100 and glycerol removal each step.	91%	30 h (91%) 100 h (77%)	Chen et al. (2009)
Restaurant grease and ethanol	*Burkholderia cepacia* immobilized lipase in Phyllosilicate sol-gel	1 bioreactor operated in batch mode ($\tau_{RT} = 48$ h, $T = 50$ °C) with 4:1 alcohol/grease molar ratio. Cosolvent-free system	96%	72 h	Hsu et al. (2004)
Soy bean oil and ethanol	*Burkholderia cepacia*—lyophilized and delipidated fermented solid	1 bioreactor operated in batch mode ($\tau_{RT} = 46$ h, $T = 50$ °C) with 2 stepwise additions of alcohol (3:1 alcohol/oil molar ratio). Cosolvent-free system	95%	140 h	Salum et al. (2010)
Soybean oil and methanol	*Rhizopus oryzae* whole cell immobilized in polyurethane foam	1 bioreactor operated in batch mode ($\tau_{RT} = 50$ h, room temperature) with 3 stepwise additions of alcohol. Cosolvent-free system, water/oil emulsification	80-90%	600 h	Hama et al. (2007)

τ_{RT}, reaction time; τ_R, residential time; Novozyme 435, *Candida antarctica* immobilized lipase.

of the biocatalyst, reduced mass transfer resistance, and increased the overall reaction rate. The same authors also studied the agitation rate and method, finding that these variables were extremely important for obtaining high conversions and good enzyme stability. Sanches and Vasudevan (2006) also used Novozyme 435 in a BSTR (60 °C, 100 rpm). They observed a slight drop-off after the initial activity level, but even so it did not drop under 95% during the first five batches, and remained above 70% after as many as eight batch cycles.

The use of continuous and batch-stirred tank reactors has been investigated in some studies and patents, which indicate that the more efficient use of the enzyme preparation as the main advantage, in view of the fact that the tanks can operate with enzymes of different ages/activities and units can be installed between the reactors to separate out the glycerol formed during the reaction. Bassheer et al. (2009) patented an enzymatic biodiesel production system using two continuous-stirred-tank reactors (CSTRs), with a bottom sintered glass filter, operating in series. The authors used a multi-enzyme preparation (enzymes immobilized in different microorganisms) in a cosolvent-free system. Separation equipment was installed between the reactors to remove the glycerol and excess water formed during the reaction. The system operated at high conversions (98%) over a short space of time (4 h), and the enzymes were reused in over 100 consecutive batch cycles.

It can therefore be concluded that the mechanical resistance of the support, the set-up of the reactor (agitation, use of separators), the process conditions (temperature, type of alcohol/oil, use of solvents) and the way the process is conducted are the main points that must be assessed when a reactor is being selected. The immiscibility of the lipid and alcohol phases causes mass transfer problems in PBRs and CSTRs. This should be addressed when the design and optimization of the method for producing biodiesel is being prepared.

One potential option that is not yet economically feasible would be to use bioreactors coupled to membranes to simultaneously separate out the product and recover the biocatalyst. The choice of the most suitable membrane would depend on what kind of biocatalyst was being used (free or immobilized).

5.1 Larger-Scale Reactors

Many articles have been published about enzymatic biodiesel production on a bench scale, yet pilot-scale operations are fundamental for developing and consolidating the enzyme-based technology for producing this commodity.

Brenneis et al. (2004) described biodiesel production from used cooking oil and 2-ethyl-1-hexanol using a liquid preparation and a commercially available thermostable lipase from *Candida antarctica* (Lipase A) called Novozym 735. They used a 3000-L STR at 500 rpm (disk-type agitator) and 50-57 °C. The authors reported that alcoholysis was completed after about 7-10 h, when they used a TAG and 2-ethyl-1-hexanol solution (molar ratio of 1:3-1:3.1) and a lipase solution (1 wt% in relation to TAG).

Park et al. (2008) produced methyl biodiesel on a pilot scale (50 L) by the transesterification of a waste material (activated bleaching earth) from the oil refining industry. They used a BSTR equipped with a filter press for separation of the FAME and solvent mixture. The biocatalyst used was a *C. cylindracea* lipase (added as a powder). Diesel oil was used as a cosolvent with the aim of obtaining a mixture with the biodiesel formed during the transesterification reaction. This mixture can be used to make biodiesel fuel, if it is blended with diesel oil at an appropriate ratio. They obtained 97% conversion after 12 h at 25 °C and 30 rpm, when 1% (w/w) lipase was added to the waste. The two biggest drawbacks noted by the authors were the difficulty of recovering the biodiesel from the waste, given that the cake of vegetable oil-free waste activated bleaching earth contained approximately 14% FAME and 16% solvent on a weight basis. To recover 100% FAME from the waste, the activated bleaching earth would require additional processing, that is, extraction using *n*-hexane. Another disadvantage was that it was impossible to separate the lipase from the final filter cake.

In their review, Tan et al. (2010) cited the operation of two industrial-scale biodiesel production plants in China. In 2007, Lvming Co. Ltd. established an enzymatic production line with 10,000-ton capacity in Shanghai, with immobilized *Candida sp.* 99-125 lipase as a catalyst. The plant uses very acidic used cooking oil and methanol as substrates. The process is conducted in STRs and a centrifuge is used to separate out the glycerol and the water produced during the reaction. The authors reported 90% yields under optimal conditions.

6 ECONOMIC EVALUATION OF ENZYMATIC BIODIESEL PRODUCTION

It seems to be a consensus in the literature that the cost of enzymes will have to fall before the process will be economically feasible. Alternatively, very high yields will have to be achievable, as already obtained by some authors (Chen and Wu, 2003; Shimada et al., 2002; Watanabe et al., 2002a), in which case the lipase can be recycled in a batch system or a continuous-flow process. Nielsen et al. (2008) analyzed studies from the literature to calculate the minimum yield in terms of kg biodiesel to kg enzyme. They calculated the maximum cost of the lipase, assuming that it should be the same as that of a chemical catalyst (25 USD/ton biodiesel). They found that enzymes costing 12-185 USD/kg could be feasible, depending on the process productivity. Fijerbaek et al. (2009) also calculated productivity from studies in the literature in order to compare them against an alkaline catalyst (NaOH; 1 wt% based on the mass of oil and complete conversion), presenting a yield of around 100 kg biodiesel per kilo of catalyst. According to their calculations, the lipases obtained yields that were up to 74 times higher. The average purchase price of 1000 US$ per kg for Novozym 435, compared to just 0.62 US$ (Haas et al., 2006) for NaOH, when offset against their respective yields, puts the cost of the enzyme at 0.14 US$ per kg of ester as against 0.006 US$ per kg of ester for NaOH. If the acquisition cost of the enzyme dropped to 44 US$ per kg or the enzyme could be reused for around 6 years, enzymes would become economically feasible from the perspective of process productivity alone.

Nevertheless, it is no easy task to give a precise answer as to how cheap enzymes would have to be to compete with chemical catalysts. They are hard to compare because the chemical and enzyme processes are so different. Clearly, it is unsatisfactory to merely compare the cost of the catalysts in isolation, but a full economic analysis of enzyme versus chemical catalysts for biodiesel production would require whole host of assessments, including cost of oil (the use of low-cost oils with a high FFA concentration can have a major impact on overall process costs); cost of alcohol; cost of pretreatment stages; process yield; cost of waste treatment; commercial value of glycerol; and cost of downstream stages. Enzyme technology has a positive impact on several of these factors: its feasibility for use with raw material of varying quality; process with fewer stages; better quality glycerol; better phase separation (with no emulsion caused by soap formation); less energy consumption; and less wastewater production (Nielsen et al., 2008).

Sotoft et al. (2010) simulated the processes used in different enzymatic biodiesel production plants and evaluated them economically, using data taken from experiments by Shimada et al. (1999) and Li et al. (2006), who did excellent studies into solvent-free systems and systems using tert-butanol, respectively. They assessed continuous-flow biodiesel production plants that used high-quality rapeseed oil and methanol. They investigated two production scales (8 and 200 M.kg of biodiesel/year) and two enzyme prices (current prices of 762.71 €/kg enzyme and a lower future price of 7.627 €/kg enzyme). The economic analysis showed that the process that used solvent was more expensive to the point of being unfeasible, while the solvent-free process was found to be feasible on a larger production scale (200 M.kg of biodiesel/year) at today's lipase price. At the projected future price of the enzyme, the small- and large-scale production processes were found to be feasible using a solvent-free medium.

The total capital investment (TCI) was found to be lower for the solvent-free system than for the system using a solvent on both the scales studied. The equipment cost was cheaper for the plant using a solvent, but when the cost of installing the solvent recovery column was added, the total cost was higher, even taking into account the extra reactors and settling tanks required for the solvent-free set-up (Sotoft et al., 2010).

As for production costs, the main contributory factors in all the scenarios studied were raw material costs and the sale price of the byproduct. The biggest single factor to affect raw material costs in the solvent-free system was the cost of the enzyme; its influence was less in the system using a solvent because the lipase was more productive in this system. The cost of the solvent, tert-butanol, was not significant, as it is reused, while the oil was the most expensive single element in the system using a solvent. The sale of the glycerol was found to be equally important in all the scenarios. The cost of utility bills was found to be very significant in the operation of the plants using a solvent because of the amount of energy required; indeed, this was one of the factors that made this process economically unfeasible. Meanwhile, the electricity costs of the solvent-free process were low (Sotoft et al., 2010).

The economic feasibility study showed that at current lipase prices, the only plant that would be cost effective was the large-scale solvent-free plant, with a very short payback period of 0.25 year (assuming 1.12 €/kg as the sale price of the biodiesel) or a minimum product price of 0.73 €/kg. For the other plants, the minimum product price stood at 1.49 €/kg for the small-scale solvent-free plant, 2.38 €/kg for the small-scale plant using solvent, and 1.70 €/kg for the large-scale plant using solvent, all of which put the product price higher than 1.12 €/kg. Even when the projected future price of lipase was used, the plants using solvents were not deemed cost effective on either scale. The large solvent-free plant was considered very feasible, with a payback period of 0.09 year and a minimum product price of 0.05 €/kg. The product price is low because of the sale of the glycerol. The small-scale solvent-free plant gave a minimum product price of 0.75 €/kg, which is lower than the market price, but its payback period would be 3.59 years, which is borderline for projects of this kind. Generally speaking, the payback period for feasible processes should be under 2 years for high-risk projects; anything over 4 years is considered unfeasible (Seider et al., 2004 cited by Sotoft et al., 2010). Given the uncertainties inherent to the new and as yet volatile biofuel market, this kind of project is inherently high risk. The results of the economic feasibility analysis are very promising and enzymatic biodiesel production seems to be bordering on becoming a truly feasible industrial-scale option. A comparison with feasibility studies from the literature of processes using traditional catalysts shows that enzymatic biodiesel production is more expensive, but if the lifespan and yield of the lipases can be improved, plus the major improvement in environmental impacts when this technology is used, then the enzymatic production of biodiesel is sure to become a very attractive prospect (Sotoft et al., 2010).

7 CONCLUSIONS

There are a few process conditions that should be taken into account before enzymatic technology can be feasibly designed for producing a commodity like biodiesel: (i) correlation between enzyme and raw material types and costs; mass transfer and reaction conditions; and product recovery when choosing whether to use a solvent; (ii) the choice of whether to use a free or immobilized biocatalyst should be dictated by weighing the cost of the support against the biocatalyst reuse capacity; (iii) long-term continuous-flow or batch experiments should be undertaken.

References

Al-Zuhair, S., 2007. Production of biodiesel: possibilities and challenges. Biofuel Bioprod. Bior. 1, 57–66.

Antczak, M.S., Kubiak, A., Antczak, T., Bielecki, S., 2009. Enzymatic biodiesel synthesis—key factors affecting efficiency of the process. Renew. Energy 34, 1185–1194.

Bajaj, A., Lohan, P., Jha, P.N., Mehrotra, R., 2010. Biodiesel production through lipase catalyzed transesterification: an overview. J. Mol. Catal. B Enzym. 62, 9–14.

Balat, M., Balat, H., 2010. Progress in biodiesel processing. Appl. Energy 87, 1815–1835.

Ban, K., Kaieda, M., Matsumoto, T., Kondo, A., Fukuda, H., 2001. Whole cell biocatalyst for biodiesel fuel production utilizing *Rhizopus oryzae* cells immobilized within biomass support particles. Biochem. Eng. J. 8, 39–43.

Ban, K., Hama, S., Nishizuka, K., Kaieda, M., Matsumoto, T., Kondo, A., et al., 2002. Repeated use of whole-cell biocatalysts immobilized within biomass support particles for biodiesel fuel production. J. Mol. Catal. B Enzym. 17, 157–165.

Bassheer, S., Haj, M., Kaiyal, M.A., 2009. Robust multi-enzyme preparation for the synthesis of fatty acid alkyl ester. Patent WO 2009/O69116, PCT/IL2008/001497.

Bisen, P.S., Sanodiya, B.S., Thakur, G.S., Baghel, R.K., Prasad, G.B.K.S., 2010. Biodiesel production with special emphasis on lipase-catalyzed transesterification. Biotechnol. Lett. 32, 1019–1030.

Brenneis, R., Baeck, B., Kley, G., 2004. Alcoholysis of waste fats with 2-ethyl-1-hexanol using Candida antarctica lipase A in large-scale tests. Eur. J. Lipid Sci. Technol. 106, 809–814.

Cavalcanti-Oliveira, E.D., Silva, P.R.R., Ramos, A.P., Aranda, D.A.G., Freire, D.M.G., 2011. Study of soybean oil hydrolysis catalyzed by *Thermomyces lanuginosus* lipase and its application to biodiesel production via hydroesterification. Enzym. Res. in press. 2011, 1–8.

Chang, C., Chen, J.H., Chang, C.M.J., Wu, T.T., Shieh, C.J., 2009. Optimization of lipase-catalyzed biodiesel by isopropanolysis in a continuous packed-bed reactor using response surface methodology. New Biotechnol. 26, 187–192.

Chen, J.W., Wu, W.T., 2003. Regeneration of immobilized Candida antarctica lipase for transesterification. J. Biosci. Bioeng. 95, 466–469.

Chen, G., Ying, M., Li, W., 2006. Enzymatic conversion of waste cooking oils into alternative fuel-biodiesel. Appl. Biochem. Biotechnol. 132, 911–921.

Chen, Y., Xiao, B., Chang, J., Fu, Y., Lv, P., Wang, X., 2009. Synthesis of biodiesel from waste cooking oil using immobilized lipase in fixed bed reactor. Energy Convers. Manage. 50, 668–673.

Demirbas, A., 2007. Progress and recent trends in biofuels. Progress in Energy and Combustion Science. 33, 1–18.

Deng, L., Tan, T., Wang, F., Xu, X., 2003. Enzymatic production of fatty acid alkyl esters with a lipase preparation from Candida sp. 99-125. Eur. J. Lipid Sci. Technol. 105, 727–734.

Deng, L., Xu, X., Haraldsson, G.G., Tan, T., Wang, F., 2005. Enzymatic production of alkyl esters through alcoholysis: a critical evaluation of lipases and alcohols. J. Am. Oil Chem. Soc. 82, 341–347.

Dizge, N., Keskinler, B., Tanriseven, A., 2009. Biodiesel production from canola oil by using lipase immobilized onto hydrophobic microporous styrene-divinylbenzene copolymer. Biochem. Eng. J. 44, 220–225.

Du, W., Wang, L., Liu, D., 2007. Improved methanol tolerance during Novozym 435-mediated methanolysis of SODD for biodiesel production. Green Chem. 9, 173–176.

Fijerbaek, L., Cristhensen, K.V., Norddahl, B., 2009. A review of the current state of biodiesel production using enzymatic transesterification. Biotechnol. Bioeng. 102, 1298–1315.

Gan, Q., Rahmat, H., Weatherley, L.R., 1998. Simultaneous reaction and separation in enzymatic hydrolysis of high oleate sun flower oil evaluation of ultrafiltration performance and process synergy. Chem. Eng. J. 71, 87–96.

Ganesan, D., Rajendran, A., Thangavelu, V., 2009. An overview on the recent advances in the transesterification of vegetable oils for biodiesel production using chemical and biocatalysts. Rev. Environ. Sci. Biotechnol. 8, 367–394.

Gorman, L.S., Dordick, J.S., 1991. Organic solvents strip water off enzymes. Biotechnol. Bioeng. 39, 392–397.

Gunstone, F.D., Herslöf, B.G., 2004. Lipid Glossary, second ed. The Oily Press Bridgwater Vol. 12.

Haas, M.J., McAloon, A.J., Yee, W.C., Foglia, T.A., 2006. A process model to estimate biodiesel production costs. Bioresour. Technol. 97, 671–678.

Hajar, M., Shokrollahzadeh, S., Vahabzadeh, F., Monazzami, A., 2009. Solvent-free methanolysis of canola oil in a packed-bed reactor with use of Novozym 435 plus loofa. Enzyme. Microb. Technol. 45, 188–194.

Halim, S.F.A., Kamaruddin, A.H., Fernando, W.J.N., 2009. Continuous biosynthesis of biodiesel from waste cooking palm oil in a packed bed reactor: Optimization using response surface methodology (RSM) and mass transfer studies. Bioresource Technology. 100, 710–716.

Hama, S., Yamaji, H., Fukumizo, T., Numata, T., Tamalampudi, S., Kondo, A., et al., 2007. Biodiesel fuel production in a packed-bed reactor using lipase-producing *Rhizopus oryzae* cells immobilized within biomass support particles. Biochem. Eng. J. 34, 273–278.

Hsu, A.F., Jones, K., Marmer, W.N., Foglia, T.A., 2001. Production of alkyl esters from tallow and grease using lipase immobilized in a phyllosilicate sol-gel. J. Am. Oil Chem. Soc. 78, 585–588.

Hsu, A.F., Jones, K.C., Foglia, T.A., Marmer, W.N., 2004. Continuous production of ethyl esters of grease using an immobilized lipase. J. Am. Oil Chem. Soc. 81, 749–752.

Iso, M., Chen, B., Eguchi, M., Kudo, T., Shrestha, S., 2001. Production of biodiesel fuel from triglycerides and alcohol using immobilized lipase. J. Mol. Catal. B Enzym. 16, 53–58.

Jachmanián, I., Dobroyán, M., Moltini, M., Segura, N., Irigaray, B., Veira, J.P., et al., 2010. Continuous Lipase-Catalyzed Alcoholysis of Sunflower Oil:Effect of Phase-Equilibrium on Process Efficiency. J. Am. Oil Chem. Soc. 87, 45–53.

Jegannathan, K.R., Abang, S., Poncelet, D., Chan, E.S., Ravindra, P., 2008. Production of biodiesel using immobilized lipase—a critical review. Crit. Rev. Biotechnol. 28, 253–264.

Jothiramalingam, R., Wang, M.K., 2009. Review of recent developments in solid acid, base, and enzyme catalysts (heterogeneous) for biodiesel production via transesterification. Ind. Eng. Chem. Res. 48, 6162–6172.

Kaieda, M., Samukawa, T., Matsumoto, T., Ban, K., Kondo, A., Shimada, Y., et al., 1999. Biodiesel fuel production from plant oil catalyzed by *Rhizopus oryzae* lipase in a water-containing system without an organic solvent. J. Biosci. Bioeng. 88, 627–631.

Kaieda, M., Samukawa, T., Kondo, A., Fukuda, H., 2001. Effect of methanol and water contents on production of biodiesel fuel from plant oil catalyzed by various lipases in a solvent-free system. J. Biosci. Bioeng. 91, 12–15.

Koike, S., Yokoo, M., Nabetani, H., Nakajimna, M., 1992. Membrane separation of fats and oils in organic solvents. Progr. Bioseparation Eng. (Japan Soc. Chem. Eng.) 33, 84–87.

Kojima, S., Du, D., Sato, M., Park, E.Y., 2004. Efficient production of fatty acid methyl ester from waste activated bleaching earth using diesel oil as organic solvent. J. Biosci. Bioeng. 98, 420–424.

Kumari, V., Shah, S., Gupta, M.N., 2007. Preparation of biodiesel by lipase-catalyzed transesterification of high free fatty acid containing oil from madhuca indica. Energy Fuels 21, 368–372.

Lai, C.C., Zullaikah, S., Vali, S.R., Ju, Y.H., 2005. Lipase-catalyzed production of biodiesel from rice bran oil. J. Chem. Technol. Biotechnol. 80, 331–337.

Lara, P.V., Park, E.Y., 2004. Potential application of waste activated bleaching earth on the production of fatty acid alkyl esters using Candida cylindracea lipase in organic solvent system. Enzyme. Microb. Technol. 34, 270–277.

Li, L., Du, W., Liu, D., Wang, L., Li, Z., 2006. Lipase-catalyzed transesterification of rapeseed oils for biodiesel production with a novel organic solvent as the reaction medium. J. Mol. Catal. B Enzym. 43, 58–62.

Li, L., Du, W., Liu, D., 2007a. Optimization of whole cell-catalyzed methanolysis of soybean oil for biodiesel production using response surface methodology. J. Mol. Catal. B Enzym. 45, 122–127.

Li, L., Du, W., Liu, D., 2007b. *Rhizopus oryzae* IFO 4697 whole cell catalyzed methanolysis of crude and acidified rapeseed oils for biodiesel production in tert-butanol system. Process Biochem. 42, 1481–1485.

Li, N.W., Zong, M.H., Wu, H., 2009. Highly efficient transformation of waste oil to biodiesel by immobilized lipase from *Penicillium expansum*. Process Biochem. 44, 685–688.

Lu, J., Nie, K., Xie, F., Wang, F., Tan, T., 2007. Enzymatic synthesis of fatty acid methyl esters from lard with immobilized Candida sp. 99-125. Process Biochem. 42, 1367–1370.

Marchetti, J.M., Miguel, V.U., Errazu, A.F., 2007. Possible methods for biodiesel production. Renew. Sust. Energy Rev. 11, 1300–1311.

Marchetti, J.M., Miguel, V.U., Errazu, A.F., 2008. Techno-economic study of different alternatives for biodiesel production. Fuel Proc. Technol. 89, 740–748.

Matsumoto, T., Takahashi, S., Kaieda, M., Ueda, M., Tanaka, A., Fukuda, H., et al., 2001. Yeast whole-cell biocatalyst constructed by intracellular overproduction of *Rhizopus oryzae* lipase is applicable to biodiesel fuel production. Appl. Microbiol. Biotechnol. 57, 515–520.

Matsumoto, T., Fukuda, H., Ueda, M., Tanaka, A., Kondo, A., 2002. Construction of yeast strains with high cell surface lipase activity by using novel display systems based on the flo1p flocculation functional domain. Appl Environ Microbiol 68, 4517–4522.

Mittelbach, M., 1990. Lipase catalyzed alcoholysis of sunflower oil. J. Am. Oil Chemi. Soc. 67, 168–170.

Nelson, L.A., Foglia, T.A., Marmer, W.N., 1996. Lipase-catalyzed production of biodiesel. J. Am. Oil Chem. Soc. 73, 1191–1195.

Nie, K., Xie, F., Wang, F., Tan, T., 2006. Lipase catalyzed methanolysis to produce biodiesel: optimization of the biodiesel production. J. Mol. Catal. B Enzym. 43, 142–147.

Nielsen, P.M., Brask, J., Fjerbaek, L., 2008. Enzymatic biodiesel production: technical and economical Considerations. Eur. J. Lipid Sci. Technol. 110, 692–700.

Noor, I.M., Hasan, M., Ramachandran, K.B., 2003. Effect of operating variables on the hydrolysis rate of palm oil by lipase. Process Biochem. 39, 13–20.

Noureddini, H., Gao, X., Philkana, R.S., 2005. Immobilized *Pseudomonas cepacia* lipase for biodiesel fuel production from soybean oil. Bioresour. Technol. 96, 769–777.

Ognjanovic, N., Bezbradica, D., Knezevic-Jugovic, Z., 2009. Enzymatic conversion of sunflower oil to biodiesel in a solvent-free system: process optimization and the immobilized system stability. Bioresour. Technol. 100, 5146–5154.

Park, E.Y., Sato, M., Kojima, S., 2008. Lipase-catalyzed biodiesel production from waste activated bleaching earth as raw material in a pilot plant. Bioresour. Technol. 99, 3130–3135.

Pizarro, A.V.L., Park, E.Y., 2003. Lipase-catalyzed production of biodiesel fuel from vegetable oils contained in waste activated bleaching earth. Process Biochem. 38, 1077–1082.

Pugazhenthi, G., Kumar, A., 2004. Enzyme membrane reactor for hydrolysis of olive oil using lipase immobilized on modified PMMA composite membrane. J. Membr. Sci. 228, 187–197.

Qin, H., Yan, X., Yun, T., Dong, W., 2008. Biodiesel production catalyzed by whole-cell lipase from rhizopus chinensis. Chin. J. Catalysis 29, 41–46.

Raman, L.P., Cheryan, M., Rajagopalan, N., 1996. Deacidification of soybean oil by membrane technology. J. Am. Oil Chem. Soc. 73, 219–224.

Ranganathan, S.V., Narasimhan, S.J., Muthukumar, K., 2008. An overview of enzymatic production of biodiesel. Bioresour. Technol. 99, 3975–3981.

Reddy, K.K., Nakajima, M., Snape, J.B., Kawakatsu, K., Ichikawa, S., Nabetani, H., 1996. Membrane decolourisation of crude oil. In: Int. Congr. Memb. Proc., Yokohama, Japan, paper 894.

Rosa, C.D., Morandim, M.B., Ninow, J.L., Oliveira, D., Treichel, H., Oliveira, J.V., 2009. Continuous lipase-catalyzed production of fatty acid ethyl esters from soybean oil in compressed fluids. Bioresour. Technol. 100, 5818–5826.

Roy, I., Gupta, M.N., 2004. Freeze-drying of proteins: some emerging concerns. Biotechnol. Appl. Biochem. 39, 165–177.

Royon, D., Daz, M., Ellenrieder, G., Locatelli, S., 2007. Enzymatic production of biodiesel from cotton seed oil using t-butanol as a solvent. Bioresour. Technol. 98, 648–653.

Saifuddin, N., Raziah, A.Z., Nor farah, H., 2009. Production of biodiesel from high acid value waste cooking oil using an optimized lipase enzyme/acid-catalyzed hybrid process. J. Chem. 6, 485–495.

Salis, A., Pinna, M., Monduzzi, M., Solinas, V., 2008. Comparison among immobilised lipases on macroporous polypropylene toward biodiesel synthesis. J. Mol. Catal. B Enzym. 54, 19–26.

Salum, T.F.C., Villeneuve, P., Barea, B., Yamamoto, C.I., Côcco, L.C., Mitchell, D.A., et al., 2010. Synthesis of biodiesel in column fixed-bed bioreactor using the fermented solid produced by *Burkholderia cepacia* LTEB11. Process Biochem. 45, 1348–1354.

Sanchez, F., Vasudevan, T., 2006. Enzyme catalyzed production of biodiesel from olive oil. Applied Biochemistry and Biotechnology. 135, 1–14.

Seider, W.D., Seader, J.D., Lewin, D.R., 2004. Product and Process Design Principles, second ed. John Wiley & Sons, Inc., New York.

Shah, S., Gupta, M.N., 2007. Lipase catalyzed preparation of biodiesel from Jatropha oil in a solvent free system. Process Biochem. 42, 409–414.

Shah, S., Sharma, S., Gupta, M.N., 2003. Enzymatic transesterification for biodiesel production. Indian Journal of Biochemistry & Biophysics. 40, 392–399.

Shah, S., Sharma, S., Gupta, M.N., 2004. Biodiesel preparation by lipasecatalyzed transesterification of jatropha oil. Energy Fuels 18, 154–159.

Shaw, J.F., Chang, S.W., Lin, S.C., Wu, T.T., Ju, H.Y., Akoh, C.C., et al., 2008. Continuous enzymatic synthesis of biodiesel with Novozym 435. Energy Fuels 22, 840–844.

Shimada, Y., Watanabe, Y., Samukawa, T., Sugihara, A., Noda, H., Fukuda, H., et al., 1999. Conversion of vegetable oil to biodiesel using immobilized Candida antarctica lipase. J. Am. Oil Chem. Soc. 76, 788–793.

Shimada, Y., Watanabe, H., Sugihara, A., Tominaga, Y., 2002. Enzymatic alcoholysis for biodiesel fuel production and application of the reaction to oil processing. J. Mol. Catal. B Enzym. 17, 133–142.

Sotoft, L.F., Rong, B.G., Christensen, K.V., Norddahl, B., 2010. Process simulation and economical evaluation of enzymatic biodiesel production plant. Bioresour. Technol. 101, 5266–5274.

Soumanou, M.M., Bornscheuer, U.T., 2003. Improvement in lipase-catalyzed synthesis of fatty acid methyl esters from sunflower oil. Enzyme. Microb. Technol. 33, 97–103.

Sousa, J.S., Cavalcanti-Oliveira, E.D., Aranda, D.A.G., Freire, D.M.G., 2010. Application of lipase from the physic nut (Jatropha curcas L.) to a new hybrid (enzyme/chemical) hydroesterification process for biodiesel production. J. Mol. Catal. B Enzym. 65, 133–137.

Talukder, M.d.M.R., Wu, J.C., Fen, N.M., Melissa, Y.L.S., 2010a. Two-step lipase catalysis for production of biodiesel. Biochemical. Eng. J. 49, 207–212.

Talukder, M.d.M.R., Wu, J.C., Chua, L.P.L., 2010b. Conversion of waste cooking oil to biodiesel via enzymatic hydrolysis followed by chemical esterification. Energy Fuels 24, 2016–2019.

Tamalampudi, S., Talukder, M.R., Hama, S., Numata, T., Kondo, A., Fukuda, H., 2008. Enzymatic production of biodiesel from Jatropha oil: a comparative study of immobilized-whole cell and commercial lipases as a biocatalyst. Biochem. Eng. J. 39, 185–189.

Tan, T., Lu, J., Nie, K., Deng, L.A., Wang, F., 2010. Biodiesel production with immobilized lipase: a review. Biotechnol. Adv. 28, 628–634.

Ting, W.J., Huang, C.M., Giridhar, N., Wu, W.T., 2008. An enzymatic/acid-catalyzed hybrid process for biodiesel production from soybean oil. J. Chin. Inst. Chem. Eng. 39, 203–210.

Torres, M., Loscos, V., Sanahuja, V., Canela, R., 2003. Reactive extraction of acylglycerides using aspergillus flavus resting cells. J. Am. Oil Chem. Soc. 80, 347–351.

Triantafyllou, A.O., Wehtje, E., Adlercreutz, P., Mattiasson, B., 1995. Effects of sorbitol addition on the action of free and immobilized hydrolytic enzymes in organic media. Biotechnol. Bioeng. 45, 406–414.

Vavra, C., Koseuglu, S.S., 1994. Catalyst removal from hydrogenated oils using membrane technology. In: Yango, T., Natsuno, R., Nakamuja, K. (Eds.), Developments in Food Engineering. Blackie Academic and Professional, Glasgow, pp. 683–685.

Wang, L., Du, W., Liu, D., Li, L., Dai, N., 2006. Lipase-catalyzed biodiesel production from soybean oil deodorizer distillate with absorbent present in tert-butanol system. J. Mol. Catal. B Enzym. 43, 29–32.

Watanabe, Y., Shimada, Y., Sugihara, A., Noda, H., Fukuda, H., Tominaga, Y., 2000. Continuous production of biodiesel fuel from vegetable oil using immobilized Candida antarctica lipase. J. Am. Oil Chem. Soc. 77, 355–360.

Watanabe, Y., Shimada, Y., Sugihara, A., Tominaga, Y., 2001. Enzymatic conversion of waste edible oil to biodiesel fuel in a fixed-bed bioreactor. J. Am. Oil Chem. Soc. 78, 703–707.

Watanabe, Y., Shimada, Y., Baba, T., Ohyagi, N., Moryiama, S., Terai, T., et al., 2002a. Methyl esterification of waste fatty acids with immobilized Candida antarctica lipase. J. oleo Sci. 51, 655–661.

Watanabe, Y., Shimada, Y., Sugihara, A., Tominaga, T., 2002b. Conversion of degummed soybean oil to biodiesel fuel with immobilized Candida antarctica lipase. J. Mol. Catal. B Enzym. 17, 151–155.

Watanabe, Y., Pinsirodom, P., Nagao, T., Kobayashi, T., Nishida, Y., Takagi, Y., et al., 2005. Production of FAME from acid oil model using immobilized Candida antarctica lipase. J. Am. Oil Chem. Soc. 82, 825–831.

Watanabe, Y., Pinsirodom, P., Nagao, T., Yamauchi, A., Kobayashi, T., Nishida, Y., et al., 2007a. Conversion of acid oil by-produced in vegetable oil refining to biodiesel fuel by immobilized Candida antarctica lipase. J. Mol. Catal. B Enzym. 44, 99–105.

Watanabe, Y., Nagao, T., Nishida, Y., Takagi, Y., Shimada, Y., 2007b. Enzymatic production of fatty acid methyl esters by hydrolysis of acid oil followed by esterification. J. Am. Oil Chem. Soc. 84, 1015–1021.

Ying, M., Chen, G., 2007. Study on the production of biodiesel by magnetic cell biocatalyst based on lipase-producing *Bacillus subtilis*. Appl. Biochem. Biotechnol. 136–140, 793–803.

Zeng, J., Du, W., Liu, X., Liu, D., Dai, L., 2006. Study on the effect of cultivation parameters and pretreatment on *Rhizopus oryzae* cell-catalyzed transesterification of vegetable oils for biodiesel production. J. Mol. Catal. B Enzym. 43, 15–18.

Zhang, Y., Dubé, M.A., McLean, D.D., Kates, M., 2003a. Biodiesel production from waste cooking oil: 1. Process design and technological assessment. Bioresour. Technol. 89, 1–16.

Zhang, Y., Dubé, M.A., McLean, D.D., Kates, M., 2003b. Biodiesel production from waste cooking oil: 2. Economic assessment and sensitivity analysis. Bioresour. Technology. 90, 229–240.

Zheng, Y., Quan, J., Ning, X., Zhu, L.M., Jiang, B., He, Z.Y., 2009. Lipase-catalyzed transesterification of soybean oil for biodiesel production in tert-amyl alcohol. World J. Microbiol. Biotechnol. 25, 41–46.

List of Abbreviations

NaOH Sodium Hydroxide
KOH Potassium Hydroxide
TAG Triacylglycerols
DAG Diacylglycerols
MAG Monoacylglycerols
FFA Free Fatty Acid
FAME Fatty Acid Methyl Esters
FAEE Fatty Acid Ethyl Esters
CLEA Cross-linked Enzyme Aggregates
PCMC Protein-coated Microcrystals
BSTR Batch-Stirred-Tank Reactor
CSTR Continuous-Stirred-Tank Reactor
PBR Packed-Bed Reactor
MKg 10^6.Kg

CHAPTER

14

Biodiesel Production in Supercritical Fluids

Kok Tat Tan, Keat Teong Lee*

School of Chemical Engineering, Universiti Sains Malaysia, Engineering Campus, Seri Ampangan, 14300 Nibong Tebal, Pulau Pinang, Malaysia

***Corresponding author: E-mail: koktat@hotmail.com**

1 INTRODUCTION

The demand for fossil fuels such as petroleum, natural gas, and coal has been escalating for the past few decades owing to rapid development and urbanization occurring throughout the world. Furthermore, the demand for these energy sources is projected to be mounting significantly in the future. Consequently, the costs of these non-renewable sources of energy have increased substantially in recent years due to high demand and limited supply in the world market. However, these fossil fuels are non-renewable and will be depleted in the future which prompted concerns of energy security and sustainability. Apart from that, employment of these exhaustible energy sources also caused environmental degradation with the emission of greenhouse gases (GHG) which include carbon monoxide (CO), carbon dioxide (CO_2), nitrogen oxide (NO), nitrogen dioxide (NO_2), and sulfur dioxide (SO_2). The release of GHG to the atmosphere would trap enormous amount of heat which leads to environmental catastrophes such as greenhouse effect, global warming, and acid rain. Hence, the escalating utilization of non-renewable fuels throughout the world implies that excessive GHG are being emitted globally and a collective effort at international level to address this global issue is inevitable. Therefore, there is an urgent need to find alternative energy source which is renewable, economical, and environmental-friendly to solve these global problems of energy security and environmental degradation.

Currently, extensive researches have been carried out worldwide to produce renewable energy which could address these issues. Generally, renewable energy is produced from infinite sources such as biomass, sunlight, or wind. Besides, utilization of renewable energy sources does not release harmful GHG gases to the atmosphere. Hence, it could contribute toward climate change mitigation and solve the environmental degradation crisis. For instance, biodiesel, one of the most researched renewable energy sources, is produced from biomass, particularly crops such as rapeseed, palm, and soybean. As they grow, these crops absorb carbon dioxide from the atmosphere and accumulate the carbon as biomass. Subsequently, during the combustion of biodiesel, the carbon will be released and returned to the atmosphere. Therefore, biodiesel is a carbon "neutral" source of renewable energy which does not emit additional carbon to the environment. In addition, biodiesel is superior to petroleum-derived diesel in terms of biodegradability, flash point, and sulfur content. Apart from that, liquid biodiesel also offers a promising solution for energy security and sustainability. Currently, the demand for liquid fuels comprises more than 40% of the total energy consumption in the world. However, other sources of renewable energy such as solar, wind, and hydrothermal are only able to provide renewable energy in the form of electricity or thermal energy. In this context, biodiesel is superior to other renewable energy sources as it could accommodate the demand of liquid fuels in the world market, particularly in transportation sector. In terms of application, biodiesel and diesel have similar physico-chemical properties, implying that no modification in existing diesel engine is required. Furthermore, biodiesel and diesel could be blended and commercially employed as transportation fuel as well. Collectively, biodiesel is environmental-friendly and has the potential to replace fossil fuels as the main source of energy in the future.

Fatty acid alkyl esters (FAAE), or the commonly known biodiesel, is produced from transesterification reaction involving triglycerides and alcohol. This reaction is similar to hydrolysis but instead of water molecule, alcohol molecule acts as acyl acceptor to produce FAAE and glycerol. In this reversible reaction, 1 mol of triglycerides reacts with 3 mol of alcohol, producing 3 mol of FAAE and 1 mol of glycerol as shown in Figure 1. Generally, methanol or ethanol is employed as the source of alcohol in transesterification reaction. If methanol is utilized, fatty acid methyl esters (FAME) will be produced while fatty acid ethyl esters (FAEE) are formed with the presence of ethanol. Both FAME and FAEE are also known as biodiesel. On the other hand, the triglycerides are acquired from crops such as rapeseed, palm, soybean, and jatropha and these oil-bearing crops produce huge amount of oil per ton of biomass. Hence, these crops are suitable to be employed as the source of triglycerides in biodiesel production.

Due to immiscibility between triglycerides and alcohol, the rate of reaction in transesterification reaction is extremely slow. Hence, catalysts are usually introduced in the reaction medium to enhance the reaction rate. Transesterification reaction can be catalyzed by homogeneous or heterogeneous catalysts. In addition, the catalysts can be either acidic or alkaline-based compounds. Currently, the most common technology employed in the industries involves homogeneous catalysts such as sulfuric acid, hydrochloric acid, sodium hydroxide, and potassium hydroxide. These homogeneous catalysts are cheap and easily introduce inside the reaction medium. However, it was found that separation and purification of products and catalysts required complicated procedures due to the homogenous phase of the mixture. Consequently,

$$
\begin{array}{lccccc}
CH_2-OCOR_1 & & & R_1COOR & & CH_2-OH \\
| & & & & & | \\
CH-OCOR_2 & +\,3\,ROH & \leftrightarrow & R_2COOR & + & CH-OH \\
| & & & & & | \\
CH_2-OCOR_3 & & & R_3COOR & & CH_2-OH \\
\text{Triglycerides} & \text{Alcohol} & & \text{FAAE} & & \text{Glycerol}
\end{array}
$$

FIGURE 1 General transesterification between triglycerides and alcohol to produce fatty acid alkyl esters (FAAE) and glycerol.

the production cost and energy consumption in the process become unattractive and impractical from economic considerations. Apart from that, base catalyst will react with free fatty acids (FFA) normally found in oils and subsequently produces unwanted side product such as soap. Furthermore, homogeneous catalytic reaction is sensitive to the presence of impurities such as water molecule which prompted the utilization of expensive refined oils. Consequently, the total production costs of biodiesel via homogeneous reaction become uneconomical.

Subsequently, a new technology in transesterification reaction emerged with the development of heterogeneous catalytic reaction. Similar to homogeneous catalysts, these catalysts can be either acidic or alkaline-based compounds. In heterogeneous catalytic reaction, the catalyst is generally in solid phase which is different from the liquid reactants in the reaction. Therefore, application of heterogeneous catalysts simplifies separation and purification of products since the products are in different phase from the catalysts. Furthermore, it was reported that solid catalysts are not sensitive to the presence of impurities (FFA and water molecule) which allows the employment of cheap sources of triglycerides such as waste cooking oil. Therefore, utilization of inexpensive feedstock in heterogeneous catalytic reaction would reduced the total processing costs of biodiesel substantially as the cost of feedstock comprises more than 70% of the total production costs. However, heterogeneous catalytic reaction suffers from lower yield and longer reaction period compared to homogeneous reaction. In heterogeneous reaction, the reaction rate is limited significantly by diffusion factor which leads to longer reaction time. Furthermore, solid catalysts are more expensive compared to homogeneous catalysts which increases the production cost of biodiesel.

2 SUPERCRITICAL FLUID REACTION

Due to limitations and weaknesses of catalytic reactions in biodiesel production, there are numerous alternative technologies that have been proposed which could overcome these issues. One of them which have been widely reported is by employing non-catalytic supercritical fluid technology. In this method, the reactants are subjected to supercritical conditions of solvent/reactant (i.e., alcohol) without the presence of any catalysts. During supercritical conditions, the properties of the solvent do not fulfil the definition of neither liquid nor gas but in between these two phases. Hence, supercritical fluid possesses unique properties such as solubility parameter, diffusion coefficient and density. The critical properties of selected solvents are shown in Table 1. During subcritical state of solvent, the reactants form two layers of oil phase and solvent phase due to immiscibility between these two compounds. Subsequently, increase in reaction temperature will enhance the solubility of solvent in oil phase due to decrease in solubility parameter of solvent. Solubility parameter is defined as

TABLE 1 Critical Properties of Selected Solvents

Solvent	Critical Temperature, T_C (°C)	Critical Pressure, P_C (MPa)
Methanol	239	8.09
Methyl Acetate	234	4.69
Dimethyl Carbonate	275	4.63

the square root of cohesive density of a liquid and only those with similar values could form a homogeneous phase. For instance, the solubility parameter for methanol is 29.7 $(MPa)^{1/2}$ while for oil it is approximately 18 $(MPa)^{1/2}$. Increment in reaction temperature decreases the solubility parameter of methanol to a value similar to oil which leads to formation of homogeneous methanol-oil mixture. Consequently, transesterification can proceed even without the presence of catalysts in supercritical fluid reaction.

Furthermore, due to the absence of catalyst, separation and purification processes in supercritical reaction become simpler and cost-effective compared to catalytic reaction. For instance, biodiesel can be separated easily without intervention from catalyst and no huge amount of waste will be produced. In supercritical fluid reaction, there are four important parameters which influence the yield of biodiesel significantly which are reaction temperature, reaction pressure, reaction time, and molar ratio of solvent to oil. The reaction temperature and pressure employed in the reaction must be above the critical points of the solvent to ensure that supercritical conditions are achieved. The yield of biodiesel is highly dependent on the reaction temperature and pressure which influence the reaction rate of transesterification substantially. On the other hand, it was reported that supercritical fluid reaction could achieve high yield of biodiesel in shorter amount of reaction time compared to conventional catalytic reaction which makes this process more economical. Besides, due to the absence of catalysts, a high molar ratio of solvent to oil is commonly employed in supercritical reaction to push the reversible transesterification reaction toward producing more biodiesel.

3 BIODIESEL PRODUCTION IN NON-CATALYTIC SUPERCRITICAL FLUID REACTION

3.1 Supercritical Alcohol (SCA) Reaction

Application of SCA in biodiesel production has been reported by several researchers including Saka and Kusdiana (2001). In their study, rapeseed oil was used as the source of triglycerides and the source of alcohol was methanol. The supercritical methanol (SCM) reaction was carried out in a batch 5 ml reaction vessel made of Inconel-625, a special material used to sustain the high temperature and pressure needed in this supercritical reaction. The set-up includes a pressure and temperature controller and monitoring system covering up to 200 MPa and 550 °C respectively. It was found that optimum conditions of 350 °C and molar ratio of methanol to oil of 42:1 and 4 min of supercritical treatment of methanol was sufficient to achieve more than 95% conversion of triglycerides into methyl esters. Compared to catalytic reaction which generally requires hours of reaction period, SCM treatment requires shorter reaction time which could reduce the processing cost of biodiesel substantially. Besides, simpler separation and purification of biodiesel from side product (glycerol) was reported due to the absence of catalyst in the reaction medium and the glycerol produced was also found to be of high purity.

Due to dissimilarity of composition in oils, it is vital to investigate the influence of oils in biodiesel yield. Demirbas (2002) carried out a comprehensive study to investigate the yield of biodiesel from various refined oils by using SCM technology. The oils that were investigated include cottonseed, hazelnut kernel, poppyseed, rapeseed, safflowerseed, and sunflowerseed.

In this study, all experimental works were conducted in a 100-ml cylindrical autoclave made from stainless steel 316 where the pressure and temperature were covered up to 100 MPa and 577 °C, respectively. The effects of methanol to oil molar ratio, reaction temperature, and reaction time on the yield of biodiesel were investigated. It was found that the biodiesel produced from SCM reaction has similar value of viscosity, ranging from 2.8 to 3.5 mm^2/s which is comparable with conventional diesel of 2.7 mm^2/s. Hence, it proves that biodiesel produced from SCM treatment is compatible with conventional diesel and suitable to be employed in existing diesel engine. In addition, it was reported that optimum yield of 95% was achieved by using hazelnut kernel oil with operating conditions of 250 °C, 41:1 molar ratio (methanol:oil), and 200 s of reaction time.

Apart from methanol, production of biodiesel from supercritical ethanol (SCE) reaction was also reported by several researchers. The justification to utilize ethanol instead of methanol is mainly because the latter is derived from fossil fuels such as petroleum and natural gas, implying that biodiesel from methanol-based reaction is not entirely renewable. On the other hand, ethanol can be derived from biomass via fermentation process and thus ensuring that biodiesel produced from SCE reaction is completely renewable (Gui et al., 2009). Balat (2008) reported SCE reactions with several refined oils such as rapeseed, sunflower, and cottonseed oils and the effects of reaction temperature, reaction time, and molar ratio of ethanol to oil were examined in a single-factor experimental design. It was reported that the viscosities of FAEE are higher ranging from 3.9 to 5.1 mm^2/s compared to FAME. Apart from that, it was found that increasing the reaction temperature and molar ratio of ethanol to oil enhances the yield of FAEE gradually until optimum yield is obtained. In addition, optimum yield of 85% was achieved with optimum conditions of 244 °C, 40:1 molar ratio of ethanol to oil, and 250 s of reaction time.

On the other hand, Gui et al. (2009) carried out optimization of SCE reaction by employing response surface methodology (RSM) design. RSM is useful in developing and optimizing processes by using data obtained from experiments in order to solve multivariable parameters simultaneously. Apart from that, RSM analysis allows a more comprehensive analysis on the interactions between experimental variables than single-factor experimental design which leads to better understanding and knowledge of the process. In this work, refined palm oil was utilized as the source of triglycerides and important process parameters such as reaction temperature, reaction time, and ethanol to oil molar ratio were investigated. In the results, it was reported that interactions between parameters were significant in determining the yield of ethyl esters. For instance, interaction term of reaction time and molar ratio implies that the influence of reaction time is substantially prominent at high molar ratio (40:1 mol/mol) compared to low molar ratio (16:1 mol/mol). In this context, at high molar ratio the yield of FAEE increased rapidly with the augmentation of reaction time while the increment in yield is slow at low molar ratio. This trend demonstrates that reaction rate of SCE enhances significantly with the presence of excessive concentration of ethanol which will shift transesterification reaction forward to produce higher yield of FAAE. In addition, it was reported that optimum conditions of 349 °C reaction temperature, 30 min reaction time, and molar ratio ethanol to oil of 33:1 could produce optimum biodiesel yield of more than 79%.

The effect of alcohol in supercritical reaction is an important parameter which needs to be investigated. Differences in chemical properties of the alcohol employed could influence the yield of biodiesel significantly. Hence, Tan et al. (2010a) carried out a comparative study between SCM and SCE reactions by employing RSM analysis to examine the influence of alcohol on optimum biodiesel yield. In this study, SCM reaction was conducted by employing

TABLE 2 Optimum Conditions and Yields of SCM and SCE Reactions[a] (Tan et al., 2010a)

	SCM	SCE
Reaction time (min)	16	29
Reaction temperature (°C)	372	349
Molar ratio (mol/mol)	40	33
Predicted yield (%)	84.1	83.1
Experimental yield (%)	81.5	79.2

[a]*Republished with permission from Elsevier.*

refined palm oil as the source of triglycerides and the optimum conditions and yields of biodiesel in SCM were compared with reported SCE results by Gui et al. (2009). In terms of optimum conditions, SCM and SCE reactions showed their own characteristics as shown in Table 2. For instance, SCM reaction required a shorter amount of reaction time (16 min) compared to 29 min needed in SCE reaction to achieve optimum yields of 81.5% and 79.2%, respectively. However, SCM reaction required higher reaction temperature and molar ratio compared to SCE treatment. In this context, the effect of reaction temperature on biodiesel yield is substantially higher in SCM compared to SCE reaction. The discrepancy in behavior of the reactions is mainly attributed by the difference in solubility parameter of the solvents. The solubility parameter of methanol and ethanol are 29.7 and 26.2 $(MPa)^{1/2}$ respectively while for oil it is approximately 18 $(MPa)^{1/2}$. As mentioned previously, the high temperature and pressure employed in supercritical reaction reduces the dielectric constant and subsequently the solubility parameters of alcohol to a value similar to oil which allow the formation of a homogeneous phase of alcohol-oil. Hence, ethanol which has lower value of solubility parameter would achieve homogeneous phase at relatively lower temperature compared to methanol. Consequently, SCE reaction suffered a substantial negative effect of decomposition and subsequently produced lower yield during high temperature while SCM did not show substantial reduction in biodiesel yield. Besides, due to elevated reaction temperature in SCM reaction, the yield is highly sensitive to long reaction time as well. Warabi et al. (2004) reported that methanol has higher reactivity than ethanol in supercritical reaction which allows the reaction to be completed in shorter reaction period. Hence, prolonging supercritical treatment in elevated temperature will induce decomposition of FAME in SCM reaction. On the other hand, extending the reaction period in SCE reaction will not affect the yield significantly due to inferior reactivity of ethanol and lower optimum temperature compared to SCM process. Hence, it can be concluded that reaction temperature is the most important parameter in SCM reaction while for SCE process, reaction time is the most significant variable affecting the yield of biodiesel.

3.2 Supercritical Methyl Acetate (SCMA) Reaction

Conventional route of producing biodiesel with alcohol produces glycerol as side product which leads to oversupply and devaluation in the world market. Furthermore, biodiesel is yet to be commercialized comprehensively worldwide due to high processing costs and expensive feedstock. Hence, the current conventional biodiesel production process is costly and unattractive economically. In addition, the poor performance of biodiesel at low temperature in terms of viscosity, pour point and oxidation stability are also some of the contributing factors toward its limited commercial application. Hence, biodiesel additives are commonly utilized to improve the properties of biodiesel to accommodate the demand in cold climate countries. Therefore, the quest to produce biodiesel additive that can improve the quality of biodiesel and revalorizing glycerol to value-added products is vital to ensure that the total processing costs of biodiesel is economical and competitive. Currently, there are numerous studies reporting the conversion of glycerol into biodiesel additives, which not only solve the problem of glycerol glut in the market but has the potential to improve the quality of biodiesel. One of the possible methods is to produce triacetylglycerol or commonly known as triacetin from glycerol and acetic acid via acetylation reaction. Triacetin is a valuable biodiesel additive which could improve the properties of biodiesel in terms of pour point, cloud point, and viscosity. However, the total costs to produce FAME and triacetin independently are enormous. Hence, it is promising to produce FAME and triacetin simultaneously in a single-step reaction which will minimize the cost of producing biodiesel additive and improve the quality of biodiesel substantially. This reaction is made possible by transesterification reaction between triglycerides and methyl acetate (MA) to produce FAME and triacetin as the side product instead of glycerol as shown in Figure 2. Furthermore, only simplified separation procedures are needed since the mixture of FAME and triacetin could be employed as biodiesel, instead of FAME only.

The first attempt to employ non-catalytic SCMA process to produce FAME and triacetin simultaneously from rapeseed oil was reported by Saka and Isayama (2009). In this study, rapeseed oil was utilized as the source of triglycerides and a single-factor experiment design was carried out to explore the effects of reaction temperature and reaction time on biodiesel yield. Furthermore, mixture of FAME and triacetin could be employed as biodiesel, rather than FAME only as in conventional alcohol-based transesterification. Since the molar ratio of FAME/triacetin in product mixture is 3:1 (mol/mol) which is equivalent to 4:1 in mass ratio (w/w), the theoretical weight of biodiesel (FAME + triacetin) is 125%, instead of 100% (FAME only). Results from the study found that optimum yield of 105% was achieved when reaction temperature of 350 °C, 45 min of reaction period, and molar ratio of MA/oil of 42:1 were employed. Moreover, the presence of triacetin in the biodiesel mixture improved the cold flow properties of the biodiesel substantially which is vital to accommodate the demand of biodiesel in cold climate countries. Hence, this study has testified the potential of SCMA reaction in producing high quality biodiesel with the presence of triacetin as side product.

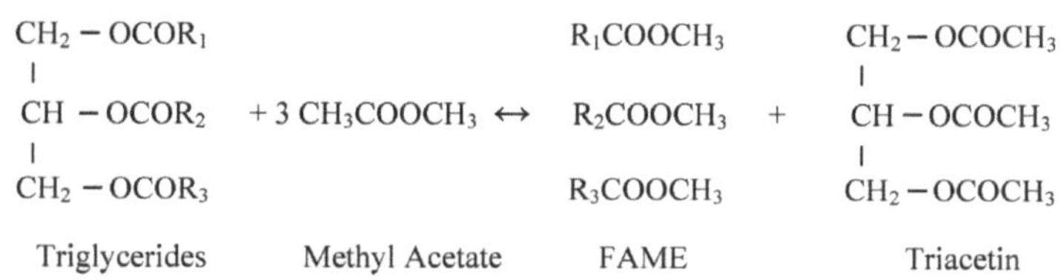

FIGURE 2 Transesterification reaction between triglycerides and methyl acetate to produce fatty acid methyl esters (FAME) and triacetin.

Apart from refined vegetable oil, Campanelli et al. (2010) also carried out SCMA reaction by employing non-edible and waste cooking oil as the source of triglycerides. In this study, refined soybean and sunflower oils were utilized as edible oils while *Jatropha curcas* oil was used as non-edible oil. In addition, these sources of triglycerides were subjected to conventional alkaline transesterification with methanol for comparison purposes with SCMA reaction. The influences of reaction temperature, reaction pressure and molar ratio of reactants on the yield of biodiesel were studied as well. Results from the study showed that oil composition does not affect the yield substantially as all the oils achieved high conversion (103-106%) after 50 min of reaction period, 20 MPa of reaction pressure, 345 °C reaction temperature, and 42:1 molar ratio of MA/oil. Furthermore, the high yield achieved in SCMA reaction with waste cooking oil which commonly contained high percentage of FFA demonstrates that the influence of FFA on biodiesel yield is minimal. Apart from that, thermal stability of triacetin was examined as well by subjecting triacetin under SCMA operating conditions. It was found that triacetin is vulnerable to thermal decomposition with substantial reduction in content after 50 min of exposure time. This observation could be the main factor that the maximum theoretical yield of 125% was not achieved in SCMA reaction.

Optimization study is important for scale-up and commercialization of SCMA process. Hence, Tan et al. (2010c) carried out optimization study of SCMA reaction involving refined palm oil to obtain optimum yield of biodiesel by employing RSM analysis. Besides, interaction effects between parameters such as reaction temperature, reaction time, and molar ratio of MA/oil were investigated as well. Results showed that mathematical model developed by RSM analysis was found to be adequate and statistically significant to predict the optimum yield. Furthermore, interaction effect between reaction temperature and molar ratio of MA/oil demonstrates that the yield of biodiesel increased gradually with increment of reaction temperature at any designated molar ratio from 30:1 to 50:1 mol/mol. On the other hand, the yield decreased steadily when the molar ratio was augmented from 30:1 to 50:1 mol/mol at any constant reaction temperature within the range of 360-400 °C. In this context, increasing the reaction temperature enhanced the reaction rate of transesterification which leads to high yield of biodiesel. However, the same trend is not applicable for molar ratio of MA/oil. Although increment in molar ratio will push the reversible transesterification to produce more FAME and triacetin, the limitation of reaction equilibrium and difficulties in separating and purifying excessive MA from FAME and triacetin have greater influences which lead to lower yield of biodiesel. Furthermore, optimum conditions were found to be 399 °C for reaction temperature, 30:1 mol/mol of MA/oil, and reaction time of 59 min to achieve optimum yield of 97.6%.

3.3 Supercritical Dimethyl Carbonate (SCDMC) Reaction

Apart from triacetin, glycerol can also be revalorized into other value-added compounds such as glycerol carbonate (GC) which is a versatile compound with enormous applications. It is useful in producing polymers such as polyesters, polyurethanes and polyamides which have higher market value than glycerol. Apart from that, GC is also a valuable compound for the production of glycidol which is widely utilized in pharmaceutical, cosmetics, and plastics industries. In addition, GC is a potential renewable substitute for petroleum-based chemicals such as ethylene carbonate or propylene carbonate which are novel components in synthesizing CO_2 separation membrane. Simultaneous production of FAME and GC via single-step transesterification has been reported by Fabbri et al. (2007). However, the reaction suffered from long reaction time and the employment of homogeneous base catalyst in the study required tedious separation and purification procedures. Consequently, it is interesting to produce FAME and GC simultaneously via single-step non-catalytic transesterification reaction. The absence of catalyst will simplify the process significantly which is vital to make it viable for commercialization purposes.

Ilham and Saka (2009) conducted a study to produce biodiesel and GC by employing SCDMC transesterification treatment without the presence of any catalysts. In this work, optimization of biodiesel yield from rapeseed oil was carried out by single-factor experimental design and fixed molar ratio of DMC/oil (42:1). Apart from that, potential of DMC to esterify FFA was also conducted by utilizing oleic acid in fixed molar ratio of 14:1 (DMC/oleic acid). Results from the study showed that optimum FAME yield of 94% could be achieved with conditions of 350 °C reaction temperature, 20 MPa reaction pressure, and 12 min of reaction time. In addition, the valuable by-product (GC) could be separated easily from FAME. Furthermore, FFA could be esterified as well to produce FAME with glyoxal and water molecule formed as side products.

Apart from that, Ilham and Saka (2010) also carried out two-step process involving subcritical water treatment and subsequently SCDMC reaction to produce FAME. In the first step, oil was mixed with water and subjected to subcritical water conditions of 270 °C reaction temperature, 27 MPA of reaction pressure, and 25 min of reaction period to hydrolyze the oil into FFA and glycerol. Subsequently, the FFA is treated with SCDMC reaction at conditions of 300 °C reaction temperature, 9 MPa reaction pressure and 15 min of reaction period for esterification reaction to produce FAME and glyoxal. Results showed that yield of 97% could be obtained with the employment of two-step procedures instead of conventional single-step reaction. Furthermore, this new route only requires milder operating conditions with lower reaction temperature and pressure compared to previously reported single-step SCDMC reaction. In addition, the mild operating conditions also allow the employment of feedstock with high amount of FFA such as crude *J. curcas* oil. Furthermore, the FAME produced in the study was also found to comply with international standards of biodiesel fuel.

On the other hand, Tan et al. (2010d) reported optimization study of SCDMC process by employing RSM analysis to obtain optimum yield of biodiesel. In this optimization study, the effects of important parameters including reaction temperature, molar ratio of DMC to oil and reaction time on the yield were examined. Interaction terms between the parameters revealed that reaction temperature and molar ratio of DMC/oil have the most significant influence on the yield. For instance, at low molar ratio (30:1 mol/mol), the yield increased substantially when the reaction temperature was increased within the range of 340-380 °C. However, the yield only augmented steadily at high molar ratio (50:1 mol/mol) within similar range of reaction temperature. These observations showed that reactivity between DMC and oil is usually low at low temperature and increment in reaction temperature induced the yield to enhance proportionally. However, the effect of reaction temperature is more prominent at low molar ratio compared to high molar ratio conditions. In this context, the high temperature promotes greater reactivity with the formation of homogeneous phase during supercritical fluid conditions, leading to insignificant effect of escalating DMC concentration in the reaction medium. In addition, optimization study found out that optimum yield of 91% of FAME could be achieved with optimum conditions of 380 °C reaction temperature, 39:1 mol/mol of DMC/oil and 30 min of reaction time.

3.4 Stability of FAME in Supercritical Fluid Reaction

Stability of FAME produced in supercritical fluid reaction is vital due to employment of high reaction temperature and pressure in the reaction. The severe operating conditions could affect the molecular structure and induced decomposition of FAME which leads to lower yield of biodiesel. Hence, Imahara et al. (2008) carried out investigation on thermal stability of FAME produced in SCM reaction to examine the influence of high temperature and pressure. Furthermore, the effect of thermal degradation on cold flow properties of biodiesel was studied as well. In the results, it was reported that saturated FAME such as methyl palmitate and methyl stearate are stable at conditions of 300 °C/19 MPa or below and when the conditions increased to 350 °C/43 MPa, there was a slight decomposition after exposure period of 60 min in supercritical conditions. On the other hand, there were substantial reductions in yield for unsaturated FAME at elevated reaction temperature. For instance, unsaturated FAME such as methyl oleate, methyl linoleate, and methyl linolenate are only stable at conditions of 270 °C/17 MPa while increment in operating conditions to 350 °C/43 MPa showed significant decreased in biodiesel yield particularly for methyl linolenate. The difference in behavior between saturated and unsaturated FAME could be explained by isomerization phenomenon of *cis*-type to *trans*-type double bond in unsaturated FAME. The absence of double bond in saturated FAME makes them more stable even at severe operating conditions of 350 °C/43 MPa while the poly-unsaturated FAME like methyl linolenate is vulnerable to high operating conditions and induces decomposition from *cis*-type to *trans*-type. In addition, the isomerization phenomenon also causes marginal adverse effect in cold flow properties of pour point and cold point of unsaturated FAME when the operating conditions were increased from 270 °C/17 MPa to 350 °C/43 MPa. In contrast, the cold flow properties of saturated FAME were unaffected even at high operating conditions of 350 °C/43 MPa. Hence, it can be concluded that saturated FAME are relatively stable at high operating conditions while unsaturated FAME, particularly poly-unsaturated compounds such as methyl linoleate and methyl linolenate are vulnerable at elevated conditions.

Apart from thermal stability, oxidation stability is also one of the most important parameters which need to be investigated in biodiesel production. Xin et al. (2008) reported oxidation stability of biodiesel produced from various refined oils including safflower, rapeseed, and palm. It was found that oxidation stability of biodiesel depended significantly on the degree of saturation. For instance, safflower oil which contained high percentage of poly-unsaturated fatty acids has the lowest oxidation stability among the oils while palm oil, with high percentage of saturated fatty acids showed substantially high oxidation stability. On the other hand, exposure of biodiesel to supercritical treatment of 270 °C/17 MPa for 30 min revealed that the content of tocopherols decreased slightly. Tocopherols are natural antioxidant in vegetable oils which could contribute to oxidation stability of biodiesel. In addition, it was reported that supercritical treatment could reduce the peroxide value of biodiesel efficiently compared to conventional alkali-based reaction. Waste oils commonly contained high peroxide value due to the presence of hydroperoxide, leading to poor oxidation stability of biodiesel derived from these sources. Therefore, biodiesel produced from supercritical treatment has lower value of peroxide and greater oxidation stability which is important for long term storage of biodiesel.

3.5 Effects of Water and FFA Content

Conventional biodiesel production by catalytic reaction suffers from low tolerance toward impurities such as water and FFA compounds which are common in waste oils/fats. Consequently, expensive refined oils must be employed as feedstock in catalytic reaction to avoid unwanted side reactions which could reduce the yield of FAME substantially. However, the cost of feedstock consists of more than 70% of the total production costs, leading to uneconomical production of biodiesel. Hence, in order to reduce the cost of biodiesel, waste or low-quality oils/fats which are inexpensive and abundantly available could be utilized as feedstock. Therefore, it is vital to investigate the performance of supercritical reaction with oils containing high percentage of water and FFA. Kusdiana and Saka (2004) reported a study to examine the effects of water and FFA content in SCM reaction with rapeseed oil. Furthermore, the performance of SCM reaction was compared with homogeneous alkaline- and acid-catalyzed reactions. It was found that the presence of water did not adversely affect the yield in SCM reaction regardless of the concentration of water in the reaction medium. In contrast, the yield increased marginally with the augmentation of water concentration. This observation can be best explained by a two-step process which is hydrolysis of triglycerides and esterification of FFA reactions, instead of the conventional transesterification reaction between triglycerides and methanol. In SCM reaction, the presence of water in the reaction mixture induces the hydrolysis of triglycerides which produces FFA and glycerol as shown in Figure 3. Subsequently, the FFA will be esterified with methanol to produce FAME and water as side product as illustrated in Figure 4. Therefore, the yield is not adversely affected but instead increases slightly due to simultaneous reactions of transesterification, hydrolysis, and esterification in SCM process. On the other hand, the yields in homogeneous alkaline- and acid-catalyzed reactions showed significant reduction with increment of water concentration due to side reactions between water and catalysts. For the effect of FFA content, similar trend was observed in SCM reaction with no significant changes in yield with the increment

CH_2-OCOR_1			R_1COOH		CH_2-OH
|					|
$CH-OCOR_2$	$+ 3\ H_2O$	$\leftrightarrow$	R_2COOH	+	$CH-OH$
|					|
CH_2-OCOR_3			R_3COOH		CH_2-OH
Triglycerides	Water		Fatty Acids		Glycerol

FIGURE 3 Hydrolysis of triglycerides to produce fatty acids and glycerol.

R_1COOH				R_1COOCH_3		
R_2COOH	+	$3CH_3OH$	$\leftrightarrow$	R_2COOCH_3	+	$3H_2O$
R_3COOH				R_3COOCH_3		
Fatty acids		Methanol		FAME		Water

FIGURE 4 Esterification of fatty acids to produce fatty acid methyl esters and water molecule.

of FFA concentration. In SCM reaction, the addition of FFA will not produce any undesirable effects as it can be esterified with methanol to produce FAME as shown earlier in Figure 4. Furthermore, esterification reaction also produces water molecule as side product which helps to hydrolyze the triglycerides and subsequently increases the reaction yield as mentioned previously. For alkaline-catalyzed reaction, the presence of FFA will deactivate alkaline catalyst and leads to formation of soap and emulsion, resulting in lower yield and complicated downstream processes. On the other hand, esterification of FFA could proceed in acid-catalyzed reaction but the water molecule produced as side product in esterification reaction will reduce the efficiency of acid catalysts and thus reduces the yield significantly with the enhancement of FFA concentration.

Apart from that, Tan et al. (2010b) also investigated the effects of water and FFA in SCM reaction but with different source of triglycerides. In this study, refined palm oil was employed to examine the influence of water and FFA on biodiesel yield and subsequently compared with heterogeneous catalyst Montmorillonite KSF. Furthermore, the performance of palm oil in this study could be compared with previously reported rapeseed oil which contains different composition of fatty acids. The results showed that no adverse effect was observed when the water content was increased within the range of 0-25 wt%. Similar with the trend reported by Kusdiana and Saka (2004), the yield increases steadily with increasing amount of water concentration. As discussed previously, the yield increased due to hydrolysis of triglycerides to FFA which was subsequently esterified to FAME. On the other hand, the yield of Montmorillonite KSF reaction suffered a significant reduction with the augmentation of water content. This observation was due to inhibition of acidic Montmorillonite KSF activities by water molecule which has strong affinity for acidic compounds such as sulfuric acid in Montmorillonite KSF catalysts. Consequently, leaching phenomenon occurred and the efficiency of Montmorillonite KSF was severely affected and resulted in low yield of FAME. As far as FFA content is concerned, increment in FFA content in reaction medium did not adversely affect both SCM and Montmorillonite KSF reactions. Instead, both reactions showed a gradual increase in yield with the enhancement of FFA concentration. For SCM reaction, the addition of FFA could be esterified with methanol to produce higher yield of FAME as discussed previously. On the other hand, for reaction catalyzed by Montmorillonite KSF, this acidic heterogeneous catalyst is not sensitive to the presence of FFA as well and it could esterify the FFA to FAME. Unlike homogeneous catalyzed reaction, there was no formation of soap or emulsion as the catalysts and reactants were in different phase and no base compounds were present. Therefore, it can be concluded that SCM reaction has high tolerance toward high concentration of FFA and water content which allow the utilization of inexpensive feedstock such as waste oils/fats in biodiesel production.

3.6 Application of Co-solvent in Supercritical Fluid Reaction

Although supercritical fluid reaction has been shown to have advantages in terms of reaction time and yield, the severe operating conditions required is not feasible for industrial application. Hence, it is vital to reduce them to milder operating conditions without compromising on the advantages of supercritical-based reaction. One of the possible methods is to introduce co-solvent into the reaction medium as reported by Han et al. (2005). In this study, CO_2 was employed as co-solvent in SCM reaction with refined soybean oil. It was found that the addition of co-solvent decreased the extreme conditions usually required in supercritical reaction. For instance, it was shown that the optimum reaction temperature was reduced substantially to 280 °C to produce 98% yield with the addition of CO_2 to methanol molar ratio of 1:10. Furthermore, the optimum yield was achieved in 10 min reaction period and reaction pressure of 14 MPa. On the other hand, without the presence of co-solvent, the reaction did not achieve optimum yield even above the temperature of 320 °C. In this study, the presence of CO_2 as co-solvent increased the mutual solubility between methanol and soybean oil under supercritical conditions. Furthermore, CO_2 is a good solvent for vegetable oil, leading to formation of homogeneous phase between oil and methanol at lower reaction temperature and pressure. Therefore, only mild operating conditions were required such as lower molar ratio of methanol to oil and lower supercritical conditions in SCM reaction with co-solvent.

Apart from that, Yin et al. (2008) also conducted similar study to investigate the potential of hexane as co-solvent in SCM reaction. In this work, SCM reaction was conducted by employing soybean oil and without any co-solvent, the FAME yield was only 67% for reaction conducted at 300 °C with constant shaking for 30 min of the reaction period at 200 rpm. However, with the addition of 2.5 wt% of hexane as co-solvent into the reaction medium, the yield was enhanced to 85%. Similarly, Tan et al. (2010b) employed heptane as feasible co-solvent in SCM reaction with refined palm oil. Without co-solvent, the optimum conditions were found to be 360 °C of reaction temperature and 22 MPa of reaction pressure with FAME yield of 80%. However, when a small amount of 0.2 molar ratio of heptane to methanol was added, yield of 66% could be obtained even at mild supercritical conditions of 280 °C/15 MPa. Hydrocarbon such as hexane and heptane are excellent solvents for non-polar compounds such as triglycerides. Hence, the introduction of hydrocarbon as co-solvent allows the formation of homogeneous phase between triglycerides and methanol even under mild supercritical conditions. Furthermore, the critical point of the mixture was reduced with the presence of co-solvent and supercritical conditions can be achieved at lower temperature and pressure.

4 CONCLUSION

Supercritical fluid reaction has been shown to have several advantages compared to conventional catalytic reaction in biodiesel production. The absence of catalysts in supercritical-based reaction simplifies the reaction route and downstream processes significantly. Furthermore, the high yield achieved in short reaction period makes it an attractive technology for commercialization purposes. In addition, supercritical reaction has been proven to have high tolerance towards impurities in oils/fats such as FFA and water content and thus allowing the utilization of inexpensive feedstock such as waste oils/fats. Although supercritical reaction required severe conditions, the introduction of co-solvent has been shown to have significant potential to reduce them to mild operating conditions. Therefore, it can be concluded that supercritical fluid reaction has a huge potential to be the major technology for biodiesel processing in the near future.

Acknowledgment

The authors would like to acknowledge Universiti Sains Malaysia and Elsevier for the contribution toward this chapter.

References

Balat, M., 2008. Biodiesel fuel production from vegetable oils via supercritical ethanol transesterification. Energy Source A Recov. Util. Environ. Eff. 30, 429–440.

Campanelli, P., Banchero, M., Manna, L., 2010. Synthesis of biodiesel from edible, non-edible and waste cooking oils via supercritical methyl acetate transesterification. Fuel 89, 3675–3682.

Demirbas, A., 2002. Biodiesel from vegetable oils via transesterification in supercritical methanol. Energy Convers. Manage. 43, 2349–2356.

Fabbri, D., Bevoni, V., Notari, M., Rivetti, F., 2007. Properties of a potential biofuel obtained from soybean oil by transmethylation with dimethyl carbonate. Fuel 86, 690–697.

Gui, M.M., Lee, K.T., Bhatia, S., 2009. Supercritical ethanol technology for the production of biodiesel: process optimization studies. J. Supercrit. Fluids 49, 286–292.

Han, H., Cao, W., Zhang, J., 2005. Preparation of biodiesel from soybean oil using supercritical methanol and CO_2 as co-solvent. Proc. Biochem. 40, 3148–3151.

Ilham, Z., Saka, S., 2009. Dimethyl carbonate as potential reactant in non-catalytic biodiesel production by supercritical method. Bioresour. Technol. 100, 1793–1796.

Ilham, Z., Saka, S., 2010. Two-step supercritical dimethyl carbonate method for biodiesel production from *Jatropha curcas* oil. Bioresour. Technol. 101, 2735–2740.

Imahara, H., Minami, E., Hari, S., Saka, S., 2008. Thermal stability of biodiesel in supercritical methanol. Fuel 87, 1–6.

Kusdiana, D., Saka, S., 2004. Effects of water on biodiesel fuel production by supercritical methanol treatment. Bioresour. Technol. 91, 289–295.

Saka, S., Isayama, Y., 2009. A new process for catalyst-free production of biodiesel using supercritical methyl acetate. Fuel 88, 1307–1313.

Saka, S., Kusdiana, D., 2001. Biodiesel fuel from rapeseed oil as prepared in supercritical methanol. Fuel 80, 225–231.

Tan, K.T., Gui, M.M., Lee, K.T., Mohamed, A.R., 2010a. An optimized study of methanol and ethanol in supercritical alcohol technology for biodiesel production. J. Supercrit. Fluids 53, 82–87.

Tan, K.T., Lee, K.T., Mohamed, A.R., 2010b. Effects of free fatty acids, water content and co-solvent on biodiesel production by supercritical methanol reaction. J. Supercrit. Fluids 53, 88–91.

Tan, K.T., Lee, K.T., Mohamed, A.R., 2010c. A glycerol-free process to produce biodiesel by supercritical methyl acetate technology: an optimization study via Response Surface Methodology. Bioresour. Technol. 101, 965–969.

Tan, K.T., Lee, K.T., Mohamed, A.R., 2010d. Optimization of supercritical dimethyl carbonate (SCDMC) technology for the production of biodiesel and value-added glycerol carbonate. Fuel 89, 3833–3839.

Warabi, Y., Kusdiana, D., Saka, S., 2004. Reactivity of triglycerides and fatty acids of rapeseed oil in supercritical alcohols. Bioresour. Technol. 91, 283–287.

Xin, J., Imahara, H., Saka, S., 2008. Oxidation stability of biodiesel fuel as prepared by supercritical methanol. Fuel 87, 1807–1813.

Yin, J.Z., Xiao, M., Song, J.B., 2008. Biodiesel from soybean oil in supercritical methanol with co-solvent. Energy Convers. Manage. 49, 908–912.

CHAPTER

15

Production of Biodiesel Using Palm Oil

*Man Kee Lam, Keat Teong Lee**

School of Chemical Engineering, Universiti Sains Malaysia, Engineering Campus, Seri Ampangan, 14300 Nibong Tebal, Pulau Pinang, Malaysia.

***Corresponding author: E-mail: chktlee@eng.usm.my**

1 INTRODUCTION

The world is gradually heading toward severe energy crisis due to limited availability of fossil fuels, such as petroleum oil, natural gas, and coal. These fossil fuels are categorized as nonrenewable energy resources that cannot be replaced in a relatively short time after being utilized. Nevertheless, it is an undeniable fact that man is still heavily dependent on fossil fuels for electricity generation, transportation, and development. In addition to that, over-exploiting the usage of fossil fuels by human beings has raised severe environmental issues and directly caused negative impacts on the Earth. One of the most critical examples is climate change due to excessive emission of green house gases caused by the burning of fossil fuels. Global warming and extreme weather changes such as sudden drought, flash flood, windstorms, and heat waves are the evidences of climate change. Therefore, the search of an alternative and renewable energy source has emerged as one of the key challenges in this century in order to protect the environment and creating a sustainable world for future generation.

There are indeed a lot of renewable energy sources that have been explored, including solar, hydropower, wind, wave, geothermal, and nuclear energy. However, most of these options are not economically feasible due to the requirement of high capital and operating cost that has limited its usage in many countries over the world that would likely to diversify their energy sources. Furthermore, availability of those renewable energies is highly dependent on regional or local condition that can be very unpredictable and inconsistence. For example, solar collector would require clear sky and plenty of sunshine to generate a sufficient amount of energy and, therefore, it is certainly not an appropriate choice for temperate countries. However, a hybrid energy conversion system can be recommended to overcome the problem and to achieve satisfactory energy conversion efficiency. Nevertheless, developing a hybrid system is not an easy task as the technology know-how to integrate the operations of the whole process is still at an infancy stage. As a consequence, it is not feasible to introduce the renewable energy integrated systems in the third world and underdeveloped countries.

Recently, biodiesel has emerged as a spark of hope in the field of renewable energy. This is because biodiesel has close similarity with conventional fossil diesel in terms of chemical structure and energy content. Apart from that, modification of a diesel engine is not required as biodiesel is compatible with existing engine and has been commercially blended with diesel as transportation fuel in many European countries (Lam et al., 2009b). Besides, significant reduction in greenhouse gases emission has been proven by burning biodiesel, and this result directly reflects the unique benefit of using biodiesel (Basha et al., 2009). Furthermore, biodiesel is a nontoxic alternative fuel and easily biodegradable in freshwater and soil, making it unquestionably good for the environment (Pasqualino et al., 2006). In general, biodiesel can be produced through transesterification reaction, in which triglyceride from vegetable oil is reacted with short-chain alcohol (e.g., methanol) in the presence of catalyst as shown in Equation (1). Soybean, rapeseed, sunflower, and palm oils are among the common vegetable oils that are used in biodiesel production. However, since these oils are edible resources, many nongovernment organizations in the world have raised the "food versus fuel" feud and, therefore, biodiesel production has shifted to other alternative feedstock such as waste frying oil (WFO) and nonedible oil (e.g., jatropha curcas, karanja, pongamia pinnata, and microalgae). The use of WFO and nonedible oil has its fair share of problem, mainly due to the exceptional high free fatty acid (FFA) content that complicates the overall biodiesel processing steps. Soap is easily formed (saponification reaction) if a base catalyst is used and consequently increases the difficulty in final product purification.

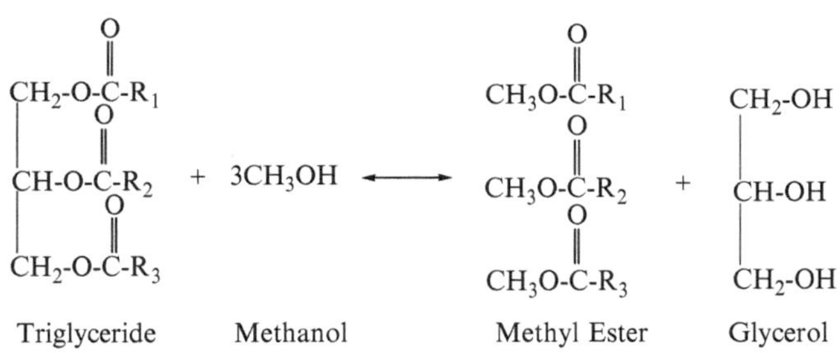

(1)

In this chapter, focus will be given toward biodiesel production from palm oil. Lately, oil palm plantation has been criticized to cause several serious environmental issues such as deforestation and habitat destruction of endangered species (specifically *orangutan*). Fortunately, with various researches and scientific findings, these accusations were found to be baseless (Lam et al., 2009b). Up to date, oil palm still remains as the most efficient edible oil-producing crop as shown in Table 1 (Malaysian Palm Oil Council (MPOC)). Oil palm plantation area only accounted for less than 5% of the world's agriculture land in year 2007, but yet it is able to supply 25% of the global oils and fats (Lam et al., 2009b). Hence, if the intention is to optimize land usage to meet the food and fuel demand simultaneously, oil palm will be the outstanding option as large quantity of oil can be produced with minimum land requirement. In addition to that, new breeds of oil palm cloned by Applied Agricultural Resources Sdn. Bhd. are able to produce 10.6 tonne/ha/year of crude palm oil (CPO), almost double of

TABLE 1 Oil Yield and World Plantation Area for Major Edible Oils

Oil Crop	Average Oil Yield (Tonne/Ha/Year)	Planted Area (Million Hectare)	% of Total Planted Area
Soybean	0.4	94.15	42.52
Sunflower	0.46	23.91	10.8
Rapeseed	0.68	27.22	12.29
Oil palm	3.62	10.55	4.76

the current yield (Lam et al., 2009b). Apart from that, palm oil production has the highest energy efficiency factor (energy output to energy input) of 9.6 compared to rapeseed of 3.0 and soybean of 2.5 (Lam et al., 2009b). This is because less fertilizer and diesel (machinery and agrochemical usage) are required to produce 1 tonne of palm oil. Apart from the positive contributions toward the environment, sustainable oil palm plantation program can also leverage poverty by helping the poor farmers and rural dwellers to improve their living standards. The successful story of Malaysian palm oil industries in transforming the rural communities to have access to their basic needs for a healthy life reflects the significant outputs of the strategy. In fact, even the Food and Agriculture Organization (FAO) does agree that new demand for biofuels production from sustainable agricultural feedstock can indeed generate a new income opportunity for farmers, leading to increased food production and poverty eradication.

2 PALM BIODIESEL CONVERSION TECHNOLOGY

2.1 Overview on the Existing Process and Technology

Currently, commercial-scale palm biodiesel production is usually carried out in a batch-type continuous stirred tank reactor. Initially, CPO is pretreated to increase its oxidative stability and to minimize the FFA content in the oil. A series of pretreatment steps are adopted such as degumming, neutralization by caustic soda, pigment removal using bleaching earth and, finally, high-temperature vacuum deodorization (Lim and Teong, 2010). The refined, bleached, and deodorized (RBD) palm oil in the presence of excess methanol and base catalyst is then heated to certain reaction temperature to produce biodiesel. Normally, multistage batch reactors are used in series to drive the reaction toward completion (Lim and Teong, 2010). After each stage of reactions, glycerol (byproduct) is withdrawn to push the reaction forward to attain higher biodiesel conversion within a minimum reaction time (Lipochem (M) Sdn Bhd and MPOB). After the reaction is completed, excess methanol is recovered through flashing in a flash vessel and further purified in a structured packing distillation column (Lipochem (M) Sdn Bhd and MPOB). The purified methanol can be recycled and use as reactant in the subsequent reactions. Apart from that, glycerol will also go through a few purification steps and is stored in a storage tank as crude glycerol. Meanwhile, the crude biodiesel is subjected to water-washing stages in cyclones to remove the remaining catalyst as well as to purify the biodiesel. Finally, the water is discharged at the bottom of the cyclone as wastewater, and the washed biodiesel is dried under vacuum condition to reduce its water content within the specified limits of biodiesel standards. Figure 1 illustrated the overall process involved in palm biodiesel production.

Biodiesel derived from palm oil has been reported to have similar fuel properties to petroleum diesel as shown in Table 2 (Lim and Teong, 2010). In addition, the palm biodiesel meets the international biodiesel specification as underlined by EN 14214 and ASTM D 6751. It was reported that pure palm biodiesel (without blending with petroleum diesel) can be directly used as fuel in a diesel engine without prior modification (Lipochem (M) Sdn Bhd and MPOB). Alternatively, it can also be blended with petroleum diesel at any proportion to initiate the implementation of biodiesel at national level and subsequently promote the advantages of using biodiesel toward environmental sustainability. Exhaustive test on the performance of palm biodiesel as an alternative fuel on diesel engine has also been conducted, including on 36 Mercedes Benz engines mounted onto passenger buses (Choo et al., 2005).

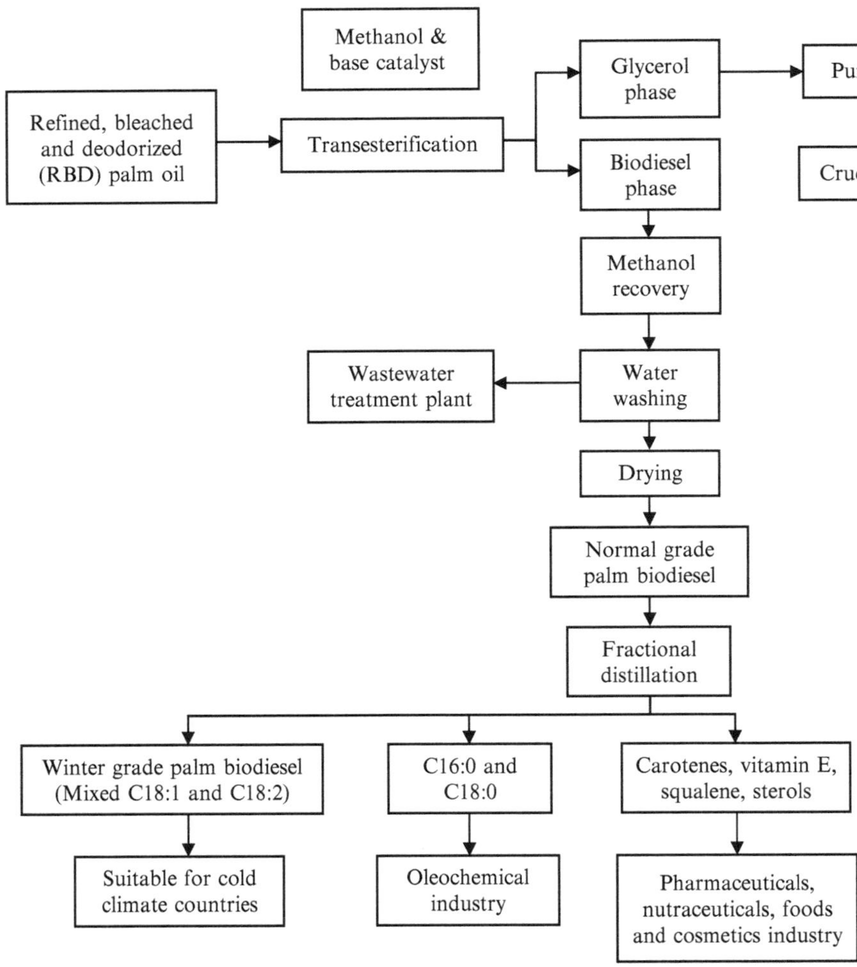

FIGURE 1 Overview on the existing palm biodiesel production process

TABLE 2 Properties of Palm Biodiesel (Normal and Winter Grade)

			Palm Diesel			
Property	**Unit**	**Petroleum Diesel**	**Normal Grade**	**Winter Grade**	**EN 14214**	**ASTM D6751**
Ester content	% mass	–	98.5	98.0-99.5	96.5 (min)	–
Free glycerol	% mass	–	<0.02	<0.02	0.02 (max)	0.02 (max)
Total glycerol	% mass	–	<0.25	<0.025	0.25 (max)	0.24 (max)
Density at 15 °C	kg/L	0.853	0.878	0.87-0.89	0.86-0.89	–
Viscosity at 40 °C	cSt	4	4.4	4.0-5.0	3.5-5.0	1.9-6.0
Flash point	°C	98	182	150-200	120 (min)	130 (min)
Cloud point	°C	–	15.2	−18 to 0	–	–
Pour point	°C	15	15	−21 to 0	–	–
Cold filter plugging point	°C	–	15	−18 to 3	–	–
Sulfur content	% mass	0.1	<0.001	<0.001	0.001 (max)	0.0015
Carbon residue	% mass	0.14	0.02	0.02-0.03	0.3 (max)	0.05 (max)
Cetane index		53	58.3	53.0-59.0	51 (min)	47 (min)
Acid value	mg KOH/g	–	0.08	<0.3	0.5 (max)	0.8 (max)
Copper strip corrosion	3 h at 50 °C	–	1a	1a	1	3 (max)
Gross heat of combustion	kJ/kg	45,800	40,135	39,160	–	–

The engines were able to successfully complete for over 30,000 km mileage, the expected performance of the engine. Apart from that, no technical problem was reported throughout the trial period, provided the engines were maintained according to their service manual (Choo et al., 2005).

On the other hand, the fluidity of fuel in an engine is a crucial factor to ensure its efficient performance. When starting up an engine especially during cold weather, it is vital that the fuel can be pumped into the engine and mechanical parts are able to move freely. Otherwise, this may result in engine malfunctioning associated with long-term use (Choo et al., 2002b). In this regard, unfortunately, palm biodiesel has a relatively high pour point of +15 °C which limits its usage in cold climate countries. Pour point is defined as the lowest temperature for an oil to pour or flow freely under a specified condition (Lee et al., 1995). If the surrounding temperature approaches or becomes lower than the pour point temperature of palm biodiesel, the fuel will be solidified and cause cold flow-related problems such as blockage to the flow pipes and filters (Chen et al., 2010). One of the possible ways to overcome this limitation is adding chemical additives such as pour point depressants, flow improvers, paraffin inhibitors, or wax modifiers to the biodiesel (Soriano et al., 2006). However, it was reported that these commercially available chemical additives which were developed for petroleum diesel may not be suitable for biodiesel application and the results are not satisfactory (Soriano et al., 2006).

Table 3 shows the composition of fatty acid in palm oil (Ma and Hanna, 1999). From the table, saturated fatty acid such as lauric (C12:0), myristic (C14:0), palmitic (C16:0), and stearic (C18:0) contributed nearly 50% of the overall palm oil fatty acid composition. Because of this, the pour point of palm biodiesel is relatively higher than biodiesel derived from other feedstock such as soybean (0 °C) and canola oil (−9 °C). However, it should be noted that palm oil constitutes various high-valued phytonutrients, namely, carotenes, vitamin E, squalene, and sterols (Harrison Lau et al., 2009). These phytonutrients bring great benefits, especially to pharmaceuticals, nutraceuticals, foods and cosmetics industry, as well as an ample health source for human consumption. After years of research and development, Malaysian Palm Oil Council (MPOC) had designed an integrated technology to recover these phytonutrients and to produce winter grade biodiesel simultaneously that was filed under Malaysian Patent PI 20021157 (Choo et al., 2002a). The integrated process starts with esterification and transesterification of CPO to produce biodiesel. Under mild reaction conditions, the phytonutrients are not completely destroyed and thus can be recovered before the palm biodiesel is burnt as fuel. After purification with water washing and drying, the palm biodiesel is further processed using short path distillation to produce three product streams—distilled palm biodiesel, carotene (vitamin A), and vitamin E. The distilled palm biodiesel will then be fed into a fractional distillation column to separate saturated methyl ester (C16 and C18) and to produce winter grade biodiesel (mixed C18:1 and C18:2); carotene will be subsequently concentrated to obtain a carotene concentrate; and finally vitamin E will be further processed, polished and solvent fractionation to obtain a concentrate vitamin E (Toh and Koh, 2008). It was reported that the expected recovery rate of carotene and vitamin E concentrate is 50 and 100 kg/day (Toh and Koh, 2008). Consequently, with the attractive market price for carotene at RM 760/kg ($ 217/kg) and vitamin E at nearly RM 1900/kg ($ 543/kg), this definitely offers a great opportunity for international business investors to gain valuable monetary return (Harrison Lau et al., 2009). Apart from that, the separated C16 and C18 methyl ester can

TABLE 3 Fatty Acid Composition of Palm Oil

Fatty Acid	Composition (%)
Lauric (12:0)	0.1
Myristic (C14:0)	1.0
Palmitic (C16:0)	42.8
Stearic (C18:0)	4.5
Oleic (C18:1)	40.5
Linoleic (C18:2)	10.1
Others	1
Total	100

be used as oleochemical feedstock for the production of white soap as well as used as active ingredients in detergent formulations (Choo et al., 2002c). The production of all these diversified byproducts simultaneously with palm biodiesel certainly enhances the economic viability of using palm oil as feedstock for biodiesel production. Specifically, the relatively higher prices of biodiesel compared to petroleum diesel may be offset by the revenue obtained from selling those phytonutrients byproducts.

2.2 Catalysis Process for Palm Biodiesel Conversion

2.2.1 *Homogeneous Base Catalyst*

Base catalysts such as sodium hydroxide (NaOH) and potassium hydroxide (KOH) are the most commonly used catalysts in industrial biodiesel production plant. The reasons for this are: (1) relatively low cost compared to heterogeneous and enzymatic catalysts, (2) easily available in the market and (3) able to accelerate transesterification effectively under a mild reaction condition. Furthermore, base-catalyzed transesterification was 4000 times faster than acidic catalysts (Fukuda et al., 2001; Kulkarni and Dalai, 2006). NaOH and KOH are available in pellet form but highly soluble in alcohol. Therefore, these catalysts are normally premixed with alcohols before transesterification reaction takes place. However, homogeneous base catalysts suffer a serious drawback in the biodiesel industry due to their high sensitivity toward FFA content in oil. FFA content in oils needs to be kept as low as possible (0.5-1%) to hinder saponification reaction from occurring because it will react with the base catalyst to produce soap as the byproduct. Excessive soap formation inhibits the biodiesel-glycerol phase separation and thus reduces biodiesel yield drastically.

Generally, RBD palm oil is used to produce biodiesel due to the low FFA content (0.1-0.5%) and thus minimize the impact of saponification reaction. Darnoko and Cheryan (2000) studied the transesterification of RBD palm oil with methanol catalyzed by KOH. From the study, the highest biodiesel concentration attained was 90% with the following reaction conditions: reaction temperature of 60 °C, methanol to oil molar ratio of 6, catalyst concentration of 1%, and reaction time of 40-60 min. The result was in accordance to other feedstock catalyzed by KOH or NaOH, such as soybean (Dias et al., 2008), sunflower (Rashid et al., 2008), and rapeseed oil (Jeong et al., 2004). However, since a series of refining processes are required to convert CPO to RBD, the additional processing cost has increased the overall RBD palm biodiesel production cost and making the whole process not economic viable. Feedstock cost has been reported to contribute the most in the whole biodiesel production chain, nearly 80% of the overall biodiesel production cost (Lam et al., 2009b). Although CPO is cheaper than RBD palm oil, it has high FFA contents, ranging from 3% to 6.5% (Che Man et al., 1999). Normally, pre-esterification step is required to reduce the FFA content in the CPO before base-catalyzed transesterification reaction takes place.

2.2.2 *Homogeneous Acid Catalyst*

Apparently, homogeneous acid catalysts are preferred for feedstock that contains high FFA in biodiesel production. Sulfuric acid (H_2SO_4) and hydrochloric acid (HCl) are the most widely used due to their strong acidic properties and low cost. Nevertheless, it was reported that H_2SO_4 can give better performance than HCl in transesterification of waste frying palm oil (Al-Widyan and Al-Shyoukh, 2002). Apart from that, the advantage of using acidic catalysts is insensitivity to FFA content in the oil and thus elimination of side reaction such as saponification (Kulkarni and Dalai, 2006). Furthermore, acidic catalysts are able to perform esterification and transesterification simultaneously. Esterification occurs when FFA reacts with alcohol in the presence of acidic catalysts to form ester as the reaction product. This is the most common method to reduce the FFA content in jatropha (Lu et al., 2009), waste cooking oil (Wang et al., 2006), pongamia pinnata (Sharma et al., 2010), and kusum (Sharma and Singh, 2010). However, high alcohol to oil molar ratio is required to accelerate acid-catalyzed transesterifcation (e.g., molar ratio of methanol: oil = 20-30:1) with reaction temperature ranging from 65 to 99 °C, H_2SO_4 loading ranging from 1 to 4 wt% (referred to weight of oil) and reaction time ranging from 20 to 70 h (Freedman et al., 1984; Wang et al., 2006).

On the other hand, homogeneous acid catalysts posed several disadvantages, such as (1) strong acidic properties caused serious corrosion to reactor wall, pipelines, and valves, (2) slow reaction rate, (3) difficulty in catalyst separation. Therefore, homogeneous acid catalysts are not favored for commercial biodiesel production, but they appear as a suitable choice in esterification reaction rather than transesterification due to the simple molecular structure of FFA as compared to triglycerides (Wang et al., 2006). It was recommended that homogeneous acid catalysts are used initially to reduce the FFA content in the oil to a lower content and then only followed by transesterification reaction catalyzed by homogeneous base catalysts (Canakci and Van Gerpen, 2003). This combined two-step process gives better advantages than the individual single step, such as relatively less energy requirement, minimizing saponification reaction, and resulting in easy separation of biodiesel and glycerol. Perhaps, the main problem associated with this combined process is the difficulty in catalyst separation that requires multiple water washing steps, resulting to huge amount of wastewater that is not environmental friendly. Up to now, research on this two-step process is still limited for palm biodiesel. Nevertheless, it holds an important key to be easily incorporated into the existing palm biodiesel plant (only homogeneous base catalysts are utilized) when there is a need to change the feedstock from RBD palm oil to CPO.

2.2.3 *Heterogeneous Catalysts*

Due to the severe difficulty in separating homogeneous catalyst after reaction and also the huge amount of wastewater generated, heterogeneous catalysts appear as an excellent solution to this problem. Heterogeneous catalysts in the form of powder or pellet can be easily separated out after the reaction is completed, and the catalysts have the potential to be recycled, regenerated, and reused. This approach is more environment friendly and indirectly reduces the overall biodiesel production cost. In fact, the use of heterogeneous catalysts in transesterification is not a new phenomenon as in the past few years; extensive researches have been carried out to explore its potentials. However, high reaction temperature, high alcohol to oil molar ratio, and long reaction time are generally required due to mass transfer limitation of oil-alcohol-heterogeneous catalyst (three-phase system) in the initial stage of the reaction. Therefore, utilization of heterogeneous catalysts for commercial biodiesel production is still not attractive. The following sections depict various types of heterogeneous base and acid catalysts used in transesterification of palm oil.

2.2.3.1 HETEROGENEOUS BASE CATALYSTS

Heterogeneous base catalysts such as basic zeolites, alkaline earth metal oxides, and hydrotalcites which exhibited alkaline properties on its surface are being identified as a good option in replacement of current homogeneous base catalysts for biodiesel production. Nevertheless, it should be noted that heterogeneous base catalysts are still very sensitive to FFA content in the oil; only oil with less than 1 wt% of FFA is favored. Thus, the major challenges facing the development of heterogeneous base catalysts are their ability to withstand high FFA oil at a mild reaction condition and their reusability.

2.2.3.1.1 CALCIUM OXIDE (CAO) CaO is one of the most favorable heterogeneous base catalysts due to their relatively high basic sites, nontoxic, low solubility in methanol and can be prepared from cheap resources like lime stone and calcium hydroxide. The catalytic activity among alkaline earth metal oxides catalyst in transesterification is according to the following order: BaO > SrO > CaO > MgO (Yan et al., 2008), which also represents the order for the amount of basic sites. Although BaO and SrO have the highest number of basic sites and catalytic activity, it was found that both BaO and SrO are soluble in methanol. Therefore, BaO and SrO are more inclined toward homogeneous system rather than heterogeneous system. However, BaO and SrO are not used as homogeneous catalyst in the industry as NaOH or KOH is much cheaper with higher efficiency. Thus, among heterogeneous base catalysts, only CaO had been extensively tested for transesterification of various vegetable oils such as soybean, rapeseed, and sunflower oils. The results reported were indeed promising with high biodiesel yield at mild reaction conditions. Nevertheless, the active sites on CaO can be easily poisoned by contact with air due to adsorption of CO_2 and H_2O on the surface of the catalyst as carbonates and hydroxyl groups (Hattori, 1995). Therefore, activation of CaO by calcinations at >700 °C is generally required to revert CO_2 poisoning (Granados et al., 2007). The activated CaO shows complete decarbonation in which all $CaCO_3$ are converted to CaO. Apart from that, activated CaO is also covered with several layers of $Ca(OH)_2$ and thus minimizing H_2O adsorption on its surface. Nevertheless, care must be taken on the activated CaO to avoid further contact with ambient air that can cause reoccurrence of carbonation and hydration especially if exposed for long period.

A recent study on the use of CaO as the catalyst in transesterification of palm oil was reported by Yoosuk et al. (2010). It was revealed that the activated CaO can be subjected to hydration-dehydration method in order to further improve its physical and chemical properties such as surface area, pore volume, number of basic sites, and basic strength. Consequently, the catalyst was used in transesterification of palm olein and the optimum yield attained was 93.9% at the following reaction conditions: reaction temperature of 60 °C, methanol to oil molar ratio of 15, catalyst loading of 7 wt%, and reaction time of 45 min. Furthermore, the catalyst can be reused for up to five cycles with minimum drop in biodiesel yield. The decrease in the activity of the catalyst was attributed to active site blockage by adsorbed impurities or product species (monoglyceride, diglyceride, triglyceride, and glycerol) and leaching of active sites into the reaction media.

Apart from hydration-dehydration method, catalytic activity of CaO can also be enhanced by the addition of catalyst support. Among various catalyst supports available in the market, alumina (Al_2O_3) has been identified as a cheap and effective support for various catalytic chemical reactions such as steam reforming and hydrogenation. This is because Al_2O_3 has high specific surface area, large pore volume, mesopore size, high thermal stability, and mechanical strength (Arzamendi et al., 2007; Komintarachat and Chuepeng, 2009; Zabeti et al., 2010). These physical characteristics are important in heterogeneous catalysts that are to be used in transesterification reaction in order to minimize mass transfer limitation. Zabeti et al. (2010) optimized the reaction parameters in transesterification of palm oil using CaO supported with Al_2O_3 as the catalyst. Respond surface methodology (RSM) coupled with Central Composite Design (CCD) was used to identify the correlation and interaction between the reaction parameters. It was found that the optimum biodiesel yield attained was 98.6% at the following reaction conditions: 65 °C of reaction temperature, methanol to oil molar ratio of 12:1, catalyst loading of 6 wt%, and reaction time of 5 h. In addition, the catalyst was reused for two cycles with sustained catalytic activity.

Although CaO appears to have the potential to replace current homogeneous base catalysts, several important issues must still be addressed. One of them is the loss of active sites that can be leached out during transesterification reaction (Granados et al., 2007; Kouzu et al., 2008a,b). This does not only cause catalyst deactivation but also results in product contamination and thus extra purification steps (water washing) are required. Due to the additional purification step, most of the active sites will be lost and therefore limit catalyst recovery. Apart from that, sensitivity of CaO toward FFA content in oil is another problem that needs to be underlined. This is because FFA will react with the basic sites of CaO to form soap and cause serious difficulty in product separation (Kouzu et al., 2008a,b). In the case of using unrefined palm oil as biodiesel feedstock, the FFA content needs to be reduced before CaO can be used as the heterogeneous catalyst.

2.2.3.2 OTHER METAL OXIDES

Apart from CaO, there are several other alkaline metal oxides reported to have good performance in transesterification reaction. Bo et al. (2007) revealed the potential of alumina-supported potassium fluoride (KF/Al_2O_3) as an alternative heterogeneous base catalyst to produce biodiesel. KF/Al_2O_3 is regarded as a low-cost, commercially available, reusable, and environment-friendly catalyst in various organic processes, such as preparation of amides from nitriles, conversion of aldehydes to nitriles, and hydrothilation of alkynes (Bo et al., 2007; Zare et al., 2009). In their study, KF/Al_2O_3 was prepared through impregnation method with 0.33 as the optimum load ratio of KF to Al_2O_3. In order to achieve a good interaction between KF and Al_2O_3, the impregnated catalyst was subjected to calcinations at 600 °C for 3 h. The resulted catalyst was applied for transesterification of palm oil and the optimum reaction condition were as follows: reaction temperature of 65 °C, methanol to oil molar ratio of 12:1, catalyst loading of 4 wt%, and reaction time of 3 h. The optimum biodiesel yield attained was almost 90%. Nevertheless, the active sites of the catalyst were lost after the first cycle of reaction and regeneration study is promptly required to further strengthen the feasibility of the catalyst to be applied in the industry.

Apart from that, KF loaded to ZnO was also identified as an active and promising heterogeneous base catalyst in transesterification. When 15 wt% KF is loaded on ZnO and calcined at 600 °C for 5 h, the resulting catalyst contains a very high number of basic sites (1.47 mmol/g; Xie and Huang, 2006). It was observed that when calcination temperature was increased further, the KF active sites were decomposed and consequently lower down the number of basic sites. Application of KF/ZnO in biodiesel production from palm oil was reported by Hameed et al., (2009). From the study, optimum yield of biodiesel attained was 89.2% at reaction temperature of 65 °C, methanol to oil molar ratio of 11.4, catalyst loading of 5.5%, and reaction time of 9.7 h. However, further improvement in catalyst preparation steps is required to obtain a higher number of basic sites on KF/ZnO and to reduce the catalyst deactivation rate.

Besides, direct impregnation of potassium (K) onto SBA-15 also exhibited high catalytic activity in transesterification. SBA-15 is selected as a good support due to the following reasons: (1) high surface area (600-1000 m^2/g), (2) tunable mesopore size (5-30 nm), and (3) high thermal stability (Abdullah et al., 2009; El Berrichi et al., 2007). Generally, microporous structure is preferred for a heterogeneous catalyst in transesterification reaction since triglycerides are categorized as large molecules with average molecules size of 2 nm (Lam et al., 2010). Therefore, SBA-15 posed a superior advantage compared to zeolite (microporous solid) as support in which mass transfer resistance and diffusion limitation can be significantly reduced. Abdullah et al. (2009) synthesized high catalytic activity of K/SBA-15 through impregnation method and calcined at 350 °C for 3 h. The resulted catalyst exhibited high surface area of 539 m^2/g and average pore diameter of 5.63 nm. Subsequently, optimization on the transesterification reaction variables using palm oil and the synthesized catalyst were carried out using design of experiment (DOE). The optimum biodiesel yield attained was 87.3% at the following optimum reaction conditions: reaction temperature of 70 °C, methanol to oil molar ratio of 11.6:1, catalyst loading of 3.91 wt%, and reaction time of 5 h.

2.2.3.3 WASTE MATERIAL

In line with the world's sustainability concept, reutilization of waste material has emerged as a new trend in order to reduce the accumulation of waste and to protect the environment. As such, researchers are now looking for potential waste materials to be converted to catalyst for various applications. One of the potential sources of waste material is agricultural waste such as biomass from oil palm industries in Malaysia. In the year of 2005 alone, it was reported that Malaysia produces about 55.73 million tonnes of oil palm waste biomass in the form of empty fruit bunches (EFBs), shell, fiber, palm kernel, frond, and trunk (Shuit et al., 2009). The synthesis of catalyst from oil palm waste biomass for transesterification reaction has been reported. Initially, oil palm biomass such as palm shell must be converted to activated carbons through pyrolysis and steam activation process. Then, active compounds are impregnated on the surface of the activated carbon. Activated carbon produced from oil palm biomass has numerous applications typically in wastewater treatment as an effective adsorbent (Hameed et al., 2008, 2009; Tan et al., 2008). High adsorption capacities attributed by oil palm-activated carbons are always correlated to their physical properties, such as high surface area, pore volume, and pore diameter. Baroutian et al. (2010) reported that by depositing KOH on palm shell-activated carbon, the activated carbon can act like a catalyst for the transesterification of palm oil. Optimum biodiesel yield of 98% was attained at the following reaction condition: reaction temperature of 64 °C, methanol to oil molar ratio of 24:1, catalyst loading of 30.3 wt%, and reaction time of 1 h. However, leaching of the active sites into the reaction media was observed, but at minimum level with 0.98 and 0.80 ppm, respectively, in the first and second cycles of reaction. Nevertheless, the presence of KOH (due to leaching) in the product mixture does not affect the quality of the biodiesel as it still meets the basic standard of biodiesel in which concentration of mineral matter should be below 200 ppm (Kouzu et al., 2008a,b).

More study on the utilization of activated carbon (derived from oil palm biomass) as the catalyst for transesterification could be found in the study by Chin et al. (2009).

Besides oil palm biomass, waste eggshells have also been identified as a low-cost catalyst for biodiesel production. It was reported that chicken eggshells consist high content of calcium carbonate ($CaCO_3$) which can be converted to CaO after calcinations at 700 °C (Cho and Seo, 2010; Wei et al., 2009). The product after the calcinations process can act as a catalyst for heterogeneous transesterification due to their high number of basic sites. Cho and Seo (2010) reported that by introducing hydrochloric acid (HCl, weak acid) solution to quail eggshells, this can further enhance its catalytic activity in transesterification of palm oil. The purpose of the weak acid treatment is to remove the dense cuticle outer layer of the eggshells which are not porous and therefore facilitating the diffusion of triglycerides to the porous palisade middle layer of the eggshell (Cho and Seo, 2010). The acid-treated quail eggshell catalyst was able to maintain its catalytic activity with 98% palm biodiesel conversion even after five cycles of reaction at temperature of 65 °C, methanol to oil molar ratio of 12:1, catalyst loading of 1.5 wt%, and reaction time of 2 h. Similar reports on utilization of waste mud crab shell for transesterification of palm oil could be found in the study by Boey et al. (2009).

2.2.4 Heterogeneous Acid Catalysts

The development of different types of heterogeneous acid catalysts lately has widened the choice of feedstock for biodiesel production including CPO that has high FFA. The advantages of using heterogeneous acid catalysts in transesterification are: (1) insensitive to FFA content in the oil, (2) catalyzed esterification and transesterification simultaneously, (3) easy catalyst recovery from reaction media, (4) have potential to be recycled and regenerated, (5) minimize the number of washing steps required and (6) less corrosion toward reactors wall, pipelines and valves compared to homogeneous acid catalyst (Jitputti et al., 2006; Kulkarni and Dalai, 2006; Suarez et al., 2007). The potential and performance of various heterogeneous acid catalysts in transesterification of different oil sources have been explored extensively in the past few years such as sulfated zirconium oxide (SO_4^{2-}/ZrO_2; Furuta et al., 2004; Jitputti et al., 2006; Park et al., 2008), sulfated titanium oxide (SO_4^{2-}/TiO_2; de Almeida et al., 2008; Peng et al., 2008), sulfated tin oxide (SO_4^{2-}/SnO_2; Furuta et al., 2004; Lam et al., 2009a), sulfonic ion-exchange resin (Dos Reis et al., 2005; Heidekum et al., 1999), sulfonated carbon-based catalyst (Lou et al., 2008; Takagaki et al., 2006), and heteropolyacids (Cao et al., 2008; Zhang et al., 2009). Nevertheless, the main challenges of commercializing these catalysts are their relatively high cost; complicated synthesis procedures and extreme reaction conditions are generally required for transesterification (e.g., high reaction temperature).

Jitputti et al. (2006) reported that SO_4^{2-}/ZrO_2 has high catalytic activity in the transesterification of palm kernel oil (Jitputti et al., 2006). The catalyst exhibited extremely strong acid strength on its surface and therefore it is suitable to be used in transesterification of oil with high FFA. The palm kernel oil used in the experiments contains FFA value of 1% (as lauric acid) which is not favorable for base catalysts. The yield of palm kernel biodiesel attained was more than 90% at reaction temperature of 200 °C, methanol to oil molar ratio of 6:1, and catalyst loading of 3 wt% in a nitrogen-pressurized reactor at 50 bars. They also studied the recycling and regeneration of the SO_4^{2-}/ZrO_2 catalyst. After the first cycle of reaction, the catalyst was recovered, dried at 100 °C, and used in the subsequent reaction. It was found that the catalytic activity of SO_4^{2-}/ZrO_2 dropped tremendously to give only 28% biodiesel yield.

This may be due to several factors such as catalyst leaching and active sites blockage by reactants or products. Nevertheless, the catalyst can be easily regenerated and gave the same activity as fresh catalyst. More research studies on palm oil conversion to biodiesel using heterogeneous acid catalysts can be found in the studies by Yee et al. (2010), Melero et al. (2010), Aderemi and Hameed (2009), and Kansedo et al. (2009).

2.2.5 *Enzymatic Catalyst*

More recently, enzymatic catalyst (especially lipase) has revealed its potential as a catalyst in transesetrification reaction to produce biodiesel. Different from chemical catalyst, enzymatic catalyst offers several advantages such as: (1) not energy intensive because reactions normally occur at room temperature, (2) insensitive to FFA content, (3) easy recovery of catalyst and glycerol, and (4) minimize water-washing step that consequently reduce wastewater treatment cost. Mucor miehei (Lipozym IM 60), Pseudomonas cepacia (PS 30), C. antarctica (Novozym 435), and Bacillus subtilis are the examples of enzymes that have shown good catalytic activity in transesterification. In addition to that, enzymatic catalyst has attained another significant milestone with the introduction of immobilization technology. The purpose of immobilization is to provide a more rigid external backbone for lipase so that it can maintain its high stability, easily recycle and reuse for the subsequent reactions (Jegannathan et al., 2008; Knezevic et al., 1998). Lipase can be immobilized into ion-exchange resin, photocrosslinkable resin, silica beads, alumina and activated carbon through adsorption, covalent bonding, entrapment, encapsulation, and cross-linking (Tan et al., 2010). However, the main limitation of using enzymatic catalyst in commercial scale is the high cost of lipase and slow reaction rate.

In transesterification using enzymatic catalyst, solvent is added into the reaction media to ensure homogeneous phase between oil and alcohol (reducing mass transfer limitation) and thus enhancing lipase catalytic activity (Fjerbaek et al., 2009). Generally, hexane is preferred due to low cost and easily availability in the market. However, it was found that solubility of methanol and glycerol in hexane is low and resulted to lipase deactivation (poisoned by methanol or glycerol; Royon et al., 2007). After years of research, *tert*-butanol was discovered as a superior solvent than hexane. Methanol and glycerol are easily soluble in *tert*-butanol which minimizes the poisoning rate caused by methanol and also reduces the heavy deposition of glycerol on the immobilized lipase (Nielsen et al., 2008; Watanabe et al., 2000). The importance of *tert*-butanol in enzymatic transesterification using unrefined palm oil is more prevalent as the oil contains a high level of phospholipids (major components of oil gum), which increases viscosity and mass transfer limitation (Talukder et al., 2009). Talukder et al. (2009) reported that the catalytic activity of Novozym 435 increased nearly 358% due to the positive effect of adding *tert*-butanol into the reaction mixture. A similar observation was also discovered by Halim and Harun Kamaruddin (2008) in transesterification of waste cooking palm oil using Novozym 435. Nevertheless, several issues must still be addressed due to the use of solvent in transesterification, such as: (1) extra reactor volume is required to accommodate the additional volume of solvent, (2) plant safety requirement (toxicity of solvent), (3) extra production cost (extra solvent recovery steps and the loss of solvent; Nielsen et al., 2008).

Unlike chemical catalyst, enzymatic transesterification requires certain amount of water in the reaction media to maintain the enzyme catalytic activity. Generally, the availability of interfacial area is one of the factors that influence enzyme activity. The presence of water will facilitate the formation of oil-water droplet which will increase the interfacial area between the reactants that will subsequently accelerate the transesterification reaction rate (Noureddini et al., 2005). However, it should be noted that if too much water is added, this could cause the hydrolysis of oil that leads to low biodiesel conversion. In addition, it is also important to ensure the support used to immobilize lipase does not adsorb water or otherwise it could inhibit the penetration of oil to the lipase layer (Samukawa et al., 2000; Talukder et al., 2009). The minimum and also optimum amount of water added into a specific reaction mixture is strongly dependent on the type of lipase used. For example, lipase PS (Burkholderia cepacia) immobilized within κ-carrageenam (biopolymer) requires an optimum water to palm oil volumetric ratio of 0.085:1 (v/v) (Jegannathan et al., 2010), whereas *C. rugosa* requires 1:1 (v/v) (Talukder et al., 2010). However, if the feedstocks used for biodiesel do contain water such as CPO and waste cooking palm oil, this important factor must be considered in determining the optimum amount of water to be used during transesterification reaction.

Theoretically, in transesterification, 3 mol of alcohol is required to produce 3 mol of biodiesel and 1 mol of glycerol. However, since the reaction is reversible, excess alcohol is generally preferred to drive the reaction toward completion within a minimum reaction time. In homogeneous and heterogeneous base catalysts, methanol to oil molar ratio of 6-15:1 is generally used. However, the same ratio cannot be applied to enzymatic catalysis as this will cause the poisoning of lipase and results in exceptionally low biodiesel yield. Therefore, in enzymatic catalysis, alcohol is normally added to the reaction mixture in stepwise order to minimize lipase poisoning by the alcohol and thus prolong its durability (Watanabe et al., 2001). Using this strategy, high yield of biodiesel can be attained (Chen et al., 2006; Ying and Chen, 2007) and the lipase has a higher possibility to be reused in the subsequent reaction. Nevertheless, there is very limited research being carried out on the development of stepwise addition of alcohol on transesterification of palm oil.

Apart from the effect of water and alcohol concentration in the reaction mixture, reaction temperature also plays a significant role in enzymatic transesterification. It is well reported that in transesterification, high reaction temperature will reduce mass transfer limitation due to the decrease in oil viscosity and therefore accelerates the transesterification rate. Nevertheless, this is not true for enzymatic catalysis. Enzyme is extremely sensitive to the surrounding temperature; once the surrounding temperature exceeds a certain limit, the lipase will deactivate immediately and perhaps permanently. Therefore, the control of temperature in enzymatic catalysis is a very important factor that does not only affect the biodiesel yield, but also indirectly determines the survival of lipase and its reusability. On the other hand, immobilized enzyme has a higher temperature resistance compared to free enzyme due to the binding of the enzyme within a carrier material that gives it a higher stability and therefore decreases the effect of thermal deactivation (Fjerbaek et al., 2009). In general, reaction temperature between 30 and 40 °C is favorable, depending on the type of lipase used in the reaction (Sim et al., 2010a,b; Talukder et al., 2009). Sim et al. (2010b) had reported on the effect of temperature in enzymatic catalysis toward biodiesel production from palm oil (Sim et al., 2010a). In that study, Lipozyme TL IM was used as catalyst and CPO as feedstock. The optimum biodiesel yield attained was 85% at the following reaction conditions: reaction temperature of 30 °C, enzyme loading of 6.67 wt%, agitation speed of 150 rpm, and reaction time of 6 h. Thus, from the result of this study, it proves that it is indeed a plausible option to produce palm biodiesel at room temperature and hence reduce energy requirement in biodiesel production.

2.3 Other Technologies

2.3.1 *Noncatalytic Transesterification: Supercritical Alcohol Method*

Supercritical fluid technology has opened up a new dimension in various reaction, extraction and separation processes. Under supercritical condition, thermophysical properties of fluid such as dielectric constant, viscosity, and specific polarity are drastically changed depending on the temperature and pressure (Lee and Saka, 2010). For transesterification reaction, supercritical alcohol (typically methanol) has been introduced as an alternative means to produce biodiesel without the need of a catalyst. More importantly, supercritical alcohol technology poses several advantages over catalysis method, such as: (1) no further purification step (water washing) required and indirectly reduce the wastewater treatment cost, (2) insensitive to water and FFA level in the oil and (3) fast reaction rate and thus, has a higher possibility to produce biodiesel on a continuous basis. During transesterification reaction under supercritical conditions, triglyceride becomes miscible in alcohol due to the decrease in dielectric constant of alcohol. Consequently, the reaction mixture becomes a single phase rather than a heterogeneous system, and this seems to be the key factor that accelerates transesterification reaction. Up to now, supercritical alcohol technology has attracted considerable attention by researchers in this field, but the true potential of this technology in the commercial scale is yet to be revealed.

On the other hand, although supercritical alcohol enjoys enormous advantages over the catalysis method, there are a few imperative issues that need to be addressed urgently upon scaling the technology up. Because high temperature and pressure are the major factors to reach the supercritical state, the energy output by burning biodiesel may not counter the extensive energy input. A proper life-cycle assessment (LCA) should be studied to further justify the possibility of using supercritical alcohol technology in biodiesel production. Despite the overall energy balance, the decomposition of biodiesel was found to be the most severe problem. When the reaction temperature was over a certain limit, the decomposition of unsaturated fatty acids was observed, which deteriorated the biodiesel conversion seriously (Imahara et al., 2008). Apart from thermal decomposition, isomerization of polyunsaturated methyl ester from cis-type carbon bonding into trans-type carbon bonding was detected through Fourier transform infrared spectrometry (FT-IR; Imahara et al., 2008). Trans-type fatty acids are naturally unstable, which degrades the cold flow properties of biodiesel (Gui et al., 2009; Imahara et al., 2008). Furthermore, the ultra high alcohol to oil molar ratio in the supercritical method is also a crucial factor in determining biodiesel conversion. Consequently, the energy consumption for alcohol separation from biodiesel and glycerol (alcohol recycling process) is expected to be exceptionally high at the industrial scale.

The transesterification of palm oil using supercritical methanol technology was investigated by Song et al. (2008). From the study, more than 90% of the biodiesel content was obtained at the following optimum reaction conditions: a reaction temperature and pressure of 350 °C and 40 MPa, respectively, a methanol to oil molar ratio of 30, and a reaction time of 5 min. The reaction rate with supercritical methanol was much faster than the reaction catalyzed by a homogeneous base catalyst. This promising result has escalated further development and research using palm oil as the feedstock to produce biodiesel. Continuous biodiesel production through supercritical methanol, which was carried out by Bunyakiat et al. (2006), indicated the special advantage of the technology (Bunyakiat et al., 2006). Palm kernel oil was tested as one of the feedstocks with the optimum conversion of 95% at the following reaction conditions: a reaction temperature and pressure of 350 °C and 19 MPa, respectively, and a molar ratio of methanol to oil of 42 at a time of 400 s. The result attained was similar to that of Song et al. (2008), although the FFA level in the palm kernel oil was exceptionally high, 31 mg KOH/g, which is far beyond the limit if a homogeneous base catalyst is utilized. Because supercritical methanol is highly tolerant to the FFA content in the oil, unrefined palm oil and waste frying palm oil are an attractive choice to further reduce the overall biodiesel production cost. However, care should be taken that the supercritical temperature should not exceed 375 °C as low biodiesel content was observed due to thermal decomposition (Song et al., 2008).

Overall, biodiesel is not a totally green biofuel if methanol is used as one of the reactants because the methanol available on the market is generally derived from fossil fuels, such as petroleum and natural gas. Therefore, ethanol has emerged as an alternative to methanol because ethanol can be derived from agricultural renewable biomass through pretreatment, hydrolysis, and fermentation processes. In fact, ethanol is more soluble in oil compared to methanol and, consequently, minimizes the mass transfer limitation of the transesterification reaction. Nevertheless, findings from various studies have proven that ethanol has lower reactivity than methanol (Meneghetti et al., 2006) due to (1) steric hindrance effects of the larger alkyl chains in ethanol (Suwannakarn et al., 2008) and (2) the stable emulsification created by ethanol during transesterification, resulting in the difficulty of ethyl ester separation from glycerol (Encinar et al., 2007). The potential of supercritical ethanol transesterification from palm oil was first explored by Gui et al. (2009). It was possible to produce palm biodiesel through supercritical ethanol; however, relatively lower yield of biodiesel was attained (79.2%) with a long reaction time (30 min). This finding may be attributed to the type of oil used because palm oil contains high-saturated fatty acids compared to other types of oil reported, and the saturated fatty acid indirectly increases the viscosity of the oil. Nonetheless, the environmental advantages of using ethanol to produce greener biodiesel should not be ignored and should be explored through intensive research.

2.3.2 Ultrasonic-Assisted Transesterification

Currently, ultrasonic technology is on the frontier of improving the mass transfer rate between the immiscible liquid-liquid phase within a heterogeneous system (Ji et al., 2006). Ultrasound is defined as sound with a frequency beyond the response of the human ear. The normal sound frequency that can be detected by the human ear lies between 16 and 18 kHz, but the frequency for ultrasound generally lies between 20 kHz and 100 MHz (Vyas et al., 2010). This high-frequency sound wave compresses and stretches the molecular spacing of a medium through which it passes. Thus, molecules are continuously vibrated, and cavities are created. As a result, microfine bubbles are formed through the sudden expansion and collapse, generating energy for chemical and mechanical effects (Colucci et al., 2005). Furthermore, the collapsed bubbles disrupt the phase boundary and impinging of the liquids to create microjets, leading to intensively emulsification of the system (Ji et al., 2006). Subsequently, the mixing effect due to emulsification increases the interfacial area between the immiscible reactants and thus facilitated reaction kinetics (Kalva et al., 2009).

The positive effect of introducing ultrasound in transesterification has been reported recently. Transesterification is well recognized as a slow reaction process; generally, 40-60 min of reaction are required for the reaction to go to completion if a homogeneous base catalyst is used under favorable reaction conditions because alcohol (typically methanol) has a very low solubility value in oil at room temperature. Therefore, a two-phase system is created in the initial reaction. This two-phase system has a limited mass transfer rate between the reactants and, thus, slows down the overall reaction rate. This phenomenon becomes even worse if heterogeneous catalysts are used. A three-phase system is created and directly interferes with the reaction rate significantly. Ultrasound technology is relatively a new method to improve the mass transfer rate in transesterification rather than mechanical agitation. A shorter reaction time and better energy efficiency for transesterification were observed with ultrasound (Colucci et al., 2005; Ji et al., 2006; Singh et al., 2007). In addition, the reaction rate constants were enhanced by a factor of 3-5 higher than those for the mechanically agitated process (Kalva et al., 2009).

Currently, research on using ultrasound technology in transesterification of palm oil is relatively limited compared to the research on other vegetable oils such as rapeseed, soybean, and sunflower oils. Inevitably, there is a knowledge gap between the effects of ultrasound on the palm biodiesel conversion efficiency under various reaction conditions and different types of catalysts. Mootabadi et al. (2010) demonstrated a significant improvement in the palm oil transesterification reaction rate catalyzed by heterogeneous base catalysts via ultrasound technology(Mootabadi et al., 2010). Refined palm oil was used in the study, and high biodiesel yields were attained after performing optimization. The reaction catalyzed by BaO and enhanced with ultrasound attained the highest biodiesel yield, which was 95.2%. In comparison, only a 67.3% biodiesel yield was attained if the reaction was enhanced with magnetic stirring (mechanical mixing). Furthermore, lower methanol and catalyst concentrations were used in the reaction due to the ultrasound effect. All of the evidence proves the efficiency of ultrasound technology in enhancing the transesterification of palm oil. In addition, Deshmane et al. (2009) demonstrated the positive effect of ultrasound in the esterification of palm fatty acid distillate (PFAD). Normally, PFAD is generated as a byproduct during the refinement of palm oil and has a lower market value compared to palm oil. Due to the high FFA value of PFAD, acid catalysts are preferred. Nevertheless, acid catalysts suffer from very slow reaction rates, and long reaction times are generally required. With the introduction of ultrasound into the reaction media, the reaction time was reduced by half due to the turbulence created intensively by ultrasound through cavitation, which resulted in excellent mixing between the two phases (oil and alcohol; Deshmane et al., 2009).

3 CONCLUSIONS

Although oil palm has been severely questioned regarding its sustainability and environmental issues, the oil yield attained annually is still far superior than other edible oil bearing crops. Therefore, palm oil should be considered as an alternative and promising feedstock to further diversified the biodiesel production in the global market. In addition, palm oil contains various phytonutrients that can be separated out prior to biodiesel production. These phytonutrients have a high market value and can thus offset the overall palm biodiesel production cost. Indeed, this benefit has not been foreseen for other edible oil crops.

To date, palm biodiesel conversion technologies have been well researched, especially the catalysis method. Homogeneous base catalysts are the most common but pose severe problems when high FFA appears in CPO. Other methods, such as heterogeneous (base and acid), enzymatic and supercritical technologies, have emerged as an alternative route to produce palm biodiesel in a greener manner with excellent biodiesel yield. However, these new methods have not been readily available at the commercial scale because the catalysts are easily poisoned and deactivated, a high energy input is required, and there are safety-related issues. Extensive research is still required to produce a breakthrough for these technologies in palm biodiesel conversion.

References

Abdullah, A.Z., Razali, N., Lee, K.T., 2009. Optimization of mesoporous K/SBA-15 catalyzed transesterification of palm oil using response surface methodology. Fuel Proc. Technol. 90, 958–964.

Aderemi, B.O., Hameed, B.H., 2009. Alum as a heterogeneous catalyst for the transesterification of palm oil. Appl. Catal. A 370, 54–58.

Al-Widyan, M.I., Al-Shyoukh, A.O., 2002. Experimental evaluation of the transesterification of waste palm oil into biodiesel. Bioresour. Technol. 85, 253–256.

Arzamendi, G., Campo, I., Arguiñaren, E., Sánchez, M., Montes, M., Gandía, L.M., 2007. Synthesis of biodiesel with heterogeneous NaOH/alumina catalysts: comparison with homogeneous NaOH. Chem. Eng. J. 134, 123–130.

Baroutian, S., Aroua, M.K., Raman, A.A.A., Sulaiman, N.M.N., 2010. Potassium hydroxide catalyst supported on palm shell activated carbon for transesterification of palm oil. Fuel Proc. Technol. 91, 1378–1385.

Basha, S.A., Gopal, K.R., Jebaraj, S., 2009. A review on biodiesel production, combustion, emissions and performance. Renew. Sustain. Energy Rev. 13, 1628–1634.

Bo, X., Guomin, X., Lingfeng, C., Ruiping, W., Lijing, G., 2007. Transesterification of palm oil with methanol to biodiesel over a KF/Al_2O_3 heterogeneous base catalyst. Energy Fuels 21, 3109–3112.

Boey, P.L., Maniam, G.P., Hamid, S.A., 2009. Biodiesel production via transesterification of palm olein using waste mud crab (Scylla serrata) shell as a heterogeneous catalyst. Bioresour. Technol. 100, 6362–6368.

Bunyakiat, K., Makmee, S., Sawangkeaw, R., Ngamprasertsith, S., 2006. Continuous production of biodiesel via transesterification from vegetable oils in supercritical methanol. Energy Fuels 20, 812–817.

Canakci, M., Van Gerpen, J., 2003. A pilot plant to produce biodiesel from high free fatty acid feedstocks. Trans. Am. Soc. Agric. Eng 46, 945–954.

Cao, F., Chen, Y., Zhai, F., Li, J., Wang, J., Wang, X., et al., 2008. Biodiesel production from high acid value waste frying oil catalyzed by superacid heteropolyacid. Biotechnol. Bioeng 101, 93–100.

Che Man, Y.B., Moh, M.H., Van De Voort, F.R., 1999. Determination of free fatty acids in crude palm oil and refined-bleached-deodorized palm olein using Fourier transform infrared spectroscopy. J. Am. Oil. Chem. Soc. 76, 485–490.

Chen, G., Ying, M., Li, W., 2006. Enzymatic conversion of waste cooking oils into alternative fuel—biodiesel. Appl. Biochem. Biotechnol 132, 911–921.

Chen, W., Zhao, Z., Yin, C., 2010. The interaction of waxes with pour point depressants. Fuel 89, 1127–1132.

Chin, L.H., Hameed, B.H., Ahmad, A.L., 2009. Process optimization for biodiesel production from waste cooking palm oil (elaeis guineensis) using response surface methodology. Energy Fuels 23, 1040–1044.

Cho, Y.B., Seo, G., 2010. High activity of acid-treated quail eggshell catalysts in the transesterification of palm oil with methanol. Bioresour. Technol. 101, 8515–8519.

Choo, Y.M., Cheng, S.F., Yung, C.L., Harrison Lau, L.N., Ma, A.N., Yusof, B., 2002a. Low pour point palm diesel. Malaysian Patent No. PI 20021157.

Choo, Y.M., Cheng, S.F., Yung, C.L., Harrison Lau, L.N., Ma, A.N., Yusof, B., 2002b. Palm diesel with low pour point for cold climate countries. MPOB information Series No. 163.

Choo, Y.M., Cheng, S.F., Yung, C.L., Harrison Lau, L.N., Ma, A.N., Yusof, B., 2002c. Production of C16 and C18 mixed methyl esters. MPOB Information Series No. 166.

Choo, Y.M., Ma, A.N., Chan, K.W., Yusof, B., 2005. Palm diesel: an option for greenhouse gas mitigation in the energy sector. J. Oil Palm Res. 17, 47–52.

Colucci, J.A., Borrero, E.E., Alape, F., 2005. Biodiesel from an alkaline transesterification reaction of soybean oil using ultrasonic mixing. J. Am. Oil. Chem. Soc. 82, 525–530.

Darnoko, D., Cheryan, M., 2000. Kinetics of palm oil transesterification in a batch reactor. J. Am. Oil. Chem. Soc. 77, 1263–1267.

de Almeida, R.M., Noda, L.K., Gonçalves, N.S., Meneghetti, S.M.P., Meneghetti, M.R., 2008. Transesterification reaction of vegetable oils, using superacid sulfated TiO_2-base catalysts. Appl. Catal. A 347, 100–105.

Deshmane, V.G., Gogate, P.R., Pandit, A.B., 2009. Ultrasound-assisted synthesis of biodiesel from palm fatty acid distillate. Ind. Eng. Chem. Res. 48, 7923–7927.

Dias, J.M., Alvim-Ferraz, M.C.M., Almeida, M.F., 2008. Comparison of the performance of different homogeneous alkali catalysts during transesterification of waste and virgin oils and evaluation of biodiesel quality. Fuel 87, 3572–3578.

Dos Reis, S.C.M., Lachter, E.R., Nascimento, R.S.V., Rodrigues, J.A., Reid, M.G., 2005. Transesterification of Brazilian vegetable oils with methanol over ion-exchange resins. J. Am. Oil. Chem. Soc. 82, 661–665.

El Berrichi, Z., Louis, B., Tessonnier, J.P., Ersen, O., Cherif, L., Ledoux, M.J., et al., 2007. One-pot synthesis of Ga-SBA-15: activity comparison with Ga-post-treated SBA-15 catalysts. Appl. Catal. A 316, 219–225.

Encinar, J.M., González, J.F., Rodríguez-Reinares, A., 2007. Ethanolysis of used frying oil. Biodiesel preparation and characterization. Fuel Proc. Technol. 88, 513–522.

Fjerbaek, L., Christensen, K.V., Norddahl, B., 2009. A review of the current state of biodiesel production using enzymatic transesterification. Biotechnol. Bioeng. 102, 1298–1315.

Freedman, B., Pryde, E.H., Mounts, T.L., 1984. Variables affecting the yields of fatty esters from transesterified vegetable oils. J. Am. Oil. Chem. Soc. 61, 1638–1643.

Fukuda, H., Kondo, A., Noda, H., 2001. Biodiesel fuel production by transesterification of oils. J. Biosci. Bioeng. 92, 405–416.

Furuta, S., Matsuhashi, H., Arata, K., 2004. Biodiesel fuel production with solid superacid catalysis in fixed bed reactor under atmospheric pressure. Catal. Commun. 5, 721–723.

Granados, M.L., Poves, M.D.Z., Alonso, D.M., Mariscal, R., Galisteo, F.C., Moreno-Tost, R., et al., 2007. Biodiesel from sunflower oil by using activated calcium oxide. Appl. Catal. B 73, 317–326.

Gui, M.M., Lee, K.T., Bhatia, S., 2009. Supercritical ethanol technology for the production of biodiesel: process optimization studies. J. Supercrit. Fluids 49, 286–292.

Halim, S.F.A., Harun Kamaruddin, A., 2008. Catalytic studies of lipase on FAME production from waste cooking palm oil in a tert-butanol system. Proc. Biochem. 43, 1436–1439.

Hameed, B.H., Tan, I.A.W., Ahmad, A.L., 2008. Optimization of basic dye removal by oil palm fibre-based activated carbon using response surface methodology. J. Hazard. Mater. 158, 324–332.

Hameed, B.H., Tan, I.A.W., Ahmad, A.L., 2009. Preparation of oil palm empty fruit bunch-based activated carbon for removal of 2,4,6-trichlorophenol: optimization using response surface methodology. J. Hazard. Mater. 164, 1316–1324.

Harrison Lau, L.N., Choo, Y.M., Ng, M.H., Puah, C.W., Yahaya, H., Ma, A.N., 2009. Palm phytonutrients: opportunity beyond palm biodiesel. MPOB e-journal-Palm Oil Devel. 50, 9–11.

Hattori, H., 1995. Heterogeneous basic catalysis. Chem. Rev 95, 537–558.

Heidekum, A., Harmer, M.A., Hoelderich, W.F., 1999. Addition of carboxylic acids to cyclic olefins catalyzed by strong acidic ion-exchange resins. J. Catal. 181, 217–222.

Imahara, H., Minami, E., Hari, S., Saka, S., 2008. Thermal stability of biodiesel in supercritical methanol. Fuel 87, 1–6.

Jegannathan, K.R., Abang, S., Poncelet, D., Chan, E.S., Ravindra, P., 2008. Production of biodiesel using immobilized lipase—a critical review. Crit. Rev. Biotechnol. 28, 253–264.

Jegannathan, K.R., Jun-Yee, L., Chan, E.S., Ravindra, P., 2010. Production of biodiesel from palm oil using liquid core lipase encapsulated in κ-carrageenan. Fuel 89, 2272–2277.

Jeong, G.T., Park, D.H., Kang, C.H., Lee, W.T., Sunwoo, C.S., Yoon, C.H., et al., 2004. Production of biodiesel fuel by transesterification of rapeseed oil. Appl. Biochem. Biotechnol. Enzym Eng. Biotechnol. 114, 747–758.

Ji, J., Wang, J., Li, Y., Yu, Y., Xu, Z., 2006. Preparation of biodiesel with the help of ultrasonic and hydrodynamic cavitation. Ultrasonics 44, e411–e414.

Jitputti, J., Kitiyanan, B., Rangsunvigit, P., Bunyakiat, K., Attanatho, L., Jenvanitpanjakul, P., 2006. Transesterification of crude palm kernel oil and crude coconut oil by different solid catalysts. Chem. Eng. J. 116, 61–66.

Kalva, A., Sivasankar, T., Moholkar, V.S., 2009. Physical mechanism of ultrasound-assisted synthesis of biodiesel. Ind. Eng. Chem. Res. 48, 534–544.

Kansedo, J., Lee, K.T., Bhatia, S., 2009. Biodiesel production from palm oil via heterogeneous transesterification. Biomass Bioenergy 33, 271–276.

Knezevic, Z.D., Siler-Marinkovic, S.S., Mojovic, L.V., 1998. Kinetics of lipase-catalyzed hydrolysis of palm oil in lecithin/izooctane reversed micelles. Appl. Microbiol. Biotechnol. 49, 267–271.

Komintarachat, C., Chuepeng, S., 2009. Solid acid catalyst for biodiesel production from waste used cooking oils. Ind. Eng. Chem. Res. 48, 9350–9353.

Kouzu, M., Kasuno, T., Tajika, M., Sugimoto, Y., Yamanaka, S., Hidaka, J., 2008a. Calcium oxide as a solid base catalyst for transesterification of soybean oil and its application to biodiesel production. Fuel 87, 2798–2806.

Kouzu, M., Kasuno, T., Tajika, M., Yamanaka, S., Hidaka, J., 2008b. Active phase of calcium oxide used as solid base catalyst for transesterification of soybean oil with refluxing methanol. Appl. Catal. A 334, 357–365.

Kulkarni, M.G., Dalai, A.K., 2006. Waste cooking oil—an economical source for biodiesel: a review. Ind. Eng. Chem. Res. 45, 2901–2913.

Lam, M.K., Lee, K.T., Mohamed, A.R., 2009a. Sulfated tin oxide as solid superacid catalyst for transesterification of waste cooking oil: an optimization study. Appl. Catal. B 93, 134–139.

Lam, M.K., Tan, K.T., Lee, K.T., Mohamed, A.R., 2009b. Malaysian palm oil: surviving the food versus fuel dispute for a sustainable future. Renew. Sustain. Energy Rev. 13, 1456–1464.

Lam, M.K., Lee, K.T., Mohamed, A.R., 2010. Homogeneous, heterogeneous and enzymatic catalysis for transesterification of high free fatty acid oil (waste cooking oil) to biodiesel: a review. Biotechnol. Adv. 28, 500–518.

Lee, J.S., Saka, S., 2010. Biodiesel production by heterogeneous catalysts and supercritical technologies. Bioresour. Technol. 101, 7191–7200.

Lee, I., Johnson, L.A., Hammond, E.G., 1995. Use of branched-chain esters to reduce the crystallization temperature of biodiesel. J. Am. Oil Chem. Soc. 72, 1155–1160.

Lim, S., Teong, L.K., 2010. Recent trends, opportunities and challenges of biodiesel in Malaysia: an overview. Renew. Sustain. Energy Rev. 14, 938–954.

Lipochem (M) Sdn Bhd, MPOB, Fuel of the future. Available from: http://www.lipochem.com/pdf/Fuel%20of%20The%20Future%20.%20English.pdf (accessed October 2010).

Lou, W.Y., Zong, M.H., Duan, Z.Q., 2008. Efficient production of biodiesel from high free fatty acid-containing waste oils using various carbohydrate-derived solid acid catalysts. Bioresour. Technol. 99, 8752–8758.

Lu, H., Liu, Y., Zhou, H., Yang, Y., Chen, M., Liang, B., 2009. Production of biodiesel from Jatropha curcas L. oil. Comput. Chem. Eng. 33, 1091–1096.

Ma, F., Hanna, M.A., 1999. Biodiesel production: a review. Bioresour. Technol. 70, 1–15.

Malaysian Palm Oil Council (MPOC), Available from: http://www.mpoc.org.my/ (accessed October 2010).

Melero, J.A., Bautista, L.F., Morales, G., Iglesias, J., Sánchez-Vázquez, R., 2010. Biodiesel production from crude palm oil using sulfonic acid-modified mesostructured catalysts. Chem. Eng. J. 161, 323–331.

Meneghetti, S.M.P., Meneghetti, M.R., Wolf, C.R., Silva, E.C., Lima, G.E.S., Silva, L.d.L., et al., 2006. Biodiesel from castor oil: a comparison of ethanolysis versus methanolysis. Energy Fuels 20, 2262–2265.

Mootabadi, H., Salamatinia, B., Bhatia, S., Abdullah, A.Z., 2010. Ultrasonic-assisted biodiesel production process from palm oil using alkaline earth metal oxides as the heterogeneous catalysts. Fuel 89, 1818–1825.

Nielsen, P.M., Brask, J., Fjerbaek, L., 2008. Enzymatic biodiesel production: technical and economical considerations. Eur. J. Lipid. Sci. Technol. 110, 692–700.

Noureddini, H., Gao, X., Philkana, R.S., 2005. Immobilized Pseudomonas cepacia lipase for biodiesel fuel production from soybean oil. Bioresour. Technol. 96, 769–777.

Park, Y.M., Lee, D.W., Kim, D.K., Lee, J.S., Lee, K.Y., 2008. The heterogeneous catalyst system for the continuous conversion of free fatty acids in used vegetable oils for the production of biodiesel. Catal. Today 131, 238–243.

Pasqualino, J.C., Montanè, D., Salvadoì, J., 2006. Synergic effects of biodiesel in the biodegradability of fossil-derived fuels. Biomass. Bioenergy. 30, 874–879.

Peng, B.X., Shu, Q., Wang, J.F., Wang, G.R., Wang, D.Z., Han, M.H., 2008. Biodiesel production from waste oil feedstocks by solid acid catalysis. Proc. Saf. Environ. Prot. 86, 441–447.

Rashid, U., Anwar, F., Moser, B.R., Ashraf, S., 2008. Production of sunflower oil methyl esters by optimized alkali-catalyzed methanolysis. Biomass. Bioenergy. 32, 1202–1205.

Royon, D., Daz, M., Ellenrieder, G., Locatelli, S., 2007. Enzymatic production of biodiesel from cotton seed oil using t-butanol as a solvent. Bioresour. Technol. 98, 648–653.

Samukawa, T., Kaieda, M., Matsumoto, T., Ban, K., Kondo, A., Shimada, Y., et al., 2000. Pretreatment of immobilized Candida antarctica lipase for biodiesel fuel production from plant oil. J. Biosci. Bioeng. 90, 180–183.

Sharma, Y.C., Singh, B., 2010. An ideal feedstock, kusum (Schleichera triguga) for preparation of biodiesel: optimization of parameters. Fuel 89, 1470–1474.

Sharma, Y.C., Singh, B., Korstad, J., 2010. High yield and conversion of biodiesel from a nonedible feedstock (Pongamia pinnata). J. Agric. Food. Chem 58, 242–247.

Shuit, S.H., Tan, K.T., Lee, K.T., Kamaruddin, A.H., 2009. Oil palm biomass as a sustainable energy source: a Malaysian case study. Energy 34, 1225–1235.

Sim, J.H., Kamaruddin, A.H., Bhatia, S., 2010a. The feasibility study of crude palm oil transesterification at 30°C operation. Bioresour. Technol. 101, 8948–8954.

Sim, J.H., Kamaruddin, A.H., Bhatia, S., 2010b. Biodiesel (FAME) productivity, catalytic efficiency and thermal stability of lipozyme TL im for crude palm oil transesterification with methanol. J. Am. Oil. Chem. Soc. 87, 1027–1034.

Singh, A.K., Fernando, S.D., Hernandez, R., 2007. Base-catalyzed fast transesterification of soybean oil using ultrasonication. Energy Fuels 21, 1161–1164.

Song, E.S., Lim, J.w., Lee, H.S., Lee, Y.W., 2008. Transesterification of RBD palm oil using supercritical methanol. J. Supercrit. Fluids 44, 356–363.

Soriano, J.N.U., Migo, V.P., Matsumura, M., 2006. Ozonized vegetable oil as pour point depressant for neat biodiesel. Fuel 85, 25–31.

Suarez, P.A.Z., Plentz Meneghetti, S.M., Meneghetti, M.R., Wolf, C.R., 2007. Transformation of triglycerides into fuels, polymers and chemicals: some applications of catalysis in oleochemistry. Quim. Nova 30, 667–676.

Suwannakarn, K., Lotero, E., Goodwin, J.G., Lu, C., 2008. Stability of sulfated zirconia and the nature of the catalytically active species in the transesterification of triglycerides. J. Catal. 255, 279–286.

Takagaki, A., Toda, M., Okamura, M., Kondo, J.N., Hayashi, S., Domen, K., et al., 2006. Esterification of higher fatty acids by a novel strong solid acid. Catal. Today 116, 157–161.

Talukder, M.M.R., Wu, J.C., Van Nguyen, T.B., Fen, N.M., Melissa, Y.L.S., 2009. Novozym 435 for production of biodiesel from unrefined palm oil: comparison of methanolysis methods. J. Mol. Catal. B Enzym. 60, 106–112.

Talukder, M.M.R., Wu, J.C., Fen, N.M., Melissa, Y.L.S., 2010. Two-step lipase catalysis for production of biodiesel. Biochem. Eng. J. 49, 207–212.

Tan, I.A.W., Ahmad, A.L., Hameed, B.H., 2008. Adsorption of basic dye using activated carbon prepared from oil palm shell: batch and fixed bed studies. Desalination 225, 13–28.

Tan, T., Lu, J., Nie, K., Deng, L., Wang, F., 2010. Biodiesel production with immobilized lipase: a review. Biotechnol. Adv. 28, 628–634.

Toh, T.S., Koh, P.M., 2008. Building plants for biodiesel and co-products. Int. News Fats Oils Rel. Mater. 19, 302–305.

Vyas, A.P., Verma, J.L., Subrahmanyam, N., 2010. A review on FAME production processes. Fuel 89, 1–9.

Wang, Y., Ou, S., Liu, P., Xue, F., Tang, S., 2006. Comparison of two different processes to synthesize biodiesel by waste cooking oil. J. Mol. Catal. A Chem. 252, 107–112.

Watanabe, Y., Shimada, Y., Sugihara, A., Noda, H., Fukuda, H., Tominaga, Y., 2000. Continuous production of biodiesel fuel from vegetable oil using immobilized Candida antarctica lipase. J. Am. Oil. Chem. Soc. 77, 355–360.

Watanabe, Y., Shimada, Y., Sugihara, A., Tominaga, Y., 2001. Enzymatic conversion of waste edible oil to biodiesel fuel in a fixed-bed bioreactor. J. Am. Oil. Chem. Soc. 78, 703–707.

Wei, Z., Xu, C., Li, B., 2009. Application of waste eggshell as low-cost solid catalyst for biodiesel production. Bioresour. Technol. 100, 2883–2885.

Xie, W., Huang, X., 2006. Synthesis of biodiesel from soybean oil using heterogeneous KF/ZnO catalyst. Catal. Lett 107, 53–59.

Yan, S., Lu, H., Liang, B., 2008. Supported CaO catalysts used in the transesterification of rapeseed oil for the purpose of biodiesel production. Energy Fuels 22, 646–651.

Yee, K.F., Kansedo, J., Lee, K.T., 2010. Biodiesel production from palm oil via heterogeneous transesterification: optimization study. Chem. Eng. Commun. 197, 1597–1611.

Ying, M., Chen, G., 2007. Study on the production of biodiesel by magnetic cell biocatalyst based on lipase-producing Bacillus subtilis. Appl. Biochem. Biotechnol. 137-140, 793-803.

Yoosuk, B., Udomsap, P., Puttasawat, B., Krasae, P., 2010. Modification of calcite by hydration-dehydration method for heterogeneous biodiesel production process: the effects of water on properties and activity. Chem. Eng. J. 162, 135–141.

Zabeti, M., Daud, W.M.A.W., Aroua, M.K., 2010. Biodiesel production using alumina-supported calcium oxide: an optimization study. Fuel Proc. Technol. 91, 243–248.

Zare, A., Hasaninejad, A., Beyzavi, M.H., Zare, A.R.M., Khalafi-Nezhad, A., Asadi, F., et al., 2009. KF/Al_2O_3 as a highly efficient, green, heterogeneous, and reusable catalytic system for the solvent-free synthesis of carboacyclic nucleosides via Michael addition reaction. Synth. Commun. 39, 139–157.

Zhang, X., Li, J., Chen, Y., Wang, J., Feng, L., Wang, X., et al., 2009. Heteropolyacid nanoreactor with double acid sites as a highly efficient and reusable catalyst for the transesterification of waste cooking oil. Energy Fuels 23, 4640–4646.

CHAPTER

16

Biodiesel Production from Waste Oils

Lien-Huong Huynh[1,2], *Novy S. Kasim*[1], *Yi-Hsu Ju*[1,*]

[1]Department of Chemical Engineering, National Taiwan University of Science and Technology, 43 sec. 4, Keelung Road, Taipei 10607, Taiwan.

[2]Department of Chemical Engineering, Can Tho University, 3-2 Street, Cantho City, Vietnam

*Corresponding author: E-mail: yhju@mail.ntust.edu.tw

1 INTRODUCTION

With the depletion of fossil-based diesel and rising environmental concerns, biodiesel (BD) has been receiving special attention recently. Between 2004 and 2006, BD consumption increased from roughly 25-250 million gallons (http://www.livescience.com/environment/081215-energy-debates-biodiesel.html). However, expensive feedstocks have become the major barrier for the commercialization of BD. It has been estimated that the cost of feedstock can account for 75-95% of the production cost of BD (Chhetri et al., 2008). It is of great importance to search for abundant and cheap feedstock to cut down the production cost of BD.

Waste oil is one of the potential candidates for the production of low-cost BD. Waste oil can be obtained from cooking oil, animal fat, yellow or brown grease, and sludge oil or soapstock from the refining process of vegetable oil. The most important source is the waste oil derived from households and industrial waste cooking oil. The use of waste cooking oil for BD production can help in solving the problem of waste oil disposal. The cost of waste oil mainly arises from the costs in collection, transportation, and pretreatment.

Large amounts of waste cooking oil are available throughout the world. In China the amount is 4-8 million tons (Mt)/year (Fu et al., 2009), whereas 1.4 Mt/year and 150,000 tons/year are available in the USA and Canada, respectively (Chhetri et al., 2008). In the EU countries, about 700,000 to 1 Mt/year of waste cooking oil is produced (Jacobson et al., 2008). The UK and Japan create around 200,000 tons and 400,000 to 1 Mt of waste cooking oil per year, respectively (Chhetri et al., 2008; Phan and Phan, 2008).

This chapter is intended to cover BD production from waste oil, including the physical and chemical properties of waste oil, methods for preparation of BD from waste oil, and the characterization, environmental, and economic aspects of converting waste oil to BD.

2 BIODIESEL

The energy sector accounts for 83% of global anthropogenic greenhouse-gas (GHG) emissions, in which CO_2, CH_4, and N_2O constitute approximately 94%, 5%, and 1%, respectively (UNFCCC, 2006). BD/FAME (Fatty Acid Methyl Ester) is a mixture of mono-alkyl esters of long-chain fatty acids derived from vegetable oils or animal fats, and is nontoxic, biodegradable, and sulfur- and aroma free. It works in any kind of diesel engine with few or no modifications and offers similar horsepower and torque when compared to petroleum-diesel. BD (B100) has fewer harmful emissions than petroleum diesel: total unburned hydrocarbons (−67%), CO (−48%), particulate matter (−47%), NO_x (+10%), sulfates (−100%), polycyclic aromatic hydrocarbon (PAHs, −80%), and nitrated PAHs (−90%). Edible oils-derived BD has recently become an economic competitor to edible oils. Waste oils have been considered one of the promising sources in BD production and subjected to intensive research in the BD synthesis.

The properties of the BD produced from waste oils depend on the properties and composition of the feedstock (Encinar et al., 2002; Kulkarni and Dalai, 2006; Lang et al., 2001). Details of BD properties produced from waste cooking oil are explained in Section 5.2.1.

3 WASTE OILS

3.1 Sources

The major sources of waste oil which have been used to study the production of BD were obtained from (Kemp, 2006): (i) waste cooking oil; (ii) animal fats; (iii) yellow grease; (iv) brown grease obtained from highly oxidized yellow grease or recovered waste grease from plumbing trap; and (v) waste sludge or soapstock from the vegetable oil refining process.

3.2 Chemical and Physical Properties

Acylglycerol (AG) constitution of used cooking oil or animal fat is similar to that of vegetable oil. However, differences in fatty acid profiles of animal fat and changes of oil composition occurring during frying have great influences on the quality of BD produced from these oils as compared to that from refined vegetable oil.

Animal fats are generally solid or highly viscous at ambient temperature. Higher contents of long-chain saturated fatty acids, nitrogen, or sulfur in animal fats cause BD produced from them to have poor cold-flow properties and SO_2 emission. The short supply of low-cost animal fats as raw materials also limits its industrial application.

Waste cooking oils have experienced particular physical and chemical changes in the frying process. Frying oil is repeatedly exposed to heat at 160-200 °C in open air and in the presence of light for long periods. Multiple subjection of oil to frying is the major causes for (i) an increase in free fatty acids (FFAs) content; (ii) a change in oil color to dark brown or red; (iii) an increase in viscosity and specific heat; (iv) a change in surface tension; and (v) an increase in the tendency of fat to form foam (Kulkarni and Dalai, 2006). According to some authors (Cvengros and Cvengrosová, 2004; Frankel, 1980; Kulkarni and Dalai, 2006;

Mittelbach and Enzelsberger, 1999; Nawar, 1984), three main reactions occur during frying: thermolytic or cracking, hydrolytic, and oxidative reactions.

Under high temperature and low oxygen, glycerides and saturated fatty acids are cracked into smaller molecules such as *n*-alkanes, alkenes, lower fatty acids, etc. For unsaturated fatty acids, most products are dimeric compounds such as dehydrodimers, saturated dimmers, polycyclic compounds as well as polymeric acids (Kulkarni and Dalai, 2006).

Steam produced during heating will enter the fat and fried food. The presence of water gives rise to the hydrolysis of triglycerides to form FFAs, glycerols, mono-, and di-glycerides.

Autoxidation of unsaturated fat takes place as soon as oxygen from air is dissolved into the fats. The susceptibility of unsaturated AGs to autoxidation depends on the availability of their allylic hydrogens for reactions with peroxy radicals (Frankel, 1980). Hydroperoxides and isomeric hydroperoxides formed as primary products during the reaction will further decompose via the hemolytic of oxygen-oxygen bond to yield alkoxy and hydroxyl radicals (Frankel, 1980). Scission of alkoxy at C–C bond or C–O bond leads to various products such as aldehydes, ketones, hydrocarbons, lactones, alcohols, acids, and esters (Cvengros and Cvengrosová, 2004; Kulkarni and Dalai, 2006). Excess oxygen can cause the formation of dimeric and oligomeric compounds (Nawar, 1984).

4 TECHNICAL ASPECT OF BD PRODUCTION FROM WASTE OILS

4.1 Alkali-Catalyzed BD Production

In oil with high FFA content, alkali catalyst will react with FFAs and form soap, and thus reduce the BD yield and add difficulties to downstream product separation. High water content will lead to the hydrolysis of AGs to FFAs, followed by saponification of the FFAs and thus reduce the concentration of the alkali (Komers et al., 2001).

For low FFA waste oils, alkali catalyst still exhibits good performance and can attain high BD yield. Unfortunately, waste cooking oil usually has FFA content of not less than 5 wt%. Treatment of waste oils to reduce FFAs and water content can significantly reduce the amount of alkali required to obtain high BD yield. Table 1 summarizes the homo- and heterogeneous alkali-catalyzed process reported in literatures.

Both homo- and heterogeneous alkaline catalysts can give high BD yield. The quantity required by heterogeneous alkali is greater than that required by the homogeneous one (see Table 1). The heterogeneous catalyst does have the advantage of catalyst reusability over the homogeneous catalyst. Both catalysts still require additional cost either in removing homogeneous catalyst, or in regenerating the heterogeneous catalyst to recover its activity.

The yield of BD depends on factors such as alkali type/strength, molar ratio of alcohol to oil, temperature, and mixing strength. KOH and NaOH are commonly used in transesterification; both can be incorporated to form heterogeneous catalyst. The other heterogeneous alkali that has recently gained researchers' attention is CaO. CaO can be obtained easily from egg shells and the shells of marine animals (Viriya-empikul et al., 2010; Wei et al., 2009). It is a common practice to use more alcohol than the theoretical molar ratio of alcohol to oil of 3 in

TABLE 1 Summary of Recent Alkali-Catalyzed BD Production from Waste Oils

Ref.	Oil	Optimum Conditions	BD Yield (%)
Homogeneous alkali catalyst			
Issariyakul et al. (2008)	Canola oil/used cooking oil (60/40)	KOH (1 wt%); $T = 50$ °C for 2 h; MeOH/oil molar ratio = 6	98
Hossain and Boyce (2009)	Waste sunflower cooking oil	KOH (1 wt%); $T = 40$ °C for 3 h; MeOH/oil volume ratio = 6	97-98
Felizardo et al. (2006)	Waste frying oils	NaOH (0.6 wt%); $T = 65$ °C for 1 h; MeOH/oil molar ratio = 4.8	90
Dias et al. (2008)	Waste frying oils	NaOH or CH_3ONa (0.4 wt%); $T = 60$ °C for 1 h; MeOH/oil molar ratio = 6	92
Tomasevic and Siler-Marinkovic (2003)	Used sunflower frying oil	KOH (1 wt%); $T = 25$ °C for 30 min; MeOH/oil molar ratio = 6	~100
Muniyappa et al. (1996)	Beef tallow	NaOH (0.42 wt%); $T = 68$-70 °C for 1.5 h; MeOH/fat molar ratio = 10	98
Sakai et al. (2009)	Waste cooking oil	1. KOH (0.6 wt%); 2. CaO (0.02 wt%); $T = 60$ °C; MeOH/oil molar ratio = 4-5	89-92
Bautista et al. (2009)	Used frying oil	KOH (2 wt%); $T = 60$ °C for 1 h; MeOH/oil molar ratio = 6	97.69
Heterogeneous alkali catalyst			
Brito et al. (2007)	Fried oil	KOH/Zeolite Y756 (2.5 wt%); $T = 65$ °C for 4.5 h; MeOH/oil molar ratio = 35; Reusability cycle = 2	96.58
McNeff et al. (2008)	1. Used soybean oil	1. Base modified zirconia; $T = 465$-466 °C for 30 min	1. 92.6
	2. Acidulated soapstock	2 & 3. Base modified titania; $T = 339$-348 °C for 30 min	2. 90.2
	3. Yellow grease	All reactions were carried out in a fixed-bed reactor in continuous mode	3. 92.6
Brito et al. (2009)	Waste frying oil	Mg-Al hydrotalcite, calcined at 450 °C for 20 h (6 wt%); $T = 140$ °C for 6 h; MeOH/oil molar ratio = 24	~100
Wen et al. (2010)	Waste cooking oil	TiO_2/MgO, calcined at 650 °C, Mg/Ti molar ratio = 1 (10 wt%); $T = 150$ °C for 6 h; MeOH/oil molar ratio = 50; Reusability cycle = 5	92.3

order to favor the formation of product. The use of temperature approaching the boiling point of alcohol is beneficial to the increasing of the reaction rate.

Yin et al. (2008) employed subcritical methodology for BD production. Without catalyst, the BD yield obtained at 250 °C, a methanol-to-oil molar ratio of 24, 300 rpm shaking and 20 min reaction time was very low (less than 5%). By 0.1 wt% KOH addition, the yield dramatically increased to 98% for reaction carried out at 160 °C while all other conditions remained the same.

4.2 Acid-Catalyzed BD Production

By using a heterogeneous catalyst, the need for BD washing to remove catalyst after the reaction is eliminated and the catalyst can be readily recycled. However, its activity usually is lower than that of the homogeneous one (Lou et al., 2008).

Acid can catalyze the esterification of FFAs with methanol effectively, but not the transesterification of AGs. The reaction rate of acid-catalyzed transesterification is only about 1/4000 of that in an alkali-catalyzed one (Lam et al., 2010). Haas (2005) identified the problem of incomplete reaction when converting acid oil (59.3% FFAs and 33.4% AGs) into BD catalyzed by H_2SO_4, which is caused by the slow reaction of acid-catalyzed transesterification. Boucher et al. (2008) found that after recovery and reuse, H_2SO_4 activity reduced to half of its initial value, while HCl could maintain its activity when more water is produced in the reaction system.

A superacid catalyst was found to have the remarkable ability of promoting the reaction rate of the transesterification reaction to a degree that is comparable to that of alkaline catalyst (see Table 2), owing to its high strength of acid sites (Furuta et al., 2004; Lam et al., 2009).

4.3 Lipase-Catalyzed BD Production

The enzyme utilized in the BD synthesis is lipase. Lipase has a unique feature of acting at the interface between aqueous and organic phases. Its activity increases with the increase of interfacial area and with the addition of water into the reaction system. However, in the presence of excess water, hydrolysis instead of alcoholysis is favored. Lipase is suitable for converting waste oils with high FFAs and high water contents into BD with a high yield.

As summarized in Table 3, *Pseudomonas* genus has been recently discovered as an important lipase-producing bacteria. Some important lipase-producing fungi or yeast are *Candida rugosa/cylindracea*, *C. antarctica*, *Thermomyces lipase*, *Rhizomucor miehei*, and *Rhizopus oryzae*. Free lipase produced by both microorganisms shows comparable activity in catalyzing the production of BD. It is generally difficult to recover and reuse free lipase. Its stability and activity must be maintained under certain conditions. One way to overcome these drawbacks is by using immobilized lipase.

The lipase-catalyzed reaction results in purer BD than those obtained by chemical interesterification as glycerol, the major byproduct, can be recovered easily and efficiently (Shimada et al., 2002). Certain biocatalysts, such as Lipozyme TL IM, are intended for use in the interesterification of waste oils and frying fats. Novozym-435 can also perform well in the interesterification of the waste oils (see Table 3).

Immobilized lipases have been inactivated by short-chain linear alcohols such as methanol and ethanol used in BD production (Chen and Wu, 2003). Immobilized lipase is inactivated by contact with methanol because the hydrophilic methanol will strip out essential water layer around the lipase, which is required to maintain lipase activity (Guo and Sun, 2004). The degree of inactivation of the lipase was found to be inversely proportional to the number of carbon atoms in linear alkyl alcohols (Chen and Wu, 2003). Treating with alcohol with three or more carbon atoms, for example, 2-butanol or *tert*-butanol, can regenerate the deactivated immobilized lipase.

TABLE 2 Summary of Recent Acid-Catalyzed BD Production from Waste Oils

Ref.	Oil	Optimum Conditions	BD Yield (%)
Homogeneous acid catalyst			
Haas (2005)	Soybean soapstock-derived acid oil	Pretreatment: complete hydrolysis of lipids and soapstock, followed by acidulation to produce high acid oil (95 wt% FFAs); FFAs/MeOH/H_2SO_4 molar ratio = 1:1.8:0.17; T = 65 °C for 14 h	90-95(purity)
Zheng et al. (2006)	Waste frying oil	H_2SO_4 (3.8 mol%); T = 70 °C for 4 h; MeOH/oil molar ratio = 245; FFAs immediately reacted after a few minutes.	99
Montefrio et al. (2010)	Grease interceptor (20% FFAs)	(1) H_2SO_4 (10 wt%) (2) $Fe_2(SO_4)_3$ (10 wt%); T = 50 °C for 24 h; MeOH/oil molar ratio = (1) 20 & (2) >26	1. 97.04 2. ~80
Santos et al. (2009)	Used soybean oil	H_2SO_4 (3.5 wt%); T = 28 °C for 1 h; MeOH/oil molar ratio = 9; AGs hydrolysis and saponification were performed prior to esterification. Ultrasound was employed to replace mechanical stirring.	99.8
Ghassan et al. (2004)	Waste animal fat	2.25 M H_2SO_4 (10 wt%); T = 50 °C for 2 h; EtOH/oil mass ratio = 2.5; EtOH is employed to lower viscosity of final FAME.	82
Heterogeneous acid catalyst			
Jacobson et al. (2008)	Waste cooking oil	Zn/Si (3 wt%); T = 200 °C for 10 h; MeOH/oil molar ratio = 18 Zn/Si is a zinc stearate immobilized on silica gel. Reusability cycle: 4	98
Lou et al. (2008)	Waste cooking oil	Starch derived catalyst (10 wt%); T = 80 °C for 8 h; MeOH/oil molar ratio = 30; Reusability cycle: 50	92-93
Gan et al. (2009)	Waste cooking oil	$Fe_2(SO_4)_3$/C catalyst (3.5 wt%); T = 95 °C for 4 h; MeOH/Oil molar ratio = 18; Reusability cycle = 4	96-98
Guan et al. (2009)	Waste cooking oil	K_3PO_4 (4 wt%) + co-solvent THF; T = 60 °C for 2 h; MeOH/oil molar ratio = 6; Reusability cycle: 2	~95
Lam et al. (2009)	Waste cooking oil	H_2SO_4/SnO_2-SiO_2 catalyst (3 wt%); T = 150 °C for 3 h; MeOH/oil molar ratio = 15	92.3
Gan et al. (2010)	Waste cooking oil	$Fe_2(SO_4)_3$ catalyst (2 wt%); T = 60 °C for 1 h; MeOH/oil molar ratio = 15	59.15
Fu et al. (2009)	Waste cooking oil	H_2SO_4/ZrO_2 catalyst (3 wt%); T = 120 °C for 4 h; MeOH/oil molar ratio = 9	93.6
Feng et al. (2010)	Waste frying oil	NKC-9 (cation-exchange resin) catalyst (18 wt%); T = 66 °C for 3 h; MeOH/oil molar ratio = 3; Reusability cycle = 10	90

TABLE 3 Summary of Recent Lipase-Catalyzed Production of BD from Waste Oils

Ref.	Oil	Optimum Conditions	BD Yield (%)	Notes
Chen et al. (2009)	Waste cooking oil	*Candida* lipase (25 wt%); $T = 40$ °C; MeOH/oil molar ratio = 1; Life time = 100 h	91.08 (purity)	Three-series fixed-bed reactors with immobilized lipase were performed.
Wang et al. (2008)	Waste oil	Lipozyme TL-IM (18 U/g oil)	93.7	MeOH was added at 0, 2, and 4 h, respectively, to the reaction system.
		$T = 35$ °C for 12 h; MeOH/oil molar ratio = 0.5, 0.5, 2.8; *t*-butanol/oil = 0.58 (v/v)		
Li et al. (2009)	Waste oil	*Penicillium expansum* (84 U/g oil); $T = 35$ °C for 7 h; *t*-amyl alcohol/oil = 20 wt%; Reusability cycle = 10	92.8	Blue silica gel (0.48 g/g oil) is the best adsorbent to control water content during the reaction.
Paola et al. (2009)	Low-quality olive husk oil	Lipozyme *Rhizomucor miehei* (8 wt%); $T = 37$ °C for 24 h; EtOH/oil molar ratio = 2	~90	
Watanabe et al. (2001)	Waste edible oil	Novozym-435 (4 wt%); $T = 30$ °C for 74 h; MeOH/oil molar ratio = 1	92.7 (conversion)	Three-step batch methanolysis of waste or vegetable oil was conducted.
Maceiras et al. (2009)	Waste frying oil	Novozym-435 (10 wt%); $T = 50$ °C for 4 h; MeOH/oil molar ratio = 25; Reusability cycle = 3	89.1	Acid value of waste frying oil = 1.35 mg KOH/g oil.
Halim et al. (2009)	Waste palm oil	Novozym-435 (~3 wt%); $T = 40$ °C for 3 h; *t*-butanol/oil = 1 (v/v); MeOH/oil molar ratio = 4	79.1	Two-series packed-bed reactor was employed.
Salis et al. (2008)	Waste frying oil	*Pseudomonas fluorescens* (6.25 wt %); $T = 40$ °C for 4 h; MeOH/oil = 8 (w/w)	27 mol%	Assume the substance is oleic acid, then the yield is *ca.* 28.3%
Yagiz et al. (2007)	Waste cooking oil	Lipozyme-TL IM (4 wt%); $T = 24$ °C for MeOH/oil molar ratio = 4; Reusability cycle = 3	92.8	Hydrocalcite was the best support material for the lipase.
Dizge et al. (2009)	Waste cooking oil	Lipozyme-TL (~1 wt%); $T = 65$ °C for 24 h; MeOH/oil molar ratio = 6; Reusability cycle = 10	90.2	Powder microporous polymeric matrix gave best performance as support material.

4.4 Two-Step BD Production

Two-step BD synthesis includes two-step alkali-catalyzed saponification-transesterification, two-step acid-catalyzed esterification-transesterification, and acid-catalyzed esterification followed by alkali-catalyzed transesterification (see Table 4). In two-step alkali-catalyzed saponification-transesterification, the alkali is fed into high FFAs-contained waste oil to remove FFAs by saponification (Predojević, 2008; Thanh et al., 2010).

TABLE 4 Summary of Recent Two-Step BD Production from Waste Oils

Ref.	Oil	Optimum Conditions of 1st Step	Optimum Conditions of 2nd Step	BDl yield (%)
Patil et al. (2010)	Waste cooking oil	$Fe_2(SO_4)_3$ (2 wt%); $T = 100$ °C for 2 h; MeOH/FFAs molar ratio = 9	KOH (~0.5 wt%); $T = 100$ °C for 1 h; MeOH/oil molar ratio = ~7	96
Liu et al. (2010)	Waste cooking oil	H_2SO_4 (3 wt%); $T = 60$-66 °C (via radio frequency) for 8 min; MeOH/FFAs molar ratio = 7	NaOH (0.91 wt%); $T = 60$-66 °C (via radio frequency) for 5 min; MeOH/oil molar ratio = 14.2	98.8
Hayyan et al. (2010)	Sludge palm oil	Toluene-4-sulfonic monohydrate acid (0.75 wt%); T = 60 °C for 1 h; MeOH/oil molar ratio = 10	KOH (1 wt%); $T = 60$ °C for 1 h; MeOH/oil molar ratio = 10	76.62
Wang et al. (2007)	Waste cooking oil	$Fe_2(SO_4)_3$ (2 wt%); $T = 95$ °C for 4 h MeOH/oil molar ratio = 10	KOH (1 wt%); $T = 65$ °C for 1 h; MeOH/oil molar ratio = 6	97.02
Berrios et al. (2010)	Spanish used frying oil	H_2SO_4 (5 wt%); $T = 60$ °C for 2 h; MeOH/oleic acid molar ratio = 120	KOH (1 wt%); $T = 60$ °C for 30 min MeOH/oil molar ratio = 6	88.3
Dias et al. (2009)	Acid waste lard	H_2SO_4 (2 wt%); $T = 65$ °C for 5 h; MeOH/lard molar ratio = 6	NaOH (1 wt%); $T = 65$ °C for 30 min; MeOH/lard molar ratio = 6	64.4
Thanh et al. (2010)	Waste cooking oil	KOH (0.7 wt%); $T = 30$-32 °C for 25 min; MeOH/oil molar ratio = 2.5	KOH (0.3 wt%); $T = 27$-29 °C for 20 min; MeOH/oil molar ratio = 1.5	93.8
		Reactions were enhanced by using ultrasonic irradiation		
García et al. (2010)	Waste vegetable oil	H_2SO_4 (1 wt%); $T = 60$ °C; MeOH/oil molar ratio = 6	NaOH (1 wt%); $T = 60$ °C; MeOH/oil molar ratio = 6	89-94
Wang et al. (2010)	Waste cooking oil	Polyferric sulfate (3 wt%); $T = 67$ for 4 h; MeOH/oil molar ratio = 30	KOH (1.2 wt%); $T = 40$ °C for 1 h; MeOH/oil molar ratio = 6	95.08
Sun et al. (2010)	Acid oil	H_2SO_4 (3 wt%); $T = 100$ °C for 7 min; MeOH/FFAs molar ratio = 30	H_2SO_4 (3 wt%); $T = 120$ °C for 5 min; MeOH/oil molar ratio = 20	99.5
Math and Irfan (2007)	Restaurant waste oil	H_2SO_4 (1 vol.%); $T = 35$ °C for 1 h; MeOH/oil molar ratio = 9	NaOH (0.3 wt%); $T = 55$ °C for 90 min; MeOH/oil molar ratio = 9	85.5
Predojević (2008)	Waste sunflower oil	KOH (1 wt%); $T = 30$ °C for 30 min; MeOH/oil molar ratio = 3	KOH (1 wt%); $T = 60$ °C for 30 min; MeOH/oil molar ratio = 3	94-96
The author performed silica gel chromatography or phosphoric acid purification to obtain 97-98% biodiesel purity.				
Meng et al. (2008)	Waste cooking oil	H_2SO_4; Optimum condition: NA	NaOH (1 wt%); $T = 50$ °C for 90 min; MeOH/oil molar ratio = 6	89.8
Kouzu et al. (2008)	Waste cooking oil	SO_4^{2-}/cation exchange resin; $T =$ NA for 4 h; MeOH/oil molar ratio = 5	CaO, calcined at 900 °C (1 wt%); $T = 65$ °C for 2 h; MeOH/oil molar ratio = 12	>99

For two-step acid-catalyzed esterification-transesterification, as mentioned in Section 4.2, only superacid catalyst can efficiently perform these two consecutive reactions (Sun et al., 2010). Shu et al. (2010) reported a heterogeneous superacid catalyst from sulfonated-carbonized-vegetable oil asphalt and succeeded in achieving *ca.* 93% total conversion.

For the acid-catalyzed esterification followed by alkali-catalyzed transesterification, FFAs are first esterified to become BD and after removing the acid catalyst, the AGs are transesterified. This process is one of the best alternatives in BD synthesis. However, tedious washing steps are required and the final product may still contain small quantities of the catalysts. A combination of heterogeneous acid-alkaline catalyst is a well-advised option. However, only a few studies related to this issue are available to shed light on the feasibility of this option (see Table 4). Kouzu et al. (2008) applied the combination of heterogeneous acid-alkali catalysts (SO_4^{2-}/resin and CaO) to convert waste cooking oil to FAMEs and was able to obtain a BD yield higher than 99%.

4.5 Catalyst-Free (Supercritical) Production of BD

Due to the poor miscibility between oil and methanol, a catalyst is required in order to achieve a high conversion in a reasonable time. If good contact between oil and methanol can be provided efficiently, reaction can proceed rapidly without the need of catalyst.

In the past decade, supercritical methodology has been touted as a promising method for producing BD. Producing BD under supercritical conditions does offer some advantages over the conventional BD production process, especially when waste cooking oil with high FFAs and high water content is used as the feedstock. Recently in Japan, a continuous pilot-scale production of FAMEs from superheated waste cooking oil and superheated methanol with no catalyst has been demonstrated by consuming 500 L/day waste cooking oil to produce 425 L/day FAMEs (Japan Chemical Web, 2008).

Under supercritical conditions, methanol can react with neutral lipid and reach very high conversion in relatively short time without the need of catalyst. Parameters such as temperature, FFAs and water content, molar ratio of methanol to oil, and cosolvent affect the yield in supercritical methanol production of BD. The critical temperature and critical pressure of methanol is 239 °C and 8.09 MPa, respectively. Note that supercritical condition for methanol is different from that of oil. One indicator to ensure the supercritical conditions for the whole system is a high BD conversion greater than 90%. Demirbas (2009) carried out the supercritical methanol production of BD using waste cooking oil at 247, 267, and 287 °C, and found that at 287 °C a yield close to 99% can be achieved in 20 min.

Since waste cooking oil usually contains significant amounts of water and FFAs, the transformation of waste cooking oil under supercritical condition offers great advantage by eliminating the need of pretreatment for removing FFAs and the need of catalyst. Tan et al. (2010a) compared the effects of FFAs and water on BD production using supercritical methanol method and a heterogeneous base (Montmorillonite KSF)-catalyzed reactions. Both methods show comparable BD yield for FFAs content up to 30%. However, water has different effect on BD yield for the two methods. Supercritical methanol method can maintain high BD yield (>80%) for water content as high as 20%. For the heterogeneous base-catalyzed reaction, the BD yield dropped from 80% to 13% when water content was increased from 5% to 15%.

Kusdiana and Saka (2004a) studied the effects of FFAs and water content not only on supercritical but also on acid- and base-catalyzed reactions of waste palm oil whose contents of FFAs and water are >20 wt% and >61 wt%, respectively. Unlike acid- and base-catalyzed reactions, the supercritical methanol reaction is much more tolerant of high FFAs and water content in oil; they can still obtain high BD yield (95.8 wt%).

Higher molar ratio of methanol to waste oil results in a higher yield in shorter time. Most literatures reported an optimum methanol-to-oil molar ratio of 24 when cosolvent was used and 42 if cosolvent was not added. A higher molar ratio of methanol to oil results in higher reaction pressure, which imposes stringent requirements on the reaction vessel. Cosolvent is usually employed to reduce pressure in the supercritical system. Propane, heptane, and CO_2 were usually chosen as the cosolvents (Tan et al., 2010a; van Kasteren and Nisworo, 2007).

Kusdiana and Saka (2004b) developed a two-step reaction in BD synthesis at very mild reaction conditions. In this method, TAGs were hydrolyzed by subcritical aqueous processing to FFAs and glycerol. After a self-separation of FFAs and glycerol, methanol was added to FFAs to perform esterification under supercritical condition. Because the reaction conditions were relatively mild (270 °C and 7 MPa, and 270 °C and 17 MPa for hydrolysis and esterification, respectively), undesirable change of unsaturated fatty acids was barely induced. This two-step method is more suitable for practical application, compared with the one-step method.

Table 5 summarizes the state of art of BD production supercritical methanol and supercritical methyl acetate. In order to replace glycerol by more valuable triacetin, methyl acetate was chosen as the solvent (Saka and Isayama, 2009; Tan et al., 2010b). Saka and Isayama (2009) investigated various fuel characterizations of FAMEs and triacetin mixture and reported that triacetin can be used as fuel additive to improve pour point, cold, and viscosity properties of FAMEs. After the reaction, there is no need to separate triacetin from FAMEs (BD) and quality of the BD produced is enhanced.

TABLE 5 Summary of Supercritical BD Production from Waste Oils

Ref.	Oil	Co-Solvent	MeOH/Oil (mol/mol)	Process Mode	°C (bars)	Time (min)	Product Yield (%)
Patil et al. (2010)	Waste cooking oil	NA	40	Batch	300 (100)	20	~80
Saka and Isayama (2009)[a]	a. Rapeseed oil b. Oleic acid	NA	a. 42 b. 14	Batch	350 (200)	a. 45 b. 20	a. 97 b. 91
Demirbas (2009)	Waste sunflower seed	NA	41	Batch	287 (NA)	~15	~98
van Kasteren and Nisworo (2007)	Waste cooking oil	Propane	24	Continuous	280 (128)	17	~100
Kusdiana and Saka (2004a)	a. Rapeseed b. Palm c. Used frying d. Waste palm	NA	42	Batch	350 (430)	4	a. 98.5 b. 98.9 c. 96.9 d. 95.8

[a]*Methyl acetate replaced methanol in supercritical reaction.*

The results reported seem to indicate that for supercritical methanol production of BD, the optimum temperature is about 350 °C without cosolvent and 280 °C with cosolvent; the optimum methanol-to-oil molar ratio 42 without cosolvent and 24 with cosolvent. The yield that can be achieved under these optimum conditions is more than 95%. Supercritical methyl acetate can be considered as an alternative to methanol to produce high-quality BD.

5 FEASIBILITY AND ECONOMIC ANALYSES ON BD PRODUCTION FROM WASTE OILS

5.1 Downstream Processing

After the reaction, the crude BD obtained is a mixture of FAMEs, excess methanol, and other impurities such as glycerol, unconverted oil, remaining catalyst, and soap formed during the reaction. Figure 1 presents generally principal steps for the separation and purification of FAMEs.

5.1.1 Separation

After a completed reaction, crude BD is left for a minimum of 8-24 h to ensure that all glycerol has settled. Phase separation occurs instantly; however, the impurities in the feedstock may lead to the formation of emulsion and slow down the settling of glycerol. Salting out and centrifugation can help breaking the emulsion and hasten the separation (Canoira et al., 2008; Enweremadu and Mbarawa, 2009).

Separation is carried out mostly in a separator funnel or in a decantation funnel (Felizardo et al., 2006; Phan and Phan, 2008). Recently, Saifuddin and Chua (2004) have employed microwave irradiation to speed up phase separation to several minutes. Additionally, sedimentation (Azocar et al., 2007; Encinar et al., 2007) and centrifugation (Wang et al., 2007) can also be utilized for removing glycerol. In some special cases when the settle of glycerol by gravity did not occur, pure glycerol was added for accelerating and completing the removal of glycerol (Issariyakul et al., 2007).

5.1.2 Alcohol Recovery

In order to obtain high BD yield, excess alcohol is required. However, the presence of large amount of alcohol would cause difficulty in phase separation. As soon as BD is separated from glycerol, alcohol is removed before being subjected to washing.

The simplest way to remove alcohol is evaporation under atmospheric pressure (Leung and Guo, 2006), or under vacuum (Issariyakul et al., 2007; Predojević, 2008; Wang et al., 2007). Another method for alcohol recovery is distillation.

5.1.3 Washing of BD

The purpose washing is to remove catalyst, soap, glycerol, and other impurities. There are three common methods for washing of BD (Enweremadu and Mbarawa, 2009): (i) stir or mix washing; (ii) mist washing, (iii) bubble washing. Stir/mix washing is the quick, effective, and most commonly used method (Al-Widyan and Al-Shyoukh, 2002; Canoira et al., 2008; Chhetri et al., 2008; Encinar et al., 2007; Leung and Guo, 2006; Meng et al., 2008; Phan and Phan, 2008;

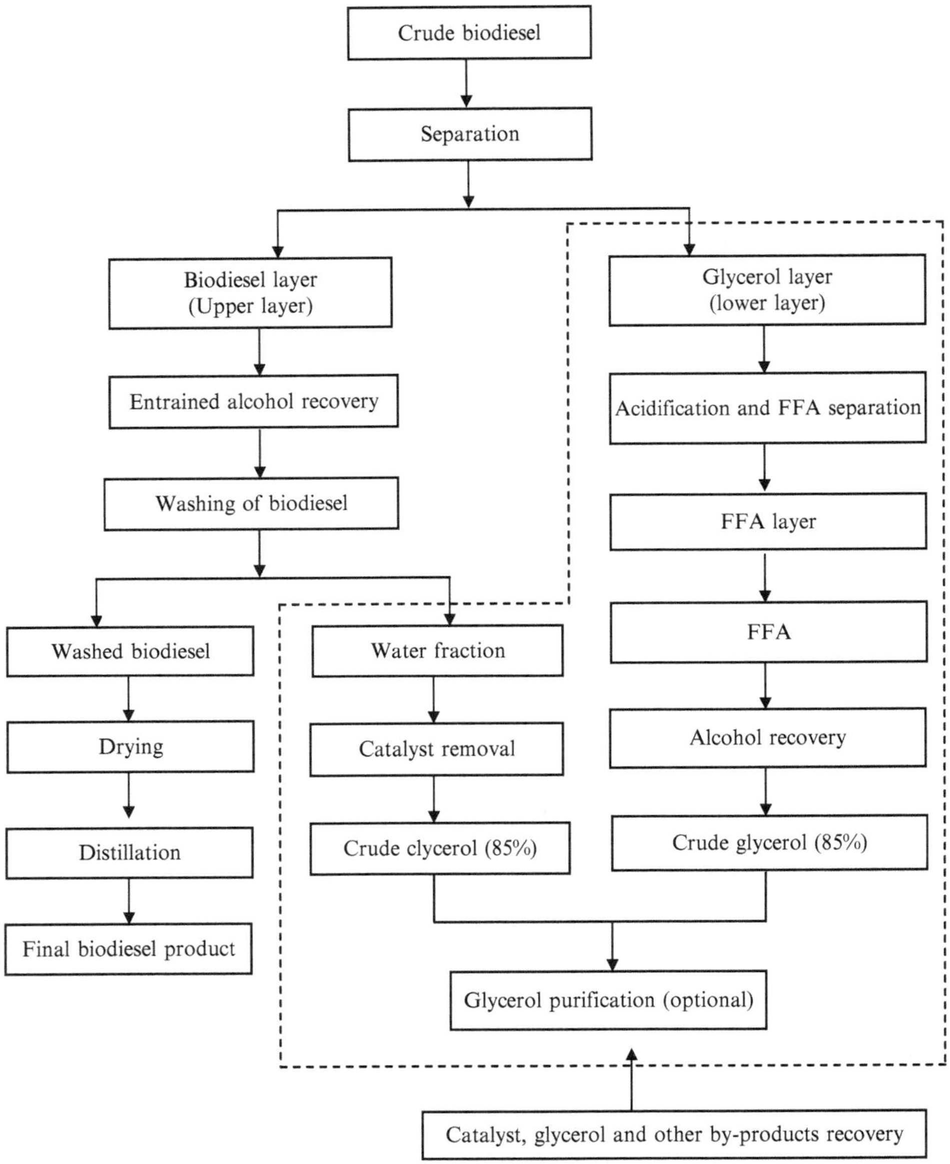

FIGURE 1 Flowchart of downstream processing of biodiesel from waste cooking oil.

Reefat et al., 2008; Sabudak and Yildiz, 2010; Tomasevic and Siler-Marinkovic, 2003; Wang et al., 2006; Zullaikah et al., 2005). BD with equal amount of water is mixed and stirred until the mixture is homogeneous. The mixture is left to settle and water is then drained to obtain clean BD. Use of hot water at 50-60 °C can help improve the quality and speed up the washing. On the other hand, mist washing and bubble washing were developed to enhance the contact between water and impurities and hence improve the quality of BD. In mist washing, water is finely sprayed on to the surface of BD, water will diffuse through BD layer and pick up all soluble impurities (Chhetri et al., 2008), whereas in bubble washing, very fine air bubbles are generated and travel from the bottom through the fuel layer, taking away all soluble impurities. On reaching the surface, the bubbles collapse and are sent back to the bottom to carry out another washing (Demirbas, 2005; Utlu and Koçak, 2008). Either individuality or combination of different washing methods can be applied to obtain clean BD. For example, Lapuerta et al. (2008) used a two-step washing: stir/mix washing followed by bubble method to increase pellucidity of BD.

Other methods can also be used to obtain a high-quality fuel. Felizardo et al. (2006) proposed an acidulated wash with water, 0.5% HCl solution, and again with water, whereas multiple washes by 50% (v/v) of a 0.2% HCl solution accompanied with 50% (v/v) of distilled water until pH of the washing water was neutral was employed in other studies (Dias et al., 2008; Lapuerta et al., 2008). In addition, Canoira et al. (2008) and Sabudak and Yildiz (2010) either used Magnesol (magnesium silicates) to absorb impurities or purified crude BD by ion-exchange resin. Predojević, 2008 suggested a silica-bed purification procedure or an acidulated wash with 5% phosphoric acid solution to obtain a higher yield compared to the common wash with distilled water at 50 °C. Demirbas (2005) recommended a two-step acidulated wash for the washing of esters. A solution of tannic acid (1 g of tannic acid/L of water) with volume percentage of 28% of oil was added and mildly agitated; and then air was carefully bubbled from the bottom of the vessel to perform a combination of bubble washing and stir washing. The process was repeated until the ester layer was clean.

5.1.4 *Drying of BD*

After washing, BD may still contain trace water. The presence of water can reduce heat of combustion of the bulk fuel or cause corrosion of vital fuel system components, gelation of residual fuel, etc. It is essential to reduce the amount of water in BD to less than 500 ppm final BD (the ASTM D6751 and EN14214 standards). When BD is clean and dry it is clear, translucent, and cloudless (Enweremadu and Mbarawa, 2009).

The oldest method for drying is by settling. For the home producer, the fuel can be dried by blowing bubbles of air from the tank bottom. Two methods that are currently commonly used are by heating and usage of chemicals (Enweremadu and Mbarawa, 2009). For the heating method, washed BD was dried by agitating at 110-120 °C for about 20 min (Encinar et al., 2007; Sabudak and Yildiz, 2010) or at a milder temperature (90-110 °C) under vacuum (5-25 mmHg) for 20 min to 1 h (Al-Widyan and Al-Shyoukh, 2002; Wang et al., 2006; Wang et al., 2007; Zullaikah et al., 2005). Heating of BD also help drive off any remaining alcohol. Drying of the fuel can also be achieved by using drying agents such as anhydrous sodium sulfate and magnesium sulfate (Canakci and Gerpen, 2001a; Dias et al., 2008; Felizardo et al., 2006; Lertsathapornsuk et al., 2008; Leung and Guo, 2006; Meng et al., 2008; Phan and Phan, 2008; Predojević, 2008; Saifuddin and Chua, 2004; Tomasevic and Siler-Marinkovic, 2003) or 4-Å molecular sieves (Canoira et al., 2008).

5.1.5 *Distillation of BD*

Distillation resulted in the purest BD available. Different fraction of distillates are collected and analyzed. Usually, fractions of FAME were collected at 90-240 °C under atmospheric or vacuum conditions.

Wang et al. (2006) carried out the distillation of FAME under vacuum (40 ± 5 mmHg). The first fraction was collected at 180 °C and the distillation was terminated when no more FAME appeared at 240 °C, 93% of the BD recovery was obtained. In the study of Zullaikah et al. (2005), distillation of FAME was performed under a lower vacuum (5 ± 1 mm Hg). Three fractions were consequently collected at 160 °C (1 h), 200 °C (30 min), and 220 °C (20 min).

5.1.6 *Catalyst, Glycerol, and Other by-Products Recovery*

The glycerol fraction obtained after the separation and washing steps contains remaining catalyst, soap, glycerol, water, and trace amount of alcohol. Salts formed after the neutralization of catalyst were removed by gravity separation (Zhang et al., 2003a). When solid catalysts were used, the catalyst was recovered by filtration and purified by solvent washing steps (Jacobson et al., 2008) or by the ashing process (Wang et al., 2006).

After removing the catalyst, mixture of glycerol, water, alcohol, and fatty acids derived from soap would pass through the distillation chamber or tower to remove any leftover water and alcohol (Encinar et al., 2007; Zhang et al., 2003a). A high-grade glycerol (purity *ca.* 92%) could be obtained when fatty acids derived from soap and entrained fatty acid esters were eliminated.

5.2 Characterization, Environmental, and Economic Aspects of BD Derived from Waste Cooking Oil

5.2.1 *Characterization of BD from Waste Cooking Oil*

Since different feedstock will greatly affect the properties of BD produced, tests of the fuel properties are required. Important parameters used to characterize BD are: (i) BD performance: cetane number, flash and combustion points, heating value, and iodine value. (ii) BD flow and cold weather properties: density, kinematic viscosity, cloud point, pour point, and cold filter plugging point. (iii) Purity of BD: Conradson carbon numbers, sulfur, water, and alcohol contents as well as amount of unreacted oil.

Table 6 lists the properties of BD derived from certain waste cooking oils and comparison of waste cooking oil-derived BD with BD standards as well as with petrodiesel and commercial BD.

As shown in Table 6, BDs from different sources show little differences in their chemical and physical properties. Almost all BDs derived from waste cooking oil meet either ASTM or European BD standards, which can be generally characterized as a diesel substitute with: (i) A high flash point: this is one of the BD advantages over diesel fuel. (ii) Similar density to diesel fuel but higher viscosity. (iii) Higher cetane number but lower heating value than diesel fuel. Its heating value is approximately 10% less than that of diesel fuel (Chhetri et al., 2008; Enweremadu and Mbarawa, 2009). (iv) Higher cloud and pour points than diesel fuel. Blending with petroleum diesel or treating with commercial petrodiesel cold flow improver additives can improve its cold weather properties. (v) High carbon residue, approximately 20% (by vol.) exceeds the PPSR standard limit (Phan and Phan, 2008), coupled with high iodine numbers: it is one of the disadvantages for considering BD as a substitute for diesel fuel.

TABLE 6 Properties of BD Derived from Certain Waste Cooking Oils and Their Comparison to Commercial BD as well as the ASTM D6751 and EN14214 Standards

Parameters	ASTM D6751	EN14214	ASTM No.2D Diesel	Dorado et al. (2003)	Phan and Phan (2008)	Chhetri et al. (2008)	Enweremadu and Mbarawa (2009)[a]	Cetinkaya and Karaosmanoglu (2004)	Georgogianni et al. (2007)	Sabudak and Yildiz (2010)	Commercial BD
FAME content (% mass)		96.5								80.8-89.9 (c) 91.0-93.3 (d) 95.6-97.2 (e)	
Density (kg/m^3) @ 15°C	860-900	860-900	836	882.3	880 (a) 850-880 (b)	870	854.8-890	882.3-887.4	826-857	882-886 (c) 880-884 (d) 882-885 (e)	882
Kinematic viscosity @ 40°C (mm^2/sec)	1.9-6.0	3.5-5.0	2.5	5.29	4.89 (a) 3.56-4.64 (b)	5.03	4.23-6.32	5.29-6.46	4.45-4.76	5.32-5.82 (c) 5.14-5.31 (d) 4.63-4.92 (e)	4.3
Cloud point (°C)	Report	–		−2	3	−1	10.7-(−2)	9	(−3)-(−4)	–	
Pour point (°C)	–	–		−6	0 (a) (−12)-(−4.5) (b)	−16	10-(−6)	−3		(−4)-(−5)	
Flash point (°C) Cetane Number	>130	>101	65	169	120 (a) 74-106.5 (b)	164	109-171	176	67-83	151-159 (c) 159-163 (d) 146-151 (e)	125
	>47	>51	50.9	58.7			47.9-62				45.8
Heating value (MJ/kg)			46.17	39.67			37.27-40.72				40.64
Iodine value		<120		78						109	128.5
Acid value		<0.5			0.43			0.289	0.5-0.8	0.29-0.41 (c) 0.26-0.39 (d) 0.23-0.38 (e)	2.76
Water content (ppm)	<500	<500	26		Trace	0		480.07		381-637 (c) 317-526 (d) 372-487 (e)	

a, fuel properties of those derived from waste cooking oil from restaurants and shops; b, properties of biodiesel blends with diesel from B5 to B75; c, d, e, properties of biodiesel from waste frying oil (WFO) with different methods: (c), one-step base-catalyzed transesterification with co-solvent of WFO (FFA value 2%); (d), two-step base-catalyzed transesterification without cosolvent of WFO (FFA value: 2%); (e), two-step acid- base-catalyzed transesterification without cosolvent of WFO (FFA value: 4.6%).
[a]*data obtained from the max and min values of selected used cooking oils.*

5.2.2 *Environmental Aspect of BD from Waste Cooking Oil*

The environmental aspect of waste cooking oil-based BD mentioned in this section is focused mainly on vehicle emission. Vehicle emission tests of FAMEs from different waste cooking oils on different vehicle engines are briefly described in Table 7.

Test results show that performance of engine on BD or blend of BD from waste cooking oil was better than that on diesel fuel. The emissions especially CO_x, SO_2, and hydrocarbon were significantly reduced. However, there was a small increase in NOx emission and a slight fried food smell (Kulkarni and Dalai, 2006).

5.2.3 *Economic Aspect of BD Production from Waste Oils*

The last aspect, but not the least important, is to assess the economic feasibility of BD production from waste oils. The economic feasibility of BD depends on the price of its competitor, petroleum-diesel. The price of petroleum-diesel is determined by its demand and supply

TABLE 7 Vehicle Emission Tests for BD Produced from Waste Cooking Oil

Sample[a]	Engine Type	Test Results (Compared to Diesel Fuel)	References
WCO-FAME	Inertia weight: 1360 Kg; 2.3 L turbocharged 4-cylinder, 4-stroke, direct injection diesel with exhaust gas recirculation (EGR) under transient operating conditions on a chassis dynamometer.	Hydrocarbon (HC) and CO emission: slightly lower; NO_x emission: a little higher (US-FTP 72) or two times higher (HWEFT); Poly Aromatic Hydrocarbon (PAH): slightly higher, but the differences are within tolerance limits.	Mittelbach and Tritthart (1988)
Blend of used palm oil FAEE and diesel fuel[b]	Single cylinder direct-injection (DI) engine	All blends give low emission, including CO and HC; however, the 50D give the best results	Al-Widyan et al. (2002)
Waste olive FAME	2500 cm^3, three cylinder, four-stroke, water-cooled, 18.5:1 compression ratio, direct injection Diesel engine Perkins AD 3-152, under several steady-state operating conditions	Significant reduction of CO, CO_2, NO, and SO_2 emission up to 58.9, 8.6, 37.5, 57.7%, respectively. Slightly increase in NO_2 emission. Exhaust gases: odorous, slight fried food smell	Dorado et al. (2003)
YGME and SME pure fuel and blends of 20% with diesel fuel	4-cyclinder turbocharged diesel engine under steady-state engine operating conditions	Significant reduction of CO and unburned HC. NO_x emission from YGME and SME increased by 11.6 and 13.1%, respectively	Canakci and Gerpen (2001b)
Used frying oil FAME	Fiat Doblo 1.9 DS, 4-cylinder, 4-stroke 46-kW-power capacity diesel engine	CO emission reduced by 8.6%, HC by 30.7%. Increase of CO_2 and NO_x emission by 2.6 and 5.0%, respectively	Ulusoy and Tekin (2004)

Reference fuel used for comparison in all studies is No. 2 Diesel, except for the first test which is US-2D.
USFTP, United State Federal Test Procedure; HWEFT, Highway Fuel Economy Test.
[a]*WCO-FAME: waste cooking oil fatty acid methyl esters; FAEE: fatty acid ethyl esters; YGME: yellow grease methyl esters; SME: soybean methyl esters.*
[b]*Ethyl esters were blended with diesel fuel with different ratio: 100O (100% esters); 25D (75:25 ester/diesel); 50D (50:50 esters/diesel); 75D (25:75 esters/diesel); 100D (100% diesel fuel).*

market. In USA, the current price of soybean oil-derived BD is nearly 20% greater than the diesel price.

BD price depends highly on the cost of feedstock. Reduction of the raw starting material cost is the first step in optimizing BD production cost (Zhang et al., 2003b). Low-priced waste oil could be considered as the raw material of choice. However, heterogeneity in lipid mixture of waste oil requires multiple steps or alternate approaches to produce BD and requires more clean-up than BD derived from refining oils. According to Zhang et al. (2003b), the main economic criteria for BD production are capital cost, manufacturing cost, and BD break-even price. The economic performance of BD can be determined once factors such as plant capacity, process technology, and raw material and chemical cost are identified. The final cost of BD comprises raw material price, manufacturing cost, government tax, transport and marketing costs (Encinar et al., 2007).

There are several works (Araujo et al., 2010; Marchetti et al., 2008; van Kasteren and Nisworo, 2007; Zhang et al., 2003b) concerning the cost estimation of industrial-scale BD production from waste oils. van Kasteren and Nisworo (2007) estimated BD price based on a supercritical process, while the analysis of Zhang et al. (2003b) was based on three process technologies, viz. alkaline-catalyzed process, acid-catalyzed process with water washing, and acid-catalyzed process with hexane washing. The economic assessment of the BD plant, carried out by van Kasteren and Nisworo (2007), showed that BD can be sold at US$ 0.17/l (125,000 tons/year), US$ 0.24/l (80,000 tons/year), and US$ 0.52/l (8000 tons/year), all of which can compete with the alkaline- and acid-catalyzed process. On the other hand, Zhang et al. (2003b) concluded that on the basis of 8000 tons/year BD capacity, acid-catalyzed process to produce BD from waste cooking oil gives lower manufacturing costs, a more attractive after-tax rate of return, and a lower BD break-even price than the alkaline-catalyzed one. Encinar et al. (2007) estimated the break-even price of BD obtained from a reaction of used frying oil with ethanol to be US$537/ton, which is lower than the range US$644 to US$884 in the study of Zhang et al. (2003b). Moreover, an economic analysis by Haas (2005) suggested that production cost of BD from a soapstock was approximately US$ 0.41/l which was 25% less than those obtained from soy oil. Marchetti et al. (2008) also suggested producing BD from a low-cost feedstock and the use of heterogeneous catalyst as a promising alternative.

It should be noted that the economic assessments were done by assuming constant feedstock price. Thompson et al. (2010) argued against overly simplifying feedstock markets by holding prices constant when considering the economics of a particular feedstock or in estimating the broader impacts of rising BD production on competing uses. Szulczyk and McCarl (2010) identified some problems with BD such as its poor cold fuel properties, and the limited amount of available animal fat as a source of BD. GHG prices have more expansionary impact on the BD prices and U.S. government subsidies have an expansionary impact on BD production, but only help expanding its market penetration by an additional 3% in 2030.

6 CONCLUDING REMARKS AND FUTURE PROSPECTS

This chapter highlights a discussion about the BD production from waste cooking oils, from the pretreatment of waste oils, technical aspect of various possible methods employed in BD synthesis, various downstream procedures that are necessary to obtain final BD adhering to the ASTM D6751 and EN14214 standards, as well as the characterization and environmental aspect of the BD produced from waste oils.

For both environmental and economic reasons, it is beneficial to produce green BD from waste cooking oils and technically it is possible. It is desirable to help solving waste oil disposal by utilizing it oils as an inexpensive starting material in BD synthesis. This process can make the BD price more competitive than that from refined vegetable oils. Therefore, this promising alternative could be implemented in both small and large scales of BD production plant. However, it does not mean we should depend on waste oils as the solely reliable starting material for producing an economically competitive BD as there are other promising sources such as nonedible oils and microorganisms.

Acknowledgment

The authors want to express their thanks to Dr. Fred Quarnstrom for his help during the course of the preparation of this book chapter.

References

Al-Widyan, M.I., Al-Shyoukh, A.O., 2002. Experimental evaluation of the transesterification of waste palm oil into biodiesel. Bioresour. Technol. 85, 253–256.

Al-Widyan, M.I., Tashtoush, G., Abu-Qudais, M., 2002. Utilization of ethyl ester of waste vegetable oils as fuel in diesel engines. Fuel Process. Technol. 76 (2), 91–103.

Araujo, V.K.W.S., Hamacher, S., Scavarda, L.F., 2010. Economic assessment of biodiesel production from waste frying oils. Bioresour. Technol. 101, 4415–4422.

Azocar, L., Scheuermann, E., Hidalgo, P., Betancourt, R., Navia, R., 2007. Biodiesel production from Rapeseed oil with waste frying oils. ISWA World Congress, Amsterdam.

Bautista, L.F., Vicente, G., Rodríguez, R., Pacheco, M., 2009. Optimisation of FAME production from waste cooking oil for biodiesel use. Biomass Bioenerg. 33, 862–872.

Berrios, M., Gutiérrez, M.C., Martín, M.A., Martín, A., 2010. Obtaining biodiesel from spanish used frying oil: issues in meeting the EN 14214 biodiesel standard. Biomass Bioenerg. 34, 312–318.

Boucher, M.B., Unker, S.A., Hawley, K.R., Wilhite, B.A., Stuart, J.D., Parnas, R.S., 2008. Variables affecting homogeneous acid catalyst recoverability and reuse after esterification of concentrated omega-9 polyunsaturated fatty acids in vegetable oil triglycerides. Green Chem. 10, 1331–1336.

Brito, A., Borges, M.E., Arvelo, R., Garcia, F., Diaz, M.C., Otero, N., 2007. Reuse of fried oil to obtain biodiesel: zeolite Y as a catalyst. Int. J. Chem. React. Eng. 5, 1–11.

Brito, A., Borges, M.E., Garín, M., Hernández, A., 2009. Biodiesel production from waste oil using Mg-Al layered double hydroxide catalyst. Energ. Fuel. 23, 2952–2958.

Canakci, M., Gerpen, J.V., 2001a. Biodiesel production from oils and fats with high free fatty acids. T ASAE 44 (6), 1429–1436.

Canakci, M., Gerpen, J.V., 2001b. The performance and emissions of a diesel engine fuelled with biodiesel from yellow grease and soybean oil. Proceedings of 2001 ASAE International Meeting Sacramento, CA. .

Canoira, L., Rodríguez-Gamero, M., Querol, E., Alcantara, R., Lapuerta, M., Oliva, F., 2008. Biodiesel from low-grade animal fat: production process assessment and biodiesel properties characterization. Ind. Eng. Chem. Res. 47, 7997–8004.

Cetinkaya, M., Karaosmanoglu, F., 2004. Optimisation of based catalyzed transesterification reaction of used cooking oil. Energ. Fuel. 18, 1888–1895.

Chen, J.W., Wu, W.T., 2003. Regeneration of immobilized *Candida antarctica* lipase for transesterification. J. Biosci. Bioeng. 95, 466–469.

Chen, Y., Xiao, B., Chang, J., Fu, Y., Lv, P., Wang, X., 2009. Synthesis of biodiesel from waste cooking oil using immobilized lipase in fixed bed reactor. Energ. Convers. Manage. 50, 668–673.

Chhetri, A.B., Watts, K.C., Islam, M.R., 2008. Waste cooking oil as an alternate feedstock for biodiesel production. Energies 1 (1), 3–18.

Cvengros, J.J., Cvengrosová, Z., 2004. Used frying oils and fats and their utilization in the production of methyl esters of higher fatty acids. Biomass Bioenerg. 27 (2), 173–181.

Demirbas, A., 2005. Biodiesel Production from vegetables oils by supercritical methanol. J. Sci. Ind. Res. 64, 858–865.

Demirbas, A., 2009. Biodiesel from waste cooking oil via base-catalytic and supercritical methanol transesterification. Energ. Convers. Manage. 50, 923–927.

Dias, J.M., Alvim-Ferraz, M.C.M., Almeida, M.F., 2008. Comparison of the performance of different homogeneous alkali catalysts during transesterification of waste and virgin oils and evaluation of biodiesel quality. Fuel 87, 3572–12578.

Dias, J.M., Alvim-Ferraz, M.C.M., Almeida, M.F., 2009. Production of biodiesel from acid waste lard. Bioresour. Technol. 100, 6355–6361.

Dizge, N., Aydiner, C., Imer, D.Y., Bayramoglu, M., Tanriseven, A., Keskinler, B., 2009. Biodiesel production from sunflower, soybean, and waste cooking oils by transesterification using lipase immobilized onto a novel microporous polymer. Bioresour. Technol. 100, 1983–1991.

Dorado, M.P., Ballesteros, E., Arnal, J.M., Gómez, J., López, F.J., 2003. Exhaust emissions from a diesel engine fueled with transesterified waste olive oil[small star, filled]. Fuel 82 (11), 1311–1315.

Encinar, J.M., González, J.F., Rodríguez, J.J., Tejedor, A., 2002. Biodiesel fuels from vegetable oils: transesterification of *Cynara cardunculus* L. oils with ethanol. Energ. Fuel. 16, 443–450.

Encinar, J.M., González, J.F., Rodríguez-Reinares, A., 2007. Ethanolysis of used frying oil. Biodiesel preparation and characterization. Fuel Process. Technol. 88, 513–522.

Enweremadu, C.C., Mbarawa, M.M., 2009. Technical aspects of production and analysis of biodiesel from used cooking oil—a review. Renew. Sust. Energ. Rev. 13 (9), 2205–2224.

Felizardo, P., Correia, M.J.N., Raposo, I., Mendes, J.F., Berkemeier, R., Bordado, J.M., 2006. Production of biodiesel from waste frying oils. Waste Manage. 26, 487–494.

Feng, Y., He, B., Cao, Y., Li, J., Liu, M., Yan, F., et al., 2010. Biodiesel production using cation-exchange resin as heterogeneous catalyst. Bioresour. Technol. 101, 1518–1521.

Frankel, E.N., 1980. Lipid Oxidation. Prog. Lipid Res. 19, 1–22.

Fu, B., Gao, L., Niu, L., Wei, R., Xiao, G., 2009. Biodiesel from waste cooking oil via heterogeneous superacid catalyst SO_4^{2-}/ZrO_2. Energ. Fuel 23, 569–572.

Furuta, S., Matsuhashi, H., Arata, K., 2004. Biodiesel fuel production with solid superacid catalysis in fixed bed reactor under atmospheric pressure. Catal. Commun. 5, 721–723.

Gan, M., Pan, D., Ma, L., Yue, E., Hong, J., 2009. The kinetics of the esterification of free fatty acids in waste cooking oil using Fe2(SO4)3/C catalyst. Chin. J. Chem. Eng. 17, 83–87.

Gan, S., Ng, H.K., Ooi, C.W., Motala, N.O., Ismail, M.A.F., 2010. Ferric sulphate catalysed esterification of free fatty acids in waste cooking oil. Bioresour. Technol. 101, 7338–7343.

García, M., Gonzalo, A., Sánchez, J.L., Arauzo, J., Peña, J.Á., 2010. Prediction of normalized biodiesel properties by simulation of multiple feedstock blends. Bioresour. Technol. 101, 4431–4439.

Georgogianni, K.G., Kontominas, M.G., Tegou, E., Avlonitis, D., Vergis, V., 2007. Biodiesel Production: reaction and process parameters of alkali catalysed transesterification of waste frying oils. Energ. Fuel. 21, 3023–3027.

Ghassan, M.T., Al-Widyan, M.I., Al-Jarrah, M.M., 2004. Experimental study on evaluation and optimization of conversion of waste animal fat into biodiesel. Energ. Convers. Manage. 45, 2697–2711.

Guan, G., Kusakabe, K., Yamasaki, S., 2009. Tri-potassium phosphate as a solid catalyst for biodiesel production from waste cooking oil. Fuel Process. Technol. 90, 520–524.

Guo, Z., Sun, Y., 2004. Characteristics of immobilized lipase on hydrophobic superparamagnetic microspheres to catalyze esterification. Biotechnol. Prog. 20, 500–506.

Haas, M.J., 2005. Improving the economics of biodiesel production through the use of low value lipids as feedstocks: vegetable oil soapstock. Fuel Process. Technol. 86, 1087–1096.

Halim, S.F.A., Kamaruddin, A.H., Fernando, W.J.N., 2009. Continuous biosynthesis of biodiesel from waste cooking palm oil in a packed bed reactor: optimization using respnse surface methodology (RSM) and mass transfer studies. Bioresour. Technol. 100, 710–716.

Hayyan, A., Alam, M.Z., Mirghani, M.E.S., Kabbashi, N.A., Hakimi, N.I.N.M.H., Siran, Y.M., et al., 2010. Sludge palm oil as a renewable raw material for biodiesel production by two-step processes. Bioresour. Technol. 101, 7804–7811.

Hossain, A.B.M.S., Boyce, A.N., 2009. Biodiesel production from waste sunflower cooking oil as an environmental recycling process and renewable energy. Bulg. J. Agric. Sci. 15, 312–317.

Issariyakul, T., Kulkarni, M.G., Dalai, A.K., Bakhshi, N.N., 2007. Production of biodiesel from waste fryer grease using mixed methanol/ethanol system. Fuel Process. Tehnol. 88 (5), 429–436.

Issariyakul, T., Kulkarni, M.G., Meher, L.C., Dalai, A.K., Bakhshi, N.N., 2008. Biodiesel production from mixtures of canola oil and used cooking oil. Chem. Eng. J. 140, 77–85.

Jacobson, K., Gopinath, R., Meher, L.C., Dalai, A.K., 2008. Solid acid catalyzed biodiesel production from waste cooking oil. Appl. Catal. B-Environ. 85, 86–91.

Japan Chemical Web, 2008. Biodiesel Made Continuously from Waste Cooking Oil without Catalyst. The Chemical Daily Co., Ltd, Tokyo.

Kemp, W.H., 2006. Biodiesel: Basic and Beyond—A Comprehensive Guide to Production and Use for Home and Farm. Aztext Press, Ontario,Canada.

Komers, K., Machek, J., Stloukal, R., 2001. Biodiesel from rapeseed oil, methanol and KOH. 2. Composition of solution of KOH in methanol as reaction partner of oil. Eur. J. Lipid Sci. Technol. 103, 359–362.

Kouzu, M., Kasuno, T., Tajika, M., Sugimoto, Y., Yamanaka, S., Hidaka, J., 2008. Calcium oxide as a solid base catalyst for transesterification of soybean oil and its application to biodiesel production. Fuel 87, 2798–2806.

Kulkarni, M.G., Dalai, A.K., 2006. Waste Cooking Oil—An Economical Source for Biodiesel: a Review. Ind. Eng. Chem. Res. 45, 2901–2913.

Kusdiana, D., Saka, S., 2004a. Effects of water on biodiesel fuel production by supercritical methanol treatment. Bioresour. Technol. 91, 289–295.

Kusdiana, D., Saka, S., 2004b. Two-step preparation for catalyst-free biodiesel fuel production: hydrolysis and methyl esterification. Appl. Biochem. Biotech. 115, 781–791.

Lam, M.K., Lee, K.T., Mohamed, A.R., 2009. Sulfated tin oxide as solid superacid catalyst for transesterification of waste cooking oil: an optimization study. Appl. Catal. B Environ. 93, 134–139.

Lam, M.K., Lee, K.T., Mohamed, A.R., 2010. Homogeneous, heterogeneous and enzymatic catalysis for transesterification of high free fatty acid oil (waste cooking oil) to biodiesel: a review. Biotechnol. Adv. 28, 500–518.

Lang, X., Dalai, A.K., Bakhshi, N.N., Reaney, M.J., Hertz, P.B., 2001. Preparation and characterization of biodiesels from various bio-oils. Bioresour. Technol. 80, 53–62.

Lapuerta, M., Herreros, J.M., Lyons, L.L., García-Contreras, R., Briceño, Y., 2008. Effect of the alcohol type used in the production of waste cooking oil biodiesel on diesel performance and emissions. Fuel 87 (15-16), 3161–3169.

Lertsathapornsuk, V., Pairintra, R., Aryusuk, K., Krisnangkura, K., 2008. Microwave assisted in continuous biodiesel production from waste frying palm oil and its performance in a 100 kW diesel generator. Fuel Process. Technol. 89 (12), 1330–1336.

Leung, D.Y.C., Guo, Y., 2006. Transesterification of neat and used frying oil: optimization for biodiesel production. Fuel Process. Technol. 87 (10), 883–890.

Li, N.W., Zong, M.H., Wu, H., 2009. Highly efficient transformation of waste oil to biodiesel by immobilized lipase from *Penicillium expansum*. Process Biochem. 44, 685–688.

Liu, S., McDonald, T., Wang, Y., 2010. Producing biodiesel from high free fatty acids waste cooking oil assisted by radio frequency heating. Fuel 89, 2735–2740.

Lou, W.Y., Zong, M.H., Duan, Z.Q., 2008. Efficient production of biodiesel from high free fatty acid-containing waste oils using various carbohydrate-derived solid acid catalysts. Bioresour. Technol. 99, 8752–8758.

Maceiras, R., Vega, M., Costa, C., Ramos, P., Márquez, M.C., 2009. Effect of methanol content on enzymatic production of biodiesel from waste frying oil. Fuel 88, 2130–2134.

Marchetti, J.M., Miguel, V.U., Errazu, A.F., 2008. Techno-economic study of different alternatives for biodiesel production. Fuel Process. Technol. 89, 740–748.

Math, M.C., Irfan, G., 2007. Optimization of restaurant waste oil methyl ester yield. J. Sci. Ind. Res. India 66, 772–776.

McNeff, C.V., McNeff, L.C., Yan, B., Nowlan, D.T., Rasmussen, M., Gyberg, A.E., et al., 2008. A continuous catalytic system for biodiesel production. Appl. Catal. A-Gen. 343, 39–48.

Meng, X., Chen, G., Wang, Y., 2008. Biodiesel production from waste cooking oil via alkali catalyst and its engine test. Fuel Process. Technol. 89, 851–857.

Mittelbach, M., Tritthart, P., 1988. Diesel fuel derived from vegetable oils, III. Emission tests using methyl esters of used frying oil. J. Am. Oil Chem. Soc. 65 (7), 1185–1187.

Mittelbach, M., Enzelsberger, H., 1999. Transesterification of heated rapeseed oil for extending diesel fuel. J. Am. Oil Chem. Soc. 76 (5), 545–550.

Montefrio, M.J., XinWen, T., Obbard, J.P., 2010. Recovery and pre-treatment of fats, oil and grease from grease interceptors for biodiesel production. Appl. Energ. 87, 3155–3161.

Muniyappa, P.R., Brammer, S.C., Noureddini, H., 1996. Improved conversion of plant oils and animal fats into biodiesel and co-product. Bioresour. Technol. 56, 19–24.

Nawar, W.W., 1984. Chemical changes in lipids produced by thermal processing. J. Chem. Ed. 61 (4), 299–302.

Paola, M.G.D.P., Calabrò, V., Curcio, S., Iorio, G., 2009. Factor analysis of transesterification reaction of waste oil for biodiesel production. Bioresour. Technol. 100, 5126–5131.

Patil, P., Deng, S., Rhodes, J.I., Lammers, P.J., 2010. Conversion of waste cooking oil to biodiesel using ferric sulfate and supercritical methanol processes. Fuel 89, 360–364.

Phan, A.N., Phan, T.M., 2008. Biodiesel production from waste cooking oils. Fuel 87 (17-18), 3490–3496.

Predojević, Z.J., 2008. The production of biodiesel from waste frying oils: a comparison of different purification steps. Fuel 87, 3522–3528.

Reefat, A.A., Attia, N.K., Sibak, H.A., El Sheltawy, S.T., El Diwani, G.I., 2008. Production optimization and quality assessment of biodiesel from waste vegetable oil. Int. J. Environ. Sci. Tech. 5 (1), 75–82.

Sabudak, T., Yildiz, M., 2010. Biodiesel production from waste frying oils and its quality control. Waste Manage. 30 (5), 799–803.

Saifuddin, N., Chua, K.H., 2004. Production of ethyl ester (biodiesel) from used frying oil: optimization of transesterification process using microwave irradiation. Malaysian J. Chem. 6 (1), 77–82.

Sakai, T., Kawashima, A., Koshikawa, T., 2009. Economic assessment of batch biodiesel production processes using homogeneous and heterogeneous alkali catalysts. Bioresour. Technol. 100, 3268–3276.

Saka, S., Isayama, Y., 2009. A new process for catalyst-free production of biodiesel using supercritical methyl acetate. Fuel 88, 1307–1313.

Salis, A., Pinna, M., Monduzzi, M., Solinas, V., 2008. Comparison among immobilised lipases on macroporous polypropylene toward biodiesel synthesis. J. Mol. Catal. B Enzym. 54, 19–26.

Santos, F.F.P., Matos, L.J.B.L., Rodrigues, S., Fernandes, F.A.N., 2009. Optimization of the production of methyl esters from soybean waste oil applying ultrasound technology. Energ. Fuel. 23, 4116–4120.

Shimada, Y., Watanabe, Y., Sugihara, A., Tominaga, Y., 2002. Enzymatic alcoholysis for biodiesel fuel production and application of the reaction to oil processing. J. Mol. Catal. B Enzym. 17, 133–142.

Shu, Q., Nawaz, Z., Gao, J., Liao, Y., Zhang, Q., Wang, D., et al., 2010. Synthesis of biodiesel from a model waste oil feedstock using a carbon-based solid acid catalyst: reaction and separation. Bioresour. Technol. 01, 5374–5384.

Sun, P., Sun, J., Yao, J., Zhang, L., Xu, N., 2010. Continuous production of biodiesel from high acid value oils in microstructured reactor by acid-catalyzed reactions. Chem. Eng. J. 162, 364–370.

Szulczyk, K.R., McCarl, B.A., 2010. Market penetration of biodiesel. Renew. Sust. Energ. Rev. 14, 2426–2433.

Tan, K.T., Lee, K.T., Mohamed, A.R., 2010a. Effects of free fatty acids, water content and co-solvent on biodiesel production by supercritical methanol reaction. J. Supercrit. Fluids 53, 88–91.

Tan, K.T., Lee, K.T., Mohamed, A.R., 2010b. A glycerol-free process to produce biodiesel by supercritical methyl acetate technology: an optimization study via response surface methodology. Bioresour. Technol. 101, 965–969.

Thanh, L.T., Okitsu, K., Sadanaga, Y., Takenaka, N., Maeda, Y., Bandow, H., 2010. A two-step continuous ultrasound assisted production of biodiesel fuel from waste cooking oils: a practical and economical approach to produce high quality biodiesel fuel. Bioresour. Technol. 101, 5394–5401.

Thompson, W., Meyer, S., Green, T., 2010. The U.S. biodiesel use mandate and biodiesel feedstock markets. Biomass Bioenerg. 34, 883–889.

Tomasevic, A.V., Siler-Marinkovic, S.S., 2003. Methanolysis of used frying oil. Fuel Process. Technol. 81, 1–6.

Ulusoy, Y., Tekin, Y., 2004. The engine tests of biodiesel from used frying oil. Energ. Source. 26, 927–932.

UNFCCC, 2006. Greenhouse Gas Data, 2006. United Nations Framework Convention on Climate Change.

Utlu, Z., Koçak, M.S., 2008. The effect of biodiesel fuel obtained from waste frying oil on direct injection diesel engine performance and exhaust emissions. Renew. Energ. 33 (8), 1936–1941.

van Kasteren, J.M.N., Nisworo, A.P., 2007. A process model to estimate the cost of industrial scale biodiesel production from waste cooking oil by supercritical transesterification. Resour. Conserv. Recy. 50, 442–458.

Viriya-empikul, N., Krasae, P., Puttasawat, B., Yoosuk, B., Chollacoop, N., Faungnawakij, K., 2010. Waste shells of mollusk and egg as biodiesel production catalysts. Bioresour. Technol. 101, 3765–3767.

Wang, Y., Ou, S., Liu, P., Xue, F., Tang, S., 2006. Comparison of two different processes to synthesize biodiesel by waste cooking oil. J. Mol. Catal. A Chem. 252, 107–112.

Wang, Y., Ou, S., Liu, P., Zhang, Z., 2007. Preparation of biodiesel from waste cooking oil via two-step catalyzed process. Energ. Convers. Manage. 48, 184–188.

Wang, Y., Wu, H., Zong, M.H., 2008. Improvement of biodiesel production by lipozyme TL IM-catalyzed methanolysis using response surface methodology and acyl migration enhancer. Bioresour. Technol. 99, 7232–7237.

Wang, Y., Nie, J., Zhao, M., Ma, S., Kuang, L., Han, X., et al., 2010. Production of biodiesel from waste cooking oil via a two-step catalyzed process and molecular distillation. Energ. Fuel. 24, 2104–2108.

Watanabe, Y., Shimada, Y., Sugihara, A., Tominaga, Y., 2001. Enzymatic conversion of waste edible oil to biodiesel fuel in a fixed-bed bioreactor. J. Am Oil Chem. Soc. 78, 703–707.

Wei, Z., Xu, C., Li, B., 2009. Application of waste eggshell as low-cost solid catalyst for biodiesel production. Bioresour. Technol. 100, 2883–2885.

Wen, Z., Yu, X., Tu, S.T., Yan, J., Dahlquist, E., 2010. Biodiesel production from waste cooking oil catalyzed by TiO_2-MgO mixed oxides. Bioresour. Technol. 101, 9570–9576.

Yagiz, F., Kazan, D., Akin, A.N., 2007. Biodiesel production from waste oils by using lipase immobilized on hydrocalcite and zeolites. Chem. Eng. J. 134, 262–267.

Yin, J.-Z., Xiao, M., Song, J.-B., 2008. Biodiesel from soybean oil in supercritical methanol with co-solvent. Energ. Convers. Manage. 49, 908–912.

Zhang, Y., Dubé, M.A., McLean, D.D., Kates, M., 2003a. Biodiesel production from waste cooking oil: 1. Process design and technological assessment. Bioresour. Technol. 89 (1), 1–16.

Zhang, Y., Dubé, M.A., McLean, D.D., Kates, M., 2003b. Biodiesel production from waste cooking oil: 2. Economic assessment and sensitivity analysis. Bioresour. Technol. 90, 229–240.

Zheng, S., Kates, M., Dubé, M.A., McLean, D.D., 2006. Acid-catalyzed production of biodiesel from waste frying oil. Biomass Bioenerg. 30, 267–272.

Zullaikah, S., Lai, C.C., Vali, S.R., Ju, Y.H., 2005. A two-step acid-catalyzed process for the production of biodiesel from rice bran oil. Bioresour. Technol. 96, 1889–1896.

SECTION IIIB

PRODUCTION OF BIOFUELS FROM ALGAE

CHAPTER

17

Production of Biodiesel from Algal Biomass: Current Perspectives and Future

*Yi-Feng Chen, Qingyu Wu**

School of Life Sciences, Tsinghua University, Beijing 100084, People's Republic of China

***Corresponding author: E-mail: qingyu@tsinghua.edu.cn**

1 INTRODUCTION

Biodiesel is defined as monoalkyl esters of plant oils or animal fats through transesterification reactions. These neutral lipids serving as energy stores in plants and animals bear a common structure of triple esters where usually three long-chain fatty acids are coupled to a glycerol. Transesterification action turns the triple esters to single ones by displacing glycerol with small alcohols (e.g., methanol). Biodiesel was initially adopted for the compression-ignition (diesel) engine invented about one century ago by a German engineer Rudolf Diesel. His first engine was run with peanut oils, and subsequent experiments confirmed that biodiesel was better than raw oils. However, biodiesel was not as popular in the world as diesel engines, replaced by petroleum-derived diesel with cheap and abundant supplies shortly after the 1920s. Recently, the rise in petroleum price and the need to reduce greenhouse gas emission reactivated interest in biodiesel. Commercial application of biodiesel started in the 1990s in the United States and European countries and now has expanded into many regions of the world including China.

Despite its acceptance worldwide and various sources of feedstocks of vegetable oils or animal fats, the current global yield of biodiesel only accounts for less than one per cent of the whole diesel market. One of the main obstacles is the supply of biodiesel feedstocks currently in usage is nonsustainable or limited to small scales. For example, soybean, a major form of feedstocks for the United States biodiesel market, is produced from arable land and also consumed as food. The feedstock competition for food and biofuel should therefore restrict the development of large-scale industrial biodiesel from soybean. Additional sources of feedstocks coming from animal fats or used oils of restaurants obviously have problems in limited and unstable supply. To these problems, more and more people through the exploration of the field believe the eventual answers might be found from microalgae due to ultrahigh biomass yields and high contents of oil in some algal species. A nationwide research activity named the Aquatic Species Program was initiated in 1978 by the Department of Energy of the United States and lasted for 18 years (Sheehan et al., 1998). One of the main tasks of the Program was to collect and identify microalgal strains with high contents of lipids and develop the cultivation systems, all of which was expected to lead to biodiesel production at commercial levels. Though the project was terminated in 1996 because of the budget difficulty, plenty of lessons and knowledge have been generated. Some of them are introduced here: (1) more than 3000 microalgal strains were collected and 10% of them turned to be most promising in oil contents and other traits; (2) the feasibility of mass cultivation of these promising strains in open ponds was confirmed and thus the culture technology was established; (3) methods to analyze algal lipids were partially established, for instance the dyer Nile red used for in vivo monitoring of lipid droplets in algal cells; (4) genetic improvement through key genes of lipid metabolism pathways was preliminarily proved effective in eukaryotic microalgae. All these successful efforts inspired continual exploration at the field, and nowadays obvious progresses have been made in both insights into algal metabolism pathways especially surrounding the model system *Chlamydomonas* and development of genetic tools.

The advantages of algal biomass as feedstock of biodiesel have been discussed in several review articles and proceedings, exemplified by Chisti (2007), Rosenberg et al. (2008), Hu et al. (2008) and a latest report of the DOE of the U.S. on the National Algal Biofuels Technology Roadmap (Ferrell and Sarisky-Reed et al., 2010). We intend to provide a brief summary of these advantages of algal feedstock below. For more details, please refer to these articles.

(1) High biomass productivity. Microalgae have been demonstrated to possess higher efficiency in photosynthesis and adaptation to stressful conditions. Many microalgal species exist in single cells or simple clusters of a few cells. These simple cellular organizations usually enable microalgae to grow fast or accumulate more lipid or starch at excessive energy. Microalgae possess multiple metabolic pathways and are interchangeable under different nutrient and sunlight situations. Algal bioreactors might be built up vertically or in a multiple-layer manner to utilize the space to maximum. On the other hand, microalgae of some fast-growing species could be harvested continuously at a daily basis. All this greatly increase algal biomass productivity per area annually.

(2) High oil contents or yields. Oleaginous microalgae contain storage oils at triacylglycerides (TAGs) more than 50% of dry cell weight. It has been known that autotrophic microalgae could accumulate more neutral lipids up on stresses such as nitrogen deficiency. More interestingly, some heterotrophic species such as *Chlorella protothecoides* that are able to grow fast meanwhile accumulate over 50% of neutral lipids, and eventually produce much higher yields of oil.

(3) Less usage of arable land and freshwater. Because microalgae will not be cultured directly in soil, theoretically any kind of lands including many nonarable lands could be exploited for algal mass cultivation. Also, algal bioreactors could be arranged in an industrial manner, which is far advanced to crops of conventional agriculture. Consequently, efficiency of the actual land usage is projectioned to be improved drastically. Microalgae can utilize various wastewaters, seawater, and other forms of produced water which cannot be introduced into the agricultural system. These features of microalgal culture obviously reduce competition for limited land and freshwater resources.

(4) In contrast to crops of conventional agriculture, microalgae of some species may utilize high doses of CO_2 present in flue gases, indicating they have the potential to reduce greenhouse gas emission. It is also interest for researchers to investigate if the high-dose CO_2 adaptation could further improve photosynthesis and thus biomass.

(5) Algal cultivation for biomass production might be compatible with biorefinery to produce a variety of fuels and value-added coproducts. In this way, all biomass components might be fully utilized.

Among these advantages of microalgae, microalgal metabolic pathways are particularly important, mainly because these pathways directly determine algal biomass formation and lipid accumulation, and the lipid yield is the most important trait for liquid transportation fuel production. We will preliminarily introduce microalgal metabolic pathways and their relevance to biomass and lipid production. More specific descriptions of their metabolic properties will be placed in the next section.

There are three metabolic pathways present in microalgae—autotrophy, heterotrophy, and mixtrotrophy. Autotrophic algae uptake light, CO_2 under simple inorganic media; their main storage products include starch and/or lipids. Under optimal light and temperature, some autotrophic algal species are able to rapidly grow at a rate of 0.2 g(DW)/l/day and eventual biomass can be formed at 2 g (DW)/l (Gouveia and Oliveira, 2009). For most of autotrophic algae, their rapid growth usually alleviates lipid accumulation and vice versa. Stress treatments like deficit in nutrients (N, P, or Si) result in lipid accumulation but meanwhile reduce growth rates. In the presence of organic carbon sources such as glucose or acetate, some species from *Chlorella*, *Chlamydomonas*, and so on can thrive with increased growth rates for example of 24 g(DW)/l/day in comparison with autotrophic conditions, and eventual maximal biomass yield could be over 100 g(DW)/l of the best strains (Wu and Shi, 2006). Among these species, a few of them can accumulate higher contents of TAG, usually over 50% (Miao and Wu, 2006), whose final lipid yields are thus extremely higher than those of autotrophic partners. When both light and organic carbons are present, some microalgae may additively make use of inorganic and organic carbons, through a mixotrophic (or named photoheterotrophic) pathway. The consequences of this pathway in biomass production are however hard to predict, maybe beyond the heterotrophy plus autotrophy (Lee, 2004), or between within the ranges of heterotrophy and autotrophy in most of cases. Less is known about lipid accumulation under mixotrophic conditions. These three metabolic pathways might be switched from one to another in certain algal species after acclimation of many generations; for example, in *C. protothecoides*, autotrophy can be reversibly changed to heterotrophy monitored by availability of glucose or light (Miao and Wu, 2006).

In this chapter, we will focus on the development of biodiesel technology based on heterotrophic microalgae since its significant progresses during the last several years in comparison to the autotrophic microalgae-based biodiesel technology. Additional progresses in regard to lipid analysis methods, improvement of algal biodiesel quality, updated transesterification reactions, and integration of biodiesel production with environmental remediation through algae (e.g., wastewater-rich nutrients or flue gas high-dose CO_2) are also introduced to try to give readers a balanced view of the current status of the field.

2 TWO APPROACHES LEADING TO LIPID ACCUMULATION IN MICROALGAE

2.1 Photoautotrophic Microalgae Accumulate Lipids upon Stresses But at Expense of Slow Growth

At present, researchers of algal biodiesel make use of two strategies to produce biodiesel, which are inherently associated with two fundamental physiological processes of microalgae—photosynthesis or aerobic respiration. Photosynthesis is the most obvious feature of all microalgae that distinguishes themselves from other microorganisms (e.g., bacteria and fungi). The energy driving photosynthesis is provided by sunlight, so fair amounts of efforts are being devoted to improve light absorbance and thus increase biomass. For this purpose, many suggestions had been made from lowering chlorophyll contents to adapt to high light, tubular photobioreactors to increase area to volume ratios, and so on, which was highlighted in Chisti (2007). There are a plenty of publications on the theme of photoautotrophic biomass. Due to the limitation of space, details as to structure, function, and regulation of microalgal photosynthetic apparatus are not mentioned, but its general characterizations are pointed out in this review. All feedstocks used in biofuel production are derived, directly or indirectly, from photosynthesis, which enables algae and plants occupy a unique position in research and program of biofuel technology. On the other hand, photosynthesis is a cheap process in nature, which might be an inherent driving power for researchers to try to integrate into industrial infrastructures.

Glucose, sucrose, and starch are the main carbohydrate products of algal photosynthesis. Under normal conditions, autotrophic algal cells accumulate few amounts of lipids, usually less than 20% of dry cell weight in eukaryotic microalgae or even less than 10% of dry cell weight in prokaryotic cyanobacteria (Hu et al., 2008). The underlying mechanism might be interpreted by the negative relationship between photosynthetic starch accumulation and lipid synthesis (Li et al., 2010). Li et al. (2010) observed that triacylglycerol (TAG) was overproduced by 10-fold in a *Chlamydomonas* starchless mutant with inactivation of ADP-glucose pyrophosphorylase. This work suggests a strategy to increase lipid production by directing more photosynthetic carbon partitioning to lipid.

On the other hand, many stress conditions ranging from nutrition deficiency (N, P, or Si) to osmotic or temperature changes induce neutral lipid accumulation within cells frequently over 50% of dry cell weight while alleviate cellular growth rates (Rodolfi et al., 2009; Zhekisheva et al., 2002). Naturally oleaginous autotrophic algae, albeit with higher contents of lipids, show slow growth, which might be the native indicator of the inherent conflict between rapid growth and lipid accumulation. This conflict might generally exist in all kinds of photosynthetic organisms from algae to higher plants.

A technology of two-phase cultivation was developed to increase lipid accumulation in autotrophic microalgae with starvation. It composes of algal growth under sufficient nutrients first for biomass formation then under nitrogen deficiency for a couple of days for the conversion to lipids. There are several shortages of the technology needed to overcome. First, the operation to switch low-cell-density algae in the nitrogen concentrations from high to low is not easy at large scale, but the reverse operation is easy to achieve by feeding certain nitrogen-containing chemicals. Second, the two-phase cultivation is discontinuous for algae harvesting and lipid production, since the first phase is dedicated to build up a proper algal biomass that is converted in the subsequent phase to lipid under starvation. This design should limit lipid yields if compared with lipid yield at a daily basis. Third, the whole cultivation process lasts for a longer period for one time production, given the lower biomass formation by autotrophic algae, which limits the technology to be scaled up for commercial application. As an example, a latest attempt was aimed to improve some of these disadvantages by Rodolfi et al. (2009). From 30 microalgal strains, *Nannochloropsis sp.* F&M-M24 was characterized with 60% lipid content after nitrogen starvation. Further experiments were performed in a 110-liter photobioreactor. Under nitrogen starvation, lipid accumulated by the species at 0.204 g (DW)/l/day with 60% final lipid content, the growth rate was 0.3 g(DW)/l/day.

2.2 Heterotrophic Microalgae Accumulate Lipids and Grow Fast, But Consume more Glucose or Other Organic Carbons

A few of microalgal species, exemplified by *C. protothecoides*, has ability to grow with organic carbon sources. These strains, after acclimation of many generations, were conferred with expanded capacities in organic carbon utilization at high efficiency in well-controlled fermenters. Under optimized conditions, their biomass formation and high cell densities are comparable even to those of bacteria or yeasts which are typical industrial microorganisms widely employed for food and drug production at large scales. Both bacteria and yeasts have been recently adopted for biofuel production of substances such as bioethanol and long-chain alcohols (Peralta-Yahya and Keasling, 2010) based on mature industrial technologies. The same situation might apply to heterotrophic microalgae according to their apparent similarities to other microorganisms of aerobic fermentation. Investigations during the past several years from our group and other laboratories over the world have confirmed that heterotrophic microalgae may fit well into the infrastructure of mature fermentation industry to express the high potential in biomass productivity as well as lipid productivity (Alabi et al., 2009; Li et al., 2007; Rosenberg et al., 2008).

Heterotrophic microalgae display another advantage, that is, there is a coordinate growth with lipid accumulation under normal growth conditions. Lipid content in *C. protothecoides* in the presence of glucose was about 4-fold of that under the light and without glucose (Miao and Wu, 2006). This phenotype is impressively different from the situation in autotrophic microalgae where fast growth and lipid accumulation usually conflict. As a consequence, the coordination of these two fundamental characters confers heterotrophic microalgae with extremely high efficiency in lipid accumulation for biodiesel production. The underlying mechanism was preliminarily exploited in heterotrophic *C. protothecoides* through metabolic flux analysis (Xiong et al., 2010b). A metabolic network composed of the glycolysis, the pentose phosphate pathway, and the tricarboxylic acid cycle was revealed, which might be involved in biomass formation and lipid accumulation. It was confirmed that excessive NADPH requirements for lipid biosynthesis, indicated by increased relative activity of the pentose phosphate pathway to the glycolysis under heterotrophic conditions interfered with nitrogen limitation. The nitrogen limitation experiment further revealed that although the stress altered the growth rate and cellular oil content and thus absolute fluxes, relative global flux distribution in the species remained stable, suggesting that the heterotrophic alga possesses high capacity of metabolic buffering up on variable environments, which might ensure the alga with stable and high metabolic output for neutral lipid accumulation. These results may also arouse additional exploration into the autotrophic partner in regard to growth and lipid biosynthesis under nitrogen limitation in the near future.

A series of explorations were made to improve the cultivation of heterotrophic microalgae. One attempt was made to modify the preliminary fed-batch culture by accurately maintaining glucose feeding rate, pH, temperature, dissolved oxygen through computer. The supplement of glucose was limited no more than 24 g/l (Xiong and Wu, 2008). It was tested with heterotrophic *C. protothecoides* in a 5-liter fermenter. By using this strategy, the algal biomass increased from 16.8 in preliminary fed-batch culture to 51.2 g/l in the modified culture within about 1 week of fermentation. Other important issues relevant to the cultivation of heterotrophic algae will be addressed in subsequent separate sections, including cost reduction by displacing glucose with cheap or even waste organic carbon sources as well as mass cultivation in a 11,000-liter fermenter to examine the scalability of the technology.

As mentioned earlier, some microalgal species are able to run a mixotrophic pathway when faced by both sunlight and organic carbon nutrients of the surroundings. At present, little is known about characterizations and mechanisms of the pathway in microalgae. Apparently, a few of microalgal species might attain benefits of both autotrophy under light and heterotrophy under organic carbons, an additive phenotype under mixotrophic conditions. Xiong et al. (2010a) successfully observed the similar results in the PFM (the photosynthesis-fermentation model) system they developed. In the PFM, the same *Chlorella* species first grows under light to develop functional photosynthetic apparatus to capture solar energy then to be switched to organic carbons to perform heterotrophic growth. Algal biomass and lipid accumulation in the PFM system was further improved in comparison to sole heterotrophic growth, which implies an approach based on the new metabolic mechanism might developed, which provides a novel strategy for the cost reduction in addition to selection of cheap alternatives of glucose or the scaleup in mass cultivation.

3 EFFORTS TO FURTHER REDUCE COSTS OF MASS CULTURE OF HETEROTROPHIC MICROALGAE IN SEARCH FOR CHEAP SUBSTITUTES OF GLUCOSE

Because glucose supplement accounts for most of the medium cost of algal cultivation (it was estimated up to 80%), the cost reduction by seeking alternatives of cheap organic carbon sources becomes the priority task of the technological improvement, in addition to other approaches like the scaleup for cost dilution. The price of starch from products such as corn is half of that of glucose, and the medium cost could be reduced to 40% when glucose is replaced with starch. Corn powder hydrolysate has been used to replace glucose for developing a cheap medium for heterotrophic *C. protothecoides* (Xu and Wu, 2006). To further reduce the medium cost, many additional options of organic carbon sources were screened and tested for feasibility instead of glucose and starch, including juices from sugar cane (Cheng et al., 2009a) or sweet sorghum (Gao et al., 2010). Even for a single carbon source like starch, there is still some room to reduce cost by selecting cheaper commercial starch products of various sources, exemplified by an extensive survey from Jerusalem artichoke tube (Cheng et al., 2009b) to cassava starch (Lu et al., 2010; Wei et al., 2009). However, all these organic carbons are the feedstocks widely used for food and chemical industries as well as for other biofuels like ethanol, which should arouse a competition between algal biodiesel and other purposes. Sugar degraded from lignocellulosic feedstock might provide potentially much cheaper and sustainable source and an approval-of-concept work has showed the application of semicellulose as a substrate for engineered *Escherichia coli* for liquid biofuel production (Steen et al., 2010). Yet, the realistic application awaits the breakthrough in the enzymatic degradation of cellulose at commercial scales.

Industrial or municipal wastewaters with abundant organic carbon contents might be ideal alternatives of glucose to support heterotrophic microalgae, but so far few evidence has been raised to approve the speculation. The municipal wastewater is usually introduced from the surroundings to a centered facility for the treatment and thus with massive and stable daily supplies. The rich nutrients in these wastewaters include nitrogen, phosphate, vitamins, and trace elements, in addition to organic carbon substances, variable depending on different sources. These wastewaters generate serious environmental concerns and must be treated before released out. Potentially, growing microalgae in such wastewaters may reach the effect of "one stone hits two birds." Inhibitors and toxic chemicals present in them threaten growth and even survival of many microalgae. So screening and acclimation become an essential step to acquire adequate microalgal species.

Waste molasses as the byproduct in sugar refinery is a kind of industrial wastewaters. It contains nearly 50% of the total sugar content and other nutrients necessary for microorganism growth. The baker's yeast could grow in waste molasses (Sirianuntapiboon and Prasertsong, 2008), implying that it might be suitable for algal growth too. A report from Yan et al. (2011) indicated that the biomass and lipid accumulation in *C. protothecoides* grown in treated waste molasses were comparable to those grown in glucose, which confirmed the feasibility of the innovation along the direction.

4 THE SCALEUP OF HETEROTROPHIC MICROALGAL BIOMASS PRODUCTION

Although so far no commercial oil and biodiesel have been reported from heterotrophic microalgae, there was a piece of news from U.S. Department of Energy that Solazyme, Inc. will build up a demonstration facility at the size of 12 metric tons of dry feedstock per day by using heterotrophic microalgae for biodiesel production. They aim to collect, from the facility, the data necessary to complete design of the first commercial plant.

Li et al. (2007) compared biomass yield and lipid accumulation in heterotrophic *C. protothecoides* when grown in fermenters with volumes of 5, 750, up to 11,000 liters, respectively. Within about 8 days of fed-batch culture, the algal biomass achieved 15.5, 12.8, and 14.2 g(DW)/l, respectively in the 5, 750, and 11,000 l fermenters. The lipid contents reached 46.1, 48.7, and 44.3% of dry cell weight accordingly. With up to 98.15% of the conversion rate of transesterification, the biodiesel production rates were 7.02, 6.12, and 6.24 g/l, respectively, which confirmed quite well the stability of the technology based on heterotrophic microalgae during the scaleup.

A theoretic prediction for the commercial scale and cost reduction was made based on the demands of the European diesel market (Wijffels and Barbosa, 2010). If the annual need for nearly 0.4 billion m^3 of transportation fuels was replaced by algal biodiesel, and given the algal biodiesel productivity was 40,000 l per hectare annually based on a 3% photosynthetic efficiency and 50% oil content, 9.25 million hectare of land would therefore be required for the biodiesel supply. It was proposed that the biodiesel production scale needs to leap at least three orders of magnitude and the cost has to be decreased by a factor of 10 (Wijffels and Barbosa, 2010). The prediction was based on the autotrophic microalgae. When the heterotrophic microalgae with higher efficiency in biomass and oil production produced under concentrated chemical energy supplies and in well-controlled fermentation facility and other advantages were considered into the framework of the prediction, less than 9.25 million hectare of land would be expected to achieve the similar goal.

5 PROGRESSES IN LIPID ANALYSIS

General requirements in lipid analysis include high-throughput screening, on-site and *in vivo* monitoring, small samples, less time, and of course accuracy. During the past several years, a lot of efforts have been made to boost the improvements in lipid analysis. Only several cases were adopted herein as snapshots of the area due to the limited space of the article.

The Soxhlet method was proved efficient for total lipid extraction and gravimetric quantification from microalgae. It is frequently adopted as a standard to evaluate other newly developed methods. However, it has two obvious shortages. The conventional organic solvents such as n-hexane are costly and also not environmentally friendly. It takes considerable time-consuming steps and also consumes samples at large amounts, for example with minimum requirement of at least 100-ml cultures. Wawrik and Harriman (2010) developed a rapid and colorimetric method for the quantification of algal lipid from 1 ml of cultures. Algal lipids are saponified to fatty acids and then mixed with a copper reagent and a color developer diethyldithiocarbamate. The formed yellow product is then colorimetrically measured. Fatty acids with chain lengths of C10:0 to C16:0 fell into the ranges of linear responses, below C10 or beyond C16 caused underestimation. But the method could be formatted in microcentrifuge tubes and an analysis of 30 samples could be finished in less than 2 h which was confirmed by monitoring dynamic total lipid contents in *Phaeodactylum tricornutum* and *Chlorella vulgaris*, indicating the method might be suitable for fast monitoring or screening of species in terms of lipid contents.

Su et al. (2008) proposed a rapid method for nondestructive detection of chlorophyll a and lipid contents of microalgae by using RGB model of two linear correlation functions based on the brightness values of the three primary colors (red, green, and blue). The prediction results from the model were agreeable with experimental results and the reliability of the model was also verified in a photosynthetic microalga *Nannochloropsis oculata*. It is curious that if this simple approach could be tested in a broad ranges of algal species, for example to estimate lipid contents in yellow heterotrophic algal cells.

Under optimized conditions, the dyer Nile red specifically binds to nonpolar macromolecules like neutral lipids. By using this feature, the Nile red staining method was established to determine neutral lipids in a broad spectrum of organisms including microalgae in about half a century ago. The method is fast, sensitive, and requires just a few of *in vivo* algal cells, and therefore has the potential for the high-throughput screening. However, the staining signals are affected by the thickness of algal cell walls that varies among species, which limits its application for the large-scale screen. Chen et al. (2009) introduced DMSO (dimethyl sulfoxide) with mild heating to speed up the penetration of the dyer throughout algal cell walls. They claimed this modification can buffer the differences of staining signals caused by differences in cell wall thickness. With such, the modified method could be fitted into 96-well plates on a fluorescent spectrometer that serves for quantification or screening of large-scale microalgal samples. They also showed that microwave pretreatment within less than 1 min could further improve the measurement of *in vivo* neutral lipids stained with Nile red (Chen et al., 2010).

Gao et al. (2008) have applied time-domain nuclear magnetic resonance (TD-NMR) to quantify lipid contents in *C. protothecoides*. The method was found simple, quick, and less expensive, but still with desired accuracy when compared with ordinary NMR. In the method, spin-echo NMR pulse sequence is employed to separate the lipid hydrogen nuclei signal from other hydrogen nuclei signals, and after calibration lipid contents in microalgae can be measured accurately.

As a nondestructive and rapid analytic technique, the near-infrared spectroscopy has many advantages in comparison to standard and traditional techniques such as the chromatographic method. Near-infrared reflectance spectroscopy (NIRS) has been used to determine the oil content and fatty acid composition in intact seeds (Kim et al., 2007). To our knowledge, a similar approach is being transplanted to analyze lipid levels and category in several microalgal species by Dr. Al Darzins's group and others.

6 THE CONVERSION OF ALGAL BIOMASS TO BIODIESEL

There are two general strategies for the conversion of algal biomass to biodiesel—lipid extracted from algae is transesterified to biodiesel or the whole algal biomass is decomposed via physical processing to bio-oil which in turn is refined to biodiesel. The past several years have seen obvious progresses in both directions(Farrell et al., 2010).

The conversion of lipid extracts is the typical mode of biodiesel production from algae, especially from highly dense heterotrophic microalgae that offer high yields of oil and turn the extraction to easy and low cost processing. The transesterification reactions are applied to convert algal triacylglycerols to FAMEs (fatty acid methyl esters), a displacement process of glycerol by mono-alcohols (e.g., methanol or ethanol). The technique is well developed to maturity and used as a standard in the conversion of vegetable oils into biodiesel. The reactions could be completed chemically with inorganic catalysts, or biochemically with the enzyme lipase (Farrell et al., 2010).

The transesterification reaction could be achieved under bases or acids. The base-catalyzed route is preferred though the acid-catalyzed one is an attractive alternative in certain situations, for example less sensitivity to water presence or reduction of saponification and emulsification (Wahlen et al., 2008). However, in general, acid catalysis possesses lower activity over that of base catalysis. As a result, higher temperatures and longer reaction times are required in these alkaline or acid-catalyzed processes. It means the chemical transesterification is energy intensive, and also requires the separation of glycerol and alkaline catalyst from biodiesel as well as the treatment of alkaline wastewater. Several variants of the transesterification reaction were developed to address these problems, including fast heating aided by microwave (Refaat and El Sheltawy, 2008), or improved mixing and heating by ultrasonic treatments (Kalva et al., 2008). Ultrasound-aided transesterification can also run inline which is superior to batch reaction in traditional methods. Particularly as the core of the technique, various catalysts of new types and catalysis modes are being developed. For example, heteropolyacids (HPA, e.g., $H_3PW_{12}O_{40}/Nb_2O_5$) have been shown to lower the required temperatures and decrease the reaction times (Alsalme et al., 2008; Cao et al., 2008). Mild Lewis acid catalysts such as AlC_{13} or ZnC_{12} are extremely efficient, in the presence of a cosolvent such as tetrahydrofuran, to convert triacylglycerols into fatty acid methyl esters up to 98% (Soriano et al., 2009). The third example is catalysts derived from the titanium compound (e.g., $HTiNbO_3$) or vanadate metal compounds (e.g., $TiVO_4$). These hydrophobic catalysts are insensitive to or reducing free fatty acid concentration, and used to achieve FAME and glycerol yields over 90% under moderate temperature and pressure (e.g., 200 °C and 35 bar; Cozzolino et al., 2006). Due to their insolubility to all other substances of the reaction system as well as their stability, they may offer an ideal option toward commercial transesterification reaction. Cheaper alternatives of this type might be identified from MgO, CaO, or Al_2O_3.

Lipases as biocatalysts are characterized with extremely specificity of transesterification reaction under quite mild conditions. The enzymes are also more attractive in environmental compatibility than classic or other inorganic catalysts. However, currently high prices and short lifetime of the enzymes limit their application at large scales. Biochemical engineering is required to improve their stability and lower the cost. Alternatively, by offering the interface for heterogeneous catalysis, immobilized lipases could reach high conversion rate up to 98%, which was demonstrated in microalgal oils (Xiong and Wu, 2008). The immobilization may also extend the lifetime of lipases and therefore improve the economy of enzymatic transesterification reaction.

By using supercritical alcohols (methanol or ethanol), oil extraction and transesterification reaction might be simplified to one-step conversion from whole wet biomass to biodiesel, in which the alcohols may function as both extractor of lipids and stimulator of transesterification reaction. The new processing has been confirmed in vegetable oils (Demirbas, 2006, 2009). The advantages of the approach are immediately foreseeable such as (1) convenient—all processing being done at one spot, (2) selective—multiple extractions of different components at high purity and concentration from algal biomass by selecting supercritical fluids, (3) "green" processing due to only alcohols being used and catalyst-free, (4) high quality and stability of biodiesel and other value-added compounds attainable at modest conditions (e.g., less than 50 °C), (5) efficient—whole native algae being employed without dewatering and oil extraction. Yet, the supercritical manipulation with methanol or ethanol may still decompose algal biomass that may potentially reduce biodiesel yield (Hawash et al., 2009; Vieitez et al., 2009). Further study is required to detail the operating conditions in the microalgal system (Patil et al., 2010).

When algal biomass contains lower amounts of lipids or when the secondary conversion from the remnants of oleaginous microalgae should be performed after lipid extraction, the physical processing such as pyrolysis or hydrothermal liquefaction might be suitable to turn these kinds of algal biomasses to biodiesel. Pyrolysis is used to decompose condensed biomass by extremely fast heating (e.g., a couple of seconds) in the absence of oxygen. For example, heating to 350-500 °C is completed within less than 2 s in fast pyrolysis in which finely ground particles of biomass should be provided. This created an advantageous situation for microalgal biomass because microalgae exist as single cells with micrometer-scale sizes (e.g., 10-30 μm in diameter), and therefore the grinding processing is not necessary for microalgal particles. Oxygen contents in different sources of biomass might affect the quality of bio-oil, the product of pyrolysis. Our previous observation showed that bio-oil from heterotrophic microalgae contained lower oxygen over that from autotrophic microalgae of the same species (Miao and Wu, 2004), implying the metabolic pathways might alternate oxygen contents in algal biomass and presenting a practical biological approach to monitor the quality of biofuels. The bio-oil may directly enter the biorefinery stream for production of useable biodiesel. The research in this direction is still in its infancy.

Wet biomass is suitable to supply to a hydrothermal liquefaction processing where water is held in a liquid state above 100 °C under pressure (called subcritical water). Biocrude, the main product of the processing, contains smaller molecules of high energy density that might be comparable to fossil diesel. The biocrude may contain additional valuable compounds and need further upgrade. Except a couple of old reports that indicated bio-oil yields of 37-64% could be generated at 300 °C and 10 MPa from *Botryococcus braunii* (Sawayama et al., 1995) or *Dunaliella tertiolecta* (Minowa et al., 1995), quite few reports on the topic could be identified from the recent literature. Since the processing may mimic the natural geological processes associated with petroleum formation and it is believed microalgae like diatom might have a major contribution to the evolution of fossil fuels, more researches are interestingly awaited in the near future.

7 THE QUALITY AND ECONOMIC ANALYSIS OF ALGAL BIODIESEL

Based on the ASTM biodiesel standard (e.g., ASTM International, 2009a,b), Miao and Wu (2006) examined properties of biodiesel from microalgal oil. The parameters range from density, viscosity, flash point, cold filter plugging point, solidifying point, to heating value. Most of the parameters comply with the ASTM standard. The properties of algal biodiesel were also comparable to those of fossil diesel, indicating that algal biodiesel is probably able to blend with fossil diesel.

Three culture modes—autotrophic algae in raceway open pond or closed photobioreactor, and heterotrophic algae in fermenter—were compared in terms of economic parameters from biomass production and harvest, oil extraction, capital, labor, to operational costs based on conditions in British Columbia (Alabi et al., 2009). Total costs for these three modes could be evaluated by using a thermodynamic model into which local light and temperature values as well as various cost parameters were integrated. The results showed that the base case costs for the three biomass production systems were \$14.44 (raceway), \$24.60 (photobioreactor), and \$2.58 (fermenter) per liter of algal oil, respectively. As a reference, the cost for per liter of canola oil was \$0.88. All these biological oils currently cannot compete with petroleum for fuel production. An immediate conclusion from this analysis could be heterotrophic fermentation appears to be the most promising approach for biofuel production, while autotrophic algae in photobioreactor are unaffordable economically. Further analysis indicated that the cost of fermentation was mainly composed of the power and organic carbon substrate. As addressed before, the two critical issues could be solved through cheap alternatives of current organic carbons and through large-scale production. The cost for oil conversion to biodiesel and the comparison between autotrophic and heterotrophic modes are not yet available, but a case study in vegetable oil showed that the conversion costs were \$0.26 or \$0.51 gal^{-1} when supercritical or traditional alkaline transesterification was employed (Anitescu et al., 2008). The supercritical transesterification showed lower cost over the conventional one because it could be finished in one step using alcohols. Taken together, these precommercial analyses favor the heterotrophic fermentation route might lead to commercial production of algal biodiesel.

8 CONCLUDING REMARK AND FUTURE PERSPECTIVES

Through the examination of major aspects of algal biodiesel production from upstream algal culture to downstream oil extraction and conversion, this review has seen significant progresses in algal biodiesel production, particularly from heterotrophic microalgae, were made during the past several years. Along with continual efforts imposed into the field, by focusing on lowering the costs of the whole process through technical improvement, comprehensive utilization of algal biomass components, and scaleup in production, it is hoped that the innovative algal biodiesel technology will probably find its application in commercial plants at large scale.

Specific tasks aiming to further yield algal biodiesel in quantity and quality are proposed and might be the focus of the next phase explorations in the field, in addition to general concerns on algal biofuels addressed by Wijffels and Barbosa (2010) and many others.

(a) In-depth insights are required about lipid metabolism in either autotrophic or heterotrophic microalgae. Particularly, the knowledge of lipid droplet biology ranging from dynamic structure, storage or degradation, to regulation will directly benefit the technological development involved in biodiesel production. For autotrophic microalgae, by using genetic engineering in capture, storage, and conversion of solar energy through photosynthesis, either biomass yield of oleaginous algae or oil yield of fast-growing algae might be enhanced. The underlying mechanisms of stress-induced accumulation of neutral lipid need to be elucidated, from which new approaches might be figured out to attenuate adverse effects of stress treatments on growth. For heterotrophic microalgae, more deep insights into the conversion from glucose to lipid might be generated that will help in seeking substitutes of costly glucose—either cheaper nutritional or regulatory substances. In addition, mixotrophic pathway may provide a unique strategy for algae to utilize both sunlight and chemical energy. The details of mixotrophic features should be characterized in the near future.

(b) A systems approach is required to study the whole chain of the process from algal strain development to utilization of all biomass components for the reduction of the costs. Based on the fast progresses that have been made in heterotrophic microalgae that could be fitted into the infrastructure of mature fermentation industry with high efficiency, heterotrophic microalgae might be placed on the priority list of the algal biodiesel production.

(c) Integration and optimization of the whole processing of algal biomass to acquire maximal benefits and increasing competiveness over other feedstock-based processing. Coproduct extraction might be coupled in parallel or in tandem with oil extraction from heterotrophic microalgae. After extraction, algal remnants might be turned to biogas through the anaerobic fermentation, or directly used as feeds and fertilizer. As a long-term goal, the remnants might be subject to the physical treating like pyrolysis or liquefaction to produce bio-oils which in turn are refined to various biofuels when the biorefining is applied to commercial production.

Acknowledgments

This study was supported by the projects 30970224 and 41030210 from the NSF of China, the MOST 863 projects 2009AA064401 and 2010AA101601, the MOST supporting project 2011BAD14B05. We apologize for those progresses made in the field that could not be cited owing to the limited space assigned to the article.

References

Alabi, A.O., Tampier, M., Bibeau, E., 2009. Microalgae technologies & progresses for biofuels/bioenergy production in British Columbia: current technology, suitability & barriers to implementation. January 2009.

Alsalme, A., Kozhevnikova, E.F., Kozhevnikov, I.V., 2008. Heteropoly acids as catalysts for liquid-phase esterification and transesterification. Appl. Catal. A Gen. 349 (1-2), 170–176.

Anitescu, G., Deshpande, A., Tavlarides, L.L., 2008. Integrated technology for supercritical biodiesel production and power cogeneration. Energy Fuels 22 (2), 1391–1399.

ASTM International, 2009a. Standard Specification for Diesel Fuel Oils. ASTM International, West Conshohocken, PA. Available at: http://www.astm.org/Standards/D975.htm.

ASTM International, 2009b. Standard Specification for Biodiesel Fuel Blend Stock (B100) for Middle Distillate Fuels. ASTM International, West Conshohocken, PA. Available at: http://astm.org/Standards/D6751.htm.

Cao, F., Chen, Y., Zhai, F., Li, J., Wang, J., Wang, X., et al., 2008. Biodiesel production from high acid value waste frying oil catalyzed by superacid heteropolyacid. Biotechnol. Bioeng. 101 (1), 93–100.

Chen, W., Zhang, C., Song, L., Sommerfeld, M., Hu, Q., 2009. A high throughput Nile red method for quantitative measurement of neutral lipids in microalgae. J. Microbiol. Methods 77, 41–47.

Chen, W., Sommerfeld, M., Hu, Q., 2010. Microwave-assisted Nile red method for in vivo quantification of neutral lipids in microalgae. Bioresour. Technol. 102, 135–141.

Cheng, Y., Yue Lu, Y., Gao, C., Wu, Q., 2009a. Alga-based biodiesel production and optimization using sugarcane as feedstock. Energy Fuels 23, 4166–4173.

Cheng, Y., Zhou, W., Gao, C., Lan, K., Gao, Y., Wu, Q., 2009b. Biodiesel production from Jerusalem artichoke (*Helianthus Tuberosus* L.) tuber by heterotrophic microalgae *Chlorella protothecoides*. J. Chem. Technol. Biotechnol. 84, 777–781.

Chisti, Y., 2007. Biodiesel from microalgae. Biotechnol. Adv. 25, 294–306.

Cozzolino, M., Tesser, R., Di Serio, M., Gaigneaux, E.M., Eloy, P., Santacesaria, E., 2006. Supported vanadium oxide nanoparticles: effect of preparation method, support and type of precursor on the catalytic performances in the ODH of methanol to formaldehyde. Scientific Bases for the Preparation of Heterogeneous Catalysts Proc. 9th Int. Symposium, Louvain-la-Neuve, Belgium, September 10-14, 2006. In: Gaigneaux, E.M., Devillers, M., De Vos, D.E., Hermans, S., Jacobs, P.A., Martens, J.A., Ruiz, P. (Eds.), Studies in Surface Science and Catalysis, vol. 162. 697–704.

Demirbas, A., 2006. Oily products from mosses and algae via pyrolysis. Energy Sourc. A Recov. Util. Environ. Eff. 28 (10), 933–940.

Demirbas, A., 2009. Biodiesel from waste cooking oil via base-catalytic and supercritical methanol transesterification. Energy Convers. Manag. 50 (4), 923–927.

Farrell, J., Sarisky-Reed, V., et al., 2010. National Algal Biofuels Technology Roadmap. U.S. Department of Energy, Office of Energy Efficiency and Renewable Energy, Biomass Program. May 2010. Available at: http://biomass.energy.gov.

Gao, C., Xiong, W., Zhang, Y., Yuan, W., Wu, Q., 2008. Rapid quantitation of lipid in microalgae by time-domain NMR in alga-based biodiesel production. J. Microbiol. Methods 75, 437–440.

Gao, C., Zhai, Y., Wu, Q., 2010. Application of sweet sorghum for biodiesel production by heterotrophic microalga *Chlorella protothecoides*. Appl. Energy 87, 756–761.

Gouveia, L., Oliveira, A.C., 2009. Microalgae as a raw material for biofuels production. J. Ind. Microbiol. Biotechnol. 36 (2), 269–274.

Hawash, S., Kamal, N., Zaher, F., Kenawi, O., El Diwani, G., 2009. Biodiesel fuel from Jatropha oil via non-catalytic supercritical methanol transesterification. Fuel 88 (3), 579–582.

Hu, Q., Sommerfeld, M., Jarvis, E., Ghirardi, M., Posewitz, M., Seibert, M., et al., 2008. Microalgal triacylglycerols as feedstocks for biofuel production: perspectives and advances. Plant J. 54 (4), 621–639.

Kalva, A., Sivasankar, T., Moholkar, V.S., 2008. Physical mechanism of ultrasound-assisted synthesis of biodiesel. Ind. Eng. Chem. Res. 48 (1), 534–544.

Kim, K.S., Park, S.H., Choung, M.G., 2007. Nondestructive Determination of Oil Content and Fatty Acid Composition in Perilla Seeds by Near-Infrared Spectroscopy. J. Agric. Food Chem. 55, 1679–1685.

Lee, Y.K., 2004. Algal nutrition: heterotrophic carbon nutrition. In: Richmond, A. (Ed.), Handbook of Microalgal Culture: Biotechnology and Applied Phycology. Blackwell Publishing, pp. 116–124.

Li, X., Xu, H., Wu, Q., 2007. Large-scale biodiesel production from microalga Chlorella protothecoids through heterotrophic cultivation in bioreactors. Biotechnol. Bioeng. 98 (4), 764–771.

Li, Y., Han, D., Hu, G., Dauvillee, D., Sommerfeld, M., Ball, S., et al., 2010. Chlamydomonas starchless mutant defective in ADP-glucose pyrophosphorylase hyper-accumulates triacylglycerol. Metab. Eng. 12 (4), 387–391.

Lu, Y., Zhai, Y., Liu, M., Wu, Q., 2010. Biodiesel production from algal oil using cassava (*Manihot esculenta* Crantz) as feedstock. J. Appl. Phycol. 22, 573–758.

Miao, X., Wu, Q., 2004. High yield bio-oil production from Fast pyrolysis by metabolic controlling of *Chlorella protothecoides*. J. Biotechnol. 110 (1), 85–93.

Miao, X., Wu, Q., 2006. Biodiesel production from heterotrophic microalgal oil. Bioresour. Technol. 97, 841–846.

Minowa, T., Yokoyama, S.Y., Kishimoto, M., Okakura, T., 1995. Oil production from algal cells of *Dunaliella tertiolecta* by direct thermochemical liquefaction. Fuel 74, 1735–1738.

Patil, P.D., Gude, V.G., Mannarswamy, A., Deng, S., Cooke, P., Munson-McGee, S., et al., 2010. Optimization of direct conversion of wet algae to biodiesel under supercritical methanol conditions. Bioresour. Technol. 102, 118–122.

Peralta-Yahya, P.P., Keasling, J.D., 2010. Advanced biofuel production in microbes. Biotechnol. J. 5, 147–162.

Refaat, A.A., El Sheltawy, S.T., 2008. Time factor in microwave-enhanced biodiesel production. WSEAS Trans. Environ. Devel. 4 (4), 279–288.

Rodolfi, L., Chini Zittelli, G.C., Bassi, N., Padovani, G., Biondi, N., Bonini, G., et al., 2009. Microalgae for oil: strain selection, induction of lipid synthesis and outdoor mass cultivation in a low-cost photobioreactor. Biotechnol. Bioeng. 102, 100–112.

Rosenberg, J.N., Oyler, G.A., Wilkinson, L., Betenbaugh, M.J., 2008. A green light for engineered algae: redirecting metabolism to fuel. Curr. Opin. Biotechnol. 19, 430–436.

Sawayama, S., Inoue, S., Dote, Y., Yokoyama, S.Y., 1995. CO2 fixation and oil production Through microalga. Energy Convers. Manag. 36, 729–731.

Sheehan, J., Dunahay, T., Benemann, J., Roessler, P., (1998). A Look Back at the U.S. Department of Energy's Aquatic Species Program: Biodiesel from Algae. .

Sirianuntapiboon, S., Prasertsong, K., 2008. Treatment of molasses wastewater by acetogenic bacteria BP103 in sequencing batch reactor (SBR) system. Bioresour. Technol. 99, 1806–1815.

Soriano Jr., N.U., Venditti, R., Argyropoulos, D.S., 2009. Biodiesel synthesis via homogeneous Lewis acid-catalyzed transesterification. Fuel 88 (3), 560–565.

Steen, E.J., Kang, Y., Bokinsky, G., Hu, Z., Schirmer, A., McClure, A., et al., 2010. Microbial production of fatty-acid-derived fuels and chemicals from plant biomass. Nature 463, 559–562.

Su, C.H., Fu, C.C., Chang, Y.C., Nair, G.R., Ye, J.L., Chu, I.M., et al., 2008. Simultaneous estimation of chlorophyll a and lipid contents in microalgae by three-color analysis. Biotechnol. Bioeng. 99, 1034–1039.

Vieitez, I., Da Silva, C., Alckmin, I., Borges, G.R., Corazza, F.C., Oliveira, J.V., et al., 2009. Effect of temperature on the continuous synthesis of soybean esters under supercritical ethanol. Energy Fuels 23 (1), 558–563.

Wahlen, B.D., Barney, B.M., Seefeldt, L.C., 2008. Synthesis of biodiesel from mixed feedstocks and longer chain alcohols using an acid-catalyzed method. Energy Fuels 22 (6), 4223–4228.

Wawrik, B., Harriman, B.H., 2010. Rapid, colorimetric quantification of lipid from algal cultures. J. Microbiol. Methods 80 (3), 262–266.

Wei, A., Zhang, W., Wei, D., Chen, C., Wu, Q., Yang, S.T., 2009. Effects of cassava starch hydrolysate on cell growth and lipid accumulation of the heterotrophic microalgae *Chlorella protothecoides*. J. Ind. Microbiol. Biotechnol. 36, 1383–1389.

Wijffels, R.H., Barbosa, M.J., 2010. An outlook on microalgal biofuels. Science 329, 796–799.

Wu, Z.Y., Shi, X.M., 2006. Optimization for high-density cultivation of heterotrophic Chlorella based on a hybrid neural network model. Lett. Appl. Microbiol. 44, 13–18.

Xiong, W., Wu, Q., 2008. High-density fermentation of microalga *Chlorella protothecoides* in bioreactor for microbiodiesel production. Appl. Microbiol. Biotechnol. 78, 29–36.

Xiong, W., Gao, C., Yan, D., Chao, W.u., Wu, Q., 2010a. Double CO_2 fixation in photosynthesis-fermentation model enhances algal lipid synthesis for biodiesel production. Bioresour. Technol. 101, 2287–2293.

Xiong, W., Liu, L., Wu, C., Yang, C., Wu, Q., 2010b. 13C-Tracer and gas chromatography-mass spectrometry analyses reveal metabolic flux distribution in the oleaginous microalga *Chlorella protothecoides*. Plant Physiol. 154, 1–12.

Xu, H., Wu, Q., 2006. High quality biodiesel production from a microalgae *Chlorella protothecoides* by heterotrophic growth in fermenters. J. Biotechnol. 126, 499–507.

Yan, D., Lu, Y., Chen, Y.F., Wu, Q., 2011. Waste molasses alone displaces glucose-based medium for microalgal fermentation towards cost-saving biodiesel production. Bioresour. Technol. (in press).

Zhekisheva, M., Boussiba, S., Khozin-Goldberg, I., Zarka, A., Cohen, Z., 2002. Accumulation of oleic acid in *haematococcus pluvialis* (chlorophyceae) under nitrogen starvation or high light is correlated with that of astaxanthin esters. J. Phycol. 38 (2), 325–331.

CHAPTER

18

Overview and Assessment of Algal Biofuels Production Technologies

*Ganti S. Murthy**

Biological and Ecological Engineering, Oregon State University

***Corresponding author: E-mail: murthy@engr.orst.edu**

1 INTRODUCTION

World economy is critically dependent on fossil fuels. Fossil fuels, in addition to being limited non-renewable resources, are concentrated in unstable regions of the world constituting a national security threat. Global climate change has been shown to be a direct consequence of anthropogenic carbon dioxide in atmosphere over the last two centuries. As the world looks for for alternatives to fossil fuels, first-generation biofuels such as corn ethanol and soybean biodiesel were an initial effort to produce transportation fuels domestically and mitigate green house gas emissions. However, due to production capacity limitations, first-generation biofuels cannot meet the transportation fuel needs. Additionally, some of the issues such as food versus fuel debate, intensive use of agricultural inputs such as fertilizers and pesticides, low net energy balance, and uncertain environmental impacts have led to investigation of second-generation biofuels.

Second-generation biofuels such as ethanol from grass straws, corn stover, switch grass, and other herbaceous crops do not directly compete with food sources and have higher net energy than corn ethanol. However, even the second-generation biofuels do not completely eliminate the need for fertilizers, pesticides, arable land, and fresh water in their production. Recalcitrance of cellulosic feedstocks poses additional technological challenges in conversion of these feedstocks to ethanol. Due to the challenges in production, harvesting, and processing technologies, ethanol from cellulosic feedstocks is yet (as in 2010) to be produced in significant quantities. Algae biofuels have the potential to be a sustainable alternative to produce transportation fuels without concerns about food versus fuel or use of valuable agricultural lands (Table 1). This has led to research interest in third-generation biofuels such as biodiesel and ethanol from algae.

TABLE 1 Comparison of Different Generations of Biofuel Alternatives

	First Gen. (Corn Ethanol)	Second Gen. (Cellulosic Ethanol)	Third Gen. (Algal Biofuels)
Enhance energy security	☑	☑	☑
Minimum impact on current transportation infrastructure	☑	☑	☑
Food versus fuel debate	×	☑	☑
Limited capacity to meet the fuel needs	×	☑×	☑
Intensive use of agricultural inputs	×	☑	☑
Net energy balance studies	× (1.16-1.67)	☑ (3-5)	☑ (9-10?)
Impact on environment	×	×	×
Need for arable lands	×	×	☑
Need for fresh water	×	×	☑

This increased attention to algal biofuels can be described more as a reawakening than start of a novel paradigm as in last 50 years, algae have been investigated as sources of food, feed, and fuels. Perhaps most famous of these efforts is the US DOE's Aquatic Species Program that started in late 1970s and continued till 1995. The closeout report from this program (Sheehan et al., 1998) contains an extensive summary of the efforts to convert algae biomass into biofuels. Similar efforts were also underway in Japan during the same timeframe.

Current production of micro- and macroalgae around the world is around 10,000 tons/year. Macroalgae have been traditionally used mostly for food applications. While micro and macroalgae have been investigated for other applications such as production of pigments and neutraceuticals, most of the algal biofuels research has been focused on microalgae. This chapter mostly focuses on production technologies, resource use, energetic, and life-cycle assessment of utilizing microalgae biofuels production.

Microalgae are microscopic plants that can grow in diverse environmental conditions. Some strains of algae accumulate lipids up to 60% of their body weight, while other strains accumulate starch. Compared to oil yield of 455 L/ha (48 gal/acre) from soybeans and 5685 L/ha (600 gal/acre) from oil palms grown in tropical regions, algae have been shown to yield upto 7760 L/ha (819 gal/acre). Algae have a potential to produce up to 40,000 L/ha (4222 gal/acre) of biodiesel (Weyer et al., 2010). Due to their higher productivities (~8 times compared to soybeans) and higher lipid content, their potential for biodiesel production has been extensively investigated. Aquatic species program of the U.S. Department of Energy identified, among others, cost of production as the principal obstacle in the adoption of algae technology on a large scale (Sheehan et al., 1998).

In order to achieve a sustainable biofuels production using algae, three critical aspects of algae system need to be mastered: production, recovery, and processing. Efficient production of algae biomass by leveraging waste water treatment, use of flue gases, and waste heat will reduce the production cost of algae as a feedstock for fuels and chemicals. Second aspect of an integrated system is the separation/recovery of algae from the medium. The third aspect is the production of fuels and chemicals from algae. A brief overview of the challenges and opportunities in each of the three aspects is discussed below.

2 AUTOTROPHIC PRODUCTION TECHNOLOGIES

Algal growth in autotrophic production systems occurs through photosynthesis converting light energy into chemical energy. Open ponds and closed photobioreactor (PBR) systems are the two main types of autotrophic cultivation systems that have been investigated. Both types of systems have different advantages and disadvantages. Therefore, hybrid systems with features of both open pond and closed PBR systems have also been proposed.

2.1 Open Ponds

Open pond cultivation systems generally consist of shallow raceways (0.15-0.45 m depth) constructed as concrete, clay, or plastic-lined ponds. Capital costs for construction of open raceway ponds are lower compared to closed PBRs (Benemann and Oswald, 1996). Paddle wheels are used to provide the necessary mixing to maintain algae culture in suspension. While the mixing energy requirement of paddle wheels is relatively low, efficiency of gas transfer is also lower. Sometimes, aerators may be used to supplement CO_2 for maximizing the algae growth. Pond temperature is not generally controlled and the light intensity is dependent on incoming solar insolation. Hence, efficiency of the open ponds is dependent on the local diurnal variations in temperature and solar insolation. While evaporative cooling of open ponds partially regulates the temperatures of open ponds, it also leads to significant loss of water (DOE, 2010).

It is difficult to maintain algae monocultures in open ponds due to contamination with native algae and algae grazers. Many strategies such as operating at higher salinity, pH, or temperatures have been proposed to maintain an exotic microenvironment in the open ponds that minimizes contamination with other strains. Successful commercial cultivation of *Spiriluna* monocultures in open ponds has been made possible by maintaining the open ponds at high pH (9.0-11.0). Similarly, β-carotene production from *Dulaliella* sp. in open ponds is possible due to high salinity levels in the open ponds. Although other systems such as circular ponds have been utilized for production of high-value products, most common open pond configuration for algal biofuels production is the open raceway ponds (Figure 1).

2.2 Closed PBRs

PBRs are closed bioreactors that permit exchange of light and energy without material exchange from surroundings. Over the years, PBRs in tubular, flat plate, column, spiral, plastic bag, vertical bag configurations have been proposed.

Growing algae in PBRs confers many advantages such as reduced land area requirement, greater productivity at higher culture densities, reduced contamination, lower evaporative water losses, and better control compared to open pond systems. However, PBR scalability is one of the challenges that still has to be addressed comprehensively. Temperature regulation

FIGURE 1 Lab-scale open pond.

of PBRs is a second challenge due to absence of evaporative cooling. Light penetration, fouling/biofilm formation on the PBR walls, gradients of pH, dissolved oxygen and CO_2 within the PBR, and hydrodynamic stresses are some of the additional challenges that have to be addressed for algae cultivation in PBR (Brennan and Owende, 2010; DOE, 2010; Pulz, 2001).

2.3 Hybrid and Novel Systems

Since cost of production, land area requirement, evaporative loss of water, and contamination are some of the critical aspects of economically viable algal biomass production, combinations of open ponds and low-cost PBRs have been proposed. In these proposed systems, first stage usually consists of closed PBRs to culture the initial inoculum with robust growth characteristics and minimum contamination. This inoculum is transferred during a second stage to an open pond for maximizing the biomass growth and lipid accumulation. Additional system configurations such as open ponds covered with transparent plastic sheets to minimize contamination and evaporative losses, low-cost PBRs utilizing plastic bags supported on vertical stands to maximize space and light utilization efficiency have also been proposed. Presently, most of the PBR systems are experimental- or laboratory-scale devices with very few examples of pilot-scale systems in continuous operation.

As any type of land-based algae cultivation system is restricted by land availability, open coastal/offshore systems for production of macro- and microalgae have been proposed. Large-scale systems for commercial production of macroalgae in offshore/near shore systems exist in China (WSA, 2009) and South Korea. One of the novel systems for microalgae production is offshore membrane enclosures for growing algae (OMEGA) (WSA, 2009).

3 HETEROTROPHIC AND MIXOTROPHIC PRODUCTION

Heterotrophic growth of algae, using organic carbon substrates such as sugars, has been successfully demonstrated for production of algae biomass and other metabolites. Since the algae do not depend on photosynthesis during heterotrophic growth phase, algae can be cultured using standard industrial fermenters to high culture densities. With greater control of culture conditions, it is possible to obtain desired product profiles such as higher lipid content during heterotrophic growth. Higher cell densities also result in lower capital and operating costs. However, long-term sustainability of such systems needs to be evaluated carefully as production of organic carbon substrate such as sugars can be energy intensive and has similar limitations as first- and second-generation biofuels.

Mixotrophic growth of the algae for production of algal biofuels and high-value products has been proposed. Some of the proposed systems supplement the initial inoculum preparation with organic carbon substrates to achieve higher cell densities before introduction to open ponds for final increase in biomass and contamination reduction by outcompeting the other algae. The main advantage of such systems is increasing productivities by utilizing both autotrophic and heterotrophic growth phases. However, as with heterotrophic systems, long-term sustainability of utilizing organic carbon substrates for production of algal biomass needs to be evaluated.

4 HARVESTING AND PROCESSING OF ALGAL BIOMASS

4.1 Harvesting/Dewatering Technologies

Algae after cultivation in open raceway ponds or closed PBRs exist as dilute solution of algae (0.1-10 g/L). Recovering algae biomass from such dilute solutions poses many challenges, especially for open pond cultivation systems. Therefore, development of efficient processes to recover algae is critical for economic viability of algal biodiesel.

Over the years, many of the technologies such as bio/chemical flocculation, sedimentation, dissolved air floatation; various types of centrifuges and filtration systems, hydrocyclones, vacuum filters have been proposed and utilized (Mohn, 1980; Molina Grima et al., 2003). Electroflocculation and ultrasonic-assisted algae concentration techniques have also been proposed recently (DOE, 2010). A summary of the performance of harvesting/dewatering technologies is presented in Table 2.

Most of the proposed strategies to recover algae from the growth media such as centrifuges, screens, and bio/chemical flocculation are expensive or unreliable in a continuous large-scale operation. Although some of these processes are used in commercial production of *Spiriluna,* the economics of operations are different economics as it is sold as human food. Therefore, to produce biodiesel from algae, it is critical to use simple, reliable, and low-cost algae recovery processes. Suitability of a particular harvesting technology is critically dependent on the strain of the algae. Therefore, pilot-scale tests must be conducted before any decision regarding the optimum harvesting technology can be made.

4.2 Processing Technologies

Over the years, many processing technologies (Figure 2) have been investigated to convert algal biomass into useful products. A brief overview of the technologies is provided below.

TABLE 2 Harvesting and Dewatering Technologies (Data from Mohn, 1980; Molina Grima et al., 2003)

Harvesting	Suspended Solids (%)	Concentration Factor	Energy Use (kWh/m^3)	Continuous?
Sedimentation tank	1.5	15	0.1	D
Lamella separator	1.6	16	0.1	D
Self-cleaning disk stack	12	120	1	C
Nozzle discharge	2-15	20-150	0.9	D
Decanter bowl	22	11	8	C
Hydrocyclones	0.4	4	0.3	C
Netzsch chamber filter	22-27	245	0.88	D
Netzsch belt filter	18	180	0.5	C
Suction filter	16	160	-	D
Cylindrical sieve rotators	7.5	75	0.3	C
Filter basket	5	50	0.2	D
Nonprecoat vacuum filter	18	180	5.9	C
potato starch precoat vacuum filter	37	2-18.5	-	C
Vacuum Suction filter	8	80	0.1	D
Vacuum belt filter	9.5	95	0.45	C
Filter thickener	5-7	50-70	1.6	D
Nonprecoat vacuum filter	18	180	5.9	C
Solar drying	85	8.5	0.01	D
Thermal drying	90	9	0.627 Kwh/kg H_2O	C

4.2.1 Anaerobic Digestion

One of the most direct approaches for utilization of algae biomass is methane production in anaerobic digester. This approach has the advantage of utilizing wet algae without the need for additional drying. Many researchers had tried anaerobic digestion of both micro- and macroalgae biomass since 1950s. In addition to production of a clean combustible gas, sludge from the anaerobic digestion can be used as a nitrogen-rich organic fertilizer. Anaerobic digestion can be integrated in waste water treatment systems. While anaerobic digestion has many advantages, the process cannot be used to produce liquid transportation fuels and takes comparatively longer time for algae processing. Additionally building industrial-scale anaerobic digestion facilities that can handle millions of tons of biomass annually can pose significant challenges.

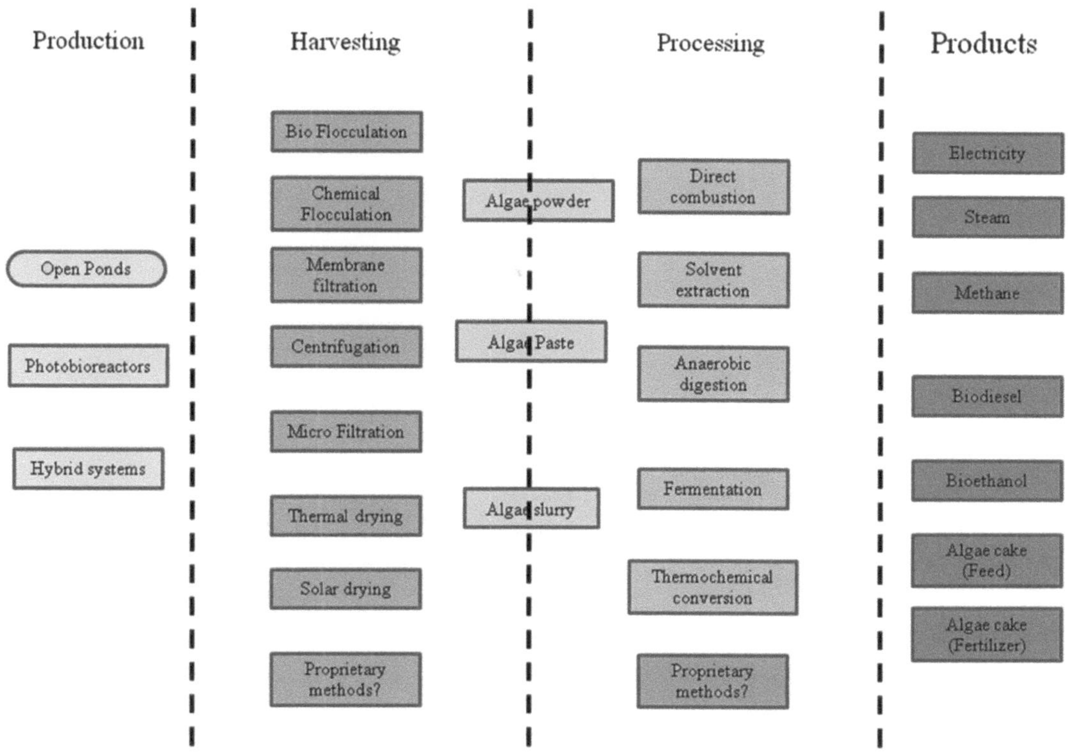

FIGURE 2 Processing technologies for algal biomass.

4.2.2 *Thermochemical Conversion*

Various types of thermochemical processes including direct combustion, fast pyrolysis, gasification, and pyrolysis have been suggested for utilizing algae biomass. Main advantage of the thermochemical conversion processes is that algae or any other biomass with varying composition can be used. However, thermochemical processes also result in a wide range of products and require additional processing to produce usable fuels. Different temperatures ranges and oxygen levels differentiate various thermochemical conversion processes (Figure 3). Direct combustion is combustion of biomass in excess of oxygen and mainly leads to heat energy which can be utilized for steam and/or electricity production. Gasification involves combustion of the biomass in presence of limited amount of oxygen and produces syngas and char as the main products (Balat et al., 2009a). Pyrolysis, on the other hand, involves heating of dry biomass in complete absence of oxygen and produces syngas, bio-oil, and bio char as the main products (Balat et al., 2009b). Direct hydrothermal liquefaction is very similar to pyrolysis except that it occurs at much higher feedstock moisture contents. Product composition can be controlled in these processes by controlling the biomass composition, catalysts, operating temperatures, and pressures. Syngas produced from these technologies can be further upgraded to liquid fuels through Fischer-Tropcsh process.

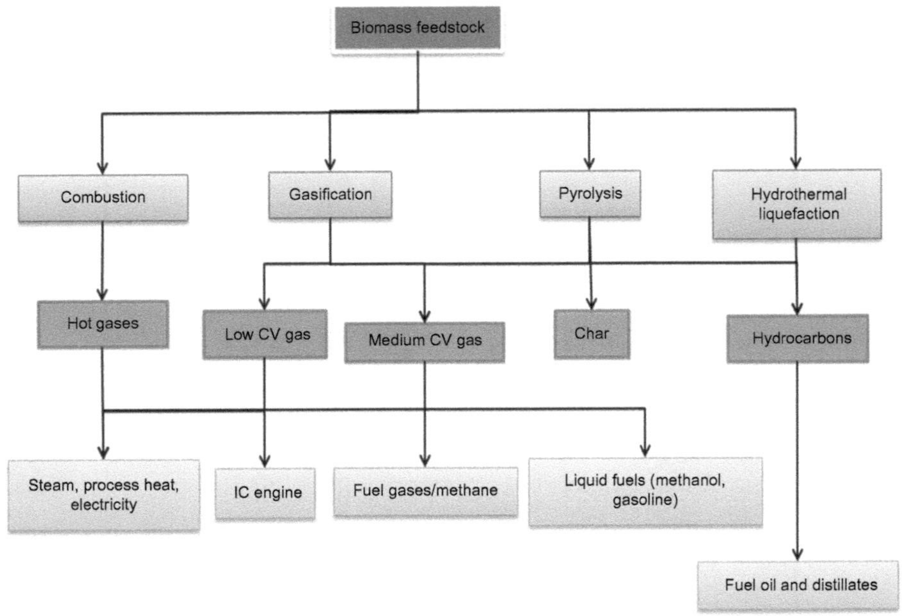

FIGURE 3 Overview of thermochemical conversion processes.

One of the simplest approaches for utilization of algae biomass is combustion for combined heat and power production. However, energy costs for drying the biomass may be prohibitive. However, with the advent of new types of fluidized bed burners, even wet biomass with 50% moisture can be directly combusted. Cofiring of coal-algae mixture in coal-fired power plants has also been suggested and could result in lower green house gas emissions.

Gasification of algae has been extensively studied. A novel system of low-temperature gasification that converts all the nitrogen in algae into recoverable ammonia has been proposed (Minowa and Sawayama, 1999). Hirano et al. (1998) studied the gasification of *Spiriluna* and obtained theoretical yields of methane (64% biomass) with an energy balance of 1.1 (ratio of methanol produced to total energy required).

Pyrolysis has been used to produce bio-oil from algae and other biomass. Bio-oil can be refined in existing petroleum refineries after some additional processing such as hydrotreating and hydrocracking (DOE, 2010). Three types of pyrolysis have been reported in literature and are defined based on the temperatures, residence time, and heating rates (Balat et al., 2009b; McKendry, 2002). In general, liquid product yields are favored by faster heating rates, shorter residence times, and moderate temperatures. Since efficiency of the pyrolysis is dependent on the feedstock particle size, microalgae have an advantage over lignocellulosic feedstocks. However, since pyrolysis can be performed only on low moisture (<15%) feedstocks, the need to dry the algae before pyrolysis would offset any particle size advantage.

Direct hydrothermal liquefaction process can be used to produce liquid fuels from wet algal biomass (Aresta et al., 2005; Minowa et al., 1995). This process is conducted at moderate temperatures (300-350°C) and pressures (5-20 MPa) in presence of a catalyst to yield bio-oil (Brennan and Owende, 2010). Use of microwave heating has been suggested for efficient heating of wet biomass. Subcritical water reacts with the biomass and results in formation of energy dense biocrude with lower heating values of 30-35 Mk/Kg (Patil et al., 2008). Biocrude yields up to 37% of the biomass have been reported (Minowa et al., 1995) with properties similar to heavy oil (Amin, 2009). Thermal efficiency of this process has been claimed to be up to 75%. Since wet biomass can be utilized, hydrothermal liquefaction process has much more favorable energetic balance compared to gasification and pyrolysis.

4.2.3 *Solvent Extraction*

For liquid transportation fuels production, algae cells must be disrupted and oil present extracted. Most common technology for lipid recovery from dried algae is solvent extraction using one of the many polar solvents such as hexane, chloroform, petroleum ether, butanol and methanol. Hexane extraction is the most common solvent due to nontoxicity of hexane compared to other solvents. Solvent extraction has been successfully used for extraction of β-carotene and asthaxanthan from algae. In addition to the traditional solvent extraction of dead algae, 'milking' of live algae cells using decane and dodecane as solvents has also been proposed (DOE, 2010). Using this approach, the live algae cells can be stripped of their lipids and returned to production system for continued accumulation of triglycerides.

Thermal pretreatment of the algae cells has been recently demonstrated to enhance solvent recovery (Kita et al., 2010). Among the different methods for cell disruption such as autoclaving, bead-beating, sonication and microwave heating prior to solvent extraction using chloroform and methanol, microwave heating method was identified as the most effective and simple method of cell disruption (Lee et al., 2010).

While conversion of oil in algae to biodiesel can be accomplished using standard commercially available technology, extraction of oil is one of the key challenges. Remaining solids rich in protein and carbohydrates can be used as animal feed.

4.2.4 *Fermentation*

Significant fraction of algal biomass is carbohydrates. Carbohydrate fraction of the algae biomass can be hydrolyzed and fermented using yeast to produce ethanol. The carbohydrates in algae can exist as starch, cellulose, and other structural. Hydrolysis of starch is an established technology and widely used in production of corn ethanol; similar extension to algae starch hydrolysis should not pose significant difficulties. Hydrogen-bonding pattern differences in cellulose (O'Sullivan, 1997) from algae (1_α) and higher plants (1_β) could result in their different response to cellulases and thus different hydrolysis yields. With a modest pretreatment process, the hydrolysis efficiencies can be increased. Algae cells contain other structural carbohydrates such as pectin and chitin. There is a need for additional research to identify enzymes and pretreatments for effective hydrolysis of algae carbohydrates.

5 CHALLENGES IN LARGE-SCALE CULTIVATION OF ALGAE

Many of the open ponds and closed PBR systems have been demonstrated at laboratory and pilot scale. A larger commercial scale facility would face additional challenges in culture mixing, gas exchange, contamination control, and process management. These challenges had been identified in the Aquatic species program (Sheehan et al., 1998). Some of these challenges are only beginning to be addressed in a comprehensive way. In a recent report, the US DOE has identified that stability of large-scale cultures, system productivity, access to nutrient sources, water management and recycling, control of systems, and coproducts will play a crucial role in the success of any algal biofuels project (DOE, 2010).

6 RESOURCE CONSTRAINTS FOR MASS PRODUCTION OF MICROALGAE

One of the biggest challenges for large-scale production of algal biofuels is in the sustainable use of resources. Although, algae can be produced theoretically at higher productivities than terrestrial plants, it must be noted that any large-scale production must address the challenges in land, water, and nutrient availability.

6.1 Nutrients

Algae are photosynthetic organisms and when produced under autotrophic conditions require light, nitrogen, phosphorous, CO_2, and trace metals for their growth. Although sunlight is virtually free, access to solar insolation is not, as distribution of sunlight varies with latitude and time of the year.

Nitrogen is one of the essential macronutrients for algae growth. Nitrogen can be supplied as chemical fertilizers or from other waste streams such as municipal waste water. Production of urea and other nitrogen fertilizers is an energy-intensive process that uses large amounts of natural gas. About 42 MJeq of natural gas is required and 1346 Kg CO_2 are released for every kg of urea produced (GREET, 2010). Therefore, integration of waste water treatment and algae production systems has been proposed to simultaneously treat the waste water and supply nitrogen for growth of algae.

Advances in modern agriculture are dependent upon application of chemical fertilizers. Theoretically, sufficient nitrogenous fertilizers can be produced if there is access to large amount of low-cost energy source. Of much greater concern is the use of phosphorous fertilizers as most of the phosphorous used in modern agriculture is mined from few locations around the world (Cordell et al., 2009). Once the phosphorous mines are depleted, entire modern agriculture will be in jeopardy. This has led to some authors to term this imminent crisis as peak phosphorous (Figure 4). Currently, over 90% of the demand for phosphate fertilizer is in food production and with large-scale increase in biofuels production, concerns could be raised about food versus fuel in the context of phosphorous use. In view of long-term sustainability, it is important to integrate waste water treatment systems for utilization of nitrogen and phosphorous.

Algae growth has been demonstrated to be faster when supplemented with CO_2. Utilization of CO_2 from power plants, cement plants, ethanol plants, petroleum refineries, and other large emitters of CO_2 was a driver for earlier research in microalgae. Many authors have reported use of flue gases, concentrated CO_2 streams for enhancing the algae growth (Benemann, 1997; Brown, 1996; Maeda et al., 1995). Significant barriers for CO_2 use from large stationary sources of CO_2 emissions such as power plants are optimizing pumping

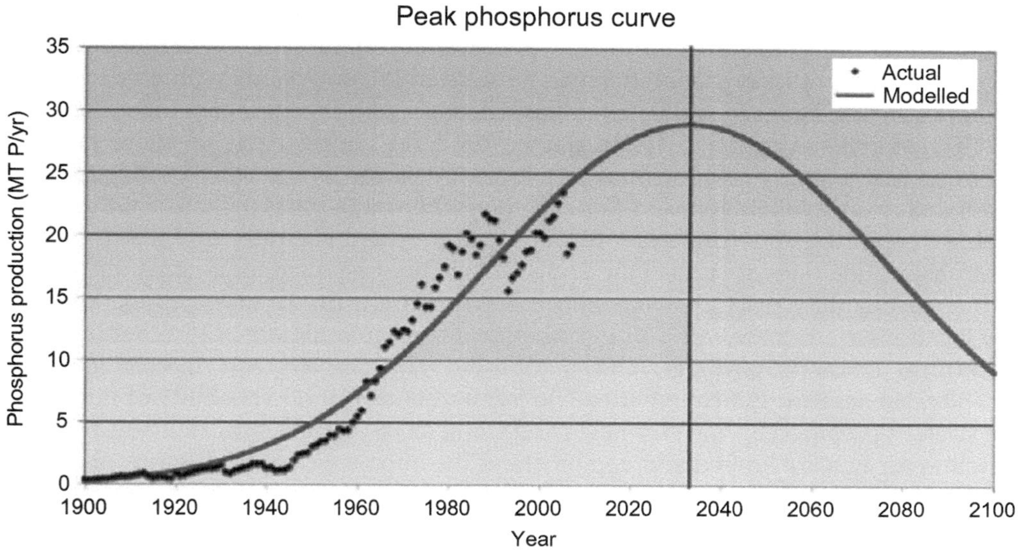

FIGURE 4 Issue of peak phosphorous *(figure from Cordell et al., 2009).*

costs, transportation logistics, and efficiencies. Maximum utilization efficiency is limited to 20-30% due to mass transfer limitations in open ponds and diurnal light availability variation even in an optimistic scenario. Additionally, availability of land and water sources around the CO_2 sources is a challenge that needs site-specific solutions.

6.2 Climate

Solar insolation, mean temperature, temperature variations, amount and distribution of rainfall, relative humidity, and cloud cover are some of the climatic variables that affect the algae productivity. Range and mean values of temperature are one of the important variables and van Harmelen and Oonk (2006) in a preliminary analysis indicated the areas with mean annual temperatures of $\geq$15°C to be suitable for algae production. However, as identified by Lundquist et al. (2010), it is important to consider the variations in temperatures along with the mean temperature while assessing the suitability of a location for algae production. It must be recognized that site-specific analyses are needed to estimate actual productivity and the aforementioned guidelines are only indicators of productivity. For example, a typical power plant operates at 40% efficiency thus producing significant quantities of waste heat which could be used for heating algae ponds in locations with lower mean annual temperatures.

6.3 Land

On a global scale in year 2000, about 12% (12.2-17.1 million sq. Km) of the earth's ice-free land surface was used for agriculture, while 28% (23.6-30.0 million sq. Km) was used as pastures (Ramankutty et al., 2008). Producing algae biofuels on large scale requires large tracts of land. Land slope, permeability of the soils, alternative uses of the land and economic value are important factors in determination of the land suitability for algae biomass production. Since the cost of land preparation can be substantial, only lands with moderate slopes (<5%) are suitable for large-scale algae production.

While nonarable marginally lands can be used for construction of algae production facilities, in reality many of these lands are located in areas where other challenges such as nutrient and water availability exist. For example, while large tracts of lands are theoretically available in southwestern United States for algae production, water availability and high evapotranspiration rates are two of the important challenges. Additional constraints in terms of nutrient availability and CO_2 availability could limit the areas for mass production of algae to few locations around the world. A global map of favorable climate (<15°C annual average temperature), lands (<5% slopes, <500 m altitude, <250 persons/Km^2 indicating low land cost, >50 persons/Km^2 indicating adequate infrastructure), and availability of nutrient-rich waste streams prepared by van Harmelen and Oonk (2006) indicates that suitable microalgae production regions are limited and demonstrates the importance of considering the resource constraints.

6.4 Water

Water is used in production and processing of algae. Primary sources of fresh water are surface waters (rivers, lakes, streams, and lakes) and ground water. Water use is generally categorized as consumptive and nonconsumptive use. Consumptive water use refers to water that is not returned to the original source such as the water lost in due to evaporation in algae ponds. Evapotranspiration losses represent the most important consumptive water use in agriculture and any potential algae production system in open ponds. Closed PBR minimize evapotranspiration losses and thus can be beneficial in areas with high evapotranspiration and lower water availability. Depending on the climatic factors such as temperature, relative humidity, and wind speed, evapotranspiration losses can be significant up to 2.7 m/year (Lundquist et al., 2010).

Increasingly, overuse of ground water and surface waters for use in agriculture, industries, and other uses has resulted in lowering of water table in many areas (IWMI, 2006; Pate et al., 2007; WEC, 2010). Due to uneven availability and allocation of fresh water sources around the world, many areas will face increasing water stress (Figure 5) and projected global water availability scenario is a cause of concern (IWMI, 2006; WEC, 2010). One of the advantages of algae over terrestrial crops is that many strains of algae can be grown in water with different salinity levels such as waste water, brackish or sea water. These sources of water do not impact the use or allocation of fresh water, and therefore algae production using these sources of water could aid in sustainable use of water resources.

7 ENERGY ANALYSIS

Multiple pathways for production, harvesting, and processing algae biomass have been proposed (Figure 2). Suitability of a pathway is dependent on various factors including location, availability of resources, and end use for the algae biomass. A technoeconomic analysis

FIGURE 5 Overview of world water stress. (Figure from Report: Comprehensive Assessment of Water Management in Agriculture, International Water Management Institute, 2006 (IWMI, 2006).) Physical water scarcity: Red, >75% river flows already in use; Light Red, >60% rivers flows already used; Orange: Economic Water scarcity, <25% used due to economic reasons; Blue: Water resources available, <25% is with drawn for human purposes. (For interpretation of the references to color in this figure legend, the reader is referred to the Web version of this chapter.)

for each of these pathways must be performed before the viability of the pathways can be commented upon. For example, most common pathway of algae production in open ponds, harvesting using a combination of flocculation and centrifugation, thermal drying followed by solvent extraction to produce algae oil and cake may not be feasible in terms of energy returns as the thermal energy input exceeds the recoverable energy from algae oil and coproducts (Sander and Murthy, 2010). However, if the algae can be dried using solar energy or waste heat, the energetics may be more favorable. Therefore, it is important to conduct energetic and sustainability assessment for each of the algae biomass utilization pathways before large implementation.

7.1 Theoretical Production Estimates

Determining theoretical maximum productivity limits based on thermodynamics and other laws of nature is important for realistic assessment of the algal biofuels potential. One of the methods for estimating theoretical maximum algal biomass and lipid productivity was discussed by Weyer et al. (2010). The model proposed by Weyer et al. (2010) consists of a number of factors such as full-spectrum solar energy (FSSE, MJ/m^2-year), photosynthetically active radiation (PAR), photon energy (PE, MJ/mol), efficiencies of photon transmission (η_{PT}) and utilization (η_{PU}), quantum requirement for conversion of PE into chemical energy (QR, mol photons/mol carbohydrate), energy in carbohydrates (CE, KJ/mol), biomass accumulation efficiency ($\eta_{Biomass}$), biomass energy content estimate (BE, KJ/g), cell oil content (CO, %), and cell oil density (CD, Kg/m^3).

Basic photosynthesis equation used to estimate the quantum requirement for conversion of PE into chemical energy (QR, mol photons/mol carbohydrate) is:

$$CO_2 + H_2O + 8 \text{ Photons} \rightarrow CH_2O + O_2.$$

Energy content of the biomass is estimated based on the proximate analysis of the dry algae as:

$$\text{BE}(\text{KJ g}^{-1}) = 15.7(\text{KJ g}^{-1}) \text{ Protein} + 16.7(\text{KJ g}^{-1}) \text{ Carbohydrate} \\ +37.6(\text{KJ g}^{-1}) \text{ Lipids}.$$

Using these assumptions, the maximum daily productivity can be estimated as:

$$\text{Maxmimum daily growth}(\text{g m}^2 \text{ day}^{-1}) = \frac{\text{FSSE}\cdot\text{PAR}\cdot\eta_{\text{PT}}\cdot\eta_{\text{PU}}\cdot\text{CE}\cdot\eta_{\text{Biomass}}}{\text{PE}\cdot\text{QR}\cdot\text{BE}\cdot 365}.$$

Similarly, annual lipid production can be estimated from the maximum daily growth as:

$$\text{Anual oil production}(\text{L Ha year}^{-1}) = \frac{\text{Maxmimum daily growth}(\text{g m}^2 \text{ day}^{-1})\text{CO}\cdot 365 \times 10{,}000}{\text{CD}}.$$

An estimate of the theoretical and best case algal lipid production scenarios along with estimates for open raceway ponds and PBRs is provided in Tables 3 and 4. These estimates

TABLE 3 Production Model (Based on Production Model from Weyer et al., 2010).

				Open Ponds		PBR	
Description	**Units**	**Th. Max**	**Best**	**Min**	**Max**	**Min**	**Max**
Full-spectrum solar energy	MJ/sq m-year	11,616.0000	5623.0000	4700	4700	4700	5621
PAR	-	0.4580	0.4580	0.458	0.458	0.458	0.458
Photon energy	MJ/mol	0.2253	0.2253	0.2253	0.2253	0.2253	0.2253
$\eta_{\text{Photon transmission}}$	-	1.0000	0.9500	0.8	0.9	0.95	0.95
$\eta_{\text{Photon utilization}}$	-	1.0000	0.5000	0.5	0.5	0.7	0.7
Quantum requirement	-	8.0000	8.0000	8	8	8	8
Carbohydrate energy content	KJ/mol	482.5000	482.5000	482.5	482.5	482.5	482.5
$\eta_{\text{Biomass accumulation}}$	-	1.0000	0.5000	0.5	0.7	0.6	0.8
Biomass energy content	KJ/g	21.9000	21.9000	21.9	21.9	21.9	21.9
Cell oil content	-	0.5000	0.5000	0.1	0.15	0.15	0.25
Oil density	Kg/m^3	918.0000	918.0000	918	918	918	918

TABLE 4 Production Estimates

Production metric	Units	Th. Max	Best	Open Ponds		PBR	
				Min	Max	Min	Max
Maximum daily growth	g/sq m-day	178.2	20.5	14.4	22.7	28.8	45.9
Annual oil production	L/ha year	354,202	40,722	5733	13,543	17,155	45,592
Annual oil production	gal/ac year	37,383	4298	605	1429	1811	4812

provide a preliminary guide to productivities, and more accurate results would be obtained using site- and strain-specific parameters in the model.

7.2 Energy Use in Processing Algae Biomass

Similar to the importance of algae production estimates, energy used in processing the algae biomass is critical in assessing the energy use in the production of biofuels. As there are no large-scale algal biofuel production facilities in operation, there is presently little information available on the production processes. Nevertheless, an initial estimate for the energy used in processed can be useful in identifying the energy intensive unit operations. A framework for reporting the algal biofuels production has been recently suggested (Beal et al., 2010). Some of the estimated energy use for algae harvesting and dewatering technologies based on lab-scale data are presented in Table 2. Similarly, energy contained in some of the products from different processing technologies with estimated process efficiencies is presented in Table 5. Using the production estimates, energy used in harvesting and processing steps, an initial energetic assessment of an algal biofuel production pathway can be conducted.

A systematic energy assessment of a biofuels pathway has to incorporate the energy used in all subprocesses, energy embodied in the fuels and the coproducts. In evaluating the energetic assessment of biofuels, solar energy is not considered as an energy input. Thus, the total energy input only includes the nonrenewable energy sources used in the production. These

TABLE 5 Products from Different Processing Technologies

Processing	Fuel	Max yield (Kg or L/Kg Biomass)	Efficiency	HHV (MJ/L or MJ/kg)
Direct combustion	Biomass	1.0 of biomass	80	18.15
Solvent extraction	Biodiesel	1.0 of lipid content	80	35.7
Anaerobic digestion	Biogas (62% CH_4)	475.8 L/Kg of biomass	95	2.375×10^{-2}
Fermentation	Ethanol	0.51 of carbohydrate	85	23.4
Thermochemical conversion (fast pyrolysis)	Bio-oil	0.553 of biomass	90	33.64

values are used to calculate total, net energy input, net energy balance, and net energy ratio as follows:

$$\text{Total energy input} = \sum \text{Sub process energy inputs},$$

$$\text{Net energy input} = \text{Total energy input} - \text{Coproduct allocation},$$

$$\text{Net energy balance} = \text{Net energy input} - \text{Energy in functional unit},$$

$$\text{Net energy ratio} = \frac{\text{Net energy input}}{\text{Energy in functional unit}}$$

$$\text{Net energy return on energy invested (EROEI)} = \frac{\text{Energy in functional unit}}{\text{Net energy input}}.$$

Use of net energy as the sole metric has been debated and is of limited use in complete assessment of biofuels advantages (Dale, 2008). While a positive net energy is essential, it is important to note that biofuels were not produced just for energetic advantages. Therefore, a complete life-cycle assessment is needed for a comprehensive assessment of any biofuels technology.

8 LIFE-CYCLE ASSESSMENT

One of the important aspects of long-term algal biofuels production is the sustainability of the resource use. Although sustainability is a more abstract concept and is difficult to quantify, a widely accepted definition of sustainability was given by Brundlandt (1987) as "Sustainable development is a social development which fulfils the needs of present generations without endangering the possibilities of fulfillment of the needs of future generations." The first basic condition for a sustainable process is that resource consumption rate should be slower than the resource regeneration rate. For example, a sustainable process would release less carbon dioxide than it would consume during the production, use, and recycle processes. The second condition relates to the emissions during the life cycle of a product: the emissions must not harm the environment or at the minimum be lower than the assimilative capacity of ecosystem (Dewulf et al., 2000).

Life-Cycle Assessment (LCA) is a tool to assess impact of products, processes, and services on the environment. All elements of the process are considered starting with the acquisition of raw materials, production, manufacturing, and disposal. LCA studies were first developed by the U.S. Environmental Protection Agency (EPA) to estimate the resource use in production of different materials. Researchers at Argonne National Laboratory have developed a transportation sector-specific version called Well to Wheels LCA. GREET, MS Excel-based software developed by Argonne National Laboratory researchers, has been used for conducting LCA analysis for transportation fuels (Wang, 2005).

LCA consists of four iterative steps (Figure 6):

1. Goal definition and scoping: Defining the functional unit, scope of the LCA, and the system boundary identification.
2. Inventory analysis: Compilation of the mass and energy flows for all the inputs, outputs entering/leaving the system boundary.

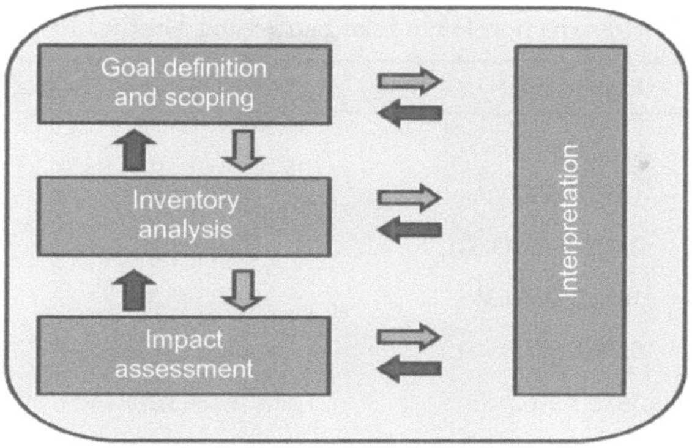

FIGURE 6 Steps in life-cycle assessment *(from SAIC, 2006)*

3. Impact assessment: Assess the environmental, human and societal effects of the materials use and emissions due to mass and energy flows into/out of the system boundary.
4. Interpretation: Interpret the results of the impact assessment in the context of functional unit, assumptions, quality of data source, and uncertainty in the data.

Since impact assessment and interpretation involve many more qualitative assumptions, many LCA are performed only until Step 2. Such LCA are called Life-Cycle Inventories (LCI).

While research in algal biofuels area is rapidly expanding, there are relatively few studies on the sustainability and life-cycle assessment of the algae pathways (Aresta et al., 2005; Batan et al., 2010; Campbell et al., 2011; Clarens et al., 2010; Lardon et al, 2009; Luo et al., 2010; Sander and Murthy, 2010; Stephenson et al., 2010).

Lardon et al. (2009) performed a comparative LCA study of two different culture conditions (nominal nitrogen or nitrogen starvation) and two different extraction options (wet or dry extraction technology) for a virtual facility. They concluded that while microalgae could have lower impacts than corn ethanol, it is imperative to lower energy and fertilizer consumption. They suggest using low nitrogen input production and wet extraction technology to accomplish these goals.

Clarens et al. (2010) compared the environmental impacts of producing microalgae, corn, switch grass, and canola. Contrary to Lardon et al. (2009), they concluded that microalgae production has higher energy use, green house gas emissions, and water consumption regardless of cultivation location. Algae production had lower impacts in the eutrophication and land area requirement categories. However, as indicated by the authors, most of these results can be attributed to energy use in fertilizer production and CO_2 delivery. Additionally, processing of algae biomass into fuels and coproducts was not considered in this LCA.

Sander and Murthy (2010) performed a complete well to pump LCA incorporating an objective system boundary definition. Alternative technology pathways including carrying composition of algae, filter press/centrifuges, solar/thermal drying were considered. Their analysis (Table 6.) indicated that algae biodiesel could result in lower GHG emissions and positive net energy. These results were critically dependent on the large coproducts credits and the harvesting technology. Only carbohydrate fraction was considered for coproducts credits, and the potential uses of algal protein for animal feed or organic fertilizers were

TABLE 6 Energy and CO_2 Emissions for Algal Biodiesel Production (Data from Sander and Murthy, 2010)

Functional Unit: 1000 MJ of Algal Biodiesel	Energy (MJ)	CO_2 Emissions (Kg)
Filter press (centrifuge) primary dewatering		
Growth	15.43 (15.43)	0.00 (0.00)
Harvest	2915.27 (5743.32)	241.87 (398.48)
Separation	165.03 (165.03)	6.33 (6.33)
Transportation	8.79 (8.79)	0.65 (0.65)
Biodiesel conv.	36.02 (36.02)	3.18 (3.18)
MeOH prod. and transport	72.24 (72.24)	0.06 (0.06)
Biodiesel transport and dist.	9.66 (9.66)	0.66 (0.66)
Natural gas prod.	69.52 (69.52)	0.00 (0.00)
Coproduct allocation	−9971.77 (−9971.77)	−273.60 (−273.60)
Total	−6679.81 (−3777.63)	−20.90 (135.71)

not included in the LCA. They identified water loss due to evapotranspiration as a significant sink of water and demonstrated that thermal dewatering technologies were not suitable for sustainable algae biofuels production.

Batan et al. (2010) performed a "well to pump" LCA for a low-cost PBR system using GREET 1.8c model as the basis. The CO_2 was not purchased in their modeled production system although fertilizers were used as input for providing nitrogen and phosphorous. They reported an NER of 0.93 MJ consumed/MJ produced and 75 g avoided CO_2/MJ of energy produced.

Stephenson et al. (2010) compared the performance of raceways and air-lift tubular bioreactors for production of microalgal biodiesel in the UK. They considered a two-stage production system with a first nitrogen-sufficient stage to accumulate biomass followed by a nitrogen-starvation phase to accumulate lipids. They concluded that open pond cultivation has lower energy consumption compared to air-lift tubular PBRs. The results were sensitive to oil productivity, energy use in circulation, recycling of culture media, and concentration of CO_2 in flue gas.

Campbell et al. (2011) performed LCA for an open pond system in Australia under three different CO_2 supplementation and two production scenarios. They concluded that algae have reduced green house gas emissions when the algae production facilities are co-located next to power plants or ammonia plants. They identified the algae productivity as one of the key variables that influence the LCA results.

Luo et al. (2010) conducted LCA for direct ethanol production from blue-green microalgae through intracellular photosynthesis-fermentation. In their analysis, direct ethanol secretion from live algal cells was assumed to reach concentrations between 0.5% and 5.0% in the culture media consisting of sea/brackish water in a flexible-film PBR. The culture broth was distilled to produce ethanol. They reported NER as 0.55-0.2 MJ/MJ ethanol and

29.8-12.3 CO_{2Eq}/MJ ethanol. The CO_2 values represented 67-87% reduction in carbon footprint compared to gasoline on an energy-equivalent basis.

Aresta et al. (2005) developed COMPUBIO software to conduct LCA for conversion of macroalgae into biofuels. The modeled system consisted of CO_2-supplemented macroalgae production in sea shore facilities using nutrients from effluent waters, harvesting of the biomass, and user-defined conversion technology. In the simulated best case scenario, macroalgae had higher net energy (11,000 MJ/ton dry algae) compared to microalgae gasification (9500 MJ/ton dry algae).

The difference in the results from many of these LCAs and even contrasting conclusions is not surprising. Such contrasting results were also obtained in the LCAs for corn ethanol and have been subject of intense debates (Wang, 2005).

Some of the factors that can lead to differences in the LCAs are:

1. System boundaries: Unit processes, inputs included in the system definition.
2. Data quality, accuracy: Age of data, sources of data, geographical context
3. Allocation of coproducts: Mass, energy, and displacement methods
4. Nitrous oxide emissions: Varying fertilizer use for different crops
5. Indirect land use change (ILUC): Variation in ILUC among first-, second-, and third-generation fuels.
6. Reported units, for example, GHG emissions/MJ of fuel or GHG emissions/Ha of land.

While LCA methodology can be used to assess the impact on environment and thus infer sustainability of a technology, some limitations are inherent in the LCA. Some of the limitations and of LCA are

1. Economic or risk assessment is not performed.
2. Energy quality is not considered.
3. Natural resource (e.g., water) use in LCA.
4. Local versus Global impacts is not characterized well.
5. Potential direct/ILUC is difficult to study.
6. Policy driven change is not incorporated.
7. Cannot predict the influence of game-changing technologies.

In recent years, many methods have been proposed to improve the LCA methodology to overcome some of its limitations. Some of the proposed improvements are:

1. Systematic boundary definition (Raynolds et al., 2000),
2. Enhancing LCA inventories using thermodynamics (Hau et al., 2007),
3. Dynamic LCA (Pehnt, 2005),
4. Accounting for water use (Koehler, 2008),
5. Incorporating alternative technology pathways.

8.1 Water Use in LCA

Water is one of the most important resources used in algal production. Current LCA framework does not incorporate the effect of water use. While this may not be very relevant for fossil fuel LCAs, it is very important to consider the impacts of water use in any biofuels production system. This is mainly due to the intensive water use in both production and processing of first, second, or third generation of biofuels. Due to large-scale production of biofuels, understanding the interdependencies between energy and water will be crucial for sustainable use of water resources (Pate et al., 2007).

One of the approaches for accounting the water use in different life-cycle assessment was proposed by Mulder et al. (2010). Mulder et al. (2010) developed a generalized indicator for water use intensity of different energy production technologies by defining energy return on water invested (EROWI) similar to energy return on energy invested (EROEI). A metric, Net EROWI, was proposed as a function of EROWI and EROEI to estimate water use in different fuel production systems as follows:

$$\omega = \frac{\text{EROEI}}{\text{EROEI} - 1},$$

$$\text{Net EROWI} = \frac{\text{Gross EROWI}}{\omega}.$$

Based on the water consumption data, Mulder et al. (2010) estimated the EROEI and EROWI for different fuels (Table 7). Similar analysis or algal biofuels indicates that the EROWI for algae using fresh water is similar to other biofuels crops and hence would lead to significant water scarcity issues if produced on large scale. However, by utilizing waste water/sea water for algae production, net EROWI can be improved significantly (Table 8).

TABLE 7 EROEI and EROWI for Different Fuels

	Water usage (L/MJ)	EROWI (MJ/L)	EROEI (MJ/MJ)	Net EROWI
Nuclear electric	1.162(0.145)	0.861(1.517)	10	0.775 (1.137)
Coal electric	0.560(0.488)	1.786 (2.049)	-	-
Conv. diesel	0.0035	285.3	5.01	228.4
		Biodiesel		
Rapeseed	100-175	0.010-0.0057	2.33	0.0057-0.0033
Algae (ponds)	20.142[a]	0.004965[a]	3.33[a]	0.03475[a]
		Ethanol		
Sugarcane	38-156	0.026-0.0065	8.3	0.023-0.0057
Corn	73-346	0.014-0.0029	1.38	0.0039-0.00081
		Lignocellulosic Crops		
Ethanol	11-171	0.091-0.0058	4.55	0.0071-0.0045
Hydrogen	15-129	0.067-0.0078	4.67	0.053-0.0062
Electricity	13-195	0.077-0.0051	5.0	0.062-0.0041

[a] *Calculated based on data from Sander and Murthy (2010). 1000 MJ ~27.89 L (7.36 gal) biodiesel.*
(Data from Mulder et al., 2010).

TABLE 8 EROEI and EROWI for Algal Fuels: Alternative Scenarios (Murthy, 2010)

Scenario	Energy Output (MJ)	Energy Input (MJ)	Net Energy	EROEI (NER)	Water Usage[c] (L/MJ)	EROWI[c]	Net EROWI[c]
Base case[a]	10,971.77	3291.96	7679.81	3.33	20.142 (0.403)	0.05 (2.48)	0.035 (1.74)
Improved harvesting[b]	10,971.77	1105.51	9866.26	9.925	20.142 (0.403)	0.05 (2.48)	0.045 (1.74)
Without coproduct credits	1000	3291.96	−2291.96	0.308	20.142 (0.403)	0.05 (2.48)	−0.114 (−5.69)
Waste Water/Sea Water	10,971.77	1105.51	9866.26	9.925	0.1	10	9.25

[a] *Base case as in Sander and Murthy (2010). 1000 MJ ~27.89 L (7.36 gal) biodiesel.*
[b] *Improved harvesting assumes a 75% reduction in energy to harvest and drying algae.*
[c] *Numbers in parentheses indicate photobioreactor case, assuming a 50 times lower water consumption than an equivalent open pond.*

9 FUTURE PERSPECTIVES: CHALLENGES AND OPPORTUNITIES

Algal biofuels due to their higher productivity compared to traditional biofuels hold great promise for sustainable biofuels production. Due to their versatility, algae can be grown in many locations around the world and can use diverse water sources that are not suitable for crop production or other industrial uses. While many researchers have demonstrated the technical feasibility of producing algal biofuels, some of the challenges for large-scale production of algal biofuels remain.

Challenges in production technologies still need research and the engineering challenge of managing large-scale ponds has to be addressed. Control systems for automated nutrient supply, harvesting, and contamination control of large ponds need to be developed. Development of low-cost PBRs that have higher productivity than open ponds and yet do not suffer from some of their disadvantages will reduce the algal biomass production cost.

There is a critical need to develop wet processing technologies for converting wet algal biomass into biofuels. Although the importance of coproducts utilization has been demonstrated, research into utilization of algae carbohydrate and protein fractions needs to be performed.

Resource availability and constraints for algae production need to be assessed in a comprehensive way. Presently, there are a large number of disparate sources and there is a need to develop unified analysis of resource availability using GIS-based tools. Comprehensive LCA using standard boundary definitions, incorporating water use can be the first steps in assessing various technologies for algal biomass processing.

In view of the challenges that the world faces in terms of limited fossil fuels, limited resources such as land, fresh water access, and global climate change, the opportunities that algal biofuels provide for a long-term sustainable solution are much greater than the technological challenges.

References

Amin, S., 2009. Review on biofuel oil and gas production processes from microalgae. Energy Convers. Manag. 50, 1834–1840.

Aresta, M., Dibenedetto, A., Barberio, G., 2005. Utilization of macro-algae for enhanced CO_2 fixation and biofuels production: development of a computing software for an LCA study. Fuel Proc. Technol. 86, 1679–1693.

Balat, M., Balat, M., Kirtay, E., Balat, H., 2009a. Main routes for the thermo-conversion of biomass into fuels and chemicals. Part 2: gasification systems. Energy Convers. Manag. 50, 3158–3168.

Balat, M., Balat, M., Kirtay, E., Balat, H., 2009b. Main routes for the thermo-conversion of biomass into fuels and chemicals. Part 1: pyrolysis systems. Energy Convers. Manag. 50, 3147–3157.

Batan, L., Quinn, J., Willson, B., Bradley, T., 2010. Net energy and greenhouse gas emission evaluation of biodiesel derived from microalgae. Environ. Sci. Technol. 44, 7975–7980.

Beal, C.M., Smith, C.H., Webber, M.E., Ruoff, R.S., Hebner, R.E., 2010. A framework to report the production of renewable diesel from algae. Bioenerg. Res. 4, 36–60.

Benemann, J.R., 1997. CO_2 mitigation with microalgae systems. Energy Convers. Manag. 38, S475–S479.

Benemann, J.R., Oswald, W.J., 1996. Systems and Economic Analysis of Microalgae Ponds for Conversion of CO_2 to Biomass. US Department of Energy, Pittsburgh Energy Technology Centre.

Brennan, L., Owende, P., 2010. Biofuels from microalgae—a review of technologies for production, processing, and extractions of biofuels and co-products. Renew. Sust. Energy Rev. 14, 557–577.

Brown, L.M., 1996. Uptake of carbon dioxide from flue gas by microalgae. Energy Convers. Manag. 37, 1363–1367.

Brundlandt, G.H., 1987. Report of the world commission on environment and development: Our common future. United Nations.

Campbell, P.K., Beer, T., Batten, D., 2011. Life cycle assessment of biodiesel production from microalgae in ponds. Bioresour. Technol. 102, 50–56.

Clarens, A., Resurreccion, E., White, M., Colosi, L., 2010. Environmental life cycle comparison of algae to other bioenergy feedstocks. Environ. Sci. Technol. 44, 1813–1819.

Cordell, D., Dranget, J.O., White, S., 2009. The story of phosphorus: global food security and food for thought. Glob. Environ. Change 19, 292–305.

Dale, B., 2008. Net energy: still a (mostly) irrelevant, misleading and dangerous metric. Biofuels Bioprod. Bioref. 2, 495–496.

Dewulf, J., Langenhove, H.V., Mulder, J., van den Berg, M.M.D., van der Kooi, H.J., de Swaan Arons, J., 2000. Illustrations towards quantifying the sustainability of Technology. Green Chem. 2, 108–114.

DOE, 2010. National Algal Biofuels Technology Roadmap. U.S. Department of Energy Efficiency and Renewable Energy, Biomass Program.

GREET, 2010. The greenhouse gases, regulated emissions and energy use in transportation model. Tech. Rep., Argonne National Laboratory, U.S. Department of Energy. Available at: http://www.transportation.anl.gov/modeling_simulation/GREET/.Version 1.8d. Accessed Dec 2010.

Hau, J.L., Yi, H., Bakshi, B.R., 2007. Enhancing Life-Cycle Inventories via reconciliation with the laws of thermodynamics. J. Ind. Ecol. 4, 5–25.

Hirano, A., Hon-Nami, K., Kunito, S., Hada, M., Ogushi, Y., 1998. Temperature effect on continuous gasification of microalgal biomass: theoretical yield of methanol production and its energy balance. Catal. Today 45, 399–404.

IWMI, 2006. Water for food, water for life: insights from the comprehensive assessment of water management in agriculture. Stockholm World Water Week. Available at: http://www.iwmi.cgiar.org/assessment/files_new/publications/Discussion%20Paper/InsightsBook_Stockholm2006.pdf. Accessed Dec 2010.

Kita, K., Okada, S., Sekino, H., Imou, K., Yokoyama, S., Amano, T., 2010. Thermal pre-treatment of wet microalgae harvest for efficient hydrocarbon recovery. App. Energy 87, 2420–2423.

Koehler, A., 2008. Water use in LCA: managing the planet's freshwater resources. Int. J. LCA 13, 451–455.

Lardon, L., Helias, A., Sialve, B., Stayer, J.P., Bernard, O., 2009. Lifecycle assessment of biodiesel production from microalgae. Environ. Sci. Technol. 43, 6475–6481.

Lee, J.Y., Yoo, C., Jun, S.Y., Ahn, C.Y., Oh, H.M., 2010. Comparison of several methods for effective lipid extraction from microalgae. Bioresour. Technol. 101, S75–S77.

Lundquist, T.J., Woertz, I.C., Quinn, N.W.T., Benemann, J., 2010. A realistic technology and engineering assessment for algal biofuels production. Energy Biosciences Institute, Univ. of California, Berkeley, CA.

Luo, D., Hu, Z., Choi, D.G., Thomas, V.M., Realff, M.J., Chance, R.R., 2010. Life Cycle energy and greenhouse gas emissions for an ethanol production process based on blue-green algae. Environ. Sci. Technol. 44, 8670–8677.

Maeda, K., Owada, M., Kimura, N., Omata, K., Karube, I., 1995. CO_2 fixation from the flue gas on coal-fired thermal power plant by microalgae. Energy Convers. Manag. 36, 717–720.

McKendry, P., 2002. Energy production from biomass (part 2):conversion technologies. Bioresour. Technol. 83, 47–54.

Minowa, T., Sawayama, S., 1999. A novel microalgal system for energy production with nitrogen cycling. Fuel 78, 1213–1215.

Minowa, T., Yokoyama, S.Y., Kishimoto, M., Okakura, T., 1995. Oil production from algal cells of *Dunaliella tertiolecta* by direct thermochemical liquefaction. Fuel 74, 1735–1738.

Mohn, F.H., 1980. Experiences and strategies in the recovery of biomass in mass culture of microalgae. In: Shelef, G., Soeder, C.J. (Eds.), Elsevier, Algal biomass. Amsterdam, pp. 547–571.

Molina Grima, E., Belarbi, E.H., Acien Fernandez, F.G., Robles Medina, A., Chisti, Y., 2003. Recovery of microalgal biomass and metabolites: process options and economics. Biotechnol. Adv. 20, 491–515.

Mulder, K., Hagen, N., Fisher, B., 2010. Burning water: a comparative analysis of the energy return on water invested. Ambio 39, 30–39.

Murthy, G.S., 2010. Energetic and environmental assessment of algae biofuels. In: Panel: Enhancing Algae Biotechnology and Biofuels in Water Limited Western States. Pacific Rim Summit on Industrial Biotechnology and Bioenergy, Honolulu, HI.

O'Sullivan, A.C., 1997. Cellulose: the structure slowly unravels. Cellulose 4, 173–207.

Pate, R., Hightower, M., Cameron, C., Einfeld, W., 2007. Overview of energy-water interdependencies and the emerging energy demands on water resources. Report: SAND 2007-1349C.

Patil, V., Tran, K.Q., Giselrad, H.R., 2008. Towards sustainable production of biofuels from microalgae. Int. J Mol. Sci. 9, 1188–1195.

Pehnt, M., 2005. Dynamic life cycle assessment (LCA) of renewable energy technologies. Renew Energy 31, 55–71.

Pulz, O., 2001. Photobioreactors: production systems for phototrophic microorganisms. App. Microbiol. Biotechnol. 57, 287–293.

Ramankutty, N., Evan, A.T., Monfreda, C., Foley, J.A., 2008. Farming the planet: 1. geographic distribution of global agriculture lands in the year 2000. Global Biogeochem. Cycles GB1003 22, 19.

Raynolds, M., et al., 2000. The Relative Mass-Energy-Economic (RMEE) method for system boundary selection. Part 1: a means to systematically and quantitatively select LCA boundaries. Int. J. LCA 5, 37–46.

SAIC, 2006. Life cycle assessment: principles and practice. Report No: EPA/600/R-06/060. In: Scientific applications international corporation (SAIC), Reston, VA.

Sander, K., Murthy, G.S., 2010. Life cycle analysis of algae biodiesel. Intl. J. LCA 15, 704–714.

Sheehan, J., Dunahay, T., Benemann, J., Roessler, P., 1998. A look back at the US Department of Energy's aquatic species program-biodiesel from algae. National Renewable Energy Laboratory, Golden CO, Report: NREL/TP-580-24,190.

Stephenson, A.L., Kazamia, E., Dennis, J.S., Howe, C.J., Scott, S.A., 2010. Life-cycle assessment of potential algal biodiesel production in the United Kingdom: a comparison of raceways and air-lift tubular bioreactors. Energy Fuels 24, 4062–4077.

van Harmelen, T., Oonk, H., 2006. Microalgae Biofixation Processes: Applications and Potential Contributions to Greenhouse Gas Mitigation Options. TNO Built Environment and Geosciences. Available at: http://www.fluxfarm.com/uploads/3/1/6/8/3168871/biofixation.pdf. Accessed Dec, 2010.

Wang, M., 2005. Updated energy and greenhouse gas emission results of fuel ethanol. In: The 15th International Symposium on Alcohol Fuels San Diego, CA, USA.

WEC, 2010. Water for Energy. World Energy Council. http://www.worldenergy.org/documents/water_energy_1.pdf. Accessed: Dec, 2010.

Weyer, K.A., Bush, D.R., Darzins, A., Wilson, B.D., 2010. Theoretical maximum algal oil production. Bioenerg. Res. 3, 204–213.

WSA, 2009. Wind sea algae: workshop proceedings. In: Trent, J.D. (Ed.), International workshop on offshore algae cultivation. Lolland, Denmark. Available at: http://wind-sea-algae.org. Accessed: 25, Dec, 2010.

CHAPTER

19

Cultivation of Algae in Photobioreactors for Biodiesel Production

*J. Pruvost**

GEPEA, Université de Nantes, CNRS, UMR6144, boulevard de l'Université, CRTT – BP 406, 44602 Saint-Nazaire Cedex, France

***Corresponding author: E-mail: jeremy.pruvost@univ-nantes.fr**

1 INTRODUCTION

Photosynthetic microorganisms such as microalgae and cyanobacteria (named for convenience "microalgae" in what follows, except when cited) have a high potential in biofuel production. Their main advantages are solar production with higher surface productivities than plants, simultaneous consumption of inorganic carbon, allowing a null carbon balance exploitation, and possible production in closed systems, offering several advantages including an intensified, controlled production with very low environmental impact (no fertilizer is released and water can be reused). The high biodiversity of microalgae means that a variety of energy-rich substances can be produced, such as hydrogen by water photolysis, lipids for biodiesel or biokerosene production, and sugars for biomass fermentation (methane) or gasification (Benemann, 2004; Chisti, 2007; Degrenne et al., 2010; Ghirardi et al., 2000; Hu et al., 2008; Melis, 2002; Rodolfi et al., 2009; Schlegel et al., 2004; Scragg et al., 2002; Spolaore et al., 2006; Tsukahara and Sawayama, 2005). However, using microalgae for biofuels introduces several constraints, in particular the need to set up mass-scale, cost-effective, and sustainable plant. This last constraint implies, for example, achieving a positive energy balance, which is not straightforward considering the different steps required to obtain usable biofuel (production, harvesting, and downstream processing of biomass into biofuel). Mass-scale production of microalgae has proved feasible for several decades, but in domains other than biofuel production (Richmond, 2004a). Significant research and development efforts are still needed to define an integrated, efficient production system meeting the specific constraints of the energy market.

This chapter is devoted to a key step in using algae for energy production purposes, namely, the biomass production system. Specific technology is required. Microalgae cultivation possesses features common to bioreactors in general, such as thermal regulation, nutrient feeding procedures, pH regulation, and mixing for heat and mass transfer enhancement. However, a light supply is necessary for photosynthetic growth, with several consequences, in particular the need for a dedicated cultivation system emphasizing large illuminated areas. Unlike other more classical bioprocesses where mixing tanks display standard geometries, cultivation systems for microalgae are characterized by a broad diversity, ranging from open ponds (open systems) to photobioreactor technology (closed systems). A detailed description of existing geometries can be found in the literature (Carvalho et al., 2006; Lehr and Posten, 2009; Richmond, 2004a; Ugwu et al., 2008). This chapter will present only a brief overview. Photobioreactor technology will be highlighted as it offers several advantages of special interest to biofuel production. However, as is well known, it also leads to more complex and costly processes, and is difficult to scale up for mass production on large land areas. Engineering breakthroughs are thus still needed before suitable systems could be set up. Recent scientific work has brought new insights into how such systems might be achieved, especially by clarifying the parameters governing photobioreactor productivities and establishing engineering bases to optimize and scale them. These aspects will be presented here in the specific context of solar production.

2 BASIC CONCEPTS OF PHOTOBIOREACTOR ENGINEERING

2.1 General Description

Photosynthetic growth in standard autotrophic conditions is based on the assimilation, under illumination, of inorganic carbon and mineral nutrients dissolved in the medium. The cultivation of photosynthetic microorganisms will thus require:

- A light supply (solar or artificial source, with an appropriate light spectrum in the photosynthetic active radiation (PAR) range, usually 0.4-0.7μm),
- An inorganic carbon source (such as dissolved CO_2),
- Mineral nutrients (major nutrients such as N, S, P sources and micronutrients such as Mg, Ca, Mn, Cu, Fe, etc.),
- Set culture conditions (pH, temperature).

Growth medium composition depends on the species cultivated. For a given species, mineral requirements can be ascertained using various methods, for example, direct measurement of their consumption or elemental composition analysis. This is easy for major nutrients (a detailed explanation can be found in Pruvost et al., 2009), but can be very difficult for micronutrients, which may require specific analytical methods (see Cogne et al., 2003). Mineral requirements can be expressed in the form of a stoichiometric equation that can be used to prevent mineral limitation by adapting nutrient concentration as a function of biomass concentration achieved in the cultivation system (Roels, 1983). Following are two examples for the fresh water species *Chlamydomonas reinhardtii* (Equation 1) and *Neochloris oleaobundans* (Equation 2):

$$CO_2 + 0.593\ H_2O + 0.176\ NH_4^+ + 0.007\ SO_4^{2-} + 0.018\ PO_4^{3-} \rightarrow CH_{1.781}O_{0.437}N_{0.176}S_{0.007}P_{0.018} + 0.108\ H^+ + 1.127\ O_2 \quad (1)$$

$$CO_2 + 0.751\ H_2O + 0.148\ NO_3^- + 0.014\ SO_4^{2-} + 0.012\ PO_4^{3-} + 0.212H^+ \rightarrow CH_{1.715}O_{0.427}N_{0.148}S_{0.014}P_{0.012} + 1.437\ O_2 \quad (2)$$

Equations (1) and (2) show the high biological requirement for CO_2. As an acid, CO_2 has a direct influence on pH. Its uptake leads to a progressive but significant basification of the medium (Chiu et al., 2008). Equations (1) and (2) also emphasize the difference due to the nitrogen source (ammonium for *Chlamydomonas reinhardtii* vs. nitrate for *Neochloris oleaobundans*). Oxygen release, for example, differs by 25-30%. In addition, ammonium consumption tends to lower the pH (H^+ release), while nitrate consumption tends to raise it (H^+ consumption). Special attention must therefore be paid to the nitrogen source when pH regulation is applied. In any case, pH will be affected by growth, with a significant influence of the carbon uptake due to its high consumption during photosynthetic growth. pH regulation will then be necessary to maintain an optimal value during cultivation (especially in the case of high volumetric productivities involving high nutrient consumption).

Most of the problems described previously (design of the medium composition, influence of biological uptake on physical and chemical characteristics of the medium) are common to all classical bioprocesses. Light energy supply, however, is highly specific. Unlike dissolved nutrients, which can be assumed to be homogeneous in well-mixed conditions, light energy is heterogeneously distributed in the culture due to absorption and scattering by cells, independent of the mixing conditions (Figure 1). As light is the principal energy source of photosynthesis, this simple fact makes microalgae cultivation systems different from other classical bioprocesses: specific approaches are thus needed for the design, optimization, and control of the cultivation system. Photosynthetic activity (P) is directly related to the light received. This is usually represented as the light response curve as given in Figure 1. This curve is characterized by a progressive saturation of photosynthesis with irradiance G up to an irradiance of saturation G_s. For higher irradiances, photoinhibition phenomena can occur with a negative influence on growth (Vonshak and Torzillo, 2004). We also note that a threshold value of irradiance is needed to obtain positive growth. This value is termed irradiance of compensation G_C (corresponding to the "compensation point of photosynthesis"). In cultivation systems, this nonlinear, complex response of photosynthesis has to be considered in combination with the light attenuation conditions. In extreme cases of high light illumination and high light attenuation (high biomass concentration), cells in different physiological states will co-occur: some may be photoinhibited (close to the light source) and some will receive no light (deep in the culture). Ideally, the control of the system would require taking all these processes into account, a far from trivial task.

2.2 Characterization of the Incident PFD

The light energy received by the cultivation system is represented by the hemispherical incident light flux density q, or photon flux density (PFD) as it is commonly termed in microalgae studies. For any light source, the PFD has to be expressed in the range of PAR,

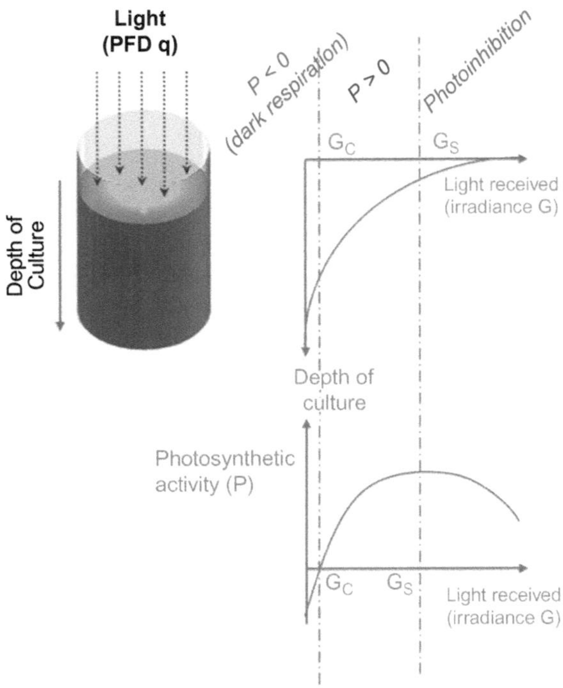

FIGURE 1 Relation between light attenuation and photosynthetic growth in microalgal cultivation systems.

in most cases in the 0.4-0.7 μm bandwidth. For example, the whole solar spectrum at ground level covers the range 0.26-3 μm. The PAR range thus corresponds to almost 43% of the full solar energy spectrum.

As light is converted inside the culture volume, it is also necessary to add to PFD determination a rigorous treatment of radiative transfer inside the culture. This enables us, for example, to couple the resulting irradiance field with photosynthetic conversion of the algal suspension to simulate light-limited growth. However, this determination requires certain information. In addition to the PFD value, light source positioning with respect to the optical transparent surface of the cultivation system is important, as light penetration inside a turbid medium is affected by the incident polar angle θ of the radiation on the illuminated surface (Figure 2). Ideally, beam and diffuse components of radiation should be considered separately. By definition, the direction of a beam of radiation, which represents direct radiation received from the light source, will define the incident polar angle θ with the illuminated surface. By contrast, diffuse radiation cannot be defined by a single incident angle, but has an angular distribution over the illuminated surface (on a 2π solid angle for a plane). We note that isotropic angular distribution is usually assumed, although an anisotropic distribution should ideally be considered because of the dependency of radiative transfer inside the culture volume on the angular nature of incident diffuse PFD. Both the incident angle and the degree of collimation of the light flux can be difficult to characterize. However, in most artificial light cultivation systems, normal incidence is usually chosen as the most effective way to transfer light into the culture volume (less reflection on optical surfaces and better light

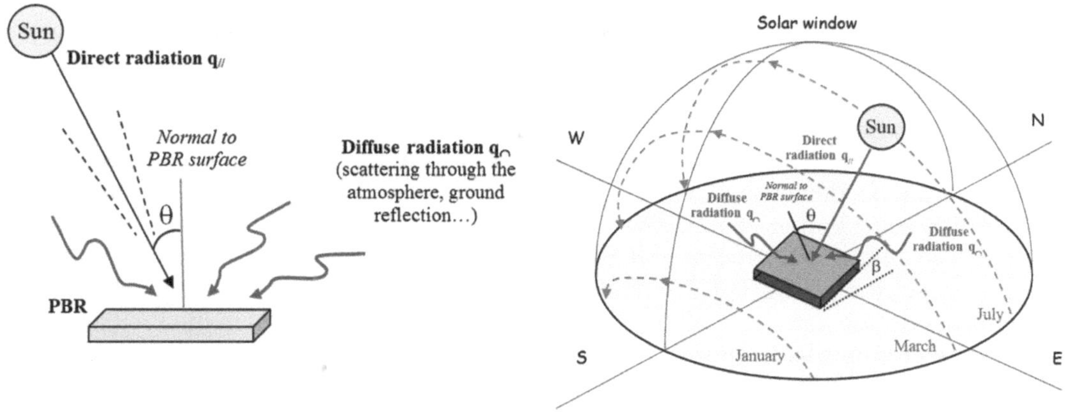

FIGURE 2 Solar radiation on a microalgal cultivation system: incident angle and diffuse-beam radiations (left), evolution of solar sky path during the year in France (right).

penetration in the culture bulk). The PFD can also in most cases be assumed to be quasicollimated (so we can consider the PFD as beam radiation only). However, these characteristics cannot be assumed in solar technology. The sun's displacement makes the incident angle time dependent and so non-normal incidence conditions will be encountered. Sunlight can also present a large proportion of diffuse radiation due to scattering through the atmosphere or by reflection from various surfaces, such as the ground. A detailed description of the respective consequences of neglecting incidence angle and direct/diffuse distribution effects in solar cultivation systems was recently published (Pruvost et al., in press). It was shown that each assumption led to an overestimation of 10-20% in biomass productivity. When the two assumptions were combined (the simplest case of radiative transfer representation), an overestimation of up to 50% was obtained, emphasizing the relevance of an accurate consideration of the incident angle and direct/diffuse distribution in the radiative transfer modeling when applied to the solar case.

The PFD can be measured using a cosine quantum sensor (LI-190-SA, LI-COR, Lincoln, NE) with multipoint measurements to obtain an average over the illuminated surface (Janssen et al., 2000b; Pottier et al., 2005; Sanchez Miron et al., 2003). The accuracy will closely depend on the average procedure, especially if the PFD is unevenly distributed. Actinometry could also be used for accurate characterization, as this is sensitive to all photons absorbed in the reaction volume. A detailed example of the experimental procedure in artificial light can be found in Pottier et al. (2005). In the case of sunlight, measurement is obviously also possible, but mathematical relations are also available to determine radiation conditions on a collecting surface as a function of the Earth's location, year period, and surface geometry (Duffie and Beckman, 2006). An example was recently given by Sierra et al. (2008) for a solar photobioreactor. Some commercial software packages integrating solar models are also available (METEONORM 6.0 software; www.meteonorm.com). These allow easy determination of irradiation conditions on a given surface. Such an approach is thus of particular interest in the case of solar production and was applied in Pruvost et al. (in press).

2.3 Light Attenuation in the Culture Bulk

Owing to absorption and scattering by cells, light distribution in microalgae cultures is highly heterogeneous. This light distribution directly influences the light received by cells (termed the irradiance G) and thus process efficiency. Light attenuation in a given cultivation system geometry depends on the optical properties and concentration of cells. Optical properties can be determined either experimentally or theoretically (Berberoglu et al., 2008; Cornet, 2007; Pottier et al., 2005).

For a given culture, the irradiance field can be obtained experimentally using an underwater spherical sensor (US-SQS/A, Heinz Walz, LI-COR, Effeltrich, Germany). Such a quantum sensor measures the light from all incoming directions (4π solid angle) in the PAR and has a small diameter (3 mm) allowing the photon fluence rate measured to be taken as the irradiance (Pottier et al., 2005). However, as any modification in cell concentration will modify light attenuation, this is of little interest. Radiative transfer models are to be preferred, of which further details are given in the next section.

3 MODELING OF MICROALGAE CULTIVATION SYSTEMS

3.1 Mass Balance

The mass balance relates concentration in the cultivation system to kinetic rates of biological production (biomass, O_2) or consumption (nutrients, CO_2) and system input and output. For a continuous system assuming perfectly mixed conditions, the biomass concentration C_x is then given by (Cornet et al., 2003; Pruvost et al., 2008; 2011):

$$\frac{dC_x}{dt} = \langle r_x \rangle - \frac{C_x}{\tau} = \langle r_x \rangle - DC_x, \qquad (3)$$

with C_x the biomass concentration, $\langle r_x \rangle$ the mean biomass volumetric growth rate in the system, and τ the residence time resulting from the liquid flow rate of the feed (fresh medium) (with $\tau = 1/D$, where D is the dilution rate).

3.2 Kinetic Modeling of Photosynthetic Growth

Solving Equation (3) involves determining the mean volumetric growth rate $\langle r_x \rangle$. This rate is linked to all possible limitations that can occur in the cultivation system. As will be shown later, light-limited conditions allow the best productivity to be obtained, and they will be retained here as an example. With appropriate kinetic relations, other limitations can be considered (growth limitation by inorganic carbon or mineral nutrient concentration, temperature influence, etc.). The interested reader can refer to Fouchard et al. (2009), where both light and nutrient limitations were modeled in the particular case of sulfur deprivation, which leads to hydrogen production by *Chlamydomonas reinhardtii*.

There are numerous kinetic models linking photosynthetic microorganism growth to the light received (Aiba, 1982; Muller-Feuga, 1998). For example, the following equations were applied for the cyanobacterium *Arthrospira platensis* (Cornet and Dussap, 2009; Equation 4) and the microalga *Neochloris oleoabundans* (Pruvost et al., 2011; Equation 5), respectively:

$$r_x = \rho\overline{\varphi}A = \rho_M \frac{K}{K+G}\overline{\varphi}E_a G C_x, \tag{4}$$

$$r_x = \rho\overline{\varphi}A - \mu_s C_x = \rho_M \frac{K}{K+G}\overline{\varphi}E_a G C_x - \mu_s C_x. \tag{5}$$

where G is the irradiance, ρ_M the maximum energy yield for photon conversion, $\overline{\phi}$ the mass quantum yield for the Z-scheme of photosynthesis, K the half saturation constant for photosynthesis, E_a the mass absorption coefficient, and μ_s a specific respiration rate.

Both equations link the photosynthetic growth rate to the local radiant light power density absorbed A and so to the local value of irradiance G inside the culture bulk ($A = E_a G C_x$). As a prokaryotic cell, with therefore a common electron carrier chain for photosynthesis and respiration, *Arthrospira platensis* displays no respiration in light (Gonzalez de la Vara and Gomez-Lojero, 1986). This is not the case for microalgae, growth in light being the result of the biomass increase caused by photosynthesis in chloroplasts (anabolism) and its partial degradation by respiration in mitochondria (catabolism). It is thus necessary to introduce a catabolism respiration term, expressed here as a function of a constant specific respiration rate μ_s. This formulation is certainly oversimplified, as chloroplast and mitochondrial activities are not independent (Kliphuis, 2010). It was, however, shown to be sufficient in the case of *Neochloris oleoabundans* and could be retained in a first assumption at least for algae presenting a low respiration activity in light.

3.3 Radiative Transfer Modeling

Light attenuation conditions can be represented using radiative transfer models. Several examples can be found in the literature (Cornet et al., 1998; Csogör et al., 2001; Pruvost et al., 2002a; Tredici and Chini Zittelli, 1998; Yun and Park, 2003). These models introduce assumptions in the radiative transfer equation, the solution of which requires complex numerical tools and long calculation times. However, several cultivation systems come under the so-called one-dimensional hypothesis, where light attenuation occurs mainly along a single direction perpendicular to the illuminated surface, termed the depth of culture z (like a rectangular photobioreactor illuminated on one or both sides, cylindrical or spherical geometry with radial illumination). In this case, simple radiative models can be applied with relative accuracy. The simplest one is the Lambert-Beer law, but because of the scattering generated by cells, its use for microalgae is not recommended, especially when working in full light attenuation conditions (Aiba, 1982; Cornet et al., 1992a,b, 1994, 1995; Pottier et al., 2005). The two-flux model offers a useful compromise, often giving a sufficiently accurate prediction of the radiation field in the context of photosynthetic microorganism cultivation (Cornet and Dussap, 2009; Cornet et al., 1998; Cornet, 2010; Pottier et al., 2005) with analytical solutions that facilitate further coupling with kinetic growth models. If geometries do not allow the one-dimensional hypothesis to be applied, numerical approaches will be required, entailing a significant computational effort (Cornet, 2007).

An example of an analytical solution for the irradiance field determination is given below using the two-flux model. This example is given here for the solar case taking into account non-normal incidence (thus introducing the incident angle θ) with a separate treatment of the direct and diffuse parts of the radiation (Pruvost et al., in press). The PFD q is thus divided into $q_{//}$ and $q_{\cap}(q = q_{//} + q_{\cap})$, the direct and diffuse parts of the PFD, respectively. The solution can be easily adapted for collimated radiation (diffuse radiation is then null) and normal incidence ($\theta = 0$). They are expressed here in Cartesian coordinates. For other geometries such as cylindrical ones, solutions can be adapted from works of Loubiere et al. (2009) and Takache et al. (2010). We also note that with an increase in the computational effort, the irradiance field can be solved spectrally, taking into account the spectral distributions of PFD and of optical properties of photosynthetic microorganisms. This has already been applied for artificial light (see again Pottier et al., 2005 and Farges et al., 2009).

The irradiance field for collimated radiation is given by:

$$\frac{G_{col}}{q_{//}} = \frac{2}{\cos\theta} \frac{(1+\alpha)\exp[-\delta_{col}(z-L)] - (1-\alpha)\exp[\delta_{col}(z-L)]}{(1+\alpha)^2\exp[\delta_{col}L] - (1-\alpha)^2\exp[-\delta_{col}L]} \quad (6)$$

with $\delta_{col}(\alpha C_X/\cos\theta) = (E_a + 2bE_s)$ the two-flux collimated extinction coefficient and $\alpha = \sqrt{E_a/(E_a 2bE_s)}$ the linear scattering modulus. E_a and E_s are the mean (spectrally averaged over the PAR) mass absorption and scattering coefficients, respectively, for the cultivated photosynthetic microorganism, b the backward scattering fraction, and C_x the biomass concentration in the culture medium.

For diffuse radiation, the following equation is obtained:

$$\frac{G_{dif}}{q_{\cap}} = 4\frac{(1+\alpha)\exp[-\delta_{dif}(z-L)] - (1-\alpha)\exp[\delta_{dif}(z-L)]}{(1+\alpha)^2\exp[\delta_{dif}L] - (1-\alpha)^2\exp[-\delta_{dif}L]} \quad (7)$$

with $\delta_{dif} = 2\alpha C_x(E_a + 2bE_s)$ the two-flux diffuse extinction coefficient.

The total irradiance (representing the amount of light impinging on algae) is finally given by simply summing the collimated and diffuse components:

$$G(z) = G_{col}(z) + G_{dif}(z), \quad (8)$$

Equations (6) and (7) show that penetrations of collimated and diffuse radiation inside the culture volume are markedly different. This will be especially important in solar conditions where the diffuse component of the radiation is non-negligible. We also note the influence of the incident angle θ on the collimated part, with a decrease in light penetration with the increase in the incident angle (the diffuse radiation is here assumed to have an isotropic angular distribution on the illuminated surface). Like the degree of collimation of the radiation, this will influence cultivation system efficiency. An example of light attenuation profile is given in Figure 3 for an incident angle $\theta = 30°$ for typical values of beam-diffuse repartition ($q_{//} = 2q_{\cap}$). Both collimated and diffuse radiations contribute to the resulting irradiance-field (for a more detailed description see again Pruvost et al., in press).

3.4 The Working Illuminated Fraction γ

The irradiance distribution allows a significant parameter to be determined: the illuminated fraction γ (Cornet and Dussap, 2009; Cornet et al., 1992a,b; Degrenne et al., 2010; Takache et al., 2010). Schematically, the culture bulk can be delimited into two zones,

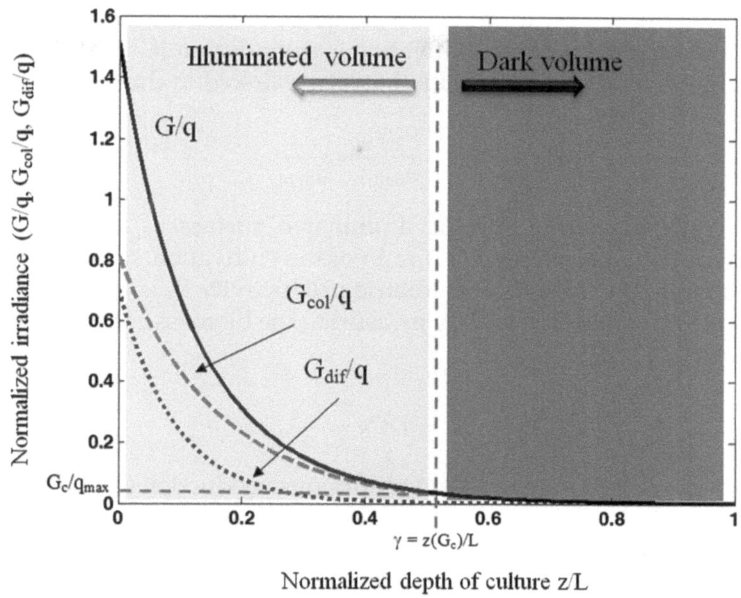

FIGURE 3 Example of irradiance field in bulk culture (solar radiation, full light absorption).

an illuminated zone and a dark zone (Figure 3). Partitioning is obtained by the compensation irradiance value G_c corresponding to the minimum value of radiant energy required to obtain a positive photosynthetic growth rate. For example, compensation irradiances $G_c = 1.5$ μmole m^{-2} s^{-1} (Cornet and Dussap, 2009) and $G_c = 10$ μmole m^{-2} s^{-1} (Takache et al., 2010) were found for *Arthrospira platensis* and *Chlamydomonas reinhardtii*, respectively. The illuminated fraction γ is then given by the depth of the culture z_c where the irradiance of compensation $G(z_c) = G_c$ is obtained (Figure 3).

In the case of cultivation systems with one-dimensional light attenuation, we have:

$$\gamma = \frac{V_i}{V_r} = \frac{Z_c}{L}, \tag{9}$$

where V_i and V_r are the illuminated and total culture volumes, respectively.

Values of γ below 1 indicate that all the available light for photosynthesis received is absorbed by the culture. Conversely, when the illuminated fraction is greater than 1, some of the light is transmitted (kinetic regime). This value has been shown to directly influence the performance of any light-limited biomass production (Cornet and Dussap, 2009; Takache et al., 2010). Because it does not allow full absorption of the light captured, the kinetic regime always leads to a loss of efficiency ($\gamma > 1$). Full light absorption is thus to be preferred ($\gamma \leq 1$), with, however, a negative influence of the dark zone for microorganisms presenting respiration in light, such as microalgae (see below).

3.5 Biomass Productivity Determination

Biomass productivity P_x is usually expressed in terms of volumetric productivity (kg m^{-3} h^{-1}). In the context of mass-scale production, surface productivity (S_x, kg m^{-2} h^{-1}) is also a useful variable to extrapolate to land area production. It has also been shown that maximal performance of a cultivation system (in light-limited conditions) when expressed on a surface basis is independent of the cultivation system design (Cornet, 2010; Pruvost et al., 2011). Both volumetric and surface productivities are linked in the following relation:

$$S_X = \frac{P_X V_r}{S_{\text{light}}} = \frac{P_X}{a_{\text{light}}}. \tag{10}$$

This equation introduces the specific illuminated surface a_{light}, which represents the illuminated surface (S_{light}) to volume (V_r) ratio of the cultivation system.

In continuous mode, the biomass volumetric productivity P_x is obtained for a given residence time τ (or dilution rate $D = 1/\tau$) by measuring the biomass concentration C_x inside the cultivation system:

$$P_x = DC_X = \frac{C_X}{\tau}. \tag{11}$$

We note that in the case of a steady-state continuous production ($dC_x/dt = 0$, see Eq.3), the biomass volumetric productivity is equal to the mean biomass volumetric growth rate in the cultivation system ($P_x = \langle r_x \rangle$).

For a batch culture, the mean biomass volumetric productivity can be estimated from the culture duration t_c before harvesting:

$$\overline{P_X} = \frac{C_X - C_{X_0}}{t_c}, \tag{12}$$

where C_{X_0} is the initial biomass concentration.

Biomass productivity can be obtained experimentally by direct measurement of the biomass concentration (Takache et al., 2010), or theoretically by solving Equation (3) (here in light-limited conditions) in combination with an appropriate formulation of kinetic growth (Equations 4 and 5) and radiative transfer in the culture bulk (Equations 6-8). This involves integrating the volumetric growth rate r_x over the reactor volume, because the heterogeneous distribution of the irradiance field makes growth rate a local value. This integration enables us to determine the mean volumetric growth rate $\langle r_x \rangle$ to solve Equation (3):

$$\langle r_x \rangle = \frac{1}{V_r} \iiint_{V_r} r_x \, dV. \tag{13}$$

For a cultivation system with one-dimensional light attenuation, this consists in a simple integration along the depth of culture z:

$$\langle r_x \rangle = \frac{1}{L} \int_0^L r_x \, dz, \tag{14}$$

where L is the photobioreactor depth.

For a given species (characterized by its optical properties and kinetic growth parameters), biomass productivity will be a function of cultivation system engineering (especially the depth of culture) and operating parameters such as the dilution rate D (or residence time τ)

or incident PFD. As a result, biomass productivity is difficult to predict in a simple manner. This makes the theoretical approach of prime relevance to predicting productivity evolution as a function of these key parameters and thus to photobioreactor optimization.

4 PRODUCTIVITY OF MICROALGAL CULTIVATION SYSTEMS

4.1 Main Limiting Parameters Affecting Productivity

Assuming that culture conditions (pH and temperature) are kept optimal, light, carbon, and mineral nutrient supplies are the main variables liable to limit photosynthetic growth and thereby reduce the productivity of cultivation systems (assuming there are no predatory contaminations). As discussed below, nutrient and CO_2 limitations can be avoided, but not light limitation, because of light attenuation in culture and of the high light requirement for photosynthesis. This simple but important observation is central to the optimization of microalgal cultivation systems. One major consequence will be the need to develop specific geometries maximizing light supply to the culture.

4.2 Nutrient and Carbon Source Limitation

To prevent mineral limitation, the growth medium must contain all the necessary nutrients (macro and micro) in sufficient quantities, and must therefore be adjusted according to the biomass concentration planned. Stoichiometric equations (Equations 1 and 2) can be used for this purpose or, more simply, concentration monitoring during cultivation. The reader can also refer to studies in which the method has been applied to various species (Pruvost et al., 2009, 2011). In specific cases, it would also be of interest to apply mineral limitation to induce specific metabolic responses, such as lipid accumulation (nitrogen source deprivation) or hydrogen production (sulfur deprivation). Stoichiometric equations can obviously be used for this purpose. As this chapter is devoted to microalgal biomass production, these aspects will not be detailed further (the interested reader can see Pruvost et al., 2011, for example).

Because the inorganic carbon source comes ideally from CO_2 dissolved in the culture medium, preventing carbon limitation is more problematic. It depends on the gas-liquid mass transfer rate and the dissolved carbon concentration obtained. CO_2 dissolution also affects the pH value (acidification), which in turn influences the amount and form of dissolved carbon obtained (CO_2, HCO_3^- or CO_3^{2-}). As stated above, nutrient consumption can also interact with pH evolution. Keeping an optimal pH value for growth while averting limitation by the carbon source may therefore not be trivial. However, in most cases, simple CO_2 bubbling is usually found to suffice in the first instance for both pH regulation (acidification) and carbon feeding. In specific cases, however, such as when using an ammonium source (the consumption of which also leads to acidification), this could be more difficult. The dissolved carbon concentration can always be monitored experimentally to forestall limitation (Degrenne et al., 2010).

4.3 Achieving Maximal Productivities

The growth of photosynthetic microorganisms depends on various parameters. Culture conditions (pH and temperature) can be kept optimal by appropriate regulation, although at large scale and in external solar conditions this can be very difficult. Chemical nutrients (inorganic dissolved carbon and mineral nutrients) can be supplied, while avoiding limiting or toxic concentrations. If all parameters are kept at their optimal value and nutrients are given in adequate quantities, light-limited conditions where light alone limits growth will be achieved. By definition, this will allow maximal biomass performance.

As recently discussed and clarified elsewhere, the light-limited regime is not sufficient to obtain maximal biomass productivities. This implies additionally controlling the radiative transfer conditions inside the culture, as represented by the γ parameter (Cornet and Dussap, 2009; Pruvost et al., 2011; Takache et al., 2010). If the biomass concentration is too low, some of the light is transmitted through the culture (low absorption, favoring the "kinetic" regime). Conversely, if the biomass is too high, a dark zone appears deep in the culture (favoring the light-limitation regime). A distinction must be made here between eukaryotic (microalgae) and prokaryotic (cyanobacteria) cells. In the case of cyanobacteria cultivation, having common electron carrier chains and no short-time respiration in the dark (Gonzalez de la Vara and Gomez-Lojero, 1986), a dark zone will be sufficient ($\gamma \leq 1$) to guarantee maximal productivity (Cornet and Dussap, 2009; Cornet, 2010). For eukaryotic cells presenting respiration in the light (microalgae), a dark zone in the culture volume where respiration is predominant will result in a loss of productivity due to biomass catabolism. Maximal productivity will then require the γ fraction to meet the exact condition $\gamma = 1$ (the "luminostat" regime), corresponding to a full absorption of the light received, but without a dark zone in the culture volume (Takache et al., 2010).

In practice, maintaining an optimal value of the γ parameter is not easy, especially in the case of microalgae (which implies meeting the condition $\gamma = 1$). Some illustrations are given below for both batch and continuous production modes. Because it does not allow full absorption of the light captured, the kinetic regime always leads to a loss of efficiency ($\gamma > 1$). This regime is, however, usually encountered at the beginning of a batch production run (Figure 4). Because of the biomass growth, attenuation conditions will continuously evolve and the γ value will progressively decrease down to a value below 1. For prokaryotic cells (Figure 4, left), as soon as full absorption is obtained, the maximal value of the mean volumetric growth rate will be achieved and then remain constant (until a large dark zone is formed, inducing a shift in the cell metabolism). For eukaryotic cells, the $\gamma = 1$ condition, and so the maximal value of the mean volumetric growth rate will only be transitorily satisfied (mean volumetric growth rate being represented by the slope of $C_x(t)$, see Equation 3). The increase in the dark volume will then progressively lower the mean volumetric growth rate (Figure 4, right).

In continuous mode, light attenuation conditions can be controlled by modifying the dilution rate to adjust biomass concentration in the system. For cyanobacteria (Figure 5), there will be an optimal range of biomass concentrations to meet the condition $\gamma \leq 1$. For microalgae, the $\gamma = 1$ condition will require an optimal biomass concentration ($C_{x\ opt}$) corresponding exactly to the occurrence of the physical limitation by light, with all light absorbed but no dark zone (as shown in Takache et al., 2010, a deviation of the γ value in the range $\gamma = 1 \pm 15\%$ could be tolerated).

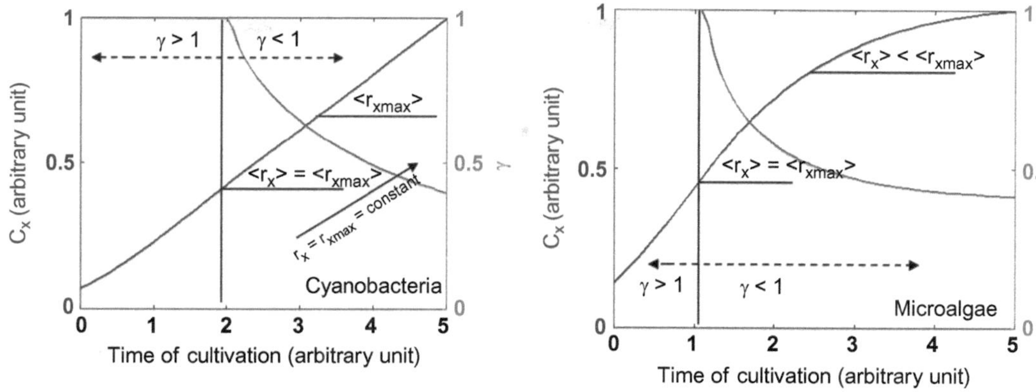

FIGURE 4 Typical evolution of biomass concentration during a batch cultivation of cyanobacteria (left) and microalgae (right) (light-limited conditions).

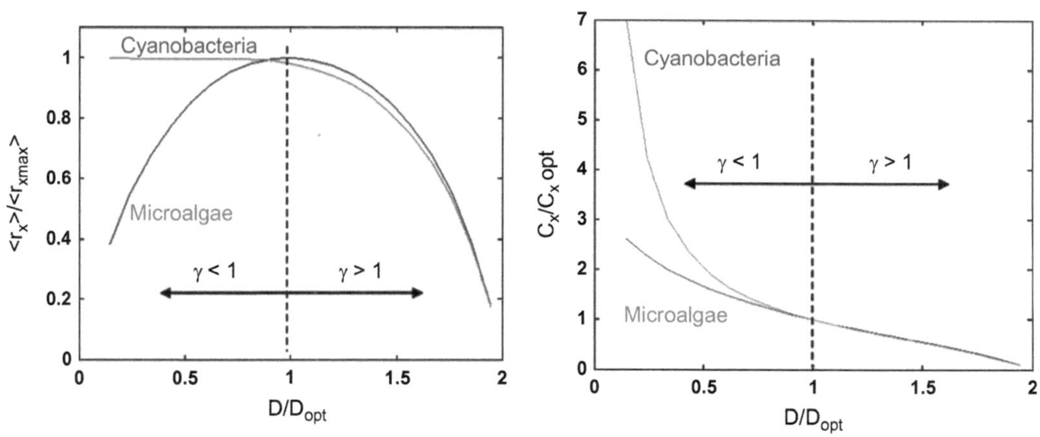

FIGURE 5 Typical evolution of biomass volumetric productivity (left) and biomass concentration (right) as a function of the dilution rate for both cyanobacteria and microalgae (continuous production in light-limited conditions).

Whichever the production mode (continuous or batch), the control of the illuminated fraction in light limited-conditions (with $\gamma \leq 1$ for cyanobacteria and $\gamma = 1 \pm 15\%$ for microalgae) will enable us to obtain maximum biomass productivity of the cultivation system in light-limited conditions (volume and surface). If radiative transfer conditions are known (using a radiative transfer model, as already described), then the optimal biomass concentration can be sought theoretically. But experimental determination is also possible simply by varying the dilution rate and measuring corresponding biomass concentration and productivity (Takache et al., 2010).

5 ENGINEERING PARAMETERS GOVERNING PHOTOBIOREACTOR PRODUCTIVITY

5.1 Optimization of the Light Supply

It is well established that cultivation of photosynthetic microorganisms is highly dependent on the light supply, especially in light-limitation conditions. The light supply can be increased either by increasing the PFD or by increasing the specific illuminated surface a_{light} (illuminated surface to culture volume ratio). Working in light-limited conditions with full light absorption is again important when increasing the PFD. The mutual shading of cells, in combination with adequate mixing conditions, will largely prevent photoinhibition effects (Richmond, 2004a,b). This enables us to work up to very high PFD (1000 μmole/m^2 s and above, see Takache et al., 2010) significantly higher than the maximum value that can be supported in dilute culture, as usually represented by the irradiance of saturation G_S (usually in the range 200-500 μmole/m^2 s).

The relation between biomass productivities and PFD was recently introduced by Cornet and Dussap (2009), who proposed a simple relation. This relation was determined for cultivation systems working in light-limited conditions meeting the condition $\gamma = 1$ (luminostat regime) and for geometries coming under the one-dimensional hypothesis (flat panel geometries, open ponds, and cylindrical and tubular photobioreactors with radial illumination). This relation has since been validated on a large number of photobioreactor geometries and species, including microalgae and cyanobacteria (Cornet and Dussap, 2009; Pruvost et al., 2011; Takache et al., 2010).

The equation for calculating maximal biomass volumetric productivity in light-limited conditions is:

$$P_{x\,\max} = \langle r_X \rangle_{\max} = \rho_M \overline{\varphi} \frac{2\alpha}{1+\alpha} a_{\text{light}} K \ln\left[1 + \frac{q}{K}\right]. \tag{15}$$

All the parameters can be determined predictively for any species or cultivation systems geometry (for details, see Cornet and Dussap, 2009). The parameters ρ_m, $\overline{\phi}$, K, and α (linear scattering modulus $\alpha = \sqrt{(E_a/(E_a + 2\,bE_s))}$ are species dependent. The specific illuminated surface a_{light} and the PFD q are engineering parameters.

Due to the progressive saturation of photosynthesis with respect to light received, increasing the PFD increases volumetric productivity, but with a progressive decrease in yield of light conversion into biomass. This results in a logarithmic relation between the productivity and the PFD. Biomass volumetric productivity is by contrast found to increase proportionally with the specific illuminated surface a_{light}, emphasizing the utility of maximizing illuminated surface with respect to culture volume. For example, for Cartesian geometries (this includes flat panel photobioreactors, but also open ponds), the a_{light} value is directly related to the depth of culture L_z by the simple relation $a_{\text{light}} = S_{\text{light}}/V_r = 1/L_z$. Very high volumetric productivities will thus be obtained for technologies with very short light paths. Depths of culture are usually in the range of 0.1 m (with depths up to 0.5 m for open ponds), but values below 0.01 m can be also encountered. Considering Equation (15), biomass volumetric productivity will then be increased 100-fold. In practice, however, very narrow light paths induce specific constraints, such as difficulty maintaining adequate heat and mass transfer conditions, or possible biofilm formation.

Equation (15) can also be expressed in terms of maximum surface biomass productivity:

$$\langle S_x \rangle_{\max} = \frac{\langle r_{x\max} \rangle}{a_{\text{light}}} = \rho_M \overline{\phi} \frac{2\alpha}{1+\alpha} K \ln\left[1 + \frac{q}{K}\right]. \tag{16}$$

We observe that surface productivity is independent of the specific illuminated surface. This is also an important conclusion. Because the specific illuminated surface is fully dependent on cultivation system geometry, surface productivity is useful for comparing efficiencies of different cultivation systems. More interestingly, it emphasizes the fact that volumetric productivity can be increased while keeping surface productivity constant (assuming that the system remains in light-limited conditions). This conclusion is of particular interest in the context of solar production of biomass for energy production uses, where surface productivity is crucial and so has to be kept maximal.

5.2 Influence of Mixing Conditions

Except for immobilized cells (not discussed here), culture mixing will be necessary not only for mass (nutrients, gas-liquid transfer) and heat transfers (temperature homogenization), but also to prevent sedimentation and biofilm formation (Muller-Feuga et al., 2003a,b; Pruvost et al., 2002a). In addition to these classical features of any bioreactor, mixing conditions also result in what are known as light-dark (L/D) cycle effects widely described in the literature (Janssen et al., 2000a,b; Perner-Nochta and Posten, 2007; Pruvost et al., 2008; Richmond, 2004a,b; Rosello Sastre et al., 2007). Cells moving in the heterogeneous radiation field experience a particular history with respect to the light they absorb, composed of variations from high irradiance level (in the vicinity of the light source) to low or quasinull values (deep in the culture) if biomass concentration is high (Figure 6).

The exact effects on the resulting growth remain to be researched. Photosynthetic conversion is indeed a dynamic process, and the fluctuating light history induced by flow can modify instantaneous conversion rates of absorbed light. However, it is very difficult to investigate those effects experimentally in cultivation systems, because of various mixing

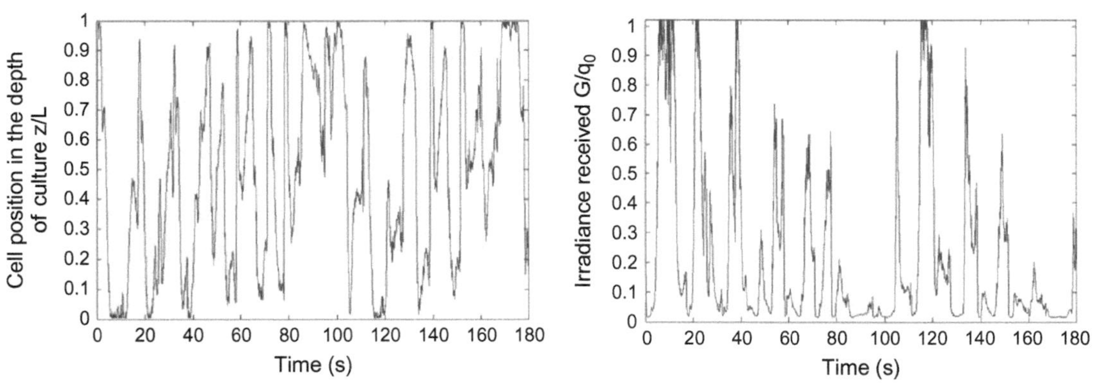

FIGURE 6 Example of cell displacement along the light gradient (left) and corresponding Light/Dark cycles (right).

effects, such as transfer enhancement (positive effect) or shear-stress generation (negative effect). Separating the coupling between the flow field and the light use from other possible mixing effects is difficult to achieve experimentally (Merchuk et al., 1998). In addition, L/D cycle effects are fully dependent on the light regime, and thus on cycle frequencies and magnitudes. In cultivation systems such values are rarely known, cell history with respect to light resulting from both flow and radiative fields, each determination being a problem on its own. Some examples can be found in the literature on the characterization of light regimes in photobioreactors. Firstly, cells trajectories are determined by using either a schematic representation of the flow (Janssen et al. 2003; Wu and Merchuk, 2002, 2004), by experimental measurement with radiative particle tracking (Luo and Al-Dahhan, 2004; Luo et al. 2003), or by a Lagrangian simulation (Pruvost et al. 2002a,b). Light regime is next obtained by introducing the light attenuation model. As shown in Pruvost et al. (2008), attention must, however, be paid to the formulation of the coupling. Mixing can influence the spatial distribution of particles participating in radiative transfer, resulting in a modification of the radiation field (Cassano et al. 1995). The calculation method for the radiative transfer has thus to be modified to take into account the effect of mixing conditions. An oversimplified formulation (as usually produced), where cell trajectories and radiative transfer are solved independently, results in a false representation of light availability in the reactor. This can lead for example to a significant overestimation of the L/D cycle effects (Pruvost et al., 2008).

In addition to the difficulty in accurately determining L/D cycle regimes experienced by flowing cells, the corresponding biological response still remains to be clarified. It is difficult to measure and the results depend on the species and the light fluctuation magnitude and frequencies applied (Janssen et al., 1999, 2000a,b). To characterize the effect of given mixing conditions on the system efficiency, an appropriate model has also to be formulated and then be associated with the L/D cycle prediction. Some attempts can be found in literature (Camacho et al., 2003; Eilers and Peeters, 1993; Luo and Al-Dahhan, 2004; Pahl-Wostl, 1992; Wu and Merchuk, 2001; 2002; Wu and Merchuk, 2004; Yoshimoto et al., 2005), but a research effort is still needed to develop robust and generalizable dynamic models that are able to represent effects of the wide range of L/D cycles encountered in microalgal cultivation system.

More vigorous mixing conditions may also have a negative effect due to the resulting hydrodynamic shear stress. Numerous species are shear stress sensitive, with various responses, ranging from modified cell response (secretion of exopolysaccharides) to cell impairment and death (Jaouen et al., 1999). A compromise has thus to be found when mixing rate is increased (Barbosa et al., 2004). However, again, very few quantitative data are available, and mixing conditions, despite their influence on cultivation systems, are usually managed empirically.

In conclusion, although many studies have shown the relevance of mixing conditions, knowledge is still insufficient and useful engineering rules have yet to be found to determine optimal conditions for a given species and cultivation system. Hydrodynamics conditions have indeed several impacts that have *in fine* to be related. For example, fast L/D cycles with frequencies higher than 1 Hz are known to have a positive effect. Such frequencies can be reached in specifically designed cultivation devices (Janssen et al., 2000b; Rosello Sastre et al., 2007). However, this improvement will also increase energy consumption and induce shear stress. Ideally, each effect has thus to be taken into account for global optimization. But this requires a significant research effort to set up the appropriate theoretical framework necessary for systematic optimization. Actually, only general rules can be currently used to guide mixing conditions in microalgae cultivation systems. Their general objectives will be to prevent cell sedimentation, guarantee medium homogenization (temperature, pH, nutrients), promote L/D cycles by generating cell displacement along the light gradient, and keep shear stress below cell fragility thresholds.

6 EXISTING TECHNOLOGY

6.1 Specific Features of Solar Cultivation Systems

Microalgal cultivation systems can use artificial or natural (sun) light sources. Obviously, for practical, economic, and environmental reasons, natural sunlight is to be preferred for mass-scale production of biomass for energy production purposes. This case will be explored here. Solar production adds a degree of complexity to the optimization and control of the cultivation system, compared with the artificial illumination case. The process is fully dynamic and driven by an uncontrolled input: the solar incident flux. Sunlight is highly variable in time (day-night cycles, season, and clouds) and space (Earth location, orientation of the cultivation system with respect to the sun path, shading by surrounding buildings, trees, or others cultivation systems, etc.). All these features already affect more classical solar processes such as photovoltaic panels, solar thermal concentrated conversion, or photocatalysis (Duffie and Beckman, 2006). Microalgae production involves in addition specific features such as the need to keep growth conditions in an acceptable range (thermal regulation will thus be of prime relevance), or the complex biological response to light (e.g., saturation or photoinhibition effects) and dark (biomass catabolism at night). Unlike processes based only on surface conversion (e.g., photovoltaic panels), optimizing the amount of light collected on the microalgal cultivation system surface is thus not sufficient. Light conversion by photosynthetic microorganisms occurs within the culture bulk: the transfer of the collected light flux inside the bulk has thus to be taken into account. As a consequence, as for any light-driven process, cultivation systems will be highly dependent on the light collected on the illuminated surface, but light transfer conditions and thus productivity will be also influenced by the variation of incident angle or direct/diffuse distribution of sunlight flux density. All these aspects formed the subject of a recent paper (Pruvost et al., in press) introducing a generic model to represent light-limited growth in solar photobioreactors (based on the theoretical framework presented in previous sections, drawn from many years of investigation in artificial light systems). The model was associated in particular with a solar database to predict surface productivity as a function of the system location or its ability to intercept solar radiation (as influenced by system inclination or season, for example).

One main consequence of working in solar conditions is the dynamic regime imposed by radiation variations (e.g., day-night cycle). Transient behavior is obtained as a result of a complex interaction between physical (light) and biological (growth) kinetics, with a specific role of night, which induces biomass catabolism. The marked, steep changes in radiation conditions during the day hinder the overall optimization of the process. Whereas in artificial light the PFD can be kept constant allowing an optimized biomass concentration to obtain maximal performance from the cultivation system, in sunlight, nonideal illumination conditions prevail most of the time due to the low growth rate of algae: at noon, biomass concentration will not increase sufficiently fast to ensure full absorption of impinging light and, at the beginning and end of day, dark zones will appear in the culture bulk (Pruvost et al., in press). All these features are characterized by the illuminated fraction, which always varies in solar conditions (Figure 7). Control strategies can be devised to optimize light use during day-night cycles, such as with the harvesting procedure to optimize biomass concentration in the system and thus in the illuminated fraction for given period of the day (an example is given in Pruvost et al., in press). However, the high variability of sunlight makes this very difficult (besides day-night cycles, weather conditions have also to be allowed for) and the species cultivated will also greatly influence the strategy (especially when cultivating eukaryotic microorganisms that are sensitive to dark zones).

Similar conclusions can be drawn for other growth parameters, such as temperature. Some 50% of the energy content of solar radiation lies in the infrared spectrum (higher than PAR wavelengths). Solar technology, and especially closed systems, thus tends toward overheating (or evaporation of water in open systems) under high light flux (depending obviously on the ambient conditions). An example is given in Figure 8 for a flat panel photobioreactor operated without thermal regulation in the South of France in the month of July (unpublished results). A temperature of 340K is reached here (67°C), obviously incompatible with a microalgal cultivation. The control of temperature is thus a further challenge for mass-scale production, especially in the case of an energy-production end use. The energy balance of the process being of prime relevance, energy consumption for thermal regulation

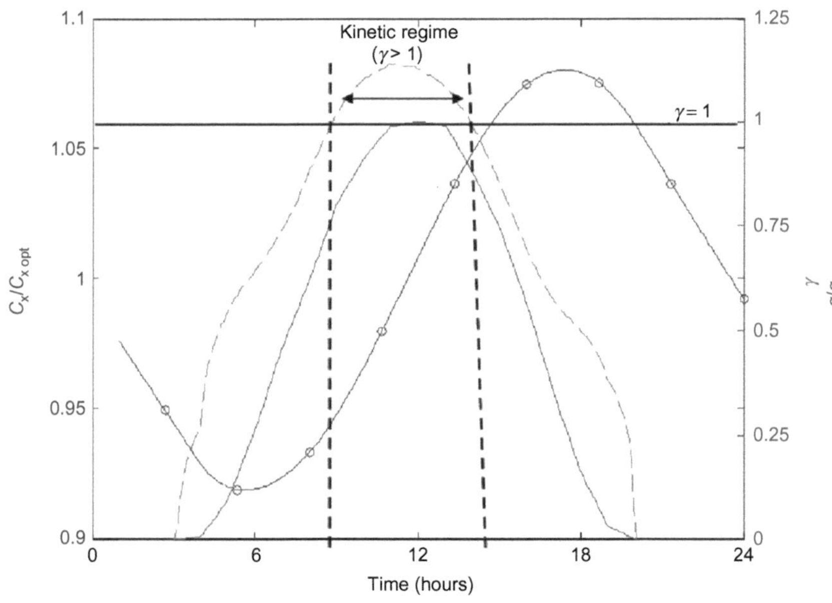

FIGURE 7 Typical day-night variation of biomass concentration (circles added on lines) and illuminated fraction (dashed line) in a surface-lightened photobioreactor during a summer day. The normalized PFD (solid line) is also given.

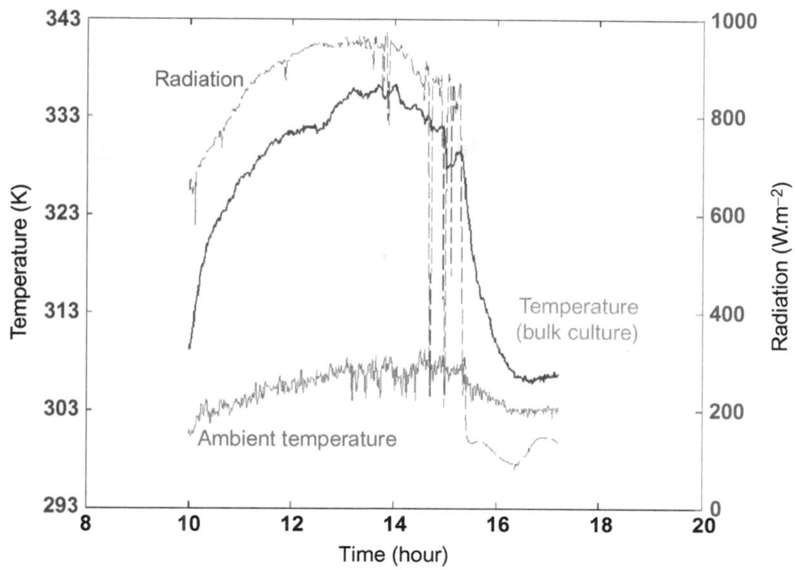

FIGURE 8 Typical thermal behavior of a flat panel system during a sunny day in France (Perpignan, July 2010). To emphasize the overheating, the system was operated without thermal regulation (water solution and black dye).

has thus to be minimized and kept in an acceptable range (at least below the energy recovered in biofuel). This implies appropriate engineering of the system but, again, the problem is not trivial, the thermal behavior (depending on the light flux) directly influencing the biomass growth.

Whatever the operating parameter (light, temperature, pH, etc.), mathematical modeling of the solar production case is certainly at least as important as technological development. Biofuel production will certainly aim at operation throughout the entire year. Hence, advanced control strategies or engineering optimization procedures are crucial to having systems operate close to their maximal performance. The utility of this approach has already been demonstrated for artificially lightened photobioreactors (Cornet et al., 2001). It will, however, require an adequate theoretical framework for the solar case.

6.2 Surface and Volumetrically Lightened Systems

Light energy can be supplied in two general ways, by direct illumination of the cultivation system (surface-lightened systems) or by inserting light sources inside the culture volume (volumetrically lightened systems). Most cultivation systems belong to the simpler surface-lightened group (Richmond, 2004a; Ugwu et al., 2008). A wide variety of geometries are encountered, from open ponds to tubular or flat panel photobioreactors. An extensive literature can thus be found on these systems, showing that all of them have advantages and limits as regards control of culture conditions, culture confinement, resulting hydrodynamics conditions, ease of upscaling, construction cost, etc. However, whatever the concept, light supply and its use by the culture will always govern the productivity of the cultivation system, so that PFD and the specific illuminated surface will be the main engineering parameters. In solar conditions, the PFD is defined by the ability of the system to collect light. As for any solar processes, various positioning options are found, with systems positioned horizontally (Acién Fernández et al., 2001; Molina et al., 2001; Oswald, 1988), vertically (Chini Zitelli et al., 2000; Chini Zittelli et al., 2006; Pulz, 2001), or in few cases tilted (Doucha and Livansky, 2006; Richmond and Cheng-Wu, 2001). Maximizing light interception is not trivial, however. It naturally depends on the system location on the Earth and on the day or year period. For example, horizontal systems are best suited to locations close to the Equator (latitude 0°). For higher latitudes, it will be necessary to increase the titled angle to maximize light collection (roughly speaking, the best inclination angle for a given position on the Earth is equal to the latitude of the location). Although maximizing light intercepted must be a basic principle of any microalgal cultivation system (as for any solar process), other constraints have also to be considered. For example, using the airlift principle for mixing will preclude horizontal geometries, and shading will have to be considered when arranging vertical or tilted systems on a given land area. Again, optimizing a solar cultivation system thus proves more complex than for other classical solar processes, such as photovoltaic panels, where light intercepted is the only parameter (of a given panel technology).

Volumetrically lightened systems lead to more complex technologies, but allow a further optimization of the light use in the culture. Firstly, insertion of light sources in the culture bulk guarantees a maximal use of emitted photons. For surface-lightened systems, and especially for artificially lightened systems, it is very difficult to collimate all the emitted light onto the optical surface of a photobioreactor. Secondly and more interestingly, internal lighting allows light to be diluted. As discussed previously, increasing light leads to higher volumetric productivity, but with a progressive decrease in the conversion yield, due to photosynthesis saturation. By diluting light received in the culture volume, a high yield can be maintained. This is of particular interest for solar use and energy production applications. In this case, solar light is collected on a given surface (using for example a parabolic device) and then transmitted to the culture (using for example optical fibers). Because of the high PFD usual in solar conditions, a significant increase in surface productivity can be obtained. Furthermore, the dilution principle can be combined with a solar tracking system, giving an additional possibility of optimization by maximizing light intercepted during the sun's travel. By combining these advantages in systems with high specific illuminated surfaces, the most efficient system of light conversion into biomass can be obtained, with both high volumetric and surface productivities. A full description of such a principle is described by Cornet (2010) with a volumetrically lightened photobioreactor based on the "DiCoFluV" concept (see publication for details). Theoretically, such technology allows the highest biomass productivities permitted with algae. The author presents maximal productivities that could be achieved for both surface and volumetrically lightened systems, assuming ideal sunlight conditions when located at the Equator. For surface-lightened systems (fixed horizontal photobioreactor), a mean daily ideal value of the PFD around 1000 μmole m^{-2} s^{-1} was harnessed, leading to a surface biomass productivity of 100 t ha^{-1} y^{-1} with an exergetic yield of the photobioreactor of 6%. For volumetrically lightened systems with a sun tracking system ("DiCoFluV" concept), the daily averaged PFD was increased to 1400 μmole m^{-2} s^{-1} (same equatorial location). In combination with the dilution principle, a surface biomass productivity of 400 t ha^{-1} y^{-1} was obtained, corresponding to a maximum energy yield of the photobioreactor of 17%. Because all the calculations were conducted here for an ideal case (solar radiation, growth kinetics, photobioreactor design, and light use), this corresponds to the upper limit of productivity that may be achieved on the Earth with photosynthetic microorganisms. Despite their promise, only a few examples of volumetrically lightened photobioreactors can be found (Cornet, 2010; Csogör et al., 2001; Hsieh and Wu, 2009; Ogbonna et al., 1996; Zijffers et al., 2008). This is mainly explained by the increase in technological complexity, and by the difficulty scaling up to large areas. Further technological developments are still needed.

6.3 Open Systems and Photobioreactors

Several recent reviews on existing technologies for microalgal production can be found in the literature and only the main aspects will be given here. The cheapest systems that can be easily extended today to a large scale are open systems. These systems have been used for many decades at an industrial scale, but for applications other than biofuels (Richmond, 2004a). Technologically, such systems could, however, be used for that purpose. The two main groups of open systems are natural ponds and raceways. The main difference between them is in the mixing regime. Open ponds are unmixed (except naturally, e.g., by wind), unlike raceways, where paddle wheels are used to circulate the culture in a loop configuration. The best productivities are obtained in these last systems. The main limitations of open systems are inherent to their operating principle. Owing to the direct contact with the atmosphere, there is a high risk of biological contamination (other microalgae species, bacteria, predators, etc.). Only resistant species can thus be long term cultivated in such systems. Because there is a large interface between the culture and the atmosphere, their control is also difficult, for example, to maintain optimal temperature (although open systems are less subject to overheating than closed systems). In addition, the gas-liquid equilibrium with the rather low atmospheric CO_2 content generally results in a limiting concentration of dissolved carbon, insufficient to meet the needs of photosynthetic microorganisms in the case of intensive production. A carbon supply can be added (CO_2 or carbonate), but a significant part of this will inevitably be degassed into the atmosphere, making carbon limitation difficult to prevent entirely.

Closed geometries reduce risks of external contamination and a better control of culture conditions is obtained. The higher partial pressure allowed in the gas phase will also prevent carbon limitation. All these advantages will allow the light-limitation condition to be obtained and, as already discussed, productivity will then be limited only by the solar energy entering the cultivation system and by its use either by direct illumination or by diluting light in the culture bulk. However, closed geometries suffer from several limitations, also inherent to their operating principles. Culture confinement increases the risk of biofilm formation, leads to oxygen accumulation in the culture (toxic effects), and overheating can occur (especially in solar conditions due to the large amount of infrared light collected). Unlike open systems where the only way to prevent external contamination or carbon limitation is to close the systems (working then with photobioreactor technology), the limits of closed geometries can be at least partly overcome by appropriate engineering of photobioreactors (e.g., by adapting mixing conditions to increase heat and gas-liquid mass transfer or to prevent biofilm formation). However, this most often also results in increased cost and complexity. As the photobioreactor is the only system that allows maximal productivity to be obtained (by working in light-limited conditions), great efforts are currently being made to develop new technology devoted to mass-scale production. Current mass-scale production comes mainly from extensive open systems easier to build and operate, that is, open ponds or raceway systems. However, it may well be that in the near future a suitable closed technology will be devised that meets the criteria for mass-scale intensified production of photosynthetic microorganisms.

Figure 9 (top) gives a rough estimate of the maximal surface productivity that could be achieved with the different technologies (all for an ideal case, as defined by Cornet, 2010). The lower surface productivity of open systems is assumed here, considering the lower control of culture conditions and effect of carbon limitation, with raceways presenting higher productivities than open ponds due to the mixing optimization they permit. Higher surface productivities are obtained with volumetrically lightened photobioreactors allowing light dilution in the culture bulk to prevent from adverse effects of photosynthesis saturation to light, as encountered in surface-lightened photobioreactors having thus lower surface productivity. Figure 9 (bottom) gives an overview of volumetric productivity of microalgal cultivation systems that is highly linked to their specific illuminated surface and thus culture depth. Raceway depths are usually about 0.2 m, while photobioreactor depths can be as low as a few centimeters or even less. Using Equation (15) and considering also that open systems are usually submitted to other limitations than light, the volumetric productivity of photobioreactor can be thus higher by two orders of magnitude. As already discussed, surface productivity is relevant for its direct impact on land areas required for a given

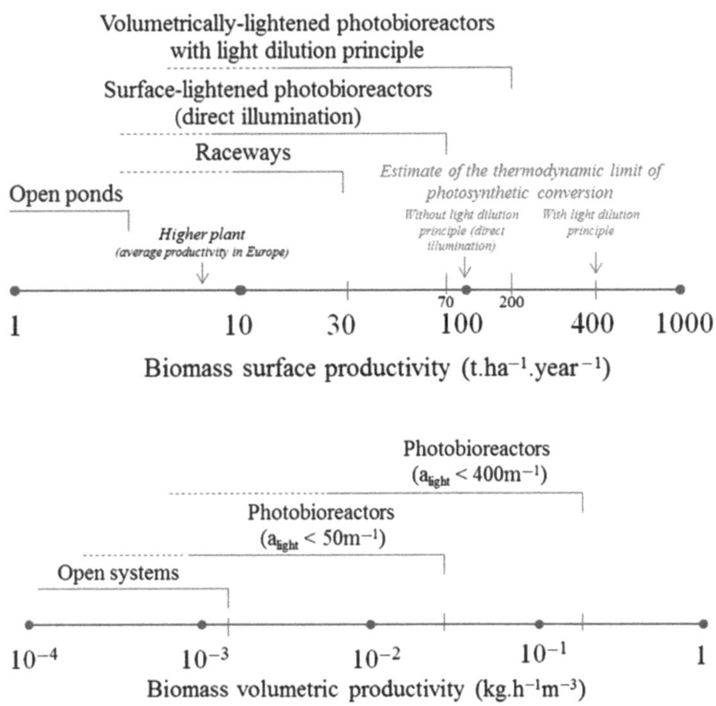

FIGURE 9 Estimate of the maximal surface (top) and volumetric (bottom) productivities that could be achieved with different microalgal cultivation systems

production. Volumetric productivity would have also a decisive impact on the global biofuel production process. Increasing volumetric productivity will indeed allow high biomass concentration and thus lower harvesting cost, and will also lower the culture volume to be managed and so energy consumption for mixing. All these aspects will contribute to a positive energy balance at the overall process level. Considering in addition that only closed systems allow carbon limitation to be prevented when working at high biomass volumetric productivity, thus leading to a higher surface productivity than in open systems, photobioreactors clearly offer the highest potential. Maximal areal productivity can be sought while increasing volumetric productivity. Limits are here mainly in engineering aspects making the development of specific cultivation systems for mass-scale production of algae at an acceptable cost as one of the main current challenges to the global use of photosynthetic microorganisms for energy production.

Acknowledgments

This work was supported by the French national research agency for bioenergy production (ANR-PNRB), and is part of the French "BIOSOLIS" research program on developing photobioreactor technology for mass-scale solar production (http://www.biosolis.org/). This book chapter is also the result of many years of collaboration with two remarkable scientists, Prof. Jack Legrand and Prof. Jean-François Cornet, to whom the author is especially grateful. Thanks also go to Vincent Goetz for his help on all solar aspects and to François Le Borgne for solar experiments.

References

Acién Fernández, F.G., Fernández Sevilla, J.M., Sánchez Pérez, J.A., Molina Grima, E., Chisti, Y., 2001. Airlift-driven external-loop tubular photobioreactors for outdoor production of microalgae: assessment of design and performance. Chem. Eng. Sci. 56, 2721–2732.

Aiba, S., 1982. Growth kinetics of photosynthetic microorganisms. Adv. Biochem. Eng. Biotechnol. 23, 85–156.

Barbosa, M.J., Hadiyanto, Wijffels, R.H., 2004. Overcoming shear stress of microalgae cultures in sparged photobioreactors. Biotechnol. Bioeng. 85, 78–85.

Benemann, J.R., 2004. Hydrogen and methane production by microalgae. In: Ltd, B.S. (Ed.), Handbook of Microalgal Culture: Biotechnology and Applied Technology. (A. Richmond).

Berberoglu, H., Pilon, L., Melis, A., 2008. Radiation characteristics of *Chlamydomonas reinhardtii* CC125 and its truncated chlorophyll antenna transformants tla1, tlaX and tla1-CW+. Int. J. Hydrogen Energy 33, 6467–6483.

Camacho, F.R., Camacho, F.G., Sevilla, F.J.M., Chisti, Y., Molina Grima, E., 2003. A mechanistic model of photosynthesis in microalgae. Biotechnol. Bioeng. 81, 459–473.

Carvalho, A.P., Meireles, L.A., Malcata, F.X., 2006. Microalgal reactors: a review of enclosed system designs and performances. Biotechnol. Prog. 22, 1490–1506.

Cassano, A.E., Martin, C.A., Brandi, R.J., Alfano, O.M., 1995. Photoreactor analysis and design: fundamentals and applications. Industrial and Engineering Chemistry Research 34 (7), 2155–2201.

Chini Zitelli, G.C., Pastorelli, R., Tredici, M.R., 2000. A Modular Flat Panel Photobioreactor (MFPP) for indoor mass cultivation of Nannochloropsis sp. under artificial illumination. J. Appl. Phycol. 12, 521–526.

Chini Zittelli, G., Rodolfi, L., Biondi, N., Tredici, M.R., 2006. Productivity and photosynthetic efficiency of outdoor cultures of *Tetraselmis suecica* in annular columns. Aquaculture 261, 932–943.

Chisti, Y., 2007. Biodiesel form microalgae. Biotechnol. Adv. 25, 294–306.

Chiu, S.Y., Kao, C.Y., Chen, C.H., Kuan, T.C., Ong, S.C., Lin, C.S., 2008. Reduction of CO_2 by a high-density culture of *Chlorella sp.* in a semicontinuous photobioreactor. Bioresour. Technol. 99, 3389–3396.

Cogne, G., Lehmann, B., Dussap, C.G., Gros, J.B., 2003. Uptake of macrominerals and trace elements by the cyanobacterium Spirulina platensis (Arthrospira platensis PCC 8005) under photoautotrophic conditions: culture medium optimization. Biotechnol. Bioeng. 81, 588–593.

Cornet, J.F., 2007. Procédés limités par le transfert de rayonnement en milieu hétérogène. Etude des couplages cinétiques et énergétiques dans les photobioréacteurs par une approche thermodynamique. Université Blaise Pascal - Clermont-Ferrand, n° d'ordre 236.

Cornet, J.F., 2010. Calculation of optimal design and ideal productivities of volumetrically-lightened photobioreactors using the constructal approach. Chem. Eng. Sci. 65, 985–998.

Cornet, J.F., Dussap, C.G., 2009. A simple and reliable formula for assessment of maximum volumetric productivities in photobioreactors. Biotechnol. Prog. 25, 424–435.

Cornet, J.F., Dussap, C.G., Cluzel, P., Dubertret, G., 1992a. A structured model for simulation of cultures of the cyanobacterium Spirulina platensis in photobioreactors. 1. Coupling between light transfer and growth kinetics. Biotechnol. Bioeng. 40, 817–825.

Cornet, J.F., Dussap, C.G., Cluzel, P., Dubertret, G., 1992b. A structured model for simulation of cultures of the cyanobacterium Spirulina platensis in photobioreactors. 2. Identification of kinetic parameters under light and mineral limitations. Biotechnol. Bioeng. 40, 826–834.

Cornet, J.F., Dussap, C.G., Gros, J.B., 1994. Conversion of radiant light energy in photobioreactors. AIChE J. 40, 1055–1066.

Cornet, J.F., Dussap, C.G., Gros, J.B., 1995. A simplified monodimensional approach for modeling coupling between radiant light transfer and growth kinetics in photobioreactors. Chem. Eng. Sci. 50, 1489–1500.

Cornet, J.F., Dussap, C.G., Gros, J.B., 1998. Kinetics and energetics of photosynthetic micro-organisms in photobioreactors: application to Spirulina growth. Adv. Biochem. Eng. Biotechnol. 59, 155–224.

Cornet, J.F., Dussap, C.G., Leclercq, J.J., 2001. Simulation, design and model based predictive control of photobioreactors. In: Thonart, M.H.a.P. (Ed.), Focus on Biotechnology. Engineering and manufacturing for biotechnology, vol. 4. Kluwer Academic Publishers, Dordrecht, pp. 227–238.

Cornet, J.F., Favier, L., Dussap, C.G., 2003. Modeling stability of photoheterotrophic continuous cultures in photobioreactors. Biotechnol. Prog. 19, 1216–1227.

Csogör, Z., Herrenbauer, M., Schmidt, K., Posten, C., 2001. Light distribution in a novel photobioreactor—modelling for optimization. J. Appl. Phycol. 13, 325–333.

Degrenne, B., Pruvost, J., Christophe, G., Cornet, J.F., Cogne, G., Legrand, J., 2010. Investigation of the combined effects of acetate and photobioreactor illuminated fraction in the induction of anoxia for hydrogen production by Chlamydomonas reinhardtii. Int. J. of Hydrogen Energy 35, 10741–10749.

Doucha, J., Livansky, K., 2006. Productivity, CO2/O2 exchange and hydraulics in outdoor open high density microalgal (Chlorella sp.) photobioreactors operated in a Middle and Southern European climate. J. Appl. Phycol. 18, 811–826.

Duffie, J.A., Beckman, W.A., 2006. Solar Engineering of Thermal Processes, third ed John Wiley & Sons, New York.

Eilers, P.H.C., Peeters, J.C.H., 1993. Dynamic behaviour of a model for photosynthesis and photoinhibition. Ecol. Model. 69, 113–133.

Farges, B., Laroche, C., Cornet, J.F., Dussap, C.G., 2009. Spectral kinetic modeling and long-term behavior assessment of Arthrospira platensis growth in photobioreactor under red (620 nm) light illumination. Biotechnol. Prog. 25, 151–162.

Fouchard, S., Pruvost, J., Degrenne, B., Titica, M., Legrand, J., 2009. Kinetic modeling of light limitation and sulphur deprivation effects in the induction of hydrogen production with *Chlamydomonas reinhardtii* Part I: model description and parameters determination. Biotechnol. Bioeng. 102, 132–147.

Ghirardi, M.L., Zhang, L., Lee, J.W., Timothy Flynn, T., Seibert, S., Greenbaum, E., Melis, A., 2000. Microalgae: a green source of renewable H2. Trends. Biotechnol. 18, 506–511.

Gonzalez de la Vara, L., Gomez-Lojero, C., 1986. Participation of plastoquinone, cytochrome c553 and ferredoxin-NADP + oxido reductase in both photosynthesis and respiration in *Spirulina maxima*. Photosynth. Res. 8, 65–78.

Hsieh, C.H., Wu, W.T., 2009. A novel photobioreactor with transparent rectangular chambers for cultivation of microalgae. Biochem. Eng. J. 46, 300–305.

Hu, Q., Sommerfeld, M., Jarvis, E., Ghirardi, M.L., Posewitz, M., Seibert, M., Darzins, A., 2008. Microalgal triacylglycerols as feedstocks for biofuel production: perspectives and advances. Plant J. 54, 621–639.

Janssen, M.G.J., Kuijpers, T.C., Veldhoen, B., Ternbach, M.B., Tramper, J., Mur, L.R., Wijffels, R.H., 1999. Specific growth rate of *Chlamydomonas reinhardtii* and *Chlorella sorokiniana* under medium duration light/dark cycles 13-87 s. J. Biotechnol. 70, 323–333.

Janssen, M., De Bresser, L., Baijens, B., Tramper, J., Mur, L.R., Snel, J., Wijffels, R.H., 2000a. Scale-up aspects of photobioreactors: effects of mixing-induced light/dark cycles. J. Appl. Phycol. 12, 225–237.

Janssen, M., Janssen, M.G.J., De Winter, M., Tramper, J., Mur, L.R., Snel, J., Wijffels, R.H., 2000b. Efficiency of light utilization of *Chlamydomonas reinhardtii* under medium-duration light/dark cycles. J. Biotechnol. 78, 123–137.

Janssen, M., Tramper, J., Mur, L.R., Wijffels, R.H., 2003. Enclosed outdoor photobioreactors: Light regime, photosynthetic efficiency, scale-up, and future prospects. Biotechnology and Bioengineering 81 (2), 193–210.

Jaouen, P., Vandanjon, L., Quéméneur, F., 1999. The shear stress of microalgal cell suspensions (*Tetraselmis suecica*) in tangential flow filtration systems: the role of pumps. Bioresour. Technol. 68, 149–154.

Kliphuis, A., 2010. Modeling of microalgal metabolism. PhD Thesis, Wageningen University, Wageningen, The Netherlands, p. 200.

Lehr, F., Posten, C., 2009. Closed photo-bioreactors as tools for biofuel production. Curr. Opin. Biotechnol. 20, 280–285.

Loubiere, K., Olivo, E., Bougaran, G., Pruvost, J., Robert, R., Legrand, J., 2009. A new photobioreactor for continuous microalgal production in hatcheries based on external-loop airlift and swirling flow. Biotechnol. Bioeng. 102, 132–147.

Luo, H.P., Al-Dahhan, M.H., 2004. Analyzing and modeling of photobioreactors by combining first principles of physiology and hydrodynamics. Biotechnol. Bioeng. 85, 382–393.

Melis, A., 2002. Green alga hydrogen production: progress, challenges and prospects. Int. J. Hydrogen Energy 27, 1217–1228.

Merchuk, J.C., Ronen, M., Giris, S., Arad, S., 1998. Light/dark cycles in the growth of the red microalga Porphyridium sp. Biotechnol. Bioeng. 59, 705–713.

Molina, E., Fernández, J., Acién, F.G., Chisti, Y., 2001. Tubular photobioreactor design for algal cultures. J. Biotechnol. 92, 113–131.

Muller-Feuga, A., 1998. Growth as a function of rationing: a model applicable to fish and microalgae. J. Exp. Mar. Biol. Ecol. 236, 1–13.

Muller-Feuga, A., Le Guedes, R., Pruvost, J., 2003a. Benefits and limitations of modeling for optimization of *Porphyridium cruentum* cultures in an annular photobioreactor. J. Biotechnol. 103, 153–163.

Muller-Feuga, A., Pruvost, J., Le Guedes, R., Le Dean, L., Legentilhomme, P., Legrand, J., 2003b. Swirling flow implementation in a photobioreactor for batch and continuous cultures of *Porphyridium cruentum*. Biotechnol. Bioeng. 84, 544–551.

Ogbonna, J.C., Yada, H., Masui, H., Tanaka, H., 1996. A novel internally illuminated stirred tank photobioreactor for large-scale cultivation of photosynthetic cells. J. Ferment. Bioeng. 82, 61–67.

Oswald, W.J., 1988. Large-scale algal culture systems (engineering aspects). In: Borowitska, M. (Ed.), Microalgal Biotechnology. Cambridge University Press, Cambridge, pp. 357–394.

Pahl-Wostl, C., 1992. Dynamic versus static models for photosynthesis. Hydrobiologia 238, 189–196.

Perner-Nochta, I., Posten, C., 2007. Simulations of light intensity variation in photobioreactors. J. Biotechnol. 131, 276–285.

Pottier, L., Pruvost, J., Deremetz, J., Cornet, J.F., Legrand, J., Dussap, C.G., 2005. A fully predictive model for one-dimensional light attenuation by *Chlamydomonas reinhardtii* in a torus reactor. Biotechnol. Bioeng. 91, 569–582.

Pruvost, J., Legrand, J., Legentilhomme, P., Muller-Feuga, A., 2002a. Simulation of microalgae growth in limiting light conditions—flow effect. AIChE J. 48, 1109–1120.

Pruvost, J., Legrand, J., Legentilhomme, P., Muller-Feuga, A., 2002b. Trajectory Lagrangian Model for Turbulent Swirling Flow in an Annular Cell. Comparison with RTD Measurements. Chem. Eng. Sci. 57 (7), 1205–1215.

Pruvost, J., Cornet, J.F., Legrand, J., 2008. Hydrodynamics influence on light conversion in photobioreactors: an energetically consistent analysis. Chem. Eng. Sci. 63, 3679–3694.

Pruvost, J., Van Vooren, G., Cogne, G., Legrand, J., 2009. Investigation of biomass and lipids production with *Neochloris oleoabundans* in photobioreactor. Bioresour. Technol. 100, 5988–5995.

Pruvost, J., Van Vooren, G., Le Gouic, B., Couzinet-Mossion, A., Legrand, J., 2011. Systematic investigation of biomass and lipid productivity by microalgae in photobioreactors for biodiesel application. Bioresour. Technol. 102, 150–158.

Pruvost, J., Cornet, J.F., Goetz, V., Legrand, J., in press. Modeling dynamic functioning of rectangular photobioreactors in solar conditions. AIChE J.

Pulz, O., 2001. Photobioreactors: production systems for phototrophic microorganisms. Appl. Microbiol. Biotechnol. 57, 287–293.

Richmond, A., 2004a. Handbook of Microalgal Culture: Biotechnology and Applied Phycology. Blackwell Sciences Ltd, Oxford, UK.

Richmond, A., 2004b. Principles for attaining maximal microalgal productivity in photobioreactors: an overview. Hydrobiologia 512, 33–37.

Richmond, A., Cheng-Wu, Z., 2001. Optimization of a flat plate glass reactor for mass production of Nannochloropsis sp. outdoors. J. Biotechnol. 85, 259–269.

Rodolfi, L., Chini Zittelli, G., Bassi, N., Padovani, G., Biondi, N., Bonini, G., Tredici, M., 2009. Microalgae for oil: strain selection, induction of lipid synthesis and outdoor mass cultivation in a low-cost photobioreactor. Biotechnol. Bioeng. 102, 100–112.

Roels, J.A., 1983. Energetics and Kinetics in Biotechnology. Elsevier biomedical press, Amsterdam.

Rosello Sastre, R., Csögör, Z., Perner-Nochta, I., Fleck-Schneider, P., Posten, C., 2007. Scale-down of microalgae cultivations in tubular photo-bioreactors—a conceptual approach. J. Biotechnol. 132, 127–133.

Sanchez Miron, A., Ceron Garcia, M.C., Contreras Gomez, A., Garcia Camacho, F., Molina Grima, E., Chisti, Y., 2003. Shear stress tolerance and biochemical characterization of Phaeodactylum tricornutum in quasi steady-state continuous culture in outdoor photobioreactors. Biochem. Eng. J. 16, 287–297.

Schlegel, G.O., Burkholder, F.W., Klein, S.A., Beckman, W.A., Wood, B.D., Muhs, J.D., 2004. Analysis of a full spectrum hybrid lighting system. Sol. Energy 46, .

Scragg, A.H., Illman, A.M., Carden, A., Shales, S.W., 2002. Growth of microalgae with increased calorific values in a tubular bioreactor. Biomass Bioenergy 23, 67–73.

Sierra, E., Acien, F.G., Fernandez, J.M., Garcia, J.L., Gonzales, C., Molina, E., 2008. Characterization of a flat plate photobioreactor for the production of microalgae. Chem. Eng. J. 138, 136–147.

Spolaore, P., Joannis-Cassan, C., Duran, E., Isambert, A., 2006. Commercial applications of microalgae. J. Biosci. Bioeng. 101, 87–96.

Takache, H., Christophe, G., Cornet, J.F., Pruvost, J., 2010. Experimental and theoretical assessment of maximum productivities for the micro-algae *Chlamydomonas reinhardtii* in two different geometries of photobioreactors. Biotechnol. Prog. 26, 431–440.

Tredici, M.R., Chini Zittelli, G., 1998. Efficiency of sunlight utilization: tubular versus flat photobioreactors. Biotechnol. Bioeng. 57, 187–197.

Tsukahara, K., Sawayama, S., 2005. Liquid fuel production using microalgae. J. Jpn. Petrol. Inst. 48, 251–259.

Ugwu, C.U., Aoyagia, H., Uchiyamaa, H., 2008. Photobioreactors for mass cultivation of algae. Bioresour. Technol. 99, 4021–4028.

Vonshak, A., Torzillo, G., 2004. Environmental stress physiology. In: Richmond, A. (Ed.), Handbook of Microalgal Culture: Biotechnology and Applied Phycology. Blackwell Sciences Ltd, Oxford, UK, pp. 57–82.

Wu, X., Merchuk, J.C., 2001. A model integrating fluid dynamics in photosynthesis and photoinhibition processes. Chem. Eng. Sci. 56, 3527–3538.

Wu, X., Merchuk, J.C., 2002. Simulation of algae growth in an bench-scale column reactor. Biotechnol. Bioeng. 80, 156–168.

Wu, X., Merchuk, J.C., 2004. Simulation of algae growth in an bench-scale internal loop airlift reactor. Chem. Eng. Sci. 59, 2999–12912.

Yoshimoto, N., Sato, T., Kondo, Y., 2005. Dynamic discrete model of flashing light effect in photosynthesis of microalgae. J. Appl. Phycol. 17, 207–214.

Yun, Y.S., Park, J.M., 2003. Kinetic modeling of the light-dependent photosynthetic activity of the green microalga Chlorella vulgaris. Biotechnol. Bioeng. 83, 303–311.

Zijffers, J.W., Janssen, M., Tramper, J., Wijffels, R.H., 2008. Design process of an area-efficient photobioreactor. Marine Biotechnol. 10, 404–415.

SECTION IV

PRODUCTION OF BIOHYDROGEN

CHAPTER

20

Production of Biohydrogen: Current Perspectives and Future Prospects

Kuan-Yeow Show[1], *Duu-Jong Lee*[2,*], *Zhen-Peng Zhang*[3]

[1]Department of Environmental Engineering, Faculty of Engineering and Green Technology, Universiti Tunku Abdul Rahman, Jalan University, Bandar Barat, 31900 Kampar, Perak, Malaysia

[2]Department of Chemical Engineering, National Taiwan University, Taipei, Taiwan

[3]Beijing Enterprises Water Group Limited, BLK 25, No. 3 Minzhuang Rd, Beijing, China

***Corresponding author: E-mail: djlee@ntu.edu.tw**

1 INTRODUCTION

Hydrogen has been identified as one of the most promising fuels for the future. It is a promising alternative to conventional fossil fuels because it has the potential to eliminate all the problems that fossil fuels create. Molecular hydrogen has the highest calorific value per unit mass (143 GJ/ton) among the known gaseous fuels (Boyles, 1984) and may release energy explosively in heat engines or generate electricity quietly in fuel cells while producing water as the only by-product. Hydrogen is also the raw material for the synthesis of ammonia, alcohols, and aldehydes, and for the hydrogenation of various petroleum and edible oils, coal, and shale oil (Fang et al., 2002). Hydrogen gas is proposed as the ultimate transport fuel for cars, trucks, and buses because of its nonpolluting characteristics and because it enables the use of highly efficient fuel cells to convert chemical energy to electricity (Forsberg, 2007). It has been known that the USA, EU, and Japan have already established hydrogen fuel stations, and car manufacturers have also invested in the development of hydrogen fuel-powered cars. This review seemed to indicate that hydrogen plays an important role in energy supply for the transport sector.

A major skepticism, however, with hydrogen as a clean energy alternative is the way it is being produced. Most of the hydrogen at the present technological development is generated from fossil fuels through thermochemical processes, such as hydrocarbon reforming, coal gasification, and partial oxidation of heavier hydrocarbons (Das and Veziroglu, 2001; Levin et al., 2004).

While the current process is based on the use of carbon-based nonrenewable resources, the cost of production poses a major challenge for the development of conventional hydrogen production.

Biohydrogen production is considered a development vital to a sustainable global clean energy supply and a promising alternative to conventional fossil fuels, as it has the potential to eliminate most of the problems that fossil fuels create, if not all. The major advantage of energy from biohydrogen is the avoidance of greenhouse gas emissions as the conversion of hydrogen to energy, either via combustion or via fuel cells, results only in pure water (Claassen et al., 1999). With the use of appropriate technologies, biohydrogen would be the desired clean product of the microbial process.

Biohydrogen production has attracted worldwide attention because of its potential to become an inexhaustible, low-cost, and renewable source of clean energy. This review provides an overview of the state of the art and perspectives of biohydrogen production research. This review focuses on recent developments in hydrogen production via different feedstock. Economics, perspectives, and prospects of biobiohydrogen production are also outlined.

2 FEEDSTOCK

For hydrogen to be renewable, it must come from renewable feedstock. Apart from water, potential resources in feedstock include agricultural crops and their waste by-products, lignocellulosic products such as wood and wood waste, waste from food processing and aquatic plants and algae, and effluents produced in human habitats. Energy from water-containing biomass such as sewage sludge, agricultural and livestock effluents, and animal excreta is recovered mainly by microbial fermentation. Biomass as energy source is characterized in the form of both flow and stock. If these resources were used under appropriate control, they would become the major source of energy in the future.

Biomass has the potential to become a significant source of renewable hydrogen. In addition, many of the processes that produce hydrogen from biomass are complementary to those that produce biomaterials. Therefore, countries with large agricultural economies have potential for significant economic development through incorporation of bioenergy into bioindustry. While renewable hydrogen technologies that use low-value waste biomass as feedstock have great potential to become cost-competitive, it is currently more expensive to produce hydrogen from biomass than it is to derive it from natural gas. Independent of the source of hydrogen, there are many logistical and market challenges that must also be overcome before a hydrogen economy can become a reality:

$$\text{Biomass} + O_2 \rightarrow CO + H_2 + CO_2 + \text{Energy}.$$

As reported by many studies of biohydrogen fermentative processes, carbohydrates are the main source of hydrogen. Thus, wastes and biomass rich in sugars and/or complex carbohydrates appear to be suitable feedstock for biohydrogen generation (Ntaikou et al., 2010). The major criteria that have to be met for the selection of substrates suitable for fermentative biohydrogen production are availability, cost, carbohydrate content, and biodegradability. Simple sugars such as glucose, sucrose, and lactose are readily biodegradable and thus preferred as model substrates for hydrogen production. However, pure carbohydrate sources are expensive raw materials for real-scale hydrogen production, which can only be viable when based on renewable and low-cost sources (Das and Veziroglu, 2001).

2.1 Pure Carbohydrates

It has been reported that biological hydrogen can be produced from a wide spectrum of carbohydrates, but most studies have been limited to using food products such as pure monosaccharides (glucose; Horiuchi et al., 2001; Zhang et al., 2007), disaccharides (sucrose; Lee et al., 2003; Yu and Mu, 2006), and polysaccharides (starch, cellulose, and hemicellulose; Czernik et al., 1999; Van Ginkel et al., 2001; Yokoi et al., 1998; Zhang et al., 2003). Simple carbohydrates such as xylose, lactose, maltose (Ferchichi et al., 2005b), and cellobiose (Kumar and Das, 2000) have also been used as feedstock for hydrogen production. Glucose and sucrose are easily degraded, and thus a short period of fermentation time or process HRT is required in batch and continuous hydrogen production experiments, respectively. On the contrary, it takes a relatively longer fermentation time to degrade starch and cellulose, since they have to be hydrolyzed into monosaccharides before being used for hydrogen production. The maximum hydrogen yields obtained from these pure carbohydrates vary from 2.40 mol H_2/mol hexose using cellulose (Ueno et al., 1995) to 3.33 mol H_2/mol hexose from starch (Kanai et al., 2005) and glucose or sucrose (van Niel et al., 2002), indicating that these carbohydrates are suitable as feedstock for dark hydrogen fermentation. However, it seems obvious that hydrogen conversion capability from different substrates is related to the microbial species as the hydrogen yields by the same cultures obtained under the similar operating conditions are rather inconsistent. For example, Ferchichi et al. (2005b) found that *Clostridium saccharoperbutylacetonicum* ATCC 27021 grown on disaccharides (lactose, sucrose, and maltose) produces on average more than twice (2.81 mol H_2/mol sugar) as much hydrogen as monosaccharides (1.29 mol H_2/mol hexose). Kumar and Das (2000) working with immobilized *Enterobacter cloacae* IIT-BT08 suggested that hydrogen yields using this microorganism vary from substrate to substrate (2.2 mol H_2/mol glucose, 6 mol H_2/mol sucrose and 5.4 mol H_2/mol cellobiose). On the contrary, Logan et al. (2002) found the maximum hydrogen yields obtained from glucose, sucrose, and potato starch by a mixed culture to be 0.9 mol H_2/mol glucose, 1.8 mol H_2/mol sucrose, and 0.59 mol H_2/mol starch, respectively. Oh et al. (2002) found that the H_2 production rate with disaccharides (lactose, sucrose) and starch was much slower than that with monosaccharides (glucose, galactose, fructose). The maximal H_2 yield and H_2 production rate were estimated to be 2.76 mol H_2/mol glucose and 29.9 mmol H_2/g cell h, respectively.

2.2 Food-Related Wastes and Wastewater

Recently, considerable research activity on fermentative hydrogen production has been focused on the conversion of reproducible biomass resources to hydrogen by microbes (Fan et al., 2006; Han and Shin, 2004; Shin et al., 2004; Wu and Lin, 2004). The renewable biomass is mainly derived from food-related wastes, that is, food industry wastes/wastewater and food residues in municipal solid wastes. The food manufacturing industry produces high-strength organic wastes or wastewater. Highly concentrated carbohydrates, mainly in terms of single sugars, starch, and cellulose in the food manufacturing by-products, make such wastes/wastewater a potential and important feedstock for biological hydrogen production. The biotransformation of wastes and wastewater to hydrogen can be considered quite appealing from both the environmental and the economic standpoint (Ntaikou et al., 2010).

Various studies have been conducted to investigate the feasibility of biological hydrogen recovery from processing by-products, including rice and wheat bran (Noike and Mizuno, 2000), wheat bran (Noike and Mizuno, 2000), bean curd manufacturing waste (Mizuno et al., 2000), noodle manufacturing waste (Mizuno et al., 2000), apple processing and potato processing wastewater (Van Ginkel et al., 2005), cheese whey (Ferchichi et al., 2005a), sugar factory wastewater (Ueno et al., 1996), rice winery wastewater (Yu et al., 2002, 2003), molasses (Tanisho and Ishiwata, 1995), sugar beet wastewater (Hussy et al., 2005), dehydrated brewery mixture (Fan and Chen, 2004), starch manufacturing wastes (Yokoi et al., 2002, 2001), and organic wastewater (Show et al., 2010). The hydrogen yields obtained from food processing wastes or wastewater range from 0.68 to 2.70 mol H_2/mol hexose, which are comparable to those from pure carbohydrates. Of the yields reported, the highest yield was demonstrated by Yokoi et al. (2002) while fermenting sweet potato starch residue with a mixed continuous culture, whereas the lowest one was demonstrated by Fan and Chen (2004) in a mixed batch culture fed with dehydrated brewery solid waste.

Complex solid wastes, such as wastes from kitchen, food processing, mixed wastes, and municipal wastes, have also been tested as feedstock for fermentative hydrogen production (Ntaikou et al., 2010). Apart from carbohydrates, such wastes usually have quite high contents of proteins and fats, and thus their biotransformation to hydrogen is comparatively lower than those obtained from carbohydrate-based wastewater. The hydrogen production potential of carbohydrate-based wastes was reported as 20 times higher than that of fat-based and protein-based wastes (Lay et al., 2003). This observation was partially attributed to the consumption of hydrogen to form ammonium using nitrogen generated from protein biodegradation.

Among the different types of wastes, organic fraction of municipal solid wastes (OFMSW) can be considered an important feedstock for hydrogen production. From an environmental viewpoint, there is an urgent need for appropriate management of municipal solid wastes. On average, almost 50% of the municipal solid wastes of underdeveloped countries consist of a fermentable and biodegradable fraction (Valdez-Vazquez et al., 2005). Because of their high carbohydrate and protein contents, food wastes constitute a major OFMSW. The OFMSW has significant potential of biological hydrogen production, but it is heavily dependent on its composition of. Among the OFMSW, Okamoto et al. (2000) reported that hydrogen production potentials of individual carbohydrates varied from 26.3-61.7 (cabbage) and 44.9-70.7 (carrot) to 19.3-96.0 mL/g VS (rice), and carbohydrates produced the most hydrogen compared with proteins (egg: 2.60-7.07 mL/g VS; lean meat: 2.47-7.68 mL/g VS) or lipids (chicken skin: 3.56-10.2 mL/g VS; fat: 4.41-11.14 mL/g VS). Later, a similar observation reported by Lay et al. (2003) supported the fact that the hydrogen production potential of carbohydrate-rich high-solid organic wastes (HSOW) (rice and potato) is approximately 20 times larger than that of fat-rich HSOW (fat meat and chicken skin) and protein-rich HSOW (egg and lean meat).

For comparison purposes, Table 1 briefly summarizes the maximum hydrogen yields obtained from different substrates as reported in literature. This, together with the studies

TABLE 1 Comparison of Maximum Hydrogen Yields of Different Substrates

Substrate	Constituents	Seed Sludge	Hydrogen Yield Conversion (mL H_2/g VS)	Conversion Efficiency (%)	References
Carbohydrates	Pure glucose	*Thermotoga maritima*	497.8 ds	100	Schröder et al. (1994)
	Cellulose	Sludge compost	298.7[a] ds	60	Ueno et al. (1995)
	Starch	*Thermococcus kodakaraensis* KOD1	414.4[b] ds	83.25	Kanai et al. (2005)
Proteins	Peptone	UASB sludge inhibited by chloroform	33.6 ds	6.75	Liang et al. (2001)
	Egg	Digested sludge	7.07 vs	1.42	Okamoto et al. (2000)
	Lean meat	Digested sludge	7.68 vs	1.54	Okamoto et al. (2000)
Lipids	Chicken skin	Digested sludge	10.2 vs	2.05	Okamoto et al. (2000)
	Fat	Digested sludge	11.1 vs	2.23	Okamoto et al. (2000)

[a]*Calculated from the reported value of 2.40 mol H_2/mol hexose.*
[b]*Calculated from the reported value 3.33 mol H_2/mol hexose, starch is calculated as hexose [$(C_6H_{10}O_5)_n$].*

mentioned above, supports the theory that most carbohydrate-rich materials are a suitable feedstock for dark hydrogen fermentation, while protein- and lipid-rich substances are probably less suitable or even unsuited. Although the yield of hydrogen from protein-rich substrates is reported to be in a low range, generally of 7.07-33.6 mL H_2/g DS (Lay et al., 2003; Liang et al., 2001; Okamoto et al., 2000), supplementing proteins to carbohydrate-rich feedstock could enhance overall hydrogen production. Kim et al. (2004) found that the addition of sewage sludge to food waste up to 13-19% increased culture alkalinity and thus enhanced overall hydrogen production potential due to the high protein content in sewage sludge. This is largely attributed to the fact that production of ammonia from the proteinaceous substrates such as peptone can avoid culture pH drops by neutralizing fatty acids produced from glucose fermentation (Cheng et al., 2002).

A large yield fluctuation is commonly noted in the OFMSW fermentation as the collected samples are heterogeneous in nature. For example, Lay et al. (1999) demonstrated a hydrogen yield of 180 mL H_2/g TVS by preheated digested sludge; Kim et al. (2004) found that 60.1 mL hydrogen was produced from 1 g TVS of a mixture of OFMSW and sewage sludge; and Fang et al. (2006) obtained a hydrogen yield of 346 mL/g carbohydrate, corresponding to a 2.50 mol H_2/mol hexose while fermenting rice slurry with a mixed culture. Notably, the highest hydrogen yield obtained from OFMSW is 3.20 mol H_2/mol hexose by thermophiles (Valdez-Vazquez et al., 2005), which is similar to the work with pure cultures of extreme thermophilic microorganisms fermenting glucose or sucrose where a hydrogen yield of 3.33 mol H_2/mol hexose is obtained (van Niel et al., 2002). This implies that OFMSW has a great potential to be used as substrate for biological hydrogen production. The fact that the mixed culture using OFMSW as substrate in solid substrate fermentation mode achieved similar hydrogen yields to pure cultures fermenting easily degradable carbohydrates in liquid cultures is very encouraging.

2.3 Wastewater Sludge

Wasted sludge produced during wastewater treatment processes is mainly composed of microorganism bodies, and its organic substance composition is therefore different from carbohydrate-rich substrates such as glucose or starch. Cai et al. (2004) suggested that proteins are the largest constituent of the waste-activated sludge, accounting for about 32-41% of total dry solid, while Kim et al. (2004) indicated a much higher proportion of proteins in sewage sludge (0.20 g carbohydrate-COD/g VS and 0.73 g protein-COD/g VS). Lower hydrogen yields from such wastewater sludge would be expected, because of the large protein proportion. In fact, Cai et al. (2004) reported a hydrogen yield of aerobic-activated sludge at 9.1 mL H_2/g dry solids (DS) and found that hydrogen yield could be enhanced to 16.6 mL H_2/g DS if the raw sludge was subjected to alkaline treatment. Wang et al. (2003) obtained a hydrogen yield of 20.2 mL H_2/g DS, equivalent to 0.6 mmol H_2/g COD of the inoculum and sludge while fermenting wastewater sludge with *Clostridium bifermentan*. Pretreatment of the original sludge through freezing and thawing and sterilization markedly increased the hydrogen yield from 20.2 to 49.3-71.7 mL H_2/g DS (Wang et al., 2003). However, the enhanced yields are little higher than those obtained from protein-rich OFMSW, and rather lower compared to the reported yields from carbohydrate-rich OFMSW (Lay et al., 2003; Okamoto et al., 2000). These studies indicate that it might be uneconomical to recover hydrogen from wastewater sludge because of its high protein proportion. Moreover, the hydrogen yield by anaerobically fermenting domestic wastewater is rather low. Van Ginkel et al. (2005) obtained a hydrogen yield of 1.8 mmol H_2/g COD from the 25 times concentrated domestic wastewater with a total COD of 6.2 g/L. This value is close to that of wastewater sludge, but a low strength of organic substances appears to limit hydrogen production efficiently from domestic wastewater.

2.4 Future Feedstock

In the above sections, the range of organic compounds that can be utilized for dark hydrogen production has been reviewed. In spite of the limited investigation that has been carried out, the studies conducted enable a conclusion to be made, that is, suitable feedstock for dark hydrogen fermentation would be carbohydrate-rich materials. Figure 1 depicts the potential and sustainable biomass sources serving as feedstock for hydrogen fermentation. Biomass is basically a stored source of solar energy initially collected by plants during the process of photosynthesis whereby carbon dioxide is captured and converted to plant materials mainly in the form of cellulose, hemicellulose, and lignin. The term "biomass" therefore covers a range of organic materials recently produced from plants, and animals that feed on the plants. The biomass includes crop residues (primary residues), forest and wood process residues (secondary residues), animal wastes, including human sewage, organic municipal solid waste, food processing wastes (secondary and tertiary residues), purpose grown energy crops, and short rotation forests (IEA, 2008). However, as stated earlier, the feedstock currently consists of food-related substance, such as some pure chemicals (food products) and some food-related

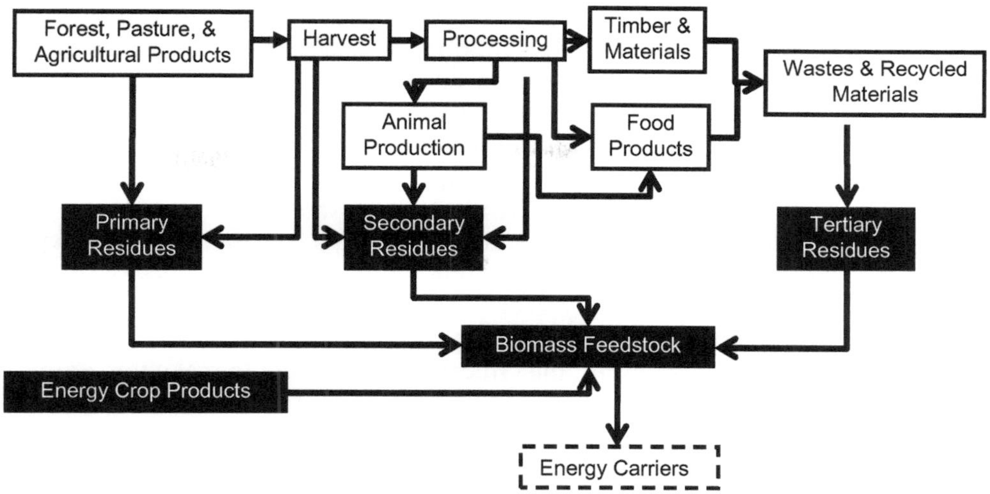

FIGURE 1 Potential sources of feedstock for dark hydrogen fermentation (IEA, 2008).

secondary and tertiary residues. The technology would have a rather limited scope if the feedstock range and sources cannot be expanded extensively.

Apart from the more easily degradable substances such as simple saccharides, starch, and cellulose, the main components of future feedstock will, most probably, be derived, to a large extent, from primary resides and energy crops, totally called "green wastes." This is largely due to a high proportion of microbial convertible carbohydrates in the green wastes. For example, the woody waste consists of about 70% (dried weight percentage) polysaccharides (cellulose and hemicellulose) and 25-35% lignin, and lignin covers the polysaccharides that are substrate convertible into fuel materials by enzymes or/and microorganisms (Take et al., 2006). However, it is difficult to directly convert green wastes into biohydrogen gas by anaerobic fermentative microorganisms because of their complex composition and polymeric structure. To the author's knowledge, only a few studies fermenting green wastes into hydrogen are available in the literature (de Vrije et al., 2002; Fan et al., 2006; Guo et al., 2010; Ntaikou et al., 2010; Ren et al., 2009). Fan et al. (2006) investigated the feasibility of obtaining biological hydrogen from agricultural wastes, that is, wheat straw wastes with cow dung compost. The maximum hydrogen yield of 68.1 mL H_2/g TVS was obtained as the raw wheat straw was pretreated with HCl and microwave heating, which is comparable to some of reported values for carbohydrate-rich wastes (Fan and Chen, 2004; Kim et al., 2004; Okamoto et al., 2000). This value is about 136-fold as compared with that of fermenting raw wheat straw wastes, indicating that appropriate pretreatment is necessary for microbial hydrogen fermentation of complex agricultural wastes.

In general, the green waste is a complex polymer consisting of tightly bound lignin, cellulose, and hemicellulose (de Vrije and Claasen, 2003; Ntaikou et al., 2010; Ren et al., 2009). Cellulose and hemicellulose can be converted into hydrogen by anaerobic fermentation, but lignin is not degraded under anaerobic conditions. Moreover, lignin strongly hampers the utilization of cellulose and hemicellulose because the bonding in lignocellulose resists mobilization and chemically degraded lignin is often inhibitory to microbial growth (de Vrije and Claasen, 2003). Therefore, it is necessary to destroy the polymeric bonding structure of lignin in the green wastes for effective production of dark hydrogen feedstock through various pretreatment methods. Delignification of green waste would be the crucial step in dark hydrogen fermentation. The methods most common applied are physicochemical treatment (steaming explosion and acidification) and enzymatic treatment. de Vrije et al. (2002) demonstrated that high delignification values were obtained by the combination of mechanical (extrusion or milling) and chemical pretreatment (sodium hydroxide) on miscanthus, a lignocellulosic biomass. An optimized process consisted of a one-step extrusion-NaOH pretreatment at moderate temperature (70 °C); a mass balance of this process in combination with enzymatic hydrolysis showed the following: pretreatment resulted in 77% delignification, a cellulose yield of more than 95% and 44% hydrolysis of hemicellulose. No matter which method is employed, the critical factors needed to make this process economically viable are to optimize sugar recovery and minimize process cost and environmental impact.

Exploitation of energy crops as feedstock for energy production and biotransformation to biofuels has been escalating in recent years. In general, the sustainability of such processes can be assured only if (a) the crops are produced at low cost, thus with minimum nutrient and water requirements; (b) they are resistant to environmental stresses, and (c) they are highly biomass yielding. Furthermore, such plants, in order to be suitable for hydrogen production via dark fermentation, should also have high sugar and/or carbohydrates content and low lignin content (Hawkes et al., 2002). According to their chemical composition, energy crops used for fermentative hydrogen production can be divided into sugar-based crops (e.g., sweet sorghum, sugarcane, and sugar beet), starch-based crops (e.g., corn and wheat), and lignocellulose-based crops including herbaceous (e.g., switch grass and fodder grass) and woody (e.g., miscanthus and poplar).

However, the continuously rising food prices, the sustainability doubts, and the energy-equation challenges have led to a backlash against the use of energy crops as feedstock for biofuel generation. This backlash was centered on the food-vs.-fuel debate, as in many countries huge agricultural areas have been turned into feedstock farms for the production of biofuels. The main arguments against the use of energy plants is that crops that could support human dietary needs either directly or indirectly through farm animals are diverted to the production of biofuels. Even in the case of nonfood crop cultivation, the sustainability issue is under question. As an answer to those issues, the production of second-generation biofuels is proposed, that is, biofuels produced by feedstock such as wastes and residues that are not a competition to edible crops (Ntaikou et al., 2010).

3 ECONOMICS OF BIOHYDROGEN PRODUCTION

While there are many reports in the literature about biohydrogen production, only a few of them deal with the economic analyses of the biohydrogen production. Benemann (1997) estimated an initial cost for an indirect microalgal biophotolysis system consisting of open ponds (140 ha) and photobioreactors (14 ha). The plant was assumed to generate 1.2 million GJ/yr at 90% plant capacity, with estimated total capital costs for the system at USD 43 million and annual operating costs at USD 12 million. Overall, total hydrogen production costs were estimated at USD 10/GJ. The capital costs were almost 90% of total costs at a 25% annual capital charge (Akkerman et al., 2003). The algal ponds were estimated at a cost of USD 6 per m^2, while the photobioreactors with assumed costs of USD 100 per m^2 were the major capital and operating cost factors. The costs of gas handling were not estimated but were presumed a significant cost factor.

An initial cost for a large-scale (>100 ha) single-stage algal or cyanobacterial biophotolysis process in a near-horizontal tubular reactor system was analyzed (Tredici and Zittelli, 1998). The main objective of the analysis was to determine if the proposed photobioreactor design could meet the cost requirements for hydrogen production through single-stage biophotolysis. The tubular photobioreactor offers superior features for biohydrogen production due to the internal gas exchange and the effective water spray cooling. Based on a 10% solar energy conversion efficiency, the costs of the tubular photobioreactor were estimated at USD 50 per m^2. The analysis did not include costs for gas handling and assumed a relatively low annual capital charge at 17%. The capital fixed costs were estimated at 80% of total costs, with the tubular material for the photobioreactor as the major cost. The hydrogen production costs were estimated at USD 15/GJ, which are comparable to the costs projected for hydrogen produced in a two-stage process from biomass residues projected at Euro 19/GJ (Tredici and Zittelli, 1998).

The economic analyses discussed above indicates that photobiohydrogens could be produced at a cost between USD 10/GJ and USD 20/GJ (Akkerman et al., 2003). This is a reasonable maximal cost target for renewable hydrogen fuel according to Benemann (2000). It should be noted that the economic analyses were based on optimistic assumptions and are highly presumptive, and were intended predominantly to ascertain the major cost drivers for photobiological hydrogen production. Currently, biologically produced hydrogen is more costly than other fuel alternatives. There is no doubt that many technical and engineering challenges have to be resolved before economic barriers can be meaningfully considered. It is clear from the economic analyses that the development of low-cost photobioreactors and the optimization of photosynthetic efficiency are the major R&D challenges.

4 FUTURE PROSPECTS AND CHALLENGE

In general, the molar yield of hydrogen and the cost of the feedstock are the two main barriers for fermentation technology. The main challenge in fermentative production of hydrogen is that less than 15% of the energy from the organic source can typically be obtained in the form of hydrogen (Logan, 2004). Consequently, it is not surprising that major efforts are directed to substantially increase the hydrogen yield. The U.S. Department of Energy (2007) program goal for fermentation technology is to realize yields of 4 and 6 mol H_2/mol of glucose by 2013 and 2018, respectively, as well as to achieve 3 and 6 months of continuous operation for the same years. Additionally, some integrated strategies are now under development, such as the two-step fermentation process (acidogenic + photobiological or acidogenic + methanogenic processes) or the use of modified microbial fuel cells (de Vrije and Claasen, 2003; Logan and Regan, 2006; Ueno et al., 2001). Through these coupled processes, more hydrogen or energy per mole of substrate can be achieved in the second stage.

Enhancement in hydrogen yield may be possible by using a suitable microbial strain, process modification, efficient bioreactor design, and genetic and molecular engineering techniques, to redirect the metabolic pathway. At this moment, the acceptability of genetically modified microorganisms is a challenge, because of the possible risk of horizontal transference of genetic material. However, this can be ruled out by chromosomal integration and the elimination of plasmids containing antibiotic markers with the available molecular tools (Datsenko and Wanner, 2000). Moreover, the improvement of hydrogen production by gene manipulation is focused mainly on the disruption of endogenous genes and not introducing new activities in the microorganisms. New pathways must be discovered to directly take full advantage of the 12 mol H_2 available in a mole of hexose.

Hydrogen is currently more expensive than other fuel options, so it is likely to play a major role in the economy in the long run, if technology improvements succeed in bringing down costs. Biohydrogen production employing renewable biomass may be a potential answer to overcome some of the economic constraints to fulfill many of our energy needs. There is scope to use sugarcane juice, molasses, or distillery effluent as substrates, because they contain sugar in significant quantities. Therefore, production as well as unit energy cost of biohydrogen would be reduced drastically. However, a rigorous technoeconomic analysis is necessary to draw a cost-effective comparison between biologically produced hydrogen and conventional fossil fuels.

Another challenge of the hydrogen fermentation is unstable hydrogen production. Unstable hydrogen production is possibly attributed to the metabolic shift of hydrogen-producing bacteria, and this could be minimized by in-depth microbial study. A major technical prerequisite for stable and efficient hydrogen fermentations is the maintenance of low hydrogen partial pressures through continuous removal of hydrogen from the fermentation broth.

5 CONCLUSIONS

Biohydrogen is believed to be one of the biofuels of the future, combining its ability to potentially reduce the dependence on foreign oil and contribute to lower the GHG emissions. The future role of hydrogen as a clean fuel for fuel cells producing near-zero emissions and as an intermediate energy carrier for storage and transport of renewable energy is increasingly recognized worldwide. Expert groups of various disciplines throughout the world are focusing on the production and application of biohydrogen, as well as the societal impacts of implementation. The R&D in the field of hydrogen and related technologies will therefore be intensified. In order to accelerate the technological development and to generate critical mass for the development of a hydrogen-based economy, international knowledge exchange and cooperation are required.

References

Akkerman, I., Janssen, M., Rocha, J.M.S., Reith, J.H., Wijffels, R.H., 2003. Photobiological hydrogen production: photochemical efficiency and bioreactor design. In: Reith, J.H., Wijffels, R.H., Barten, H. (Eds.), Biomethane and Biohydrogen: Status and Perspectives of Biological Methane and Hydrogen Production. Dutch Biological Hydrogen Foundation, Hague, the Netherlands.

Benemann, J.R., 1997. Feasibility analysis of photobiological hydrogen production. Int. J. Hydrogen Energy 22 (10-11), 979–987.

Benemann, J.R., 2000. Hydrogen production by microalgae. J. Appl. Phycol. 12 (3-5), 291–300.

Boyles, D., 1984. Bioenergy Technology—Thermodynamics and Costs. Wiley, NewYork.

Cai, M.L., Liu, J.X., Wei, Y.S., 2004. Enhanced biohydrogen production from sewage sludge with alkaline pretreatment. Environ. Sci. Technol. 38, 3195–3202.

Cheng, S.S., Chang, S.M., Chen, S.T., 2002. Effects of volatile fatty acids on a thermophilic anaerobic hydrogen fermentation process degrading peptone. Water Sci. Technol. 46, 209–214.

Claassen, P.A.M., van Lier, J.B., Contreras, A.M.L., van Niel, E.W.J., Sijtsma, L., Stams, A.J.M., et al., 1999. Utilisation of biomass for the supply of energy carriers. Appl. Microbiol. Biotechnol. 52, 741–755.

Czernik, S., French, R., Feik, C., Chornet, E., 1999. Production of hydrogen from hemicellulose-rich fractions. Abstr. Pap. Am. Chem. Soc. 217, U808–U809.

Das, D., Veziroglu, T.N., 2001. Hydrogen production by biological processes: a survey of literature. Int. J. Hydrogen Energy 26, 13–28.

Datsenko, K.A., Wanner, B.L., 2000. One-step inactivation of chromosomal genes in Escherichia coli K-12 using PCR products. Proc. Natl. Acad. Sci. USA 97, 6640–6645.

de Vrije, T., Claasen, P.A.M., 2003. Dark hydrogen fermentation. In: Reith, J.H., Wijffels, R.H., Barten, H. (Eds.), Bio-methane and Bio-hydrogen. Dutch Biological hydrogen foundation, The Hague, the Netherlands, pp. 103–123.

de Vrije, T., de Haas, G.G., Tan, G.B., Keijsers, E.R.P., Claassen, P.A.M., 2002. Pretreatment of Miscanthus for hydrogen production by *Thermotoga elfii*. Int. J. Hydrogen Energy 27, 1381–1390.

Fan, K.S., Chen, Y.Y., 2004. H-2 production through anaerobic mixed culture: effect of batch S-0/X-0 and shock loading in CSTR. Chemosphere 57, 1059–1068.

Fan, Y.T., Zhang, Y.H., Zhang, S.F., Hou, H.W., Ren, B.Z., 2006. Efficient conversion of wheat straw wastes into biohydrogen gas by cow dung compost. Bioresour. Technol. 97, 500–505.

Fang, H.H.P., Zhang, T., Liu, H., 2002. Microbial diversity of a mesophilic hydrogen-producing sludge. Appl. Microbiol. Biotechnol. 58, 112–118.

Fang, H.H.P., Li, C.L., Zhang, T., 2006. Acidophilic biohydrogen production from rice slurry. Int. J. Hydrogen Energy 31, 683–692.

Ferchichi, M., Crabbe, E., Gil, G.H., Hintz, W., Almadidy, A., 2005a. Influence of initial pH on hydrogen production from cheese whey. J. Biotechnol. 120, 402–409.

Ferchichi, M., Crabbe, E., Hintz, W., Gil, G.H., Almadidy, A., 2005b. Influence of culture parameters on biological hydrogen production by *Clostridium saccharoperbutylacetonicum* ATCC 27021. World. J. Microbiol. Biotechnol. 21, 855–862.

Forsberg, C.W., 2007. Future hydrogen markets for large-scale hydrogen production systems. Int. J. Hydrogen Energy 32, 431–439.

Guo, X.M., Trably, E., Latrille, E., Carrère, H., Steyer, J.P., 2010. Hydrogen production from agricultural waste by dark fermentation: a review. Int. J. Hydrogen Energy doi:10.1016/j.ijhydene.2010.03.008.

Han, S.K., Shin, H.S., 2004. Biohydrogen production by anaerobic fermentation of food waste. Int. J. Hydrogen Energy 29, 569–577.

Hawkes, F.R., Dinsdale, R., Hawkes, D.L., Hussy, I., 2002. Sustainable fermentative hydrogen production: challenges for process optimisation. Int. J. of Hydrogen Energy 27, 1339–1347.

Horiuchi, J., Kikuchi, S., Kobayashi, M., Kanno, T., Shimizu, T., 2001. Modeling of pH response in continuous anaerobic acidogenesis by an artificial neural network. Biochemical. Engineering. Journal. 9, 199–204.

Hussy, I., Hawkes, F.R., Dinsdale, R., Hawkes, D.L., 2005. Continuous fermentative hydrogen production from sucrose and sugarbeet. Int. J. Hydrogen Energy 30, 471–483.

IEA, 2008. World Energy Outlook 2008. In: OECD/IEA, Paris, p. 578.

Kanai, T., Imanaka, H., Nakajima, A., Uwamori, K., Omori, Y., Fukui, T., et al., 2005. Continuous hydrogen production by the hyperthermophilic archaeon, *Thermococcus kodakaraensis* KOD1. J. Biotechnol. 116, 271–282.

Kim, S.H., Han, S.K., Shin, H.S.H.S., 2004. Feasibility of biohydrogen production by anaerobic co-digestion of food waste and sewage sludge. Int. J. Hydrogen Energy 29, 1607–1616.

Kumar, N., Das, D., 2000. Enhancement of hydrogen production by *Enterobacter cloacae* IIT-BT 08. Process Biochem. 35, 589–593.

Lay, J.J., Lee, Y.J., Noike, T., 1999. Feasibility of biological hydrogen production from organic fraction of municipal solid waste. Water Res. 33, 2579–2586.

Lay, J.J., Fan, K.S., Chang, J., Ku, C.H., 2003. Influence of chemical nature of organic wastes on their conversion to hydrogen by heat-shock digested sludge. Int. J. Hydrogen Energy 28, 1361–1367.

Lee, K.S., Lo, Y.S., Lo, Y.C., Lin, P.J., Chang, J.S., 2003. H-2 production with anaerobic sludge using activated-carbon supported packed-bed bioreactors. Biotechnol. Lett. 25, 133–138.

Levin, D.B., Pitt, L., Love, M., 2004. Biohydrogen production: prospects and limitations to practical application. Int. J. Hydrogen Energy 29, 173–185.

Liang, T.M., Wu, K.L., Cheng, S.S., 2001. Hydrogen Production of Chloroform Inhibited Sludge Granular Sludge. Fukuoka, Japan.

Logan, B.E., 2004. Extracting hydrogen and electricity from renewable resources. Environ. Sci. Technol. 38, 160A–167A.

Logan, B.E., Regan, J.M., 2006. Microbial fuel cells-challenges and applications. Environ. Sci. Technol. 40, 5172–5180.

Logan, B.E., Oh, S.E., Kim, I.S., Van Ginkel, S., 2002. Biological hydrogen production measured in batch anaerobic respirometers. Environ. Sci. Technol. 37, 2530–2535.

Mizuno, O., Ohara, T., Shinya, M., Noike, T., 2000. Characteristics of hydrogen production from bean curd manufacturing waste by anaerobic microflora. Water Sci. Technol. 42, 345–350.

Noike, T., Mizuno, O., 2000. Hydrogen fermentation of organic municipal wastes. Water Sci. Technol. 42 (12), 155–162.

Ntaikou, I., Antonopoulou, G., Lyberatos, G., 2010. Biohydrogen production from biomass and wastes via dark fermentation: a review. Waste Biomass Valorization 1 (1), 21–39.

Oh, Y.K., Seol, E.H., Lee, E.Y., Park, S., 2002. Fermentative hydrogen production by a new chemoheterotrophic bacterium *Rhodopseudomonas Palustris P4*. Int. J. Hydrogen Energy 27, 1373–1379.

Okamoto, M., Miyahara, T., Mizuno, O., Noike, T., 2000. Biological hydrogen potential of materials characteristic of the organic fraction of municipal solid wastes. Water Sci. Technol. 41, 25–32.

Ren, N., Wang, A., Cao, G., Xu, J., Gao, L., 2009. Bioconversion of lignocellulosic biomass to biohydrogen: potential and challenges. Biotechnol. Adv. 27, 1051–1060.

Schröder, C., Selig, M., Schönheit, P., 1994. Glucose fermentation to acetate, CO_2 and H_2 in the anaerobic hyperthermophilic eubacterium *Thermotoga maritima*—involvement of the Embden-Meyerhof pathway. Arch. Microbiol. 161, 460–470.

Shin, H.S., Youn, J.H., Kim, S.H., 2004. Hydrogen production from food waste in anaerobic mesophilic and thermophilic acidogenesis. Int. J. Hydrogen Energy 29, 1355–1363.

Show, K.Y., Zhang, Z.P., Tay, J.H., Liang, T., Lee, D.J., Ren, N.Q., et al., 2010. Critical assessment of anaerobic processes for continuous biohydrogen production from organic wastewater. Int. J. Hydrogen Energy 35, 13350–13355.

Take, H., Andou, Y., Nakamura, Y., Kobayashi, F., Kurimoto, Y., Kuwahara, M., 2006. Production of methane gas from Japanese cedar chips pretreated by various delignification methods. Biochem. Eng. J. 28, 30–35.

Tanisho, S., Ishiwata, Y., 1995. Continuous hydrogen production from molasses by fermentation using urethane foam as a support of flocks. Int. J. Hydrogen Energy 20, 541–545.

Tredici, M.R., Zittelli, G.C., 1998. Efficiency of sunlight utilization: tubular versus flat photobioreactors. Biotechnol. Bioeng. 57, 187–197.

Ueno, Y., Kawai, T., Sato, S., Otsuka, S., Morimoto, M., 1995. Biological production of hydrogen from cellulose by natural anaerobic microflora. J. Ferment. Bioeng. 79, 395–397.

Ueno, Y., Otsuka, S., Morimoto, M., 1996. Hydrogen production from industrial wastewater by anaerobic microflora in chemostat culture. J. Ferment. Bioeng. 82, 194–197.

Ueno, Y., Haruta, S., Ishii, M., Igarashi, Y., 2001. Microbial community in anaerobic hydrogen-producing microflora enriched from sludge compost. Appl. Microbiol. Biotechnol. 57, 555–562.

U.S. Department of Energy, 2007. Hydrogen, Fuel Cells and Infrastructure Technologies Program, Multi-Year Research. Development and Demonstration Plan, U.S. Department of Energy.

Valdez-Vazquez, I., Rios-Leal, E., Esparza-Garcia, F., Cecchi, F., Poggi-Varaldo, H.A., 2005. Semi-continuous solid substrate anaerobic reactors for H-2 production from organic waste: mesophilic versus thermophilic regime. Int. J. Hydrogen Energy 30, 1383–1391.

Van Ginkel, S., Sung, S.W., Lay, J.J., 2001. Biohydrogen production as a function of pH and substrate concentration. Environ. Sci. Technol. 35, 4726–4730.

Van Ginkel, S.W., Oh, S.E., Logan, B.E., 2005. Biohydrogen gas production from food processing and domestic wastewaters. Int. J. Hydrogen Energy 30, 1535–1542.

van Niel, E.W.J., Budde, M.A.W., de Haas, G.G., van der Wal, F.J., Claasen, P.A.M., Stams, A.J.M., 2002. Distinctive properties of high hydrogen producing extreme thermophiles, *Caldicellulosiruptor saccharolyticus* and *Thermotoga elfii*. Int. J. Hydrogen Energy 27, 1391–1398.

Wang, C.C., Chang, C.W., Chu, C.P., Lee, D.J., Chang, B.V., Liao, C.S., 2003. Producing hydrogen from wastewater sludge by *Clostridium bifermentans*. J. Biotechnol. 102, 83–92.

Wu, J.H., Lin, C.Y., 2004. Biohydrogen production by mesophilic fermentation of food wastewater. Water Sci. Technol. 49, 223–228.

Yokoi, H., Tokushige, T., Hirose, J., Hayashi, S., Takasaki, Y., 1998. H2 production from starch by a mixed culture of *Clostridium butyricum* and *Enterobacter aerogenes*. Biotechnol. Lett. 20, 143–147.

Yokoi, H., Saitsu, A., Uchida, H., Hirose, J., Hayashi, S., Takasaki, Y., 2001. Microbial hydrogen production from sweet potato starch residue. J. Biosci. Bioeng. 91, 58–63.

Yokoi, H., Maki, R., Hirose, J., Hayashi, S., 2002. Microbial production of hydrogen from starch-manufacturing wastes. Biomass Bioenergy 22, 389–395.

Yu, H.Q., Mu, Y., 2006. Biological hydrogen production in a UASB reactor with granules. II: reactor performance in 3-year operation. Biotechnol. Bioeng. 94, 988–995.

Yu, H., Zhu, Z., Hu, W., Zhang, H., 2002. Hydrogen production from rice winery wastewater in an upflow anaerobic reactor by using mixed anaerobic cultures. Int. J. Hydrogen Energy 27, 1359–1365.

Yu, H.Q., Hu, Z.H., Hong, T.Q., 2003. Hydrogen production from rice winery wastewater by using a continuously-stirred reactor. J. Chem. Eng. Jpn. 36, 1147–1151.

Zhang, T., Liu, H., Fang, H.H.P., 2003. Biohydrogen production from starch in wastewater under thermophilic condition. J. Environ. Manage. 69, 149–156.

Zhang, Z.P., Tay, J.H., Show, K.Y., Yan, R., Liang, D.T., Lee, D.J., et al., 2007. Biohydrogen production in a granular activated carbon anaerobic fluidized bed reactor. Int. J. Hydrogen Energy 32, 185–191.

CHAPTER

21

Biohydrogen Production from Bio-oil

*Amit Kumar**, *Susanjib Sarkar*

Department of Mechanical Engineering, University of Alberta, Edmonton, Alberta, Canada T6G 2G8

***Corresponding author: E-mail: amit.kumar@ualberta.ca**

1 INTRODUCTION

Interest in utilization of renewable energy sources for production of energy and intermediate products has increased in recent time. There are two key drivers behind this. First one is to reduce dependence on import of fossil fuel and, second is to reduce the emission of greenhouse gases (GHGs). Fossil fuel utilization for production of energy contributes significantly to the emission of GHGs. A lot of effort is being put into reduction of the emission of GHGs. Broadly, there are three categories for mitigation of the GHGs. These include improvement of efficiency of energy utilization; sequestration of emitted GHGs; and, switching to renewable energy sources. Renewable energy sources offer attractive opportunities and address both the drivers behind their utilization. Among different renewable energy technologies, biomass-based energy technologies have high potential and are the only renewable energy source from which liquid fuels can be produced directly. Biomass could be used to produce heat, power, liquid fuels, and chemicals through different conversion processes. For example, heat and power can be produced through direct combustion or gasification of biomass sources (Bain et al., 1996; Cameron et al., 2006; Kumar et al., 2003). Similarly, liquid fuels such as bioethanol (Aden et al., 2002), biodiesel (Haas et al., 2006; Kinast, 2003), and bio-oil (Dynamotive Energy Systems, 2010; Ensyn, 2010; Pootakham and Kumar, 2010a) could be produced from different conversion processes using different types of biomass feedstocks. Chemicals such as ammonia and hydrogen (Larson et al., 2005; Sarkar and Kumar, 2009, 2010a,b; Spath et al., 2005) can also be produced from biomass.

There is a range of biomass feedstocks which can be used to produce various biomass-based end products. The focus here is on the lignocellulosic biomass utilization for fuels and chemicals. These lignocellulosic biomass feedstocks can be based on forest or agricultural resources.

Forest biomass could include whole-tree biomass, forest harvest residues, mill residues, and hybrid species' biomass (i.e., hybrid poplar, willow). Forest harvest residues are generated during logging operations. In some jurisdictions around the world in about 80% of the logging operations, trees are cut in the stand, skidded to the roadside, and delimbed at the roadside. These residues are then forwarded and piled and burnt to prevent forest fires (Kumar et al., 2003; Sarkar and Kumar, 2010a). Most of the mill residues today at the pulp and lumber mills are already in use for the production of energy (Kumar et al., 2003). Agricultural biomass could include straw, corn stover, and energy crops (such as switchgrass, miscanthus). In some jurisdictions, these are used for the production of energy such as in Europe (Caddet Renewable Energy, 1988a,b,c; Larsen, 1999), but in others, these are just left in the field to rot and could be used for the production of fuels and chemicals (Kumar et al., 2003; Sultana et al., 2010). With the transition and diversification of forest industry, whole-tree biomass also provides an attractive option.

There are two main characteristics of lignocellulosic biomass feedstocks which have significant impacts on the production of fuels/chemicals. First, the field/forest sourced biomass has low yield per unit area (dry tonnes of biomass/ha) as compared to the fossil fuels. This means that for a particular size of fuel production facility, the area required to supply field/forest sourced biomass is much larger as compared to supply area of the fossil fuels (Pootakham and Kumar, 2010a,b; Sarkar and Kumar, 2010a,b). As a result of this, the average transportation distance of biomass to the production facility is longer, and this contributes to higher transportation cost and hence, the high cost of production of fuels/chemicals from biomass. Several studies have reported that the transportation cost of biomass is between 25% and 50% of the delivered cost of biomass (Aden et al., 2002; Atchison and Hettenhaus, 2003; Glassner et al., 1998; Kumar et al., 2005; Perlack and Turhollow, 2002) for ethanol production from agricultural biomass. Similar numbers are reported for power generation facilities based on biomass. Second, the energy density of biomass (GJ/m^3) is lower as compared to fossil fuels. This also contributes to high cost of energy production from biomass (Pootakham and Kumar, 2010a; Sarkar and Kumar, 2010a,b). These two characteristics need to be improved to increase the competitiveness of biomass-based fuels and chemicals. Production of bio-oil could help in addressing these issues.

1.1 Bio-oil Production Process, Characteristics, and Status

Conversion of biomass to bio-oil helps in increasing the energy density of biomass and also helps in converting it to a liquid form. The energy density of bio-oil is about 6-7 times that of "as received" agricultural biomass from field (Badger and Fransham, 2006; Pootakham and Kumar, 2010a; Sarkar and Kumar, 2010a,b). Bio-oil is a thick dark liquid produced from the fast pyrolysis of biomass (Bridgwater, 2004; Dynamotive Energy Systems, 2010; Ensyn, 2010). Fast pyrolysis is the process of rapid heating of biomass in the absence of air in a temperature range of 400-500°C. This results in production of organic vapors which are further condensed to produce bio-oil. The process also produces char and water. The yield of bio-oil, char, and water depends on the types of biomass and the process of conversion (Dynamotive Energy Systems, 2010; Ensyn, 2010; Hassan et al., 2009; Mohan et al., 2006; Pootakham and Kumar, 2010a; Pootakham, 2009; Sarkar and Kumar, 2010a,b).

Bio-oil is a highly viscous liquid similar to fuel oil grade #2 and is dark brown in color. It has a pH of 2.8. The low pH has an impact on the type of material for the equipment required for its storage and handling (Darmstadt et al., 2004; Diebold, 2000). Low pH of bio-oil is due to the presence of acidic components (Yaman, 2004). Bio-oil has a density of 1,200 kg/m^3 and higher heating value of 20 MJ/m^3. The lower hearing value depends on the water content of the oil. Another key characteristic of bio-oil is that it behaves like crude oil in a temperature range of 30-40°C (Bridgwater, 2003; Pootakham and Kumar, 2010a; Thamburaj, 2000). This could help in transportation of bio-oil in the form of a liquid on a large scale through pipeline (Pootakham and Kumar, 2010a). The high energy and mass densities of bio-oil as compared to "as received" biomass from field/forest are the key reasons for increased interest in the bio-oil utilization for the production of fuels and chemicals.

Bio-oil can be used for the production of heat and power. It can be directly combusted to produce heat or electricity (Dynamotive Energy Systems, 2010). Few companies are working on using it for heat and power production (Badger and Fransham, 2006; Brammer and Bridgwater, 2002; Bridgwater, 1999; Yaman, 2004). Bio-oil can also be used to produce fuels and chemicals (Bridgwater, 1999; IEA, 2007; Shaw, 2006; Yaman, 2004). It can be gasified to produce intermediate product, that is, syngas which can be converted into fuels and chemicals (Henrich et al., 2009). One of these products could be hydrogen. This will be discussed further in this chapter. Several companies are producing bio-oil to use it for the production of fuel and chemicals. Dynamotive Energy Systems Inc. has a plant using 100 dry tonnes/day of biomass in Canada. Bio-oil produced from this plant would be used for electricity production (Dynamotive Energy Systems, 2010). Other prominent companies which produce bio-oils are ENSYN Systems and ROI (Bridgwater, 2003; IEA, 2007). ENSYN is focused on the production of specialty chemicals.

1.2 Biohydrogen Production from Biomass and Bio-oil

Biohydrogen can be produced through thermochemical conversion, electrohydrogenesis, and biological conversion of biomass. Details on electrohydrogenesis and biological conversion can be found in literature elsewhere (Cheng and Logan, 2007; Cortright et al., 2002; Deluga et al., 2004; Salge et al., 2006; Zhang et al., 2007). Thermochemical conversion of biomass for production of biohydrogen is the focus of this chapter. Thermochemical conversion process includes gasification of biomass and bio-oil. Gasification of biomass in limited supply of oxygen results in the formation of syngas which is a mixture of carbon monoxide, carbon dioxide, methane, and hydrogen. One of the direct pathways is by converting syngas to biohydrogen. Syngas can be converted to biohydrogen by steam reforming (Czernik et al., 2007; Larson et al., 2005; Ruyck et al., 2007; Spath et al., 2005). Another method of production of biohydrogen from biomass is through the production of bio-oil by fast pyrolysis and reforming of bio-oil to biohydrogen. Figure 1 shows the detailed unit operations for the production of biohydrogen from biomass and bio-oil. Bio-oil normally does not have the inorganics as these are mostly present in the char during its production; hence, the produced syngas is much cleaner than if produced by the gasification of biomass. Compared to "as received" biomass, bio-oil is in liquid form and has higher energy and mass density. If biomass is transported in the form of bio-oil, it has lower transportation cost and hence helps in reducing the overall processing cost (Pootakham and Kumar, 2010a,b).

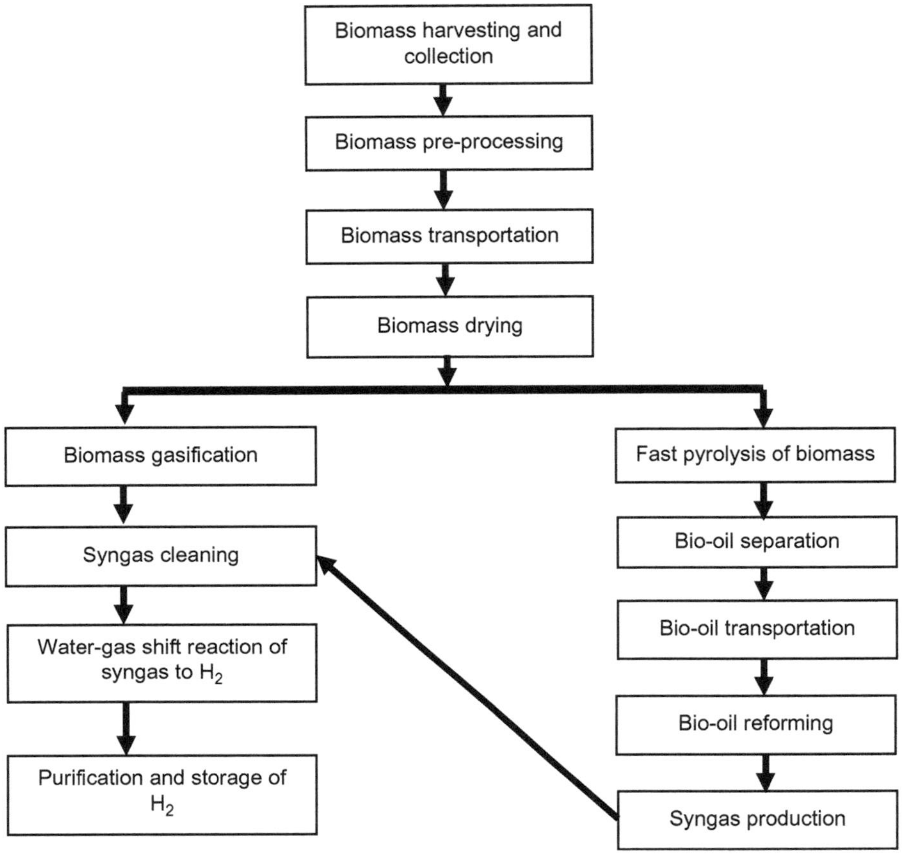

FIGURE 1 Biohydrogen production from biomass (Sarkar and Kumar, 2009, 2010a,b).

Some earlier works have been done on the production of syngas from bio-oil through gasification, but there is a scarcity of data on the production of biohydrogen from bio-oil (Bimbela et al., 2007; Davidian et al., 2008a,b; Domine et al., 2008; Rioche et al., 2005). Production of biohydrogen from biomass has been studied earlier, but these are mostly based on gasification of biomass. Syngas produced during gasification is reformed in the presence of steam. This is followed up by the water-gas shift reaction of syngas to increase the concentration of biohydrogen (DOE, 2003; McHugh, 2005; Sarkar and Kumar, 2009, 2010a,b).

The production of biohydrogen from bio-oil includes two key steps. First step is the production of bio-oil by fast pyrolysis of biomass and the second step is reforming of bio-oil to biohydrogen. If biohydrogen production plant is integrated with the bio-oil production plant, there is no need to transport bio-oil. If the biohydrogen production plant is located far away from the bio-oil production plant, bio-oil needs to be transported by truck or pipeline to the biohydrogen plant (Pootakham and Kumar, 2010a,b). The analysis presented in this chapter assumes that the two plants are located near to each other, and hence there is no cost of transportation of bio-oil. Although bio-oil could be produced from a range of biomass feedstocks including forest- and agriculture-based biomass, the focus here is on forest biomass.

Bio-oil Production

Forest biomass in the form of chips, delivered at the bio-oil production plant, is reduced to size of 2-3 mm and dried to a moisture content of less than 10% (Bridgwater, 2003, 2004; Mohan et al., 2006). This dried biomass feedstock is fast pyrolyzed. In fast pyrolysis, dried and ground biomass is heated to a temperature in the range of 450-500°C in the absence of oxygen. This rapid heating of biomass results in the formation of volatile vapors. These volatile vapors are cooled down to about 20°C to produce bio-oil (Mohan et al., 2006; Sarkar and Kumar, 2010a,b). The heat recovery from the high-temperature vapor can be used for raising steam. There are some commercial-scale bio-oil production plants around the world with minor variation in technology (e.g., Dynamotive Energy Systems, 2010; Ensyn, 2010).

Yield of bio-oil from biomass depends on the type of feedstocks and the process of conversion. This yield could vary from 42 to 75 wt%. The variation in the yield of bio-oil is due the variation in the content of fixed carbon content and ash. Average yield of bio-oil from wood chips from whole-tree biomass is about 60-70 wt%. Along with bio-oil, fast pyrolysis of biomass also produces char and incondensable fuel gas which account for about 30-40% (Bridgwater, 1999; Brow, 2009; Liu et al., 2010; Ringer et al., 2006; Sarkar and Kumar, 2010a; Tsai et al., 2007; Yaman, 2004).

Biohydrogen Production from Bio-oil

Bio-oil produced from the fast pyrolysis of biomass consists of organic acids, aldehydes, ketones, alcohol, and water (Jung et al., 2008; Oasmaa and Peacocke, 2001; Sarkar and Kumar, 2010b; Tsai et al., 2007). To stabilize water-insoluble fraction and reduce its viscosity, methanol is added to it. Some experimental work has been done on the mixing of various percentages of methanol to bio-oil to study its impacts. Ten percent blending is considered to be sufficient for its stability (Domine et al., 2008; Lu et al., 2008; Oasmaa et al., 2004; Yu et al., 2007).

2 REFORMING OF BIO-OIL

In the bio-oil reforming process, steam is provided in the presence of a noble metal catalyst with heat and results in the formation of syngas which is predominantly a mixture of CO and H_2. Several studies have been conducted on reforming of biomass using catalyst such as Rhodium (Rh), Platinum (Pt), and Nickel (Ni) with varying success (Basagiannis and Verykios, 2007; Bimbela et al., 2007; Davidian et al., 2008b; Domine et al., 2008; Iojoiu et al., 2007; Magrini-Bair et al., 2002; Rioche et al., 2005; Vagia and Lemonidou, 2008). One of the problems that reforming of bio-oil has is that the process generates carbon on catalyst surface, and this results in the need for regeneration of catalyst. Research is currently underway in developing these catalysts and development of more control on carbon deposition. Some earlier work have been done on reforming of bio-oil, but continuous production of bio-oil has not yet been demonstrated (Sarkar and Kumar, 2010b).

3 WATER-GAS SHIFT REACTION OF SYNGAS

Syngas produced by reforming of bio-oil further goes through water-gas shift reaction with steam. This helps in increasing the concentration of biohydrogen. During this reaction, CO reacts with steam to form CO_2 and biohydrogen. Typically, two types of reactor are used for water-gas shift reaction. These include low-temperature shift reactor and high-temperature shift reactor. Water-gas shift reaction is an exothermic reaction; the release of heat results in increased temperature of the reactor and results in reduced amount of production of biohydrogen. In order to control this, oxides of iron and chromium are used in high-temperature shift reactor, and oxides of copper and zinc are used in low-temperature shift reactor (Chen et al., 2008; Sarkar and Kumar, 2010b).

4 PURIFICATION OF BIOHYDROGEN AND STORAGE

After the water-gas shift reaction, a mixture of gases with high content of biohydrogen is obtained. Further purification of biohydrogen is required before it can be used as a fuel or chemical. The purification of biohydrogen is carried out through the pressure swing absorption system. A pressure swing absorption unit consists of a series of columns where successive pressurization and depressurization of the gas mixture take place. This results in the separation of biohydrogen from other gases with an efficiency of about 80% (Sarkar and Kumar, 2010a,b; Sircar and Golden, 2000). The pure biohydrogen can be further stored for use.

5 STEAM REQUIREMENT FOR BIOHYDROGEN PRODUCTION

Steam is required in the biohydrogen production process in two key steps. First, steam is required during bio-oil reforming. The steam required for this step is mainly for providing heat for the reforming process as it is an endothermic process. The amount of steam required dictates the extent of biohydrogen production. Steam is also required in the water-gas shift reaction of syngas to biohydrogen. Steam for these two steps could be generated by cooling of the organic vapors produced during the fast pyrolysis of biomass to produce bio-oil. This steam could be generated in a heat recovery steam generation unit (Sarkar and Kumar, 2010a). The steam required for these two steps could also be produced from an external source either by combusting biomass, produced bio-oil or any other fossil fuel.

5.1 Cost of Biohydrogen Production from Bio-oil

The process of production of biohydrogen from bio-oil is shown in Figure 1. The whole process involves a series of unit operations. The total cost of production of biohydrogen depends on the cost incurred at each of the unit operations and its characteristics. Given below is a description of each of the unit operations and also its characteristics. Although bio-oil could be produced from a range of biomass feedstocks including forest- and agriculture-based biomass, the focus here is on forest biomass. Overall scope of the cost estimation includes harvesting of the whole-tree biomass from the forest, chipping of the biomass on the roadside in the forest, transportation of the chips from the roadside to the bio-oil production plant by truck, fast pyrolysis of the wood chips for the production of bio-oil in the bio-oil production plant, reforming of bio-oil to produce syngas, and further conversion of this to biohydrogen through water-gas shift reaction and finally, purification and storage of produced biohydrogen. Several studies have discussed the size of bio-oil production plants

(Curtis et al., 2003; Mullaney et al., 2002; Phillips et al., 2007). Currently, bio-oil is produced in plants utilizing 100 dry tonnes or less of biomass per day (Sarkar and Kumar, 2010b; Svoboda et al., 2009). Here, the focus is on a bio-oil production plant using 500 dry tonnes of biomass/day, and all the cost estimates are developed at this size and are results of the technoeconomic modeling of the process. The cost estimates are in 2008 U.S. dollars.

6 HARVESTING AND PROCESSING OF FOREST BIOMASS

This step includes the felling of trees in the forest, skidding of the trees to the roadside, and chipping of the whole trees on the roadside, and transportation of chips to the bio-oil production plant by truck. The details on harvesting, processing, and transportation of biomass could be found in earlier studies (Kumar et al., 2003; Kumar, 2009; Sarkar and Kumar, 2009; Sultana et al., 2010). The total cost of delivered biomass for western Canada is $35 per dry tonne including harvesting cost of $10.7 per dry tonne, chipping cost of $3.84 dry tonne, and transportation cost of $8.8 per dry tonne (Sarkar and Kumar, 2010a) at a yield of 84 dry tonnes/ha from whole-forest. The transportation costs are based on chips transportation by B-train chip van. These costs can vary depending on the yield of the biomass (dry tonnes/ha), transportation costs, and scale of processing in different jurisdiction (Kumar et al., 2003).

7 CAPITAL COST OF BIO-OIL PRODUCTION PLANT

Table 1 gives the capital cost of bio-oil production plant at a size of 500 dry tonnes/day. The capital cost of production plants consists of fuel handing and drying system, fast pyrolysis unit, heat recovery and oil quenching unit, bio-oil recovery and storage unit, steam and power generation with cooling tower unit and others.

TABLE 1 Capital Cost of Bio-Oil Production Plant (Derived from Sarkar and Kumar, 2010b)

Components	Costs (In Thousand Dollars)	Comments/Remarks
Fuel handling and drying system	8,053	This unit has a scale factor of 0.70 and if the size of the plant is different, this scale factor can be used to estimate the cost at a particular size.
Fast pyrolysis unit	4,335	This cost is based on a fluidized-bed reactor with scale factor of 0.65-0.70 (Hamelinck and Faaij, 2002; Jones et al., 2009; Phillips et al., 2007; Spath et al., 2005).
Heat recovery and quenching	3,406	This cost includes capital cost of heat recovery and bio-oil quenching unit and has a scale factor of 0.65-0.70.
Bio-oil recovery and storage unit	885	It has a scale factor of 0.60.
Steam and power generation unit with cooling tower unit and others	8,483	This cost consists of steam and power generation units with cooling tower unit and other utilities. The scale factor is in the range of 0.70-0.78.

In addition to the costs shown in Table 1, after factoring in contingency, warehouse cost, site development cost, indirect cost and site development cost, the total cost is $58 million (details given in Sarkar and Kumar, 2010b). This cost could vary depending on the location of the plant around the world. This capital cost is very much representative of North America. A further penalty of 10% should be attributed if the location of the plant is remote as in the whole-forest biomass case.

8 OPERATING COST OF BIO-OIL PLANT

One of the key components of operating cost of a bio-oil plant is the cost of electricity for running the auxiliaries and grinder for grinding biomass. Typically, char produced during the pyrolysis process is combusted to generate steam and power. This power is sufficient to run the plant and often surplus and can be sold to the grid (Mullaney et al., 2002; Sarkar and Kumar, 2010b). The amount of electricity required for the compression of noncondensable gases is about 125 kWh/dry tonne and feedstock grinding process (to about 3 mm) requires about 65 kWh/dry tonne (Ringer et al., 2006). The estimated yearly operating cost of bio-oil plant excluding the feedstock cost and ash disposal cost is about 1.2% of the capital cost. The cost of bio-oil production is estimated at a forest biomass-based feedstock cost of $35 per dry tonne. Also, the cost estimates consider an ash disposal cost which consists of ash transportation cost and ash spreading cost. The cost of ash transportation is $0.18 per dry tonne/km and ash spreading cost of $25.22 per dry tonne/ha (Kumar et al., 2003; Sarkar and Kumar, 2009, 2010a; Zundel et al., 1996). Typically, the cost of storage of biomass up to 3 months is required as during early spring the forestry roads are impassable (especially in North America and Scandinavian countries). This cost is about $4.06 per dry tonne (Hamelinck et al., 2005; Sarkar and Kumar, 2010b). Further to these costs, administrative staff cost of $64 per h and labor cost of $40 per h is considered (Sarkar and Kumar, 2009). Maintenance cost per year is 2% of the capital cost (Kumar et al., 2003; Spath et al., 2005). Purchase price for methanol is $0.25/liter for this study (Haas et al., 2006).

9 PRODUCTION COST OF BIO-OIL

Once all the cost components and input parameters are determined for the bio-oil production plant, the total cost of production of bio-oil can be determined through the development of technoeconomic model and is shown in Table 2. The cost of production of bio-oil is about 20 cents/kg at a plant size of 500 dry tonnes/day. The largest component of the production cost is the capital cost at about 38%. Figure 2 shows the cost of production of bio-oil with the increased size of the bio-oil production plant. The cost of production of bio-oil decreases with the increase in capacity of the production plant. This trend continues up to a plant size of 1500 dry tonnes/day plant ($0.15 per kg). This is due to the benefit of economy of scale in the capital cost of the production plant. Although transportation cost increases with the size of the plant, it is less than the economy of scale benefits. The cost of production of bio-oil from a plant having capacity larger than 1500 dry tonnes/day does not increase significantly and is almost flat. In this region, the benefit due to economy of scale is nearly the same as the increase in the transportation cost due to increase in size. Cost of transportation of biomass

TABLE 2 Cost of Production of Bio-Oil (Derived from Sarkar and Kumar, 2010b)

Cost Components	Values (cents/kg)	Comments/Remarks
Capital cost	7.6	Determined based on data given in Table 1.
Operating cost	0.8	Determined based on operating cost parameters discussed in this section.
Maintenance cost	1.2	Determined based on operating cost parameters discussed in this section.
Administrative cost	4.4	Determined based on operating cost parameters discussed in this section.
Harvesting cost	1.8	Determined based on feedstock production cost parameters discussed in this section.
Transportation cost	1.4	Determined based on feedstock production cost parameters discussed in this section.
Feedstock storage cost	0.2	Determined based on operating cost parameters discussed in this section.
Road and infrastructure cost	1.4	This refers to the cost of construction of road and details are given in Kumar et al. (2003).
Silviculture cost	0.4	This refers to the cost of preparation of forest for replanting and details are given in Kumar et al. (2003).
Royalty cost	0.8	This is based on assumption of a royalty payment of $5 per dry tonne (Kumar et al., 2003; Sarkar and Kumar, 2009).
Ash disposal cost	0	Negligible.
Total production cost	20	Estimated based on developed technoeconomic model.

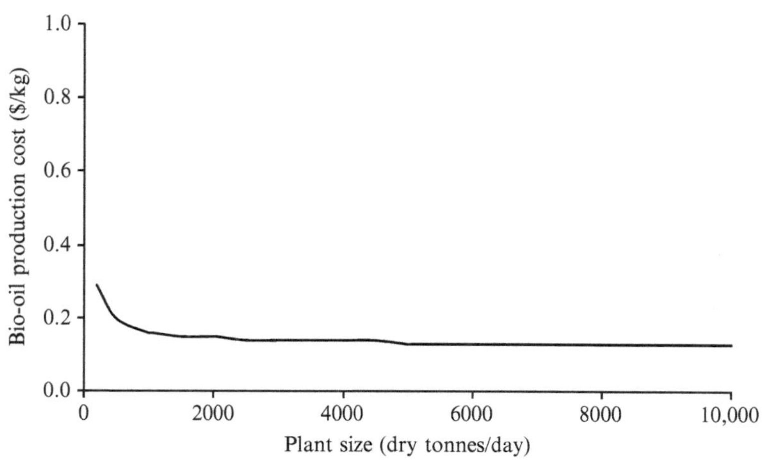

FIGURE 2 Cost versus capacity curve for bio-oil production plant.

increases as the area for collection of biomass increases with the increase in the size of the plant. Optimum sizes of other biomass conversion facilities are given elsewhere (Kumar et al., 2003; Sarkar and Kumar, 2009, 2010a).

10 COST OF BIO-OIL REFORMING

The next step in the cost analysis is determination of the cost of bio-oil reforming. Bio-oil produced from the forest biomass can be reformed to produce biohydrogen. Based on an assumption that the bio-oil yield from the forest biomass is about 60 wt% (Sarkar and Kumar, 2010b), a bio-oil production plant consuming 2,000 dry tonnes of biomass/day would produce about 1,200 tonnes of bio-oil/day. Using a scale factor of 0.70, the capital cost of bio-oil reforming plant is $155 million (Sarkar and Kumar, 2010b). The total capital cost is estimated by developing the size of different equipment using Aspen Plus modeling. These equipment are assumed to be similar to the equipment used in the bio-oil production. The capital costs of the equipment include storage units, product gas production and clean up unit, gas compression unit, steam generation unit, and pressure swing absorption unit. The details on the cost of different equipment are obtained from literature and in consultation with the experts (Corradetti and Desideri, 2007; Hamelinck and Faaij, 2002; Kreutz et al, 2005; Larson et al., 2005; Spath et al., 2005). Blending of methanol with bio-oil helps in stabilization of bio-oil and 10% (mass basis) blending is sufficient. Other operating parameters are also based on Aspen Plus modeling of the process. The key operating costs include power requirement for the gas compression and air separation units, natural gas consumption for the production of steam for reforming reaction and water-gas shift reaction, cost of reforming catalyst, water consumption, and cost of water treatment. Table 3 gives the key costs for bio-oil reforming. Once all these cost parameters and characteristics of the plants are developed, a technoeconomic model could be developed to estimate the cost of production of biohydrogen from reforming of bio-oil.

TABLE 3 Cost Parameters for Bio-Oil Reforming for a 1200 Dry Tonnes/Day Bio-Oil Plant

Items	Costs	Comments/Remarks
Capital cost	$155 million	Developed for a plant using 1,200 dry tonnes of bio-oil with an additional 10% methanol blend.
Cost of reforming catalyst	$13,540 per kg	Based on an estimated requirement of 1,890 kg of Rhodium catalyst for a 1,500 kg/day bio-oil reforming plant (Evans and Steward, 2007). Life of the catalyst is assumed to be 5 years.
Cost of water treatment	$0.78 per m^3	Based on water consumption of 0.02 m^3 per kg of H_2 (Phillips et al., 2007; Spath et al., 2005). The steam to carbon ratio is assumed to be 3.
Cost of natural gas	$20.20 million	Natural gas is used for the production of steam which is used for water-gas shift reaction and at a natural gas price of $5 per GJ with a total consumption of 4.04 million GJ.
Cost of electricity	$0.07 per kWh	This cost is used to calculate the power requirement of 39.16 MW of electricity.

12 BIO-OIL-BASED BIOHYDROGEN COST

All the cost parameters and characteristics are used to develop a technoeconomic model to calculate the cost of biohydrogen from bio-oil. Table 4 gives the biohydrogen production cost breakdown for a production plant using bio-oil produced by a plant of size 2000 dry tonnes of biomass/day. Bio-oil production cost and the operating cost for biohydrogen plant are the two largest contributors to the total cost of production of biohydrogen from bio-oil. These are 34% and 31% of the total biohydrogen production cost, respectively. Main components of the operating cost are the electricity cost for compressors, pumps, air separation unit, and electric motors. Reforming catalyst contributes to about 11% of the total operating cost. Cost of water treatment is only 2% of the total operating cost. Capital cost of biohydrogen production plant is about 12% of the total cost. Maintenance and administrative costs are minor components of the total biohydrogen production cost.

The variation in the cost of biohydrogen production with the size of the whole bio-oil reforming plant is given in Figure 3. The cost of biohydrogen production decreases rapidly for size of the bio-oil reforming plant smaller than 2000 dry tonnes/day. This capacity represents a plant using 2000 dry tonnes/day of forest biomass and producing bio-oil at this scale. This bio-oil is then passed through steam reforming process to produce biohydrogen. Below a plant size of 10,000 dry tonnes/day, there isn't any size of the plant after which the cost of biohydrogen production starts increasing with the increase in the size of the plant. It is critical to understand the nature of this curve. For plants with capacities below 2000 dry tonnes/day, the capital cost of the biohydrogen plant/unit output decreases with the increase in the capacity because of the economy of scale benefits. On the other hand, the cost of transportation of biomass increases with the increase in the size of the plant due to larger area required to supply biomass to the plant. This is reflected in the total bio-oil production cost. The trade-off between capital cost of the biohydrogen production plant and the biomass transportation cost, reflected through bio-oil production cost, results in the unique nature of the curve. For plants with capacities larger than 2000 dry tonnes/day, the curve is relatively flat. This indicates that the benefits due to economy of scale are not significantly higher as

TABLE 4 Cost of Biohydrogen Production from Bio-Oil (Derived from Sarkar and Kumar, 2010b)

Cost Components	Values ($/kg of H_2)	Comments/Remarks
Capital cost	0.28	Determined based on data given in Table 3.
Operating cost	0.74	Determined based on operating cost parameters discussed in this section.
Maintenance cost	0.03	Determined based on operating cost parameters discussed in this section.
Administrative cost	0.06	Determined based on operating cost parameters discussed in this section.
Bio-oil production cost	0.83	Determined based on bio-oil production cost estimated earlier.
Methanol purchase cost	0.20	Determined based on 10 wt% addition of methanol to bio-oil for stability.
Total production cost	2.14	Estimated through development of technoeconomic models.

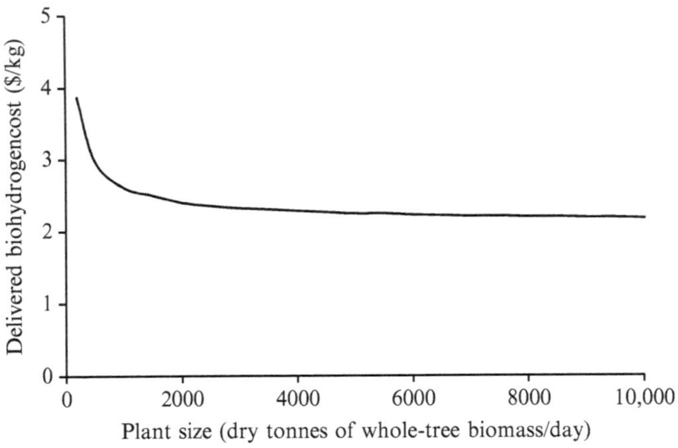

FIGURE 3 Cost versus capacity curve for biohydrogen production by reforming of bio-oil.

compared to the increased transportation cost of biomass reflected in the bio-oil production cost. Based on Figure 3, a plant using 2000 dry tonnes/day of biomass could be built with low risk and at this size most of the benefits of the economy of scale are realized. For jurisdictions where the availability of biomass is an issue, the size could be decided accordingly. If this cost of production of biohydrogen is compared with the cost of hydrogen produced from natural gas, the biohydrogen cost is higher. The cost of natural gas-based hydrogen is in the range of $0.96-$3.5 per kg (Balat, 2008; Chen and Elnashaie, 2006; Sarkar and Kumar, 2009, 2010b). The variability in the natural gas-based hydrogen cost is due to the size of the plant, the cost of natural gas, and other variables.

The biohydrogen cost from bio-oil depends on a number of factors. The key factors include the capital cost of bio-oil plant, the operating cost of bio-oil plant, capital cost of the biohydrogen production plant, operating cost of biohydrogen plant, natural gas cost, and electricity cost. Figure 4 shows the variation in the overall biohydrogen production cost with

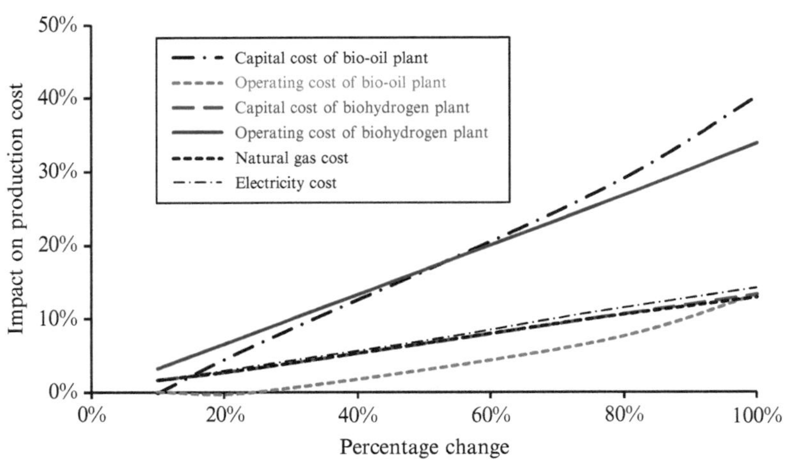

FIGURE 4 Sensitivity analysis of key input parameters on the biohydrogen production cost.

the change in the key input parameters. Capital cost of bio-oil production plant influences the overall cost of biohydrogen significantly. A 100% change in the capital cost of bio-oil production plant changes the biohydrogen production cost by 40%. Operating cost of biohydrogen plant also has a significant impact on the biohydrogen production cost. Operating cost of the bio-oil plant has least impact on the biohydrogen production cost.

12.2 Future of Biohydrogen Production from Bio-oil

Biohydrogen produced from biomass through bio-oil reforming is expensive as compared to the natural gas-based hydrogen. The main driver for using biomass to produce hydrogen is the emphasis on greenhouse gas mitigation and independence from import of fossil feedstocks such as natural gas for its production. The carbon footprint for the production of biohydrogen has been studied earlier (Sarkar and Kumar, 2010a,b). Table 5 gives the life-cycle emissions from a biohydrogen production plant based on bio-oil reforming.

Based on the total emissions of about 0.653 kg of CO_2 eq. over the life cycle of the biohydrogen production plant, the carbon credit value required for this to be competitive with the natural gas-based hydrogen is about $133 per tonne of CO_2 mitigated. This is based on a life-cycle emission of GHG in production of biohydrogen from natural gas of 11.88 kg of CO_2 per kg of H_2 produced (Spath and Mann, 2001) and a natural gas price of $5 per GJ (Sarkar and Kumar, 2010a,b). Also, a major contributor to this is the high production cost of biohydrogen from bio-oil. Technology improvement is required to bring down the cost of biohydrogen production to make it competitive with natural gas-based hydrogen.

TABLE 5 Greenhouse Gas Emissions in Biohydrogen Production from Bio-Oil Produced in a Plant Having Capacity of 2000 Dry tonnes/Day

Unit Operations	Values (g of CO_2 eq./kg of H_2)	Comments/Remarks
Production of biomass	327	Based on a bio-oil yield of about 60% and a hydrogen yield of 14.72% from bio-oil and methanol blend. The emission value includes all the unit operations of harvesting and processing of biomass (Sarkar and Kumar, 2010a,b).
Transportation of biomass	73	Transportation of biomass to a plant at an average distance of 18 km (Kumar et al., 2003).
Construction and decommissioning of bio-oil production plant	185	Based on Kumar et al. (2003).
Construction and decommissioning of bio-oil reforming plant	68	Based on estimates from Spath and Mann (2001).
Total emissions	653	

References

Aden, A., Ruth, M., Ibsen, K., Jechura, J., Neeves, K., Sheehan, J., et al., 2002. Lignocellulosic Biomass to Ethanol Process Design and Economics Utilizing Co-Current Dilute Acid Prehydrolysis and Enzymatic Hydrolysis for Corn Stover. NREL/TP-510-32438. National Renewable Energy Laboratory, Golden, CO. Available from: www.dtic.mil/cgi-bin/GetTRDoc?AD=ADA436469&Location=U2&doc=GetTRDoc.pdf.

Atchison, J.E., Hettenhaus, J.R., 2003. Innovative Methods for Corn Stover Collecting, Handling, Storing and Transporting. Prepared for National Renewable Energy Laboratory, Colorado. Report No. ACO-1-31042-01. Available from: http://www.afdc.doe.gov/pdfs/7241.pdf.

Badger, P.C., Fransham, P., 2006. Use of mobile fast pyrolysis plants to densify biomass and reduce biomass handling costs—a preliminary assessment. Biomass Bioenergy 30 (4), 321–325.

Bain, R.L., Overend, R.P., Craig, K.R., 1996. Biomass-fired power generation. In: Conference Paper: Biomass usage for utility and industrial power, Snowbird Resort and Conference Center. Snowbird, UT April 29-May 3, 1996, Engineering Foundation, NY.

Balat, M., 2008. Potential importance of hydrogen as a future solution to environmental and transportation problems. Int. J. Hydrogen Energy 33 (15), 4013–4029.

Basagiannis, A.C., Verykios, X.E., 2007. Steam reforming of the aqueous fraction of bio-oil over structured Ru/MgO/Al_2O_3 catalysts. Catal. Today 127 (1-4), 256–264.

Bimbela, F., Oliva, M., Ruiz, J., García, L., Arauzo, J., 2007. Hydrogen production by catalytic steam reforming of acetic acid, a model compound of biomass pyrolysis liquids. J. Anal. Appl. Pyrol. 79 (1-2), 112–120.

Brammer, J.G., Bridgwater, A.V., 2002. The influence of feedstock drying on the performance and economics of a biomass gasifier-engine CHP system. Biomass Bioenergy 22 (4), 271–281.

Bridgwater, A.V., 1999. Principles and practice of biomass fast pyrolysis processes for liquids. J. Anal. Appl. Pyrol. 51 (1-2), 3–22.

Bridgwater, A.V., 2003. Renewable fuels and chemicals by thermal processing of biomass. Chem. Eng. J. 91 (2), 87–102.

Bridgwater, A.V., 2004. Biomass fast pyrolysis. Therm. Sci. 8 (2), 21–49.

Caddet Renewable Energy, 1988a. The Masnedo CHP plant—using indigenous CO_2 neutral fuels. Technical Brochure No. 74, International Energy Agency (IEA).

Caddet Renewable Energy, 1988b. Straw-fired CHP plant in Rudkobing—providing environmentally. Technical Brochure No. 95, IEA.

Caddet Renewable Energy, 1988c. The Sabro straw-fired district heating plant. Technical Brochure No. 88, IEA.

Cameron, J.B., Kumar, A., Flynn, P.C., 2006. The impact of feedstock cost on technology selection and optimum size. Appl. Biochem. Biotechnol. 129 (1-3), 71–87.

Chen, Z.X., Elnashaie, S.S.E.H., 2006. Development of a novel autothermal reforming process and its economics for clean hydrogen production. Asia Pac. J. Chem. Eng. 1 (1-2), 5–12.

Chen, W.H., Hsieh, T.C., Jiang, T.L., 2008. An experimental study on carbon monoxide conversion and hydrogen generation from water gas shift reaction. Energy Convers. Manage. 49 (10), 2801–2808.

Cheng, S., Logan, B.E., 2007. Sustainable and efficient biohydrogen production via electrohydrogenesis. Proc. Natl. Acad. Sci. USA 104 (47), 18871–18873.

Corradetti, A., Desideri, U., 2007. Should biomass be used for power generation or hydrogen production? J. Eng. Gas Turb. Power 129 (3), 629–636.

Cortright, R.D., Davda, R.R., Dumesic, J.A., 2002. Hydrogen from catalytic reforming of biomass-derived hydrocarbons in liquid water. Nature 418 (6901), 964–967.

Curtis, W., Ferland, C., McKissick, J., Barnes, W., 2003. The Feasibility of Generating Electricity from Biomass Fuel Sources in Georgia. The University of Georgia, Athens, Georgia. Available from: www.caed.uga.edu/publications/2003/pdf/FR-03-06.pdf.

Czernik, S., Evans, R., French, R., 2007. Hydrogen from biomass-production by steam reforming of biomass pyrolysis oil. Catal. Today 129 (3-4), 265–268.

Darmstadt, H., Garcia-Perez, M., Adnot, A., Chaala, A., Kretschmer, D., Roy, C., 2004. Corrosion of metals by bio-oil obtained by vacuum pyrolysis of softwood bark residues. An X-ray photoelectron spectroscopy and auger electron spectroscopy study. Energy & Fuels 18 (5), 1291–1301.

Davidian, T., Guilhaume, N., Daniel, C., Mirodatos, C., 2008a. Continuous hydrogen production by sequential catalytic cracking of acetic acid Part I. Investigation of reaction conditions and application to two parallel reactors operated cyclically. Appl. Catal. Gen. 335 (1), 64–73.

Davidian, T., Guilhaume, N., Provendier, H., Mirodatos, C., 2008b. Continuous hydrogen production by sequential catalytic cracking of acetic acid: part II. Mechanistic features and characterisation of catalysts under redox cycling. Appl. Catal. Gen. 337 (2), 111–120.

Deluga, G., Salge, J., Schmidt, L., Verykios, X., 2004. Renewable hydrogen from ethanol by autothermal reforming. Science 303 (5660), 993–997.

Diebold, J.P., 2000. A Review of the Chemical and Physical Mechanisms of the Storage Stability of Fast Pyrolysis Bio-Oils. National Renewable Energy Laboratory. Golden, CO.

DOE, 2003. Office of Fossil Energy-Hydrogen Program Plan, Hydrogen from Natural Gas and Coal: The Road to a Sustainable Energy Future. Hydrogen Coordination Group, U.S. Department of Energy, Washington, DC. Available from: www.netl.doe.gov/technologies/hydrogen_clean_fuels/refshelf/pubs/fehydrogenplan2003.pdf.

Domine, M.E., Iojoiu, E.E., Davidian, T., Guilhaume, N., 2008. Hydrogen production from biomass-derived oil over monolithic Pt- and Rh-based catalysts using steam reforming and sequential cracking processes. Catal. Today 133-135 (April-June), 565–573.

Dynamotive Energy Systems, 2010. Technology. Available from: www.dynamotive.com/technology (accessed 09.04.10).

Ensyn, 2010. Technology. Available from: www.ensyn.com/tech.htm (accessed 09.04.10).

Evans, R.J., Steward, D.M., 2007. Distributed reforming of biomass pyrolysis oils. Available from: www1.eere.energy.gov/hydrogenandfuelcells/pdfs/06_nrel_distributed_reforming_biomass_pyrolysis_oils.pdf (accessed 05.01.08).

Glassner, D., Hettenhaus, J., Schechinger, T., 1998. Corn stover collection project. In: Bioenergy 98—Expanding Bioenergy Partnerships: Proceedings, vol. 2, Madison, WI. pp. 1100–1110. Available from: http://www.ceassist.com/bio98paper.pdf.

Haas, M.J., McAloon, A.J., Yee, W.C., Foglia, T.A., 2006. A process model to estimate biodiesel production costs. Bioresour. Technol. 97 (4), 671–678.

Hamelinck, C.N., Faaij, A.P.C., 2002. Future prospects for production of methanol and hydrogen from biomass. J. Power Source 111 (1), 1–22.

Hamelinck, C.N., Suurs, R.A.A., Faaij, A.P.C., 2005. International bioenergy transport costs and energy balance. Biomass Bioenergy 29 (2), 114–134.

Hassan, E.M., Yu, F., Ingram, L., Steele, P., 2009. The potential use of whole-tree biomass for bio-oil fuels. Energy Sourc A Recov. Util. Environ. Eff. 31 (20), 1829–1839.

Henrich, E., Dahmen, N., Dinjus, E., 2009. Cost estimate for biosynfuel production via biosyncrude gasification. Biofuels Bioprod. Bioref. 3, 28–41.

IEA, 2007. Biomass for Power Generation and CHP. International Energy Agency, Paris, France.

Iojoiu, E.E., Domine, M.E., Davidian, T., Guilhaume, N., Mirodatos, C., 2007. Hydrogen production by sequential cracking of biomass-derived pyrolysis oil over noble metal catalysts supported on ceria-zirconia. Appl. Catal. Gen. 323 (April) 147–161.

Jones, S., Holladay, J., Valkenburg, C., Stevens, D., Walton, C., Kinchin, C., et al., 2009. Production of Gasoline and Diesel from Biomass Via Fast Pyrolysis, Hydrotreating and Hydrocracking: A Design Case. Pacific Northwest National Laboratory, Richland, WA. Available from: www.pnl.gov/main/publications/external/technical_reports/PNNL-18284.pdf.

Jung, S.H., Kang, B.S., Kim, J.S., 2008. Production of bio-oil from rice straw and bamboo sawdust under various reaction conditions in a fast pyrolysis plant equipped with a fluidized bed and a char separation system. J. Anal. Appl. Pyrol. 82 (2), 240–247.

Kinast, J.A., 2003. Production of Biodiesels from Multiple Feedstocks and Properties of Biodiesels andBiodiesel/Diesel Blends Final Report; Report 1 in a Series of 6. National Renewable Energy Laboratory, Golden, CO.

Kreutz, T., Williams, R., Consonni, S., Chiesa, P., 2005. Co-production of hydrogen, electricity and CO_2 from coal with commercially ready technology. Part B: economic analysis. Int. J. Hydrogen Energy 30 (7), 769–784.

Kumar, A., 2009. A conceptual comparison of bioenergy options for using mountain pine beetle infested wood in Western Canada. Bioresour. Technol. 100 (1), 387–399.

Kumar, A., Cameron, J.B., Flynn, P.C., 2003. Biomass power cost and optimum plant size in western Canada. Biomass Bioenergy 24 (6), 445–464.

Kumar, A., Cameron, J.B., Flynn, P.C., 2005. Pipeline transport and simultaneous saccharification of corn stover. Bioresour. Technol. 96 (7), 819–829.

Larsen, J.B., 1999. The world's largest straw-fired power plant. Available from: www.nemesis.at/publication/gpi 99 1/articles/27.html.

Larson, E.D., Jin, H., Celik, F.E., 2005. Gasification-Based Fuels and Electricity Production from Biomass, Without and with Carbon Capture and Storage. Princeton Environmental Institute, Princeton University, Princeton, NJ. Available from: www.princeton.edu/pei/energy/publications/texts/LarsonJinCelik-Biofuels-October-2005.pdf.

Liu, R., Deng, C., Wang, J., 2010. Fast pyrolysis of corn straw for bio-oil production in a bench-scale fluidized bed reactor. Energy Sources A Recov. Util. Environ. Eff. 32 (1), 10–19.

Lu, Q., Yang, X., Zhu, X., 2008. Analysis on chemical and physical properties of bio-oil pyrolyzed from rice husk. J. Anal. Appl. Pyrol. 82 (2), 191–198.

Magrini-Bair, K., Czernik, S., French, R., Parent, Y., Ritland, M., Chorne, E., 2002. Fluidizable Catalysts for Producing Hydrogen by Steam Reforming Biomass Pyrolysis Liquids. Proceedings of the 2002 U.S. DOE Hydrogen Program Review. National Renewable Energy Laboratory, Golden, CO.

McHugh, K., 2005. Hydrogen Production Methods. MPR Associates, Inc., Alexandria, VA.

Mohan, D., Pittman, C.U.J., Steele, P.H., 2006. Pyrolysis of wood/biomass for bio-oil: a critical review. Energy & Fuels 20 (3), 848–889.

Mullaney, H., Farag, I.H., LaClaire, C.E., Barrett, C.J., Hall, K., 2002. Technical, Environmental and Economic Feasibility of Bio-Oil in New Hampshire's North Country. New Hampshire Industrial Research Center (NHIRC), Durham, NH. Available from: www.unh.edu/p2/biooil/bounhif.pdf.

Oasmaa, A., Peacocke, C., 2001. A Guide to Physical Property Characterization of Biomass-Derived Fast Pyrolysis Liquid. VTT Technical Research Centre of Finland, Finland. Available from: www.vtt.fi/inf/pdf/publications/2001/P450.pdf.

Oasmaa, A., Kuoppala, E., Selin, J.F., Gust, S., Solantausta, Y., 2004. Fast pyrolysis of forestry residue and pine. 4. Improvement of the product quality by solvent addition. Energy & Fuels 18 (5), 1578–1583.

Perlack, R.D., Turhollow, A.F., 2002. Assessment of options for the collection, handling, and transport of corn stover. Report no. ORNL/TM-2002/44. Available from: http://bioenergy.ornl.gov/pdfs/ornltm-200244.pdf.

Phillips, S., Aden, A., Jechura, J., Dayton, D., Eggman, T., 2007. Thermochemical Ethanol via Indirect Gasification and Mixed Alcohol Synthesis of Lignocellulosic Biomass. NREL/TP-41168. National Renewable Energy Laboratory, Golden, CO. Available from: www.nrel.gov/docs/fy07osti/41168.pdf.

Pootakham, T., 2009. Bio-oil transportation by pipeline. M.Sc. thesis, Department of Mechanical Engineering, University of Alberta, Edmonton, Alberta, p. 152.

Pootakham, T., Kumar, A., 2010a. A comparison of pipeline versus truck transport of bio-oil. Bioresour. Technol. 101 (1), 414–421.

Pootakham, T., Kumar, A., 2010b. Bio-oil transport by pipeline: a techno-economic assessment. Bioresour. Technol. 101 (18), 7137–7143.

Ringer, M., Putsche, V., Scahill, J., 2006. Large-Scale Pyrolysis Oil Production: A Technology Assessment and Economic Analysis. NREL/TP-37779. Golden, CO, National Renewable Energy Laboratory. Available from: www.nrel.gov/docs/fy07osti/37779.pdf.

Rioche, C., Kulkarni, S., Meunier, F.C., Breen, J.P., Burch, R., 2005. Steam reforming of model compounds and fast pyrolysis bio-oil on supported noble metal catalysts. Appl. Catal. B Environ. 61 (1-2), 130–139.

Ruyck, J.D., Delattin, F., Bram, S., 2007. Co-utilization of biomass and natural gas in combined cycles through primary steam reforming of the natural gas. Energy 32 (4), 371–377.

Salge, J.R., Dreyer, B.J., Dauenhauer, P.J., Schmidt, L.D., 2006. Renewable hydrogen from nonvolatile fuels by reactive flash volatilization. Science 314 (5800), 801–804.

Sarkar, S., Kumar, A., 2009. Techno-economic assessment of biohydrogen production from forest biomass in Western Canada. Trans. ASABE 52 (2), 519–530.

Sarkar, S., Kumar, A., 2010a. Biohydrogen production from forest and agricultural residues for upgrading of bitumen from oil sands. Energy 35 (2), 582–591.

Sarkar, S., Kumar, A., 2010b. Large-scale biohydrogen production from bio-oil. Bioresour. Technol. 101 (19), 7350–7361.

Shaw, M., 2006. Pyrolyis Lignocellulosic Biomass to Maximize Bio-Oil Yield: An Overview. CSBE/SCGAB Paper No. 06-105. CSBE/SCGAB, Edmonton, Alberta.

Sircar, S., Golden, T.C., 2000. Purification of hydrogen by pressure swing adsorption. Separ. Sci. Technol. 35 (5), 667–687.

Spath, P.L., Aden, A., Eggeman, T., Ringer, M., Wallace, B., Jechura, J., 2005. Biomass to Hydrogen Production Detailed Design and Economics Utilizing the Battelle Columbus Laboratory Indirectly-Heated Gasifier. NREL/TP-37408. National Renewable Energy Laboratory, Golden, CO. Available from: www.nrel.gov/docs/fy05osti/37408.pdf.

Spath, P.L., Mann, M.K., 2001. Life Cycle Assessment of Hydrogen Production via Natural Gas Steam Reforming. NREL/TP-570-27637. National Renewable Energy Laboratory, Golden, CO. Available from: www.nrel.gov/docs/fy01osti/27637.pdf.

Sultana, A., Kumar, A., Harfield, D., 2010. Development of agri-pellet production cost and optimum size. Bioresour. Technol. doi:10.1016/j.biortech.2010.02.011.

Svoboda, K., Pohořelý, M., Hartman, M., Martinec, J., 2009. Pretreatment and feeding of biomass for pressurized entrained flow gasification. Fuel Proc. Technol. 90 (5), 629–635.

Thamburaj, R., 2000. Dynamotive engineering. Fast pyrolysis of biomass for green power generation. Available from: www.dynamotive.com (accessed 20.06.06.).

Tsai, W.T., Lee, M.K., Chang, Y.M., 2007. Fast pyrolysis of rice husk: product yields and compositions. Bioresour. Technol. 98 (1), 22–28.

Vagia, E.C., Lemonidou, A.A., 2008. Hydrogen production via steam reforming of bio-oil components over calcium aluminate supported nickel and noble metal catalysts. Appl. Catal. Gen. 351 (1), 111–121.

Yaman, S., 2004. Pyrolysis of biomass to produce fuels and chemical feedstocks. Energy Convers. Manage. 45 (5), 651–671.

Yu, F., Deng, S., Chen, P., Liu, Y., Wan, Y., Olson, A., et al., 2007. Physical and chemical properties of bio-oils from microwave pyrolysis of corn stover. Appl. Biochem. Biotechnol. 137-140 (1-12), 957–970.

Zhang, Y.H.P., Evans, B.R., Mielenz, J.R., Hopkins, R.C., Adams, M.W.W., 2007. High-yield hydrogen production from starch and water by a synthetic enzymatic pathway. PLoS ONE 2 (5), e456.

Zundel, P.E., Hovingh, A.J., Wuest, L., MacElveney, D., Needham, T.D., 1996. Silviculture Systems for the Production of Energy Biomass in Conventional Operations in Atlantic Canada. Fredericton, New Brunswick. Available from: www.unbf.ca/forestry/centers/biomass.htm (accessed 21.04.07).

CHAPTER

22

Biohydrogen Production from Industrial Effluents

S. Venkata Mohan, G. Mohanakrishna, S. Srikanth*

Bioengineering and Environmental Centre (BEEC), Indian Institute of Chemical Technology (IICT), Hyderabad-500607, India

***Corresponding author: E-mail: vmohan_s@yahoo.com; svmohan@iict.res.in**

1 INTRODUCTION

The advent of the new century has witnessed an unchecked, overexploited use, and depletion of fossil fuels which has resulted in alarming environmental pollution and a steep rise in global warming causing an unusual increase in surface temperatures. Thus, the recognition and rapid development of alternative, renewable, carbon-neutral, and eco-friendly fuels is of paramount importance to fulfill the burgeoning energy demands. This has instigated more than ever rapid development of bioenergy to solve the looming energy crisis as well as to save the planet from the brink of an environmental catastrophe. Hydrogen (H_2) has been considered as a sustainable energy carrier as it is clean (does not emit any toxic by-product or greenhouse gases), efficient (with a high-energy yield of 142 kJ/g; 2.75-fold greater than that of methane), and renewable. Currently, H_2 is being produced mainly from fossil fuels, biomass, and water. H_2 production through biological routes is considered as one of the opportunistic and sustainable ways to meet the future energy demand and to prevent fossil fuel-based environmental impacts. Biological approaches for producing H_2 also facilitate the conversion of negative-value organic waste (Venkata Mohan, 2010, 2009).

2 BIOLOGICAL ROUTES OF H_2 PRODUCTION

Broadly, biological H_2 can be produced through two main mechanisms: photosynthesis and dark fermentation. Photosynthesis is a light-dependent process, while dark fermentation (anaerobic) is a light-independent catabolic process (Venkata Mohan, 2010, 2009). Most of the biological H_2 production processes are operated at ambient temperatures and pressures, regarded as less energy intensive and therefore considered as a potential alternative to the conventional physical/chemical methods usually opted for H_2 production.

2.1 Photobiological Process

Photobiological H_2 production can be classified into direct biophotolysis, indirect biophotolysis, and photofermentation. Direct biophotolysis converts water into H_2 and O_2 in presence of sun light during photosynthesis (Figure 1). However, H_2 production by photolysis of water is very slow due to the inhibition of hydrogenase activity by the oxygen released during photosynthesis (Miyake et al., 1999). In indirect photolysis, H_2 generates through the biochemical reduction of organic compounds during Calvin-cycle. Photofermentation is a process where the organic acids/volatile fatty acids (VFA) such as acetate, butyrate, propionate, etc., are consumed as e^- donors for H_2 production using light energy (Figure 1). Although hydrogenase activity is important nitrogenase activity is also crucial in this aspect. Light energy is not required for water oxidation in this case, and hence the conversion efficiency is higher compared to the photolysis (Miyake et al., 1999).

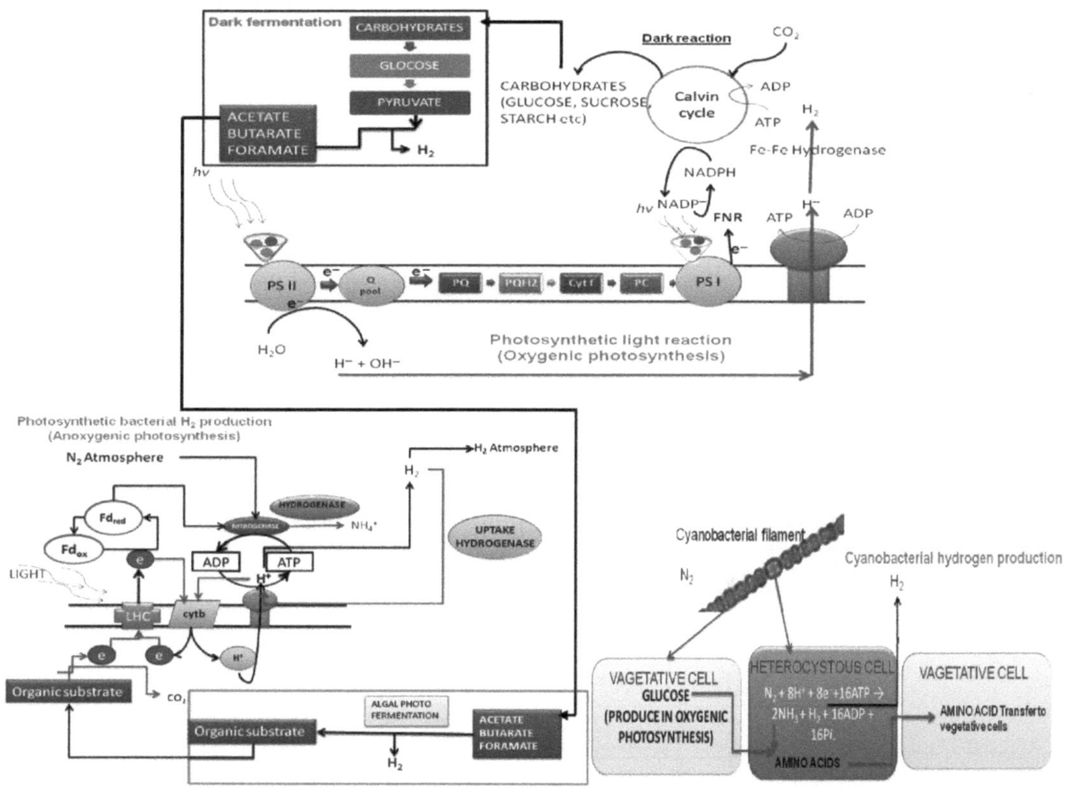

FIGURE 1 Schematic illustration of oxygenic and anoxygenic photosynthetic processes producing H_2 in algae, cyanobacteria, and photosynthetic bacteria.

Oxygenic photosynthesis occurs in algae and cyanobacteria, while anoxygenic photosynthesis is feasible in photosynthetic bacteria like purple sulfur (PSB) and purple nonsulfur bacteria (PNSB; Figure 1). During oxygenic process, photosystem II drives the first stage of the process by splitting water molecule into H^+, e^-, and oxygen. Electrons get transferred to the photosystem I through a cascade of proteins and are finally used in the Calvin cycle to fix CO_2 into biosynthetic intermediates and storage compounds (Donald, 1995). Protons gets driven through ATPase and hydrogenase and reduced with the e^- from reduced ferridoxin (Fd) to produce H_2 (Donald, 1995). In dark phase of process, the stored compounds are oxidized to form organic acids during respiration for energy generation. Hydrogenase catalyzes H_2 production in both algae and cyanobacteria, while nitrogenase catalyzes the H_2 production only in cyanobacteria (Donald, 1995). The function of hydrogenases in photofermentation is similar like acidogenic (dark) fermentation. However, both the hydrogenase and nitrogenase are sensitive to O_2, and the inhibitory effect prevents the H_2 production (Donald, 1995). In the case of cyanobacteria, oxygenic photosynthesis takes place in vegetative cells, while the nitrogen fixation along with H_2 production takes place in heterocysts. Heterocysts have leghemoglobin which scavenge the O_2 released during oxygenic photosynthesis and maintain low O_2 concentration inside the cell, which increases the hydrogenase and nitrogenase catalytic activity for higher H_2 production (Donald, 1995). PNSB has the advantage of being able to utilize various carbon sources, especially the short-chain fatty acids during anoxygenic photosynthesis (Shi and Yu, 2006) and are the most intensively studied anoxygenic phototrophs that produce H_2. Anoxygenic photosynthetic bacteria obtain e^- from organic substrate but not from water and, therefore, the inhibitory effects of O_2 on H_2-producing enzymes can be avoided. Since the organic substrates were originally derived from CO_2 fixed by green plants, anoxygenic photo-H_2 production is carbon neutral. PNSB mainly depends on the nitrogenase, which is known for N_2 fixation to NH_3. In the absence of N_2, nitrogenase acts as an ATP-powered hydrogenase, producing H_2 exclusively, without feedback inhibition (Donald, 1995). The ATP required for this can be generated with a single e^- repeatedly energized through cyclic photophosphorylation to maintain H^+ gradient and thereby ATP levels. H_2 production via nitrogenase has a specific activity but lower than [Ni–Fe] hydrogenase. Mo-nitrogenase has higher specific activity than the Fe-nitrogenase. Nitrogenase irreversibly catalyzes the reduction of molecular nitrogen to ammonium (nitrogen fixation) by consuming reducing power (e^- mediated by ferredoxin, NAD^+, etc.) and ATP. H_2 production catalyzed by nitrogenase is a side reaction at a rate of one third to one fourth that of nitrogen fixation. Nitrogen-fixing cyanobacteria are potential candidates for H_2 production by nitrogenase, but it is an energy-consuming process due to breakdown of many ATP molecules (Donald, 1995).

2.2 Dark Fermentation Process

Anaerobic conversion requires a series of four interrelated steps and five physiologically distinct groups of microorganisms to convert hydrocarbons from complex to simple molecules through H_2 and acid as intermediates finally to carbon dioxide (CO_2) and methane (CH_4; Figure 2). Obligatory H_2-producing acidogenic bacteria (AB) oxidize fermentation products to acid intermediates and H_2, which also include acetate production from H_2 and CO_2 by acetogens and homoacetogens (Angenent et al., 2004; Venkata Mohan, 2009, 2010). Fermentative process starts with the conversion of glucose to pyruvate through glycolysis by both

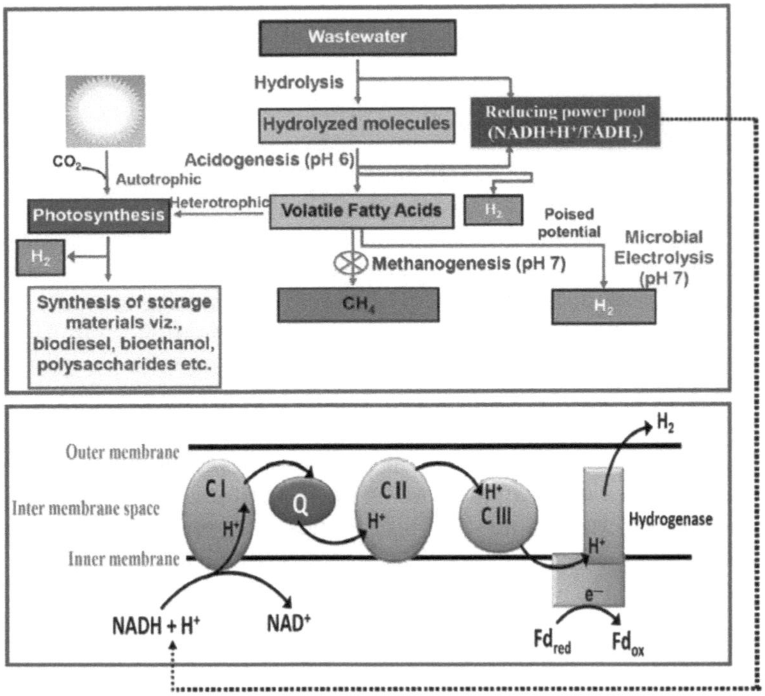

FIGURE 2 Schematic illustration of substrate conversion and H_2 production mechanism during dark fermentation.

obligate and facultative anaerobic bacteria. Facultative anaerobes convert pyruvate to acetyl-CoA and formate catalyzed by pyruvate formate lyase (PFL) where H_2 is produced from formate by the formate hydrogen lyase (FHL) complex. In obligate anaerobes, pyruvate is converted to acetyl-CoA and CO_2 through pyruvate ferredoxin oxidoreductase (PFOR), and this oxidation requires reduced ferredoxin (Fd) which again depends on the redox condition (Venkata Mohan, 2010).

Proton shuttling takes place between metabolic intermediates with the help of various redox mediators. The H^+ from the redox mediator is detached by a specific dehydrogenase (NADH-dehydrogenase) and gets reduced to generate H_2 in presence of the hydrogenase with the help of e^- donated by reduced ferredoxin (co-factor) (Venkata Mohan, 2010). Dehydrogenase enzymes are involved in the interconversion of metabolites and the transfer of H^+ between metabolic intermediates through redox reactions using several mediators (NAD^+, FAD^+, etc.; Venkata Mohan et al., 2010a; Venkata Mohan, 2009, 2010). Both dehydrogenase and hydrogenase activities are crucial to maintain H^+ equilibrium in the cell and to reduce them to H_2.

The basic functional differences between photo and dark fermentation processes are outlined in Table 1. During the initial phase of biohydrogen research, much attention was paid to the photobiological routes using specific strains and defined medium. Low rates of H_2 production was observed because of the complex reaction system, inhibitory effect of O_2 on hydrogenase and nitrogenase (Winkler et al., 2002), and lower utilization of waste (Srikanth et al., 2008, 2009; Venkata Mohan et al., 2008c) are some of the inherent

TABLE 1 Functional Differences Between Dark and Photo Fermentation Processes as a Function of H_2 Production

Dark Fermentation	Photofermentation
Light is not required	Light is required
H_2 production occurs during substrate degradation	H_2 production occurs during substrate synthesis
Utilizes wastewater as substrate	High strength and toxic wastewater cannot serve as substrate
Acidophilic pH favors H_2 production	Near-neutral pH is favorable
CO_2 releases along with H_2 generation	CO_2 gets fixed to carbon source along with H_2 generation
Hydrogenases class of enzymes involve in H_2 production	Both hydrogenases and nitrogenases involve in H_2 production
Absence of O_2 favors hydrogenases activity	O_2 generated during photolysis shows inhibitory effect on both hydrogenase and nitrogenase activity
Volatile fatty acids (VFA) are generated during acidogenesis along with H_2	VFA can be a good substrate for H_2 production
Reduced NAD^+ acts as proton source	Reduced $NADP^+$ acts as proton source
H_2 partial pressure is reduced through the reversible hydrogenases	H_2 partial pressure is reduced through the uptake hydrogenase

disadvantages linked with the photobiological process. High activation energy to drive hydrogenase and the low solar conversion efficiencies are also considered as major limitations for this process (He et al., 2005). Dark fermentation process gained importance due to its feasibility of utilizing wastewater as substrate and using mixed cultures as biocatalyst. Dark fermentative process does not rely on the light, utilizes a variety of carbon sources such as organic compounds, wastes, wastewaters, or insoluble cellulosic materials, requires less energy, technically much simpler, requires low operating costs, and is more stable (Angenent et al., 2004; Venkata Mohan, 2009, 2010). Simplicity, efficiency, and lesser foot prints are some of the strong features of the dark fermentation process which makes it practically more feasible for the mass production of H_2. Much of the literature reported on biohydrogen production concerned to wastewater utility was on dark fermentation process. On the contrary, photosynthetic bacteria can readily utilize the organic acids generated form dark fermentation process to produce additional H_2.

3 BIOCATALYST

Selection of appropriate biocatalyst or inoculum significantly influences the fermentation end-product formation. Initial reports on biohydrogen production are confined more towards the usage of pure cultures as biocatalyst with defined substrate. Diverse groups of microorganisms, viz. anaerobic (facultative and obligate), photoheterotrophic and microalgae are capable of producing H_2 through the degradation of organic substances. Bacteria capable of producing H_2 are reported to distribute across 10 of the 35 bacterial groups (Nandi and Sengupta, 1998) and widely exist in natural environments such as soil, cow dung, anaerobic sludge, municipal sludge, compost, etc. which can be used as inoculum. Obligate anaerobes, thermophiles, rumen bacteria, methanogens, and few types of facultative anaerobes were reported to produce H_2. Production rate of H_2 was comparatively lower in the case of facultative anaerobes (Chong et al., 2009). When aerobic condition prevails, facultative anaerobes consume O_2 and then switch to anaerobic fermentation which favors H_2 production (Chong et al., 2009). Obligate anaerobes, categorized as mesophiles and thermophiles, are able to produce H_2 in the pH range of 4-7 (Chong et al., 2009). Species of *Clostridium sp, Thermoanaerobacterium sp, Caldicellulosiruptor sp, Actinomyces sp., Porphyromonos sp,* etc. are obligate anaerobes, whereas the species like *Escherichia coli, Enterobacter, Citrobacter, Klebsiella,* etc. come under facultative group were reported to produce H_2 (Chong et al., 2009). H_2 production from wastewater has more significance where mixed cultures are used as biocatalyst with diverse biochemical functions. Recently, much focus is observed on the use of anaerobic microflora enriched from various sources. From the view of engineering aspects, mixed cultures are usually preferred because of low cost, operational flexibility, diverse biochemical functions, stability, and possibility of using of wider range of substrates (Wang and Wan, 2009). Moreover, mixed culture operation restricts the sterile requirement (Angenent et al., 2004; Venkata Mohan, 2010).

3.1 Pretreatment of Biocatalyst

Shifting or regulating the metabolic pathway towards acidogenesis and inhibiting methanogenesis facilitates higher H_2 yields (Srikanth et al., 2010b; Venkata Mohan, 2009, 2010). Pretreatment of biocatalyst plays a vital role in the selective enrichment of mixed consortia for the metabolic shift towards acidogenesis (Venkata Mohan et al., 2008e, 2009b; Venkata Mohan, 2009, 2010; Zhu and Beland, 2006). Typical anaerobic mixed cultures are unable to produce higher H_2 as it gets consumed by the H_2-consuming methanogenic bacteria (MB). Effective way to enhance H_2 production from anaerobic culture is to restrict or to terminate methanogenesis by allowing H_2 to become a metabolic end product. Unique function and physiological difference between AB and MB forms the main criterion for the preparation of the biocatalyst (Venkata Mohan et al., 2008e; Zhu and Beland, 2006). H_2-producing bacteria can form spores which protect them, in adverse environmental conditions (high temperature, extreme acidity and alkalinity), but methanogens lack such capability (Venkata Mohan et al., 2008e; Zhu and Beland, 2006). Various methods for the biocatalyst pretreatment are reported (Venkata Mohan et al., 2007b; Venkata Mohan, 2008, 2009, 2010; Wang and Wan, 2008a). Some of the reported pretreatment methods used and their functional property on microorganism are listed in Table 2. Different pretreatment methods have different functional property and comparison of such pretreatment methods helps to obtain an efficient pretreatment method (Wang and Wan, 2008a). Combining different pretreatment methods also showed a positive effect on H_2 production process (Srikanth et al., 2010b; Venkata Mohan et al., 2007d, 2008f; Venkata Mohan et al., 2008e, 2009b). Untreated consortia have higher bacterial population with a wide variety of biochemical functions facilitating diverse metabolic activities. On the contrary, pretreatment facilitates the selective enrichment of bacterial population leading to less diversity in their biochemical functions towards acidogenesis

TABLE 2 Pretreatment Methods Used to Enrich Biocatalyst for H_2 Production

Pretreatment Method	Conditions	Function
Acid	Extreme acidic microenvironment (pH < 4)	Enrich spore-forming bacteria by specifically suppressing the MB.
Alkaline	Alkaline microenvironment (pH > 9)	Nonspecific inhibition of MB
Heat shock	Extreme temperature (>80 °C)	Suppress non-spore-forming bacteria and helps to harvest spore-forming bacteria.
Load shock	In the presence of higher substrate concentration	Leads to the accumulation of high organic acids which prevents MB growth
Oxygen shock	In the presence of oxygen/air (0.5 mg/l)	MB are obligative anaerobes, exposure to oxygen lowers their adenylate charge and leads to death
Chemical	*2-bromoethanesulfonic acid* (*BESA*,100 mmol)	Inhibits coenzyme-M reductase complex (chief component for methanogenesis)
	Iodopropane	Iodopropane is a corrinoid antagonist that prevents the functioning of the B12 enzymes as a methyl group carrier
	Acetylene	Nonspecific inhibition of MB
	KNO_3 (10 mmol/l)	Nonspecific inhibition of MB
Combined	Combination of two or more pretreatment methods	To achieve more specific enrichment of AB

(Srikanth et al., 2010b). Untreated consortia support H^+ reduction during methanogenesis rather than the interconversion of metabolites, which is presumed to be necessary for H_2 production (Srikanth et al., 2010b). In spite of the improved H_2 production, marked reduction in the substrate degradation was observed after using the pretreated cultures (Srikanth et al., 2010b), which can be attributed to the inhibition of MB.

4 RENEWABLE WASTEWATER

Wastewater is being produced continuously and has been increased in volume with time due to industrialization. Remediation of wastewater being energy intensive adds up the expensiveness and increases the economic burden on the industry. Reducing the treatment cost and finding ways to produce useful products from wastes such as H_2 have been gaining importance. The biodegradable organic fraction present in wastewater associated with inherent net positive energy makes it as an ideal candidate for H_2 generation. Generation of bioenergy from renewable wastewater with simultaneous treatment reduces the overall cost and makes the whole process environmentally sustainable (Venkata Mohan, 2009, 2010). Technical feasibility, simplicity, economics, societal needs, and political priorities are some of the vital aspects considered to choose the bioprocess that will be used to treat wastes in the future (Angenent et al., 2004). Availability of large quantities of wastewater, presence of degradable carbon material, cost and the need for their treatment make them a potential substrate for producing H_2 (Venkata Mohan, 2009, 2010). Table 3 depicts some of the reported studies where wastewater was used as substrate to generate H_2 by biological routes.

TABLE 3 Various Types of Wastewater Used as Substrates for H_2 Production

Type of Fermentation	Industry Category	Type of Industrial Wastewater	References
Dark	Food processing industry	Food processing wastewater	Van Ginkel et al. (2005), Şentürk et al. (2010), Zhu et al. (2009a)
		Coffee manufacturing wastewater	Jung et al. (2010)
		Tofu wastewater	Kim and Lee (2010), Kim et al. (2010)
		Starch-based wastewater	Chen et al. (2008), Zhang et al. (2003)
		Citric acid wastewater	Yang et al. (2006)
		Slaughterhouse waste	Gómez et al. (2006)
		Sweet sorghum syrup/extract	Saraphirom and Reungsang, (2010), Antonopoulou et al. (2008, 2010)
		Liquid swine manure	Wu et al. (2010), Zhu et al. (2009b)
	Dairy-based industries	Dairy processing wastewater	Venkata Mohan et al. (2007a, 2008e), Gustavo et al. (2008), Ren et al. (2007)
		Dairy waste permeate/ waste lactose	Banks et al. (2010)
		Cheese processing wastewater	Ferchichi et al. (2005), Yang et al. (2007)
		Cattle wastewater	Tang et al. (2008)
	Alcohol-based industries	Brewery wastewater	Chang et al. (2008), Shi et al. (2010)
		Wine process wastewater	Yu et al. (2002), Froylán et al. (2009)
		Molasses-based wastewater	Venkata Mohan et al. (2008f), Vatsala et al. (2008), Lay et al. (2010)
		Vinasse	Buitrón and Carvajal (2010), Fernandes et al. (2010)
	Plant/ Agricultural-based waste	Vegetable-based market waste	Mohanakrishna et al. (2010a)
		Paper mill waste	Idania et al. (2005), Lakshmidevi and Muthukumar (2010)
		Lignocellulose-derived organic acids	Zhu et al. (2010)
		Wheat straw hydrolysate	Kongjan and Angelidaki (2010)
		Cassava stillage	Luo et al. (2010a,b), Sreethawong et al. (2010)
		Citrus peelings	Venkata Mohan et al. (2009a)

TABLE 3 Various Types of Wastewater Used as Substrates for H_2 Production—Cont'd

Type of Fermentation	Industry Category	Type of Industrial Wastewater	References
	Organic-based industries	Phenol-containing wastewater	Tai et al. (2010)
		Organic wastewater	Show et al. (2010)
		Chemical wastewater	Venkata Mohan et al. (2007b,c,d, 2010d), Vijaya Bhaskar et al. (2008)
	Oil-based industries	Glycerin from biodiesel production	Fernandes et al. (2010), Liu and Fang (2007)
		Palm oil mill effluent (POME)	Vijayaraghavan and Ahmad (2006), Wu et al. (2009)
		Olive mill wastewater	Ntaikou et al. (2009)
	Others	Landfill leachate	Liu et al. (2010), Hafez et al. (2010)
		Filtrate of activated sludge	Guo et al. (2010)
		Probiotic wastewater	Sivaramakrishna et al. (2009)
Photo		Olive mill waste	Ena et al. (2010), Eroglu et al. (2010)
		Potato steam peels hydrolysate	Afsar et al. (2010)
		Sugar refinery wastewater	Yetis et al. (2000)
		Dairy wastewater	Venkata Mohan et al. (2008), Srikanth et al. (2008), Seifert et al. (2010)
		Tofu wastewater	Zheng et al. (2010), Zhu et al. (2002)
		Ground wheat solution	Argun et al. (2009)
		Starch fermentation	Laurinavichene et al. (2008)

The composition of the microbial community survived in the long-term operated anaerobic sequencing biofilm reactor producing H_2 from the treatment of various types of wastewaters showed significant diversity (Venkata Mohan et al., 2010d; Figure 3). Major nucleotide sequences were affiliated to Class Clostridia followed by Bacteroidetes, Deltaproteobacteria, and Flavobacteria. Long time operation with diverse operating conditions might have resulted in the survival of robust and selectively enriched bacteria which are capable of producing H_2 under acidogenic conditions. Almost all the microbial community structure analysis showed the presence of clostridium as dominant group. Clostridia (gram positive, rod shaped) associated with Firmicutes group are obligate anaerobes capable of producing endospores and also able to survive at nutritional limiting conditions and even at higher temperatures.

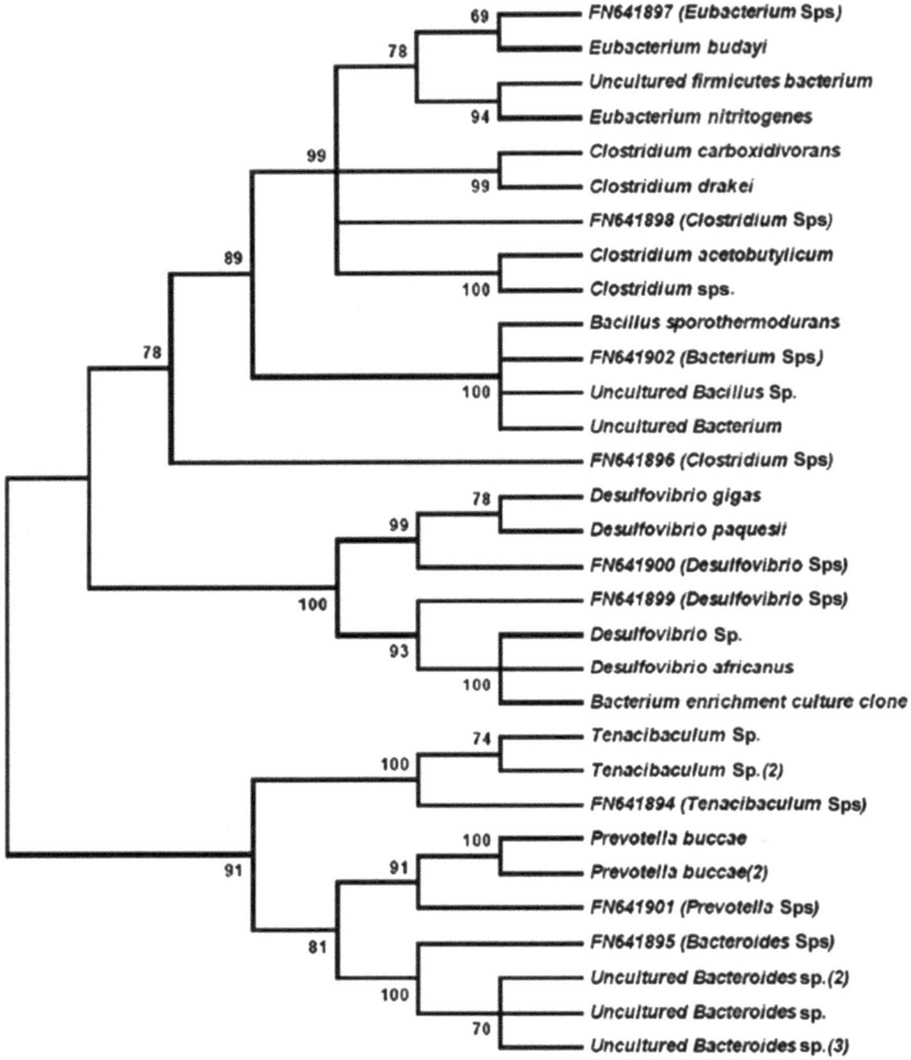

FIGURE 3 Neighbor-joining tree representing microbial diversity of long-term operated anaerobic reactor producing H_2 (constructed using Mega 4.0) showing closely related phylogenetic relationships of 16S rDNA from Gene Bank (Venkata Mohan et al., 2010d).

5 FACTORS INFLUENCING H_2 PRODUCTION

5.1 Redox Condition

Redox condition is especially important for fermentative H_2 production process where the activity of AB is considered to be crucial and rate limiting (Fan et al., 2006; Venkata Mohan et al., 2008a; Venkata Mohan, 2009, 2010). Based on the organisms and their growth conditions, changes in external pH can bring about alterations in several primary physiological parameters, including the internal pH, concentration of other ions, membrane potential, and proton-motive force. Redox condition also influences the efficiency of substrate metabolism, protein and storage materials synthesis, and release of metabolic byproducts. Under acidic conditions, pyruvate converts into VFA along with H_2 by AB. Neutral operation leads to the formation of CH_4 and CO_2 by MB, while basic operation leads to solventogenesis. Repression of MB indirectly promotes H_2 producers within the system (Venkata Mohan et al., 2007d, 2008e; Zhu and Beland, 2006). AB functions well below pH 6, while for MB optimum range is between 6.0 and 7.5 (Venkata Mohan, 2010, 2009). The pH range of 5.5-6.0 is reported to be ideal to avoid both methanogenesis and solventogenesis (Venkata Mohan et al., 2007d, 2010a; Venkata Mohan, 2009, 2010). Good H_2 yield was observed by maintaining pH in and around 6 compared to near-neutral pH (van Ginkel et al., 2001; Venkata Mohan et al., 2007d; Venkata Mohan, 2010). However, highly acidic pH (<4.5) is detrimental to H_2 production as it inactivates AB (Venkata Mohan, 2010; Zhu and Beland, 2006). Hydrogenase enzyme activity gets inhibited by maintaining low or high pH beyond optimum range (Fan et al., 2006; Venkata Mohan, 2010).

Dehydrogenase was the other important enzyme that maintains H^+ concentration in the cell through H^+ shuttling leading to increased chances of H_2 generation (Venkata Mohan et al. 2010a). Dehydrogenase-catalyzed redox reactions are higher under acidophilic operation, which enhances the H^+ shuttling between metabolic intermediates and redox mediators instead of getting reduced to end products. This H^+ shuttling provides higher availability of H^+ to make H_2, while neutral operation leads to the H^+ reduction to methane (Venkata Mohan et al., 2008a, 2010a). Alkaline operation also provides higher redox reactions, where the H^+ shuttling between metabolic intermediates results in the formation of reduced compounds such as aldehydes, alcohols, and reducing sugars (Venkata Mohan et al., 2010a). Under acidophilic operation, relatively higher accumulation of acid metabolites was observed over the corresponding neutral and basic redox condition which corroborates well with the H_2 production (Venkata Mohan et al., 2008a, 2010a). The activity of hydrogenase is greater at acidic pH whereas in the neutral to alkaline range metabolic pathway proceeds to the next step (methanogenesis) where H^+ is reduced by combining with CO_2 to form CH_4 or might lead to solventogenesis forming ethanol via acetaldehyde-reducing H^+ (at basic pH). Maintenance of acidophilic conditions in association with pretreatment was observed to be effective for H_2 production during the treatment of various wastewaters (Venkata Mohan et al., 2008a, 2010a). Therefore, sustaining the system redox condition in optimum range is essential for effective H_2 production and pH can be considered as a manipulated variable for the process control.

5.2 Temperature

Temperature affects H_2 production, metabolites distribution, substrate degradation, and bacterial growth. H_2 production was reported under ambient (15-27 °C), mesophilic (30-45 °C), and moderate thermophilic (50-60 °C) temperatures, and a few studies were reported even under extreme thermophilic conditions (over 60 °C; Venkata Mohan, 2010; Yokoyama et al., 2009). The optimal temperature for H_2 production widely varied based on the nature of the biocatalyst and the type of wastewater used. For pure cultures, the optimal temperatures were reported to be in the range of 37-45 °C, whereas for mixed microflora diverse optimum temperatures were reported (Tang et al., 2008). Both mesophilic and thermophilic temperatures were observed to be optimal for fermentative H_2 production processes.

Thermophilic conditions were reportedly advantageous due to its thermodynamics (Tang et al., 2008) which gives higher reaction rate with better process performance and decreased problems with contaminating H_2-consuming microorganisms. Even though higher temperatures favor reaction kinetics, rapid changes in the system pH may inhibit H_2-producing bacteria (Venkata Mohan, 2010). The changes in soluble metabolite composition were also observed with change in the operating temperature which resulted in metabolic shifts to dominant bacterial function at that particular temperature (Wang and Wan, 2009). Temperature control might not be a feasible option for process control at all instances (Venkata Mohan, 2010).

5.3 Fermentation Time

Fermentation time or hydraulic retention time (HRT) is one of the important and readily manipulated process control variables which can significantly influence the H_2 generation by way of restricting methanogenic activity. Maximum H_2 yields are reported between 0 and 14 h in various experiments performed under batch mode using dairy, chemical, and distillery wastewaters as substrates (Venkata Mohan et al., 2007a,c; Venkata Mohan et al., 2008f; Venkata Mohan, 2010; Vijaya Bhaskar et al., 2008). Longer HRTs induce a metabolic shift from the acidogenic process to the methanogenic process which is unfavorable for H_2 production. Therefore, many of the researchers have used short HRTs which help to restrict the MB growth (Hawkes et al., 2007). HRT mostly depends on the nature and composition of the substrate, nature of biocatalyst, loading rate, and redox condition (Venkata Mohan, 2010). Optimum HRTs between 8.0 and 14 h were reported for effective H_2 production (Hawkes et al., 2007).

5.4 Reactor Configuration and Mode of Operation

Reactor configuration in association with its operation mode governs the performance of an open engineering system and influences the microenvironment in the reactor, hydrodynamic behavior, wastewater-biocatalyst contact, survivability of microbial population, etc (Venkata Mohan, 2009, 2010). Diverse reactor configuration, viz. suspended growth, biofilm/packed-bed/ fixed bed, fluidized bed, expanded bed, upflow anaerobic sludge blanket (UASB), granular sludge, membrane-based systems, immobilized systems, etc., were reported in biohydrogen research. Biofilm/attached-growth systems are generally robust to shockloads and provide resilience and resistance to changes in the process parameters (Lalit Babu et al., 2009; Venkata Mohan et al., 2007b). They are well suited for treating highly variable wastewater which facilitates improved reaction potential leading to stable and robust system. The influence of self-immobilization of enriched acidogenic mixed consortia on H_2 production was studied on different supporting materials [SBA-15 (mesoporous) and activated carbon (granular; GAC and powder; PAC)] in comparison with suspended growth of culture (Venkata Mohan et al., 2008d). Attached growth showed higher H_2 yield (ninefolds) and substrate degradation efficiency, particularly at higher loading rates. Batch, fed-batch, semibatch/continuous, periodic discontinuous batch (sequencing batch operation), and continuous modes of reactor operation have been reported for H_2 production. Fed-batch operation reduces the accumulation of soluble metabolic intermediates due to fill-draw mode operation (Venkata Mohan et al., 2007a,b; Venkata Mohan et al., 2008a,f). Poor biomass retention/cell washout can be reduced to some extent in batch-mode operation (Yokoi et al., 1997). Batch mode operation coupled with biofilm configuration combines the dual operational advantages of both the systems and helps to maintain stable and robust cultures suitable for treating highly variable wastewater (Lalit Babu et al., 2009; Luo et al., 2010a; Venkata Mohan et al., 2007b,c,d; Venkata Mohan et al., 2008a,b,d; Yokoi et al., 1997).

6 COMBINED PROCESS EFFICIENCY

When using wastewater as substrate for H_2 production, the substrate degradation efficiency is also important when process efficiency is considered (Venkata Mohan et al., 2009b). There exists a tradeoff between technical efficiency based on H_2 production and substrate removal in association with operating conditions (Venkata Mohan, 2010). Neutral pH is ideal for wastewater treatment, while acidic pH is useful for effective H_2 production (Venkata Mohan et al., 2007b,d). Balancing conditions for combined performance are especially important to sustain the economic viability and environmental acceptability of the process. Data enveloping analysis (DEA) was used to study the role of some important factors, viz. nature of inoculum, pretreatment method, pH, cosubstrate addition, and feed composition on combined process efficiency (Venkata Mohan et al., 2008c, 2009b). Analysis showed that the untreated anaerobic inoculum under acidic conditions utilizing simple wastewater as substrate showed combined process efficiency. Taguchi's design of experimental (DOE) methodology was employed to enumerate the role of selected factors on both H_2 production and substrate degradation (Venkata Mohan et al., 2009b; Venkata Mohan, 2010). Taguchi's approach helped to identify the influence and contribution of individual selected factors on both the processes.

7 PROCESS LIMITATIONS

In spite of the striking advantages, the main challenges encountered with fermentative H_2 production process are low substrate conversion efficiency and residual substrate present in acid-rich wastewater generated from the acidogenic process (Venkata Mohan, 2010, 2009). The persistent accumulation of acidogenic byproducts causes a sharp drop in the pH resulting in the inhibition of AB activity thereby reducing H_2 production (Venkata Mohan, 2008; Wang and Wan, 2009). The undissociated soluble metabolites can permeate through the cell membrane of AB and then dissociate in to the cell leading to physiological imbalance and particularly at a high concentration the increase in ionic strength results in the cell lysis (Wang and Wan, 2009). Thus, some maintenance energy is required to restore the physiological balance in the cell, which reduces the energy required for bacterial growth and also inhibits the bacterial growth. The metabolic end products and the resultant H_2 yields vary based on the environmental conditions even within the same bacterium (Hawkes et al., 2007). H_2 yield is lower when more reduced organic compounds, such as lactic acid, propionic acid, and ethanol, are produced as fermentation products, because these represent the end products of metabolic pathways that bypass the major H_2-producing reaction (Angenent et al., 2004). Biological limitations such as H_2-end-product inhibition and acid or solvent accumulation limit the molar yield. Even under optimal operating conditions about 60-70% residual organic carbon remains in the effluent after-dark fermentation (Das and Veziroglu, 2001; Venkata Mohan et al., 2008b; Venkata Mohan, 2010), and it requires further treatment prior to disposal.

8 STRATEGIES TO ENHANCE PROCESS EFFICIENCY

8.1 Process Integration

Effective utilization of the residual organic fraction can be made possible by integrating with the other energy-producing processes. Figure 4 depicts some of the multiple process integration routes reported along with their futuristic approaches. Integration of an acidogenic process with a terminal photofermentation or acidogenic processes for additional H_2 (Srikanth et al., 2008, 2009; Venkata Mohan et al., 2009b) or methanogenic processes for methane; Venkata Mohan et al., 2008b) or microbial fuel cell (MFC) for bioelectricity generation (Mohanakrishna et al., 2010a,b) was reported along with enhanced substrate degradation (Figure 5). Photosynthetic bacteria can produce H_2 by consuming organic acids present in the effluents generated from acidogenic H_2 fermentation processes (Srikanth et al., 2008, 2009;

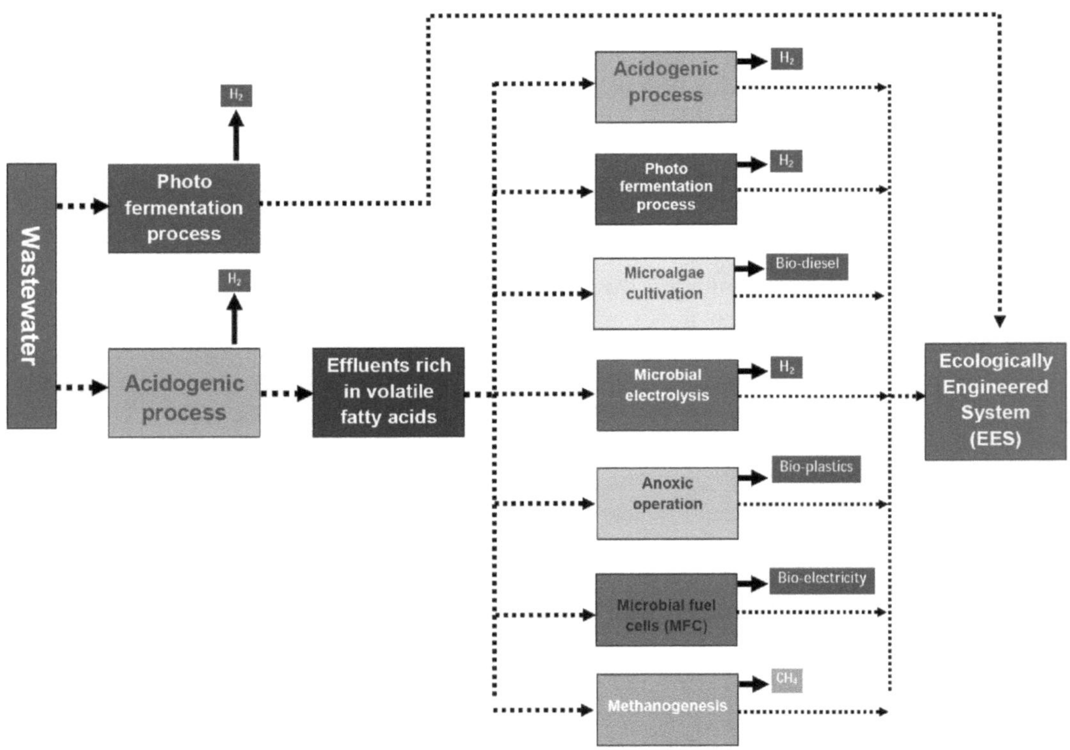

FIGURE 4 Schematic representation of various reported and possible integration processes with dark fermentation for additional bioenergy generation and treatment.

Venkata Mohan, 2008). Some PNSB can utilize metabolic intermediates of dark fermentation such as short-chain volatile fatty acid (Donald, 1995). Henceforth, integration of anoxygenic photo fermentation with dark fermentation will have dual advantages of increased H_2 production along with substrate removal (Srikanth et al., 2008, 2009; Venkata Mohan et al., 2008c; Venkata Mohan, 2010). Green algae such as *Chlorella* also have the capability to utilize organic acids produced during the dark fermentation for H_2 production, especially acetate as a substrate (Amutha and Murugesan, 2011). Some PNSB can also use sugars, aromatic compounds (*Rhodopseudomonas palustris* can use lignin monomers) and even they can oxidize inorganic substrates such as $S_2O_3^{2-}$, H_2S, or Fe^{2+} to get electrons, which are further utilized in H_2 production (Donald, 1995). Theoretically, the maximum H_2 yield (12 mol H_2/mole glucose) may be obtained when glucose is converted to acetate as the terminal product through dark fermentation, then subsequently converted into H_2 through photofermentation (Hawkes et al., 2007; Srikanth et al., 2008; Venkata Mohan et al., 2008b; Venkata Mohan, 2010). However, the efficiency of both H_2 production and substrate degradation was found to depend on the process used in the first stage along with the composition of the substrate (Venkata Mohan et al., 2008b).

Bioelectricity production was observed using acid-rich effluents from acidogenic process in MFC with 90% of substrate degradation efficiency (Mohanakrishna et al., 2010b). Simple organic acids or by-products of dark fermentation processes were more suitable substrates for bioelectricity generation (Mohanakrishna et al., 2010b). Multistage process can also be used to maximize H_2 production. The integration facilitates the economic viability, commercialization of the process and helps for environmental restoration by complete stabilization of the wastewater. Ecologically engineered system (EES) consisted of diverse biota, viz. aquatic macrophytes, submerged plants, emergent plants, and filter feeders was designed to mimic the natural cleansing functions of wetlands was successfully used to treat acidogenic wastewater from H_2 production process (Venkata Mohan et al., 2010c).

8.2 Hydrogen Partial Pressure

The partial pressure of H_2 affects H_2 production pathways and at higher concentration they act for end-product inhibition (Nath and Das, 2004). At higher H_2 concentrations, metabolic pathway shifts towards the production of more reduced substrates, such as lactate, ethanol, acetone, butanol, alanine, or methane decreasing the H_2 production (Nath and Das, 2004). H_2 production becomes thermodynamically unfavorable at H_2 partial pressures greater than 60 Pa (Angenent et al., 2004; Venkata Mohan, 2010). Operating bioreactors at low H_2 partial pressure by stripping H_2 from the solution as it is generated (Hawkes et al., 2007) accomplishes both efforts simultaneously (Angenent et al., 2004; Venkata Mohan, 2010). Sparging of inert gases like nitrogen and argon lowers the dissolved H_2 concentration resulting in an increase of H_2 yield (Mizuno et al., 2000; Tanisho et al., 1998). Low agitation and high partial pressures favor butanol production (Doremus et al., 1985), which upon stirring improved the H_2 transfer into the gas phase and helped to reduce the inhibition of H_2/acetate formation. Gas-stripping and membrane-absorption technologies are also studied for removal of H_2 from the reactor (van Groenestijn et al., 2002).

8.3 Buffering Capacity

The generation and accumulation of acid metabolites or VFA during fermentation process causes a sharp drop in the system pH and gradually reduces the buffering capacity of the system thereby inhibiting the H_2 production (Lin and Lay, 2004b; Prathima Devi et al., 2010). VFA and pH express the acid-base condition of anaerobic microenvironment and provide the information pertaining to the balance between two of the most important microbial groups, viz. AB and MB. If the pH of the system is not maintained in an optimum range, it results in the cessation of H_2 production along with a marked shift in the microbial metabolism (Prathima Devi et al., 2010; van Ginkel et al., 2001). System pH can be controlled by increasing the buffering capacity. Alkalinity indicates the buffering capacity of a system in association with redox microenvironment and VFA concentration. Buffering capability maintains stable ionic strength against changes in acidic and basic constituents. Higher alkalinity indicates more stability in the redox condition. Maintaining stable pH can be made possible with increased buffering capacity of the system. Supplementation of buffer in different concentrations at varying pHs increased the H_2 production time by maintaining the favorable pH conditions in the system (Zhu et al., 2009a). CO_2 released from the biogas was reused in the bioreactor in order to improve the system-buffering capacity during H_2 production (Prathima Devi et al., 2010). After supplementing with CO_2, the reactor showed significant improvement in the system-buffering capacity which positively influenced both H_2 production and substrate degradation. Prevailing alkaline condition helps to build up buffering nature, which resist the fluctuations in pH even at higher VFA concentrations.

8.4 Bioplastics Production

Biohydrogen can also be viewed as energy source and an intermediate towards the production of VFA which can be further transformed to polyhydroxyalkanoates (PHAs), or can be used for biohydrogenation of fatty acids into alcohols (Venkata Mohan et al., 2010b). VFA present in these effluents generated from H_2 producing reactor can be further transformed to PHAs or can be used for biohydrogenation of fatty acids into alcohols. PHA is a group of biologically derived biopolyester of hydroxyalkanoates that accumulate as carbon/energy or reducing-power storage materials in microbial cells (Van Loosdrecht et al., 1997). When a carbon source is available in excess and other nutrients are growth limiting, biopolyesters are deposited as water-insoluble cytoplasmic nonsized inclusions by Eubacteria and Archaea (Rehm, 2007). VFA under low oxygen/anoxic microenvironment gets converted to poly(β-OH)butyrate (PHB) via various biochemical routes catalyzed by the action of different enzymes. So far, many efforts have been made to produce PHA from commercial-grade substrate using pure cultures, which are expensive due to their high substrate and production costs. Compared to carbohydrates, lipids, and amino acids, VFA has simple structure with lower number of carbon atoms [acetate (2C), propionate (3C), and butyrate (4C)], which facilitates its easy synthesis to PHA without the involvement of glycolysis and b-oxidation pathways by using less number of metabolic reactions and enzymes. The feasibility of PHB was successfully reported with individual VFA and acid-rich effluents from a biohydrogen-producing reactor as primary substrates employing aerobic consortia as biocatalyst under anoxic microenvironment (Venkata Mohan et al.,

2010b). Butyrate showed higher PHB productivity followed by acetate, acids mixture, and propionate. Acid-rich effluents resulted higher PHB yield especially at low substrate load. Neutral operation showed effective PHB production compared to acidic and basic conditions due to the associated higher metabolic activity of biocatalyst. The integrated approach helped to treat additional COD from acid-rich HBR effluents apart from byproduct recovery. PHA represents a potentially sustainable replacement for fossil fuel-based conventional thermoplastics due to their easy biodegradability and capability to generate from renewable resources.

8.5 Bioaugmentation

Bioaugmentation is a strategy usually applied to improve the start up of a bioreactor, to accelerate process efficiency, to protect the existing microbial community against adverse effects, or to compensate for organic or hydraulic overloading (Venkata Mohan et al., 2005). For augmenting purpose, indigenous or allochthonous wild-type or genetically modified organisms are used (Venkata Mohan et al., 2005). Bioaugmentation strategy was applied to operating anaerobic reactor (producing CH_4) to metabolically shift from methanogenesis to acidogenesis so as to produce H_2 as end product (Venkata Mohan et al., 2007c). Selectively enriched AB (in immobilized form) was used as the augmenting inoculum. After augmenting, H_2 production rate improved significantly. This strategy can be implemented to the existing full-scale anaerobic reactors producing CH_4 in order to shift towards H_2 production. This process is an effective method to improve the process efficiency in short period of time. Bioaugmentation with cocultures *Clostridium acetobutylicum* X_9 and *Ethanoigenens harbinense* B_{49} showed to improve cellulose hydrolysis and subsequent H_2 production rates from carboxymethyl cellulose fermentation (Ren et al., 2008). The effect of constructed microbial consortium (*Enterobacter cloacae* IIT-BT 08: *Citrobacter freundii* IIT-BT L139: *Bacillus coagulans* IIT-BT S1) into sewage sludge was reported to improve the H_2 production (Kotay and Das, 2010).

8.6 Activators

Some metal ions, organic compounds, and nutrients concentration generally have a stimulating effect on the enzymatic activity pertaining to H_2 production if added at optimal concentrations (Table 4). Iron and nickel are the components of the most crucial enzyme, hydrogenases. The [Ni–Fe] and [Fe–Fe] centers in the active site of hydrogenases play significant roles to catalyze H_2 evolution and uptake (Frey, 2002). [Ni–Fe] hydrogenases have a higher substrate affinity among the hydrogenases, while [Fe–Fe] hydrogenases have higher conversion efficiency and hence are considered as most potent class of hydrogenases. The electron donors and acceptors vary in a narrow range for different hydrogenases, viz., cytochrome c_3 as electron donors and acceptor for [Ni–Fe] hydrogenase, and cytochrome c_3 and c_6 act as physiological electron donors or acceptors for [Fe–Fe] hydrogenases (Donald, 1995; Nelson and Cox, 2004). Iron was studied in detail till date as it is an important component for hydrogenase function (Donald, 1995; Nelson and Cox, 2004), and few other metals also have significant roles in fermentative H_2 production, viz. nickel, manganese, magnesium, and zinc (Nelson and Cox, 2004). Magnesium plays a vital role in the substrate

TABLE 4 Details of Some of Activators Which can Stimulate H_2 Production

Activator	Functional Role	Reference
Fe^{+3}	Acts as micronutrient and mediates between hydrogenase and nicotinamide adenine dinucleotide (NADH)-ferredoxin reductase as electron carrier. Major component in the active site of all hydrogenases and central atom of ferridoxin (Fd) which access electrons to carry towards hydrogenase	Wang and Wan (2009), Frey (2002)
Mg^{+2}	Forms complex with ATP (Mg-ATP) and involves in the substrate activation through phosphorylation. Activates the α-subunit of catalytic site of E1 of pyruvate dehydrogenase complex (PDC) leads to acetate pathway. Also activates membrane components such as cytochromes for increased electron transfer	Nelson and Cox (2004)
Ni^{+2}	Metal ion in the central complex of [Ni–Fe] hydrogenase	Wang and Wan (2008b, 2009)
Mn^{+}	Mimics the function of Mg^{+2} in its absence. Activates the Ni–Fe hydrogenases.	Nelson and Cox (2004)
Zn^{+}	Active site component of several enzymes and functions as catalyst for some reactions	Wang and Wan (2009), Nelson and Cox (2004)
Nitrogen	Constituent of all amino acids, nucleic acids, and redox powers (NAD, FAD, FMN, etc.). Involves in the growth and metabolic functions.	Wang et al. (2009)
Phosphorous	Helps in activation of substrate molecules through phosphorylation. Forms high-energy compounds (ATP), a cell energy reservoir. Provides internal cell buffering capacity.	Wang and Wan (2009)

activation prior to undergoing metabolism, where each molecule of substrate should first get activated with phosphate component (Nelson and Cox, 2004). Substrate phosphorylation is made easy through Mg-ATP complex instead ATP individually, which leads to the substrate availability to the enzyme. Manganese will do the similar function like magnesium, when its availability is more, of course with less efficiency. It is also active site component of some enzymes (Nelson and Cox, 2004). Zinc has a function as the active site component for different enzymes and showed positive influence on fermentative H_2 production (Wang and Wan, 2009). Metal ions like chromium and cobalt showed toxic effects on H_2 production (Wang and Wan, 2009).

Nitrogen and phosphorus are the important constituents of amino acids, nucleic acids, reducing powers and enzymes, and they play a major role in growth (Wang et al., 2009). However, at higher concentrations, they affect the biochemical function of the cell. Nitrogen at elevated levels could inhibit the process by affecting the intracellular pH and might lead the metabolic path towards ammonification, where the protons get consumed instead of forming H_2 (Salerno et al., 2006). Optimum C/N ratio helps in the bacterial growth as well as their function for the substrate conversion to end products (Lin and Lay, 2004a). The major role of phosphate is energy generation to the cell in the form of ATP, a high-energy nucleotide. Phosphates also help in maintaining the system-buffering capacity alternative to carbonates during fermentative process (Lin and Lay, 2004b; Wang and Wan, 2009). Comparatively phosphates can acts as better buffering agent than carbonates, since carbonate dissolution increases the CO_2 fraction in biogas (Lin and Lay, 2004b).

8.7 Microbial Electrolysis

Microbial mediated electrolysis cells (MEC), also called bioelectrochemically assisted microbial reactor (BEAMR) or electrohydrogenesis (Figure 5), provide a new approach which facilitates convertion of biodegradable material into H_2 in the presence of applied external voltages (Cheng and Logan, 2007). MEC process is mostly related to MFC technology where exoelectrogenic bacteria was employed in both the processes, but the difference is the requirement of small input of external potential for MEC in order to get H_2. Based on thermodynamics, the potential greater than 0.11 V in addition to that generated by bacteria (-0.3 V) will yield H_2 at the cathode (Cheng and Logan, 2007). This provides a route for extending H_2 production past the endothermic barrier imposed by the microbial formation of fermentation end products, such as acetic acid (Cheng and Logan, 2007). This process requires low potentials compared to theoretically applied voltage of 1.23 V for water electrolysis (Rozendal et al., 2007), and MEC can therefore be considered as a potential option for H_2 production with simultaneous treatment of wastewater. Usually, a voltage of 0.6 V or more was used for an efficient H_2 production (Cheng and Logan, 2007). Compared to the dark fermentation process where 33% of energy recovery is generally possible, MEC can achieve more than 90% of H_2 recovery (Cheng and Logan, 2007).

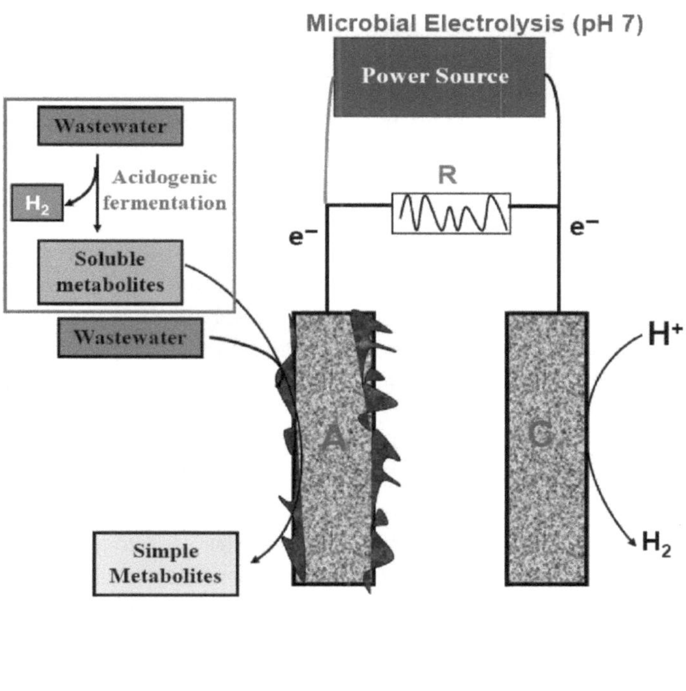

FIGURE 5 Schematic illustration of microbial electrolysis process and the utilization of acid rich effluents from acidogenic fermentation as substrates for additional H_2 production in MEC.

During the initial studies, MEC was operated in the double chambers (Cheng and Logan, 2007; Rozendal et al., 2007), and membranes are used which separates anode and cathode chambers. In the recent studies, MECs were designed to operate without membranes and the elimination of membrane not only simplifies and economies the process operation but also can decrease the internal resistance and thus increasing the H_2 production rate. Single chamber MEC was operated without membrane using wastewater as substrate and mixed consortia as biocatalyst (Venkata Mohan et al., 2011). Variation in the dehydrogenase activity in response to the poised potential and its significant role in carrying out the redox reactions for generating H_2 was also evaluated in detail during operation. Regression analysis of the data obtained supported the best performance at 0.6 V of poised potential which was reported to be ideal for the enrichment of exoelectrogenic bacteria. One more important point to be considered during the operation of MEC is to control the formation of methane which affects the H_2 production rate. Methanogenesis can be controlled by operating the experimental cycle with short HRT and by using higher applied potentials (>0.6 V) (Srikanth et al., 2010a). Hydrogen was successfully produced from various substrates such as cellulose, glucose, VFA, proteins, swine, waste water, and also with the effluents collected from the ethanol-H_2 fermentative reactor (Cheng and Logan, 2007). This process makes possible to generate H_2-utilizing effluents generated from acidogenic fermentation (Wagner et al., 2009). Multielectrode MEC was also reported for the continuous production of biogas (Rader and Logan, 2010).

9 FUTURE OUTLOOK

Nonutilized residual organic fraction remaining as a soluble fermentation product after acidogenic process is one of the most important aspects to be paid significant attention. Various routes to utilize this residual organic fraction of acidogenic process as substrate need to be explored. Multiple process integration approaches towards the utilization of wastewater effectively and completely with simultaneous bioenergy recovery are to be evaluated for economic viability of the process and commercialization. Controlling the system redox conditions (buffering capacity) leads to protest from the persistent acidophilic microenvironment due to soluble acid intermediates and thus resulting in increased H_2 yields. Photobiological processes especially with wastewater as substrate have not yet been fully exploited and are relatively less studied. Photosynthetic culture has the advantage of high substrate conversion efficiency to H_2 because of its ability to mineralize glucose to CO_2. The potential of photosynthetic culture in utilizing wastewater directly as well as the dark fermentation effluents should be evaluated. Photosynthetic sulfur and nonsulfur bacteria can be considered and evaluated with sulfur-containing wastewater where they can utilize different forms of sulfur as electron donor. Process engineering, understanding biochemistry and microbiology aspects based on functional role of membrane components and mechanism of proton reduction, community analysis, culture development aspects, design and development of bioreactor systems for dark and photofermentation operations are some of the key areas where considerable focus is required. Metabolic engineering is one of the promising areas which can be advantageously used to enhance H_2 production rate using recombinant DNA technology. Developing a process to convert existing/operating anaerobic reactors producing methane to H_2 production will pave way for large-scale implementation of this technology. Biohydrogen technology requires multidisciplinary research to make the process environmentally sustainable and economically viable.

Acknowledgments

Author thanks the Director, IICT, for his encouragement and acknowledges the inputs of A Kiran Kumar, S Veer Raghavulu, G Velvizhi, M Prathima Devi, M Lenin Babu, R Kannaiah Goud, M Venkateswar Reddy, G Venkata Subhash, K Chandra Sekhar, Rashmi Chandra, and P Chiranjeevi. This work was supported by Ministry of New and Renewable Energy (MNRE), Government of India in the form of National Mission Mode project on Biohydrogen production [No. 103/131/2008-NT].

References

Afsar, N., Özgür, E., Gürgan, M., Akköse, S., Yücel, M., Gündüz, U., et al., 2011. Hydrogen productivity of photosynthetic bacteria on dark fermenter effluent of potato steam peels hydrolysate. Int. J. Hydrogen Energy 36, 432–438.

Amutha, K.B., Murugesan, A.G., 2011. Biological hydrogen production by the algal biomass Chlorella vulgaris MSU 01 strain isolated from pond sediment. Bioresour. Technol. 102, 194–199.

Angenent, L.T., Karim, K., Al-Dahhan, M.H., Wrenn, B.A., Domíguez-Espinosa, R., 2004. Production of bioenergy and biochemicals from industrial and agricultural wastewater. Trends Biotechnol. 22, 477–485.

Antonopoulou, G., Gavala, H.N., Skiadas, I.V., Angelopoulos, K., Lyberatos, G., 2008. Biofuels generation from sweet sorghum: Fermentative hydrogen production and anaerobic digestion of the remaining biomass. Bioresour. Technol. 99, 110–119.

Antonopoulou, G., Gavala, H.N., Skiadas, I.V., Lyberatos, G., 2010. Influence of pH on fermentative hydrogen production from sweet sorghum extract. Int. J. Hydrogen Energy 35, 1921–1928.

Argun, H., Kargi, F., Kapdan, I.K., 2009. Hydrogen production by combined dark and light fermentation of ground wheat solution. Int. J. Hydrogen Energy 34, 4305–4311.

Banks, C.J., Zotova, E.A., Heaven, S., 2010. Biphasic production of hydrogen and methane from waste lactose in cyclic-batch reactors. J. Cleaner. Prod. 18, S95–S104.

Buitrón, G., Carvajal, C., 2010. Biohydrogen production from Tequila vinasses in an anaerobic sequencing batch reactor: effect of initial substrate concentration, temperature and hydraulic retention time. Bioresour. Technol. 101 (23), 9071–9077.

Chang, J., Chou, C., Ho, C., Chen, W., Lay, J., Huang, C., 2008. Syntrophic co-culture of aerobic Bacillus and anaerobic Clostridium for bio-fuels and bio-hydrogen production. Int. J. Hydrogen Energy 33, 5137–5146.

Chen, S.D., Lee, K.S., Lo, Y.C., Chen, W.M., Wu, J.F., Lin, C.Y., et al., 2008. Batch and continuous biohydrogen production from starch hydrolysate by *Clostridium* species. Int. J. Hydrogen Energy 33 (7), 1803–1812.

Cheng, S., Logan, B.E., 2007. Sustainable and efficient biohydrogen production via electrohydrogenesis. Proc. Nat. Acad. Sci. 104, 18875–118873.

Chong, L., Sabaratnam, V., Shirai, Y., Hassan, M.A., 2009. Biohydrogen production from biomass and industrial wastes by dark fermentation. Int. J. Hydrogen Energy 34, 3277–3287.

Das, D., Veziroglu, T.N., 2001. Hydrogen production by biological process: a survey of literature. Int. J. Hydrogen Energy 26, 13–28.

Donald, A.B., 1995. Molecular Biology of Cyanobacteria. Kluwer Academic Publisher, Dordrecht, The Netherland.

Doremus, M.G., Linden, J.C., Moreira, A.R., 1985. Agitation and pressure effects on acetone-butanol fermentation. Biotechnol. Bioeng. 27, 852–860.

Ena, A., Pintucci, C., Carlozzi, P., 2010. Production of bioH_2 by Rhodopseudomonas palustris (strain 6A) grown in pre-treated olive mill waste, under batch or semi-continuous regime. J. Biotechnol. 150 (1), 180.

Eroglu, E., Gunduz, U., Yucel, M., Eroglu, I., 2010. Photosynthetic bacterial growth and productivity under continuous illumination or diurnal cycles with olive mill waste water as feedstock. Int. J. Hydrogen Energy 35, 5293–5300.

Fan, Y., Zhang, G., Guo, X., Xing, Y., Fan, M., 2006. Biohydrogen-production from beer lees biomass by cow dung compost. Biomass. Bioenergy. 30, 493–496.

Ferchichi, M., Crabbe, V., Gil, G.H., Hintz, W., Almadidy, A., 2005. Influence of initial pH on hydrogen production from cheese whey. J. Biotechnol. 120, 402–409.

Fernandes, B.S., Peixoto, G., Albrecht, F.R., Katia, N., Aguila, S.D., Zaiat, M., 2010. Potential to produce biohydrogen from various wastewaters. Energy Sus. Deve. 14 (2), 143–148.

Frey, M., 2002. Hydrogenases: hydrogen-activating enzymes. Chem. Biochem. 3, 153–160.

Froylán, M.E.E., Carlos, P.O., Jose, N.C., Yolanda, G.G., André, B., Humberto, G.P., 2009. Anaerobic digestion of the vinasses from the fermentation of Agave tequilana Weber to tequila: the effect of pH, temperature and hydraulic retention time on the production of hydrogen and methane. Biomass Bioenergy 33, 14–20.

Gómez, X., Morán, V., Cuetos, M.J., Sánchez, M.E., 2006. The production of hydrogen by dark fermentation of municipal solid wastes and slaughterhouse waste: a two-phase process. J. Power Sources 157, 727–732.

Guo, L., Li, X.M., Zeng, G.M., Zhou, Y., 2010. Effective hydrogen production using waste sludge and its filtrate. Int. J. Hydrogen Energy 35 (9), 3557–3562.

Gustavo, D.V., Felipe, A.M., Antonio de, L.R., Elías, R.F., 2008. Fermentative hydrogen production in batch experiments using lactose, cheese whey and glucose: influence of initial substrate concentration and pH. Int. J. Hydrogen Energy 33, 4989–4997.

Hafez, H., George, N.G., Naggar, H.E., 2010. An integrated system for hydrogen and methane production during landfill leachate treatment. Int. J. Hydrogen Energy 35 (10), 5010–5014.

Hawkes, F.R., Hussy, I., Kyazze, G., Dinsdale, R., Hawkes, D.L., 2007. Continuous dark fermentative hydrogen production by mesophilic microflora: principles and progress. Int. J. Hydrogen Energy 32, 172–184.

He, D., Bultel, Y., Magnin, J.P., Roux, C., Willison, J.C., 2005. Hydrogen photosynthesis by Rhodobacter capsulatus and its coupling to PEM fuel cell. J. Power Sources 141, 19–23.

Idania, V.V., Sparling, R., Risbey, D., Noemi, R.S., Hécto, M.P.V., 2005. Hydrogen generation via anaerobic fermentation of paper mill wastes. Bioresour. Technol. 96, 1907–1913.

Jung, K.W., Kim, D.H., Shin, H.K., 2010. Continuous fermentative hydrogen production from coffee drink manufacturing wastewater by applying. UASB reactor. Int. J. Hydrogen Energy 35 (24), 13370–13378.

Kim, M.S., Lee, D.Y., 2010. Fermentative hydrogen production from tofu-processing waste and anaerobic digester sludge using microbial consortium. Bioresour. Technol. 101, S48–S52.

Kim, M.S., Lee, D.Y., Kim, D.H., 2010. Continuous hydrogen production from tofu processing waste using anaerobic mixed microflora under thermophilic conditions. Int. J. Hydrogen Energy doi:10.1016/j.ijhydene.2010.06.040.

Kongjan, P., Angelidaki, I., 2010. Extreme thermophilic biohydrogen production from wheat straw hydrolysate using mixed culture fermentation: effect of reactor configuration. Bioresour. Technol. 101 (20), 7789–7796.

Kotay, S.M., Das, D., 2010. Microbial hydrogen production from sewage sludge bioaugmented with a constructed microbial consortium. Int. J. Hydrogen Energy 35, 10653–10659.

Lakshmidevi, R., Muthukumar, K., 2010. Enzymatic saccharification and fermentation of paper and pulp industry effluent for biohydrogen production. Int. J. Hydrogen Energy 35 (8), 3389–3400.

Lalit Babu, V., Venkata Mohan, S., Sarma, P.N., 2009. Influence of reactor configuration on fermentative hydrogen production during wastewater. Int. J. Hydrogen Energy 34, 3305–3312.

Laurinavichene, T.V., Tekucheva, D.N., Laurinavichius, K.S., Ghirardi, M.L., Seibert, M., Tsygankov, A.A., 2008. Towards the integration of dark and photo fermentative waste treatment. 1. Hydrogen photoproduction by purple bacterium Rhodobacter capsulatus using potential products of starch fermentation. Int. J. Hydrogen Energy 33, 7020–7026.

Lay, C.H., Wu, J.H., Hsiao, C.L., Chang, J.J., Chen, C.C., Lin, C.Y., 2010. Biohydrogen production from soluble condensed molasses fermentation using anaerobic fermentation. Int. J. Hydrogen Energy 35 (24), 13445–13451.

Lin, C.Y., Lay, C.H., 2004a. Carbon/nitrogen-ratio effect on fermentative hydrogen production by mixed microflora. Int. J. Hydrogen Energy 29, 41–45.

Lin, C.Y., Lay, C.H., 2004b. Effects of carbonate and phosphate concentrations on hydrogen production using anaerobic sewage sludge microflora. Int. J. Hydrogen Energy 29, 275–281.

Liu, F., Fang, B., 2007. Optimization of biohydrogen production from biodiesel wastes by Klebsiella pneumoniae. Biotech. J. 2, 374–380.

Liu, Q., Zhang, X., Yu, L., Zhao, A., Tai, J., Liu, J., et al., 2011. Fermentative hydrogen production from fresh leachate in batch and continuous bioreactors. Bioresour. Technol. 102, 5411–5417.

Luo, G., Xie, L., Zou, Z., Wang, W., Zhou, Q., Shim, H., 2010a. Anaerobic treatment of cassava stillage for hydrogen and methane production in continuously stirred tank reactor (CSTR) under high organic loading rate (OLR). Int. J. Hydrogen Energy 35, 11733–11737.

Luo, G., Xie, L., Zou, Z., Wang, W., Zhou, Q., 2010b. Exploring optimal conditions for thermophilic fermentative hydrogen production from cassava stillage. Int. J. Hydrogen Energy 35 (12), 6161–6169.

Miyake, J., Wakayama, T., Schnackenberg, J.S., Arai, T., Asada, Y., 1999. Simulation of daily sunlight illumination pattern for bacterial photo-hydrogen production. J. Bioscience. Bioeng. 88, 659–663.

Mizuno, O., Dinsdale, R., Hawkes, F.R., Hawkes, D.L., Noike, T., 2000. Enhancement of hydrogen production from glucose by nitrogen gas sparging. Bioresour. Technol. 73, 59–65.

Mohanakrishna, G., Kannaiah Goud, R., Venkata Mohan, S., Sarma, P.N., 2010a. Enhancing biohydrogen production through sewage supplementation of composite vegetable based market waste. Int. J. Hydrogen Energy 35 (2), 533–541.

Mohanakrishna, G., Venkata Mohan, S., Sarma, P.N., 2010b. Utilizing acid-rich effluents of fermentative hydrogen production process as substrate for harnessing bioelectricity: an integrative approach. Int. J. Hydrogen Energy 35, 3440–3449.

Nandi, R., Sengupta, S., 1998. Microbial production of hydrogen-an overview. Crit. Rev. Microbiol. 24, 61–84.

Nath, K., Das, D., 2004. Improvement of fermentative hydrogen production: various approaches. Appl. Microbiol. Biotechnol. 65, 520–529.

Nelson, D.L., Cox, M.M., 2004. Lehninger Principles of Biochemistry, fourth ed WH Freeman & Co.

Ntaikou, I., Kourmentza, C., Koutrouli, E.C., Stamatelatou, K., Zampraka, A., Kornaros, M., et al., 2009. Exploitation of olive oil mill wastewater for combined biohydrogen and biopolymers production. Bioresour. Technol. 100, 3724–3730.

Prathima Devi, M., Venkata Mohan, S., Mohanakrishna, G., Sarma, P.N., 2010. Regulatory influence of CO_2 sparging on fermentative hydrogen production. Int. J. Hydrogen Energy 35, 10701–10709.

Rader, G.K., Logan, B.E., 2010. Multi-electrode continuous flow microbial electrolysis cell for biogas production from acetate. Int. J. Hydrogen Energy 35, 8848–8854.

Rehm, B.H.A., 2007. Biogenesis of microbial polyhydroxyalkanoate granules: a platform technology for the production of tailor-made bioparticles. Curr. Iss. Mol. Biol. 9, 41–62.

Ren, N.Q., Chua, H., Chan, S.Y., Tsang, Y.F., Wang, Y.J., Sin, N., 2007. Assessing optimal fermentation type for biohydrogen production in continuous flow acidogenic reactors. Bioresour. Technol. 98, 1774–1780.

Ren, N., Wang, A., Gao, L., Xin, L., Lee, D., Su, A., 2008. Bioaugmented hydrogen production from carboxymethyl cellulose and partially delignified corn stalks using isolated cultures. Int. J. Hydrogen Energy 33, 5250–5255.

Rozendal, R.A., Hamelers, H.V.M., Molenkamp, R.J., Buisman, C.J.N., 2007. Performance of single chamber biocatalyzed electrolysis with different types of ion exchange membranes. Water. Res. 41 (9), 1984–1994.

Salerno, M.B., Park, W., Zuo, Y., Logan, B.E., 2006. Inhibition of biohydrogen production by ammonia. Wat. Res. 40, 1167–1172.

Saraphirom, P., Reungsang, A., 2010b. Optimization of biohydrogen production from sweet sorghum syrup using statistical methods. Int. J. Hydrogen Energy 35 (24), 13435–13444.

Seifert, K., Waligorska, M., Laniecki, M., 2010. Hydrogen generation in photobiological process from dairy wastewater. Int. J. Hydrogen Energy 35, 9624–9629.

Shi, X.Y., Yu, H.Q., 2006. Continuous production of hydrogen from mixed volatile fatty acids with Rhodopseudomonas capsulata. Int. J. Hydrogen Energy 31, 1641–1647.

Shi, X.Y., Jin, D.W., Sun, Q.Y., Li, W.W., 2010. Optimization of conditions for hydrogen production from brewery wastewater by anaerobic sludge using desirability function approach. Int. J. Hydrogen Energy 35 (7), 1493–1498.

Show, K.Y., Zhang, Z.P., Tay, J.H., Liang, D.T., Lee, D.J., Ren, N., et al., 2010. Critical assessment of anaerobic processes for continuous biohydrogen production from organic wastewater. Int. J. Hydrogen Energy 35 (24), 13350–13355.

Sivaramakrishna, D., Sreekanth, D., Himabindu, V., Anjaneyulu, Y., 2009. Biological hydrogen production from probiotic wastewater as substrate by selectively enriched anaerobic mixed microflora. Renew. Energy 34, 937–940.

Sreethawong, T., Chatsiriwatana, S., Rangsunvigit, P., Chavadej, S., 2010. Hydrogen production from cassava wastewater using an anaerobic sequencing batch reactor: effects of operational parameters, COD:N ratio, and organic acid composition. Int. J. Hydrogen Energy 35 (9), 4092–4102.

Srikanth, S., Venkata Mohan, S., Prathima Devi, M., Lenin Babu, M., Sarma, P.N., 2008. Effluents with soluble metabolites generated from acidogenic and methanogenic processes as substrate for additional hydrogen production through photo-biological process. Int. J. Hydrogen Energy 34, 1771–1779.

Srikanth, S., Venkata Mohan, S., Prathima Devi, M., Dinikar, P., Sarma, P.N., 2009. Acetate and butyrate as substrates for hydrogen production through photo-fermentation using mixed culture: optimization of process parameters and combined process evaluation. Int. J. Hydrogen Energy 34, 7513–7522.

Srikanth, S., Venkata Mohan, S., Sarma, P.N., 2010a. Positive anodic poised potential regulates microbial fuel cell performance with the function of open and closed circuitry. Bioresour. Technol. 101, 5337–5344.

Srikanth, S., Venkata Mohan, S., Lalit Babu, V., Sarma, P.N., 2010b. Metabolic shift and electron discharge pattern of anaerobic consortia as a function of pretreatment method applied during fermentative hydrogen production. Int. J. Hydrogen Energy 35, 10693–10700.

Tai, J., Adav, S.S., Su, A., Lee, D.J., 2010. Biological hydrogen production from phenol-containing wastewater using Clostridium butyricum. Int. J. Hydrogen Energy 35 (24), 13345–13349.

Tang, G.L., Huang, J., Sun, Z.J., Tang, Q.Q., Yan, C.H., Liu, G.Q., et al., 2008. Biohydrogen production from cattle wastewater by enriched anaerobic mixed consortia: influence of fermentation temperature and pH. J. Biosci. Bioeng. 106, 80–87.

Tanisho, S., Kuromoto, M., Kadokura, N., 1998. Effect of CO_2 removal on hydrogen production by fermentation. Int. J. Hydrogen Energy 23, 559–563.

van Ginkel, S., Sung, S.W., Lay, J.J., 2001. Biohydrogen production as a function of pH and substrate concentration. Environ. Sci. Technol. 35, 4726–4730.

Van Ginkel, S.W., Oh, S., Logan, B.E., 2005. Biohydrogen gas production from food processing and domestic wastewaters. Int. J. Hydrogen Energy 30, 1535–1542.

van Groenestijn, J., Hazewinkel, J., Nienoord, M., Bussmann, P., 2002. Energy aspects of biological hydrogen production in high rate bioreactors operated in the thermophilic temperature range. Int. J. Hydrogen Energy 27 (11), 1141–1147.

Van Loosdrecht, M.C.M., Pot, M.A., Heijnen, J.J., 1997. Importance of bacterial storage polymers in bioprocesses. Wat. Sci. Technol. 35 (1), 41–47.

Vatsala, T.M., Mohan Raj, S., Manimaran, A., 2008. A pilot-scale study of biohydrogen production from distillery effluent using defined bacterial co-culture. Int. J. Hydrogen Energy 33, 5404–5415.

Venkata Mohan, S., 2008. Fermentative hydrogen production with simultaneous wastewater treatment: influence of pretreatment and system operating conditions. J. Scie. Ind. Res. 67 (11), 950–961.

Venkata Mohan, S., 2009. Harnessing of biohydrogen from wastewater treatment using mixed fermentative consortia: process evaluation towards optimization. Int. J. Hydrogen Energy 34, 7460–7474.

Venkata Mohan, S., 2010. Waste to renewable energy: a sustainable and green approach towards production of biohydrogen by acidogenic fermentation. In: Om, S., Steve, H. (Eds.), Sustainable Biotechnology: Renewable Resources and New Perspectives. Springer, pp. 129–164.

Venkata Mohan, S., Chandrasekhara Rao, N., Prasad, K.K., Sarma, P.N., 2005. Bioaugmentation of anaerobic sequencing batch biofilm reactor (ASBBR) with immobilized sulphate reducing bacteria (SRB) for treating sulfate bearing chemical wastewater. Proc. Biochem. 40 (8), 2849–2857.

Venkata Mohan, S., Lalit Babu, V., Sarma, P.N., 2007a. Anaerobic biohydrogen production from dairy wastewater treatment in sequencing batch reactor (AnSBR): effect of organic loading rate. Enz. Microbial. Technol. 41 (4), 506–515.

Venkata Mohan, S., Bhaskar, Y.V., Sarma, P.N., 2007b. Biohydrogen production from chemical wastewater treatment by selectively enriched anaerobic mixed consortia in biofilm configured reactor operated in periodic discontinuous batch mode. Water. Res. 41, 2652–2664.

Venkata Mohan, S., Mohanakrishna, G., Veer Raghuvulu, S., Sarma, P.N., 2007c. Enhancing biohydrogen production from chemical wastewater treatment in anaerobic sequencing batch biofilm reactor (AnSBBR) by bioaugmenting with selectively enriched kanamycin resistant anaerobic mixed consortia. Int. J. Hydrogen Energy 32, 3284–3292.

Venkata Mohan, S., Bhaskar, Y.V., Krishna, T.M., Chandrasekhara Rao, N., Lalit Babu, V., Sarma, P.N., 2007d. Biohydrogen production from chemical wastewater as substrate by selectively enriched anaerobic mixed consortia: influence of fermentation pH and substrate composition. Int. J. Hydrogen Energy 32, 2286–2295.

Venkata Mohan, S., Lalit Babu, V., Srikanth, S., Sarma, P.N., 2008a. Bio-electrochemical evaluation of fermentative hydrogen production process with the function of feeding pH. Int. J. Hydrogen Energy 33 (17), 4533–4546.

Venkata Mohan, S., Mohankrishna, G., Sarma, P.N., 2008b. Integration of acidogenic and methanogenic processes for simultaneous production of biohydrogen and methane from wastewater treatment. Int. J. Hydrogen Energy 33, 2156–2166.

Venkata Mohan, S., Srikanth, S., Dinakar, P., Sarma, P.N., 2008c. Photo-biological hydrogen production by the adopted mixed culture: data enveloping analysis. Int. J. Hydrogen Energy 33 (2), 559–569.

Venkata Mohan, S., Mohanakrishna, G., Reddy, S.S., Raju, B.D., Rao, K.S., Sarma, P.N., 2008d. Self-immobilization of acidogenic mixed consortia on mesoporous material (SBA-15) and activated carbon to enhance fermentative hydrogen production. Int. J. Hydrogen Energy 33, 6133–6142.

Venkata Mohan, S., Lalit Babu, V., Sarma, P.N., 2008e. Effect of various pretreatment methods on anaerobic mixed microflora to enhance biohydrogen production utilizing dairy wastewater as substrate. Bioresour. Technol. 99, 59–67.

Venkata Mohan, S., Mohanakrishna, G., Ramanaiah, S.V., Sarma, P.N., 2008f. Simultaneous biohydrogen production and wastewater treatment in biofilm configured anaerobic periodic discontinuous batch reactor using distillery wastewater. Int. J. Hydrogen Energy 33 (2), 550–558.

Venkata Mohan, S., Lenin Babu, M., Mohanakrishna, G., Sarma, P.N., 2009a. Harnessing of biohydrogen by acidogenic fermentation of Citrus limetta peelings: effect of extraction procedure and pretreatment of biocatalyst. Int. J. Hydrogen Energy 34 (15), 6149–6156.

Venkata Mohan, S., Veer Raghuvulu, S., Mohanakrishna, G., Srikanth, S., Sarma, P.N., 2009b. Optimization and evaluation of fermentative hydrogen production and wastewater treatment processes using data enveloping analysis (DEA) and Taguchi design of experimental (DOE) methodology. Int. J. Hydrogen Energy 34, 216–226.

Venkata Mohan, S., Srikanth, S., Lenin Babu, M., Sarma, P.N., 2010a. Insight into the dehydrogenase catalyzed redox reactions and electron discharge pattern during fermentative hydrogen production. Bioresour. Technol. 101, 1826–1833.

Venkata Mohan, S., Venkateswar Reddy, M., Venkata Subhash, G., Sarma, P.N., 2010b. Fermentative effluents from hydrogen producing bioreactor as substrate for poly (β-OH) butyrate production with simultaneous treatment: an integrated approach. Bioresour. Technol. doi:10.1016/j.biortech.2010.06, 109.

Venkata Mohan, S., Mohanakrishna, G., Chiranjeevi, P., Dinakar, P., Sarma, P.N., 2010c. Ecologically engineered system (EES) designed to integrate floating, emergent and submerged macrophytes for the treatment of domestic sewage and acid rich fermented-distillery wastewater: evaluation of long term performance. Bioresour. Technol. 101, 3363–3370.

Venkata Mohan, S., Veer Raghavulu, S., Kannaiah Goud, R., Srikanth, S., Lalit Babu, V., Sarma, P.N., 2010d. Microbial diversity analysis of long term operated biofilm configured anaerobic reactor producing hydrogen from wastewater under diverse conditions. Int. J. Hydrogen Energy 35, 12208–12215.

Venkata Mohan, S., Lenin Babu, M., 2011. Dehydrogenase activity in association with poised potential during biohydrogen production in single chamber microbial electrolysis cell. Bioresour. Technol. 102 (12), 6637–6782.

Vijaya Bhaskar, Y., Venkata Mohan, S., Sarma, P.N., 2008. Effect of substrate loading rate of chemical wastewater on fermentative biohydrogen production in biofilm configured sequencing batch reactor. Bioresour. Technol. 99, 6941–6948.

Vijayaraghavan, K., Ahmad, D., 2006. Biohydrogen generation from palm oil mill effluent using anaerobic contact filter. Int. J. Hydrogen Energy 31, 1284–1291.

Wagner, R.C., Regan, J.M., Oh, S.E., Zuo, Y., Logan, B.E., 2009. Hydrogen and methane production from swine wastewater using microbial electrolysis cells. Wat. Res. 43 (5), 1480–1488.

Wang, J.L., Wan, W., 2008a. Comparison of different pretreatment methods for enriching hydrogen-producing cultures from digested sludge. Int. J. Hydrogen Energy 33, 2934–2941.

Wang, J., Wan, W., 2008b. Influence of Ni^{2+} concentration on biohydrogen production. Bioresour. Technol. 99, 8864–8868.

Wang, J., Wan, W., 2009. Factors influencing fermentative hydrogen production: a review. Int. J. Hydrogen Energy 34, 799–811.

Wang, B., Wan, W., Wang, J., 2009. Effect of ammonia concentration on fermentative hydrogen production by mixed cultures. Biores. Technol 100, 1211–1213.

Winkler, M., Hemschemeier, A., Gotor, C., Melis, A., Happer, T., 2002. [Fe]- hydrogenases in green algae: photo-fermentation and hydrogen evolution under sulfur deprivation. Int. J. Hydrogen Energy 27, 1431–1439.

Wu, T.Y., Mohammad, A.W., Jahim, J.M.d., Anuar, N., 2009. A holistic approach to managing palm oil mill effluent (POME): biotechnological advances in the sustainable reuse of POME. Biotechnol. Adv. 27, 40–52.

Wu, X., Yao, W., Zhu, J., 2010. Effect of pH on continuous biohydrogen production from liquid swine manure with glucose supplement using an anaerobic sequencing batch reactor. Int. J. Hydrogen Energy 35 (13), 6592–6599.

Yang, H., Shao, P., Lu, T., Shen, J., Wang, D., Xu, Z., et al., 2006. Continuous bio-hydrogen production from citric acid wastewater via facultative anaerobic bacteria. Int. J. Hydrogen Energy 31, 1306–1313.

Yang, P., Zhang, R., McGarvey, J.A., Benemann, J.R., 2007. Biohydrogen production from cheese processing wastewater by anaerobic fermentation using mixed microbial communities. Int. J. Hydrogen Energy 32, 4761–4771.

Yetis, M., Gündüz, U., Eroğlu, İ., Yücel, M., Türker, L., 2000. Photoproduction of hydrogen from sugar refinery wastewater by Rhodobacter sphaeroides O.U.001. Int. J. Hydro. Energy. 25, 1035–1041.

Yokoi, H., Tokushige, T., Hirose, J., Hayashi, S., Takasaki, Y., 1997. Hydrogen production by immobilized cells of aciduric Enterobacter aerogenes strain HO-39. J. Ferment. Bioeng. 83, 481–484.

Yokoyama, H., Ohmori, H., Waki, M., Ogino, A., Tanaka, Y., 2009. Continuous hydrogen production from glucose by using extreme thermophilic anaerobic microflora. J. Biosci. Bioeng. 107, 64–66.

Yu, H., Zhu, Z., Hu, W., Zhang, H., 2002. Hydrogen production from rice winery wastewater in an upflow anaerobic reactor by using mixed anaerobic cultures. Int. J. Hydrogen Energy 27, 1359–1365.

Zhang, T., Liu, H., Fang, H.H.P., 2003. Biohydrogen production from starch in wastewater under thermophilic condition. J. Environ. Manag. 69, 149–156.

Zheng, G.H., Wang, L., Kang, Z.H., 2010. Feasibility of biohydrogen production from tofu wastewater with glutamine auxotrophic mutant of Rhodobacter sphaeroides. Ren. Energy. 35, 2910–2913.

Zhu, H., Beland, M., 2006. Evaluation of alternative methods of preparing hydrogen producing seeds from digested wastewater sludge. Int. J. Hydrogen Energy 31, 1980–1988.

Zhu, H., Ueda, S., Asada, Y., Miyake, J., 2002. Hydrogen production as a novel process of wastewater treatment—studies on tofu wastewater with entrapped R. sphaeroides and mutagenesis. Int. J. Hydrogen Energy 27, 1349–1357.

Zhu, H., Parker, W., Basnar, R., Proracki, A., Falletta, P., Béland, M., et al., 2009a. Buffer requirements for enhanced hydrogen production in acidogenic digestion of food wastes. Bioresour. Technol. 100 (21), 5097–5102.

Zhu, J., Li, Y., Wu, X., Miller, C., Chen, P., Ruan, R., 2009b. Swine manure fermentation for hydrogen production. Bioresour. Technol. 100 (22), 5472–5477.

Zhu, Z., Shi, J., Zhou, Z., Hu, F., Bao, J., 2010. Photo-fermentation of Rhodobacter sphaeroides for hydrogen production using lignocellulose-derived organic acids. Proc. Biochem. 45 (12), 1894–1898.

CHAPTER

23

Thermophilic Biohydrogen Production

Dimitar Karakashev, Irini Angelidaki*

Department of Environmental Engineering, Technical University of Denmark, Lyngby, Denmark

***Corresponding author: E-mail: dbka@env.dtu.dk**

1 BACKGROUND

In recent years, hydrogen gas is attracting widespread attention as a clean and environment-friendly fuel that produces water when combusted. The future energy economy will have an important role for hydrogen as a clean, CO_2-neutral energy carrier.

Hydrogen can be produced by both biological and nonbiological methods. The main industrial process to produce H_2 consists in steam reforming from natural gas and petroleum, a process which depends on fossil fuels and thus is not CO_2 neutral. Another source is electrolysis of seawater, which could be sustainable if electricity is generated from renewable resources, such as from windmill electricity. An alternative way to circumvent the dependence of H_2 production from fossil fuels is to utilize the potential of H_2-producing microorganisms to derive hydrogen from widely available biomass as a renewable energy source (Lee et al., 2010). Currently, hydrogen is applicable in fuel cells.

Biohydrogen production can be realized by microorganisms using carbohydrate-rich and nontoxic raw materials (Kapdan and Kargi, 2006). Among the various processes leading to biohydrogen production, direct and indirect biophotolysis, hemoheterotrophic (dark) fermentation, photoheterotrophic (light-driven) fermentation, and *in vitro* enzymatic conversion of biomass are important. Currently, dark fermentation is the most feasible process for biohydrogen production from renewable biomass due to its higher rate of hydrogen evolution in the absence of any light sources as well as the versatility of the substrates used. However, the key issue that still needs to be addressed includes the much lower hydrogen yields (up to 2.5-2.9 mol H_2/mol glucose) compared to the theoretical yield of 4 mol H_2/mol glucose for fermentation with only acetate as liquid end fermentation product. One of the reasons for the low hydrogen yield is that in many microorganisms the actual yields are reduced by hydrogen recycling mechanisms due to the presence of one or more uptake hydrogenases, which consume part of the produced hydrogen (Hallenbeck and Benemann, 2002).

To reach higher rates and yields in biohydrogen production, fermentation under high temperatures (50-80 °C) with application of thermophilic and extreme thermophilic microorganisms is a very promising alternative. Higher temperatures thermodynamically favor hydrogen production. Besides, elevated temperatures contribute to better pathogenic destruction and limit hydrogen consumption by hydrogen consumers (methanogens, homoacetogens, sulfate reducers). Furthermore, certain thermophilic bacteria have the ability to produce ethanol simultaneously with hydrogen (Zhao et al., 2009). Ethanol is also a renewable biofuel, and its coproduction can increase the benefits from future commercial exploitation of those thermophilic/extreme thermophilic organisms.

This review describes the thermophilic biohydrogen production process, focusing on thermodynamics, metabolic pathways, and the microorganisms involved. The effect of operational conditions (pH, loading rate, retention time, concentrations of dissolved hydrogen and carbon dioxide, soluble metabolic profile, and SMP) on process performance as well as some practical aspects such as feedstock used and reactor technologies applied, is also given, with respect to challenges, possibilities, and future perspectives of the process.

2 THERMODYNAMIC ASPECTS

Stoichiometrically, the maximum hydrogen yield for complete conversion of glucose to H_2 and CO_2 is 12 mol H_2/mol glucose (Equation 1). However, according to standard (25 °C) Gibbs free energy of reactions (Equations 1–4), production of 12 mol of hydrogen (Equation 1) is thermodynamically unfavorable. From the thermodynamic point of view, the most favorable conversion is with butyrate as end fermentation product (Equation 3), followed by mixed propionate-acetate-type fermentation (reaction 4) and acetate fermentation (Equation 2). Meanwhile, from the practical point of view, the most desirable is acetate fermentation (Equation 2), whereby 4 mol H_2/mol glucose can be obtained. However, this stoichiometric yield is only attainable under near equilibrium conditions, which implies very slow rates and/or at very low partial pressure of hydrogen (Hallenbeck and Benemann, 2002).

$$C_6H_{12}O_6 + 12H_2O \rightarrow 6HCO_3^- + 12H_2 + 6H^+ \Delta G^{o\prime} = 241 \text{ kJ mol}^{-1}, \quad (1)$$

$$C_6H_{12}O_6 + 4H_2O \rightarrow 2CH_3COO^- + 2HCO_3^- + 4H_2 + 4H^+ \Delta G^{o\prime} = -48 \text{ kJ mol}^{-1}, \quad (2)$$

$$C_6H_{12}O_6 + 2H_2O \rightarrow CH_3CH_2CH_2COO^- + 2HCO_3^- + 2H_2 + 3H^+ \Delta G^{o\prime} = -137 \text{ kJ mol}^{-1}, \quad (3)$$

$$C_6H_{12}O_6 + 3H_2O \rightarrow CH_3CH_2OH + CH_3COO^- + 2HCO_3^- + 2H_2 + 3H^+ \Delta G^{o\prime} = -97 \text{ kJ mol}^{-1}. \quad (4)$$

A possible strategy to increase the H_2 yield is to increase the cultivation temperature and hence decrease the Gibbs free energy of the conversion process according to the second law of thermodynamics (Equation 5):

$$\Delta G = \Delta H - T\Delta S, \quad (5)$$

where ΔG is the change in Gibbs free energy, ΔH is the change in enthalpy, T is the absolute temperature, and ΔS is the change in entropy.

3 BIOCHEMICAL PATHWAYS AND MICROBIOLOGY

The total hydrogen yields from the different anaerobic glucose degradation pathways are shown in Table 1. The biochemistry of the anaerobic glucose catabolism shows that the production of propionate, ethanol, and lactate is associated with absence of hydrogen generation. In practice, highest hydrogen yields are associated with production of acetate, or a mixture of acetate and butyrate, and low hydrogen yields—with production of different combinations of butyrate, acetate, ethanol, propionate, and lactate as the main liquid fermentation end products.

The main biochemical pathways involved in dark fermentative conversion of simplest sugar-glucose to hydrogen and different end products are presented in Figure 1.

Fermentative hydrogen production is usually carried out under mesophilic or thermophilic conditions by a wide variety of microorganisms such as strict anaerobes (clostridia, ruminococci, and archae), facultative anaerobes (*Escherichia coli* and *Enterobacter aerogenes*), and facultative bacteria (*Alcaligenes eutrophus* and *Bacillus licheniformis*), when kept under anoxic conditions. Recently, attempts for improving biohydrogen production using thermophilic (50-60 °C) and extreme thermophilic (60-80 °C) bacteria and archae have been made. Among pure cultures of thermophiles and extreme thermophiles—*Caldicellulosiruptor* sp., *Thermoanaerobacter* sp., *Thermoanaerobacterium* sp., *Thermotoga* sp., *Pyrococcus* sp., *Desulfotomaculum geothermicum*, *Clostridium thermocellum* (Bartacek et al., 2007; Kapdan and Kargi, 2006; O-Thong et al., 2008a,b,c; Van Niel et al., 2002; Willquist et al., 2009)—studied for dark fermentative hydrogen production, the extreme thermophiles (70 °C) *Caldicellulosiruptor saccharolyticus* and *Thermotoga elfii* (Van Niel et al., 2002) appear to be most promising as they display superior hydrogen yields

TABLE 1 Fermentation End Products and Hydrogen Yields from the Main Anaerobic Glucose Degradation Pathways

Liquid Fermentation End Product(s)	Equation for Anaerobic Glucose Degradation Pathway	Theoretical Hydrogen Yield (mol H_2/mol glucose)
Acetic acid CH_3COOH	$C_6H_{12}O_6 + 2H_2O \rightarrow 2CH_3COOH + 4H_2 + 2CO_2$	4
Butyric acid $CH_3CH_2CH_2COOH$	$C_6H_{12}O_6 \rightarrow CH_3CH_2CH_2COOH + 2H_2 + 2CO_2$	2
Butyric acid $CH_3CH_2CH_2COOH$ + acetic acid CH_3COOH	$4C_6H_{12}O_6 + 2H_2O \rightarrow 3CH_3CH_2CH_2COOH + 2CH_3COOH + 10H_2 + 8CO_2$	2.5
Ethanol CH_3CH_2OH + acetic acid CH_3COOH	$C_6H_{12}O_6 + 2H_2O \rightarrow CH_3CH_2OH + CH_3COOH + 2H_2 + 2CO_2$	2
Propionic acid CH_3CH_2COOH	$C_6H_{12}O_6 + H_2 \rightarrow 2CH_3CH_2COOH + 2H_2O$	0
Ethanol CH_3CH_2OH	$C_6H_{12}O_6 \rightarrow 2CH_3CH_2OH + 2CO_2$	0
Lactic acid $C_3H_3O_3$	$C_6H_{12}O_6 \rightarrow C_3H_3O_3$	0

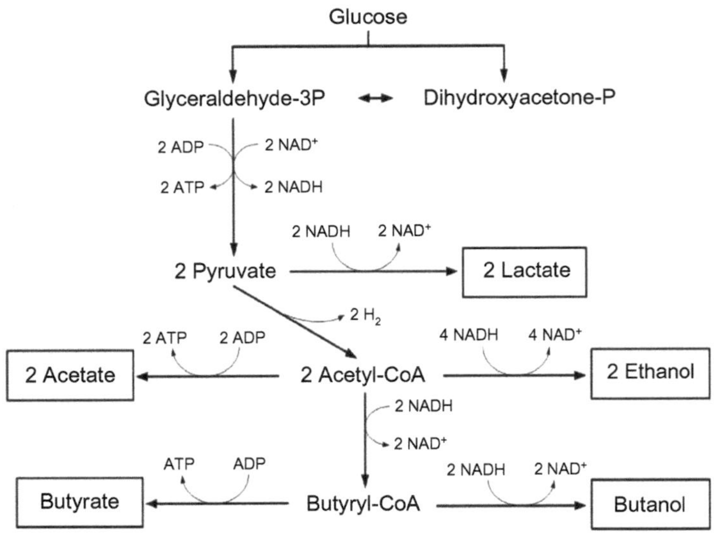

FIGURE 1 Schematic pathways for glucose conversion to hydrogen/other products via dark fermentation *(modified from Nath and Das, 2004).*

up to 3.3 mol H_2/mol hexose. In addition, these microorganisms can utilize a wide spectrum of carbon sources ranging from simple sugars (pentoses and hexoses) to more complex carbohydrates such as starch and cellulose. However, the major drawback of using these microorganisms is their large dependence on growth factor sources such as yeast extract.

Some hyperthermophiles (optimal growth at 85 °C) such as the archaeon *Thermococcus kodakaraensis* were also demonstrated to produce hydrogen (Kanai et al., 2005). Recently, mixed thermophilic cultures were also investigated with respect to their hydrogen-producing potential (Karakashev et al., 2009; Kongjan et al., 2010; Zhao et al., 2009, 2010). However, the results obtained with those mixed cultures are not comparable with those obtained by *C. saccharolyticus* or *T. elfii* (with respect to hydrogen yield). Additional attempts are required to optimize the dark fermentative thermophilic biohydrogen process with mixed cultures to make them commercially attractive.

4 EFFECT OF PROCESS CONDITIONS

The major factors influencing dark fermentative biohydrogen production are organic loading rate (OLR), pH, hydraulic retention time (HRT), dissolved hydrogen and dissolved carbon dioxide concentrations, and SMP (Table 2).

pH is a factor affecting fermentative hydrogen production to a great extent. There are many contradictory reports regarding the optimal pH for the biohydrogen production process. In general, the optimal pH for growth of fermentative hydrogen producers is around 7. However, due to fact that hydrogen producers face strong competition from hydrogen consumers (mainly methanogens) growing best at pH 7, the optimal pH of fermentative

TABLE 2 Main Factors Affecting Thermophilic Biohydrogen Production

Factor	Effect(s)	References
OLR	Affects cell metabolism High OLR results in substrate inhibition	Kraemer and Bagley (2007) and Ren et al. (2007)
pH	Affects cell metabolism-cell membrane charge, membrane transport, enzymatic activities; extreme pH stimulates spore formation	Lee et al. (2002) and Ren et al. (2007)
HRT	Too low HRT-microbial washout and substrate inhibition	Zhang et al. (2008a,b) and Hawkes et al. (2002)
Partial pressure of H_2 (pH_2)	Affects fermentation metabolism; increased pH_2 inhibits hydrogen production and leads to formation of reduced end products (ethanol, lactate)	Kengen et al. (2009) and Soboh et al. (2004)
Partial pressure of CO_2 (pCO_2)	Affects the activity of homoacetogens and methanogens	Willquist et al. (2009) and Kim et al. (2006)
Soluble metabolic profile (SMP)	Affects fermentation metabolism-end products inhibition, high organic acid concentration can result in cell lysis	Chin et al. (2003) and van Ginkel and Logan (2005)

hydrogen production process was considered to be between 5.0 and 6.0 (Cai et al., 2004). However, other studies (Lee et al., 2002; Pikuta and Hooevr, 2004) reported an unusual optimal pH for the fermentation process (around 9.00-9.5). Those findings suggest that optimal pH value largely depends on the community composition of the original microbial consortium used as inoculum—whether it is enriched by acidophilic/acidotolerant or alkalophylic/alkalotolerant hydrogen producers.

HRT and OLR are the main optimization parameters for thermophilic biohydrogen production (Hawkes et al., 2002; Kraemer and Bagley, 2007). Generally, short HRT was considered to facilitate fermentative biohydrogen production with optimal HRT around 0.25 d^{-1} due to the fact that some hydrogen consumers (slowly growing methanogens) are washed out under those conditions. On the other hand, high OLR (low HRT) can result in substrate inhibition. "Shock" loading reduces hydrogen production through accumulation of organic acids (pH decrease), metabolic inhibition, and/or increased dissolved hydrogen concentration.

Dissolved hydrogen concentration is another factor influencing thermophilic biohydrogen production. Since dissolved hydrogen concentration is difficult to monitor, often H_2 partial pressure (pH_2) is used as a parameter to approximate the dissolved hydrogen concentration. This is, however, often inaccurate, as the process is usually not in equilibrium (Van Niel et al., 2003). Since hydrogen is known to have an inhibitory effect on growth and its own production in a variety of thermophiles (Kengen et al., 2009; Soboh et al., 2004), maximizing fermentative hydrogen yield is only possible when pH_2 is kept sufficiently low in the closed fermentation system. Normally, this can be achieved by flushing off the produced hydrogen with an inert gas such as N_2 or He (Kraemer and Bagley, 2007). As the inert gases are expensive, use of CO_2 might be a cheaper alternative since it is a by-product from the fermentation process. However, stripping with CO_2 has some disadvantages as discussed later.

Carbon dioxide may also affect thermophilic hydrogen production (Willquist et al., 2009). Although stripping with carbon dioxide was employed for hydrogen removal from gas phase and subsequent increase of hydrogen yields (Kraemer and Bagley, 2007), there is one major complication with this approach. Elevated carbon dioxide partial pressure (pCO_2) might inhibit hydrogen production as it triggers homoacetogenic reaction, resulting in increased acetate levels (Equation 6) and finally acidification:

$$4H_2 + 2CO_2 \rightarrow CH_3COOH + 2H_2O. \qquad (6)$$

SMP and more specifically organic acid (acetate and butyrate) levels could have a significant effect on the fermentative metabolism. Those acids in their undissociated form can pass the cell membrane, dissociate within a cell, release a proton, and finally uncouple the proton motive force across the cell membrane, which can result in metabolic inhibition and cell lysis. At high concentrations, organic acids can decrease the cell growth rate (Chin et al., 2003) and cause a metabolic shift, from hydrogen production (acetate or butyrate pathway) to propionate or solvent (ethanol, butanol) synthesis (Van Niel et al., 2003).

5 PRACTICAL APPLICATIONS

Practical application of thermophilic biohydrogen production depends on the microorganisms employed, feedstock (substrates), and process technology (operational conditions such as temperature and pH, fermentation mode, and reactor type applied) utilized (Table 3).

In comparison to mesophilic pure cultures, utilization of pure thermophilic (f.ex. *Thermoanaerobacterium thermosaccharolyticum*) or extreme thermophilic (*T. elfii, C. saccharolyticus*) cultures for biohydrogen production is more sustainable as those cultures are more resistant to contaminations by traditional mesophilic hydrogen scavengers such as methanogens and sulfate reducers. However, application of pure cultures (although they have the highest hydrogen yields) is still limited to fermentation of defined substrates, mainly sugars that are usually sterilized. When nonsterile waste materials are used as feedstock, pure cultures would face strong competition with the complex microflora of those wastes. In such cases, application of mixed thermophilic or extreme thermophilic cultures (Table 3) will be a more appropriate choice as, generally, mixed cultures are more robust to process imbalances and stress situations inside the reactor.

The substrates (Table 3) utilized for thermophilic biohydrogen production can be divided into the following groups:

- Pure substrates (glucose, xylose, arabinose, cellulose, starch)
- Industrial wastes and wastewaters from agriculture, food, sugar, pulp, and paper-processing industry (cow waste slurry, palm oil mill effluent, rice winery wastewater, wheat straw hydrolysate, paper sludge hydrolysate)

Among the different reactor technologies studied on lab scale—anaerobic sequencing batch (ASBR), continuously stirred tank (CSTR), membrane bioreactor (MBR), and upflow anaerobic sludge blanket (UASB) reactors—CSTR and UASB were the most widely used. The highest thermophilic hydrogen production rate of 199 mmol H_2/L/d was obtained with mixed cultures in CSTR operating at 60 °C and pH 5.0 (Ueno et al., 2006). With respect to

TABLE 3 Thermophilic Biohydrogen Production Processes with Different Feedstocks: Defined and Nondefined Carbon Sources (Organic Waste Materials)

	Conditions				Hydrogen Production		
Microorganism(s)	**Fermentation Mode**	***T* (°C)**	**Initial pH**	**Feedstock**	**Rate (mmol H_2/L/h)**	**Yield (mol H_2/mol hexose)**	**Reference**
Mixed culture	Batch	55	5.5	Food waste	NA	1.8	Shin et al. (2004)
Mixed culture	Batch	70	7.0	Glucose	NA	2.4	Zhao et al. (2009)
Mixed culture	Batch	51	6.5	Glucose	NA	1.52	Karadag et al. (2009)
Caldicellulosiruptor saccharolyticus, *Thermotoga elfii*	Batch	70 (*C. saccharolyticus*), 65 (*T. elfii*)	7 (*C. saccharolyticus*), 7.4 (*T. elfii*)	Sucrose (*C. saccharolyticus*), glucose (*T. elfii*)	NA	3.3	Van Niel et al. (2002)
Mixed culture	Batch	55	NA	Palm oil mill effluent	98	1.72	Ismail et al. (2010)
Thermotoga elfii, *Caldicellulosiruptor saccharolyticus*	Batch	65 (*T. elfii*), 70 (*C. saccharolyticus*)	NA	Paper sludge hydrolysate	0.3 (*T. elfii*), 0.25 (*C. saccharolyticus*)	NA	Kadar et al. (2003)
Mixed culture	Batch	60, 75	7.0	Cow waste slurry	30.0	NA	Yokoyama et al. (2007)
Mixed culture	Continuous (UASB)	55	5.5	Rice winery wastewater	6.56	2.14[a]	Yu et al. (2002)

Continued

TABLE 3 Thermophilic Biohydrogen Production Processes with Different Feedstocks: Defined and Nondefined Carbon Sources (Organic Waste Materials)—Cont'd

	Conditions				Hydrogen Production		
Microorganism(s)	**Fermentation Mode**	***T* (°C)**	**Initial pH**	**Feedstock**	**Rate (mmol H_2/L/h)**	**Yield (mol H_2/mol hexose)**	**Reference**
Thermoanaerobacterium thermosaccharolyticum	Continuous (UASB)	60	5.5	Sucrose	152	1.7	O-Thong et al. (2008b)
Mixed culture	Continuous (CSTR)	70	NA	Wheat straw hydrolysate	8.2	NA	Kongjan et al. (2010)
Mixed culture	Continuous (UASB)	70	4.5	Glucose	2.3	2.5	Kotsopoulos et al. (2006)
Mixed culture	Continuous (ASBR)	55	5.5	Palm oil mill effluent	2.6	2.24	O-Thong et al. (2007)
Mixed culture	Continuous (CSTR)	60	5.0	Artificial garbage slurry containing paper (AGSP)	199	NA	Ueno et al. (2006)
Mixed culture	Continuous (CSTR)	55	NA	Olive pulp	0.58	NA	Gavala et al. (2005)
Mixed culture	Continuous (CSTR)	55	5.5	Food waste	1.7	2.2	Shin and Youn (2005)
Mixed culture	Continuous (MBR)	Thermophilic	5.5	Glucose	48	NA	Oh et al. (2004)

NA, not available.
[a]*Calculated based on the information provided.*

hydrogen yield from thermophilic mixed cultures, the highest value of 2.5 mol H_2/mol glucose was obtained in UASB operating at 70 °C and pH 4.5 (Kotsopoulos et al., 2006).

Recently, a very promising approach for enhancement of fermentative hydrogen production based on nanoparticle addition was developed (Zhang and Shen, 2007). However, this investigation was performed only under mesophilic conditions. More detailed investigations are required to clarify the technological and economical feasibility of this approach at higher temperatures.

Although several efforts have been made to improve biohydrogen reactor performance, full-scale biohydrogen production has not been developed yet. However, some pilot-scale applications have emerged in the recent years (Ren et al., 2006). An anticipated disadvantage of large-scale hydrogen production that needs to be addressed during scale-up is the escape of hydrogen through large plastic enclosures and thin metal sheets that might occur due to the high diffusivity of hydrogen. However, as hydrogen becomes a more important fuel, full-scale applications will emerge in the future.

6 CHALLENGES, POSSIBILITIES, AND FUTURE PERSPECTIVES

The ultimate goal, and challenge, for fermentative hydrogen research and development focuses essentially on attaining higher yields of hydrogen. Significant improvement can be expected through rapid gas removal and separation, optimized bioreactor design, and genetic modifications in the microorganisms. Nonetheless, it is rather difficult to predict which of the various approaches will ultimately succeed in substantial enhancement of hydrogen yields so that the dark fermentation process of hydrogen generation becomes commercially competitive.

To increase the economic feasibility of the process, recent attempts have concentrated on further biological processing (second stage) of the organic acids-rich effluents from dark fermentation hydrogen production. Two possible second stages (Figure 2) were investigated: photoheterotrophic hydrogen production (Hawkes et al., 2007) and methanogenesis

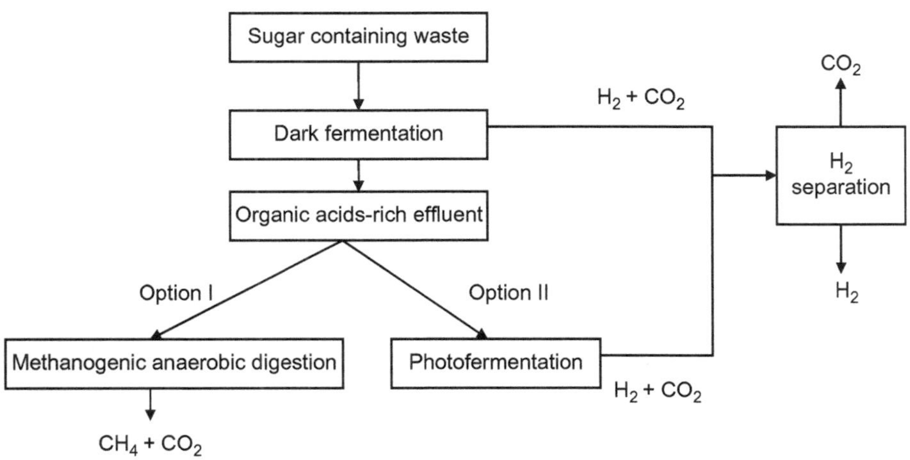

FIGURE 2 Possible treatments of the effluent from dark fermentative biohydrogen production.

(Liu et al., 2006). If the technologically effective and cost-effective photobioreactors were available, the two-stage process combining dark and light-driven hydrogen fermentation would be a very promising method as it has a theoretical maximal molar yield of 12 mol H_2/mol hexose converted in the two-stage process (Hawkes et al., 2007). However, some studies indicate that photofermentation is a very inefficient and expensive process with respect to high energy demands for light sources and requirement for elaborate photobioreactors covering large areas (Hallenbeck and Benemann, 2002). There are good indications that a two-stage process with an acidifying hydrogen-producing first stage and a methanogenic second stage gives rise to more efficient waste treatment and energy recovery than a single-stage methanogenic process (Liu et al., 2006). This process could easily be implemented in existing and new biogas plants, where an additional reactor could be added before the traditional biogas reactor for production of biohydrogen. As a powerful combustion stimulant for accelerating methane combustion, a mixture of 20% hydrogen and 80% methane known as hythane (The Hythane System) will reduce the emission of CO, CO_2, and NO_x of natural gas powered vehicles and increase the efficiency of internal combustion engines.

It is likely that the food industry and kitchen wastes will initially prove most attractive as substrates for biohydrogen production at elevated temperature conditions. Reactor capital and operating costs are likely to be similar to those already well known for anaerobic digestion. The next major challenge is to determine whether the economics and reliability of dark fermentative hydrogen production are sufficiently attractive for commercial production.

Acknowledgments

This study received support from Danish Agency for Science, Technology, and Innovation under Bio REF. Project No. 2104-06-0004 and from Danish Council for Strategic Research under Project No. 2101-09-0135 "High rate algal biomass production for food, biochemicals and biofuels."

References

Bartacek, J., Zabranska, J., Lens, P.N.L., 2007. Developments and constraints in fermentative hydrogen production. Biofuels Bioprod. Bioref. 1, 201–214.

Cai, M., Liu, J., Wei, Y., 2004. Enhanced biohydrogen production from sewage sludge with alkaline pretreatment. Environ. Sci. Technol. 38, 3195–3202.

Chin, H.L., Chen, Z.S., Chou, C.P., 2003. Fedbatch operation using Clostridium acetobutylicum suspension culture as biocatalyst for enhancing hydrogen production. Biotechnol. Prog. 19, 383–388.

Gavala, H.N., Skiadas, I.V., Ahring, B.K., Lyberatos, G., 2005. Potential for biohydrogen and methane production from olive pulp. Water Sci. Technol. 52 (1/2), 209–215.

Hallenbeck, P.C., Benemann, J.R., 2002. Biological hydrogen production; fundamentals and limiting processes. Int. J. Hydrogen Energy 27, 1185–1193.

Hawkes, F.R., Dinsdale, R., Hawkes, D.L., Hussy, I., 2002. Sustainable fermentative hydrogen production: challenges for process optimization. Int. J. Hydrogen Energy 27, 1339–1347.

Hawkes, F.R., Hussy, I., Kyazze, G., Dinsdale, R., Hawkes, D.L., 2007. Continuous dark fermentative hydrogen production by mesophilic microflora: principles and progress. Int. J. Hydrogen Energy 32, 172–184.

Ismail, I., Hassan, M.A., Rahman, N.A.A., Soon, C.S., 2010. Thermophilic biohydrogen production from palm oil mill effluent (POME) using suspended mixed culture. Biomass Bioenergy 34, 42–47.

Kadar, Z., de Vrije, T., Budde, M.A.W., Szengyel, Z., Reczey, K., Claasen, P.A.M., 2003. Hydrogen production from paper sludge hydrolysate. Appl. Biochem. Biotechnol. 105-108, 557–566.

Kanai, T., Imanaka, H., Nakajima, A., Uwamori, K., Omori, Y., Fukui, T., et al., 2005. Continuous hydrogen production by the hyperthermophilic archaeon, *Thermococcus kodakaraensis* KOD1. J. Biotechnol. 116, 271–282.

Kapdan, I.K., Kargi, F., 2006. Biohydrogen production from waste materials. Enzyme Microb. Technol. 38, 569–582.

Karadag, D., Makinen, A.E., Efimova, E., Puhakka, J.A., 2009. Thermophilic biohydrogen production by an anaerobic heat treated-hot spring culture. Bioresour. Technol. 100, 5790–5795.

Karakashev, D., Kotay, S., Trably, E., Angelidaki, I., 2009. A strict anaerobic extreme thermophilic hydrogen—producing culture enriched from digested household waste. J. Appl. Microbiol. 106, 1041–1049.

Kengen, S.W.M., Goorissen, H.P., Verhaart, M., van Niel, E.W.J., Claasen, P.A.M., Stams, A.J.M., 2009. Biological hydrogen production by anaerobic microorganisms. In: Soetaert, W., Vandamme, E.J. (Eds.), Biofuels. John Wiley & Sons, Chichester, pp. 197–221.

Kim, D.H., Han, S.K., Kim, S.H., Shin, H.S., 2006. Effect of gas sparging on continuous fermentative hydrogen production. Int. J. Hydrogen Energy 31, 2158–2169.

Kongjan, P., O-Thong, S., Kotay, M., Min, B., Angelidaki, I., 2010. Biohydrogen production from wheat straw hydrolysate by dark fermentation using extreme thermophilic mixed culture. Biotechnol. Bioeng. 105 (5), 899–908.

Kotsopoulos, T., Zeng, R., Angelidaki, I., 2006. Biohydrogen production in granular up-flow anaerobic sludge-blanket (UASB) reactors with mixed cultures under hyper-thermophilic temperature (70 °C). Biotechnol. Bioeng. 94, 296–302.

Kraemer, J.T., Bagley, D.M., 2007. Improving the yield from fermentative hydrogen production. Biotechnol. Lett. 29 (5), 685–695.

Lee, H.S., Vermaas, W.F.J., Rittmann, B.E., 2010. Biological hydrogen production: prospects and challenges. Trends Biotechnol. 28 (5), 262–270.

Lee, Y., Miyahara, T., Noike, T., 2002. Effect of pH on microbial hydrogen fermentation. J. Chem. Technol. Biotechnol. 77, 694–698.

Liu, D., Liu, D., Zheng, R., Angelidaki, I., 2006. Hydrogen and methane production from household solid waste in the two-stage fermentation process. Water Res. 40 (11), 2230–2236.

Nath, K., Das, D., 2004. Improvement of fermentative hydrogen production: various approaches. Appl. Microb. Biotechnol. 65, 520–529.

Oh, S.E., Lyer, P., Bruns, M.A., Logan, B.E., 2004. Biological hydrogen production using a membrane bioreactor. Biotechnol. Bioeng. 87, 119–127.

O-Thong, S., Prasertsan, P., Intrasungkha, N., Dhamwichukorn, S., Birkeland, N.K., 2007. Improvement of biohydrogen production and treatment efficiency on palm oil mill effluent with nutrient supplementation at thermophilic condition using an anaerobic sequencing batch reactor. Enzyme Microb. Technol. 41, 583–590.

O-Thong, S., Prasertsan, P., Intrasungkha, N., Dhamwichukorn, S., Birkeland, N.K., 2008a. Optimization of simultaneous thermophilic fermentative hydrogen production and COD reduction from palm oil mill effluent by *Thermoanaerobacterium*-rich sludge. Int. J. Hydrogen Energy 33, 1221–1231.

O-Thong, S., Prasertsan, P., Karakashev, D., Angelidaki, I., 2008b. High-rate continuous hydrogen production by *Thermoanaerobacterium thermosaccharolyticum* PSU-2 immobilized on heat-pretreated methanogenic granules. Int. J. Hydrogen Energy 33, 6498–6508.

O-Thong, S., Prasertsan, P., Karakashev, D., Angelidaki, I., 2008c. Thermophilic fermentative hydrogen production by the newly isolated *Thermoanerobacterium thermosaccharolyticum* PSU-2. Int. J. Hydrogen Energy 33, 1204–1214.

Pikuta, E.V., Hooevr, R.B., 2004. Potential application of anaerobic extremophiles for hydrogen production. Instruments, Methods, and Missions for Astrobiology VIII. In: Hoover, R.B., Gilbert, V.L., Alexei, Y.R. (Eds.), Proc of SPIE, vol. 5555. SPIE, Belingham, WA.

Ren, N., Li, J., Li, B., Wang, Y., Liu, S., 2006. Biohydrogen production from molasses by anaerobic fermentation with a pilot-scale bioreactor system. Int. J. Hydrogen Energy 31, 2147–2157.

Ren, N.Q., Chua, H., Chan, S.Y., Tsang, Y.F., Wang, Y.J., Sin, N., 2007. Assessing optimal fermentation type for biohydrogen in continuous flow acidogenic reactors. Bioresour. Technol. 98, 1774–1780.

Shin, H.S., Youn, J.H., 2005. Conversion of food waste into hydrogen by thermophilic acidogenesis. Biodegradation 16, 33–44.

Shin, H.S., Youn, J.H., Kim, S.H., 2004. Hydrogen production from food waste in anaerobic mesophilic and thermophilic acidogenesis. Int. J. Hydrogen Energy 29, 1355–1363.

Soboh, B., Linder, D., Hedderich, R., 2004. A multisubunit membrane-bound [NiFe] hydrogenase and an NADH-dependent Fe-only hydrogenase in the fermenting bacterium *Thermoanaerobacter tengcongensis*. Microbiology 150, 2451–2463.

Ueno, Y., Sasaki, D., Fukui, H., Haruta, S., Ishii, M., Igarashi, Y., 2006. Changes in bacterial community during fermentative hydrogen and acid production from organic waste by thermophilic anaerobic microflora. J. Appl. Microbiol. 101, 331–343.

van Ginkel, S., Logan, B.E., 2005. Inhibition of biohydrogen production by undissociated acetic and butyric acids. Environ. Sci. Technol. 39, 9351–9356.

Van Niel, E.W.J, Claasen, P.A.M., Stams, A.J.M., 2003. Substrate and product inhibition of hydrogen production by the extreme thermophile, *Caldicellulosiruptor saccharolyticus*. Biotechnol. Bioeng. 81 (3), 255–262.

Van Niel, E.W.J., Budde, M.A.W., de Haas, G.G., van der Wal, F.J., Claassen, P.A.M., Stams, A.J.M., 2002. Distinctive properties of high hydrogen producing extreme thermophiles, *Caldicellulosiruptor saccharolyticus* and *Thermotoga elfii*. Int. J. Hydrogen Energy 27, 1391–1398.

Willquist, K., Claasen, P.A.M., Van Niel, E.W.J., 2009. Evaluation of the influence of CO_2 on hydrogen production by *Caldicellulosiruptor saccharolyticus*. Int. J. Hydrogen Energy 34, 4718–4726.

Yokoyama, H., Waki, M., Moriya, N., Yasuda, T., Tanaka, Y., Haga, K., 2007. Effect of fermentation temperature on hydrogen production from cow waste slurry by using anaerobic microflora within the slurry. Appl. Microbiol. Biotechnol. 74, 474–483.

Yu, H., Zhu, Z., Hu, W., Zhang, H., 2002. Hydrogen production from rice winery wastewater in an upflow anaerobic reactor by using mixed anaerobic granules. Int. J. Hydrogen Energy 27, 1359–1365.

Zhang, Y., Shen, J., 2007. Enhancement effect of gold nanoparticles on biohydrogen production from artificial wastewater. Int. J. Hydrogen Energy 32, 17–23.

Zhang, Z.P., Show, K.Y., Tay, J.H., Liang, D.T., Lee, D.J., 2008a. Biohydrogen production with anaerobic fluidized bed reactors—a comparison of biofilm based and granule-based systems. Int. J. Hydrogen Energy 33 (5), 1559–1564.

Zhang, Z.P., Show, K.Y., Tay, J.H., Liang, D.T., Lee, D.J., 2008b. Enhanced continuous biohydrogen production by immobilized anaerobic microflora. Energy Fuels 22, 87–92.

Zhao, C., Karakashev, D., Lu, W., Wang, H., Angelidaki, I., 2010. Xylose fermentation to biofuels (hydrogen and ethanol) by extreme thermophilic (70 °C) mixed culture. Int. J. Hydrogen Energy 35 (8), 3415–3422.

Zhao, C.X., O-Thong, S., Karakashev, D., Angelidaki, I., Lu, W.J., Wang, H.T., 2009. High yield simultaneous hydrogen and ethanol production under extreme-thermophilic (70 °C) mixed culture environment. Int. J. Hydrogen Energy 34, 5657–5665.

CHAPTER

24

Biohydrogen Production with High-Rate Bioreactors

*Wen-Wei Li, Han-Qing Yu**

Department of Chemistry, University of Science & Technology of China, Hefei, 230026 China

***Corresponding author: E-mail: hqyu@ustc.edu.cn**

1 INTRODUCTION

Biohydrogen production, as a promising avenue for obtaining clean and sustainable energy from wastes, has attracted increasing attention in the past decades (Hallenbeck and Ghosh, 2009; Holladay et al., 2009). Attributed to the recent advances in reactor configuration, functional microbes, and process control techniques, the biological hydrogen production rate (HPR) has significantly improved, approaching a practical level (Lee et al., 2010b). Biological hydrogen production process can be classified into four major categories, depending on the dominating microorganisms and sources: dark fermentation of organic wastes or energy crops by heterotrophic bacteria; biophotolysis of water using algae and cyanobacteria; photofermentation of organic compounds by photosynthetic bacteria (PSB); and electrochemically assisted hydrogen production using microbial electrolysis cells (MECs) with anode-respiring bacteria (ABR; Hallenbeck and Benemann, 2002; Lee et al., 2010b). Accordingly, different adaptive bioreactors have been developed, including continuously stirred tank reactors (CSTRs), anaerobic fluidized-bed reactors (AFBRs), packed-bed reactors (PBRs), and upflow anaerobic sludge blanket (UASB) reactors for dark fermentation, as well as specific photobioreactors and MECs. Among the various options of dark-fermentation reactors, CSTRs have been conventionally adopted attributed to their simplicity and ease of operation (Ren et al., 2006). However, they generally suffer from a low HPR and poor operation stability due to biomass washout. Furthermore, high energy input is usually needed to achieve a vigorous mechanical stirring. Comparatively, the AFBRs, PBRs, and UASBs enable effective biomass retention and allow for operation under higher organic loading rates (OLRs), which lead to better hydrogen production performance (Barros et al., 2010). In addition, several modifications of CSTRs have also been proposed to enhance biomass retention by incorporating into biomass-liquid separation techniques. Meanwhile, efforts have also been devoted to the development of efficient photobioreactors and MECs to ensure efficient hydrogen production.

Development of bioreactors presents one important aspect of the fast-progressing biohydrogen production technologies. It is widely accepted that the reactor configuration and operating mode can pose significant influences on the system biohydrogen production performance by directly affecting the hydrodynamics and bacteria properties. Both volumetric HPR and hydrogen yield (HY) have been considered as important indexes to evaluate the performance of a biohydrogen production system. However, it is noted that the HY, defined by the amount of hydrogen produced per substrate consumed, is more dependent upon the microbial properties, substrate type, and environmental conditions rather than the reactor configurations. In fact, for dark-fermentation processes, HYs obtained with different systems are rather inconsistent so far and mostly no greater than those achieved with CSTRs (Chang and Lin, 2004; Hallenbeck and Ghosh, 2009). Thus, there is no reason to think that the reactor type would essentially influence the HY in dark-fermentation systems, and a comparison of these reactors here shall mainly be based on their HRPs. For the phototrophic and MEC hydrogen production processes, however, the HY is usually used for reactor evaluation because of the fewer microbial species and substrates involved as well as the relatively low HPR value of such systems. In addition, the operating stability is also a critical factor to be considered for the biological hydrogen production systems. Therefore, here we offer a comprehensive introduction and systematical comparison of high-rate hydrogen-producing bioreactors based on their HPR, HY, biomass retention capability, and process stability.

In fact, there are so many variations of reactor configuration and different operating strategies, for which it is impossible to give a full list. While biohydrogen production can be achieved in both batch and continuous operation systems, a continuous process generally requires a smaller reactor size and is favorable for practical application (Bartacek et al., 2007; Cheong et al., 2007). Therefore, in this work, we focus on the recent progresses of the main high-rate bioreactors operated on continuous basis. The principles and technical foundations of these bioreactors are summarized, and their improvements to existing reactors and the remaining drawbacks are discussed. In addition, we compare these high-rate bioreactors and explore the further developing directions and possibilities to overcome the challenges and promote their large-scale application.

2 INTEGRATED CONTINUOUSLY STIRRED TANK REACTORS

One of the most frequently employed configurations of reactor in biohydrogen fermentation is CSTR, in which substrates and products are continuously added and withdrawn, and mechanical agitation is provided to achieve a full mixing of the sludge suspension within the reactor (Yu et al., 2003). However, this configuration has inherent limitations due to an identical hydraulic retention time (HRT) and solids retention time (SRT). Biomass washout and system failure are liable to occur in such systems at a short HRT; thus, a longer HRT is required in the conventional CSTRs, which decreases biohydrogen productivity (Wu et al., 2008). An HPR of below 0.2 L L^{-1} h^{-1} has been generally reported for most CSTRs (Ren et al., 2007). Therefore, to maintain adequate biomass retention in high-rate operation of CSTRs, a decoupling of SRT and HRT is necessary. This leads to the development of several integrated CSTRs with enhanced biomass-liquid separation.

2.1 CSTR+ Settler Integrated System

Adding a gravity settler behind the CSTR presents one of the simplest ways to reduce biomass washout (Figure 1a). Through partially returning the sludge from the settler, a higher biomass concentration (expressed as the mass of volatile suspended solids (VSS) in per liter of

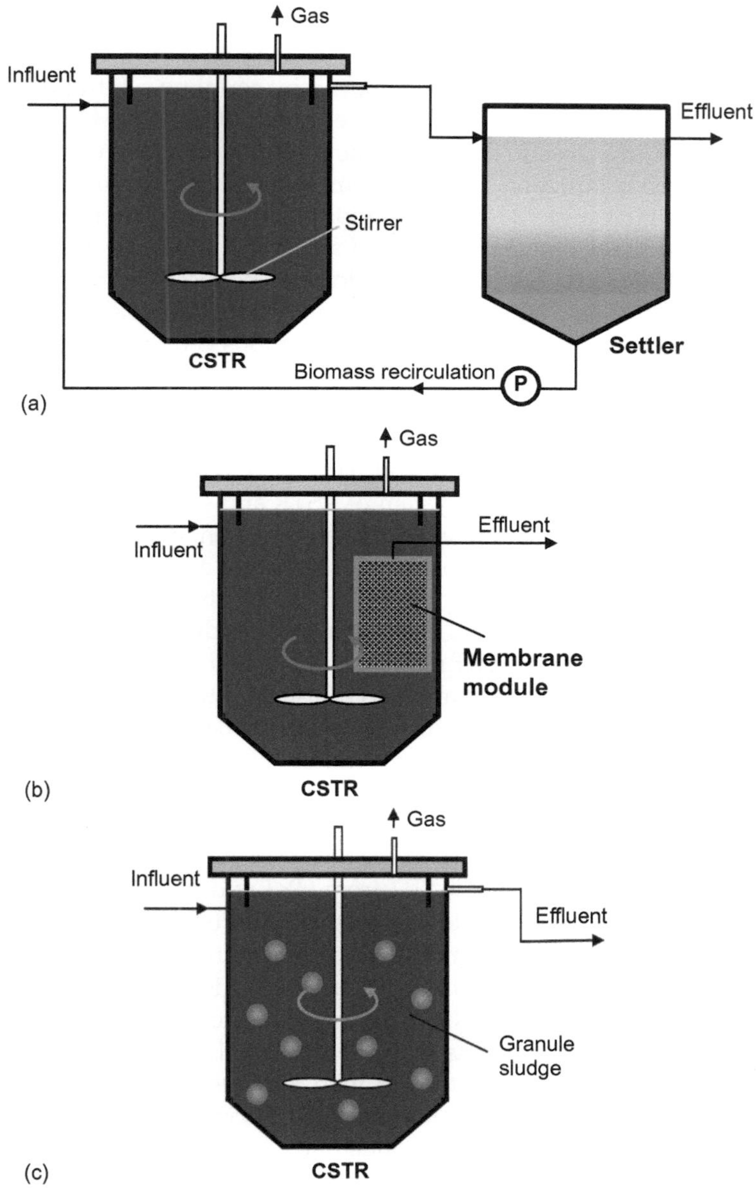

FIGURE 1 Schematic diagram of integrated CSTR: (a) CSTR+ clarifier integrated system *(adapted from Hafez et al., 2010)*; (b) CSTR+ membrane integrated system *(Lee et al., 2009a)*; (c) granule-based CSTR *(Ren et al., 2010)*.

fluid) can be maintained in the reactor (Hafez et al., 2009, 2010). In a CSTR+ settler integrated system, a remarkably high biomass concentration of 18.4 g-VSS L^{-1} was achieved at 8 h HRT, compared to that of 2.3 g-VSS L^{-1} in a CSTR under identical HRT conditions (Wu et al., 2008). As a result, the HPR increased significantly from 0.02-0.08 L L^{-1} h^{-1} in the conventional CSTR to 0.10-0.40 L L^{-1} h^{-1} in the integrated system (Hafez et al., 2009).

2.2 CSTR+ Membrane Integrated System

Improved biomass retention can also be achieved by coupling a filtration membrane to a CSTR for biomass-liquid separation (Figure 1b; Lee et al., 2007, 2009a; Oh et al., 2004a). Such an integrated system using crossflow alumina membrane was first investigated by Oh et al. (2004a) for biohydrogen production. In the system, the SRT can be well controlled independent of HRT. At 3.3-h HRT and 12-h SRT, the biomass concentration reached 5.8 g-VSS L^{-1} which was nearly two times higher than that of a CSTR at the same HRT. Meanwhile, the HPR increased from 0.26 to 0.32 L L^{-1} h^{-1}. Significantly enhanced biomass retention was also reported by Lee et al. (2007) in a similar system at a extremely low HRT of 1 h, where hollow-fiber microfiltration membrane was adopted. While serious cell washout was observed in a CSTR at HRT of 2-4 h, the HPR and yield increased substantially in this system at 1 h HRT and stably maintained at 2.75 L L^{-1} h^{-1} and 1.36 mol-H_2 mol^{-1}-fructose, respectively. Compared with the crossflow (or side stream) configuration, submerged membrane offers more advantages in terms of energy demanding and bacteria activity, and is thus more preferable for large-scale applications. A coupling system of CSTR and submerged membrane was successfully applied by Lee et al. (2009a) to improve cell retention, where the biomass concentration grew steadily and reached 9.5 g-VSS L^{-1} after 3 days at an HRT of 9 h and SRT of 90 d. In comparison, the biomass concentration in a CSTR under the same HRT decreased rapidly from 0.93 to 0.24 g L^{-1} within this period, leading to poor hydrogen production performance. It is worthwhile to note that a longer SRT seems to favor a higher HPR due to a raised biomass concentration, but would possibly decrease the HY ascribed to the accumulation of inert biomass and EPS and a metabolic pathway shift (Lee et al., 2010a). Therefore, a proper control of SRT is of pivotal importance for such CSTR-membrane integrated systems.

2.3 Granule-Based CSTR

Another important means to uncouple SRT from HRT and reduce cell washout is to adopt granule-based techniques (Figure 1c). It has been demonstrated that the granular sludge, formed either by self-flocculation or immobilization onto natural or synthetic carriers, generally possesses a higher settleability and enables a more efficient biomass-liquid separation than the flocs (Liu et al., 2009). In a CSTR, the microbial cells can aggregate under appropriate hydrodynamic and operating conditions (Fang et al., 2002; Liu and Fang, 2003; Zhang et al., 2007b) or grow sufficiently on the surface of inert carriers (Ren et al., 2010) to form H_2-producing granules (HPGs). The granulation of H_2-producing sludge was found to markedly improve the HPR of a CSTR treating sucrose-containing wastewater (Fang et al., 2002) or glucose wastewater (Show et al., 2007). Especially, a higher OLR derived from appropriate HRTs and substrate concentrations in such granule-based system was found to further increase the HPR to 3.26 L L^{-1} h^{-1}, in contrast to only 0.15 L L^{-1} h^{-1} in a conventional CSTR (Fang and Liu, 2002).

Substantial improvement in HPR and system stability was also demonstrated in CSTR system with bacteria attached onto inert carriers (Ren et al., 2010; Wu et al., 2006). The addition of granular-activated sludge (GAC), which served as the carrier for sludge immobilization, led to a much higher biomass concentration (15.5 g-VSS L^{-1}) than CSTR with suspended H_2-producing flocs (HPFs; 6.48 g-VSS L^{-1}) under the same conditions (Ren et al., 2010). Consequently, a substantially higher HPR (0.41 vs. 0.28 L L^{-1} h^{-1}) was obtained in such an attached-type granule-based CSTR. An even higher HPR (15 L L^{-1} h^{-1}) was reported by Wu et al. (2006) in a silicone-immobilized CSTR, where a extremely high biomass concentration of up to 35.4 g-VSS L^{-1} was successfully maintained at a 0.5-h HRT. Furthermore, such attached-growth systems also demonstrated excellent operation stability (Ren et al., 2010) and favorable bacteria community (Wu et al., 2006), showing great potential for practical application.

2.4 Limitations of Integrated CSTRs

Despite of the improvements in CSTR setup by integrating with other reactor configurations, all the CSTRs invariably suffer from high energy demand due to the continuous intensive stirring, and there are several inherent limitations for each of such reactors.

(1) A large volume of settler is required for the CSTR-settler integrated system to achieve efficient biomass-liquid separation, which increases the cost. Moreover, this system is unstable because the separation performance would deteriorate markedly in case of sludge bulking under shock loading.
(2) CSTR-membrane integrated system generally possesses compact configuration and higher operating stability. However, the high investment and operating cost as well as the membrane fouling would present serious barriers to their practical application at large-scale (Lee et al., 2010a; Shen et al., 2010).
(3) In comparison with the other two integrated systems, granule-based reactors offer better antishocking capability and enable more cost-effective operations. However, the overgrowth of methanogens, which consume hydrogen and reduce HPR, in the granules may become a big challenge especially for the treatment of real wastewaters. In addition, washout of carriers may occur at larger granule sizes. Although various pretreatment methodologies such as heat shock and acid incubation have been demonstrated effective to suppress the methodogens, they also increase operation cost and add to complexity in practical operation.

So far, studies of these integrated CSTRs are still limited at laboratory scale, and breakthrough of the remaining hurdles should be pursued in future investigations to make them technically and commercially viable.

3 ANAEROBIC FLUIDIZED-BED REACTOR

3.1 Main AFBRs

Although CSTRs are still the most frequently used reactor type for continuous biohydrogen production, a range of other reactor types based on cell-immobilization technologies have been developed in recent years, such as AFBRs, UASB, anaerobic APBR and trickling-bed reactors

(TBR; Hallenbeck and Ghosh, 2009). These immobilized-cell systems have become popular alternatives to the conventional CSTRs in continuous biohydrogen fermentation processes, mainly attributed to their superior biomass retention capabilities (Chang et al., 2002; Wu et al., 2002). Among these, AFBRs exhibit a couple of advantages.

AFBRs are similar to CSTRs in configuration, but usually with larger height to diameter (H/D) ratios and are characterized by high-velocity fluid circulation (Figure 2). In an AFBR, microorganisms are retained in the form of biofilm on carriers or self-flocculated granules and maintained in suspension by the drag force of upward fluid flow, known as fluidization (Barros et al., 2010). A high H/D ratio is favorable to minimize the liquid circulation rate needed for efficient fluidization of the granules inside the reactor. Due to the intrinsic fluid-like behavior of these granules, a better mixing and full sludge-substrate contact can be achieved in such systems than CSTRs, which allows high-rate biohydrogen production. Moreover, a stressful environment such as the high upflow rate seemed to favor a higher activity of H_2-producing bacteria in the granules compared with the floc biomass in suspended systems (Zhang et al., 2008a).

An AFBR for biohydrogen production was firstly explored by Zhang et al. (2007a) using GAC as the support medium. A high biomass concentration of 21.5 g-VSS L^{-1} and a maximum HPR of 2.36 L L^{-1} h^{-1} were achieved at 0.5 h HRT, and no methane generation was detected in the process. Although an appropriately designed three-phase separator on the top of the reactor offers effective separation of gas from the fluid, a high effluent flow may

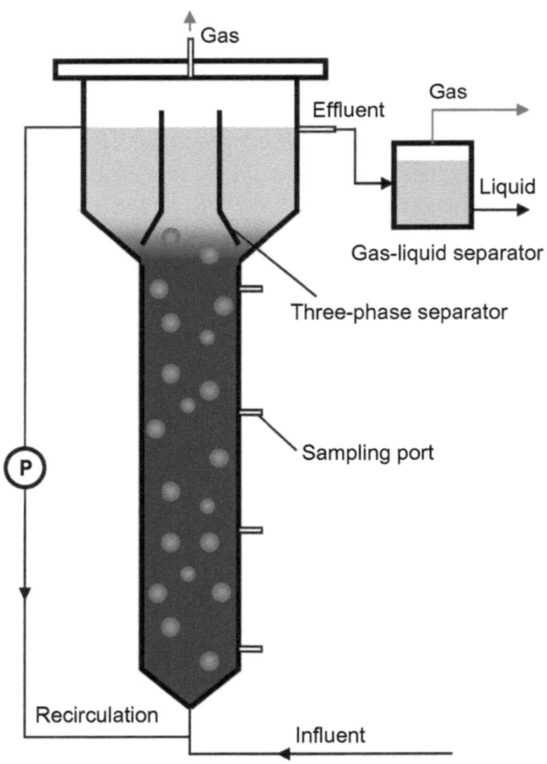

FIGURE 2 Schematic diagram of AFBR *(adapted from Barros et al., 2010; Zhang et al., 2007c).*

bring out some gas. Therefore, a gas-liquid separator was often adopted at the effluent outlet to further recover the remaining hydrogen gas (Koskinen et al., 2007; Wu et al., 2007). In order to minimize the liquid recycle rate required for fluid fluidization, an AFBR with an internal draft tube of high H/D ratio (25) was proposed by Lin et al. (2006). An HRP of 2.1 L L^{-1} h^{-1} was obtained in this novel reactor, which was 50% higher than that of a normal AFBR under identical conditions. Similar draft tube AFBR configuration was also successfully applied by Peintner et al. (2010) in thermophilic hydrogen production processes to achieve satisfactory performances. In addition to GAC, other materials such as polymeric matrix polyethylene-octene elastomer (Wu et al., 2007), polystyrene, expanded clay (Barros et al., 2010), and celite (Koskinen et al., 2007) have also been used as the support media in AFBRs for efficient hydrogen production. While inert carriers have been commonly adopted to favor a fast granulation in AFBRs, self-flocculated HPGs have also been developed in such reactors, demonstrating comparable capabilities in biomass retention (34-37 g-VSS L^{-1}) and hydrogen production (HPR = 6.6-7.6 L L^{-1} h^{-1}) to the attached growth systems (Cavalcante de Amorim et al., 2009a; Zhang et al., 2008a).

3.2 Limitations of AFBRs

AFBRs in different setups have been applied for hydrogen fermentation processes, with high biomass content, good turbulence, and excellent mass transfer performance. Indeed, recent studies have shown that high HPRs can be achieved in such reactors with self-immobilized or attached-growth HPGs (Davila-Vazquez et al., 2008; Hawkes et al., 2007). Nevertheless, they invariably suffered from several constraints in practical use.

(1) The requirement for a high upflow fluid velocity to suspend the biomass in the reactor necessitates a high fluid recirculation ratio, which means more energy consumption for pumping and a high operating cost. In addition, the pressure drop associated with high H/D ratio of reactor also increases the energy demand for fluid recirculation.

(2) A relatively high fluid velocity in AFBR is necessary to maintain the fluidization state and keep a good mechanical stability of the granules. But a violent turbulence at high fluid velocities may result in disintegration of bioaggregates and suppressed bacteria activities. The disintegrated fine suspended biomass would then enter into the effluent, which requires further separation treatment. Moreover, the wash out of fine biomass and low bacteria activity would lead to poor HPR and decrease operating stability. Thus, a proper control of the fluid velocity is important for such systems and may present a challenge in large-scale application.

4 PACKED-BED REACTORS

4.1 Anaerobic Contact Filter (ACF)

In contrast to fluidized bed where the granules exhibit a fluid-like behavior, a packed bed is composed of a fixed layer of support materials (Figure 3), which serve as the carriers for biofilm formation and favor an intimate contact between the substrate and sludge (Wu et al., 2007). In fact, AFBRs at low fluid velocity can be regarded as packed-bed ones, where the support matrix settles on the porous plate at a static state as the fluid passes through the

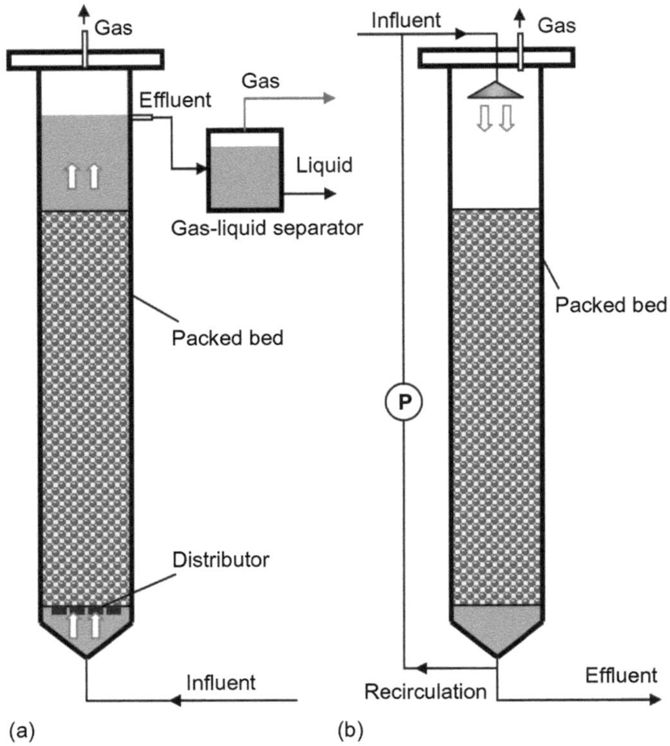

FIGURE 3 Schematic diagram of PBRs: (a) ACF *(adapted from Chang et al., 2002; Vijayaraghavan and Ahmad, 2006)* and (b) TBR *(adapted from Oh et al., 2004b).*

voids. PBRs have also been comprehensively investigated for hydrogen fermentation in the past decade (Chang et al., 2002; Lee et al., 2003; Leite et al., 2008; Wu et al., 2007). In a PBR, microorganisms exist not only in the spaces within the medium, but also attach on its surface, creating a hybrid layer of "biofilter" (Li et al., 2006). Hence, PBRs with an upflow through the packed bed was also frequently referred to as ACFs (Figure 3a; Vijayaraghavan et al., 2007). In order to avoid short-circuiting flow through the packed column, a distributor is usually set at the reactor bottom to ensure a homogeneous fluid flow. At the outlet of the reactor, a gas/liquid separator is used to separate and recover the fluid and biohydrogen gas (Chang et al., 2002).

Compared with the AFBRs, ACFs have more compacted structure and usually smaller vessel size, need no high-rate flow circulation, and is more favorable for biofilm growth (Palazzi et al., 2000). Thus, both the initial investment and operating costs can be reduced, while better biomass retention can be achieved. Like many other attached-growth systems, the performance of an ACF is highly dependant on the properties of the support materials. Among the various options, GAC has been commonly adopted attributed to its high bioadsorptivity, good mechanical stability, and low cost (Chang et al., 2002; Lee et al., 2003). The use of a GAC-packed column was found to enable a high biomass density on the carrier surface (Chang et al., 2002) and lead to high HPRs of up to 7.4 L L^{-1} h^{-1} (Lee et al., 2003). One critical parameter to be controlled in such systems is the packed-bed porosity. A higher porosity means more void space for microbial growth and thus enables a reduced pressure drop through the bed layer. In addition, a bed porosity of above 90% with GAC medium was found to effectively induce sludge granulation within an ACF. Based on this, a novel carrier-induced granular sludge bed reactor was developed by Lee et al. (2004b) in which only a small quantity of high-porosity GAC was packed on the bottom of the reactor as the supporting matrices. This reactor enabled a stable operation at an extremely low HRT of 0.5 h without biomass washout, and a rapid formation of HPGs was witnessed. As a result, a remarkably high biomass concentration of 26 g-VSS L^{-1} and HPR of 7.3 L L^{-1} h^{-1} were achieved. Nevertheless, it should be noted that a too loose packed bed and overhigh bed porosity increase the risk of short-circuiting flow and reduce biomass-substrate contact. Besides GAC, several other materials have also been frequently used as the packing medium for ACFs, such as expanded clay (Leite et al., 2008), loofah sponge (Chang et al., 2002), and polyethylene-octene elastomer (Wu et al., 2007). These systems showed slightly lower HPRs (0.36-1.3 L L^{-1} h^{-1}) compared to the GAC-packed ACFs (Chang et al., 2002; Lee et al., 2003), possibly because of a relatively lower biomass accumulation and insufficient biomass-substrate contact. To overcome this, a horizontal-flow anaerobic immobilized biomass reactor coupling with expanded-clay bead support material was recently developed (Leite et al., 2008). This horizontal configuration favors a fast gas-liquid separation and adequate biomass-substrate contact along the reactor, thus enhanced hydrogen production can be achieved.

Aside from the readily degradable substrates, ACFs also show good performance in recovering hydrogen from lignocellulosic wastes, which is highly desirable for commercialization of hydrogen production (Vijayaraghavan and Ahmad, 2006; Vijayaraghavan et al., 2007). A high-rate ACF reactor with enhanced mixing was proposed by Vijayaraghavan et al. (2007) to improve lignocellulose bioconversion, which consists of four zones, namely, bottom zone, packing zone, outlet zone, and head space for the gas accumulation. Rigid spherical plastic balls of 40-mm diameter, with a number of slotted openings on the surface, were used as the packing material. An additional recirculation pump was used to promote mixing of the digester content, which was carried out for six cycles per day. The average HYs reached 0.46 L g^{-1}-VSS destroyed at an HRT of 5 days for fruit peel waste fermentation (Vijayaraghavan et al., 2007) and 0.42 L g^{-1}-COD destroyed at 7-d HRT for palm oil mill effluent treatment (Vijayaraghavan and Ahmad, 2006).

4.2 Trickling-Bed Reactor

Another important form of PBR for hydrogen fermentation is trickling-bed reactor (TBR), also called trickling biofilter (Oh et al., 2004b). Unlike the ACFs that involve upflow of fluid within the reactor, a TBR entails the downward movement of the liquid over a packed bed under gravity, which favors a lower gas holdup and thus a more efficient gas-liquid separation (Figure 3b). This alleviates the H_2 inhibition and the severe channeling of liquid and gas flow in the packed bed. Beside, the close proximity of the substrate, bacteria, and the resulting gas phase facilitates a good biomass-substrate interaction and gas-liquid exchange during biohydrogen production (van Groenestijn et al., 2002 van Groenestijn et al., 2002).

Oh et al. (2004b) studied the thermophilic hydrogen production in a TBR using fibrous polyvinylidene dichloride particles as the support material. Influent was fed into the reactor from the headtop through a distributor. Part of the liquid broth at the outlet, after pH adjustment, was returned to the headtop and joined the influent. This process offers an easy pH control and good operation robustness. Most of all, a high biomass density of 18-24 g-VSS L^{-1} was maintained in such a system, and the high mass transfer rate at the biomass-liquid interface led to an HPR of 0.98 L L^{-1} h^{-1} at 2-h HRT (Oh et al., 2004b). This improved liquid-gas-biomass interface offers a better condition for hydrogen fermentation than a UFBR (Peintner et al., 2010). Consistent results were also reported in a TBR for mesophilic hydrogen fermentation using pure culture, where glass beads were used as the packing materials and glucose as the substrate (Zhang et al., 2006). In this system, the fluid rate was appropriately controlled to create a thin fluid film over the biofilm, which facilitated a rapid hydrogen gas evolution out of the biofilm and thus promoted hydrogen production.

4.3 Limitations of PBRs

As in any design, the PBR also has limitations that need to be considered in design and be overcome in future applications.

(1) A packed bed of carriers undoubtedly favors biofilm growth and biomass retention, but usually at the cost of a lowered mixing within the bed compared to the fluidized type of reactors like CSTRs and AFBRs;

(2) The packed bed is liable to blocking by biomass and suspended solids (SS) that passing through. Thus, this process is unsuitable for treatment of waste streams with high SS concentration.

(3) Channeling of liquid and gas flow is easy to occur in the packed bed of an ACF especially under high bed porosity, causing dead zones within the bed. In addition, biogas release out of the liquid phase can be limited in such reactors without intensive stirring, which also inhibit hydrogen production (Rachman et al., 1998).

(4) TBR is sensitive to changes in environmental conditions such as pH and temperature (Oh et al., 2004b). The microbial population and their biological activities can be significantly affected by small variations of pH and temperature. Thus, an accurate control of the operating conditions is needed, which presents another challenge for its practical application.

5 UPFLOW ANAEROBIC SLUDGE BLANKET REACTORS

5.1 Standard UASB Reactors

Granulation technologies are among the most attractive means of enhancing H_2 production by reducing washout of bacterial cells. It is well recognized that the success of HPGs relies largely on the development of UASB reactors (Lee et al., 2010b). As thus, these reactors were also frequently referred to as anaerobic granular sludge bed reactors. The scheme of a typical standard UASB is shown in Figure 4a. Simple column-type reactors with three-phase separators mounted on the top are usually adopted. The overall reactor configuration is mostly similar to an AFBR, but no fluid recirculation is applied. Therefore, unlike the AFBR which achieve fluidization and mixing under a rapid upflow of fluid, the UASB reactors take advantage of the upward motion of the produced gas (mostly H_2 and CO_2 in biohydrogen fermentation processes) to cause hydraulic turbulence without any mechanical parts.

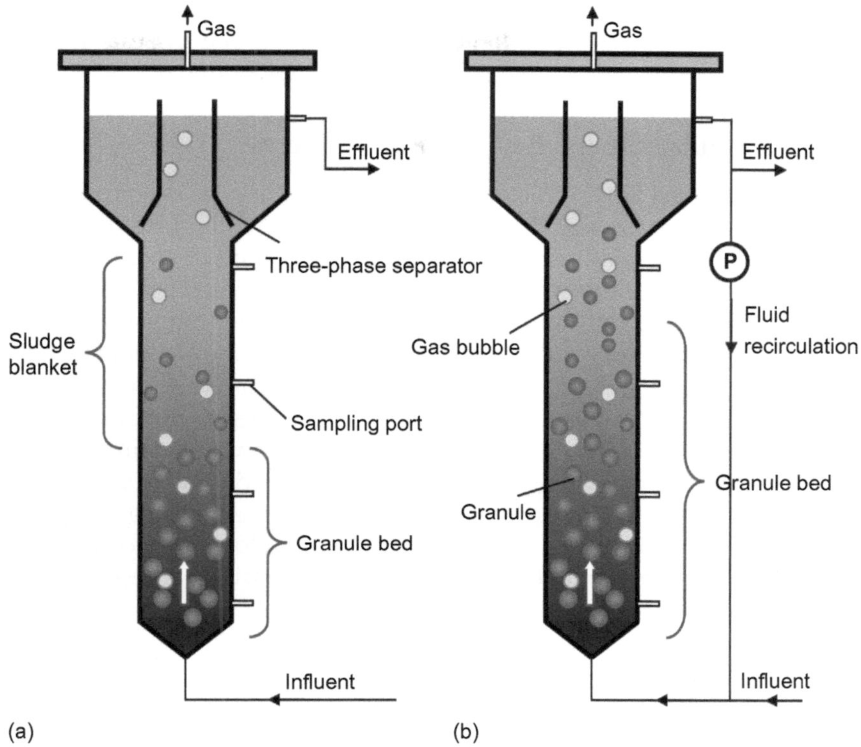

FIGURE 4 Schematic diagram of (a) UASB and EGSB reactor *(adapted from Mu and Yu, 2006).*

One most important feature of UASB reactors for biohydrogen production processes is the stratification of sludge at different height of the reactor. Generally, when the granules develop increasingly in the reaction zone, a dense sludge bed of granules and a thin sludge blanket dominated by floc sludge would come into being, and between them a clear interface can be seen (Chang and Lin, 2004; Mu and Yu, 2006). During operation, the wastewater passes upward through the suspended granule bed and sludge blanket in turn, where the bacteria in the sludge come into full contact with the substrates (Kotsopoulos et al., 2006).

The feasibility of employing a standard UASB reactor for continuous biohydrogen fermentation was firstly explored by Yu et al. (2002). By using mixed anaerobic cultures as the seed sludge and rice winery wastewater as the substrate, an HPR of 0.16 L L^{-1} h^{-1} was achieved at 2-h HRT, and no methane generation was detected. More importantly, UASB reactor provides a favorable environment for sludge granulation (Adav et al., 2008). A systematic investigation on the operation stability, HRT dependence, and sludge granulation of UASB reactors for hydrogen production was conducted by Chang and Lin (2004). Mature HPGs with an average diameter of 0.43 mm were formed after operating for 173 days, and a constant hydrogen production was maintained for a long period of 8 months. This high hydrogen production efficiency and long-term stability of UASB reactors was further demonstrated by Yu and Mu (2006), with HYs of 0.49-1.44 mol-H_2 mol^{-1}-glucose achieved during a 3-year operation despite the fluctuation of environmental parameters such as pH and temperature. A side-by-side comparison of different reactors for hydrogen production processes by Gavala et al. (2006) found that the UASB gave a significantly higher HPR than CSTR (0.43 vs. 0.19 L $L^{-1}h^{-1}$ at 2 h HRT) and showed a more stable performance under changed HRTs.

5.2 Expanded Granular Sludge-Bed (EGSB) Reactors

A variant of the standard UASB reactors, EGSB reactors, has also been recently applied for biohydrogen production processes (Guo et al., 2008b). One distinguishing feature of EGSB reactors compared with the standard UASBs is the incorporation of an effluent recirculation, which enables a higher rate of upward flow through the sludge bed and causes expansion of the granular sludge bed (Figure 4b). In addition, tall-column configuration is frequently adopted to further increase the flow velocity. This expanded bed favors efficient biomass-substrate contact and more valid segregation of small suspended particles from the sludge bed (Li et al., 2010). In an EGSB reactor with GAC carriers for treatment of molasses-containing wastewater, a high biomass concentration of 19.4 g-VSS L^{-1} was successfully maintained during the near 400-d operation, and a HPR of up to 0.71 L $L^{-1}h^{-1}$ was attained (Guo et al., 2008b). The same reactor configuration was also successfully applied to produce hydrogen from starch waste water, with a maximum HPR of 1.64 L $L^{-1}h^{-1}$ obtained at a lower pH of 4.2-4.4 (Guo et al., 2008a). The low pH operation implies that less or no alkaline addition is needed in such systems which saves the cost.

5.3 Limitations of UASB Reactors

The feasibility of employing UASB reactors for biohydrogen production has been widely demonstrated. UASB combines the advantages of AFBRs and PBRs in biomass retention and mixing, and presents a promising biosystem for hydrogen production from high-strength wastewaters. Particularly, no fluid circulation is applied in the standard UASB reactors, which enables a cost-effective operation. Despite these frequently cited advantages, the practical application of UASB reactors for biohydrogen fermentation processes is still constrained by several factors.

(1) The long start-up period for granulation has been one major challenge. A culturing period of several months is usually needed for the development of mature H2-producging granules (Chang and Lin, 2004; Mu and Yu, 2006). Although several approaches are available to accelerate this process, such as adding Ca^{2+} (Lee et al., 2004a) or carrier (Lee et al., 2004b) and acid pretreatment of seed sludge, they also raise the cost and may cause undesirable reactor corrosion. Furthermore, the carriers are prone to washout as the granule size grows up, arising risk of system instability.

(2) EGSB reactor imparts many technological advantages to standard UASB by adding a fluid recirculation and adopting taller reactor configurations. However, this would also be problematic in large-scale application when the economics becomes a vital consideration.

6 PHOTOBIOREACTORS

Dark-fermentation is no doubt the most attractive means of biohydrogen production at the present stage, which shows a couple of advantages over other biohydrogen options, such as the potential of directly using wastewater streams and organic wastes, a low energy demand

and, most of all, a higher HPR. However, for the treatment of non-sugar-based substrates like lignocellulose, the bioconversion rate and HY would become more important considerations than HPR, while this is exactly the biggest drawbacks of dark-fermentation processes. In this regard, photobiological processes show a great potential and are attracting increasing attention in recent years (Hallenbeck and Ghosh, 2009). Photobiological hydrogen production, through biophotolysis or photofermentation, are known to achieve more efficient conversion of organics into hydrogen and benefit direct removal of CO_2 (Lee et al., 2010b). Because of the special characteristics of such systems in light utilization and bacteria community, an appropriately designed reactors is thus of critical importance. Photobioreactors are differentiated from the dark-fermentation bioreactors in several unique features: transparent materials should be adopted to allow sufficient light penetration; the energy source of light is instantaneous and cannot be stored within the reactor; self-shading of cells occur in the reactor, which means that the light penetration becomes weak with the increase of biomass concentration and the amount of available light attenuates in the deeper reactor regions (Akkerman et al., 2002). So far, a number of photobioreactors with different geometries and illumination mode, such as tubular, flat-plate type, and internal-illuminated reactors, have been developed and intensively investigated (Dasgupta et al., 2010).

6.1 Tubular Reactors

Typically, CSTR-type photobioreactors of simple structure are adopted for phototrophic biohydrogen production processes, with certain degree of success (Oh et al., 2004c; Shi and Yu, 2005, 2006). However, such reactors invariably suffer from slow cell growth and low light conversion efficiency, causing poor H_2 production performance (Liao et al., 2010; Nath and Das, 2009). The restricted light penetration, low area/volume (A/V) ratio insufficient agitation, and poor gas exchange are considered the major limiting factors of such reactors. In this respect, tubular reactor presents an attractive alternative with enhanced properties.

Diverse geometries of tubular reactors have been designed, such as vertical, horizontal, and helical configurations. Vertical tubular reactors (including airlift and bubble column reactors) are among the most widely employed types of photobioreactors. A vertical tubular reactor typically consists of several vertically set transparent tubes, in which agitation is achieved usually with the help of bottom bubbling (Figure 5a). Such reactors are characterized by good mixing and gas-exchange properties, which enable efficient biomass growth (Miron et al., 2000). However, the bubbling of inert gas would dilute the stream of hydrogen and decrease hydrogen purity. Although recirculation of the produced hydrogen is suggested as a possible solution, it may cause hydrogenase activation and hydrogen feedback inhibition as well as gas holdup. Compared with vertical-type configuration, horizontal (or near horizontal) reactors enable a less-violent shear stress to cells and a low gas holdup, but at the cost of a poor temperature exchange (Gebicki et al., 2009; Tredici and Zittelli, 1998). A significantly higher A/V ratio can be achieved in a three-dimensional helical tubular reactor coupled with external gas and heat exchanger, leading to an improved HPR of 0.013 $L\ L^{-1}h^{-1}$ (Tsygankov et al., 1998). However, biofouling due to cell deposition in the inner wall seems to be a barrier for such helical tubular systems.

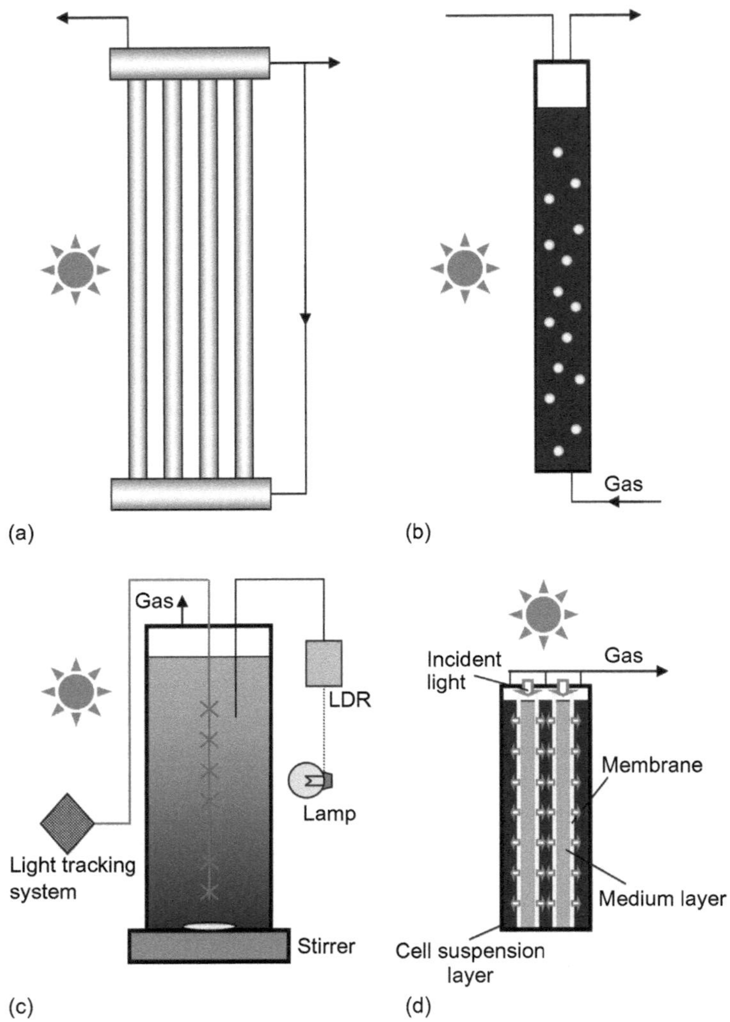

FIGURE 5 Schematic diagram of photobioreactors: (a) tubular reactor; (b) flat-plate reactor; (c) optical-fiber assisted reactor; (d) multilayered reactor *(adapted from Chen et al., 2008; Gebicki et al., 2010; Kondo et al., 2006).*

6.2 Flat-Plate Reactors

Another commonly-used photoreactor is flat-plate configuration. Flat-plate reactors are characterized by a minimized reactor thickness and an open gas-transfer area, which raises photosynthetic efficiency and enables effective control of gas pressure (Figure 5b). Furthermore, they are economically competitive with other configurations due to the highly simple structure. An outdoor flat-plate solar bioreactor using *Rhodobacter sphaeroides* has been reported to produce hydrogen at an HPR of 0.01 L $L^{-1}h^{-1}$ (Eroglu et al., 2008). However, such reactors generally face difficulties in agitation and temperature control, where mixing is achieved by recirculation of the evolved gas (Hoekema et al., 2002). Notwithstanding, better mixing was achievable in a floating-type flat-plate reactor with rocking motion (Otsuki et al., 1998). Although the achieved light conversion efficiency was still low (0.3%) in this system, it hold great promise for practical application as the motion of waves in seas or lakes can be utilized to enhance agitation and temperature control. Noticeably, an annular triple jacket reactor, which is essentially a plate-type reactor in coiled configuration, was also recently developed (Basak and Das, 2010). The axially located light source enabled superior light penetration and high-rate hydrogen production, and the maximum HPR and HY reached 0.007 L $L^{-1}h^{-1}$ and 4.5 mol-H_2 mol^{-1}-DL malic acid, respectively.

6.3 Internal-Illuminated Reactors

In conventional tubular or plate-type reactor, light is provided from outside the reactor and thus light conversion efficiency is usually low due to the self-shading effect. In contrast, internal illumination can offer more additional light energy, apart from the readily available external illumination, into the interior of the reactors and enable a higher biomass growth and hydrogen production. Such a mode can be achieved either by coupling an optical fiber into the fermentation as the internal light source (Chen et al., 2006) or through inducing a incident light into a multilayered reactor (Kondo et al., 2006).

The application of an optical-fiber-assisted illumination system was found to significantly enhance hydrogen production in a photobioreactor. In this system, a bundle of optical fibers protected in a glass tube is directly immersed into the liquid medium inside the reactor while being connected to an external light collecting point of a light-tracking system at the other end (Figure 5c). Meanwhile, direct external light illumination is also applied to surround the reactor to increase light energy utilization. With such multilight sources, a higher HPR (0.038 L $L^{-1}h^{-1}$) and HY (3.15 mol-H_2 mol^{-1}-acetate) was obtained (Chen et al., 2006). Compared with the artificial light sources that consume electricity, directly utilization of sun light is apparently more desirable for practical application. For this purpose, a solar-energy-excited optical fiber photobioreactor was also developed for hydrogen production with reduced operating cost (Chen et al., 2008). One limitation of the solar energy is the unstable and incontinuous sunlight intensity over time. Thus, a supplementary artificial light source was adopted and a light-dependent resistor (LDR) for online monitoring and irradiation intensity control was installed to ensure a stable illumination. This combinative illumination system led to a stable HPR of 0.036 L $L^{-1}h^{-1}$ and HY of 2.45 mol-H_2 mol^{-1}-acetate, which were approximate to those under sole artificial source conditions (Chen et al., 2008).

Induced and diffused internal illumination presents another alternative to enhance light utilization. It has been recognized that the light conversion efficiency of a photobioreactor can be significantly affected by the spatial distribution of light. A multilayered photobioreactor with improved light distribution was constructed by Kondo et al. (2006). In this reactor, the cell suspension layer and clear medium layer were separated by a transparent permeable membrane and alternately arranged (Figure 5d). Such a multilayered configuration enabled an efficient induction and penetration of light through the medium layers and an enhanced illumination of the cells from both sides of the suspension layers. As a result, more light energy is supplied directly to the interior of the reactor, and a more uniform distribution of light illumination as well as a higher HPR were achieved compared to the conventional reactors with solely external illumination (Kondo et al., 2006).

6.4 Limitations of Photobioreactors

While reactors for dark-fermentation process are approaching an industrial scale, photobiological reactors are still in the development phase (Dasgupta et al., 2010). Various photobioreactors have been designed and tested for hydrogen production in the past decades. Small thickness of reactors is usually adopted to increase the A/V ratio and avoid shading effect, while internal illumination and diffused-light systems are also frequently applied to enhance light utilization. Nevertheless, the achievable HPRs of the existing phototrophic biological production systems are still orders of magnitude lower than those of the dark fermentators, and several obstacles are yet to be overcome in photobioreactor design and operation.

(1) One major challenge confronting the scaling-up of photobioreactors has been the inherently limited light penetration due to a self-shading effect. Despite of the rapid development of innovative photobioreactors with enhanced light utilization features, the light conversion efficiencies in such systems are still less than 10% so far (Brentner et al., 2010).

(2) A/V ratio, agitation, and temperature and gas exchange are also limiting factors in some photobioreactors. Large illumination area can be achieved in tubular-type reactors, but the high gas holdup and high energy input for pumping present two major bottlenecks. While higher photosynthetic efficiencies and effective control of gas pressure can be achieved in flat-plate photobioreactors, difficulty arises to achieve sufficient agitation and temperature control in such systems. Therefore, development of suitable photobioreactors is still challenging.

(3) Expensive reactors with high transmittance, good impermeability, and photochemical stability are usually adopted for photobiological processes. Although less expensive open systems such as outdoor ponds can also be used for photofermentation, they generally suffer from low HPR, poor stability, and inferior controllability. Hence, development of high-rate and closed-structure photobioreactors with innovative cheap materials and optimized designs are needed to make the process viable both technologically and economically.

7 MICROBIAL ELECTROLYSIS CELLS

7.1 Two-Chamber MECs

Apart from dark fermentation, biophotolysis, and photofermentation, MEC as an emerging technology was also intensively investigated for biohydrogen production (Lee et al., 2010b). In a MEC process, bacterial metabolism and electrochemistry were combined to achieve efficient H_2 production, usually at a high HY (Rozendal et al., 2008). ARB can anaerobically oxidize organic substances and transfer the produced electrons to an anode. The electrons are then conducted through an external circuit to the cathode, where protons were reduced to generate H_2 gas. In this process, an external power supply is usually applied to provide the necessary energy for hydrogen generation. Typically, a two-chamber configuration is adopted that consists of an anodic and a cathodic chamber separated by a proton

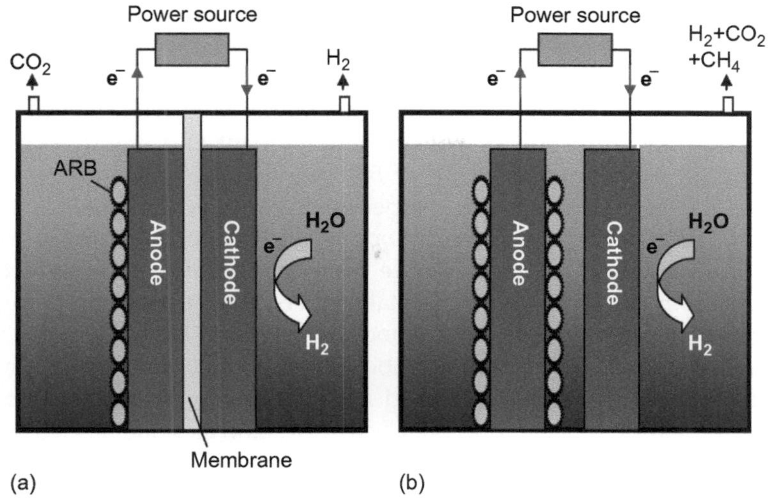

FIGURE 6 Schematic diagram of (a) two-chamber and (b) single-chamber MEC *(Call and Logan, 2008; Lee et al., 2010b; Liu et al., 2005).*

exchange membrane (PEM; Figure 6a). Two chamber MECs were firstly explored by Liu et al. (2005) using a plain carbon cloth anode, carbon paper cathode, and a maximum HY of 2.9 mol-H_2 mol^{-1}-acetate was obtained at an applied voltage of 0.25 V. In most cases, the anode reaction is a limiting step in which the electron production and transfer by ARB is constricted due to a limited anode surface. Therefore, anode surface modification is considered a useful approach to enhance H_2 production in such systems. Indeed, substantially improved H_2 production was achieved in a compact reactor system by applying a chemically modified three-dimensional graphite granule anode, where the HY reached 3.95 mol-H_2 mol^{-1}-acetate at 0.6 V (Cheng and Logan, 2007). Another inherent constraint of the two-chamber MEC is the low proton transfer rate and high energy losses caused by the PEM (Rozendal et al., 2007a). Therefore, removal of the PEM is expected to significantly reduce the system resistance and increase hydrogen production, which has led to the development of membraneless single-chamber MECs (Figure 6b; Call and Logan, 2008; Call et al., 2009; Lu et al., 2009; Tartakovsky et al., 2009).

7.2 Single-Chamber MECs

Single-chamber MECs are now gaining increasing interests attributed to its lower internal resistance and simple structure. Through the use of a membraneless system with improved graphite fiber brush anode and small electrode spacing, the pH gradient in the bulk liquid was reduced and the HPRs were significantly raised, reaching a maximum of 0.13 L L^{-1} h^{-1} at an applied voltage of 0.8 V (Call and Logan, 2008). This was significantly higher compared to most two-chamber systems (Liu et al., 2005). However, some other problems arise in a completely separator-free configuration, such as accelerated H_2 consumption by anode microorganisms and mixed gaseous products (Lee and Rittmann, 2009). As illustrated in

Figure 3b, an undesirable mixture of H_2 and other anodic metabolic gases such as CO_2, CH_4, and H_2S is usually released in such membraneless systems (Logan et al., 2008). Therefore, as a compromise, some alternative separators such as J-cloth were also tested for single-chamber MEC systems to permit ion transfer but reduce hydrogen gas diffusion (Escapa et al., 2009; Tartakovsky et al., 2009). With the employment of a J-cloth separator coupling with a novel gas-diffusion cathode, a HY of 5.4 mol-H_2 mol^{-1}-glycerol and HPR of up to 0.26 L L^{-1} h^{-1} was obtained, indicating a superior performance of such reactors compared to the conventional PEM and liquid-phase cathode systems. Other strategies to decrease anodic hydrogen consumption such as upflow-mode reactors were also recently proposed. In an upflow single-chamber MEC with a metal-catalyst-free cathode assembled on the top of the reactor and a graphite granule compacted anode, an HPR of 0.024 L L^{-1} h^{-1} was achieved at 1.0-V voltage (Lee et al., 2009b), which was about twofold higher than that of a conventional single-chamber system with platinum-coated cathode at similar voltage (Rozendal et al., 2007a). On the other hand, the development of gas-diffusion cathode has led to the initiation of membrane electrode assemblies in MEC design, which is also a single-chamber architecture (Rozendal et al., 2007a). In such a system, a membrane is integrated with the cathode that is loaded with a platinum catalyst layer at the gas collection chamber side, so that the cathode resistance can be reduced.

7.3 Self-Powered MECs

One biggest drawback of MECs is the need for external energy supply, that is, certain amount of electrical energy must be input to make the cathode potential negative enough for hydrogen generation. This increases cost and results in low net energy output. Thus, efforts are needed to reduce or even completely eliminate the energy supply by external electricity sources. In this respect, two novel MECs with interesting self-powering capabilities were recently proposed and showed great promise for practical use.

Instead of adopting an electricity power source, an MFC was utilized by Sun et al. (2008) to provide the extra energy needed for MEC processes. In this conception, the produced electricity from the MFC was in situ utilized by the MEC for hydrogen production, thus creating a situation of self-powering of the MEC-MFC coupled system. The maximum HPR and HY reached 0.62 mL L^{-1} h^{-1} and 1.6 mol-H_2 mol^{-1}-acetate respectively and no extra energy input was required in this system. Another innovative self-powering MEC design involves the in situ utilization of solar power (Chae et al., 2009). In this system, a dye-sensitized solar cell was employed to harvest light energy for powering the MEC. An HY of 3.14 mol-H_2 mol^{-1}-acetate was achieved under illuminated condition, indicating a critical role of light to drive the reaction for hydrogen evolution and a good feasibility of such an integrated strategy. However, the power efficiency of this solar-powered MEC was very low (10%), which limited the electricity supply.

Although the HPRs achieved in these novel systems are still far lower than those in normal MEC systems due to the limited achievable voltage, the rapid advances of MFC and light-conversion technologies in the future may hopefully breakthrough this barrier and significantly improve the hydrogen production efficiency to a practical level.

7.4 Limitations of MECs

Microbial electrolysis makes it possible to generate hydrogen from organics of diverse compositions such as wastewater, and opens the possibility of completely oxidizing non-fermentable substrate into CO_2 at high yield (Lee and Rittmann, 2009). However, the HPRs achieved in MECs are still very low compared to dark-fermentation processes, and there are several remaining constraints for the application of MECs.

(1) Both single-chamber and two-chamber reactors have been used for MEC processes. Although membrane elimination can help reduce the energy losses and attenuate pH gradient, a membraneless single-chamber MEC introduces risks of hydrogen consumption by methanogens or ARBs, which in turn decreases the HY and the purity of the produced gas (Lee et al., 2009b). Therefore, the limited property of the available membrane constitutes a challenge, to which the future development of separators with high conductivity and good cell retention capabilities may provide a solution. In addition, exploration of innovative electrode materials and configurations with enhanced properties is also expected to further lift the HPRs of MECs (Chae et al., 2009).

(2) High cost presents another bottleneck for the practical application of MECs, mainly due to the use of precious metal catalysts, expensive membrane, and external power source as a common practice. Recently, efforts are underway to reduce the cost and energy demand, such as adoption of biocathodes (Rozendal et al., 2007b), substitution of PEM with fabric cloth (Escapa et al., 2009), and application of self-powering systems (Chae et al., 2009; Sun et al., 2008). However, low HPR is still a major hurdle of all these systems. Cost and efficiency are two equally critical issues to be accounted for a biohydrogen production process. To achieve an efficient and cost-effective MEC process, there is still a long way to go.

8 COMPARISON OF VARIOUS BIOHYDROGEN REACTOR SYSTEMS

Various bioreactors for hydrogen production processes have been developed in the past decades with different degrees of success. Each of these reactors, including dark-fermentation reactors, photobioreactors, and MECs, has its unique advantages and inherent limitations. And the application of diverse immobilization and mixing techniques adds up to the complexity and versatility of reactors. A systematic comparison of the merits and drawbacks of different biohydrogen production systems is given in Table 1. Generally, considerably higher HPRs (ranging from 0.11 to 15.4 L $L^{-1}h^{-1}$) were found in dark-fermentation reactors compared to the biophotoreactors (0.007-0.038 L $L^{-1}h^{-1}$) and MECs (0.0006-0.26; Brentner et al., 2010; Levin et al., 2004). On the contrary, much higher HYs are commonly seen in the photobioreactors (6.75-9.45 mol-H_2 mol^{-1}-hexose) and MECs (4.8-16.56 mol-H_2 mol^{-1}-hexose). Therefore, the dark-fermentation reactors may be a superior choice for degradation of readily degradable sugar-based substrate, while the photobioreactor and MEC systems are more favorable for bioconversion of other feedstock such as practical waste streams.

Among the various options of dark-fermentation reactors, the granule-based reactors display distinctly better performances than suspended-growth HPF systems in terms of biomass

TABLE 1 Comparison of Hydrogen Production in Various High-Rate Bioreactors

Reactor	Culture	Substrate	VSS (g/L)	Maximum HPR (L L^{-1} h^{-1})	Maximum HY (mol-H_2 mol^{-1}-hexose)	Refs.
Integrated CSTR	HPF	Glucose	2.3	0.4	2.8	Hafez et al. (2009)
Integrated CSTR	HPF	Glucose	5.8	0.32	–	Oh et al. (2004a)
Integrated CSTR	HPF	Fructose	6.4	2.75	1.36	Lee et al. (2007)
Integrated CSTR	HPF	Glucose	9.5	0.11	0.96	Lee et al. (2009a)
Integrated CSTR	HPG	Sucrose	20	0.54	2.2	Fang et al. (2002)
Integrated CSTR	HPG	Glucose	7.7-16.0	3.26	1.84	Show et al. (2007)
Integrated CSTR	HPG	Molasses	15.5	0.41	–	Ren et al. (2010)
Integrated CSTR	HPG	Sucrose	35.4	15.4	3.5	Wu et al. (2006)
AFBR	HPG	Sucrose	15.0	1.32	0.57	Chang et al. (2002)
AFBR	HPG	Glucose	21.5	2.36	1.19	Zhang et al. (2007a)
AFBR	HPG	Sucrose		2.27	4.98	Lin et al. (2006)
AFBR	HPG	Glucose	29.3	0.25	2.6	Peintner et al. (2010)
AFBR	HPG	Sucrose		1.32	1.04	Wu et al. (2007)
AFBR	HPG	Glucose	34-37	7.6	1.7	Zhang et al. (2008a)
AFBR	HPG	Glucose		0.93	2.77	Wu et al. (2003)
AFBR	HPG	Glucose	–	0.97	2.49	Cavalcante de Amorim et al. (2009b)
PBR	HPG	Sucrose	15.8	1.32	–	Chang et al. (2002)
PBR	HPG	Sucrose	–	7.4	3.9	Lee et al. (2003)
PBR	HPG	Sucrose	26	7.3	3.03	Lee et al. (2004b)
PBR	HPG	Sucrose	–	–	2.48	Leite et al. (2008)
PBR	HPG	Glucose	18-24	0.98	1.11	Oh et al. (2004b)
UASB	HPG	Winery wastewater	9.4-11.7	0.16	2.14	Yu et al. (2002)
UASB	HPG	Sucrose	6.3	0.20	1.44	Yu and Mu (2006)
UASB	HPG	Glucose	–	0.43	1.6	Gavala et al. (2006)
UASB	HPG	Sucrose	19.4	0.71	3.47	Guo et al. (2008b)
UASB	HPG	Starch	8.26	1.64	0.94	Guo et al. (2008a)
Photobioreactor	Cyanobacteria	–	–	0.013	–	Tsygankov et al. (1998)
Photobioreactor	PSB	Malate	–	0.01	–	Eroglu et al. (2008)

TABLE 1 Comparison of Hydrogen Production in Various High-Rate Bioreactors—Cont'd

Reactor	Culture	Substrate	VSS (g/L)	Maximum HPR ($L\ L^{-1}\ h^{-1}$)	Maximum HY (mol-H_2 mol^{-1}-hexose)	Refs.
Photobioreactor	PSB	DL malic acid	–	0.007	6.75	Basak and Das (2010)
Photobioreactor	PSB	Acetate	–	0.038	9.45	Chen et al. (2006)
Photobioreactor	PSB	Acetate	–	0.032	7.35	Chen et al. (2008)
Photobioreactor	PSB	Lactate	–	0.021	–	Kondo et al. (2006)
MEC	ARB	Acetate	–	0.011	8.7	Liu et al. (2005)
MEC	ARB	Acetate	–	0.045	11.85	Cheng and Logan (2007)
MEC	ARB	Acetate	–	0.13		Call and Logan (2008)
MEC	ARB	Glycerol	–	0.025	10.8	Escapa et al. (2009)
MEC	ARB	Acetate	–	0.26	11.7	Tartakovsky et al. (2009)
MEC	ARB	Acetate		0.024	16.56	Lee et al. (2009b)
MEC	ARB	Acetate	–	0.0006	4.8	Sun et al. (2008)
MEC	ARB	Acetate	–		9.42	Chae et al. (2009)

retention and HPR. Although the highest HPR (15.0 $L\ L^{-1}\ h^{-1}$) reported so far was achieved in a CSTR with silicone-immobilized granules (Wu et al., 2006), the other granule-based configurations, especially UASB reactors, demonstrate higher stability and economical efficiency in long-term operation. In addition to the reactor configuration, the biohydrogen production processes can be affected by a number of other factors such as operation conditions, substrate type, and microbial species. Therefore, notwithstanding the differentiated performances of different reactor systems, it is not easy to draw a conclusion in regard to what configuration is better, even under a specific set of conditions. But it is worthwhile to put a glimpse into the characteristics of these difference systems and to find some hints for possible future improvements in reactor design and operation.

9 CHALLENGES AND FUTURE IMPLICATIONS

Although no commercial application has been established in biohydrogen production, the vast progresses in dark fermentation technology as well as the phototrophically and electrochemically assisted biotechnologies are pushing it to a near practical level. Reactor is the key to the successful application of biohydrogen production processes. For this purpose, a great number of biohydrogen reactors have been developed and extensively studied in the past decades. However, most of them were operated at a laboratory scale, and there are remaining barriers for the existing systems. Particularly, the granule-based reactors, thermophilic processes, and integrated systems hold great promise for practical application, but many challenges are yet to be addressed.

9.1 Granule-Based Reactors

Many high-rate biohydrogen systems have been developed based on granulation technologies, including the granule-based CSTR, AFBR, PBR, UASB, and even in some photobioreactors. In these systems, an enhanced biomass retention plus the superior properties of microbial granules contribute to high HPR and operating stability (Zhang et al., 2008b). However, the granule-based reactors generally suffer from constrictions in mass transfer, gas release, and carrier application.

9.1.1 Mass Transfer

One major challenge associated with granule-based systems is the limited mass transfer and the mass transfer efficiency might further decrease after a long-term operation due to the agglomeration of massive number of cells and insufficient mixing. This is especially true in UASB and PBR reactors, where insufficient mixing leads to poor substrate-biomass contact and thus low HPR. To improve the mixing and mass transfer efficiency, approaches via recycle of liquid and biogas have been universally adopted (Lee et al., 2004a). Also, the H/D ratio of the column-shaped configurations can be adjusted to further elevate the liquid upflow velocity (Lee et al., 2006). And supply of additional mechanical agitation is also helpful in lifting hydraulic turbulence and mass transfer (Lee et al., 2006). However, increasing liquid circulation or applying additional agitation would raise the energy consumption and increase the operating cost. Furthermore, a high superficial liquid velocity in the reactor may evoke washout and disruption of granules (Wu et al., 2009). Although the application of appropriate distributors in the reactor seemed to provide an effective strategy to reduce cell washout and boost H_2 production (Lo et al., 2009), it adds to reactor complexity and may increase gas holdup. Therefore, the mass transfer in granule-based reactors is a critical issue to be considered in reactor design.

9.1.2 Gas Release

Since the hydrogen concentration, specifically the dissolved hydrogen concentration, plays a significant role in hydrogen fermentation, it is essential to decrease this dissolved content and reduce hydrogen inhibition to microbes. Furthermore, an attachment of gas on the sludge particles may lead to sludge floatation and deteriorate system performance (Lin et al., 2006). Several approaches have been applied to reduce hydrogen concentrations in liquid phase through accelerate gas release, including intensive stirring (Lin et al., 2006; Wu et al., 2003), stripping with inert or recycle gases (Kim et al., 2006), and headspace pressure reduction (Mandal et al., 2006). But they invariably require considerable input of energy or even gas, causing increased operation costs or decreased hydrogen purity. Recently, another novel approach for enhanced hydrogen separation was proposed by Fritsch et al. (2008). Monoliths, a special material with high specific surface area and uneven structure, were added into the biohydrogen reactor as nucleation seeds to enhance bubble formation and increase hydrogen production. The success of this strategy implies that development of novel materials with enhanced properties for gas separation may offer another attractive avenue to promote gas release in biohydrogen systems.

9.1.3 *Biofilm Versus Self-Flocculated Granules*

Granulation technologies have been widely adopted for dark-fermentation reactors and demonstrated superior performance than suspended-growth systems. Three categories of granule systems have been applied in biohydrogen production, including surface attachment, self-flocculation, and gel entrapment approaches. All these demonstrate good capabilities in cell retention and hydrogen production. Nevertheless, the immobilized cells created by gel entrapping techniques tend to suffer from higher mass transfer resistance (Kumar and Das, 2001) and gas holdup within the gel-entrapped bioparticles (Wu et al., 2002). Comparatively, HPGs based on attached growth of bacteria and self-flocculation usually have higher porosity and better hydrogen production performances. However, regarding the benefits of carrier-based and self-flocculated granule system, controversy still remains despite of their comparable biomass retention (34-37 g-VSS L^{-1}) and hydrogen production capabilities (HPR = 6.6-7.6 L $L^{-1}h^{-1}$; Cavalcante de Amorim et al., 2009a; Davila-Vazquez et al., 2008; Zhang et al., 2008a). Kim et al. (2005) compared the characteristics of polyvinyl alcohol (PVA)-attached and self-formed HPGs in continuous hydrogen production systems, and observed that the microorganisms tended to accumulate more in the self-formed granules than the biofilm. However, other researchers demonstrated that, while a long start-up period is usually required for the self-flocculated granules, the addition of appropriate carriers can lead to a rapid formation of HPGs (Zhang et al., 2008a), and the presence of carrier can also effectively induce the formation of self-flocculated granules (Barros et al., 2010; Lee et al., 2004a, 2006). A comparison of both types of HPGs in AFBRs by Zhang et al. (2008a) revealed that carrier-based granules were liable to be washed out as the size increased and gradually replaced by self-flocculated ones during operation. Based on these studies, we can conclude that, although the washout of carrier presents a potential problem to be addressed, the addition of appropriate carriers can provide an effective approach to accelerate granulation and improve the biohydrogen production performances.

9.1.4 *Immobilized-PSB System*

To date, most previous works regarding biohydrogen production have been focusing on HPG dark fermentation, while only a few studies on PSB immobilization have been reported (Liao et al., 2010). Immobilization of PSB onto appropriate carriers was found to also increase hydrogen production in photobioreactor systems by providing a favorable environmental for PSB growth and maintaining the structural stability (Brentner et al., 2010; Tian et al., 2009). Chen and Chang (2006) added clay and silica gel into a photobioreactor for cell immobilization, and found a 67.2% increase in HPR in this enhanced system. Tian et al. (2009; 2010) reported that immobilization of *Rhodopseudomonas palustris* on PVA matrix or packed glass beads could remarkably facilitate biomass retention and hydrogen production. However, it should be noted that application of immobilization technologies may also arise difficulties in the light penetration (Tian et al., 2009). In this respect, transparent inorganic carriers with an adequate adsorptivity might provide a desirable choice. Hence, development of cheap carrier materials with high light transmittance and surface area might be an important aspect of future efforts on photobiological hydrogen production technologies.

In summary, the granule-based reactors present a preferred choice for continuous biohydrogen production and exhibit good prospect for practical applications. Although many challenges remain unsolved at the present stage, Rapid progresses in carrier materials and reactor configurations is bringing great possibilities for a breakthrough.

9.2 Thermophilic Fermentation Reactors

Biohydrogen fermentation processes are known to be influenced by a number of environmental and operating parameters (Wang and Wan, 2009), among which the temperature shows an especially significant influence. A thermophilic environment appeared to be more favorable for H_2-producting bacteria than the mesophilic conditions (Peintner et al., 2010; Yu et al., 2002). Thus, thermophilic fermentation processes have attracted increasing attention in recent years, and a variety of thermophilic fermentation reactors with sophisticated structure have been developed accordingly (Akutsu et al., 2009a; Kotsopoulos et al., 2006). One most important advantage of thermophilic processes is the suppressed methanogen activity at high temperature, which reduces hydrogen consumption (Gavala et al., 2006; Shin et al., 2004). However, there are specific constraints for such thermophilic fermentation processes and the matching reactors.

Firstly, extreme thermophilic microorganisms are known to grow slowly and generate low cell densities, which result in low hydrogen productivities. Thus, approaches for improving microbial growth like immobilization technologies need to be explored in such reactors for high-temperature operations. Secondly, high energy input is required to maintain the thermophilic conditions. One possible solution to this could be using high-temperature food wastewaters such as those from coffee processing, palm oil mill, or distillery plants (Yu et al., 2002), but this may bring about process instability due to influent fluctuations. Finally, because of the high environmental sensitivity of thermophiles (Peintner et al., 2010), the accurate control of pH and temperature is very important. All these problems, plus the cost, present challenges in reactor design and operation. Thus, to realize a stable, efficient, and cost-effective thermophilic fermentation process, a properly designed thermoreactor with heat recovery and pH control systems and incorporating immobilization techniques shall be needed, which warrants more future investigations.

9.3 Hybrid Reactors

Because each individual reactor has its limitations, combination of two or multiple reactors for hydrogen production appears to be an attractive solution. The merits of two-stage biohydrogen production processes have been long demonstrated. Through coupling the dark-fermentation process with a subsequent step such as photofermentation or SEM, not only a higher hydrogen recovery but also a nearly complete degradation of organics can be achieved (Hallenbeck and Ghosh, 2009; Lee et al., 2010b). However, the dark-photo fermentation processes so far are mostly performed in batch-transfer mode (Lu et al., 2009; Su et al., 2009). Batch operation is convenient and easily controllable though, a continuous waste utilization and hydrogen production process is more preferable for large-scale practical application. Currently, an appropriate design of reactors for the continuous-operating hybrid processes is still lacking, and several potential concerns are yet to be addressed in reactor design and operation.

One of most critical considerations is how to bridge the two reactors in a convenient and efficient way. Membrane system seems to be a viable choice. Dialysis/microfiltration membranes have been successfully applied in a two-stage system for methane and PHA production (Du and Yu, 2002). However, the feasibility of membrane separation for a dark-photo fermentation hybrid system is yet to be ascertained. Moreover, membrane fouling may become a challenge in such systems. Another aspect to be considered in reactor design is how to ensure optimized conditions of individual units without affecting each other, as the favorable conditions for the two stages may differ significantly. Finally, a hybrid-process system apparently adds to complexity of reactor design and the overall cost; therefore, the economic aspects of such systems should also be evaluated for practical application.

9.4 Reactor Optimization and Scaling Up

9.4.1 Reactor Optimization

Notwithstanding the great progresses achieved in reactor design and operation, there are still many defects in such systems which require further improvement and optimization. For example, one of the major obstacles of dark-fermentation processes has been the coexistence of hydrogen-consuming bacteria especially methanogens. To suppress these bacteria, approaches such as heat treatment, pH control, or utilization of chemical inhibitors are necessary, and these raise specific requirements on reactor design. Feedstock for biohydrogen production is another problem. Currently, most of the dark-fermentation processes use expensive readily degradable substrates such as glucose and sucrose, while only a few studies on the use of lignocellulosic biomass have been reported (Guo et al., 2010). Indeed, lignocellulosic wastes, such as agricultural biomass and food waste, present a highly desirable feedstock for sustainable biohydrogen production due to their high abundance, wide availability, and low cost. One major limitation is the low biodegradability of the lignocellulosic contents; thus, usually a longer HRT is needed for the process. While most of the existing bioreactors were designed for processing of the sugar-based substrates, optimized reactors for processing of such hard-degrading lignocellulosic wastes should be specifically developed, where longer HRT and high retention of efficient bacteria are more preferred.

For reactor optimization, numerical models can provide a useful tool, because a better understanding on how a bioreactor functions would benefit reactor design and operation. Various types of biohydrogen production reactors have been designed, but most of them are based on empirical or semiempirical studies; hence, it is difficult to achieve a truly optimized reactor design and an accurate control of the process due to a lack of quantitative information. This opens up the demand for reactor simulation, where mathematical approaches can be employed to quantitatively describe the hydrodynamics, mass transfer, and bacterial behaviors in the reactors. A variety of numerical models and tools have been utilized to simulate and optimize the biohydrogen reactors (Su et al., 2010; Wang et al., 2009). Particularly, computational fluid dynamics (CFD) has been demonstrated an effective approach to simulate the fluid flow, heat and mass transfer, and the reactions. In addition, mathematical models based on neural network and genetic algorithm have also been successfully applied to predict the hydrogen production in granule-based reactors (Mu and Yu, 2007). In another study, an optimized photobioreactor design was achieved to maximize the solar to hydrogen energy conversion efficiency based on the process kinetics (Berberoglu and Pilon, 2010).

These studies provide solid evidences that the biohydrogen reactors can be well simulated and optimized through mathematical approaches. Nevertheless, most of these studies are based on laboratory-scale experiments using synthesized wastewater, and our knowledge about the interrelations among different parameters is still limited. In light of the high complexity and significant variety in practical processes, the simulation and optimization of biohydrogen reactors for real wastewater treatment may present a challenging task.

9.4.2 *Scaling Up*

To date, none of the biohydrogen approaches is ready to meet a large-scale operation requirement, and no industrial-scale bioreactor has been set up. On the one hand, our limited knowledge on the hydrogen production processes sets barriers for reactor optimization and further scaling up. On the other, some new challenges may arise during the scaling up. For example, for dark-fermentation processes, a full mixing can be well achieved in small-size reactors, but usually present a big problem in large tanks. For phototropic hydrogen production processes, one of the biggest challenges in scaling up is the serious self-shade effects, and the manufacturing of large-scale reactors raises higher demand on material intensity and transmittance. For MECs, the requirements for extra energy input and expensive metal cathode are two major obstacles for its large-scale application, while the rapid growth and competition of the non-ARBs contained in real wastewater present another potential challenge for the practical application. A settlement of these challenges shall rely on a better understanding of the physiological, biological, and hydrodynamic aspects of various biohydrogen systems, a further development of efficient immobilization, thermophilic-fermentation and hybrid-process technologies, as well as the evolution of low-cost innovative materials with enhanced properties. All these warrant intensive future investigations.

10 CONCLUSIONS

Biohydrogen production has been a subject of intensive study in the past decades. So far, a great number of bioreactors for high-rate biohydrogen production have been developed, including integrated CSTR, AFBR, PBR, UASB, photobioreactor, and MEC. Here, we offer a systematic comparison of the characteristics and performances of these reactors in terms of HRT, HY, biomass retention, and operating stability. Dark fermentation systems generally have higher HPRs but lower HYs compared to photobioreactors and MECs, while a better biomass retention and operation stability are commonly found for the granule-based reactors than the floc systems. Nevertheless, each type of reactor has its advantages and inherent limitations, and their performance is usually dependent heavily on the specific environmental and operational conditions. Thus, it is difficult to tell which configuration is superior, and further optimization and improvement of each reactor can be expected. In fact, the recent development in granule-based reactors, thermophilic processes and integrated systems, and the progress of numerical simulation technologies have brought new opportunities for the application of biohydrogen production processes. However, several challenges remain in reactor design and scaling up and further investigations are warranted to address these issues.

References

Adav, S.S., Lee, D.J., Show, K.Y., Tay, J.H., 2008. Aerobic granular sludge: recent advances. Biotechnol. Adv. 26, 411–423.

Akkerman, I., Janssen, M., Rocha, J., Wijffels, R.H., 2002. Photobiological hydrogen production: photochemical efficiency and bioreactor design. Int. J. Hydrogen Energy 27, 1195–1208.

Akutsu, Y., Lee, D.Y., Chi, Y.Z., Li, Y.Y., Harada, H., Yu, H.Q., 2009a. Thermophilic fermentative hydrogen production from starch-wastewater with bio-granules. Int. J. Hydrogen Energy 34, 5061–5071.

Barros, A.R., Cavalcante de Amorim, E.L., Reis, C.M., Shida, G.M., Silva, E.L., 2010. Biohydrogen production in anaerobic fluidized bed reactors: effect of support material and hydraulic retention time. Int. J. Hydrogen Energy 35, 3379–3388.

Bartacek, J., Zabranska, J., Lens, P.N.L., 2007. Developments and constraints in fermentative hydrogen production. Biofuels Bioprod. Bioref. 1, 201–214.

Basak, N., Das, D., 2010. Photofermentative hydrogen production using purple non-sulfur bacteria *Rhodobacter sphaeroides* O.U.001 in an annular photobioreactor: a case study. Biomass Bioenergy 33, 911–919.

Berberoglu, H., Pilon, L., 2010. Maximizing the solar to H_2 energy conversion efficiency of outdoor photobioreactors using mixed cultures. Int. J. Hydrogen Energy 35, 500–510.

Brentner, L.B., Peccia, J., Zimmerman, J.B., 2010. Challenges in developing biohydrogen as a sustainable energy source: implications for a research agenda. Environ. Sci. Technol. 44, 2243–2254.

Call, D., Logan, B.E., 2008. Hydrogen production in a single chamber microbial electrolysis cell lacking a membrane. Environ. Sci. Technol. 42, 3401–3406.

Call, D.F., Merrill, M.D., Logan, B.E., 2009. High surface area stainless steel brushes as cathodes in microbial electrolysis cells. Environ. Sci. Technol. 43, 2179–2183.

Cavalcante de Amorim, E.L., Barros, A.R., Rissato Zamariolli Damianovic, M.H., Silva, E.L., 2009a. Anaerobic fluidized bed reactor with expanded clay as support for hydrogen production through dark fermentation of glucose. Int. J. Hydrogen Energy 34, 783–790.

Cavalcante de Amorim, E.L., Barros, A.R., Rissato, Z.D.M.H., Silva, E.L., 2009b. Anaerobic fluidized bed reactor with expanded clay as support for hydrogen production through dark fermentation of glucose. Int. J. Hydrogen Energy 34, 783–790.

Chae, K.J., Choi, M.J., Kim, K.Y., Ajayi, F.F., Chang, I.S., Kim, I.S., 2009. A solar-powered microbial electrolysis cell with a platinum catalyst-free cathode to produce hydrogen. Environ. Sci. Technol. 43, 9525–9530.

Chang, F.Y., Lin, C.Y., 2004. Biohydrogen production using an up-flow anaerobic sludge blanket reactor. Int. J. Hydrogen Energy 29, 33–39.

Chang, J.S., Lee, K.S., Lin, P.J., 2002. Biohydrogen production with fixed-bed bioreactors. Int. J. Hydrogen Energy 27, 1167–1174.

Chen, C.Y., Chang, J.S., 2006. Enhancing phototropic hydrogen production by solid-carrier assisted fermentation and internal optical-fiber illumination. Proc. Biochem. 41, 2041–2049.

Chen, C.Y., Lee, C.M., Chang, J.S., 2006. Feasibility study on bioreactor strategies for enhanced photohydrogen production from *Rhodopseudomonas palustris* WP3-5 using optical-fiber-assisted illumination systems. Int. J. Hydrogen Energy 31, 2345–2355.

Chen, C.Y., Saratale, G.D., Lee, C.M., Chen, P.C., Chang, J.S., 2008. Phototrophic hydrogen production in photobioreactors coupled with solar-energy-excited optical fibers. Int. J. Hydrogen Energy 33, 6886–6895.

Cheng, S.A., Logan, B.E., 2007. Sustainable and efficient biohydrogen production via electrohydrogenesis. PNAS 104, 18871–18873.

Cheong, D.Y., Hansen, C.L., Stevens, D.K., 2007. Production of biohydrogen by mesophilic anaerobic fermentation in an acid-phase sequencing batch reactor. Biotechnol. Bioeng. 96, 421–432.

Dasgupta, C.N., Jose Gilbert, J., Lindblad, P., Heidorn, T., Borgvang, S.A., Skjanes, K., et al., 2010. Recent trends on the development of photobiological processes and photobioreactors for the improvement of hydrogen production. Int. J. Hydrogen Energy 35, 10218–10238.

Davila-Vazquez, G., Arriaga, S., Alatriste-Mondragón, F., de León-Rodríguez, A., Rosales-Colunga, L., Razo-Flores, E., 2008. Fermentative biohydrogen production: trends and perspectives. Rev. Environ. Sci. Biotechnol. 7, 27–45.

Du, G., Yu, J., 2002. Green technology for conversion of food scraps to biodegradable thermoplastic polyhydroxyalkanoates. Environ. Sci. Technol. 36, 5511–5516.

Eroglu, I., Tabanoglu, A., Gunduz, U., Eroglu, E., Yucel, M., 2008. Hydrogen production by *Rhodobacter sphaeroides* O.U.001 in a flat plate solar bioreactor. Int. J. Hydrogen Energy 33, 531–541.

Escapa, A., Manuel, M.F., Moran, A., Gomez, X., Guiot, S.R., Tartakovsky, B., 2009. Hydrogen production from glycerol in a membraneless microbial electrolysis Cell. Energy Fuels 23, 4612–4618.

Fang, H.H.P., Liu, H., 2002. Effect of pH on hydrogen production from glucose by a mixed culture. Bioresour. Technol. 82, 87–93.

Fang, H.H.P., Liu, H., Zhang, T., 2002. Characterization of a hydrogen-producing granular sludge. Biotechnol. Bioeng. 78, 44–52.

Fritsch, M., Hartmeier, W., Chang, J.S., 2008. Enhancing hydrogen production of *Clostridium butyricum* using a column reactor with square-structured ceramic fittings. Int. J. Hydrogen Energy 33, 6549–6557.

Gavala, H.N., Skiadas, I.V., Ahring, B.K., 2006. Biological hydrogen production in suspended and attached growth anaerobic reactor systems. Int. J. Hydrogen Energy 31, 1164–1175.

Gebicki, J., Modigell, M., Schumacher, M., Burgb, J.V.D., Roebroeck, E., 2009. Development of photobioreactors for anoxygenic production of hydrogen by purple bacteria. Chem. Eng. Trans. 18, 363–366.

Gebicki, J., Modigell, M., Schumacher, M., van der Burg, J., Roebroeck, E., 2010. Comparison of two reactor concepts for anoxygenic H_2 production by Rhodobacter capsulatus. J. Clean. Prod. 18, S36–S42.

Guo, W.Q., Ren, N.Q., Chen, Z.B., Liu, B.F., Wang, X.J., Xiang, W.S., et al., 2008a. Simultaneous biohydrogen production and starch wastewater treatment in an acidogenic expanded granular sludge bed reactor by mixed culture for long-term operation. Int. J. Hydrogen Energy 33, 7397–7404.

Guo, W.Q., Ren, N.Q., Wang, X.J., Xiang, W.S., Meng, Z.H., Ding, J., et al., 2008b. Biohydrogen production from ethanol-type fermentation of molasses in an expanded granular sludge bed (EGSB) reactor. Int. J. Hydrogen Energy 33, 4981–4988.

Guo, X.M., Trably, E., Latrille, E., Carrere, H., Steyer, J.P., 2010. Hydrogen production from agricultural waste by dark fermentation: a review. Int. J. Hydrogen Energy 35, 10660–10673.

Hafez, H., Baghchehsaraee, B., Nakhla, G., Karamanev, D., Margaritis, A., El Naggar, H., 2009. Comparative assessment of decoupling of biomass and hydraulic retention times in hydrogen production bioreactors. Int. J. Hydrogen Energy 34, 7603–7611.

Hafez, H., Nakhla, G., El. Naggar, M.H., Elbeshbishy, E., Baghchehsaraee, B., 2010. Effect of organic loading on a novel hydrogen bioreactor. Int. J. Hydrogen Energy 35, 81–92.

Hallenbeck, P.C., Benemann, J.R., 2002. Biological hydrogen production; fundamentals and limiting processes. Int. J. Hydrogen Energy 27, 1185–1193.

Hallenbeck, P.C., Ghosh, D., 2009. Advances in fermentative biohydrogen production: the way forward? Trends Biotechnol. 27, 287–297.

Hawkes, F.R., Hussy, I., Kyazze, G., Dinsdale, R., Hawkes, D.L., 2007. Continuous dark fermentative hydrogen production by mesophilic microflora: principles and progress. Int. J. Hydrogen Energy 32, 172–184.

Hoekema, S., Bijmans, M., Janssen, M., Tramper, J., Wijffels, R.H., 2002. A pneumatically agitated flat-panel photobioreactor with gas re-circulation: anaerobic photoheterotrophic cultivation of a purple non-sulfur bacterium. Int. J. Hydrogen Energy 27, 1331–1338.

Holladay, J.D., Hu, J., King, D.L., Wang, Y., 2009. An overview of hydrogen production technologies. Catal. Today 139, 244–260.

Kim, J.O., Kim, Y.H., Ryu, J.Y., Song, B.K., Kim, I.H., Yeom, S.H., 2005. Immobilization methods for continuous hydrogen gas production biofilm formation versus granulation. Proc. Biochem. 40, 1331–1337.

Kim, D.H., Han, S.K., Kim, S.H., Shin, H.S., 2006. Effect of gas sparging on continuous fermentative hydrogen production. Int. J. Hydrogen Energy 31, 2158–2169.

Kondo, T., Wakayama, T., Miyake, J., 2006. Efficient hydrogen production using a multi-layered photobioreactor and a photosynthetic bacterium mutant with reduced pigment. Int. J. Hydrogen Energy 31, 1522–1526.

Koskinen, P.E.P., Kaksonen, A.H., Puhakka, J.A., 2007. The relationship between instability of H_2 production and compositions of bacterial communities within a dark fermentation fluidized-bed bioreactor. Biotechnol. Bioeng. 97, 742–758.

Kotsopoulos, T.A., Zeng, R.J., Angelidaki, I., 2006. Biohydrogen Production in Granular Up-Flow Anaerobic Sludge Blanket (UASB) Reactors with Mixed Cultures Under Hyper-Thermophilic Temperature (70 °C). In: A Wiley Company, vol. 94. Wiley Subscription Services, Inc., pp. 296–302.

Kumar, N., Das, D., 2001. Continuous hydrogen production by immobilized *Enterobacter cloacae* IIT-BT 08 using lignocellulosic materials as solid matrices. Enzyme Microb. Technol. 29, 280–287.

Lee, H.S., Rittmann, B.E., 2009. Significance of biological hydrogen oxidation in a continuous single-chamber microbial electrolysis cell. Environ. Sci. Technol. 44, 948–954.

Lee, K.S., Lo, Y.S., Lo, Y.C., Lin, P.J., Chang, J.S., 2003. H_2 production with anaerobic sludge using activated-carbon supported packed-bed bioreactors. Biotechnol. Lett. 25, 133–138.

Lee, K.S., Lo, Y.S., Lo, Y.C., Lin, P.J., Chang, J.S., 2004a. Operation strategies for biohydrogen production with a high-rate anaerobic granular sludge bed bioreactor. Enzyme Microb. Technol. 35, 605–612.

Lee, K.S., Wu, J.F., Lo, Y.S., Lo, Y.C., Lin, P.J., Chang, J.S., 2004b. Anaerobic hydrogen production with an efficient carrier-induced granular sludge bed bioreactor. Biotechnol. Bioeng. 87, 648–657.

Lee, K.S., Lo, Y.C., Lin, P.J., Chang, J.S., 2006. Improving biohydrogen production in a carrier-induced granular sludge bed by altering physical configuration and agitation pattern of the bioreactor. Int. J. Hydrogen Energy 31, 1648–1657.

Lee, K.S., Lin, P.J., Fangchiang, K., Chang, J.S., 2007. Continuous hydrogen production by anaerobic mixed microflora using a hollow-fiber microfiltration membrane bioreactor. Int. J. Hydrogen Energy 32, 950–957.

Lee, D.Y., Li, Y.Y., Noike, T., 2009a. Continuous H_2 production by anaerobic mixed microflora in membrane bioreactor. Bioresour. Technol. 100, 690–695.

Lee, H.S., Torres, C., Parameswaran, P., Rittmann, B.E., 2009b. Fate of H_2 in an upflow single-chamber microbial electrolysis cell using a metal-catalyst-free cathode. Environ. Sci. Technol. 43, 7971–7976.

Lee, D.Y., Li, Y.Y., Noike, T., 2010a. Influence of solids retention time on continuous H_2 production using membrane bioreactor. Int. J. Hydrogen Energy 35, 52–60.

Lee, H.S., Vermaas, W.F.J., Rittmann, B.E., 2010b. Biological hydrogen production: prospects and challenges. Trends Biotechnol. 28, 262–271.

Leite, J.A.C., Fernandes, B.S., Pozzi, E., Barboza, M., Zaiat, M., 2008. Application of an anaerobic packed-bed bioreactor for the production of hydrogen and organic acids. Int. J. Hydrogen Energy 33, 579–586.

Levin, D.B., Pitt, L., Love, M., 2004. Biohydrogen production: prospects and limitations to practical application. Int. J. Hydrogen Energy 29, 173–185.

Liao, Q., Wang, Y.J., Wang, Y.Z., Zhu, X., Tian, X., Li, J., 2010. Formation and hydrogen production of photosynthetic bacterial biofilm under various illumination conditions. Bioresour. Technol. 101, 5315–5324.

Li, C., Zhang, T., Fang, H.H.P., 2006. Fermentative hydrogen production in packed-bed and packing-free upflow reactors. Water Sci. Technol. 54, 95–103.

Li, Y.F., Hui, Y.M., Yao, X., Wang, L., Song, Q.W., Liu, S., et al., 2010. Continuous fermentative hydrogen production from brown sugar using EGSB reactor. Adv. Mater. Res. 113-116, 1132–1137.

Lin, C.N., Wu, S.Y., Chang, J.S., 2006. Fermentative hydrogen production with a draft tube fluidized bed reactor containing silicone-gel-immobilized anaerobic sludge. Int. J. Hydrogen Energy 31, 2200–2210.

Liu, H., Fang, H.H.P., 2003. Hydrogen production from wastewater by acidogenic granular sludge. Water Sci. Technol. 47, 153–158.

Liu, H., Grot, S., Logan, B.E., 2005. Electrochemically assisted microbial production of hydrogen from acetate. Environ. Sci. Technol. 39, 4317–4320.

Liu, X.W., Sheng, G.P., Yu, H.Q., 2009. Physicochemical characteristics of microbial granules. Biotechnol. Adv. 27, 1061–1070.

Lo, Y.C., Lee, K.S., Lin, P.J., Chang, J.S., 2009. Bioreactors configured with distributors and carriers enhance the performance of continuous dark hydrogen fermentation. Bioresour. Technol. 100, 4381–4387.

Logan, B.E., Call, D., Cheng, S., Hamelers, H.V.M., Sleutels, T.H.J.A., Jeremiasse, A.W., et al., 2008. Microbial electrolysis cells for high yield hydrogen gas production from organic matter. Environ. Sci. Technol. 42, 8630–8640.

Lu, L., Ren, N.Q., Xing, D.F., Logan, B.E., 2009. Hydrogen production with effluent from an ethanol-H-2-coproducing fermentation reactor using a single-chamber microbial electrolysis cell. Biosens Bioelectron 24, 3055–3060.

Mandal, B., Nath, K., Das, D., 2006. Improvement of biohydrogen production under decreased partial pressure of *Enterobacter cloacae*. Biotechnol. Lett. 28, 831–835.

Miron, A.S., Camacho, F.G., Gomez, A.C., Molina, G.E., Chisti, Y., 2000. Bubble column and airlift photobioreactors for algal culture. AIChE J. 46, 1872–1887.

Mu, Y., Yu, H.Q., 2006. Biological hydrogen production in a UASB reactor with granules. I: physicochemical characteristics of hydrogen-producing granules. Biotechnol. Bioeng. 94, 980–987.

Mu, Y., Yu, H.Q., 2007. Simulation of biological hydrogen production in a UASB reactor using neural network and genetic algorithm. Int. J. Hydrogen Energy 32, 3308–3314.

Nath, K., Das, D., 2009. Effect of light intensity and initial pH during hydrogen production by an integrated dark and photofermentation process. Int. J. Hydrogen Energy 34, 7497–7501.

Oh, S.E., Iyer, P., Bruns, M.A., Logan, B.E., 2004a. Biological Hydrogen Production Using a Membrane Bioreactor. In: A Wiley Company, vol. 87. Wiley Subscription Services, Inc., pp. 119–127.

Oh, Y.K., Kim, S.H., Kim, M.S., Park, S., 2004b. Thermophilic biohydrogen production from glucose with trickling biofilter. Biotechnol. Bioeng. 88, 690–698.

Oh, Y.K., Seol, E.H., Kim, M.S.M.S., Park, S., 2004c. Photoproduction of hydrogen from acetate by a chemoheterotrophic bacterium *Rhodopseudomonas palustris* P4. Int. J. Hydrogen Energy 29, 1115–1121.

Otsuki, T., Uchiyama, S., Fujiki, K., Fukunaga, S., 1998. Hydrogen production by a floating-type photobioreactor. In: Biohydrogen. Plenum Press, London, pp. 369–374.

Palazzi, E., Fabiano, B., Perego, P., 2000. Process development of continuous hydrogen production by *Enterobacter aerogenes* in a packed column reactor. Bioproc. Eng. 22, 205–213.

Peintner, C., Zeidan, A.A., Schnitzhofer, W., 2010. Bioreactor systems for thermophilic fermentative hydrogen production: evaluation and comparison of appropriate systems. J. Clean. Prod. 18, S15–S22.

Rachman, M.A., Nakashimada, Y., Kakizono, T., Nishio, N., 1998. Hydrogen production with high yield and high evolution rate by self-flocculated cells of Enterobacter aerogenes in a packed-bed reactor. Appl. Microbiol. Biotechnol. 49, 450–454.

Ren, N.Q., Li, J.Z., Li, B.Q., Wang, Y., Liu, S.R., 2006. Biohydrogen production from molasses by anaerobic fermentation with a pilot-scale bioreactor system. Int. J. Hydrogen Energy 31, 2147–2157.

Ren, N.Q., Chua, H., Chan, S.Y., Tsang, Y.F., Wang, Y.J., Sin, N., 2007. Assessing optimal fermentation type for biohydrogen production in continuous-flow acidogenic reactors. Bioresour. Technol. 98, 1774–1780.

Ren, N.Q., Tang, J., Liu, B.F., Guo, W.Q., 2010. Biological hydrogen production in continuous stirred tank reactor systems with suspended and attached microbial growth. Int. J. Hydrogen Energy 35, 2807–2813.

Rozendal, R.A., Hamelers, H.V.M., Molenkamp, R.J., Buisman, C.J.N., 2007a. Performance of single chamber biocatalyzed electrolysis with different types of ion exchange membranes. Water Res. 41, 1984–1994.

Rozendal, R.A., Jeremiasse, A.W., Hamelers, H.V.M., Buisman, C.J.N., 2007b. Hydrogen Production with a Microbial Biocathode. Environ. Sci. Technol. 42, 629–634.

Rozendal, R.A., Hamelers, H.V.M., Rabaey, K., Keller, J., Buisman, C.J.N., 2008. Towards practical implementation of bioelectrochemical wastewater treatment. Trends Biotechnol. 26, 450–459.

Shen, L., Zhou, Y., Mahendran, B., Bagley, D.M., Liss, S.N., 2010. Membrane fouling in a fermentative hydrogen producing membrane bioreactor at different organic loading rates. J. Membr. Sci. 360, 226–233.

Shin, H.S., Youn, J.H., Kim, S.H., 2004. Hydrogen production from food waste in anaerobic mesophilic and thermophilic acidogenesis. Int. J. Hydrogen Energy 29, 1355–1363.

Shi, X.Y., Yu, H.Q., 2005. Response surface analysis on the effect of cell concentration and light intensity on hydrogen production by *Rhodopseudomonas capsulata*. Proc. Biochem. 40, 2475–2481.

Shi, X.Y., Yu, H.Q., 2006. Continuous production of hydrogen from mixed volatile fatty acids with *Rhodopseudomonas capsulata*. Int. J. Hydrogen Energy 31, 1641–1647.

Show, K.Y., Zhang, Z.P., Tay, J.H., Liang, D.T., Lee, D.J., Jiang, W.J., 2007. Production of hydrogen in a granular sludge-based anaerobic continuous stirred tank reactor. Int. J. Hydrogen Energy 32, 4744–4753.

Su, H., Cheng, J., Zhou, J., Song, W., Cen, K., 2009. Combination of dark- and photo-fermentation to enhance hydrogen production and energy conversion efficiency. Int. J. Hydrogen Energy 34, 8846–8853.

Su, Z., Kang, R., Shi, S., Cong, W., Cai, Z., 2010. Study on the destabilization mixing in the flat plate photobioreactor by means of CFD. Biomass Bioenergy 34, 1879–1884.

Sun, M., Sheng, G.P., Zhang, L., Xia, C.R., Mu, Z.X., Liu, X.W., et al., 2008. An MEC-MFC-coupled system for biohydrogen production from acetate. Environ. Sci. Technol. 42, 8095–8100.

Tartakovsky, B., Manuel, M.F., Wang, H., Guiot, S.R., 2009. High rate membrane-less microbial electrolysis cell for continuous hydrogen production. Int. J. Hydrogen Energy 34, 672–677.

Tian, X., Liao, Q., Liu, W., Wang, Y.Z., Zhu, X., Li, J., et al., 2009. Photo-hydrogen production rate of a PVA-boric acid gel granule containing immobilized photosynthetic bacteria cells. Int. J. Hydrogen Energy 34, 4708–4717.

Tian, X., Liao, Q., Zhu, X., Wang, Y., Zhang, P., Li, J., et al., 2010. Characteristics of a biofilm photobioreactor as applied to photo-hydrogen production. Bioresour. Technol. 101, 977–983.

Tredici, M.R., Zittelli, G.C., 1998. Efficiency of sunlight utilization: tubular versus flat photobioreactors. Biotechnol. Bioeng. 57, 187–197.

Tsygankov, A.A., Hall, D.O., Liu, J., Rao, K.K., 1998. An automated helical photobioreactor incorporating cyanobacteria for continuous hydrogen production. In: Biohydrogen. Plenum Press, London, pp. 431–440.

van Groenestijn, J.W., Hazewinkel, J.H.O., Nienoord, M., Bussmann, P.J.T., 2002. Energy aspects of biological hydrogen production in high rate bioreactors operated in the thermophilic temperature range. Int. J. Hydrogen Energy 27, 1141–1147.

Vijayaraghavan, K., Ahmad, D., 2006. Biohydrogen generation from palm oil mill effluent using anaerobic contact filter. Int. J. Hydrogen Energy 31, 1284–1291.

Vijayaraghavan, K., Ahmad, D., Soning, C., 2007. Bio-hydrogen generation from mixed fruit peel waste using anaerobic contact filter. Int. J. Hydrogen Energy 32, 4754–4760.

Wang, J., Wan, W., 2009. Factors influencing fermentative hydrogen production: a review. Int. J. Hydrogen Energy 34, 799–811.

Wang, X., Ding, J., Ren, N.Q., Liu, B.F., Guo, W.Q., 2009. CFD simulation of an expanded granular sludge bed (EGSB) reactor for biohydrogen production. Int. J. Hydrogen Energy 34, 9686–9695.

Wu, S.Y., Lin, C.N., Chang, J.S., Lee, K.S., Lin, P.J., 2002. Microbial hydrogen production with immobilized sewage sludge. Am. Chem. Soc. 18, 921–926.

Wu, S.Y., Lin, C.N., Chang, J.S., 2003. Hydrogen production with immobilized sewage sludge in three-phase fluidized-bed bioreactors. Am. Chem. Soc. 19, 828–832.

Wu, S.Y., Hung, C.H., Lin, C.N., Chen, H.W., Lee, A.S., Chang, J.S., 2006. Fermentative hydrogen production and bacterial community structure in high-rate anaerobic bioreactors containing silicone-immobilized and self-flocculated sludge. Biotechnol. Bioeng. 93, 934–946.

Wu, K.J., Chang, C.F., Chang, J.S., 2007. Simultaneous production of biohydrogen and bioethanol with fluidized-bed and packed-bed bioreactors containing immobilized anaerobic sludge. Proc. Biochem. 42, 1165–1171.

Wu, S.Y., Hung, C.H., Lin, C.Y., Lin, P.J., Lee, K.S., Lin, C.N., et al., 2008. HRT-dependent hydrogen production and bacterial community structure of mixed anaerobic microflora in suspended, granular and immobilized sludge systems using glucose as the carbon substrate. Int. J. Hydrogen Energy 33, 1542–1549.

Wu, J., Zhou, H.M., Li, H.Z., Zhang, P.C., Jiang, J., 2009. Impacts of hydrodynamic shear force on nucleation of flocculent sludge in anaerobic reactor. Water Res. 43, 3029–3036.

Yu, H.Q., Mu, Y., 2006. Biological hydrogen production in a UASB reactor with granules. II: reactor performance in 3-year operation. Biotechnol. Bioeng. 94, 988–995.

Yu, H., Zhu, Z., Hu, W., Zhang, H., 2002. Hydrogen production from rice winery wastewater in an upflow anaerobic reactor by using mixed anaerobic cultures. Int. J. Hydrogen Energy 27, 1359–1365.

Yu, H.Q., Hu, Z.H., Hong, T.Q., 2003. Hydrogen production from rice winery wastewater by using a continuously-stirred reactor. J. Chem. Eng. Japan 36, 1147–1151.

Zhang, H., Bruns, M.A., Logan, B.E., 2006. Biological hydrogen production by Clostridium acetobutylicum in an unsaturated flow reactor. Water Res. 40, 728–734.

Zhang, Z.P., Tay, J.H., Show, K.Y., Yan, R., Tee Liang, D., Lee, D.J., et al., 2007a. Biohydrogen production in a granular activated carbon anaerobic fluidized bed reactor. Int. J. Hydrogen Energy 32, 185–191.

Zhang, Z.P., Show, K.Y., Tay, J.H., Liang, D.T., Lee, D.J., Jiang, W.J., 2007b. Rapid formation of hydrogen-producing granules in an anaerobic continuous stirred tank reactor induced by acid incubation. Biotechnol. Bioeng. 96, 1040–1050.

Zhang, Z.P., Tay, J.H., Show, K.Y., Yan, R., Liang, D.T., Lee, D.J., et al., 2007c. Biohydrogen production in a granular activated carbon anaerobic fluidized bed reactor. Int. J. Hydrogen Energy 32, 185–191.

Zhang, Z.P., Show, K.Y., Tay, J.H., Liang, D.T., Lee, D.J., 2008a. Biohydrogen production with anaerobic fluidized bed reactors—a comparison of biofilm-based and granule-based systems. Int. J. Hydrogen Energy 33, 1559–1564.

Zhang, Z.P., Show, K.Y., Tay, J.H., Liang, D.T., Lee, D.J., 2008b. Enhanced continuous biohydrogen production by immobilized anaerobic microflora. Energy Fuels 22, 87–92.

SECTION V

PRODUCTION OF BIOBUTANOL AND OTHER GREEN FUELS

CHAPTER

25

Butanol Fuel from Biomass: Revisiting ABE Fermentation

Rajeev K. Sukumaran, Lalitha Devi Gottumukkala, Kuniparambil Rajasree, Deepthy Alex, Ashok Pandey*

Centre for Biofuels, Biotechnology Division, National Institute for Interdisciplinary Science and Technology – CSIR, Industrial Estate PO, Thiruvananthapuram 695019, India

***Corresponding author: E-mail: rajeevs@niist.res.in**

1 INTRODUCTION

Increasing demand on the rapidly depleting petroleum resources, fluctuations in fuel prices, and the increasing awareness on the harmful effects of the toxic and greenhouse gases generated by burning of petroleum have resulted in an interest in alternative fuels (Lynd et al., 2008; Wyman, 2007). While bioethanol is being considered actively as a gasoline alternative or supplement, there is also a renewed interest in butanol produced using the biological route (Lee et al., 2008; Papoutsakis, 2008). Butanol is a four-carbon primary alcohol with energy density closer to gasoline. It is less volatile, less hygroscopic, and less corrosive than ethanol which makes it more desirable than the latter in terms of its better suitability to be used in current petrol engines (Durre, 2007, 2008). It is therefore considered superior to ethanol as a gasoline additive or as a fuel by itself (Durre, 2007). Indeed, vehicle tests of over 10000 km on an unmodified car using 100% butanol was claimed, proving the suitability of butanol as an effective alternative fuel and its usefulness in reducing the emissions (Butylfuel, 2011). The current applications of butanol are mainly in the manufacture of lacquers and enamels. It is also a valuable industrial solvent used in the manufacture of several compounds including vitamins, antibiotics, and hormones (Lee et al., 2008). While most of the current production of butanol is from the petrochemical base, it used to be produced mainly by microbial fermentation until the first half of the last century. Gram-positive spore-forming bacteria of the genera *Clostridium* (Specifically species of *C. acetobutylicum* and *C. beijerinckii*) were employed for butanol production using starchy substrates in what was known as Acetone-Butanol-Ethanol (ABE) fermentation (Since the primary products of fermentation were Acetone, Butanol, and Ethanol in 3:6:1 ratio) A detailed review on the history of ABE fermentation may be found in Jones and Woods (1986). The microbial route of butanol production eventually became nonprofitable due to completion from alternative petrochemical routes in early 1960s. However, scientific interest in ABE fermentation and Clostridia remained active mainly due to the efforts of select groups working on Clostridial biochemistry, genetics, and fermentations employing these organisms, and this has resulted in considerable advancement in the knowledge about this Genus. A major outcome of these efforts is the development of tools for genetic manipulations of solventogenic Clostridia that has now become indispensible for studies on these bacteria. Some of these tools were reviewed by Tummala et al. (2001) and the recent developments are described by Papoutsakis (2008).

Renewed interest in ABE fermentation has spurred an explosion of new R&D efforts in this field, and worldwide there is an active effort to make the butanol production process a commercial success. The reasons that caused the demise of commercial ABE fermentation several decades ago still prevail, and the challenge by and large is to make the process economically feasible and noncompeting with food crops for raw materials. Inherent problems associated with ABE fermentation are the high costs of substrate, low product concentrations (maximum of ~2% in batch operation), solvent toxicity, and the high cost of product recovery (Durre, 1998). ABE fermentation in Clostridia are biphasic with an initial acidogenic phase when acetate, butyrate, hydrogen, and CO_2 are produced and a later solventogenic phase during which the acids are assimilated and used in the production of Acetone, Butanol, and Ethanol (Lee et al., 2008). Acidogenic phase corresponds with the exponential growth and the switch to solventogenic phase also marks the initiation of stationary phase and sporulation. Thanks to the work done by the Papoutsakis group and others, much is known about the molecular events during sporulation and solventogenesis (Grimmler et al., 2011; Papoutsakis, 2008; Paredes et al., 2005). Nevertheless, our understanding about the signals that cause the switch from acidogenesis to solventogenesis and the exact sequence of signaling events and their culmination in terms of altered gene expression is far from complete. Strategies of overexpression of genes involved in solvent production may not result in a higher titer of ABE due to the toxicity problem. Improving the solvent tolerance is therefore imperative before attempting to improve solvent production. A detailed understanding of the metabolic networks and the intricacies of its operation are warranted to address the issue of increasing solvent production and the solvent tolerance. Until these problems are addressed, the way forward appears to be the use of improved fermentation strategies and concurrent removal of products to prevent toxicity and to achieve higher cell densities and productivity (Ezeiji et al., 2007a).

With the genome sequences of the ABE-producing bacteria *C. acetobutylicum* and *C. beijerinckii* now being available (Nölling et al., 2001, http://genome.jgi-psf.org/clobe/clobe.home.html), more advances are expected in the understanding of the Clostridial metabolism and solventogenesis which will in turn help the efforts for metabolic engineering of the organisms for better solvent production, specificity, and tolerance. Another important aspect under study is to impart the ability for cellulose hydrolysis into solventogenic Clostridia (Mingardon et al., 2005; Perret et al., 2004) with the goal of direct utilization of lignocellulosic biomass for solvent production. These along with advanced strategies of fermentation, *in situ* product separation and recovery can definitely make the ABE process more viable in the future. Parallel to these, there are important breakthroughs by Liao group in production of butanol and other higher alcohols in *Escherichia coli* utilizing its amino acid biosynthesis pathway (Atsumi et al., 2008; Zhang et al., 2008). The technology is already being commercialized (Gevo, 2010). The use of such technologies for nonfermentative production of advanced biofuels in organisms such as *E. coli* or *Saccharomyces* whose genetics is well understood opens a new path in biofuel research. In the subsequent discussion, we try to present an overview of the status of biobutanol research exploring the possibilities for rapid development of organism and processes for commercial production of butanol as a biofuel.

2 BUTANOL PRODUCTION BY CLOSTRIDIA: THE ABE FERMENTATION

The metabolic pathway leading to butanol production is shared by the solventogenic *Clostridia* (e.g., *C. acetobutylicum*, *C. beijerinckii*, *C. pasteurianum*; Zheng et al., 2009), Figure 1 gives a schematic representation of the pathway for ABE production in *C. acetobutylicum*. Clostridia are strict anaerobes that can ferment a wide variety of substrates including monosaccharides and polysaccharides (Jones and Woods, 1986).Glucose is channeled through the glycolytic (EMP) pathway generating pyruvate and 2 molecules of ATP and NADH. Solventogenesis happens in two distinct phases and in the first phase the metabolism is directed toward production of acetic and butyric acids as shown in Figure 1. This phase,

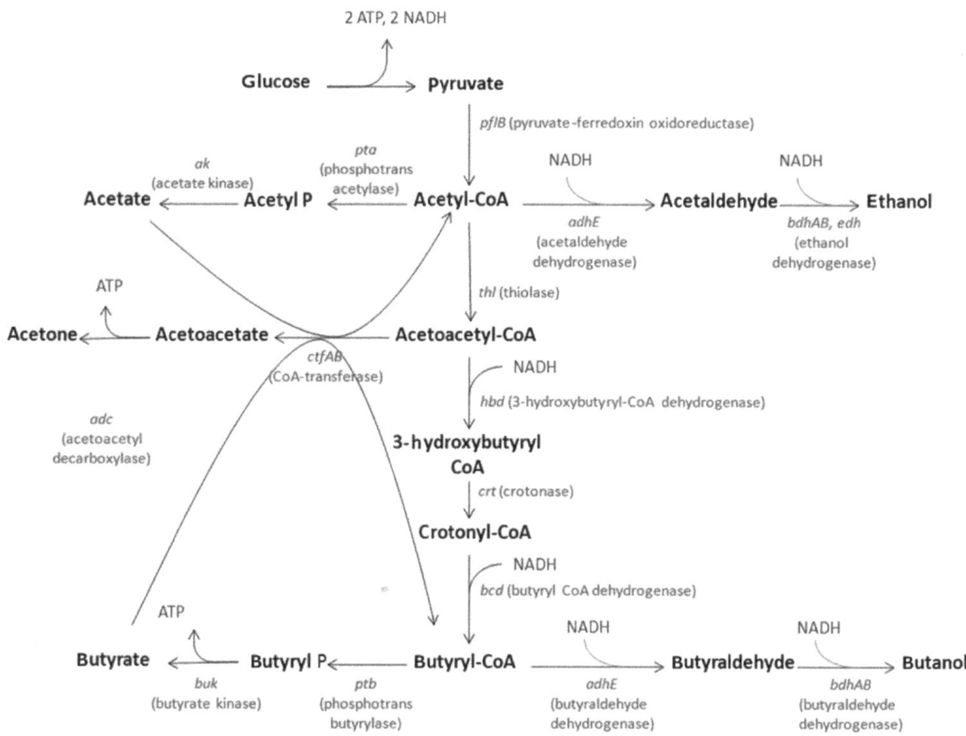

FIGURE 1 Metabolic pathways in *C. acetobutylicum*. Reactions during acidogenic phase and solventogenic phase are represented in red and blue arrows, respectively. Pale blue letters describe the genes (in italics) and enzymes involved in the reactions. (For interpretation of the references to color in this figure legend, the reader is referred to the Web version of this chapter.)

known as the acidogenic phase, corresponds with active growth of the organism and accumulation of organic acids—primarily acetate and butyrate. The synthesis of acetate and butyrate is crucial for generation of ATP needed for cell growth and metabolism (Ezeiji et al., 2010; Zheng et al., 2009). Among the acids, more butyrate is formed in comparison to acetate, since this step is where the organism can regenerate the NADH formed in glycolysis by its oxidation (Ezeiji et al., 2010).

Solventogenesis, the second phase of ABE fermentation, is believed to be triggered when the acid concentrations reach a threshold value which was proposed to be around 60 mmol/l (Maddox et al., 2000; Zheng et al., 2009). Lowering of pH and accumulation of acids is believed to shift the metabolism toward solventogenesis. During the solventogenic phase, the acids produced during the acidogenic phase are reassimilated and used for production of Acetone, Butanol, and Ethanol (Figure 1). The solventogenic phase marks only a small window between acidogenic phase and sporulation, and it is believed that the shift to butanol production and initiation of sporulation are initiated by activation of the transcriptional regulator Spo0A with pivotal role in sporulation (Papoutsakis, 2008; Paredes et al., 2005). The major challenges that lie ahead in ABE fermentation using the current strains of Clostridia include the following: (i) attaining higher biomass on cheaper substrates; (ii) improving solvent tolerance to achieve higher butanol titers; (iii) overcoming strain disintegration; (iv) delaying the initiation of sporulation to have a longer solventogenesis window; (v) improving the selectivity of ABE process to produce butanol preferentially; (vi) improving aero tolerance; and (vii) developing/improving fermentation, *in situ* solvent removal, and recovery strategies (Papoutsakis, 2008)

3 SPORULATION AND SOLVENTOGENESIS: THE SCOPE FOR DECOUPLING

In the biphasic ABE fermentation in batch mode, the solventogenic phase is marked by commitment to sporulation. Exponential growth in the acidogenic phase is followed by an accumulation of acids and resultant changes in pH which is believed to activate a solventogenic response through molecular signals that are as yet inconclusive. Nevertheless, we know that SpoA, a global transcriptional regulator, is responsible for initiating the switch to sporulation through the solventogenic phase. Solventogenesis is therefore intrinsically linked to sporulation response which limits the usability of cells grown up to this stage to a short time frame before they enter the dormant phase. A decoupling of solventogenesis from sporulation will be advantageous for full utilization of the cell biomass by enabling a longer solventogenesis window. Spo0A activation happens by its phosphorylation, but the phospho relay kinases which transfer the signal in response to environmental cues in Clostridia remain elusive. This, coupled with the findings that *buk* (butyrate kinase) knockout mutants but not *ptb* (phospho-trans-butyrylase) mutants (Desai and Papoutsakis, 1999; Harris et al., 2000) shows early and accelerated solventogenesis and sporulation, has led to the speculation that butyryl phosphate (BuP) may be the elusive signal that phosphorylates Spo0A (Paredes et al., 2005; Zhao et al., 2005). Spo0A activation is responsible for inducing the expression of solvent synthesizing enzymes. The major solventogenic genes responsible for acetone and butanol synthesis, that is, the *sol* locus genes, are organized into two major operons: *aad-ctfA-ctfB* and *adc* in Clostridia. In *C. beijerinckii*, the sol locus operons are in the bacterial chromosome, while these are located in the pSOL1 mega plasmid in *C. acetobutylicum* (Cornillot et al., 1997; Wilkinson et al., 1995) Activated Spo0A can induce the expression of these major solventogenic genes in clostridia (Harris et al., 2002; Tomas et al., 2003a). In addition to inducing the expression of solventogenic enzymes, the activation of Spo0A also triggers the early events leading to sporulation. It downregulates AbrB which is a transition state regulator (Chumsakul et al., 2010), thereby allowing the expression of σ factor σ^H. Activated Spo0A and σ^H in turn upregulates the *spoIIA*, *spoIIG*, and *ftsAZ* operons and the *spoIIE* gene (Paredes et al., 2005). While the *spoIIA* operon encodes the anti-σ factor SpoIIAA, anti-σ factor SpoIIAB and the early FORESPORE specific σ^F, the *spoIIG* operon encodes MOTHER-CELL specific σ factor σ^E and the *ftsAZ* operon encodes FtsZ, essential in formation of sporulation septum. SpoIIE is a phosphatase that activates SpoIIAA by dephosphorylation. This in turn activates a cascade of events leading to the expression of several genes culminating in sporulation. Scotcher and Bennett (2005) described an attempt to decouple solventogenesis from sporulation by antisense RNA inactivation of SpoIIE expression. Though sporulation was not stopped, there was notable delay in it without affecting solvent production. Now the attempts of such decoupling are being explored by targeting the inactivation of other genes especially those for sigma factors involved in early sporulation events (Jones et al., 2008)

4 METABOLIC ENGINEERING APPROACHES TO IMPROVE ABE FERMENTATION

The major problems associated with ABE fermentations employing clostridia are solvent toxicity, undesirable spore formation, strain degeneration, and formation of acetone and ethanol along with butanol. These problems need to be resolved and novel features such as the ability to directly ferment lignocellulose and tolerance to O_2 may be imparted to the Clostridia to make the ABE process commercially competitive. A lot of metabolic engineering attempts have been directed toward these goals which are reviewed in Papoutsakis, 2008. Central to the different challenges is the problem of solvent toxicity, and the recent developments including metabolic engineering for improving solvent resistance are reviewed by Ezeiji et al. (2010). Increasing the solvent concentration is of prime concern in ABE fermentation, and there have been attempts to improve ABE production by over expression of *adc* (aldehyde dehydrogenase) and *ctf*AB (CoA transferase) which led to an earlier induction of acetone formation and improved final concentrations of ABE (Mermelstein et al., 1993). Another important approach in enhancing ABE production was the inactivation of a putative transcriptional repressor of the *sol* locus genes called SolR. Mutants with inactivated *solR* had early induction of solventogenic genes and extended solvent production these mutants. Also, the mutant strain produced higher levels of ABE (Harris et al., 2001). Attempts were also made to inactivate the enzymes involved in butyrate formation with the objective of enhancing butanol production by redirecting carbon flow from acidogenesis to solventogenesis. Mutants with inactivated butyrate kinase (*buk*) produced considerably more amount of butanol than the parental strain (Green et al., 1996), and under optimized conditions, these mutants produced even better amounts of the solvents (Harris et al., 2000). Studies of Desai and Papoutsakis (1999) on knockdown of butyrate kinase (*buk*) and phosphor transbutyrylase (*ptb*) by antisense RNA technology had revealed interesting features on the regulation of solvent-forming enzymes and sporulation of *C. acetobutylicum*. While *buk* downregulation resulted in enhanced solventogenesis, inactivation of *ptb* had the reverse effect. These results led to the speculation that it is the accumulation of butyrate phosphate (BuP) which is an intermediate compound in the formation of butyrate from butyryl CoA (Figure 1) that is responsible for enhanced butanol production through induction of solventogenic enzymes. The regulatory role of BuP was confirmed by microarray analyses where it was observed that elevated BuP levels correspond to expression of solventogenic and stress genes (Zhao et al., 2005). Improving solvent tolerance is imperative for an enhanced solvent production by clostridia since elevated levels of the solvents especially butanol are toxic to the bacteria. Hence, solvent tolerance cannot be considered isolated from solvent production and the efforts on improving solvent production definitely have to address the issue of solvent toxicity. Solvent tolerance of Clostridia is complex issue involving induced changes in cell membrane properties and composition and the overexpression of a number of genes including molecular pumps, chaperones, transcriptional regulators and genes involved in the metabolism of fatty acids, and in sporulation (Tomas et al., 2004). Overexpression of the heat-shock proteins GroES and GroEL in *C. acetobutylicum* resulted in an improved solvent production (Tomas et al., 2003b). Borden and Papoutsakis (2007) had identified a gene (CAC1869) homologous to the "xenobiotic responsive element" (XRE) transcriptional regulators which can improve the solvent tolerance of the bacterium by 90% when transferred into *C. acetobutylicum* by screening of a genomic library. Decoupling of solventogenesis from sporulation is desired for increasing the duration of solventogenic phase before sporulation sets in and the strategies employed for decoupling of solventogenesis and sporulation are given in the preceding section. Apart from the classical strategies which included inactivation of genes responsible for onset of sporulation (Scotcher and Bennett, 2005), strategies that employ introduction of the desirable solventogenesis genes into asporogenous nonsolventogenic strains of *C. acetobutylicum* are also being attempted with considerable amount of success (e.g., Nair and Papoutsakis, 1994; Sillers et al., 2008). The major advantage of this approach is the possibility of being selective in the type of solvent being produced. The study by Sillers et al. (2008) has achieved exactly that and mutants with preferential selectivity for butanol production were generated. This is advantageous even when the total solvent production is lesser since the end product will have a comparatively purer butanol which will cut down the cost and energy for recovery and purification. Imparting cellulolytic ability into solventogenic Clostridia is another major goal of the groups working on metabolic engineering of these bacteria. Functional cellulolytic machinery will enable the bacteria to ferment lignocellulosic biomass to the solvents directly which will be tremendous practical applications in commercial biobutanol production for fuel applications. It is known that *C. acetobutylicum* contains a complete cellulosome which is a membrane-bound enzyme complex found in cellulolytic Clostridia (Nölling et al., 2001). Interested readers are directed to Bayer et al. (2004) and Fontes and Gilbert (2010) for detailed information on cellulosomes. Recent studies have concentrated on the engineering of *C. acetobutylicum* to express and assemble a functional cellulosome. Indeed, a mini cellulosome was formed in the bacterium on overexpression of the scaffolding protein CipA (Sabathe and Soucaille, 2003). Efforts were also undertaken to create heterologous mini cellulosomes in *C. acetobutylicum* utilizing components from the celluloytic strains *Clostridium celluloyticum* and *Clostridium thermocellum* (Mingardon et al., 2005; Perret et al., 2004). Though these efforts could successfully create the mini cellulosomes *in vivo*, cellulose assimilation and utilization for butanol is still far from achieved. However, it proves that it is possible in principle to assemble the celluloytic machinery in these bacteria and eventually a fully functional cellulosome and hence the cellulose to butanol conversion ability may be achieved.

5 FERMENTATION TECHNOLOGIES AND DOWNSTREAM PROCESSING FOR ABE FERMENTATION

ABE fermentation was operated largely in batch mode during the ages of its commercial production in the first half of the last century (Jones and Woods, 1986). This mode though is easier to operate and has lesser chances of contamination gives lower productivities. Continuous culture and fed-batch fermentations are the alternatives but with their own inherent drawbacks. For example, it is difficult to operate ABE fermentation in single-stage continuous cultures due to the biphasic nature of ABE fermentation. However, it may be noted that even commercial plants were run in continuous mode in the erstwhile USSR (Yarovenko, 1964). Most of the current knowledge on the ABE fermentation process has come from the studies on *C. beijerinckii* by Blaschek group (reviewed in Ezeiji et al., 2007a, 2010; Ezeji et al., 2004a). Conventional batch operation of ABE fermentation seldom yields more than 12-20 g/l solvents and the fermentation time is about 2-6 days due to solvent toxicity, and therefore the amount of sugars that can be fermented is also limited (Lee et al., 2008). Use of diluted sugar solutions necessitates the handling of large process volumes making this process uneconomical. However, recent advancements in fermentation technology coupled with newer methods for in situ product removal and recovery have revitalized the hopes on having a commercial process. While the lower productivity of batch culture can be addressed by fed batch or continuous processes, product inhibition is addressed by the use of novel solvent removal techniques Use of immobilized cell cultures and cell recycle reactor can further improve the reactor productivity even to the order of 40-50 times compared to batch rectors (Ezeji et al., 2004a). Major concerns in ABE fermentation include the availability of cheap sustainable feedstock for fermentation, achieving high cell densities, solvent toxicity, and product recovery. Some of the major work done on ABE fermentation during the last decade is presented in Table 1.

5.1 Raw Materials for ABE Fermentation

Clostridia are capable of utilizing a variety of free sugars and carbohydrates and the secretion of various enzymes by these bacteria are documented (Ezeiji et al., 2007a). The cost of substrate is a major factor affecting the economics of ABE fermentation, and hence the ability of these bacteria for utilization of mixed sugars becomes particularly relevant (Qureshi and Blaschek, 2000). Though several carbon sources such as glucose, xylose, various starches, corn, molasses, cheese whey, and even algal biomass have been employed in the past as ABE fermentation substrates (Jones and Woods, 1986), the focus has now shifted to renewable and sustainable nonfood raw materials such as agricultural waste. This feedstock is primarily lignocellulose which implies that the current ABE strains cannot directly ferment them due to want of cellulolytic machinery. So, it is imperative to use pretreatment and hydrolysis steps

TABLE 1 Methods, Carbon Sources, and Performance of ABE Fermentation

Organism	Type of Fermentation	C-Source	Method for in Situ Product Removal	Solvent Conc. (g/L)	Yield (g/g)	Productivity (g/l/h)	Reference
C. saccharoperbutylacetonicum	Batch	Glucose	Liquid-Liquid Extraction	29.8	0.4	0.55	Ishizaki et al. (1999)
C. beijerinckii	Batch	Glucose	Pervaporation	51.5	0.42	0.69	Qureshi and Blaschek (1999)
C. acetobutylicum DSM 1731	Batch	Domestic Organic Waste	Nil	9.39	ND	ND	Claassen et al. (2000)
C. beijerinckii BA101	Continuous (immobilized on clay bricks)	Glucose	Nil	7.9	0.38	15.8	Qureshi et al. (2000)
C. beijerinckii BA101	Batch	Packing peanuts	Nil	27.7	0.37	0.2	Jesse et al. (2002)
	Batch	Agro waste	Nil	14.8			
C beijerinckii BA101	Batch	Spray dried molasses	Nil	10.7	ND	ND	Qureshi et al. (2001)
C. acetobutylicum ATCC824	Continuous	Low-grade glycerol-glucose mixtures	Nil	7.9	0.29	0.33	Andrade and Vasconcelos (2003)
C. beijerinckii	Batch	Packing peanuts	Nil	18.9	0.32	0.17	Ezeji et al. (2003a)
	Continuous	Packing peanuts		8.4	8.4	0.27	
C. beijerinckii BA101	Batch	Glucose	Gas stripping	75.9	0.47	0.61	Ezeji et al. (2003b)

C. beijerinckii BA101	Continuous	Glucose	Gas stripping	460	0.4	0.91	Ezeji et al. (2004a)
C. beijerinckii BA101	Feb batch	Glucose	Gas stripping	233	0.47	1.16	Ezeji et al. (2004b)
C. acetobutylicum	Continuous (immobilized on fibrous matrix)	Glucose	Nil	5.1	0.42	0.46	Huang et al. (2003)
C. beijerinckii	Continuous (immobilized on membrane)	Corn Steep Liquor	Nil	6.2	0.3	2.01	Qureshi et al. (2004)
C. acetobutylicum P260	Batch	Corn fiber arabinoxylan	Gas stripping	24.67	0.44	0.47	Qureshi et al. (2006)
C. beijerinckii	Fed batch	Saccharified liquefied corn starch	Gas stripping	81.3	0.36	0.59	Ezeiji et al. (2007c)
C. beijerinckii P260	Batch	Wheat straw hydrolysate	Nil	20.1	0.41	0.28	Qureshi et al. (2007)
C. beijerinckii ATCC 55025	Continuous (immobilized on corn stalk)	Glucose	Nil	5.1	0.32	5.06	Zhang et al. (2009)
C. beijerinckii	Batch	Wheat bran	Nil	11.8	0.32	0.16	Liu et al. (2010)
C. saccharoperbutylacetonicum N1-4	Batch	Cassava starch	Nil	21	0.41	0.44	Thang et al. (2010)
		Cassava strips	Nil	19.4	0.38	0.4	

ND, not determined/not mentioned in the study.

for generation of fermentable sugars from the biomass. However, this process also generates a lot of compounds that can be inhibitory to the Clostridial growth and ABE production. It may be noted that the sugar degradation products furfural and hydroxy methyl furfural (HMF) are rather stimulatory to ABE fermentation, while the organic acids and lignin degradation products generated are toxic (Ezeiji et al., 2007b). During the fermentation of mixed sugars with solventogenic Clostridia, all sugars are utilized concurrently with the rates of utilization dependent on the sugar type (Ezeiji et al., 2007b). The concurrent uptake of both hexose and pentoses is considered highly advantageous (Ezeiji et al., 2007a). The major challenges in selection of carbon sources for ABE fermentation are shared with lignocellulosic ethanol production, but here the advantage is the ability of Clostridia to use pentoses. Another advantage in comparison to process development for bioethanol is the ability to use the relatively low concentration of sugars in the biomass hydrolysates as obtained by enzymatic hydrolysis and still yield better productivities. Removal of toxic compounds from hydrolysates is a major challenge though, which needs to be addressed.

5.2 Achieving High Cell Densities and High Productivity through Fermentation Process Development and Solvent Removal

In the conventional batch process for butanol production, reactor productivities are not more than 0.5 g/l/h due to reasons including low cell concentration, down time, and solvent toxicity (Ezeji et al., 2006). Batch reactors cannot achieve high cell densities due to solvent inhibition or growth and sporulation in Clostridial fermentations. On the contrary, continuous culture can attain very high productivities in comparison to the conventional batch mode of operation. Butanol productivities of 10.2 g/l/h were achieved by Lienhardt et al. (2002) using an immobilized cell biofilm reactor, whereas the same group could obtain a productivity of 15.8 g/l/h in continuous culture with cells immobilized on clay bricks (Qureshi et al., 2000). The major limitation for achieving high cell densities and hence high productivities is the toxicity of the solvents produced by the cells. Advanced technologies for continuous product removal from fermentation liquor are available now which prevents the inhibition by the solvents and hence improve the life and solvent productivity. The most important techniques employed for removal of the solvents from fermentation medium are liquid-liquid extraction, perstraction, pervaporation, and gas stripping (Figure 2).

In liquid-liquid extraction, a water-insoluble organic extractant is mixed with the fermentation broth. Since butanol is more soluble in the extractant, it gets concentrated in the organic phase. Solvents thus concentrated in the organic phase can be recovered by back extraction into another solvent or by distillation (Maddox, 1989). Oleyl alcohol which is relatively nontoxic is generally used as the extractant (Evans and Wang, 1988; Karcher et al., 2005). Disadvantages of liquid-liquid extraction include toxicity of the extractant, formation of emulsions, loss of extractant, and formation of rag layer by microbial accumulation at the liquid-liquid interface. Perstraction is similar to liquid-liquid extraction in that it recovers butanol and other solvents into an extractant, but a direct contact of the extractant and the fermentation broth is prevented by introduction of a contact membrane that separates the phases but allows diffusion of ABE (Grobben et al., 1993; Traxler et al., 1985). While Perstraction can address toxicity of the extractant and also provide some selectivity in recovery of solvents by selection of membranes, there could be serious fouling issues and the process is relatively costly to operate.

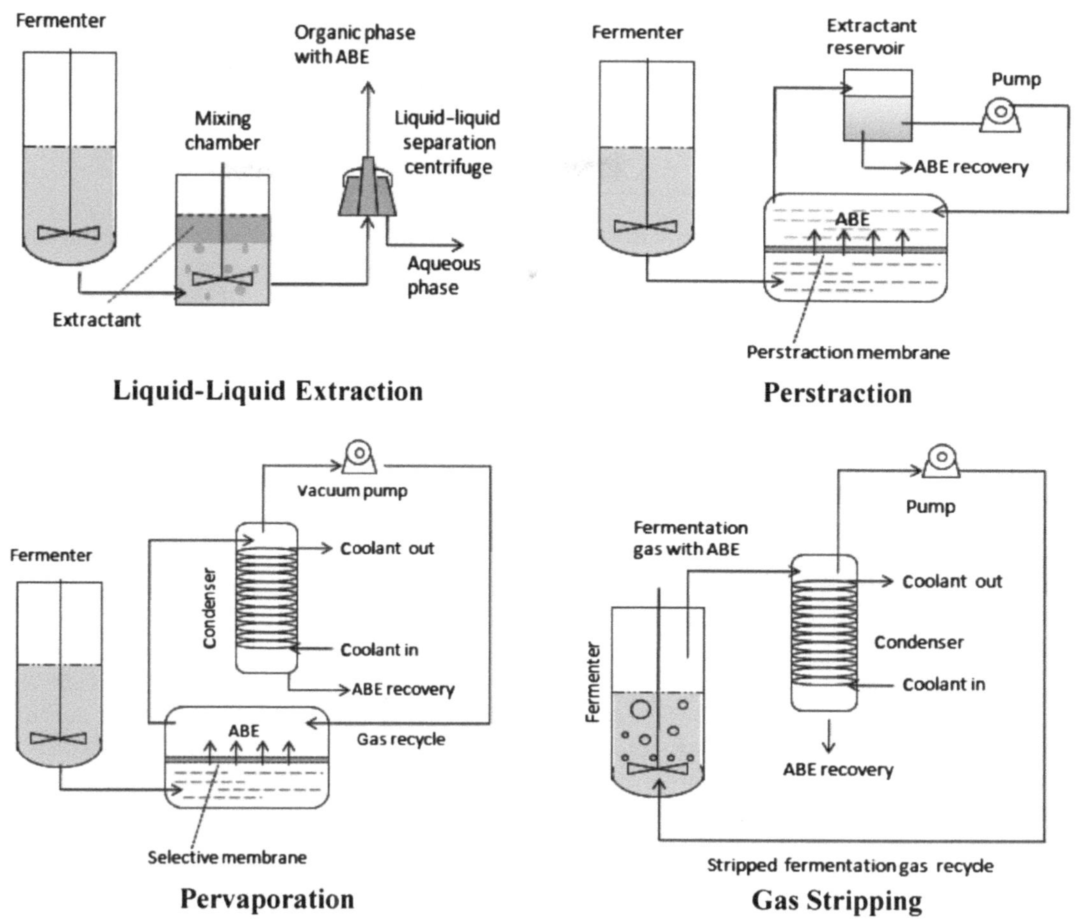

FIGURE 2 Various techniques used in ABE fermentation for *in situ* product removal.

Pervaporation allows selective removal and recovery of volatiles from fermentation broth. Here, the fermentation broth containing solvents is passed over a selective membrane in a Pervaporation module where the other side of the membrane is a gaseous phase (either vacuum or an inert sweep gas) when the volatiles are extracted into the gaseous phase from where it can be condensed and recovered (Shao and Huang, 2007). Use of Pervaporation techniques in fermentation, in particular ABE fermentation, is reviewed in Vane (2005), Ezeji et al. (2006) and Qureshi and Ezeji (2008). Membrane fouling and loss of fermentation intermediates are considered as the major drawbacks of Pervaporation (Ezeiji et al., 2010). In contrast to the membrane-based techniques for product separation, Gas stripping is a simple and cost-effective technique that can be integrated with ABE fermentation. Here, oxygen-free N_2 or the fermentation gases comprising CO_2 and H_2 is continually sparged into the reactor, and the effluent gases are channeled through a condenser where the volatized solvents are recovered by cooling it (Ezeji et al., 2003b). Gas stripping has now become one of the most promising strategies for in situ removal and recovery of ABE due to its simplicity in operation, no toxicity or removal of nutrients and intermediates from the fermentation broth. Integration of gas stripping with ABE fermentation has resulted in highly improved productivities and yield (Ezeji et al., 2004b, 2007c, Qureshi and Blaschek, 2001).

6 CONCLUSIONS

ABE fermentations were next only to ethanol fermentations and used to be a major industry until 1960s. Later, biological route for butanol production lost its importance owing to competition from petrochemical route, and today there is a renewed interest in ABE fermentation due to increased concerns over petroleum depletion and the increased pollution due to burning of petroleum fuels. Though the ABE fermentation process used to be operational decades back, the same technologies are not applicable today due to the lack of cost effectiveness and the nonavailability of conventional raw materials. The most feasible feedstock for butanol seems to be lignocellulose, but the problems plaguing bioethanol are also applicable for biobutanol. However, the future for biobutanol seems bright since the Clostridia that produce ABE are capable of utilizing a range of carbon sources for growth and solvent production and also are not inhibited by the sugar degradation products generated during biomass pretreatment are being developed. Also, nonsolventogenic Clostridia possess enzyme systems for hydrolysis of lignocellulose cellulose which in future may be incorporated into the solvent-producing species, thereby generating superbugs which can directly ferment plant biomass to produce biobutanol.

ABE fermentations are limited largely by the issues of lower cell densities and viability, smaller time window for solventogenesis, and product inhibition. Through several decades of R&D on Clostridial biochemistry and the molecular biology of ABE fermentations, lots of insights have been gained on this fermentation. Lot is now known on the molecular events leading to solventogenesis and sporulation and decoupling of these have been demonstrated at least in the laboratory. Similarly, much information has been gained on the mechanisms of solvent toxicity which might be used to engineer strains that can tolerate higher solvent concentrations. Increasing solvent production by modulation of the metabolic pathway enzymes has been demonstrated and coupled with improved solvent tolerance, this will be able to generate efficient microbes that will shift the economy of ABE fermentation toward commercial feasibility.

Meanwhile, in the short term, advanced fermentation technologies are being developed by the expert groups which tackle problems such as low cell density, viability, and solvent sensitivity by modulations in the methods of carbon feeding, mode of culture, and *in situ* removal and recovery of solvents. These efforts may be developed into commercially viable technologies.

Parallel to these, newer technologies for production of butanol and higher alcohols are also being developed utilizing non-natural routes engineered in *E. coli*. These breakthroughs might help in bypassing the organism development for ABE fermentation and may reach the market sooner. The use of such technologies for nonfermentative production of advanced biofuels in organisms such as *E. coli* or *Saccharomyces* whose genetics is well understood can herald an entirely new way of approaching the renewable fuel problem

Acknowledgments

Authors acknowledge financial support to the Centre for Biofuels (CBF) from TIFAC, DST, Govt of India.

References

Andrade, J.C., Vasconcelos, I., 2003. Continuous cultures of *Clostridium acetobutylicum*: culture stability and low-grade glycerol utilisation. Biotechnol. Lett. 25 (2), 121–125.

Atsumi, S., Hanai, T., Liao, J.C., 2008. Non-fermentative pathways for synthesis of branched-chain higher alcohols as biofuels. Nature 451 (7174), 86–89.

Bayer, E.A., Belaich, J.P., Shoham, Y., Lamed, R., 2004. The cellulosomes: multienzyme machines for degradation of plant cell wall polysaccharides. Annu. Rev. Microbiol. 58, 521–554.

Borden, J.R., Papoutsakis, E.T., 2007. Dynamics of genomic-library enrichment and identification of solvent tolerance genes for *Clostridium acetobutylicum*. Appl. Environ. Microbiol. 73, 3061–3068.

Butylfuel, 2011. Available from: website http://www.butanol.com/index.html (accessed 07.06.11).

Chumsakul, O., Takahashi, H., Oshima, T., Hishimoto, T., Kanaya, S., Ogasawara, N., et al., 2010. Genome-wide binding profiles of the *Bacillus subtilis* transition state regulator AbrB and its homolog Abh reveals their interactive role in transcriptional regulation. Nucleic Acids Res. doi: 10.1093/nar/gkq780.

Claassen, P.A., Budde, M.A., López-Contreras, A.M., 2000. Acetone, butanol and ethanol production from domestic organic waste by solventogenic clostridia. J. Mol. Microbiol. Biotechnol. 2 (1), 39–44.

Cornillot, E., Nair, R.V., Papoutsakis, E.T., Soucaille, P., 1997. The genes for butanol and acetone formation in *Clostridium acetobutylicum* ATCC 824 reside on a large plasmid whose loss leads to degeneration of the strain. J. Bacteriol. 179 (17), 5442–5447.

Desai, R.P., Papoutsakis, E.T., 1999. Antisense RNA strategies for metabolic engineering of *Clostridium acetobutylicum*. Appl. Environ. Microbiol. 65, 936–945.

Durre, P., 1998. New insights and novel developments in clostridial acetone/butanol/isopropanol fermentation. Appl. Microbiol. Biotechnol. 49, 639–648.

Durre, P., 2007. Biobutanol an attractive biofuel. Biotechnol. J. 2, 1525–1534.

Durre, P., 2008. Fermentative Butanol production, bulk chemical and biofuel. Ann. N.Y. Acad. Sci. 1125, 353–362.

Evans, P.J., Wang, H.W., 1988. Enhancement of butanol fermentation by *Clostridium cetobutylicum* in the presence of decanol-oleyl alcohol mixed extractants. Appl. Environ. Microbiol. 54, 1662–1667.

Ezeiji, T.C., Qureshi, N., Blaschek, H.P., 2007a. Bioproduction of butanol from biomass: from genes to bioreactors. Curr. Opin. Biotechnol. 18, 220–227.

Ezeiji, T.C., Qureshi, N., Blaschek, H.P., 2007b. Butanol production from agricultural residues: impact of degradation products on *Clostridium beijerinckii* growth and butanol fermentation. Biotechnol. Bioeng. 97 (6), 1460–1469.

Ezeiji, T.C., Qureshi, N., Blaschek, H.P., 2007c. Production of acetone butanol (AB) from liquefied corn starch, a commercial substrate, using *Clostridium beijerinckii* coupled with product recovery by gas stripping. J. Ind. Microbiol. Biotechnol. 34, 771–777.

Ezeiji, T.C., Milene, C., Price, N.D., Blaschek, H.P., 2010. Achievements and perspectives to overcome the poor solvent resistance in acetone and butanol producing microorganisms. Appl. Microbiol. Biotechnol. 85, 1697–1712.

Ezeji, T.C., Groberg, M., Qureshi, N., Blaschek, H.P., 2003a. Continuous production of butanol from starch-based packing peanuts. Appl. Biochem. Biotechnol. 105-108, 375–382.

Ezeji, T.C., Qureshi, N., Blaschek, H.P., 2003b. Production of butanol by *Clostridium beijerinckii* BA101 and in-situ recovery by gas stripping. World. J. Microbiol. Biotechnol. 19, 595–603.

Ezeji, T.C., Qureshi, N., Blaschek, H.P., 2004a. Butanol fermentation research: upstream and downstream manipulations. Chem. Rec. 4 (5), 305–314.

Ezeji, T.C., Qureshi, N., Blaschek, H.P., 2004b. Acetone butanol ethanol (ABE) production from concentrated substrate: reduction in substrate inhibition by fed-batch technique and product inhibition by gas stripping. Appl. Microbiol. Biotechnol. 63, 653–658.

Ezeji, T.C., Qureshi, N., Karcher, P., Blaschek, H.P., 2006. Butanol production from corn. In: Minteer, S.D. (Ed.), Alcoholic Fuels: Fuels for Today and Tomorrow. Taylor & Francis, NY, pp. 99–122.

Fontes, C.M., Gilbert, H.J., 2010. Cellulosomes: highly efficient nanomachines designed to deconstruct plant cell wall complex carbohydrates. Annu. Rev. Biochem. 79, 655–681.

Gevo, 2010. Gevo produces isobutanol, hydrocarbons and jet fuels from cellulosic biomass. Press release dated July 29, 2010 Available from: http://www.gevo.com/pr-gevo-produces-isobutanol-hydrocarbons-jet-fuel-from-cellulosic-biomass.php.

Green, E.M., Boynton, Z.L., Harris, L.M., Rudolph, F.B., Papoutsakis, E.T., Bennett, G.N., 1996. Genetic manipulation of acid formation pathways by gene inactivation in *Clostridium acetobutylicum* ATCC 824. Microbiology (UK). 142, 2079–2086.

Grimmler, C., Janssen, H., Krauβe, D., Fischer, R.J., Bahl, H., Dürre, P., et al., 2011. Genome-wide gene expression analysis of the switch between acidogenesis and solventogenesis in continuous cultures of *Clostridium acetobutylicum*. J. Mol. Microbiol. Biotechnol. 20 (1), 1–15.

Grobben, N.G., Eggink, G., Cuperus, F.P., Huizing, H.J., 1993. Production of acetone, butanol, and ethanol (ABE) from potato wastes: fermentation with integrated membrane extraction. Appl. Microbiol. Biotechnol. 39, 494–498.

Harris, L.M., Desai, R.P., Welker, N.E., Papoutsakis, E.T., 2000. Characterization of recombinant strains of the *Clostridium acetobutylicum* butyrate kinase inactivation mutant: need for new phenomenological models for solventogenesis and butanol inhibition? Biotechnol. Bioeng. 67, 1–11.

Harris, L.M., Blank, L., Desai, R.P., Welker, N.E., Papoutsakis, E.T., 2001. Fermentation characterization and flux analysis of recombinant strains of *Clostridium acerobutylicurn* with an inactivated solR gene. J. Ind. Microbiol. Biotechnol. 27, 322–328.

Harris, L.M., Welker, N.E., Papoutsakis, E.T., 2002. Northern, morphological, and fermentation analysis of spo0A inactivation and overexpression in *Clostridium acetobutylicum* ATCC 824. J. Bacteriol. 184, 3586–3597.

Huang, W.C., Ramey, D.E., Yang, S.T., 2003. Continuous production of butanol by *Clostridium acetobutylicum* immobilized in a fibrous bed bioreactor. Appl. Biochem. Biotechnol. 113-116, 887–898.

Ishizaki, A., Michiwaki, S., Crabbe, E., Kobayashi, G., Sonomoto, K., Yoshino, S., 1999. Extractive acetone-butanol-ethanol fermentation using methylated crude palm oil as extractant in batch culture of *Clostridium saccharoperbutylacetonicum* N1-4. J. Biosci. Bioeng. 87 (3), 352–356.

Jesse, T.W., Ezeji, T.C., Qureshi, N., Blaschek, H.P., 2002. Production of butanol from starch-based waste packing peanuts and agricultural waste. J. Ind. Microbiol. Biotechnol. 29 (3), 117–123.

Jones, D.T., Woods, D.R., 1986. Acetone-butanol fermentation revisited. Microbiol. Rev. 50, 484–524.

Jones, S.W., Paredes, C.J., Tracy, B., Cheng, N., Sillers, R., Senger, R.S., et al., 2008. The transcriptional program underlying the physiology of *clostridial* sporulation. Gen. Biol. 9, R114. Available form: http://genomebiology.com/2008/9/7/R114. doi: 10.1186/gb-2008-9-7-r114.

Karcher, P., Ezeji, T.C., Qureshi, N., Blaschek, H.P., 2005. Microbial production of butanol: product recovery by extraction. In: Satyanarayana, T., Johri, B.N. (Eds.), Chapter in Microbial Diversity: Current Perspectives and Potential Applications. IK International Publishing House Pvt Ltd, New Delhi, pp. 865–880.

Lee, S.Y., Park, J.H., Jang, S.H., Nielsen, L.K., Kim, J., Jung, K.S., 2008. Fermentative Butanol production by *Clostrida*. Biotechnol. Bioeng. 101 (2), 209–228.

Lienhardt, J., Schripsema, J., Qureshi, N., Blaschek, H.P., 2002. Butanol production by *Clostridium beijerinckii* BA101 in an immobilized cell biofilm reactor: increase in sugar utilization. Appl. Biochem. Biotechnol. Spring 98-100, 591–598.

Liu, Z., Ying, Y., Li, F., Ma, C., Xu, P., 2010. Butanol production by *Clostridium beijerinckii* ATCC 55025 from wheat bran. J. Ind. Microbiol. Biotechnol. 37 (5), 495–501.

Lynd, L.R., Laser, M.S., Bransby, D., Dale, B.E., Davison, B., Hamilton, R., et al., 2008. How biotech can transform biofuels. Nat. Biotechnol. 26, 169–172.

Maddox, I.S., 1989. The acetone-butanol-ethanol fermentation: recent progress in technology. Biotechnol. Genetic Eng. Rev. 7, 190–220.

Maddox, I.S., Steiner, E., Hirsch, S., Wessner, S., Gutierrez, N.A., Gapes, J.R., et al., 2000. The cause of "acid crash" and "acidogenic fermentations" during the batch acetone-butanol-ethanol (ABE-) fermentation process. J. Mol. Microbiol. Biotechnol. 2, 95–100.

Mermelstein, L.D., Papoutsakis, E.T., Petersen, D.J., Bennett, G.N., 1993. Metabolic engineering of *Clostridium acetobutylicum* ATCC 824 for increased solvent production by enhancement of acetone formation enzyme-activities using a synthetic acetone operon. Biotechnol. Bioeng. 42, 1053–1060.

Mingardon, F., Perret, S., Belaich, A., Tardif, C., Belaich, J.P., Fierobe, H.P., 2005. Heterologous production, assembly and secretion of a minicellulosome by *Clostridium acetobutylicum* ATCC 824. Appl. Environ. Microbiol. 71, 1215–1222.

Nair, R.V., Papoutsakis, E.T., 1994. Expression of plasmid-encoded aad in *Clostridium acetobutylicum* M5 restores vigorous butanol production. J. Bacteriol. 176, 5843–5846.

Nölling, J., Breton, G., Omelchenko, M.V., Makarova, K.S., Zeng, Q.D., Gibson, R., et al., 2001. Genome sequence and comparative analysis of the solvent-producing bacterium *Clostridium acetobutylicum*. J. Bacteriol. 183, 4823–4838.

Papoutsakis, E.T., 2008. Engineering solventogenic *clostridia*. Curr. Opin. Biotechnol. 19, 420–429.

Paredes, C.J., Alsaker, K.V., Papoutsakis, E.T., 2005. A comparative genomic view of clostridal sporulation and physiology. Nat. Rev. Microbiol. 3, 969–978.

Perret, S., Casalot, L., Fierobe, H.P., Tardif, C., Sabathe, F., Belaich, J.P., et al., 2004. Production of heterologous and chimeric scaffoldins by *Clostridium acetobutylicum* ATCC 824. J. Bacteriol. 186, 253–257.

Qureshi, N., Blaschek, H.P., 1999. Production of acetone butanol ethanol (ABE) by a hyper-producing mutant strain of *Clostridium beijerinckii* BA101 and recovery by pervaporation. Biotechnol. Prog. 15, 594–602.

Qureshi, N., Blaschek, H.P., 2000. Economics of butanol fermentation using hyper-butanol producing *Clostridium beijerinckii* BA101. Trans. IChemE 78 (Part C), 139–144.

Qureshi, N., Blaschek, H.P., 2001. Recovery of butanol from fermentation broth by gas stripping. Renew. Energy 22, 557–564.

Qureshi, N., Ezeji, T.C., 2008. Butanol (a superior biofuel) production from agricultural residues (renewable biomass): recent progress in technology. Biofuels Bioprod. Bioref. 2, 319–330.

Qureshi, N., Schripsema, J., Lienhardt, J., Blaschek, H.P., 2000. Continuous solvent production by *Clostridium beijerinckii* BA101 immobilized by adsorption onto brick. World. J. Microbiol. Biotechnol. 16, 377–382.

Qureshi, N., Lolas, A., Blaschek, H.P., 2001. Soy molasses as fermentation substrate for production of butanol using *Clostridium beijerinckii* BA101. J. Ind. Microbiol. Biotechnol. 26 (5), 290–295.

Qureshi, N., Karcher, P., Cotta, M., Blaschek, H.P., 2004. High-productivity continuous biofilm reactor for butanol production: effect of acetate, butyrate, and corn steep liquor on bioreactor performance. Appl. Biochem. Biotechnol. 113-116, 713–721.

Qureshi, N., Li, X.L., Hughes, S., Saha, B.C., Cotta, M.A., 2006. Butanol production from corn fiber xylan using *Clostridium acetobutylicum*. Biotechnol. Prog. 22 (3), 673–680.

Qureshi, N., Saha, B.C., Cotta, M.A., 2007. Butanol production from wheat straw hydrolysate using *Clostridium beijerinckii*. Bioproc. Biosys. Eng. 30, 419–427.

Sabathe, F., Soucaille, P., 2003. Characterization of the CipA scaffolding protein and in vivo production of a minicellulosome in *Clostridium acetobutylicum*. J. Bacteriol. 185, 1092–1096.

Scotcher, M.C., Bennett, G.N., 2005. SpoIIE regulates sporulation but does not directly affect solventogenesis in *Clostridium acetobutylicum* ATCC824. J. Bacteriol. 187, 1930–1936.

Shao, P., Huang, R.Y.M., 2007. Polymeric membrane pervaporation. J. Membr. Sci. 287, 162–179.

Sillers, R., Chow, A., Tracy, B., Papoutsakis, E.T., 2008. Metabolic engineering of the non-sporulating, non-solventogenic *Clostridium acetobutylicum* strain M5 to produce butanol without acetone demonstrate the robustness of the acid-formation pathways and the importance of the electron balance. Metab. Eng. doi:10.1016/j.ymben.2008.07.005.

Thang, V.H., Kanda, K., Kobayashi, G., 2010. Production of acetone-butanol-ethanol (ABE) in direct fermentation of cassava by *Clostridium saccharoperbutylacetonicum* N1-4. Appl. Biochem. Biotechnol. 161 (1-8), 157–170.

Tomas, C.A., Alsaker, K.V., Bonarius, H.P.J., Hendriksen, W.T., Yang, H., Beamish, J.A., et al., 2003a. DNA array-based transcriptional analysis of asporogenous, nonsolventogenic *Clostridium acetobutylicum* strains SKO1 and M5. J. Bacteriol. 185, 4539–4547.

Tomas, C.A., Welker, N.E., Papoutsakis, E.T., 2003b. Overexpression of groESL in *Clostridium acetobutylicum* results in increased solvent production and tolerance, prolonged metabolism, and changes in the cell's transcriptional program. Appl. Environ. Microbiol. 69, 4951–4965.

Tomas, C., Beamish, J., Papoutsakis, E.T., 2004. Transcriptional analysis of butanol stress and tolerance in *Clostridium acetobutylicum*. J. Bacteriol. 186, 2006–2018.

Traxler, R.W., Wood, E.M., Mayer, J., Wilson, M.P., 1985. Extractive fermentation for the production of butanol. Dev. Ind. Microbiol. 36, 519–525.

Tummala, S.B., Tomas, C.A., Harris, L.M., Welker, N.E., Rudolph, F.B., Bennett, G.N., et al., 2001. Genetic tools for solventogenic clostridia. In: Bahl, H., Durre, P. (Eds.), Clostridia—Biotechnology and Medical Applications. Wiley-VCH, pp. 105–123.

Vane, A., 2005. Review of pervaporation for product recovery from biomass fermentation processes. J. Chem. Technol. Biotechnol. 80, 603–629.

Wilkinson, S.R., Young, D.I., Morris, J.G., Young, M., 1995. Molecular genetics and the initiation of solventogenesis in *Clostridium beijerinckii* (formerly *Clostridium acetobutylicum*) NCIMB 8052. FEMS Microbiol. Rev. 17, 275–285.

Wyman, C.E., 2007. What is (and is not) vital to advancing cellulosic ethanol. Trends Biotechnol. 25, 153–157.

Yarovenko, V.L., 1964. Principles of the continuous alcohol and butanol-acetone fermentation processes. In: Malek, I. (Ed.), Continuous Cultivation of Microorganisms. 2nd Symposium, Czechoslovakian Academy of Sciences, Prague, pp. 205–217.

Zhang, K., Sawaya, M.R., Eisenberg, D.S., Liao, J.C., 2008. Expanding metabolism for biosynthesis of non natural alcohols. Proc. Natl. Acad. Sci. USA. 105 (52), 20653–20658.

Zhang, Y., Ma, Y., Yang, F., Zhang, C., 2009. Continuous acetone-butanol-ethanol production by corn stalk immobilized cells. J. Ind. Microbiol. Biotechnol. 36 (8), 1117–1121.

Zhao, Y., Tomas, C.A., Rudolph, F.B., Papoutsakis, E.T., Bennett, G.N., 2005. Intracellular butyryl phosphate and acetyl phosphate concentrations in *Clostridium acetobutylicum* and their implications for solvent formation. Appl. Environ. Microbiol. 71, 530–537.

Zheng, Y.N., Li, L.Z., Xian, M., Ma, Y.J., Yang, J.M., Xu, X., et al., 2009. Problems with the microbial production of butanol. J. Ind. Microbiol. Biotechnol. 36 (9), 1127–1138.

CHAPTER

26

Production of Green Liquid Hydrocarbon Fuels

*Anjan Ray**, *Amar Anumakonda*

UOP India Pvt Ltd, 6th floor, Building 9B, Cyber City, DLF Phase III, Gurgaon – 122002, India

UOP LLC, 25 E. Algonquin Road, Des Plaines, IL 60616, USA

***Corresponding author: E-mail: anjan.ray@uop.com**

1 INTRODUCTION

Advanced biofuels technologies are intended to progress beyond the early generations of biofuels, such as conventional biodiesel (FAME) and bioethanol, by creating bio-based hydrocarbon fuels, fuel additives, and power sources that fit seamlessly into existing supply chains and operational fuel/power retailing infrastructure.

To make this a reality, a confluence of technology, policy, and consumer education is necessary. This is not an easy task, but it can be achieved, provided several key issues are addressed by governments, international and national networks of policy makers, environmental experts, financial institutions, and technology leaders. Moreover, these efforts need to be aligned in ways that seek to move toward simultaneous solutions to a number of complex, conflicting challenges, which we shall refer to henceforth as the Four Imperatives:

1. Improved energy security to underpin economic development
2. Reduced dependence on fossil fuels, especially on substantial crude oil imports by energy-deficient nations
3. Reduced greenhouse gas (GHG) emissions
4. Minimized adverse effects on food security, water supply and quality, agricultural land and forests

1.1 The Need for Green Fuels (As Compared to Fossil Fuels)

In the context of these Four Imperatives, renewable fuels can be seen as fulfilling several important current and future needs while posing new challenges of their own.

1.1.1 *Improved Energy Security*

Energy security not only underpins economic development by providing a critical component of infrastructure, supply chains, and manufacturing processes, but also addresses indirect spinoffs into areas vital to underdeveloped and developing countries, such as healthcare, education, and safety.

A UNIDO paper (Energy, Development, and Security, 2008) stresses that "developing countries need to expand access to reliable and modern energy services to alleviate poverty and increase productivity, to enhance competitiveness and economic growth." Liquid fuel consumption forecasts reflect this trend (Figure 1).

1.1.2 *Reduced Dependence on Fossil Fuels*

Emerging economies such as Turkey and India, where fossil fuel reserves are limited but energy demand is rising, are at higher risk of economic impact from escalations in oil prices. Over a decade ago, Ediger and Kentel (1999) argued that a gradual shift from fossil fuels to renewable energy seemed to be the sole alternative for Turkey.

Such countries where crude oil imports make up a significant percentage of GDP—India being a case in point with a crude oil import bill of $79.6 billion relative to a GDP of approximately $1 trillion (for 12 months ending March 2010)—are especially vulnerable to price volatility in this commodity. Depending on the prevailing subsidy regime, as the price of crude escalates into triple digits, the economic burden on the state becomes unsustainable.

1.1.3 *Reduced GHG Emissions*

In the cycle of fuel production and consumption, the overall impact of GHG emission changes does not necessarily correspond just to the carbon footprint. Del Grosso, Adler, and Parton have pointed out (Adler et al., 2007) that nitrous oxide emissions and carbon

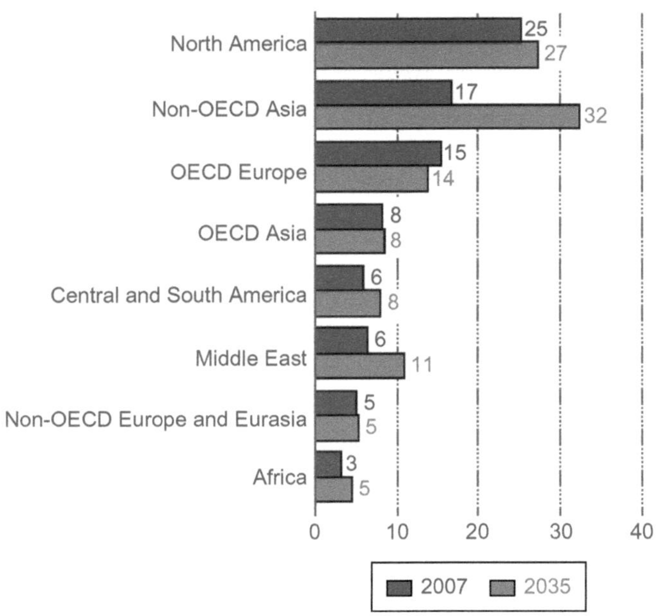

FIGURE 1 World liquids consumption by region and country group, 2007 and 2035 (million barrels per day). U.S. Energy Information Administration/International Energy Outlook (2010). Energy, Development, and Security © OECD/IEA, 2008.

dioxide emissions take place during the production of biofuels, originating from in-soil activity, farm machinery, farm inputs, agricultural processes, and fuel manufacturing processes. Yet, purpose-grown biomass species such as hybrid poplar and switch grass could nonetheless reduce net GHG emissions significantly through fixation of atmospheric carbon dioxide to organic carbon during crop growth.

1.1.4 *Minimized Adverse Effects on Natural Resources and Food Security*

An intense debate continues on whether the resources demanded for biofuels production—be it land, water, or agricultural output that can also be used for food and fodder—are being appropriately deployed. In an assessment of incentive structures facing agriculturists, refiners, and consumers in India, Srinivasan (2009) has indicated that unsustainable support process for feedstock and unviable procurement prices for finished biofuels such as bioethanol and biodiesel have hampered, rather than hastened, the gradual partial substitution of fossil fuels by biofuels that policymakers sought to bring about. The High Level Conference on World Food Security in 2008 highlighted the complex interdependence of Energy Security, Biofuels and Food Supply[a], pointing out that global policy options and strategies essentially fell into three categories:

A. A "business-as-usual" scenario in which current national activities continued unabated
B. A "moratorium" scenario where all production of biofuels is temporarily prohibited
C. An "intergovernmental consensus building on sustainable biofuels" scenario which recommends the development of an internationally agreed approach to consider the issues and develop biofuels in a sustainable manner

1.2 The Need for Hydrocarbon Fuels (As Compared to Oxygenated Fuels)

The two principal first-generation biofuels, bioethanol and biodiesel (fatty acid methyl acid, FAME), can now be considered commercially established products, albeit often with government subsidies required to ensure viability. Bioethanol has been demonstrated to work successfully in flex-fuel vehicles using ratios from 0% to 100% mixed with gasoline. FAME, on the other hand, has generally been used in blends with diesel at concentrations below 10%.

1.2.1 *What Issues Limit Widespread Adoption of These Fuels?*

The process for FAME production is very simple, involves time-tested chemistry and requires relatively inexpensive process equipment; however, it has several problems:

- It is difficult to maintain consistent quality, especially when the feedstock quality or type changes.
- The quality of biodiesel produced depends very heavily on the quality of the vegetable oil used, on which it is very difficult to simultaneously maintain control and supply security.
- Also, the final product contains oxygen and therefore offers a lower calorific value relative to petroleum-derived diesel; this implies larger capacities of storage and handling facilities for a given energy delivery than conventional diesel from refineries.

[a]Energy Security, Biofuels and Food Supply. Available from: www.unescap.org/esd/energy/theme/documents/FS7-EnergSecurityy&FoodSupply.pdf.

- The shelf life of FAME varies widely depending on feedstock and process conditions; it decays with time and typically becomes unfit for use within a few months unless stabilized with additives, an approach that can impact both economics and engine performance.
- The FAME process involves addition of excess of methanol, which is typically made from nonrenewable sources thereby reducing the green footprint of FAME Biodiesel. Additionally, issues relating to the supply chain for methanol need to be addressed in citing FAME biodiesel plants, especially in remote locations.
- FAME biodiesel production generates glycerin as a byproduct which requires energy to be transported away from the production site and sold to users in other industries, for example, pharmaceuticals. A concern exists that as FAME production grows, supply constraints could arise on ancillary feed and product streams: methanol could become more expensive and prices of output glycerin may continue to fall, impacting the overall economics for FAME production adversely.
- Most importantly, the physical and chemical characteristics for FAME necessitate dedicated handling, storage, blending, and usage infrastructure. Many vehicle manufacturers have invested millions of research dollars in adapting their diesel engines for FAME blends so as to be able to provide end-user guarantees in the face of compulsory biofuel-blending mandates.

Thus, while FAME clearly represents a beginning for biofuels in widespread use, the earlier constraints indicate the need for new technologies. Drop-in hydrocarbon replacements from renewable feedstock could eliminate a number of these issues.

Bioethanol faces similar hurdles in production and use. Like FAME biodiesel, production of bioethanol too is well established and relatively simple as it is based on fermentation of sugars. While this has allowed bioethanol to be the first and most widely used biofuel today, the rapid penetration has also unearthed some real concerns. In particular:

Energy content of Ethanol is approximately 60% of that of hydrocarbon-based gasoline it substitutes in volume. This impacts the range of the vehicle between fuel refills for a given fuel tank size.

Ethanol absorbs water from air and from dead zones in the distribution and storage infrastructure, picking up even trace amounts from the distribution system, bringing it into the delivery system, and ultimately into the power delivery circuit of the automobile, leading to corrosion and energy dilution issues.

Heavy usage of water in the production of feedstock (corn, sugarcane) is a potentially significant cost in the production of bioethanol; water is still treated as a relatively inexpensive commodity in many parts of the world but this situation is unlikely to continue.

Sugars and carbohydrates that are the source of first-generation bioethanol are in direct competition with food. While this problem is alleviated in lignocellulosic bioethanol, the as yet significantly higher costs for production of lignocellulosic bioethanol make it a technocommercially challenging option.

1.3 The Need for Liquid Fuels (As Compared to Solid and Gaseous Fuels)

The use of compressed gaseous transport fuels has emerged as a possible option to liquid fuels. As per the Society of Indian Automobile Manufacturers (www.siamindia.com), Delhi has set an example by having over 100,000 CNG (Compressed Natural Gas)-fueled vehicles, the most of any city in the world. While hydrogen-based fuel cell vehicles (FCVs) might maximize CO_2 emission reductions compared to the status quo, significant implementation barriers exist to such a solution (Hekkert et al., 2005); bridging alternatives such as CNG engines and gasoline-fuelled FCVs have meanwhile become commercial realities.

Notwithstanding these trends, we have shown earlier in this paper that liquid fuels such as gasoline and diesel dominate world transportation demand. There is as yet no solid transport fuel in commercial use of any form (barring coal for steamships), nor is there a solid or gaseous fuel readily available as of now for aviation. If liquid fuels are indeed likely to remain the leading category for the foreseeable future of transportation, it follows logically that liquids will be the most likely form of biofuels, too, for some time to come.

2 DESIRED CHARACTERISTICS OF GREEN LIQUID HYDROCARBON FUELS (GLHF)

Fuel quality plays a critical role in determining system performance, equipment life, nature, and extent of emissions as well as the overall economics of an industrial unit, residential, or commercial establishment, power plant, or transport vehicle (Figure 2). As seen in figure 2, liquid fuels find applications in a variety of energy applications, from transport to electricity generation. Hence the product quality is a critical parameter is bringing new liquid fuels to market to fill these specific needs.

GLHF, therefore, must deliver on multiple fronts relative to liquid hydrocarbon fossil fuels (LHFF):

- Deliver comparable performance in existing equipment, unless there is compelling value in equipment modification or replacement
- Reduce NO_x and SO_x emissions
- Minimize disruption of existing LHFF supply chains
- Either meet prevailing specifications and regulatory of LHFF, else be amenable to standardization such that new regulatory definitions can be instituted

In addition to this, the feedstocks and production processes for GLHF must comply with the Four Imperatives we mentioned earlier. Rigorous, consistent, and widely accepted

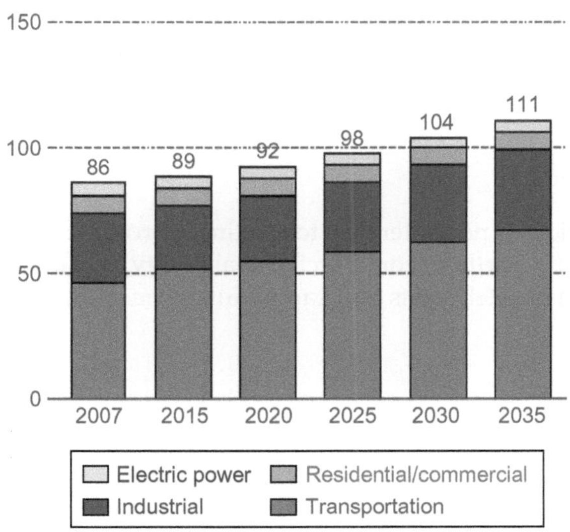

FIGURE 2 World liquids consumption by sector, 2007-2035 (million barrels per day). U.S. Energy Information Administration/International Energy Outlook (2010).

methodologies for assessment of GHG emission reduction and studies of effects on food, security, water, and land use need to be developed in parallel to ensure that the replacement of LHFF by GLHF does not create new challenges for future generations while solving problems of the present.

2.1 Automotive Fuels

Liquid hydrocarbons have remained the preferred choice for the transportation sector, driven by their high calorific value, well-established supply chain, and easy availability. However, it has been recognized that vehicular emissions from nitrogen oxides (NO_x) and sulfur oxides (SO_x) need to be controlled to limit impact on ambient air quality. Diesel fuel used in vehicles on the highway—trucks, buses, passenger cars—must ideally be low in sulfur content. Upper limits for sulfur levels are mandated in most major user countries and have progressively become stricter over the years.

CO_2 emissions from transportation rank second only to power generation as a source of GHG emissions (Figure 3). Biofuels, developed from natural feedstocks like oilseed species and algae that fix carbon dioxide through photosynthesis, thus reduce net CO_2 emissions relative to fossil fuels.

The automotive sector has been one of the early adopters of biofuels. Both major first-generation fuels, ethanol (for blending with/substitution of gasoline) and FAME biodiesel (for blending with/substitution of diesel), have received some measure of support from the industry in terms of engine development, system trials, vehicle warranties, and supply chain infrastructure. Growing rapidly from a small base over the past decade, notably in Brazil and the United States, biofuels contributed about 3% of global road-transport fuel demand in 2009. As seen in Table 1, the worldwide production of biofuels is approaching 1 million barrels per day (Mbpd) while the world demand for liquid fuels is 100-150 times that amount, as shown in Figures 1 and 2. Thus there is significant room for biofuels to take a bigger share of the world liquid fuel pool.

However, several aspects still need considerable effort (Lahaussois, 2010), such as

- Seamless use by existing and future fleet
- Infrastructure adaptation
- Customer acceptance
- Fuel filter plugging
- Injector deposits
- Material compatibility
- Fuel tank corrosion

Drop-in GLHF which are chemically identical or near-identical to existing petrodiesel and petrogasoline could address all these issues. Further, commercial availability of GLHF products could significantly reduce or eliminate R&D costs and capital investments related to such efforts.

2.2 Aviation Fuels

As might be expected, the specifications and performance requirements for aviation fuels are significantly more rigorous than for fuels used in surface and marine transport. Several issues must be considered: (a) the freezing point, which should be as low as possible to enable

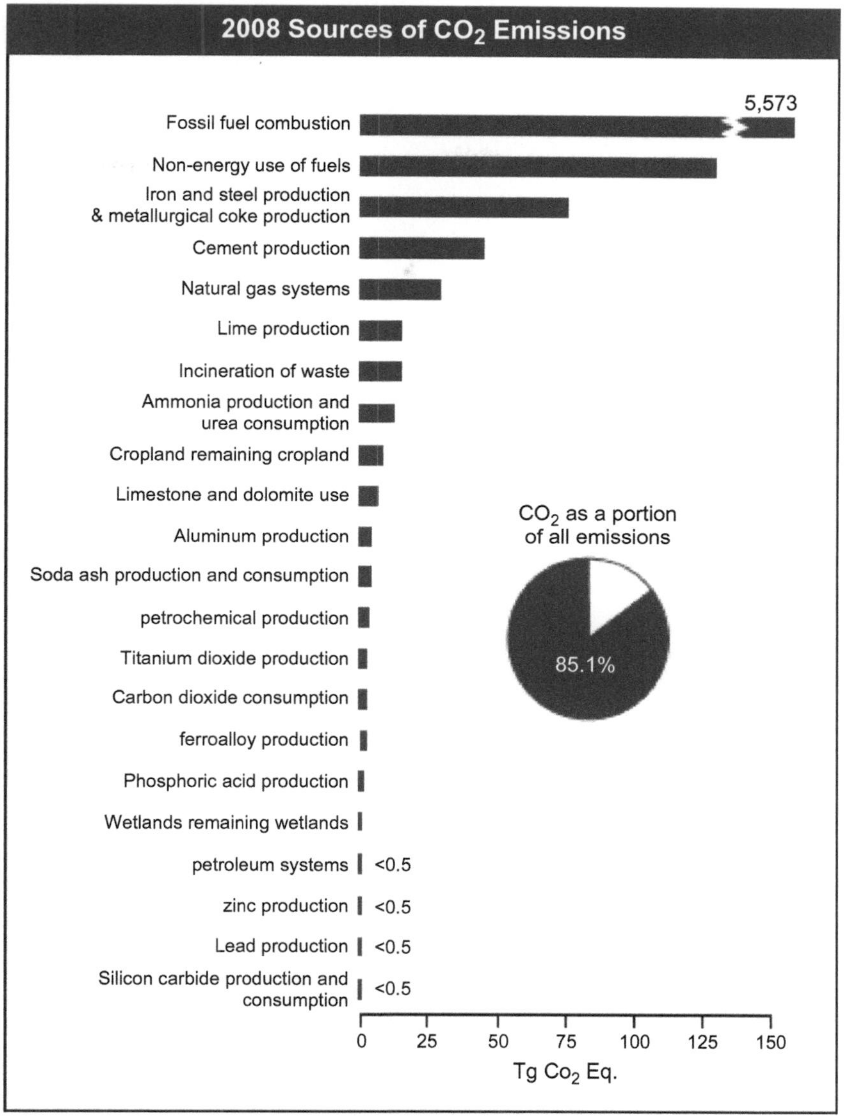

FIGURE 3 2008 Sources of CO_2 Emissions in the US (United States Environmental Protection Agency, 2010).

aircraft to fly safely at higher altitudes; (b) the flash point, ideally as high as possible to minimize fire hazard; (c) low vapor pressure for ground crew safety and (d) good thermal oxidative stability, since aviation fuel is not just a combustion energy source but also a coolant, and thus prone to oxidative degradation as fuel temperature rises[b].

As is evident (Table 2), no single fuel as yet meets all applications in the existing fleet of aircraft around the world. This implies that no single biofuel is likely to achieve this either, at

[b]Air BP website, History of Jet Fuel, http://www.bp.com/sectiongenericarticle.do?categoryId=4503664&contentId =57733.

TABLE 1 World Biofuels Production, 2009[n]

	Ethanol		Biodiesel		Total	
	Mtoe	**kb/d**	**Mtoe**	**kb/d**	**Mtoe**	**kb/d**
United States	21.5	470	1.6	33	23.1	503
Brazil	12.8	287	1.2	25	14.1	312
European Union	1.7	38	7.0	140	8.7	178
China	1.1	24	0.3	6	1.4	30
Canada	0.6	13	–	–	0.6	13
India	0.1	3	0.1	2	0.2	5
Other	0.9	20	2.7	51	3.6	72
World	38.7	855	12.9	257	51.6	1 112

TABLE 2 ASTM International Standard Specifications for Aviation Fuels (Orr, 2009)

ASTM International Standard Specification	Description	Fuel
D910-07a	Standard Specification for Aviation Gasoline	Grades 80, 91, 100LL, 100
D1655-08a	Standard Specification for Aviation Turbine Fuels	Jet A, Jet A-1
D6227-04a	Standard Specification for Grade 82 Unleaded Aviation Gasoline	Grade 82
D6615-06	Standard Specification for Jet B Wide-Cut Aviation Turbine Fuel	Jet B

least in the foreseeable future. By way of example, Jet A has a lower cost but a higher freezing point limit (-40 °C) as compared to that of Jet A-1 (-47 °C), whereby use of the latter is favored in winter conditions or on polar routes.

ASTM standard D7566-10a defines specific types of aviation turbine fuel that contain synthesized hydrocarbons for civil use in the operation and certification of aircraft and describes fuels found satisfactory for the operation of aircraft and engines. Synthetic paraffinic kerosene (SPK) may be blended with D1655 fuel; SPK thus represents a window of opportunity for replacement by GLHF.

In addition, one must also consider Aviation Biofuels for military use, where the duty requirements are further constrained by manner of use. For instance, JP-5 is a high flash point kerosene fuel that dates from the 1950s, introduced with the intent of enhancing safety on aircraft flying from aircraft carriers. JP-8, which replaced JP-4 for the U.S. Air Force by 1995, has similarities with Jet A-1 but is enhanced with additives that improve anti-icing performance, inhibit corrosion, and mitigate risks from static charge development (Agosta, 2002).

[n]World Energy Outlook 2010, International Energy Agency.

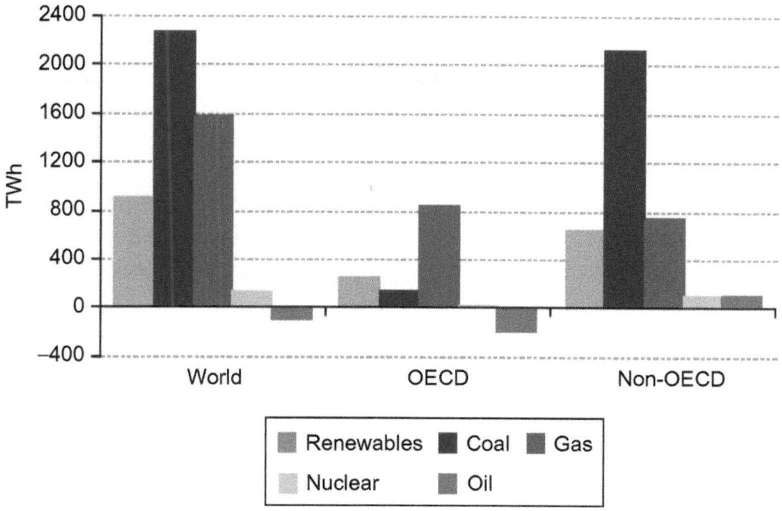

FIGURE 4 World incremental electricity generation by fuel, 2000-2008.[n]

2.3 Liquid Fuels for Power Generation

The global use of liquids for power generation is relatively small. Figure 4 shows that coal and gas are predominant, and oil use continues to decline as renewable options become less expensive and more readily available.

Given the costs of development and production of GLHF to relatively tight specifications for transportation uses, and the availability of lower cost alternatives for green power such as biomass-based generation, it is unlikely that GLHFs will be a significant fuel class for the renewable power industry, though their use as a diesel replacement in stationary applications is a possibility for regions where grid connectivity is low or erratic, and genset usage is relatively high.

2.4 Liquid Fuels for Heating Applications

Furnace and fuel oils used for residential heating and industrial thermal applications typically derive from lower value refinery cuts. As in the case of power generation earlier, it is unlikely that GLHFs will see adoption in this category any time soon.

3 TECHNOLOGIES FOR PRODUCTION OF GLHF

A number of process routes can be effectively employed to convert biofeedstock into GLHFs. These range from continuous operations such as one might find in a refinery, where triglyceride oils extracted from the oil seeds are processed in catalytic reaction systems in a

[n]World Energy Outlook 2010, International Energy Agency.

process similar to crude oil processing, to converting fatty acids using biochemical fermentation-based processes, using sugars as the biofeed source and finally to coprocessing the bio-oils along with petroleum crude oil in existing refineries. While all these processes have been demonstrated to result in production of hydrocarbon fuel, the development toward commercialization is widely different. As discussed below, each of these process routes faces unique challenges.

The biochemical processes have yet to be demonstrated to be effective at larger production scale, with issues related to enzyme availability and robustness and accompanying economic challenges of scaling up. This is a newly developing area, with companies such as Amyris Technologies, Virent Technologies, Solazyme, and LS9, among many other startup companies populating the space.

At the other end of the spectrum is coprocessing, which effectively uses existing refinery infrastructure but also at the same time, brings in new risks into existing large installations. For instance, the oxygen content and impurities occurring in natural bio-derived oils are radically different than those seen in petroleum crude oil, and as the current refinery configurations—reactors, catalysts, equipment, etc., are designed with crude oil in mind, the introduction of bio-derived oils in coprocessing could subject these investments to a degree of risk. While companies such as Conoco Phillips and other refiners have tried this approach, in practice, the coprocessing of bio-oil blends with crude oil has been limited to less than 10% of crude oil feed flow, and that too largely on an infrequent basis.

3.1 Hydroprocessing of Lipids

Hydroprocessing of triglycerides emerges as a process that is practical today, as it has both the elements of fuel flexibility on the front end and the ability to produce products that are 100% hydrocarbon and thus fully fungible with existing hydrocarbon transport fuels in use today, providing full back-end compatibility.

In effect, this process is an adaptation of a refinery operation that is hydroprocessing petroleum crude oil, wherein the catalyst and process conditions have been optimized to handle bio-derived oil feeds that have different types of organic and inorganic contaminants and contain as much as 20% oxygen in the feedstock. Companies such as Neste Oil, Honeywell's UOP, and Syntroleum have demonstrated that this approach can produce high-quality products, comparable in properties and in some ways (for instance, in cetane number) even superior to petroleum-based diesel, jet fuel, and gasoline.

A scheme for the Ecofining™ process from the renewable fuels technology portfolio of Honeywell's UOP[c] is shown in Figure 5. Here, triglycerides are treated with hydrogen in pressurized reaction systems, where specific catalytic activity and experience in refinery process integration enables high production efficiencies even at scales that are an order of magnitude smaller than current petroleum refineries. As feedstock supplies strengthen in volume and thereby become less expensive, the economies of scale afforded by such processes that fit seamlessly with the refining process industry promises to be an avenue toward eventual parity of GLHFs with petroleum-derived liquid fuels.

[c]UOP Website, http://www.uop.com/renewables/10000.html.

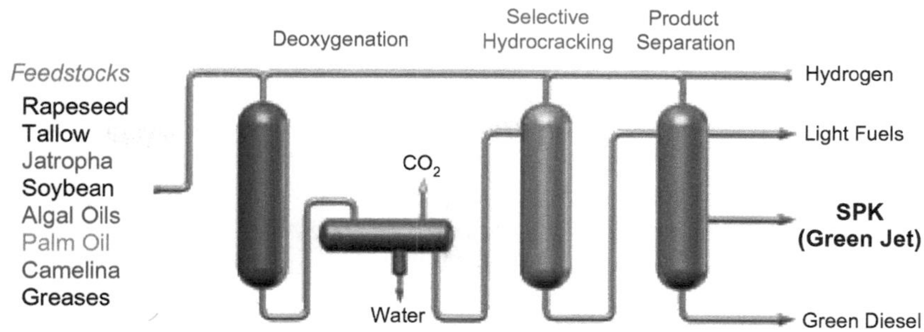

FIGURE 5 A process schematic of UOP's proprietary process for production of renewable fuels from triglycerides (Anumakonda, 2010).

The Ecofining process to produce green diesel can be effectively leveraged by locating the GLHF production unit either at or in close proximity to a petroleum refinery, affording multiple synergies:

- Refinery off-take of product at near-zero supply chain costs
- Renewable fuel distribution into the existing fuel distribution network of the refinery is possible, since the product is highly fungible with today's hydrocarbon-based transport fuels
- Integration of supplies and utilities, such as hydrogen and process water
- Reuse of Ecofining byproducts like naphtha and LPG in the petroleum refinery (National Energy Technology Laboratory, 2009)
- Leveraging the higher cetane value (75-90) of the green diesel thus produced by the refinery to upgrade inferior cuts through blending and thus expand the diesel pool

Honeywell Green Jet™ fuel is derived from similar hydroprocessing technology but incorporating an additional selective cracking step to reduce carbon chain lengths so as to decrease the freezing point. Several demonstration flights of commercial aircraft have taken place using Honeywell Green Jet™ fuel, using various blends of jatropha, camelina, and algae oils as feedstock.

3.2 Biomass Gasification and Fischer-Tropsch Catalysis

Other than hydroprocessing of lipids, a key alternative approach to GLHFs is via gasification of biomass. The resulting syngas is subsequently subjected to Fischer-Tropsch catalytic recombination of carbon and hydrogen (from CO and H_2 components of the syngas) into hydrocarbon fuel molecules.

For instance, ClearFuels Technology Inc. has developed a flexible biomass gasification technology that converts multiple rural cellulosic biomass feedstocks such as sugarcane bagasse and virgin wood waste into clean syngas suitable for integration with synthesis gas-to-liquids technologies. ClearFuels has signed an exclusive worldwide license with Rentech for the use of Rentech's patented and proprietary Fischer-Tropsch synthetic fuels technology for the production of renewable drop-in fuels from sugarcane bagasse. To facilitate the development process, a 20-ton/day ClearFuels biomass gasifier designed to produce syngas from bagasse, virgin wood waste, and other cellulosic feedstocks will be built at Rentech's Product Demonstration Unit (PDU) in Colorado. The gasifier will be integrated with Rentech's Fischer-Tropsch Process and upgrading technology from Honeywell's UOP to produce renewable drop-in synthetic jet and diesel fuel at demonstration scale.[d]

In June 2009, Neste Oil and Stora Enso inaugurated a demonstration plant at Varkaus for biomass to liquids (BtL) production utilizing forestry residues. A 50/50 joint venture, NSE Biofuels Oy, has been established first to develop technology and later to produce commercial quantities biocrude for conversion renewable diesel. The demonstration process units will cover drying of biomass, gasification, gas cleaning, and testing of Fischer-Tropsch catalysts, and will be used to develop technologies and engineering solutions for a commercial-scale plant.[e]

As may be evident from the cases mentioned, collaborations and partnerships abound in the field of GLHF because of the need for multiple domain expertise and interdisciplinary work. Another instance of this can be seen at the Iowa State University Center for Sustainable Environmental Technologies, where a partnership project with the Department of Energy and ConocoPhillips is developing a continuous operation of a biomass to FT-liquids process including the demonstration of viable cleaning technologies. Cleaning of the intermediate syngas is a critical success factor for the downstream conversion to GLHFs.[f]

Also based on a gasification-FT approach is the recent announcement of Dynamic Fuels, a 50/50 venture formed between Syntroleum and Tyson Foods. Utilizing fats and oils feedstock from Tyson, coupled with Syntroleum's Bio-Synfining™ technology, Dynamic Fuels' first plant is intended to produce 75 million gallons/year of GLHFs.[g]

The biomass gasification-FT processes face another hurdle related to the added logistics costs of collating and transporting biomass needed at the required scale. Unless the unit is sited close to an abundant and reliable biomass source, such as for the Stora Enso and Tyson Foods examples mentioned earlier, biomass collection can add significantly to the high costs of production. Further, the syngas conversion process is not very selective, resulting in a compounded problem of either additional converting steps or finding marketing avenues for a variety of high-end waxes that are coproduced with the fuel range hydrocarbons in a typical FT process.

A typical process scheme of a gasification-FT process is shown in Figure 6.

3.3 Conversion of Sugars to Hydrocarbons

Also gaining importance is an approach where sugars (derived from biosources) form an intermediate feedstock pool and these sugars are subsequently converted—either catalytically or through biochemical processes—into renewable fuels. The obvious benefit here is the utilization of intermediate sugar molecules that have multiple commercial possibilities as compared to a fuel product family alone.

[d]Rentech website, http://www.rentechinc.com/gasifier.php.

[e]Neste Oil Press Release, http://www.nesteoil.com/default.asp?path=1;41;540;1259;1260;11736;12772.

[f]Iowa State University College of Engineering web link http://www.engineering.iastate.edu/innovate/feature-stories/spring-2008/conocophillips-iowa-state-join-to-produce-synfuels-from-gasification.html.

[g]Syntroleum website, www.syntroleum.com.

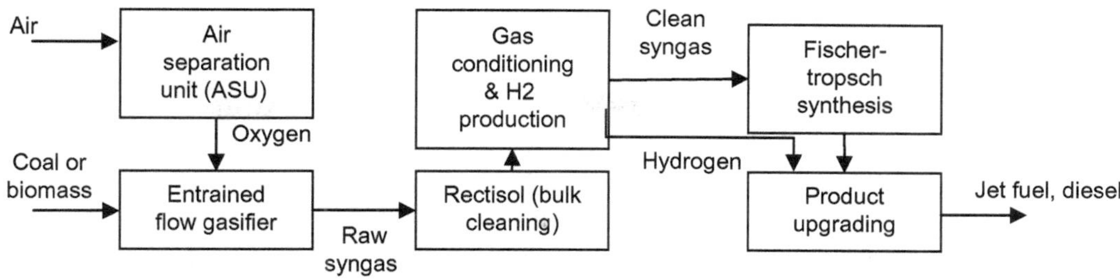

FIGURE 6 Block diagram of a typical Gas-to-Liquid scheme depicting the Fischer-Tropsch process (Boerrigter, 2006).

Virent Energy Systems has developed the BioForming® process, which is based on Virent's Aqueous Phase Reforming or APR technology. This process converts sugars derived from vegetable sources to light hydrocarbons. The BioForming process requires multiple steps to convert the light hydrocarbons to components similar to conventional transportation fuels. While the products from this process are compatible with petroleum fuels, the multiple processing steps increase the complexity and cost of a commercial process. A biogasoline demonstration plant using Virent technology in partnership with Royal Dutch Shell has commenced production.[h]

Amyris Technologies' fermentation-based process[i] uses an engineered yeast to convert sugar into isoprenoids, a useful family of compounds which includes not just fuels but also pharmaceuticals, nutraceuticals, aroma chemicals, and chemical intermediates. The initial isoprenoid Amyris focuses on a 15-carbon hydrocarbon, beta-farnesene, which easily separates from and floats on top of the aqueous fermentation broth, enabling easy recovery and purification of the hydrocarbon. Farnesene can be converted into a renewable diesel, but also into a variety of other useful chemicals, again underscoring the benefits of a fungible intermediate route.

LS9 UltraClean™ products are a family of GLHF produced by LS9 DesignerMicrobes™ created through synthetic biology,[j] starting from natural sources of sugar such as sugar cane and cellulosic biomass, LS9 claims to have developed a new means of efficiently converting fatty acid intermediates into petroleum replacement products via fermentation of renewable sugars, and has also discovered and engineered a new class of enzymes and their associated genes to efficiently convert fatty acids into hydrocarbons.

For these processes that are based on processing or fermenting sugars, the inherent assumption is that cellulosic routes to sugars will reach an economical and sustainable level. If not, commercial scaling of sugar-to-GLHF processes would be limited by competition with the food chain. The hydrocarbon products produced from sugar-based processes may also need further purification, separation, and other forms of refinement, requiring require further processing in refinery operations.

[h]Virent Energy Systems website www.virent.com.

[i]Amyris Technologies website, www.amyris.com.

[j]LS9 Inc. website, www.ls9.com.

The enzymatic and biochemical routes, whether through the gasification pathway or the sugars pathway, are at relatively early stages of research and process development today. They face the growing pains of scale-up challenges and production-scale economic hurdles. The Gasification-Fischer Tropsch (Gasification-FT) route, ironically, faces the opposite hurdle, that of scale-down. To be economical, gasification and FT process are typically mega-scale units, requiring billions of dollars (National Energy Technology Laboratory, 2009) of capital investment, as compared to hydroprocessing refining routes that need capital investments typically an order of magnitude smaller.

While each technology route has its own distinctive challenges, it is worthwhile noting that this is a wide playing field. The demand for global energy, aided by the compelling need for greener and renewable sources of energy, allows for more than one winner in the path to providing renewable fuels in the market. Secondly, renewable fuels can hope to reach economic parity only if the supply chain of the feedstocks is strong and sustainable. Volumes of bio-based feedstocks need to build up in significant quantities to achieve this and hence the spotlight is really on the nurturing and supporting the supply chain of renewable feedstocks.

By way of perspective, using the hydroprocessing approach as an illustration, the cost of fuel processing is only about 10% of the cost of the raw material—the triglyceride oil—that is converted into GLHF (estimated for a fuel production facility of roughly 100 million gallons/year, roughly an order of magnitude smaller than today's average crude oil refinery). At today's market rates, these oil feedstocks are priced in the same range as food oils in the absence of a strong and independent supply chain of inedible oils. As supply of these feed oils increase, two beneficial effects are expected to come into play. On one hand, due to supply increase, the price of inedible oils would come down, likely pegging within range of crude oil. On the other hand, larger supply quantities of feed would favor scaling up of the hydroprocessing ecorefineries, which—in turn—would reduce the fractional contribution of the production cost to even lower than 10%, tending toward the 5% range that crude oil production costs are typically in today's modern petroleum-based refineries.

On a technical level, it is premature to call any one route the winner in the race for an efficient pathway to renewable fuel production. All routes have challenges but all of them have promising avenues, and many industrious groups of researchers and companies are dedicating themselves to the task of taking them forward.

3.4 Pyrolysis and Upgrading

Gasification is one of three possible thermochemical approaches to utilize the energy contained in biomass. The other two are direct combustion (which will not be discussed here since it does not lead to liquid fuels) and pyrolysis.

Pyrolysis is a process by which higher molecular weight material is decomposed into smaller molecules by rapid heating in the absence of oxygen. Where biomass is concerned, pyrolysis can be thought of as a calorie concentrator, as the principal product is usually a liquid (pyrolysis oil or bio-oil) that has a much higher density, usually 3-4 times that of the solid feedstock.

Several types of pyrolysis reactors and technologies have been developed over the years. These include (Brown and Holmgren, 2009):

- Bubbling fluidized bed
- Vacuum Pyrolysis
- Rotating Cone Pyrolyzer
- Ablative Pyrolyzer
- Auger Reactor
- Circulating fluidized bed

Each reactor type has its own advantages and challenges. RTP™ technology is offered by Envergent Technologies LLC, a joint venture between Honeywell's UOP and Ensyn, Inc. This process, operating at atmospheric pressures and moderately high temperatures of about 500 °C, converts biomass to bio-oil within a short contact time of about 2 s in a circulating fluidized-bed reactor. Oil yields are typically in the range of 60-70%, the balance being char (12-15%) and gas. The pyrolysis oil produced contains water, acidic moieties, and significant levels of oxygenates; nonetheless, it can be used as-is in heat and power generation applications.

Various pathways for upgrading pyrolysis oil to GLHFs have been explored over the years. Dynamotive Energy Systems, Honeywell's UOP, Alphakat GmbH, and Kior, among others, have been in the news for developments in this area. Not surprisingly, hydroprocessing has been a favored approach; catalytic deoxygenation of biomass-derived pyrolysis oils *in situ* is also being worked upon.

4 FEEDSTOCK CONSIDERATIONS FOR GLHF

4.1 Lipids

Triglyceride lipids are among the most convenient sources for GLHFs.

Most vegetable oils, as well as animal fats like tallow, typically contain carbon chain distributions in a range (C16-C20) similar to those in the cuts of petrodiesel. A few natural oils, such as coconut oil, contain C12 and C14 chains, which correspond roughly to the hydrocarbon range of kerosene.

We have already shown that conversion of triglycerides to GLHFs can be achieved through catalytic hydrogenation and isomerization.

This implies that the selection of triglyceride feedstock would be determined by the target hydrocarbon carbon chain distribution and extent of branching, as also by the amenability of the feedstock to the process—particularly the specific catalyst used for a given technology, which may be sensitive to specific impurities in the feedstock, for instance heavy metals, alkali metals, alkaline earths, phosphorus, etc.

As long as the feedstock is free of such impurities, the characteristics of GLHF produced are largely independent of the feedstock carbon chain distribution, making this approach much more fungible in feedstock terms that conventional FAME biodiesel where biofuel characteristics are predicated primarily upon feedstock fatty acid composition.

As the hydrogen required in the conversion process plays an important role in the economics as well as the supply chain for a triglyceride-based GLHF, it can be inferred that highly unsaturated raw materials are less preferred compared to more saturated options.

An additional practical reason for this preference is that highly unsaturated oils oxidize more readily. The oxidation mechanism is complex and tends to generate polymeric gums and deposits, which progressively reduce the usability of the oils with time unless stored in an inert atmosphere or with antioxidants and stabilizers added. The rates of reactions in autoxidation schemes may depend on hydrocarbon structure, heteroatom concentration, heteroatom speciation, oxygen concentration, temperature, and the presence of impurities that might act as catalysts (Mushrush et al., 2000)

In addition to the degree of unsaturation, presence of unsaponifiable matter and free fatty acid (FFA) can affect the overall yield and process economics. Unlike conventional FAME biodiesel through transesterification where FFAs are severely limiting to both process and product, hydrotreatment processes for conversion of triglycerides to green liquid hydrocarbons are relatively tolerant, up to 20% FFA usually being of little consequence. Unsaponifiable matter is usually not converted in GLHF production processes and therefore does not contribute to process yield.

For the economics of a biofuel oilseed to be viable, it is imperative that the triglyceride oil yield per hectare be high and sustained at a predictable level year on year. High yields depend on multiple factors, including optimal planting density (plants per hectare), seed production (kg per plant per year), and extractable oil content (percent by weight of seed). Sowing, nutrition, irrigation, and harvesting costs must be considered, agronomic practices established, and quality of planting material needs to be ensured to avoid surprises.

While palm, soybean, mustard, and rapeseed have been successfully converted to GLHF, concerns have been raised about the use of edible oils in biofuels and the use of arable land to produce biofuel crops. The consequent potential impact on global food prices has led to a worldwide focus on oil-yielding species that grow on marginal and degraded lands, or offer the potential for rotation or intercropping with food crops.

Much of the initial attention appears to have been focused on jatropha species (particularly *Jatropha curcas*), but the oil yields per hectare under controlled conditions have not been replicated in arid and rain-fed soils in most regions. Other nonfood oilseeds that offer promise include pongamia in tropical regions (Meher et al., 2004) and camelina in temperate zones, with several others such as Ethiopian mustard (*Brassica carinata*) (Dorado et al., 2004), cuphea, nonedible safflower, and rubber seed oil being the subject of active research programs.

Jatropha oil has been shown to convert readily to GLHFs, both as a diesel drop-in replacement and bio-derived SPK blended with Jet A-1 fuel. Achten et al. (2010) have postulated that systematic breeding and domestication are essential to realize the full potential of Jatropha. They emphasize species distribution, site requirements, regeneration ecology, genetic diversity, advances in selection, development of varieties, and hybridization. Evidently, although there have been isolated success stories in commercial development of *J. curcas* as a biofuel crop, scalability and sustainability of production are still some distance away. An added factor is the time to maturity—jatropha trees take up to 4 years to attain full oil yields of 1-2 tons/ha, though fast-maturing cultivars of 1-year maturity and higher oil yields have also been reported in controlled studies.

In comparison, *Pongamia pinnata*, a hardy and perennial leguminous plant native to India, Myanmar, and Australia, can take a decade or more to mature but delivers oil yields of 3-5 tons/ha. The oil, which is bitter and otherwise inedible, is known to have value in folk medicine for the treatment of rheumatism, as well as human and animal skin diseases.

Camelina sativa is an oil seed of the mustard family (Brassicaceae), native to Europe. Cultivated since ancient times for use as lamp fuel, the seed contains 30-40% oil, can grow

TABLE 3 Indicative Composition of Various Triglycerides

	Fatty Acid Composition (wt%)									
Lipid	14:0	16:0	18:0	18:1	18:2	18:3	20:0	20:1	22:0	22:1
Camelina		7-8	2-3	16-18	20-25	28-32		11-13		1-2
Canola		4-5	1-2	55-63	20-31	9-10				1-2
Cottonseed	1-2	22-24	2-5	19	50-53					
Jatropha		11-12	16-18	10-14	45-48					
Linseed		6	3-4	13-37	5-23	26-60				
Palm	0-3	32-47	4-7	37-53	6-12					
Rapeseed	1-2	1-5	1-4	13-38	9-22	1-10				40-64
Safflower		6-7	2-3	9-14	75-81					
Pongamia		11-13	6-8	50-54	95-18		1-2		4-5	
Soybean		2-11	2-6	22-31	49-53	2-11				
Sunflower		3-7	1-6	14-43	44-69					
Tallow	3-6	25-37	14-29	26-50	1-3					

TABLE 4 Dry Matter Composition of Algae as Potential Biofuel Feedstock (Becker, 1994)

Strain	Protein	Carbohydrates	Lipids	Nucleic Acid
Scenedesmus obliquus	50-56	10-17	12-14	3-6
Scenedesmus dimorphus	8-18	21-52	16-40	–
Chlamydomonas rheinhardii	48	17	21	–
Chlorella vulgaris	51-58	12-17	14-22	4-5
Spirogyra sp.	6-20	33-64	11-21	–
Euglena gracilis	39-61	14-18	14-20	–
Prymnesium parvum	28-45	25-33	22-38	1-2
Porphyridium cruentum	28-39	40-57	9-14	–
Synechoccus sp.	63	15	11	5

on marginal land and in rotation with wheat and does not compete as a food crop (Yao, 2010). Yields, reported in the range of 1-2 metric tons/ha in different studies, appear to improve with judicious use of nitrogenous fertilizers. Unlike jatropha, this is an annual plant with a maturity of 3-4 months and therefore allows plantation owners the flexibility to produce alternative crops, unlike jatropha or Pongamia where the land is committed for several years.

It should also be noted that economics of a biofuel plantation crop will further be enhanced if there are ways to add value through coproducts and byproducts, such as medicinal extracts, nutraceuticals, or biomass deployed for fodder, fuel, or power generation uses (Tables 3 and 4).

The various carbon chain lengths are indicative of potential complex proteins and other chemical compounds that could be advantageously separated before converting the oils into fuels.

To summarize, the key technical considerations for selection of a triglyceride lipid feedstock for hydroprocessing to GLHFs are:

- Carbon chain distribution
- Degree of unsaturation
- Free fatty acid content
- Unsaponifiable matter
- Impurities that may affect the catalyst used

Commercial considerations include:

- Availability and cost of planting material
- Yield per hectare
- Time to maturity
- Dependence on irrigation
- Value addition from byproducts
- Productivity in marginal and degraded soils

4.2 Algae and Halophytes

The most significant constraint in the use of lipid glycerides as biofuels feedstock is the land area required for achieving even modest scales of production. At typical levels of 2 tons/ha/year, the demand on land-use from biofuels crops would either drive up the cost of land unreasonably beyond a point or begin to compete indirectly with food crops for nutrients, water, and labor even if marginal and degraded lands were to be used for biofuel oilseeds.

Given that the earth contains thrice as much water as land, marine and coastal sources of lipids have attracted much interest as a scalable alternative to lipid triglycerides.

4.2.1 Algae

Microalgae are a widely available source of lipids, and several algae are rich in triglycerides. They can be grown in water—often using effluent streams or atmospheric CO_2 in emission-rich zones as sources of carbon for their growth—and can be harvested daily or more often compared to tree-borne oils that are typically harvested once or twice a year.

There are five primary challenges to scaling up algae toward viability and sustainability.

- Identification of appropriate algal strains and conditions to balance their growth rates and oil content
- Adaptability of the algal production system for diurnal and seasonal variations, as also for environmental shocks and incursions of pathogens and predators
- Cost-effective dewatering of the algae and extraction of the oils on a commercial scale, given that the water content in the harvested algae is at least 95% and often higher
- High upfront capital investment of current algal cultivation systems
- Value realization from byproducts and/or from utilization of wastewater and/or carbon capture

The lipid levels of microalgae can be influenced by stress factors, for instance, nitrogen or silicon starvation. Unfortunately, this may not translate into higher oil yields per acre, as the enhanced oil production per cell may be offset by slower rates of cell growth in response to that same stress.

Genetic modification of algae to improve characteristics and deliver efficient, scalable solutions to the above challenges has attracted significant interest. Solazyme[k] has announced breakthroughs in this approach to produce algal oils, which can then be duly converted to a variety of downstream products including GLHFs.

4.2.2 *Halophytes*

The theme of land availability and appropriate use is inextricably linked with the biofuels industry. It is variously estimated that 20-25% of land on the earth is unusable for crop production due to either salinization and desertification of that land. Particularly in developing and underdeveloped regions, a significant opportunity exists for reclaiming or gainfully utilizing saline land and marshy coastal areas through the use of halophytes. These are salt-tolerant species such as Seashore Mallow (*Kosteletzkya virginica*) (Ruan et al., 2008) and Salicornia species, notably *Salicornia bigelovii* and *S. brachiata,* a succulent bushy plant found in tropical coastal areas which is rich in unsaturated fatty acids, comprising over 90% of the carbon chains in the triglyceride (Anwar et al. 2002).

Intercropping of salicornia with mangroves as part of a sustainable livelihood initiative has been carried out successfully in Eritrea (Hodges, 2010). While this can be considered proof of concept that salicornia and may be cultivated and harvested in a sustainable manner, large-scale adoption of salicornia oil as a feedstock for GLHF may still take several years.

4.3 Biomass

Biomass for heat and power production currently provides the vast majority of renewable energy consumed in the industrial sector (about 90 percent), and it is expected to remain the largest component of the industrial sector's renewable energy mix for the foreseeable future (U.S. Energy Information Administration (EIA), 2010).

However, production of liquid fuels from biomass typically requires multiple process steps, including pretreatment. As most biomass—such as agricultural byproducts—tend to be low in bulk density, extensive use of biomass for GLHF either requires locally concentrated availability in the vicinity of the production unit, or purpose-grown biomass crops on biomass plantations using dedicated land. This implies that biomass crops would compete economically and ecologically for land, water, and nutrients with food crops, forests, or land uses. Further, GHG emissions from the energy used in biomass collection and processing, including potential land use changes and rate of biomass replenishment through utilization of atmospheric CO_2, have to be evaluated carefully to ensure that any biomass-to-GLHF system genuinely reduces overall emissions.

At the other extreme, it has been argued that the efficiency of biomass conversion to energy is far lower than might be anticipated from theoretical considerations, and that replacing the 2080 W/capita energy demand derived from fossil fuels and nuclear energy today with

[k]Solzyme website, www.solazyme.com.

biomass energy could require a staggering 4000 m^2/person of biologically productive land (Burkhardt, 2006). This implies that it may be better to use byproducts of productive land use, such as agricultural residue or livestock tallow, unless the productivity of purpose-grown biomass crops is sufficiently high and can be justified through detailed analysis of impact on food security and the environment.

Along these lines, energy policy experts across continents are critically reviewing biomass availability and utility for energy uses (Dornburg et al. 2008; Luckow et al. 2010). Especially to be noted is the level of detail in the National Biomass Resource Atlas of India, which maps the availability of surplus biomass down to district levels as a guideline for potential biomass energy projects.[1]

4.4 Sugars

Sugar-rich feedstocks have been of interest to the biofuels community from an ethanol perspective for several decades, but new avenues for conversion of sugars to hydrocarbons have begun to emerge of late, as mentioned in Section 3.3 earlier. From a GLHF perspective, therefore, there is fresh interest in identification of sugar sources, especially those that do not adversely impact food security.

Sugarcane is globally the single largest crop source of fermentable sugars. Sugar beet and sweet sorghum have been researched as alternate options. In particular, the short crop duration of about 4 months, much lower water requirement in cultivation, and reduced effluent load from fermentation of sweet sorghum as compared to sugarcane have been highlighted (International Crops Research Institute for the Semi-Arid Tropics (ICRISAT), 2007).

Corn, cereals, grains, tapioca, cassava, and other starchy sources can be converted to simple sugars and subsequently to GLHF using some of the technology options shown in Section 3.3. However, direct conversion to ethanol is an existing, single-step alternative for the biofuel industry, and several commercial enzyme options already exist.

Conversion of the cellulose and hemicellulose components of lignocellulosic biomass to hexose and pentose sugars, respectively—and thereon to GLHFs—appears to be the most attractive option from the Four Imperatives standpoint. The key challenge here is to find an effective, scalable pretreatment process for the biomass, and ideally to also derive economic value from the byproduct lignin to improve prospects of commercial viability.

4.5 Municipal and Industrial Waste

Although there is considerable variance between sources, estimates of annual global municipal waste generation range between 2.5 and 4 billion tons.

Not only does this accumulated and growing mass of waste pose risks to public health and hygiene, but it also represents unproductive use of precious land by way of landfills, a significant source of GHG emissions such as methane and—of relevance to this book—a large reserve of organic matter that could be converted to biofuels if technologies were developed and appropriate supply chains established.

[1]National Biomass Resource Atlas of India. Available from: http://lab.cgpl.iisc.ernet.in/Atlas/. Combustion Gasification & Propulsion Laboratory, Indian Institute of Science, Bangalore, supported by the Ministry of New and Renewable Energy, Government of India.

Segregation of waste is the key challenge in this sector. For instance, if plastic waste could be segregated efficiently, it would be easier to scale up technologies, such as that from Ozmotech,[m] that break this plastic down into smaller hydrocarbon molecules.

5 CONCLUSION

GLHF that can fit seamlessly into existing fuel storage and distribution infrastructure appear to be both a necessity and a likely contributor to the future global fuels scenario. Some GLHFs are already available commercially today; in 2009, the U.S. Military announced a procurement program of about 600,000 gallons of green jet fuel to be delivered over 2 years. The increasing number of announcements from both established companies and startups in the field offers hope that GLHFs will be a routinely available class of commodities in the foreseeable future.

The complex matrix of feedstock sources, conversion technologies, and application areas for GLHF is such that several factors have to align for long-term success. Effective supply chains, scalable production processes, appropriate conducive regulatory frameworks, and rigorous life-cycle analyses are needed to ensure that correct choices are made in a critical area of energy security and environmental responsibility.

References

Achten, W.M.J., Nielsen, L.R., Aerts, R., Lengkeek, A.G., Kjær, E.D., Trabucco, A., et al., 2010. Towards domestication of *Jatropha curcas*. Biofuels 1 (1), 91–107.

Adapted from: Anumakonda, A., 2010. Greening global aviation, presented at MRO Americas. Phoenix, AZ, USA. (April 2010).

Adapted from: Boerrigter, H., 2006. Economy of BIOMASS-TO-LIQUIDS (BTL) plants: an engineering assessment. In: Report No. ECN-C-06-019. Energy Research Center of the Netherlands (ECN) (May 2006).

Adler, P.R., Del Grosso, S.J., Parton, W.J., 2007. Life-cycle assessment of net greenhouse-gas flux for bioenergy cropping systems. Ecol. Appl. 17, 675–691.

Agosta, A., 2002. Development of a chemical surrogate for JP-8 Aviation Fuel using a Pressurized Flow Reactor. M.S. Thesis, Department of Mechanical Engineering, Drexel University. .

Anwar, F., Bhanger, M.I., Khalil, M., Nasir, A., Ismail, S., 2002. Analytical characterization of *Salicornia bigelovii* seed oil cultivated in Pakistan. J. Agric. Food Chem. 50 (15), 4210–4214.

Becker, E.W., 1994. Baddiley, J. et al., (Ed.), Microalgae: Biotechnology and Microbiology. Cambridge Univ. Press, p. 178.

Brown, R.C., Holmgren, J., 2009. Fast Pyrolysis and Bio-oil upgrading. Available from: www.ars.usda.gov/sp2UserFiles/Program/307/biomasstoDiesel/RobertBrown&JenniferHolmgrenpresentationslides.pdf.

Burkhardt, H., 2006. Physical Limits to Large Scale Global Biomass Generation for Replacing Fossil Fuels, Round Table on Forestry organized by the Faculty of Forestry at the University of Toronto, and Science for Peace (September 2006).

Dorado, M.P., Ballesteros, E., Lopez, F.J., Mittelbach, M., 2004. Optimization of alkali-catalyzed transesterification of brassica carinata oil for biodiesel production. Energy Fuels 18, 77–83.

Dornburg, V., et al., 2008. Climate Change: Scientific Assessment and Policy Analysis—Assessment of Global Biomass Potentials and Their Links to Food, Water, Biodiversity, Energy Demand and Economy. Study carried out within the framework of the Netherlands Research Programme on Scientific Assessment and Policy Analysis for Climate Change.

[m]Ozmotech Pty Ltd website, www.ozmotech.com.au.

Ediger, V.S., Kentel, E., 1999. Renewable energy potential as an alternative to fossil fuels in turkey. Energy Conservat. Manag. 40 (7), 743–755.

Energy, Development, and Security, 2008. Energy issues in the current macroeconomic context. United Nations Industrial Development Organization.

Hekkert, M.P., Hendriks, F.H.J.F., Faaij, A.P.C., Neelis, M.L., 2005. Natural gas as an alternative to crude oil in automotive fuel chains well-to-wheel analysis and transition strategy development. Energy Pol. 33, 579–594.

Hodges, C.N., 2010. An Introduction to Integrated Seawater Agriculture Systems (ISAS): A Source of Sustainable Biofuels, 2nd International Symposium on Biofuels, organized by the Petroleum Federation of India and UOP, a Honeywell Company, New Delhi.

International Crops Research Institute for the Semi-Arid Tropics (ICRISAT), 2007. Pro-Poor Biofuels Outlook for Asia and Africa: ICRISAT's perspective (March 2007). http://test1.icrisat.org/Investors/Biofuel.pdf.

Lahaussois, D., 2010. Toyota motor Europe, presentation at World Biofuels Market, Amsterdam, 16th March 2010.

Luckow, P., Dooley, J.J., Wise, M.A., Kim, S.H., 2010. Biomass Energy for Transport and Electricity: Large Scale Utilization Under Low CO_2 Concentration Scenarios, Pacific Northwest National Laboratory (January 2010).

Meher, L.C., Naik, S.N., Das, L.M., 2004. Methanolysis of pongamia pinnata oil for production of biodiesel. J. Sci. Ind. Res. 63 (11), 913–918.

Mushrush, G.W., Beal, E.J., Hughes, J.M., Wynne, J.H., Sakran, J.V., Hardy, D.R., 2000. Biodiesel fuels: use of soy oil as a blending stock for middle distillate petroleum fuels. Ind. Eng. Chem. Resour. 39, 3945–3948.

National Energy Technology Laboratory, 2009. Affordable, low-carbon diesel fuel from domestic coal and biomass, Document # DOE/NETL-2009/1349.

Orr, M.S., FAA Aviation News July/August 2009, 24, 2009.

Ruan, C.J., Li, H., Guo, Y.Q., Qin, P., Gallagher, J.L., Seliskar, D.M., et al., 2008. *Kosteletzkya virginica*, an agroecoengineering halophytic species for alternative agricultural production in China's east coast: ecological adaptation and benefits, seed yield, oil content, fatty acid and biodiesel properties. Ecol. Eng. 32 (4), 320–328.

Srinivasan, S., 2009. The food v. fuel debate: a nuanced view of incentive structures. Renew. Energy 34, (4): 950–954.

U.S. Energy Information Administration, Annual Energy Outlook 2010, DOE/EIA-0383, 2010. (Washington, DC, July 2010), website www.eia.gov/oiaf/aeo.

United States Environmental Protection Agency, 2010. Inventory of U.S. greenhouse gas emissions and sinks: 1990–2008.

Yao, S., 2010. ARS Researching Camelina as a New Biofuel Crop. Agricultural Research Service, US Department of Agriculture. http://www.ars.usda.gov/is/pr/2010/100413.htm.

Index

A
ABE fermentation, 13, 19, 571–582
Ablative fast pyrolysis, 67
Accelerase®, 193, 194
Accelerate granulation, 559
ACEII, 180–181
Acetaldehyde, 13, 293, 295, 299, 300, 509
Acetate, 82, 83, 86, 87, 88, 131, 137, 141, 159, 210–211, 213, 217, 232, 233, 265, 297, 341, 344–346, 384, 385, 401, 500, 501, 513, 514, 515, 516, 525, 526, 527, 528, 530, 551, 553, 554, 557, 572, 574
Acetic acid, 22, 79, 82, 83, 84, 85, 86, 87, 88, 95, 137, 153, 156, 164, 166, 213, 236, 241, 257, 261, 291, 292, 296, 297, 299, 300, 303, 304, 345, 517, 527
Acetogens, 82, 85, 501
Acetone, 13, 18, 19, 85, 157, 184, 233, 319, 513, 571, 572, 574, 575
  precipitation, 184
Acetone-butanol-ethanol (ABE) fermentation, *see* ABE fermentation
Acetyl-CoA, 82–83, 86, 88, 293, 295, 502, 528
Acetylxylan esterase (EC 3.1.1.72), 205, 210–211, 213–214, 216, 219, 220, 236
ACF (anaerobic contact filter), 543–546
Acid-catalyzed
  BD production, 379, 380
  esterification, 381, 383
    -transesterification, 381, 383
  process, 391, 408
    with hexane washing, 391
    with water washing, 391
  steam pretreatment, 111
  transesterification, 379
Acidic cellulase, 188
Acidogenesis, 83, 88, 503, 504, 515, 572, 575
Acidogenic bacteria (AB), 501, 504, 505, 509, 511, 514, 515
Acidogenic phase, 572, 573, 574
Acid oil, 323, 324, 325, 379, 380, 382
Acid pretreatment, 109, 111–112, 115, 150, 153, 155–156, 160, 168, 170, 183, 256–257, 272, 273, 274, 548
Acid-rich effluents, 513, 514, 515
Acidulated wash, 387
Acrylamide, 16, 19, 244
Acrylic acid, 16, 19–20, 300
Acrylonitrile, 7, 16
cis-Acting promoter, 197
Activators, 180, 181, 515–517
Adaptation, 58, 103, 150, 207, 291, 300, 400, 401, 592, 596
Adaptive bioreactors, 537
Adsorbent, 324, 325, 363, 381
Adsorption, 72, 95, 110, 240, 241, 242, 243, 244, 246, 247, 305, 321, 327, 361, 363, 365
Aeration, 185, 186, 187
AFEX (ammonia fiber expansion), 54, 55, 141, 150, 162, 163, 168, 169, 232
Agitation, 59, 90, 92, 93, 240, 322, 330, 366, 369, 513, 538, 549, 550, 551, 552, 558
Agricultural crop residues, 113
Agricultural wastes, 363, 473, 577
Agroindustrial residues, 192, 251–279
Agroresidues, 192, 203, 204
Aircraft, 593, 594, 597
Alcohol
  dehydrogenase, 293, 295
  furfuryl, 20, 152, 157, 293, 296, 299, 300
  HMfurfuryl, 296
  oleyl, 263, 265, 580
  supercritical, 342–344, 367, 408
Alcoholysis, 317, 319, 330, 379
Aldehyde dehydrogenase, 293, 295, 575
Aldol, 13
Algae, 74, 397–461, 468, 500, 501, 513, 537, 592, 597, 604–605
  biofuels, 415, 416, 425, 432
Algal
  biodiesel production, 410, 411, 432
  biomass, 399–411, 418, 419, 421, 422, 423, 427, 435, 449, 577
  culture, 410
Aliphatic fractions, 11
Alkali
  catalyst, 315, 377, 378, 383
  -catalyzed
    BD production, 377–378
    saponification-transesterification, 381
    transesterification, 381, 383
  pretreatment, 156–157, 183, 257
Alkaline
  catalyst, 169, 316, 331, 350, 377, 379, 383, 408
  -catalyzed process, 391

Alkaline (*Continued*)
with water washing, 391
hydrolysis, 55, 109, 110, 114, 257
pretreatment, 109, 156, 157, 257, 258
Alkalinity, 471, 504, 514
Alkyds, 13
Alkylations, 12
Alkyl esters, 316, 319, 322, 340, 376
Allocation
by mass, 26
methods, 27, 28, 31, 33, 39, 41, 43–44, 45, 46, 47, 48
Allothermal, 63
Alternative biomass use, 30
Alternative fuel, 126, 199, 278, 299
Alternative land use, 30
Amano enzyme, 194
Amino acids, 15, 16, 206, 207, 211, 245, 291, 297, 300, 514, 516, 572
Ammonia and carbon dioxide pretreatment, 258–259
Ammonia fiber expansion (AFEX), *see* AFEX (ammonia fiber expansion)
Ammonia recycle percolation (ARP), 163
Amorphogenesis, 238
Anaerobes, 502, 504–505, 507, 527, 573
Anaerobic digestion, 74, 125, 129, 131, 253, 420, 429, 534
A. niger, 182, 184, 186, 189, 192, 194
Animal fats, 375–376, 380, 391, 399, 601
Animal feed biotechnology, 219
Annular triple jacket reactor, 551
Anoxygenic photosynthesis, 501
Antipilling, 188
Antisense RNA inactivation, 575
Antisense technology, 192
Applications
of cellulases, 178, 188–189, 191–192
of hemicellulases, 219–220
Applied potentials, 518
Aquatic species program, 400, 416, 424
α-Arabinofuranosidase (EC 3.2.1.55), 205, 209, 236
Arabinoxylans, 204, 209–210, 212
Arkenol process, 231
Aromatic benzene, 231
Aromatic chemicals, 9, 11
Aromatics, 6, 10, 11, 14–15, 16, 73
ARP, *see* Ammonia recycle percolation (ARP)
Artificial/engineered cellulases, 199
Artificial or natural (sun) light sources, 455
Ashing process, 388
Aspartic acid, 15
Aspartic anhydride, 16
Aspergillus, 115, 181, 182, 187, 189, 195, 196, 208, 212, 216, 235, 322
Asphalts, 6
Asporogenous nonsolventogenic, 576
ASTM, 356, 357, 389, 392, 409, 594
ASTM D6751, 357, 387, 389, 392
Attached-growth, 510, 541, 543–544
Attributional LCA, 29, 31
Auger (screw) reactor, 69
Automotive, 592
Autotrophic production, 417–418
Autotrophy, heterotrophy, and mixtrotrophy, 401
Autoxidation, 377, 602
Aviation, 591, 592–594

B

Bacillus sp., 182, 198, 209, 210, 215, 235, 245, 322, 365, 515, 527
Back mixing, 92
Back propagation neural network, 219
Bacteria, 90, 109, 110, 142, 180, 181, 182, 185, 186, 196, 205, 206, 207, 208, 210, 211, 217, 218, 219, 239, 252, 255, 257, 260, 263, 264, 271, 288, 290, 293, 299, 300, 379, 402, 403, 439, 450, 451, 452, 459, 475, 476, 500, 501, 502, 503, 504, 505, 507, 509, 510, 511, 512, 514, 516, 517, 518, 526, 527, 537, 538, 540, 541, 542, 543, 545, 546, 547, 549, 552, 556, 559, 561, 571, 572, 574, 576, 577
activities, 543
Bagasse, 54, 79, 81, 89, 103, 105, 106, 107, 113, 114, 118, 130, 146, 155, 156, 158, 159, 161, 162, 165, 184, 218, 251, 253, 254, 259, 266, 273, 304, 597, 598
Base-catalyzed, 204, 359, 383, 384, 389, 407
Baseline system, 30
Batanes/butadiene, 7
Batch
cultivation, 301, 302, 451
mode, 329, 510, 511, 574, 577, 580
operation, 131, 243, 510, 560, 572, 577
processes, 184, 301
Bench scale, 330
Beta-glucosidase (BGL), 109–110, 115, 178, 179, 180, 181, 191, 192, 193, 195, 196, 197, 198, 199, 209, 218, 234, 235, 239, 243, 244, 246, 247, 259, 262
Binders, 10
Bioaugmentation, 515
Bio-based products, 79, 82, 95
Biobutanol, 32, 75, 569–607
Biocatalyst particle sizes to, 327
Biocatalysts, 12, 21, 22, 80, 87, 315, 316, 320, 321, 322, 325, 326, 327, 330, 379, 408, 503–505, 509, 510, 514, 515, 518
Biocharacterization, 189
Biochemical, 5, 12, 16, 32–33, 55, 79, 80, 83, 85, 86, 118, 126, 130, 146, 190, 208, 209, 213, 407, 408, 500, 504, 514, 516, 527–528, 596, 598, 600
pathways, 146, 527–528

Bioconversion, 56, 80, 87, 88, 113, 188, 190–192, 196, 198, 199, 200, 203, 218, 220, 229, 291, 296, 299, 545, 549, 555
  rate, 296, 549
Biodegradable municipal waste (BMW), 117
Biodiesel
  production, 15, 43, 53, 315–333, 339–351, 353–370, 375–392, 400, 401, 403, 405, 406, 407, 410, 411, 416, 432, 439–461, 507, 590
  from the waste, 331
Bioelectricity, 253, 266–267, 277, 278, 512, 513
  production, 253, 266–267, 277, 513
Bioenergy, 33, 56, 59, 79, 103, 104, 112, 113, 140, 170, 229, 266, 468, 499, 505, 512, 518
Bioethanol, 3, 13, 15, 19, 28, 29, 31, 32, 33, 36, 38, 39, 40, 41, 43, 44, 46, 47, 56, 99–306, 403, 481, 571, 580, 582, 587, 589, 590
  production, 28, 31, 33, 38, 46, 47, 99–306
Biofilm, 93, 418, 452, 453, 459, 507, 510, 511, 542, 543, 544, 546, 559, 580
Biofuel, 3, 4, 5, 8, 16, 19, 25–48, 51–75, 79–96, 101, 103, 104, 105, 107, 115, 117, 118, 123, 127, 129, 136, 140, 149, 189–190, 191, 192, 196, 218, 220, 269, 278, 355, 368, 375, 397–461, 474, 476, 573, 582, 587, 589, 590, 591, 592, 593, 594, 598, 601, 602, 603, 604, 605, 606
  production, 16, 25, 32, 33, 52, 74, 83, 89, 94, 102, 136, 190, 192, 402, 403, 405, 410, 429, 439, 440, 457
Biogas, 116, 117, 129, 267, 269, 411, 514, 517, 518, 534, 546, 558
Biohydrogen, 465–562
  from bio-oil, 484, 486, 491, 493
  cost from bio-oil, 492
  produced from biomass through bio-oil reforming is expensive as compared to the natural gas-based hydrogen, 493
  production, 468, 474–475, 476, 481–493, 499–519, 525–534, 537–562
    from bio-oil, 481–493
    cost, 491, 492, 493
    plant, 484, 491, 492, 493
Biokerosene, 439
Biological
  detoxification, 303
  pretreatments, 55, 150, 166–167, 168
  process, 55, 160, 163, 246, 265, 500–501, 503, 518, 533, 549, 552
Biomass
  concentration, 231, 440, 441, 444, 446, 448, 449, 450, 451, 453, 455, 456, 461, 539, 540, 541, 542, 545, 548, 549
  conversion, 4, 8, 9, 21, 55, 60, 70, 72, 116, 155, 191, 194, 195, 197, 198, 199, 278, 490, 605
  -ethanol, 191
  feedstocks, 4, 7, 8, 12, 20, 53, 80, 89, 94, 95, 145, 252, 481, 482, 484, 486, 597
  -fired cogeneration, 267, 269, 277, 278
  -liquid separation, 537, 538, 540, 541, 545
  measurement, 186
  processing, 8–9, 21, 69, 193, 435
  program, 125, 126, 145
  productivity, 400, 403, 443, 447, 448, 449, 451, 453, 458
  resources, 3, 4, 13, 16, 59, 469
  retention, 510, 537, 538, 540, 542, 543, 544, 546, 548, 558, 559, 562
  -substrate interaction, 240, 545
  washout, 537, 538, 539, 545
Biomass-integrated gasification combined cycle (BIGCC), 267–268, 275, 276, 277
Bio-oil
  to biohydrogen, 483, 484
  from biomass, 485
  production cost for biohydrogen plant, 491, 493
Biophotolysis, 474, 475, 500, 525, 537, 549, 552
Bioplastics, 514–515
Bioprocess, 56, 106, 142, 183–187, 198, 199, 440, 441, 505
Bioreactor, 80, 86, 90, 185, 186, 187, 197, 218, 265, 306, 325, 326, 327, 330, 400, 402, 403, 409, 410, 432, 453, 513, 514, 515, 518, 530, 533, 537–562
  configuration, 90
Biorefineries, 4, 5, 8–9, 10, 11, 13–16, 21, 32, 33, 72, 73, 85–86, 102, 105, 146, 159, 170, 178, 193, 305, 401, 409
Biostoning, 188
Biphasic, 572, 574, 577
Birchwood xylan (BWX), 212
Botryococcus braunii, 409
Brazilian bioethanol program, 118
2-bromoethanesulfonic acid, 505
Brown grease, 375, 376
BTX (aromatic benzene, toluene, and xylene), 7, 11, 12, 14, 15
Bubble column, 84, 90, 92, 218, 549
Bubble washing, 385, 386, 387
Bubbling fluidized bed (BFB) combustion, 58
Bubbling fluidized bed (BFB) pyrolysis, 68
Buffering capacity, 186, 514, 516, 518
Bulk chemicals, 6, 7, 13–21
Butadiene, 7, 13, 15, 19, 20
Butane, 13, 19
Butanol, 13, 32, 55, 75, 82, 83, 86, 87, 88, 95, 265, 317, 320, 325, 328, 332, 365, 379, 423, 513, 528, 530, 569–607
Butyrate, 82, 87, 88, 95, 500, 514, 515, 526, 527, 528, 530, 572, 574, 575, 576
Butyrate kinase, 574, 575
Butyrate phosphate (BuP), 574, 576
Butyric acid, 86, 87, 88, 527, 573
Butyryl CoA, 88, 528, 576
Butyryl phosphate, 574

Byproducts, 12, 52, 61, 79, 89, 95, 106, 107, 109, 124, 241, 260, 270, 287, 291, 301, 359, 509, 511, 597, 603, 604, 605, 606

C

C5 (xylan), 6, 18, 20, 56, 128, 131, 181, 193, 231
C6 (glucan), 6, 13, 18, 20, 53, 56, 106, 128, 131, 136, 137, 138, 181, 193, 231, 471, 515, 526
C. acetobutylicum, 182, 571, 572, 573, 575, 576, 578, 579
Camelina, 597, 602, 603
Candida antarctica (Novozym 435), 318, 328, 329, 330
Capital, 11, 64, 73, 129, 130, 165, 167, 267, 271, 274, 275, 332, 353, 391, 409, 417, 419, 474, 475, 487–488, 489, 490, 491, 492, 493, 534, 592, 600, 604
Caprolactam, 15, 16
Carbohydrate-based, 178, 199, 200, 470
Carbohydrate-binding module (CBM), 213, 214
Carbohydrates, 4, 11, 12, 14, 18, 20, 52, 108, 109, 125, 155, 159, 162, 241, 252, 260, 289, 291, 295, 423, 427, 468, 469, 470, 471, 472, 473, 474, 514, 528, 577, 590, 604
Carbonfeeding, 582
Carbon footprint, 433, 493, 588
Carbon intensity, 31
Carbon limitation, 449, 459, 460, 461
Carbon monoxide (CO), 9, 11, 56, 61, 62, 63, 64, 65, 68, 69, 71, 80, 82, 83, 84, 85, 86, 87, 88, 89, 90, 91, 92, 93, 94, 268, 276, 339, 376, 428, 483, 485, 534, 597
Carbon-neutral, 499
Carbon sources, 101, 184, 217, 218, 219, 278, 287, 290, 297, 401, 403, 404, 440, 449, 501, 503, 514, 528, 577, 580, 582
Carboxylic acids, 13, 18, 19, 153, 160, 258, 291, 292, 296–297, 300, 306, 403
Carrier-induced granular sludge bed reactor, 545
Case studies
  biomass-integrated gasification combined cycle (BIGCC) fueled with rice hulls
    simulation procedure, 275
    simulation results, 275–277
  ethanol production from rice hulls
    process description, 270–271
    simulation procedure, 271
    simulation results, 271–272
Cassava, 105, 113, 114, 405, 506, 606
  bagasse, 113, 114
Catabolite repression, 181, 186, 187, 198, 289, 290
Catalysts
  enzymatic, 150, 316, 359, 365–366
  heterogeneous, 13, 21, 327, 340, 341, 350, 360–364, 377, 379, 391, 408
  heterogeneous acid, 364–365, 380
    -alkali, 383
    -alkaline, 383
  heterogeneous alkali, 377, 378
  heterogeneous alkaline, 377
  heterogeneous superacid, 383
  homogeneous acid, 359–360, 364, 380
  homogeneous alkali, 378
  homogeneous alkaline, 349
  homogeneous base, 347, 359, 360, 361, 362, 367, 368, 369
  microbial, 80, 85, 88, 94, 95
  new heterogeneous, 21
  superacid, 379, 383
Catalytic
  domain (CD), 179, 206, 213, 214, 235
  hydrogenation, 12, 20, 601
Catechol, 297
Categories, 10, 32, 52, 80, 104, 431, 481, 537, 559, 589
C. beijerinckii, 571, 572, 573, 574, 577, 578, 579
cbh1 promoter, 181, 195, 197
CBH (cellobiohydrolase), 109, 178, 179, 180, 181, 188, 190, 195, 197, 198, 218, 234
C. carboxydivorans, 80, 82, 83, 84, 85, 93
Cell
  concentration, 90, 302, 444, 580
  densities, 90, 196, 302, 303, 402, 403, 419, 560, 572, 577, 580–582
  fragility, 455
  membrane, 153, 288, 297, 298, 511, 529, 530, 576
  recycle reactor, 577
  recycling, 302
Cellobiohydrolase (CBH), *see* CBH (cellobiohydrolase)
Cellobiohydrolase I (CBHI) promoter, 180, 195
Cellobiose, 86, 107, 109, 141, 178, 179, 180, 187, 190, 193, 196, 197, 198, 199, 234, 235, 236, 238, 239, 242, 243, 246, 247, 259, 469
Cells trajectories, 454
Cellulase
  cocktails, 195
  market, 192–194
  production, 179, 180, 181, 182, 183–187, 192, 193, 195, 196, 197, 198, 199, 218
  systems, 179–181, 187, 190, 191, 192, 195, 196, 259
Cellulolytic
  ability, 576
  enzymes, 56, 158, 180, 181, 190, 192, 197, 203–220
  machinery, 576, 577
Cellulose
  -binding domain, 179, 213, 214, 235
  -binding module, 179, 196, 213, 214, 238
Cellulosic ethanol, 74, 103–105, 116, 137, 140, 143, 146, 170, 192, 416

Cellulosomes, 180, 196, 576
Central composite design, 219, 362
Centrifugation, 106, 233, 243, 244, 327, 385, 427
Cetane number, 101, 388, 596
Challenge, 75, 94–95, 117–118, 198–199, 423–424, 435, 475–476, 533–534, 557–562
Channeling, 545, 546
Chaperones, 576
Characterization aspects of BD, 388
Characterization of BD, 388
Characterization and environmental aspect, 392
Cheaper bioprocess technology, 199
Cheaper raw material, 192, 196, 198
Cheaper substrate, 199, 278, 574
Cheese whey, 470, 577
Chemical
  detoxification, 304
  engineering, 9, 16
  fossil bulk, 13–16
  industries, 4, 7, 8, 12, 20, 21, 52, 86, 405
  N-containing, 14, 15–16
  syngas-derived, 10, 11
  treatment, 106, 256–257, 474
Chimeric cellulosomes, 196
Chimeric scaffolding, 196
Chlamydomonas, 400, 401, 402, 441, 444, 447, 604
Chlorella
  protothecoides, 400
  vulgaris, 406, 604
Chrysosporium lucknowense, 191, 205
cipA, 180, 576
Circulating fluidized bed (CFB)
  combustion, 58
  pyrolysis, 68–69
Citrus wastes, 191, 306
Classification, 62, 116, 117, 178, 211, 234
Clean energy alternative, 467
Cleaning agents, 12
Climate change, 25, 32, 51, 103, 114, 252, 339, 353, 415
Closed geometries, 459
Closed photobioreactor (PBR), 409, 417
Closed-structure, 552
Clostridia, 95, 180, 507, 527, 572, 573, 574, 575, 576, 577, 580, 582
Clostridium acetobutylicum, 87, 515
Clostridium fimi, 198
Clostridium ljungdahlii, 80
Clostridium ragsdalei, 85
Clostridium thermoaceticum, 88
Clostridium thermocellum, 180, 527, 576
Cloud point (CP), 194, 217, 345, 357, 388, 389
C/N ratio, 516
CO_2, 3, 8, 9, 33, 34, 36, 38, 39, 42, 43, 44, 47, 52, 55, 61, 62, 63, 64, 65, 66, 69, 71, 80, 81, 82, 83, 84, 86, 87, 94, 109, 117, 141, 149, 150, 164, 168, 169, 185, 186, 192, 252, 264, 265, 268, 294, 339, 346, 350, 351, 361, 376, 384, 390, 401, 417, 424, 425, 426, 428, 431, 432, 433, 440, 441, 444, 449, 459, 468, 485, 493, 501, 502, 503, 509, 514, 517, 518, 525, 526, 529, 533, 546, 549, 553, 554, 555, 572, 581, 590, 592, 593, 604, 605
explosion, 55, 141, 150, 164, 168, 169
CoA transferase, 575
CODH enzyme, 82, 83
Coenzyme-M reductase, 505
Cofactors, 197, 293
Cold filter plugging point (CFPP), 388, 409
Cold flow improver additives, 388
Cold flow properties, 345, 348, 367, 376
Cold fuel properties, 391
Cold weather properties, 388
Combinations, 21, 22, 26, 28, 47, 70, 71, 79, 93, 115, 116, 154, 161, 164, 168, 183, 190, 191, 192, 195, 218, 219, 232, 245, 257–259, 264, 266, 275, 292, 301, 305, 320, 325, 383, 387, 418, 427, 441, 448, 452, 458, 474
Combined heat and power (CHP), 43, 59, 422
Combustion
  entrained flow, 59
  fixed-bed, 57
  fluidized-bed, 57, 58
  points, 388
Commercial value of glycerol, 331
Commodities, 4, 6, 7, 19, 25, 66, 321, 327, 330, 333, 588, 607
Compatibility, 151, 408, 592, 596
Competition, 25, 52, 74, 96, 123, 140, 144, 145, 149, 253, 399, 401, 405, 474, 528, 530, 562, 582, 590, 599
  with food, 25, 74, 96, 145, 149, 253
Composite hollow fiber membrane (CHFM), 90, 92, 93, 95, 96
Composition, 4, 5, 6, 7, 16, 53–54, 84, 107, 113, 114, 129, 130, 131, 180, 184, 185, 191, 203, 218, 236, 239, 241, 255, 266, 268, 276, 298, 300, 327, 342, 346, 350, 358, 376, 407, 421, 431, 440, 441, 470, 472, 473, 474, 507, 510, 511, 529, 576, 601, 603
Computational fluid dynamics (CFD), 561
Concentrating-solar biomass gasification (CSBG), 64
Condenser, 69, 106, 581
Conradson carbon numbers, 388
Consequential LCA, 30
Contact membrane, 580
Contamination, 114, 185, 186, 218, 239, 299, 300, 326, 362, 417, 418, 419, 423, 435, 449, 459, 530, 577
Continuous and batch-stirred tank reactors, 330
Continuous culture, 470, 577, 580
Continuous fermentation, 198, 263, 290, 300, 301, 303
Continuous-flow, 321, 327, 331, 332

Continuous mode, 184, 301, 378, 448, 450, 510, 577
Continuous operation, 184, 301, 378, 448, 450, 510, 577
Continuous process, 13, 187, 230, 263, 301, 538, 577
Continuous stirred-tank reactor (CSTR), 84, 90, 93, 330, 530, 537, 538, 539–542, 546, 548, 549, 557, 558, 562
Controllability, 552
Conventional, 9, 11, 36, 55, 60, 64, 66, 67, 73, 85, 86, 93, 95, 103, 112, 160, 232, 243, 266, 267, 275, 342, 343, 344, 345, 346, 347, 348, 349, 351, 354, 383, 400, 401, 406, 410, 467, 468, 476, 500, 515, 537, 538, 540, 542, 551, 554, 577, 580, 582, 587, 589, 599, 601, 602
Conversion
  anaerobic, 501
  biomass, 4, 8, 9, 21, 55, 60, 70, 72, 116, 155, 191, 194, 195, 197, 198, 199, 278, 490, 605
  efficiency, 88, 130, 144, 150, 248, 353–354, 369, 471, 475, 500, 511, 515, 518, 549, 551, 561
  elemental, 8
Cooking oil, 322, 326, 329, 330, 331, 341, 346, 375, 376, 377, 378, 380, 381, 382, 383, 384, 386, 388–390, 391, 392
Coprocessing, 596
Coproducts, 22, 26, 29, 30, 31, 33, 34, 39, 47, 53, 73, 88, 89, 104, 105, 112, 124, 128, 129, 141, 145, 164, 401, 411, 424, 427, 429, 431, 432, 433, 435, 526, 603
  credits, 431
Corn
  ethanol, 103, 105, 116, 415, 416, 423, 431, 433
  starch, 101, 102, 105
  stover, 6, 54, 79, 104, 107, 114, 115, 118, 125, 128, 129, 141, 155, 157, 160, 161, 162, 163, 165, 166, 253, 254, 258, 415, 482
Cosolvent, 164, 328, 329, 330, 383, 384, 385, 389, 408
Cost
  of alcohol, 331
  of biohydrogen production from bio-oil, 486–487
  of downstream stages, 331
  effective, 55, 75, 84, 94, 108, 150, 161, 164, 168, 169, 171, 184, 189, 191, 193, 196–199, 229, 267, 278, 279, 332, 342, 439, 476, 534, 541, 548, 555, 560, 581, 582, 604
  management system, 140–144
  of pretreatment stages, 331
  of production of biohydrogen from reforming of bio-oil, 490, 493
  of production of bio-oil, 488–490, 491, 492
ρ-Coumaric acid esterases (EC 3.1.12), 72, 90–93, 95, 326–331, 510–511, 537, 538, 541, 546, 548, 557, 558, 560
Covalent attachment, 321
C. pasteurianum, 573
Crabtree effect, 289
Cracking, 18, 68–69, 73, 86, 377, 597
Cradle to grave, 28
CREI, 180
Crossflow gasifiers, 62
Cross-linked enzyme aggregates, 321
Crude oil, 3, 6, 8, 67, 251, 324, 483, 587, 588, 596, 600
Crystalline cellulose, 155, 159, 178, 179, 181, 190, 191, 213, 235, 240, 242, 247, 259
Crystallinity, 54, 55, 108, 109, 110, 114, 150, 151, 154, 157, 159, 167, 169, 198, 233, 242, 255, 259
Culture
  conditions, 215, 419, 431, 440, 449, 450, 457, 459, 460
  mixing, 423, 453
  pH drops, 471
  volume, 442, 446, 447, 450, 452, 457, 458, 461
Cyanobacteria, 402, 439, 450, 451, 452, 475, 500, 501, 537

D

Dairy waste, 506, 507
Dark fermentation, 474, 499, 501–503, 512, 513, 517, 518, 525, 528, 533, 537, 538, 548, 549, 552, 555, 557, 559, 560, 561, 562
Dark zone, 447, 450, 456
Data enveloping analysis (DEA), 511
DDGS, 34, 38, 39, 40, 43, 44, 46, 47
Deactivation, 65, 240, 318, 319, 321, 362, 363, 365, 366
Decomposition, 18, 55, 62, 65, 68, 70, 154, 232, 233, 237, 257, 344, 346, 348, 367, 368
Decoupling of solventogenesis, 574, 576
Defibrillation, 155, 188
Degradation, 9, 20, 54, 55, 79, 107, 108, 109, 123, 128, 150, 151, 152, 153, 154, 156, 158, 159, 160, 161, 162, 163, 164, 165, 166, 167, 169, 183, 189–190, 196, 197, 210, 217, 230, 235, 236, 237, 241, 246, 255, 256, 257, 258, 259, 289, 291, 296, 297, 298, 300, 304, 339, 348, 402, 410, 445, 470, 503, 505, 509, 510, 511, 512, 513, 514, 527, 555, 560, 580, 582, 593
Dehydration, 12, 13, 19, 20, 21, 39, 60, 270, 271, 289, 291, 292, 295, 319, 361
Dehydrogenase, 82, 293, 295, 502, 509, 516, 518, 575
De-inking, 189
Delignification, 114, 116, 157, 158, 159, 162, 163, 166, 167, 183, 258, 305, 474
Density
  high cell, 90, 302, 303, 403, 577, 580–582
  photon flux, 441
Deoxygenation, 8, 16, 70, 597, 601
Dephosphorylation, 575
Depolymerization, 11, 16, 55, 71, 72, 163, 167, 191, 234, 237, 242
De-repression, 180, 195
Design of experimental (DOE), 511
Detergents, 188, 189, 190, 198
Detoxification, 141, 150, 151, 166, 260, 270, 271, 273, 274, 278, 299, 300, 303–305
strategies, 301, 303–305

Dewatering, 408, 419, 420, 429, 432, 604
1,2-dichloroethane, 19
DiCoFluV, 458
Diels-Alder, 15
Diesel fuel, 6, 43, 70, 388, 390, 592, 598
Differential expression, 217, 218
Diffusion, 89, 319, 322, 327, 341, 363, 364, 554, 580
Dilute acid, 109, 111, 112, 115, 116, 117, 125, 141, 156, 160, 183, 230, 231, 232, 252, 254, 257, 259–260, 270, 273, 274, 290, 292, 301, 303, 305
Dilute acid hydrolysis, 109, 116, 231, 252, 254, 259, 292
Dimethyl ether (DME), 11
Dimethylsulfoxide, 9
Direct and diffuse parts of the radiation, 446
Directed evolution, 195, 197, 245
Disintegration, 543, 574
Dispersants, 10
Dissolved air floatation, 419
Dissolved hydrogen, 526, 528, 529, 558
concentration, 529, 558
Distillation, 6, 9, 31, 39, 59, 62, 95, 106, 114, 125, 126, 129, 131, 141, 229, 232, 243, 252, 265, 271, 355, 358, 385, 387–388, 580
of BD, 387–388
Distribution, 31, 33, 59, 67, 70, 73, 85, 103, 144, 145, 149, 178, 203, 204, 218, 220, 257, 265, 403, 424, 425, 442, 443, 444, 446, 448, 454, 455, 509, 551, 590, 597, 601, 602, 604, 607
Downdraft (cocurrent) gasifiers, 62
Downregulates AbrB, 575
Drop in pressure, 327
Drying
  agents, 387
  of BD, 387
Dunaliella tertiolecta, 409
Dynamic regime, 455

E

E5, 29, 34, 36, 37, 38, 40, 41, 42, 43, 44, 45, 46, 47
E10, 34, 36, 37, 38, 40, 41, 42, 43, 44, 45, 46, 86, 102
E85, 29, 34, 36, 37, 38, 40, 41, 42, 43, 44, 45, 46, 102, 104, 116
Ecofining, 596, 597
Eco-friendly, 199, 232, 499
Ecoinvent, 33, 34, 36, 43, 46
Ecologically engineered system (EES), 513
Economic allocation, 26, 31, 32, 46, 47
Economically viable, 13, 72, 191, 418, 474
Economic analysis/analyses, 108, 244, 271, 331, 332, 385–391, 409–410, 474, 475
Economic aspect of BD, 388–391
Economic assessment, 391
Economic evaluation, 123–124, 125, 126, 127–130, 140, 141, 143, 144–145, 165, 331–333
Economic feasibility, 95, 125, 198, 199, 243, 332, 390, 533
  of BD, 390
Economic reasons, 158, 427
Economy of scale, 136, 488, 491, 492
Effluent(s), 131, 243, 263, 433, 468, 476, 499–519, 530, 531, 532, 533, 539, 542, 543, 544, 545, 547, 548, 581, 604, 606
  recirculation, 548
EG III, 189
Electricity, 4, 5, 43, 51, 59, 63, 73, 103, 116, 124, 129, 130, 131, 132, 133, 135, 136, 137, 138, 139, 145, 252–253, 266–267, 270, 271, 275, 276, 277, 278, 305, 332, 340, 353, 421, 422, 434, 467, 483, 488, 490, 491, 492, 512, 513, 525, 551, 554, 591, 595
Electrohydrogenesis, 483, 517
Electrolysis, 64, 517, 525, 537, 552–555
Electron donors, 87, 515, 518
Emissions
  greenhouse gas (GHG), 25, 26, 28, 29, 30, 36, 39, 41, 42, 43, 44, 46, 47, 59, 74, 101, 103, 116, 117, 145, 229, 376, 399, 401, 431, 433, 468, 476, 493, 587, 588–589, 592, 605, 606
  net, 36, 39, 41, 42, 43, 44, 46, 47, 103, 116, 589
Empty fruit bunches, 105, 253, 363
Emulsifiers, 10
Emulsions, 580
EN 14214, 356, 357, 387, 389, 392
Encapsulation, 303, 365
Endoglucanase (EG), 109, 141, 178, 179, 180, 181, 189, 190, 195, 197, 198, 214, 218, 234, 242, 245, 247, 259
Endoxylanase (EC 3.2.1.8), 205–208, 210, 212, 215, 219
End products, 14, 94, 110, 142, 160, 197, 208, 239, 240, 241, 242, 243, 244, 247, 258, 261, 481, 503, 504, 509, 511, 513, 515, 516, 517, 527, 529, 576
Energetic assessment, 429
Energy
  allocation, 31, 41, 47
  balances, 26, 36, 39, 47, 74, 125, 128, 130, 275, 367, 415, 416, 422, 430, 439, 456, 461
  consumption, 22, 25, 31, 32, 36, 44, 51, 108, 111, 112, 116, 117, 155, 163, 165, 168, 169, 218, 243, 271, 273, 275, 277, 278, 316, 317, 318, 332, 340, 341, 367, 432, 454, 456, 461, 543, 558
  crops, 30, 44, 52, 53, 79, 116, 472, 473, 474, 482, 537
  demand, 3, 63, 114, 128, 150, 160, 166, 169, 243, 278, 499, 534, 540, 541, 543, 548, 555, 588, 605
  efficiency, 26, 36, 65, 112, 128, 129, 165, 252, 277, 355, 369
  input, 128, 155, 166, 355, 367, 427, 429, 430, 435, 437, 537, 552, 554, 560, 562
  losses, 553, 555
  production, 56, 74, 79, 103, 104, 112, 118, 144–145, 266–269, 434, 440, 453, 455, 456, 458, 461, 474, 482

Energy (*Continued*)
using agroindustrial residues, 266–269
sources, 3, 51, 55, 59, 89, 149, 181, 278, 339, 340, 353, 429, 481
use, 8, 28, 36–38, 42, 45, 46, 47, 116, 130, 144, 267, 420, 429–430, 431, 432, 605, 606
Energy return on water invested (EROWI), 434, 435
Energy substitution efficiency, 36–38, 45–46
Engineered/artificial cellulases, 195–196
Engineering of photobioreactors, 459
Engine technology, 26
Enrichment, 214, 300, 504, 505, 518
Ensyn, 481, 482, 483, 485, 601
Entrained flow gasifiers, 63
Envergent, 601
Environment
favorabl, 547, 559
micro, 303, 417, 505, 510, 514, 518
Environmental aspect of BD, 388–391
Environmental impact, 22, 26, 31, 33, 42, 64, 74, 103, 136, 165, 333, 415, 431, 439, 474, 499
Environmental performance, 28
Environmental reasons, 455
Environmental restoration, 513
Enzyme
-based assay, 213
-based ethanol, 125
versus chemical catalysts for biodiesel production, 331
cocktail, 181, 191, 192, 199
deactivation, 318
denaturation, 240, 320
production, 125, 131, 142, 179, 181, 183, 184, 186, 192, 194, 197, 217–219, 278
reutilization, 321
Escherichia coli, 163, 405, 504, 527, 572
ESE, 45, 46
Ester, 19, 109, 204, 210, 211, 212, 236, 315, 317, 329, 331, 354, 357, 358, 360, 367, 368, 376, 387
Esterification, 13, 19, 315, 316, 323, 324, 325, 326, 347, 349, 350, 358, 360, 364, 369, 379, 381, 383, 384
Ethane, 7, 13, 19
Ethanol
first-generation, 123, 134
fuel ethanol, 26, 38, 42, 47, 102, 105–106, 149, 191, 192, 251, 253, 254, 260, 261, 262, 263, 265, 270, 274, 278
inhibition, 239, 289
lignocellulosic, 110–117, 118, 123–146, 170, 178, 191–192, 199, 229, 234, 238, 240, 241, 243, 248, 580
sugarcane ethanol, 103
thermochemical ethanol, 126
Ethanol programme cost target (EPCT), 126
Ethylene, 7, 13, 20, 21, 55, 63, 71, 80, 88, 157, 241, 325, 346, 543, 545
Eucalyptus, 125, 131, 132, 133, 134, 165
Evaporation, 57, 125, 129, 158, 186, 233, 291, 298, 304, 385, 426, 456
Evaporative, 417, 418
Evapotranspiration, 426, 432
Exoelectrogenic bacteria, 517, 518
Exoglucanase (CBH), 109, 141, 178, 179, 180, 181, 188, 190, 195, 197, 198, 199, 234, 242, 259
Expanded bed, 327, 510, 548
reactors, 327
Expensive, 11, 19, 20, 55, 56, 67, 71, 72, 80, 103, 141, 150, 231, 239, 321, 323, 332, 333, 341, 345, 349, 375, 407, 419, 468, 469, 476, 493, 514, 529, 534, 552, 555, 561, 562, 590, 595, 596
Expression cassettes, 192
External power supply, 552
Extractant, 580
Extraction
liquid-liquid, 95, 263, 265, 578, 580
oil, 322, 408, 409, 410, 411
solvent, 55, 150, 423, 427, 429
wet, 431
Extra energy input, 554, 562
Extrusion, 55, 155, 168, 474

F

Fabric-softening, 189
Facultative anaerobes, 206, 260, 502, 504, 527
FAME, *see* Fatty acid methyl ester (FAME)
Fast pyrolysis
ablative, 67
of biomass, 66, 68, 482, 484, 485, 486
rotating cone, 68
ultra, 69–70
Fatty acid ethyl ester (FAEE), 340, 343, 390
Fatty acid methyl ester (FAME), 322, 324, 325, 330, 331, 340, 343, 344, 345, 346, 347, 348, 349, 350, 351, 376, 380, 387, 388, 389, 390, 407, 408, 587, 589, 590, 592, 601, 602
Fatty acids, 82, 241, 298, 315, 323, 325, 340, 341, 345, 348, 349, 350, 354, 358, 367, 368, 369, 376, 377, 384, 388, 390, 399, 406, 407, 408, 471, 500, 501, 503, 513, 514, 576, 589, 596, 599, 601, 602, 603, 604, 605
2,5-FDCA, *see* 2,5-Furan dicarboxylic acid (FDCA)
Feasibility and economic analyses, 385–391
Fed batch
cultivation, 301, 302
cultures, 196, 404, 405
fermentations, 302, 577
Feed additive, 7, 189, 220

Feed back control, 198
Feed back inhibition, 192, 195, 196, 197, 238, 501, 549
Feeds and fertilizer, 411
Feedstock(s)
  cost, 96, 124, 132, 133, 134, 135, 137, 138, 139, 140, 141, 150, 359, 488
  nontransport costs, 133, 136, 137–140
  transport costs, 134, 136
Fermentable, 79, 106, 108, 109, 114, 116, 156, 157, 162, 167, 178, 181, 189, 190, 229, 231, 238, 248, 252, 255, 257, 260, 278, 301, 304, 305, 470, 555, 580, 606
  sugars, 79, 106, 108, 109, 156, 157, 162, 167, 178, 181, 189, 190, 229, 231, 238, 248, 252, 255, 257, 304, 580, 606
Fermentation
  broth, 83, 84, 252, 476, 580, 581, 582, 599
  industry, 403, 411, 506, 507
  mode, 306, 404, 530, 531, 532
Fermentative processes, 468, 501, 503, 517
Fermenter, 270, 403, 404, 405, 409, 419
Ferridoxin (Fd), 501, 516
Ferulic acid esterases (EC 3.1.1.73), 205, 236
FFA, *see* Free fatty acid (FFA)
Field residues, 113
Filamentous fungi, 181, 184, 186, 187, 190, 192, 195, 197, 208, 293
Filter, 88, 190, 231, 244, 270, 330, 331, 357, 388, 409, 420, 431, 432, 513, 592
Filtration membrane, 244, 540
Firmicutes, 507
First-generation
  biofuels, 53, 74, 105, 123, 415, 589
  ethanol, 123, 134
Fischer-Tropsch (FT), 11, 32, 61, 80, 597–598, 599
Fixed-bed combustion, 57
Fixed-bed gasifiers, 62, 80
Fixed operating costs, 132, 133, 134, 135, 136, 138, 139
Flash points, 340, 357, 388, 389, 409, 593, 594
Floating-type, 551
Flocculating yeast, 303
Flocculation, 419, 427
Flue gases, 58, 401, 416, 424, 432
Fluid circulation, 542, 548
Fluid film, 546
Fluidization, 542, 543, 546
Fluidized-bed combustion, 57, 58
Fluidized-bed gasifiers, 63, 80
Fluidized-bed reactors, 63, 327, 487, 537, 541–543, 601
Fluid recirculation ratio, 543
Food additives, 12
Food *versus* fuel, 354, 415, 416, 424
Food industry, 20, 189, 469
Food processing wastes, 470, 472, 506
Food-related wastes, 469–472
Food-*versus*-fuel debate, 474
Forespore, 575
Forest
  biomass, 73, 155, 482, 484, 485, 486, 487, 488, 490, 491
  management, 112
Formic acid, 152, 153, 241, 296, 297
Fossil(s)
  bulk chemicals, 13–16
  -derived, 20, 96
  fuels, 3, 8, 13, 26, 28, 29, 30, 36, 51, 52, 59, 73, 74, 104, 149, 178, 188, 190, 252, 253, 278, 339, 340, 343, 353, 368, 409, 415, 435, 467, 468, 476, 481, 482, 486, 499, 515, 525, 587–589, 591, 592, 593, 605
  refinery, 4, 14, 16, 19
Fouling, 57, 65, 418, 541, 549, 561, 580, 581
Four imperatives, 587, 591, 606
FPU activities, 190
Free biocatalysts, 320–321, 327
Free enzymes, 320, 366
Free fatty acid (FFA), 315, 316, 317, 323, 324, 325, 326, 331, 341, 346, 347, 349, 350, 351, 354, 355, 359, 360, 361, 362, 364, 365, 367, 368, 369, 370, 376, 377, 379, 380, 381, 382, 383, 384, 386, 389, 408, 602, 604
Free sugars, 109, 186, 577
Fructose, 20, 105, 260, 264, 290, 294, 469, 540, 556
Frying oil, 354, 376, 378, 380, 381, 382, 389, 390, 391
Fuel-bioethanol, 33, 36, 38, 47
Fuel blend, 26, 28, 29, 35, 36, 38, 39, 40, 41, 44, 45, 46, 47, 192, 388, 590
Fuel performance, 28, 29, 35, 37, 39, 40, 41, 44, 45, 46, 47
Fuel properties, 266, 356, 388, 389, 391
Fumaric, 12, 15, 109, 156
Fumaric acid, 15, 156
Functional food ingredients, 219
Functional genomics, 217
Functional unit, 26, 27, 28, 29, 33, 117, 430, 431, 432
Fungal fermentation, 20
Fungi, 55, 110, 118, 142, 166, 167, 180, 181, 182, 184, 185, 186, 187, 190, 192, 195, 196, 197, 199, 205, 206, 208, 210, 211, 214, 215, 217, 218, 234, 237, 260, 261, 288, 293, 303, 379, 402, 596, 597, 599, 601
2,5-Furan dicarboxylic acid (FDCA), 18, 21
Furans, 18, 20, 78, 79, 152, 160, 166, 292, 300
  compounds, 20
Furfural, 12, 18, 20, 109, 111, 124, 125, 152, 153, 156, 157, 158, 160, 163, 166, 241, 252, 255, 256, 257, 258, 270, 271, 272, 273, 291, 292–295, 296, 299, 300, 302, 304, 305, 580
Furfuryl alcohol, 20, 152, 157, 293, 296, 299, 300
Furoic acid, 153, 160, 293, 299
Future perspectives, 101–118, 196–199, 410–411, 435, 526, 533–534

G

GAC, 510, 541, 542, 543, 544, 545, 548
α-Galactosidases (EC 3.2.1.22), 205
Gamma-butyrolactone, 19
Gas chromatography (GC), 194, 346, 347, 442
Gas-diffusion cathode, 554
Gasification, 5, 9, 11, 57, 60, 61–65, 69, 71, 72, 80, 81, 84, 85, 88, 89, 94, 116, 117, 126, 252, 253, 267–268, 269, 275, 277, 421, 422, 423, 433, 439, 467, 481, 483, 484, 597–598, 600
Gasifier, 62, 63, 64, 72, 80, 81, 84, 267, 268–269, 597, 598, 599
  chamber, 267, 268–269
Gas-liquid exchange, 545
Gas-liquid mass transfer, 90, 93, 95, 96, 459
Gas-liquid mass transfer rate, 449
Gas-liquid separator, 542, 543, 544
Gasoline, 6, 11, 20, 29, 30, 34, 35, 36, 37, 38, 40, 41, 42, 43, 44, 45, 46, 47, 61, 70, 73, 86, 102, 106, 116, 117, 126, 143, 145, 191, 192, 422, 433, 571, 589, 590, 591, 592, 594, 596, 599
  alternative, 571
Gas
  release, 558–559
  stripping, 578, 579, 580, 581, 582
  transfer area, 550
  turbine, 267, 268, 269, 275, 276, 277
Gel entrapment, 559
Gene
  manipulation, 476
  regulation, 181, 191, 195, 199
Genencor, 191, 192, 193, 194
Generation
  of biofuel, 220, 434
  shear-stress, 454
Genetic(s)
  improvement, 190, 197, 400
  manipulation, 198, 572
GH3, 208, 209
GH 5, 207
GH 7, 207
GH 8, 207
GH10, 207, 208
GH11, 207
GH43, 207, 208, 209, 210, 216
GH51, 209, 210
GH54, 208, 209, 217
GH62, 209
GHG balance, 25, 26, 28, 30, 44, 46, 48, 74, 136
Gibbs free energy, 87, 526
GLHF, 591, 592, 593, 594, 595–607
Global production, 102, 130
Global warming, 3, 30, 33, 46, 51, 101, 339, 353, 499
Global warming potentials, 33
Glucanases, 178, 189, 198, 206, 207, 214, 219, 235, 246
Glucaric acid, 21
Gluconic acid, 12, 18
Glucose
  repression, 180, 181, 195
  tolerance, 199
β-Glucosidase (BG), *see* Beta-glucosidase (BGL)
α-Glucouronidase (EC 3.2.1.139), 205, 213, 236
Glutamic, 12
Glutamic acid, 15, 16
Glycerin, 5, 507, 590
Glycerols, 62, 95, 106, 107, 157, 288, 289, 290, 294, 295, 297, 315, 316, 317, 319, 320, 321, 324, 325, 326, 327, 328, 329, 330, 331, 332, 340, 342, 344, 345, 346, 347, 349, 354, 355, 356, 357, 359, 360, 361, 365, 366, 367, 368, 377, 379, 384, 385, 386, 388, 407, 408, 554, 557, 578
  removed, 324
Glycogen, 289
Glycolaldehyde, 298
Glycols, 7, 13, 20, 55, 157, 241
Goal and scope, 28
Granular anaerobic sludges, 87
GRAS, 192
Grate firings, 57–58
Green processing, 89, 94, 408
Green wastes, 473, 474
GREET, 424, 430, 432
Grid, 253, 266, 267, 277, 488, 595
Growth kinetics, 93, 459
Guaiacols, 9

H

H_2-producing granule (HPG), 540, 543, 545, 546, 547, 559
Halophyte, 604–605
Harvesting/dewatering, 419, 420, 429
Heat, 4, 5, 8, 9, 10, 29, 43, 51, 52, 56, 57, 58, 59, 61, 62, 63, 64, 65, 66, 67, 68, 69, 80, 101, 129, 131, 137, 151, 155, 160, 185, 253, 266, 267, 268, 269, 271, 275, 277, 278, 289, 292, 295, 339, 353, 357, 376, 387, 416, 421, 422, 425, 427, 440, 452, 453, 459, 467, 481, 483, 485, 486, 487, 505, 541, 549, 560, 561, 576, 601, 605
  transfer, 57, 66, 67, 68, 69, 80, 155, 160, 185, 186, 453
Heating value, 56, 57, 60, 61, 62, 67, 70, 72, 86, 128, 143, 253, 266, 388, 389, 409, 423, 483
Heat recovery steam generator (HRSG), 268, 269, 275
Height to diameter (H/D) ratios, 542
Hemicellulases, 156, 162, 177, 189, 190, 193, 195, 205, 209, 213–214, 217, 218, 219–220, 234, 262

Hemicelluloses, 4, 6, 9, 11, 20, 53, 54, 55, 72, 79, 104, 107, 108, 109, 110, 111, 113, 114, 115, 118, 128, 131, 137, 150, 151, 153, 154, 155, 156, 157, 158, 159, 160, 161, 162, 163, 164, 165, 166, 167, 168, 169, 178, 181, 183, 203–205, 209, 211, 214, 217, 218, 220, 230, 233, 234, 235, 236, 241, 242, 251, 255, 256, 257, 258, 259, 260, 264, 270, 271, 272, 278, 292, 469, 472, 473, 474, 606
  structure, 164, 203
Henry's law, 89
Heterogeneous alkali, 377, 378
Heterologous cellulases, 196
Heterotrophic, 400, 401, 403–406, 407, 409, 410, 411, 418–419, 503, 537
microalgae, 401, 403–406, 407, 409, 410, 411
Hexokinase, 293, 295
Hexoses, 54, 56, 106, 109, 125, 128, 141, 142, 152, 166, 205, 252, 260, 261, 264, 290, 295, 469, 470, 471, 476, 528, 531, 532, 534, 555, 556, 557, 580, 606
H. grisea, 182, 189
High biological requirement for CO_2, 441
High carbon residue, 388
High cell densities, 90, 302, 303, 403, 577, 580–582
High temperature, 9, 13, 18, 55, 63, 64, 65, 66, 68, 69, 80, 84, 86, 108, 111, 150, 152, 156, 164, 232, 256, 258, 288, 292, 296, 317, 321, 342, 344, 347, 348, 355, 367, 377, 485, 486, 504, 525, 560, 601
H. insolens, 182, 188, 189, 210, 217
HMF, 18, 109, 111, 152, 153, 156, 157, 158, 160, 163, 241, 255, 257, 258, 270, 271, 272, 273, 291, 292, 295, 296, 299, 300, 302, 304, 305, 580
HMFAL, 300
HMFCA, 300
HMfurfuryl alcohol, 296
HMfuroic acid, 296
Homoacetogenic microorganism, 84
Homogeneous, 7, 8, 54, 57, 64, 67, 185, 230, 232, 247, 265, 316, 319, 340, 341, 342, 344, 347, 349, 350, 351, 359–360, 361, 362, 364, 365, 366, 367, 368, 369, 377, 378, 379, 380, 386, 441, 544
Honeywell, 596, 597, 598, 601
3-HPA, *see* 3-Hydroxypropionic acid (HPA)
Humicola, 180, 181, 182, 187, 189, 198, 210, 216, 235
Hybrid catalysis, 326
Hydraulic retention time (HRT), 469, 510, 518, 528, 529, 538, 540, 541, 542, 545, 546, 547, 548, 561, 562
Hydraulic turbulence, 546, 558
Hydrocarbons (HC), 6, 55, 63, 70, 72, 80, 84, 85, 94, 351, 376, 377, 390, 422, 467, 501, 587–607
Hydroesterification, 325–326
Hydrogenase, 82, 83, 84, 85, 88, 500, 501, 502, 503, 509, 515, 516, 549
enzyme, 83, 84, 88, 509
Hydrogenation, 5, 12, 20, 21, 361, 467, 601
/reduction, 19, 21
Hydrogen (H_2)
-consuming bacteria, 561
  production, 87, 444, 449, 468, 469, 470, 471, 472, 474, 475, 476, 526, 527, 528, 529, 530, 531, 532, 533, 537, 538, 540, 543, 545, 546, 547, 548, 549, 551, 552, 553, 554, 555, 556, 557, 558, 559, 560, 561, 562
    efficiency, 547, 554
  yield, 469, 470, 471, 472, 473, 476, 493, 525, 526, 527, 528, 529, 530, 533, 538
Hydrogenolysis, 21
Hydrolases, 177, 214, 315
Hydrolysis
  acid, 54, 55, 72, 106, 108–109, 116, 124, 141, 146, 150, 230–232, 252, 254, 258, 259, 270, 291, 292, 296, 297, 298
  alkaline, 55, 109, 110, 114, 257
  dilute acid, 109, 116, 231, 252, 254, 259, 292
  enzymatic, 106, 109–110, 114, 115, 117, 125, 129, 131, 146, 150, 151, 154, 155, 156, 157, 158, 160, 161, 162, 164, 165, 167, 178, 183, 192, 198, 213, 231, 232, 234–238, 239, 240, 241–242, 243, 244–247, 256, 258, 259, 271, 278, 299, 325–326, 474, 580
Hydroprocessing, 596–597, 600, 601, 604
Hydropyrolysis, 66, 70
Hydrothermal carbonization (HTC), 60–61
Hydrothermal gasification of biomass, 65
Hydrothermal liquefaction, 70, 409, 421, 422, 423
Hydrothermal pretreatment, 109, 159, 161, 164, 170
4-hydroxybenzaldehyde, 153–154, 241, 297, 305
5-hydroxymethylfurfural (5-HMF), 152–153, 241
Hydroxymethylfurfural, 152, 153, 241, 255, 291, 295–296, 580
3-Hydroxypropionic acid (HPA), 18, 19
Hythane, 534

I

Illuminated fraction, 446–447, 451, 456
Illuminated zone, 447
Illumination mode, 549
Immobilization, 214, 243, 244, 265, 303, 321, 322, 365, 408, 510, 540, 541, 555, 559, 560, 562
Immobilized cell
  biofilm reactor, 580
  cultures, 577
  systems, 542
Immobilized enzymes, 316, 321, 322, 325, 327
Immobilized lipase, 318, 320, 321, 322, 329, 365, 379, 381, 408
Incident angle, 442, 443, 455
Incident angle θ, 446

Incorporation rate, 29, 44
Indirect land use change (ILUC), 30, 48, 433
Inducers, 180, 184, 186, 197, 208, 215, 237
Industrial
enzymes, 177, 178, 184, 194
-scale, 90, 94, 315, 317, 318, 331, 333, 391, 420, 562
Inedible, 600, 602
Inert sweep gas, 581
Infrared spectrum, 456
Inhibition, 62, 88, 94, 110, 142, 152, 192, 195, 196, 197, 198, 208, 238, 239, 240, 241, 242, 243, 246, 247, 261, 264, 265, 287, 289, 290, 292, 293, 298, 300, 301–303, 304, 305, 315, 319, 324, 327, 350, 441, 442, 452, 455, 500, 501, 505, 511, 513, 529, 530, 545, 549, 558, 577, 580, 582
Inhibitors, 88, 95, 111, 128, 131, 141, 150, 151, 152, 161, 162, 163, 164, 165, 166, 167, 169, 197, 208, 212, 215, 241–242, 243, 252, 255, 256, 257, 264, 287–306, 324, 357, 405, 561
Inorganic carbon source, 440, 449
In situ, 252, 265, 299–300, 301, 304, 554, 572, 574, 577, 578, 581, 582, 601
In situ detoxification, 299–300, 301
Integrated CSTRs, 538, 539, 541, 556, 562
Integrated upgrading system, 118
Internal illumination, 551, 552
Internal rate of return (IRR), 124, 125
Internal resistance, 518, 553
Inventory, 26, 32, 430, 431
Investment costs, 130, 132, 133, 134, 135, 136, 138, 139, 140, 142, 266, 267
In vivo, 400, 406, 407, 576
Iodine
numbers, 388
value, 388, 389
Iodopropane, 505
Ion-exchange resin, 304, 305, 364, 365, 380, 387
Ionic liquid (IL), 111, 158–159, 166, 168, 183, 230, 232, 233–234
IPCC, 26, 30, 33, 39, 43
Irradiance
distribution, 446
field, 442, 444, 446, 447, 448
Irrigation, 602, 604
ISO 14040-series, 28, 31
Itaconic acid, 18, 20

J

Jatropha, 326, 340, 346, 354, 360, 597, 602, 603
oil, 326, 602
Jerusalemartichoke tube, 404
Jet A–1, 594, 602

K

Kerosene, 6, 439, 594, 601
Kinematic viscosity, 388, 389
Kinetic models, 245–247, 271, 444–445
Koji chamber, 184, 185

L

Laboratory scale, 418, 541, 557, 562
Laccases, 150, 237–238, 300, 303, 304
Lactic acid, 13, 18, 19, 153, 265, 291, 301, 511, 527
Lactose, 180, 290, 468, 469, 506
Land, 3, 26, 27, 28, 29, 30, 33, 34, 35, 39, 40, 41, 44, 46, 47, 53, 64, 74, 103, 110, 112, 114, 118, 145, 178, 354, 399, 400, 401, 406, 415, 417, 418, 424, 425, 426, 431, 433, 440, 447, 458, 587, 589, 592, 602, 603, 604, 605, 606
Land-use change, 3, 26, 29, 30, 39, 40, 41, 44, 46, 47
Langmuir adsorption model, 246
Large-scale operation, 60, 239, 419, 562
LA, *see* Levulinic acid (LA)
Laundry, 177, 188–189, 193
Lazy learning algorithm, 219
Levelized production cost, 125, 126, 127, 130
Levulinic acid (LA), 12, 18, 20, 153, 166, 241, 296, 297
LHFF, 591, 592
Life-cycle analysis (LCA), 28, 73–74
Life-cycle assessment (LCA), 116
Life-cycle impact assessment, 32
Life-cycle inventory (LCI), 32, 33, 431
Light absorption, 447, 452
Light attenuation, 441, 442, 444, 445, 446, 447, 448, 449, 450, 454
conditions, 441, 445, 450
Light availability, 425, 454
Light collected, 455, 459
Light conversion efficiencies, 549, 551, 552
Light-dark (L/D) cycle effects, 453
Light energy, 417, 441, 457, 500, 551, 554
supply, 441
Light-limitation conditions, 452, 459
Light-limited regime, 450
Light penetration, 418, 442, 446, 549, 551, 552, 559
Light supply, 440, 449, 452–453, 457
Lignin
amphiphilic, 244
monomeric, 16
Lignocellulose, 72, 79, 103, 107, 108, 109, 111, 115, 118, 149–171, 183, 191, 192, 196, 197, 199, 214, 215, 233, 235, 237, 240, 241, 242, 243, 244, 245–247, 255, 256, 258, 259, 260, 287, 288, 291–298, 300, 305, 473, 474, 506, 545, 549, 575, 577, 582
conversion technologies, 199
Lignocellulosic
bioethanol, 101–118, 126, 140, 143, 144–145, 146, 590

biomasses, 4, 9, 14, 52, 53–55, 56, 65, 66, 68, 70, 72, 73, 74, 79, 89, 103, 104, 107, 108, 116, 137, 149, 150, 151, 156, 159, 161, 164, 169, 177–200, 203–220, 229–248, 251, 252, 253–255, 259, 262, 265, 278, 299, 474, 481, 482, 561, 572, 576, 606
ethanol, 110–117, 118, 123–146, 170, 178, 191–192, 199, 229, 234, 238, 240, 241, 243, 248, 580
feedstock, 53, 72, 101–118, 123–146, 149–171, 177–200, 203–220, 229–248, 251–279, 287–306, 405, 422
residues, 5, 118, 184, 189, 190, 191, 266
wastes, 140, 189, 196, 220, 259, 545, 561
Lignosulfonate, 112, 291
Limiting parameters, 449
Limiting step, 553
Limonene, 291, 306
Lipase
-catalyzed, 379–381
BD production, 379–381
production, 381
Lipid(s)
accumulation, 401, 402–404, 405, 418, 449
analysis, 401, 406–407
-rich substances, 471
Liquefaction, 38, 39, 60, 66, 70–71, 252, 409, 421, 422, 423
Liquid hot water (LHW), 109, 141, 150, 160, 161–162, 168, 169, 256, 257, 270, 271, 272, 273, 274, 278
Loading, 111, 112, 114, 115, 132, 150, 162, 163, 167, 183, 243, 258, 360, 361, 362, 363, 364, 366, 510, 515, 526, 528, 529, 537, 541
Low-cost PBRs, 418, 432, 435
Low FFA waste oils, 377
Low-price raw material, 316, 317, 323, 324, 391
Lubricating oils, 6
Luminostat, 450, 452
Lysine, 15

M

Maillard reaction, 291, 296
Maleic acid, 19, 20, 156
Maleic anhydride, 12, 16, 20
Malic, 12, 551, 557
Malting, 189
Mannose, 11, 54, 107, 137, 204, 211, 241, 246, 260, 261, 290, 292, 294
β-Mannosidases (EC 3.2.1.25), 205, 211, 212
Market, 4, 7, 12, 16, 19, 20, 21, 25, 31, 74, 85, 105, 112, 118, 127, 136, 140, 142, 143, 144, 145, 146, 177, 190, 192–194, 278, 332, 339, 340, 344, 345, 346, 358, 359, 361, 365, 368, 369, 391, 399, 406, 440, 468, 506, 582, 591, 600
Mass-scale intensified production, 460
Mass-scale production, 439, 447, 455, 456, 460, 461
Mass transfer, 68, 72, 80, 89–90, 91, 92, 93, 94, 95, 181, 186, 243, 319, 330, 360, 362, 363, 365, 366, 368, 369, 425, 440, 449, 452, 459, 543, 546, 558, 559, 561
Mathematical model, 245, 271, 346, 561
maximum energy, 445, 458
Mechanical comminuting, 55
Mechanical comminution, 150, 154–155, 168
Mechanical pretreatment, 155, 255
Mechanism of induction, 197
MEC-MFC coupled system, 554
Membrane
electrode assemblies, 554
fouling, 541, 561, 581
reactors, 327
separation, 244, 298, 561
Mesophilic, 83, 84, 93, 141, 198, 509, 527, 530, 533, 546, 560
Metabolic engineering, 96, 297, 300, 518, 572, 575–577
Metabolic networks, 403, 572
Metabolic response, 288, 449
Metagenomic library, 206
Metal ions, 515, 516
Methane, 3, 11, 33, 56, 63, 64, 65, 66, 80, 87–88, 96, 117, 131, 420, 422, 439, 483, 499, 501, 509, 512, 513, 518, 534, 542, 547, 561, 606
-fermenting, 87–88
Methanogenesis, 88, 504, 505, 509, 515, 518, 533
Methanogen(s), 504, 526, 528, 529, 530, 541, 555, 560, 561
activity, 560
Methanol, 11, 157, 158, 232, 233, 253, 269, 315, 316, 317, 318, 319, 320, 322, 323, 324, 325, 326, 327, 328, 329, 331, 332, 340, 341, 342, 343, 344, 346, 349, 350, 351, 354, 355, 356, 359, 360, 361, 362, 363, 364, 365, 366, 367, 368, 369, 378, 379, 381, 383, 384, 385, 399, 407, 408, 422, 423, 485, 488, 490, 491, 493, 590
Methanolysis, 319, 320, 322, 324, 381
Methodological choices, 26, 27, 28, 32, 33, 39, 41, 46, 47, 48
Methyl esters, 32, 318, 322, 324, 325, 340, 342, 345, 349, 354, 358, 367, 376, 390, 407, 408
Michaelis-Menton, 247
Microalgae, 32, 33, 354, 400, 401, 402–406, 407, 408, 409, 410, 416, 418, 422, 424, 426, 431, 432, 433, 439, 440, 441, 444–451, 452, 455, 459, 503, 604, 605
Microarray, 576
Microbial community, 507, 515
Microbial degradation, 183, 259
Microbial electrolysis, 517–518, 537, 552–555
Microbial fermentation, 80, 468, 571
Microbial fuel cell (MFC), 305, 475, 512, 513, 517, 554
Microbial genetics, 197
Microbial physiology, 199
Microbial species, 95, 261, 469, 538, 557
Microbiology, 83–84, 94, 518, 527–528

Microorganisms, 55, 82, 83, 84, 86, 87, 88, 89, 90, 91, 92, 93, 106, 108, 110, 114, 150, 151, 152, 156, 158, 159, 163, 166, 178, 181, 182, 183, 184, 186, 187, 190, 196, 197, 205, 214, 215, 217, 239, 243, 260, 261, 262, 263, 264, 265, 278, 287, 288, 289, 290, 291, 292, 293, 296, 297, 298, 299–301, 303, 304, 305, 330, 379, 402, 403, 405, 439, 440, 444, 445, 446, 447, 450, 452, 455, 456, 459, 460, 469, 471, 472, 473, 476, 501, 503, 504, 510, 525, 526, 527, 528, 530, 531, 532, 533, 537, 542, 544, 553, 559, 560
Microwave-assisted hydrothermal carbonization (MAHC), 61
Microwave pretreatment, 160–161, 407
Military, 594, 607
Mineral nutrients, 440, 444, 449, 450
Mini cellulosome, 576
Minimum ethanol selling price (MESP), 126, 127, 130
Minimum product price, 332
Mist washing, 385, 386
Mixed cultures, 90, 91, 197, 262, 264, 469, 471, 503, 504, 528, 530, 531, 532, 533
Mixed sugars, 231, 577, 580
Mixing, 58, 69, 90, 92, 95, 155, 191, 243, 368, 369, 377, 408, 417, 423, 440, 441, 452, 453–455, 458, 459, 460, 461, 485, 538, 542, 545, 546, 548, 549, 550, 551, 555, 558, 562
Mixing conditions, 441, 452, 453–455, 459
Mixotrophic, 401, 404, 410, 418–419
Modeling
  mathematical, 246, 457
  radiative transfer, 443, 445–446
Molar ratio, 86, 316, 317, 318, 323, 324, 326, 328, 329, 330, 342, 343, 344, 345, 346, 347, 351, 359, 360, 361, 362, 363, 364, 366, 367, 368, 377, 378, 380, 381, 382, 383, 384, 385
Molasses, 191, 251, 254, 278, 287, 289, 290, 291, 295, 298, 405, 470, 476, 506, 548, 556, 577, 578
Molecular pumps, 576
Molecular recombination, 195
Monoliths, 558
Mother-cell, 575
Multidisciplinary research, 519
Multifaceted approaches, 183, 192, 199
Multifunctional system, 31
Multilayered reactor, 550, 551
Multiple linear regression, 219
Multiplicity, 214–217
Multi-Year Program Plan (MYPP), 126, 127, 130
Municipal, 52, 103, 107, 115–117, 118, 253, 254, 405, 424, 469, 470, 472, 504, 606–607
Municipal solid wastes, 107, 115–117, 118, 254, 469, 470, 472
Municipal wastewater sludge, 116
Mustard, 602

N

NADH, 294, 297, 299, 300, 516, 528, 573
Nannochloropsis oculata, 406
Nannochloropsis sp., 403
Naphtha, 6, 7, 597
Near-infrared reflectance spectroscopy (NIRS), 407
Net energy ratio, 430
Net energy return on energy invested (EROEI), 430, 434, 435
Net energy use, 36, 42, 45, 46, 47
Net EROWI, 434, 435
Net production cost, 130, 132, 136, 140
Neutral/alkaline cellulases, 188
Nile red, 400, 406, 407
Nitrogen, 8, 15, 32, 53, 55, 56, 64, 70, 80, 88, 89, 94, 186, 219, 232, 242, 276, 339, 364, 376, 400, 402, 403, 404, 405, 420, 422, 424, 431, 432, 441, 449, 470, 501, 513, 516, 592, 605
Nitrogenase, 8, 15, 32, 53, 55, 56, 64, 70, 80, 88, 89, 94, 186, 219, 232, 242, 276, 339, 364, 376, 400, 402, 403, 404, 405, 420, 422, 424, 431, 432, 441, 449, 470, 500, 501, 502, 503, 513, 516, 592, 603, 605
Nitrogen oxides (NO_x), 592
4 nitrophenyl 2-O-(4-O-methyl α-D-glucuronopyrnosyl) β-D-xylopyranose, 213
4-nitrophenyl–2-O-transferuloyl α-L-arabinofuranoside, 213
4-nitrophenyl 5-O-transferuloyl α-L-arabinofuranoside, 213
Noncomplexed cellulose, 180, 195
Nonfermentable substrate, 555
Nonfermentative production, 573, 582
Nonrenewable energy use, 28
Nonrenewable primary energy, 36, 42, 45, 46
Nonselective chemistries, 11
Nonsolventogenic, 576, 582
Novozymes, 146, 191, 192, 193, 194, 321, 327, 328, 329, 330
NREL, 124
Numerical models, 561
Nutrients, 74, 83, 84, 86, 94, 106, 114, 184, 186, 219, 231, 288, 290, 297, 358, 359, 369, 400, 401, 402, 404, 405, 424, 426, 435, 440, 441, 444, 449, 450, 453, 455, 474, 514, 515, 516, 582, 604, 605
Nutrition deficiency, 402
Nylons, 7, 11, 15, 21

O

O-acetyl–4-O-methylglucuronoxylan, 204
Obligate anaerobes, 502, 504, 507
Oil
  extraction, 322, 408, 409, 410, 411
  refinery, 6, 16, 20, 67, 600
OMEGA, 418

Open-core gasifiers, 62
Open ponds, 400, 409, 417, 418, 419, 423, 425, 426, 427, 428, 429, 432, 435, 440, 452, 457, 459, 460, 474
Open raceway ponds, 417, 419, 428
Open systems, 440, 456, 459–461, 552
Operating cost for biohydrogen plant, 491
Operation
conditions, 69
robustness, 546
Operon, 574, 575
Optical fiber, 458, 550, 551
Optimization, 20, 94, 105, 192, 197, 219, 278, 330, 343, 346, 347, 363, 369, 441, 449, 452, 454, 455, 457, 458, 460, 475, 529, 561–563
Optimum sizes, 490
Organic acids, 12, 82, 86, 88, 106, 109, 156, 167, 181, 241, 485, 500, 501, 503, 505, 506, 512, 513, 529, 530, 533, 574, 580
Organic extractant, 580
Organosolv, 55, 153, 157–158, 163, 166, 167, 168, 169, 232, 254
Overheating, 185, 269, 456, 457, 459
Overliming, 150, 291, 304, 305
Oxidations, 12, 21, 54, 55, 56, 80, 150, 153, 154, 159–160, 168, 169, 237, 241, 258, 297, 299, 300, 345, 348, 377, 467, 500, 502, 514, 574, 602
Oxidative pretreatment, 159, 257, 258
Oxidative reactions, 377
Oxygen, 7, 8, 9, 14, 16, 21, 53, 56, 59, 60, 61, 63, 64, 65, 67, 70, 71, 81, 106, 157, 159, 160, 169, 186, 237, 258, 264, 268, 276, 377, 404, 409, 418, 421, 441, 459, 483, 485, 500, 501, 505, 514, 581, 589, 596, 599, 600, 602
-blown gasification, 64
Oxygenic photosynthesis, 501
Ozonolysis, 55, 150, 153, 158, 168, 169

P

Packed bed, 327, 328, 381, 510, 537, 543–546
reactors, 327, 328, 381, 537, 543–546
Palm, 253, 254, 266, 325, 328, 340, 344, 348, 350, 351, 353–370, 381, 382, 384, 390, 507, 530, 531, 532, 545, 560, 597, 602, 603
oil, 253, 254, 325, 328, 344, 348, 350, 351, 353–370, 381, 382, 390, 507, 530, 531, 532, 545, 560, 597
Palm oil mill effluent (POME), 507, 530, 531, 532, 545
Paper industry, 9, 112, 118, 184, 188, 189, 190, 196
Paper and pulp industry, 218, 219
Partially returning, 539
Partial pressure, 87, 89, 90, 459, 476, 503, 513, 526, 529, 530
Particulate matter (PM), 80, 84, 85, 88, 94, 302, 376
Payback period, 332
Penetrations of collimated and diffuse, 446
Pentoses, 54, 56, 109, 125, 128, 141, 142, 152, 163, 166, 235, 252, 261, 264, 270, 403, 528, 580, 606
fermentation, 264
Performance of fuels, 26, 36
Periodic discontinuous batch, 510
Peroxidases, 107, 236–237, 303
Perstraction, 580
Pervaporation, 578, 580, 581
Petrochemicals, 7, 11, 14, 16, 19, 145, 571, 572, 582, 593
Petroleum
refineries, 4, 8, 73, 422, 424, 596, 597
resources, 8, 25, 251, 571
pH, 73, 83, 84, 86, 87, 88, 93, 95, 106, 111, 152, 153, 157, 160, 162, 167, 184, 185, 186, 187, 191, 198, 219, 232, 234, 237, 256, 261, 264, 288, 296, 297, 304, 320, 387, 404, 417, 418, 440, 441, 449, 450, 455, 457, 471, 483, 503, 504, 505, 509, 510, 511, 514, 516, 526, 528, 529, 530, 531, 532, 533, 545, 546, 547, 548, 553, 555, 560, 561, 574
Phaeodactylum tricornutum, 406
Pharmaceuticals, 4, 12, 15, 346, 356, 358, 590, 599
Phenolic compounds, 79, 152, 153, 154, 160, 166, 237, 255, 291, 292, 296, 297–298, 303, 304, 305, 306
Phenolics, 297
Phenols, 7, 11, 14, 15, 18, 154, 158, 160, 237, 291, 297, 507
Phenylpropenyl, 9
Phosphatase, 295, 575
Phosphorus, 425, 516, 601
Phosphorylation, 114, 293, 501, 516, 574, 575
Phosphotrans-butyrylase, 574
Photobiological processes, 500–501, 503, 518, 549, 552
Photobioreactor, 402, 403, 409, 410, 417, 435, 439–461, 474, 475, 534, 537, 548–552, 555, 556, 557, 558, 559, 561, 562
Photofermentation, 500, 501, 503, 512, 513, 518, 533, 534, 537, 549, 552, 560
Photoheterotrophic, 401, 503, 525, 533
Photosynthetic activity (P), 441, 442
Photosynthetic efficiency, 406, 475, 550
Photosynthetic growth, 440, 441, 442, 444–445, 447, 449
rate, 445, 447
Physical detoxification, 304
Phytonutrients, 358, 359, 369
Pilot-scale, 85, 104, 239, 330, 383, 418, 419, 533
Plant size, 134, 136, 488, 489, 491, 492
Plasma gasification, 63–64
Platform, 5, 12, 16, 28
PNp α-L-arabinofuranoside, 212
PNP β-D xylopyranoside, 210, 213
Polybutyleneterephthalate (PBT), 21
Polyelectrolyte, 10
Polyesters, 7, 13, 16, 21, 86, 346, 514
Polyethylene terephthalate (PET), 21

Polyhydroxyalkanoate (PHA), 86, 514, 515, 561
Polyhydroxypolyamides, 21
Polymerization, 20, 21, 22, 54, 60, 109, 151, 154, 157, 165, 167, 208, 233, 247, 255, 259, 303
Polymers, 10, 11, 20, 52, 53, 54, 107, 178, 179, 203, 211, 237, 244, 251, 473
Polyols, 13
Polyurethanes, 20, 322, 327, 329, 346
Poly(β-OH)butyrate (PHB), 514, 515
Pongamia, 354, 360, 602, 603
Poplar, 6, 53, 79, 111, 125, 126, 131, 132, 133, 134, 141, 157, 158, 165, 254, 262, 474, 482, 589
Porosity, 154, 276, 544, 545, 546, 559
Possibilities, 60, 87, 161, 248, 266, 430, 526, 533–534, 538, 560, 573, 598
Pour point (PP), 345, 348, 357, 358, 384, 388, 389
Poverty, 355, 588
Power, 3, 4, 9, 10, 43, 52, 55, 59, 64, 67, 90, 95, 116, 128, 130, 151, 155, 169, 233, 253, 266, 267, 268, 269, 275, 276, 277, 278, 390, 402, 410, 422, 424, 425, 432, 445, 481, 482, 483, 487, 488, 490, 501, 514, 552, 553, 554, 555, 587, 590, 591, 592, 595, 601, 603, 605
Practical applications, 145, 384, 530–533, 538, 541, 546, 548, 551, 555, 558, 560, 561, 562, 576
Pretreatments, 9, 54–55, 67, 72, 79, 88, 105–106, 108, 109, 111, 112, 114, 115, 116, 117, 125, 128, 131, 141, 149–171, 178, 181–183, 190, 192, 232, 234, 238, 239, 241, 255–259, 264, 266, 270, 271, 272, 273, 274, 278, 287, 290, 291, 292, 299, 305, 319, 331, 355, 368, 375, 380, 383, 391, 407, 423, 472, 473, 474, 504–505, 509, 511, 541, 548, 577, 582, 605, 606
Previous land use, 26, 28, 30
Primary residues, 472, 473
Process
- conditions, 8, 20, 128, 186, 190, 330, 528–530, 590, 596
- design, 28, 125, 130, 131, 136, 137, 140, 141, 184, 187, 198
- efficiencies, 229, 248, 264, 429, 444, 511, 512–518
- integration, 262, 263, 512–513, 518, 596
- yield, 331, 602

Processing residues, 112, 113
Product inhibition, 110, 198, 239, 240, 241, 243, 247, 261, 511, 513, 577, 582
Production
- of bioethanol from agroindustrial residues as feedstocks, 251–279
- capacity, 68, 104, 105, 112, 124, 130, 132, 133, 134–136, 137, 138, 139, 275, 415
- costs, 22, 72, 103, 118, 123, 124, 125, 126, 127, 128, 129, 130–140, 141, 142, 143, 146, 150, 170, 198, 253, 254, 263, 265, 273, 274, 275, 278, 327, 332, 341, 349, 359, 360, 365, 368, 369, 375, 391, 416, 435, 475, 488–490, 491, 492, 493, 514, 600
  - of bio-oil, 488–490
- model, 428
- plants, 104–105, 277, 315, 331, 332, 359, 484, 485, 486, 487–488, 489, 490, 491, 492, 493
- process, 12, 20, 22, 31, 33, 34, 35, 105, 108, 132, 150, 151, 157, 161, 167, 170, 184, 195, 198, 270, 273, 278, 332, 345, 356, 383, 429, 461, 482–483, 486, 499, 504, 508, 509, 511, 513, 526, 528, 529, 531, 532, 537, 538, 543, 547, 548, 549, 555, 560, 562, 572, 591, 602

Productivities
- biomass, 400, 403, 443, 447, 448, 449, 451, 453, 458
- limits, 427, 461
- maximal biomass volumetric, 452
- surface, 439, 447, 448, 453, 455, 458, 460, 461
- volumetric, 167, 447, 448, 451, 452, 453, 458, 460, 461

Product value (PV), 126
Promoters of transcription elongation factors 1α and tef1, 196
1,2-propanediol, 19
2-propanol, 13
Propene, 7, 13, 19
Propene glycol, 13
Propylene, 7, 13, 346
Prospects, 75, 85, 129, 254, 333, 467–476, 560, 606
Protein engineering, 197, 198, 245, 248
Protein-rich, 470, 471, 472
Proton transfer rate, 553
PSB immobilization, 559
PSOL1 mega plasmid, 575
Pulp industry, 218, 219
Pulping, 112, 154, 170, 189, 220, 232, 263, 291
Pulsed electrical field process, 55
Purification, 11, 84, 124, 209, 213, 214, 215, 217, 233, 315, 316, 317, 322, 340, 341, 342, 347, 354, 355, 356, 358, 362, 367, 382, 385, 386, 387, 484, 486, 576, 599
Purpose-grown, 589, 605, 606
Pyrolysis, 5, 9, 55, 56, 58, 59, 61, 62, 65–70, 71, 72, 73, 269, 363, 409, 421, 422, 423, 429, 482, 483, 484, 485, 486, 487, 488, 600–601
Pyrrolydone, 16
Pyruvate dehydrogenase, 293, 295, 516

R

Raceway, 409, 417, 419, 428, 432, 459, 460
Radiative transfer inside the culture, 442
Radiative transfer models, 444, 445, 451
Rag layer, 580
Random mutagenesis, 195
Rapeseed (canola) oil, 321, 332, 342, 345, 347, 349, 350, 358, 359, 409
Rapid formation, 545, 559

Raw materials, 4–8, 9, 61, 73, 74, 85, 103, 105, 106, 112–113, 114–115, 116–117, 118, 128, 149, 150, 151, 152, 154, 156, 158, 164, 165, 166, 169, 178, 183, 192, 196, 198, 219, 254, 266, 267, 270, 271, 272, 274, 278, 287, 288, 291, 315, 316, 317, 319, 323, 324, 325, 326, 332, 376, 391, 430, 467, 469, 525, 572, 577–580, 582, 600, 601
RBB (remazol brilliant blue) dyed methyl glucuronoxylan, 212
Reaction conditions, 17, 18, 61, 65, 232, 315, 316, 317, 333, 358, 359, 361, 362, 363, 364, 366, 367, 368, 369, 384
Reaction medium, 64, 317, 318, 319, 320, 323, 324, 340, 342, 347, 349, 350
Reactor
  configurations, 72, 90–93, 95, 326–331, 510–511, 537, 538, 541, 546, 548, 557, 560
  technologies, 526, 530
  thickness, 550
Redirecting, 95, 575
Redox
  condition, 502, 508–509, 510, 514, 518
  mediators, 238, 502, 509
Reducing agents, 83, 94, 304
Reducing equivalents, 82, 83
Reduction, 13, 19, 21, 26, 28, 30, 32, 42, 43, 44, 46, 56, 57, 59, 62, 64, 65, 69, 71, 83, 89, 103, 104, 111, 112, 114, 115, 116, 117, 136, 140, 145, 151, 155, 159, 160, 163, 165, 167, 191, 192, 193, 196, 198, 229, 239, 242, 244, 252, 255, 263, 265, 270, 273, 278, 288, 293, 296, 299, 300, 320, 321, 325, 344, 346, 348, 349, 350, 354, 390, 391, 404, 406, 407, 410, 419, 433, 481, 500, 501, 505, 509, 518, 558, 590, 592
  to butanediol, 19
Reference systems, 28, 29–30, 31, 34, 35, 39, 47, 48
Refineries
  fossil, 4, 14, 16, 19
  gas, 7, 12
  oil, 6, 16, 20, 67, 600
  petroleum, 4, 8, 73, 422, 424, 596, 597
Reforming, 5, 62, 70, 361, 467, 483, 484, 485, 486, 490, 491, 492, 493, 525, 599
Refuse-derived fuel (RDF), 116, 117
Regulatory, 216, 410, 576, 591
Remediation, 401, 505
Renewable
  energy, 3, 25, 51, 52, 59, 188, 190, 339, 340, 353, 354, 476, 481, 525, 588, 605
  resource, 103, 107, 189, 196, 515, 525
Rentech, 597, 598
Replenishment, 605
Repression, 180, 181, 184, 186, 187, 192, 195, 198, 215, 289, 290, 509
Residual substrate, 511
Resins, 5, 7, 11, 20, 21, 230, 231, 303, 304, 305, 321, 364, 365, 380, 382, 383, 387
Resource availability, 435
Response surface methodology, 219
Retention time, 93, 156, 304, 510, 526, 528, 538
Reutilization, 319, 321, 363
Rice straw, 113, 114, 118, 155, 159, 161, 164, 192, 215, 217, 218, 247, 254
Rudolf Diesel, 399

S

Saccharification, 34, 35, 38, 39, 56, 110, 111, 112, 125, 126, 128, 131, 141, 142, 156, 157, 159, 160, 161, 163, 167, 191, 199, 217, 238, 239, 240, 243, 244, 247, 251, 252, 258, 259–260, 261–264, 270, 271, 304
Saccharomyces, 105, 106, 111, 159, 239, 260, 262, 288, 573, 582
Salicornia, 605
Salix, 128, 129
Salt stress, 290
Saponification, 109, 354, 359, 360, 377, 380, 381, 407
Scaffolding protein CipA, 576
Scale-down, 600
Scale-up, 12, 90, 95, 185, 346, 533, 600
Scenario, 39–41, 46, 47, 73, 75, 116, 125, 126, 136, 137, 145, 178, 192–194, 332, 425, 426, 428, 432, 433, 435, 589
Scrubbers, 85, 94
Sea water, 426, 434, 435
Secondary residues, 472, 473
Second-generation, 53, 64, 74, 107–110, 134, 140, 146, 193, 266, 415, 419, 474
Second-generation biofuels, 53, 64, 74, 107, 140, 415, 419, 474
Secretome, 217
Security, 51, 101, 103, 114, 149, 278, 339, 340, 415, 416, 587, 588, 589, 592, 606
Selective removal, 581
Self-flocculation, 540, 559
Self-immobilization, 510
Self-shade effects, 562
Self-shading, 549, 551, 552
Semicontinuous-flow processes, 327
Separate hydrolysis and fermentation (SHF), 56, 125, 238, 239, 261, 304
Separation
  biomass-liquid, 537, 538, 540, 541
  gravity, 388
  membrane, 244, 298, 561
  phase, 67, 332, 359, 385
Separation and purification of FAMEs, 385
Separator-free configuration, 553
Separators, 58, 63, 94, 330, 385, 420, 542, 543, 544, 546, 547, 553, 554, 555

Sequencing biofilm reactor, 507
Sequestrants, 10
Set aside, 25, 30, 34, 35, 40, 41, 47
Settler, 539, 540, 541
Settling, 332, 385, 387
Severity factor, 165, 234, 292
Shear stress, 218, 240, 327, 454, 455, 549
Short light paths, 452
Sigma factors, 575
Signaling events, 572
Simultaneous reaction and separation, 264–265
Simultaneous saccharification and cofermentation (SSCF), 125, 131, 157, 163, 262, 263, 264
Simultaneous saccharification and fermentation (SSF), 56, 110, 128, 160, 163, 238, 261–264, 304
Single-cell protein, 198
Site-directed mutagenesis, 198, 245
Slow pyrolysis, 66, 70
Sludge oil, 375
Sludge-substrate contact, 542
SmF, 183, 186, 187, 198, 218
Soaking aqueous ammonia (SAA), 163
Soapstock, 375, 376, 378, 380, 391
Sodium glutamate, 16
Softening, 112, 188, 189
Solar insolation, 417, 424, 425
Solar-powered, 554
Solar production, 439, 440, 443, 453, 455, 457
Solid loading, 183, 243
Solid-state fermentation (SSF), 141, 183, 184–186, 215
Solid-state fermentor, 185, 186
Sol locus, 574, 575
Solubility limit, 323
Solvent
  inhibition, 580
  tolerance, 572, 574, 576, 582
  toxicity, 265, 572, 575, 576, 577, 580, 582
Solvent-free system, 318, 322, 323, 324, 328, 329, 330, 332
Solventogenesis, 83, 88, 93, 509, 572, 573, 574–575, 576, 582
Solventogenic, 87, 572, 573, 574, 575, 576, 580, 582
Solventogenic phase, 572, 573, 574, 576
Sorbitol, 12, 13, 18, 20, 21, 260, 320
The Soxhlet method, 406
Soybean, 32, 43, 155, 323, 324, 325, 326, 328, 329, 340, 346, 350, 351, 354, 355, 358, 359, 361, 369, 378, 380, 390, 391, 399, 415, 416, 597, 602, 603
  oil, 32, 43, 323, 324, 325, 326, 328, 329, 350, 351, 378, 380, 391
Sparging, 90, 513
Specification, 67, 143, 145, 275, 319, 356, 591, 592, 594, 595
Specific illuminated surface, 448, 452, 453, 458, 460
Spectrumof hemicellulolytic enzymes, 218
SPK, 594, 597, 602
Spo0A, 574, 575
SpoA, 574
SpoIIE, 575
SPORL, 111, 112, 163, 168
Sporulation, 572, 574–575, 576, 580, 582
Spruce, 111, 128, 129, 141, 160, 163, 254, 302, 304
SSF bioreactors, 185, 197
Stability
  long-term, 547
  operating, 538, 541, 543, 558, 562
/operation time, 328
  oxidative, 355, 593
Starch, 4, 5, 101, 102, 103, 104, 105, 106, 107, 113, 126, 149, 161, 178, 191, 192, 219, 251, 252, 254, 299, 380, 400, 401, 402, 404, 405, 416, 420, 423, 469, 470, 471, 472, 473, 474, 506, 507, 528, 530, 548, 556, 571, 577, 606
Start-up, 548, 559
State of technology (SOT), 125, 142
Steam Pretreatment/Steam Explosion (ST/SE), 256
Stillage, 38, 39, 129, 141, 290, 298, 305, 506
Stir/mix washing, 385, 387
Stirred tank bioreactor, 325
Stirred-tank reactors, 84, 90, 218, 327, 330
Stoichiometry, 159
Stonewash, 188, 193
Storage, 8, 30, 58, 86, 110, 125, 130, 149, 167, 168, 348, 355, 400, 401, 410, 476, 483, 484, 486, 487, 488, 489, 490, 501, 509, 514, 589, 590
Strain disintegration, 574
Strategic cost management, 140
Strategies, 9, 25, 30, 53, 56, 150, 166, 179, 183, 191, 195, 197, 214, 217, 242, 244, 245, 287–306, 318, 355, 366, 402, 404, 407, 410, 417, 456, 512–518, 554, 558, 572, 576
Stratification, 547
Straw, 5, 34, 35, 38, 39, 40, 43, 44, 46, 47, 54, 58, 65, 74, 79, 104, 107, 113, 114, 118, 131, 132, 133, 134, 135, 136, 137, 138, 139, 140, 154, 155, 156, 157, 158, 159, 160, 161, 162, 164, 165, 167, 184, 192, 204, 215, 217, 218, 219, 247, 253, 254, 264, 415, 473, 482, 530
Stress
  genes, 576
  tolerance, 298
Structural heterogeneity, 203
Styrene-butadiene systems, 20
Subcritical, 109, 341, 347, 378, 384, 409, 423
Submerged fermentation, 141, 183, 186–187, 197
Submerged fermentation technology, 183, 197
Subsidy, 588
Substitution efficiency, 26, 36, 45–46

Substrate
degradation, 503, 505, 509, 510, 511, 512, 513, 514
metabolism, 509
swelling, 186
type, 183, 538, 557
Succinic acid, 18, 19, 20
Sucrose, 102, 105, 106, 191, 251, 260, 264, 402, 468, 469, 471, 531, 532, 539, 556, 561
Sugar
-based substrate, 549, 555, 561
inhibition, 289–290
Sugarcane
bagasse, 79, 81, 89, 103, 107, 113, 114, 118, 146, 155, 162, 165, 251, 254, 266, 273, 304, 597
ethanol, 103
residues, 130
Sulfate-reducing bacteria (SRB), 90, 91, 92
Sulfite liquor, 264
Sulfite pretreatment to overcome recalcitrance of lignocellulose (SPORL), 111, 112, 163, 168
Sulfonation, 10, 112, 163
Sunflower oils, 118, 325, 327, 329, 346, 361, 369, 382
Sun's displacement, 443
Supercritical
BD production, 384
fluids, 164, 168, 339–351, 367, 408
methanol, 342, 367, 368, 383, 384, 385
Supercritical alcohol (SCA), 342–344, 367–368, 408
Supercritical dimethyl carbonate (SCDMC), 346–347
Super critical gasification of biomass, 64–65
Supercritical methyl acetate (SCMA), 344–346, 384, 385
Support materials, 381, 543, 544, 545
Supramolecular materials, 9
Surface attachment, 559
Surface-lightened systems, 457, 458
Surfactants, 20, 21, 240–241, 244, 325
Suspended growth, 510, 555, 559
Suspended solid (SS), 131, 420, 539, 546
Sustainability, 25, 26, 32, 46, 51, 74, 86, 95, 112, 118, 142, 143, 149, 178, 199, 339, 340, 356, 363, 369, 419, 424, 427, 430, 431, 433, 474, 602, 604
Sustainable
biofuels, 25, 74, 416, 435, 589
energy, 499, 537
Sweet sorghum, 130, 254, 262, 291, 404, 474, 506, 606
Switchgrass, 6, 131, 132, 133, 134, 155, 157, 159, 161, 163, 254, 258, 482
Swollenin, 238
Syngas
-derived chemicals, 10, 11
fermentation, 79–96
quality, 80, 94
Syringaldehyde, 153, 241, 297
Syringols, 9
Systematic comparison, 555, 562
System boundaries, 28, 430, 431, 433
System definition, 27, 28–29, 34, 35, 39, 433
System expansion, 26, 27, 31, 32

T

Taguchi, 511
Target costing (TC), 140, 142, 143, 144, 146, 341, 448
Target cost of lignocellulosic ethanol, 143
Target selling price of lignocellulosic ethanol, 143
Technoeconomic analysis, 123–146, 426, 476
Technoeconomic evaluations, 123, 124, 125, 127–130, 140, 141, 143, 144, 146
Technoeconomic feasibility, 199
Technoeconomic model, 487, 488, 489, 490, 491
Technological configurations, 261, 265, 271
Technology improvement is required to bring down the cost of biohydrogen production to make it competitive with natural gas-based hydrogen, 493
Technology improvements, 476, 493
Temperature
control, 106, 185, 342, 510, 550, 551, 552
high, 9, 13, 18, 55, 63, 64, 65, 66, 68, 69, 80, 84, 86, 108, 111, 150, 152, 156, 164, 232, 256, 258, 288, 292, 296, 317, 321, 342, 344, 347, 348, 355, 367, 377, 485, 486, 504, 525, 560, 601
Terephthalic acid, 14, 21
Terpenes, 14, 15
Tert-butanol, 320, 325, 328, 332, 365, 379
Tertiary residues, 472, 473
Tetrahydrofuran, 19, 408
Textile, 7, 177, 178, 188, 189, 190, 192, 193, 329
industry, 178, 188
T. harzianum, 182, 189, 212, 215
Thermal pretreatment, 109, 159, 161, 164, 170, 255–256, 257–258, 423
Thermal pretreatment in combination with acid pretreatment, 257–258
Thermal pretreatment in combination with alkaline oxidative pretreatment, 258
Thermal pretreatment in combination with alkaline pretreatment, 258
Thermal pretreatment in combination with oxidative pretreatment, 258
Thermochemical, 5, 9, 51–75, 79, 89, 111, 118, 126, 130, 146, 268, 269, 421–423, 429, 467, 483, 600
Thermophilic, 56, 83, 84, 93, 205, 206, 215, 217, 219, 471, 509, 510, 525–534, 543, 545, 558, 560, 562
Thermophilic fermentation, 560, 562
Thickness, 186, 407, 550, 552

Thio hydroxymethylation, 10
Third-generation biofuels, 64, 74, 75, 415
Time-domain nuclear magnetic resonance (TD-NMR), 407
T. longibrachium, 191
Tolerance of lipases to alcohol, 323
Tolerant, 150, 166, 196, 306, 368, 384, 602
Toluene, 7, 12, 382
Total capital investment, 332
Toxic compounds, 93, 150, 151, 152–154, 156, 166, 168, 169, 260, 288, 291, 293, 296, 299, 300, 302, 303, 304, 580
Toxicity, 32, 95, 150, 153, 159, 265, 296, 297, 304, 305, 365, 572, 575, 576, 577, 580, 581, 582
Transcriptional regulators, 574, 576
Transcriptional regulator Spo0A, 574
Transducers, 197
Transesterification, 315, 316, 317–322, 323, 324, 327, 330, 340, 341, 342, 343, 345, 346, 347, 349, 354, 356, 358, 359, 360, 361, 362, 363, 364, 365, 366, 367–369, 377, 379, 381, 383, 389, 399, 401, 405, 407, 408, 410, 602
Transfer enhancement, 440, 454
Transglycosylation, 180, 215, 220
Transportation, 3, 4, 12, 19, 25, 47, 52, 67, 72, 73, 79, 86, 101, 110, 127, 140, 145, 149, 178, 230, 253, 269, 278, 340, 353, 354, 375, 401, 406, 415, 416, 420, 423, 425, 430, 432, 482, 483, 484, 486, 487, 488, 489, 491, 492, 493, 591, 592, 595, 599
T. reesei, 115, 179, 180, 181, 182, 183, 184, 186, 189, 190, 191, 192, 193, 194, 195, 196, 197, 198, 199, 208, 217, 218, 235, 238, 239, 245, 262, 304
T. reesei genome, 195, 199
Trehalose, 288, 289
Triacylglyceride (TAG), 315, 317, 318, 319, 322, 323, 324, 330, 400, 401, 402
Triacylglycerol, 315, 402, 407, 408
Trichoderma, 110, 131, 141, 179, 180, 181, 182, 187, 188, 189, 190, 191, 195, 197, 207, 212, 234, 259, 263, 303
Trichoderma reesei, 110, 131, 141, 179, 207, 212, 234, 259, 263, 303
Trickle-bed reactor, 92–93
Triglyceride, 340, 341, 342, 343, 344, 345, 346, 349, 350, 351, 354, 360, 361, 363, 364, 367, 377, 423, 596, 597, 600, 601, 602, 603, 604, 605
T. viride, 115, 189, 190, 194
Two-flux model, 445–446
Two-stage biohydrogen production, 560
Two stage process, 87, 230, 475, 534
Two-step acidulated wash, 387
Two-step BD production, 381–383
Two-step method, 169
Two-step washing, 387

U

Ultra filtration, 243, 244
Ultrasonic, 95, 161, 368–369, 408, 419
Ultrasonic-assisted transesterification, 368–369
Ultrasound pretreatment, 161
Underfeed stokers, 57
Unsaponifiable, 602, 604
Unsaturated fat, 367, 377, 384, 605
UOP, 68, 596, 598, 601
Updraft (countercurrent) gasifiers, 62
Upflow anaerobic sludge blanket (UASB), 471, 510, 530, 531, 532, 533, 537, 541, 546–548, 556, 557, 558, 562
Upflow-mode reactors, 554
Used cooking oil, 330, 331, 376, 378, 389
Used frying oil, 378, 382, 390, 391
Utilization phase, 26, 39, 44

V

Vacuum evaporation, 304
Vacuum pyrolysis, 66, 70, 601
Value chain, 8, 9, 16, 22, 124, 145–146
Value engineering (VE), 140–142, 144, 146
Vanillic acid, 154, 241, 291, 305
Vanillin, 9, 154, 241, 291, 297, 304
Variable operating costs, 132, 133, 134, 135, 138, 139
Vegetable oils, 32, 74, 267, 269, 315–333, 339–351, 353–370, 375–392, 399, 407, 408, 410, 589, 601
Vehicle emission test, 390
Viability, 106, 164, 243, 265, 278, 287, 293, 301, 303, 359, 419, 427, 511, 513, 518, 582, 589, 604, 606
Vinasse, 38, 106, 506
Vinyl chloride, 7, 19
Viscosity, 219, 232, 233, 243, 319, 343, 345, 365, 366, 367, 368, 376, 384, 388, 409, 485
Vitamin C, 12
Void space, 186, 544
Volatile fatty acid (VFA), 500, 503, 513
Volatiles, 56, 57, 58, 59, 62, 69, 72, 143, 158, 232, 252, 257, 276, 304, 332, 485, 500, 503, 513, 539, 571, 581
Volumetrically lightened systems, 457–459
Volumetric HPR and hydrogen yield (HY), 538
Volumetric mass transfer coefficient, 89, 91, 92
Vortex reactor, 67

W

Warranties, 592
Washes, 387
Washing, 188, 189, 233, 299, 356, 358, 360, 362, 364, 365, 367, 379, 383, 385–387, 388, 391
  of BD, 385–387
Waste
  cooking oil, 322, 326, 329, 341, 346, 360, 375, 376, 377, 378, 380, 381, 382, 383, 384, 386, 388–391

-management, 116, 117
molasses, 405
oil, 74, 323, 324, 328, 329, 348, 349, 350, 351, 375–392
palm oil, 381
sludge, 376
Wastewater(s)
food processing, 506
high-strength, 548
molasses-based, 506
sludge, 116, 472
sugar refinery, 507
Water
activity, 185, 186, 218, 288
content, 65, 67, 129, 186, 243, 266, 317, 318–319, 324, 350, 351, 355, 377, 379, 381, 383, 384, 389, 483, 604
waste, 424, 426, 434, 435, 548
treatment, 416, 420, 424
Waxes, 11, 113, 598
Well-to-Tank (WtT), 26, 27, 28, 29, 34, 36, 41, 42, 45
Well-to-Wheel (WtW), 26, 27, 29, 33, 36, 39, 41, 42, 43, 44, 45, 46, 48
Wet biomass, 64, 408, 409, 422, 423
Wet oxidation, 54, 55, 150, 153, 154, 159–160, 168, 169, 258
Wheat
grains, 34, 40, 44
straw, 34, 40, 54, 104, 113, 154, 156, 157, 158, 159, 160, 162, 165, 167, 171, 218, 253, 254, 264, 473, 506, 530, 532, 579
Whole-cell
biocatalysts, 21, 322
immobilized lipases, 320, 322
Whole-tree biomass, 482, 485, 486, 492
Wine industry, 189
Wood-Ljungdahl pathway, 82
Wood logging residues, 113
Wood processing residues, 113
Wood pulping, 189
Woody biomass, 56, 102, 110, 111, 112, 116, 156, 162, 163
1900 World Exposition, 13, 19, 20

X

Xenobiotic responsive element (XRE), 576
Xylan, 6, 55, 109, 111, 131, 137, 156, 159, 163, 193, 204, 206, 207, 208, 210, 211, 212, 213, 214, 215, 216, 217, 220, 235, 236, 255, 256, 257, 292
Xylanases, 191, 193, 205, 206, 207, 208, 209, 210, 212, 213, 214, 215, 216, 217, 218, 219, 220, 235–236, 241, 259, 261
Xylene (BTX), 7
Xylitol, 12, 13, 20, 215, 220, 255, 264, 294, 295, 301
Xylose, 11, 12, 20, 53, 54, 55, 107, 137, 155, 157, 158, 159, 163, 204, 205, 207, 208, 209, 210, 211, 212, 215, 230, 235, 236, 241, 246, 247, 251, 256, 260, 261, 262, 264, 270, 272, 290, 292, 294, 295, 469, 530, 577
β-Xylosidase (EC 3.2.1.37), 205, 208, 209, 210, 213, 215, 216, 219, 236, 259
Xylytol, 18

Y

Yeasts, 86, 106, 109, 110, 128, 129, 150, 151, 152, 166, 181, 231, 238, 239, 252, 255, 260–261, 262, 263, 264, 265, 289, 290, 293, 294, 297, 299, 302, 303, 322, 379, 403, 405, 423, 528, 599
Yellow grease, 376, 378
Yellow or brown grease, 375
Yield(s)
ethanol, 84, 86, 95, 109, 112, 118, 126, 128, 132, 133, 134, 135, 136–137, 138, 139, 153, 156, 161, 163, 165, 167, 239, 243, 258, 263, 273, 289, 293, 297, 298, 299
hydrogen, 469, 470, 471, 472, 473, 476, 493, 525, 526, 527, 528, 529, 530, 533, 538

FSC
www.fsc.org
MIX
Papier | Fördert gute Waldnutzung
FSC® C083411

Kolibri 360 GmbH
Ferdinand-Jühlke-Straße 7
99095 Erfurt, Deutschland
produktsicherheit@kolibri360.de